# Green Pesticides Handbook

## Essential Oils for Pest Control

# Green Pesticides Handbook
## Essential Oils for Pest Control

Edited by
Leo M.L. Nollet
Hamir Singh Rathore

**CRC Press**
Taylor & Francis Group
Boca Raton London New York

CRC Press is an imprint of the
Taylor & Francis Group, an **informa** business

CRC Press
Taylor & Francis Group
6000 Broken Sound Parkway NW, Suite 300
Boca Raton, FL 33487-2742

First issued in paperback 2019

© 2017 by Taylor & Francis Group, LLC
CRC Press is an imprint of Taylor & Francis Group, an Informa business

No claim to original U.S. Government works

ISBN-13: 978-1-4987-5938-0 (hbk)
ISBN-13: 978-0-367-87702-6 (pbk)

---

**Library of Congress Cataloging-in-Publication Data**

---

Names: Nollet, Leo M. L., 1948- editor. | Rathore, Hamir Singh.
Title: Green pesticides handbook : essential oils for pest control / [compiled by] Leo M.L. Nollet and Hamir
   Singh Rathore.
Description: Boca Raton : Taylor & Francis, CRC Press, 2017. | Includes bibliographical references.
Identifiers: LCCN 2016052776 | ISBN 9781498759380 (print : alk. paper)
Subjects: LCSH: Natural pesticides--Handbooks, manuals, etc. | Essences and essential oils--Handbooks,
   manuals, etc. | Green chemistry--Handbooks, manuals, etc.
Classification: LCC SB951.145.N37 G74 2017 | DDC 628.9/6--dc23
LC record available at https://lccn.loc.gov/2016052776

---

**Visit the Taylor & Francis Web site at**
**http://www.taylorandfrancis.com**

**and the CRC Press Web site at**
**http://www.crcpress.com**

# Contents

## Section I   Essential Oils as Green Pesticides

## Section II   Essential and Vegetable Oils

## Section III   Different Aspects of Essential Oils

# *Preface*

This book explores one series of biopesticides, essential oils. The number of essential oils with pesticidal possibilities or activities is large—too large to comment on all in this book. The authors made a selection of essential oils. These are discussed in depth. Other interesting properties of essential oils are also considered (in some chapters of Section III). Both editors would like to cordially thank all of the authors for their excellent work. They spent a lot of time and hard work, and the result is superb. We would also like to thank the editorial team of CRC Press/Taylor & Francis for their patience. Finally, Dr. Nollet gives a word of thanks to and expresses his appreciation for his coeditor, Professor Rathore. The idea of this book originated in his head.

Two heads are better than one.

**Leo M.L. Nollet**
**Hamir Singh Rathore**

# Editors

**Leo M.L. Nollet, PhD,** received an MS (1973) and PhD (1978) in biology from the Katholieke Universiteit Leuven, Belgium. He is an editor and associate editor of numerous books. He edited for M. Dekker, New York—now CRC Press/Taylor & Francis—the first, second, and third editions of the books entitled *Food Analysis by HPLC* and the *Handbook of Food Analysis*. The last edition is a two-volume book. He also edited the *Handbook of Water Analysis* (first, second, and third editions) and *Chromatographic Analysis of the Environment*, Third Edition (CRC Press). With F. Toldrá, he coedited two books published in 2006 and 2007: *Advanced Technologies for Meat Processing* (CRC Press) and *Advances in Food Diagnostics* (Blackwell Publishing—now Wiley). With M. Poschl, he coedited the book *Radionuclide Concentrations in Foods and the Environment*, also published in 2006 (CRC Press).

Dr. Nollet has coedited several books with Y.H. Hui and other colleagues: the *Handbook of Food Product Manufacturing* (Wiley, 2007); *Handbook of Food Science, Technology and Engineering* (CRC Press, 2005); *Food Biochemistry and Food Processing* (first and second editions; Blackwell Publishing [Wiley], 2006, 2012); and the *Handbook of Fruits and Vegetable Flavors* (Wiley, 2010). In addition, he edited the *Handbook of Meat, Poultry and Seafood Quality* (first and second editions; Blackwell Publishing [Wiley], 2007, 2012). From 2008 to 2011, he published with F. Toldrá five volumes on animal product–related books, namely, the *Handbook of Muscle Foods Analysis*, the *Handbook of Processed Meats and Poultry Analysis*, the *Handbook of Seafood and Seafood Products Analysis*, the *Handbook of Dairy Foods Analysis*, and the *Handbook of Analysis of Edible Animal By-Products*. Also in 2011 with F. Toldrá, he coedited for CRC Press two volumes: *Safety Analysis of Foods of Animal Origin* and *Sensory Analysis of Foods of Animal Origin*. In 2012, they both published the *Handbook of Analysis of Active Compounds in Functional Foods*. In a coedition with Hamir Rathore, the *Handbook of Pesticides: Methods of Pesticides Residues Analysis* was marketed in 2009, *Pesticides: Evaluation of Environmental Pollution* in 2012, and the *Biopesticides Handbook* in 2015.

Other finished book projects include *Food Allergens: Analysis, Instrumentation, and Methods* (with A. van Hengel; CRC Press, 2011) and *Analysis of Endocrine Compounds in Food* (Wiley-Blackwell, 2011). Dr. Nollet's recent projects include *Proteomics in Foods* with F. Toldrá (Springer, 2013) and *Transformation Products of Emerging Contaminants in the Environment: Analysis, Processes, Occurrence, Effects and Risks* with D. Lambropoulou (Wiley, 2014). In the series, CRC Food Analysis & Properties, Dr. Nollet edited with C. Ruiz-Capillas *Flow Injection Analysis of Food Additives* (CRC Press, 2015) and *Marine Microorganisms: Extraction and Analysis of Bioactive Compounds* (CRC Press, 2016).

**Hamir Singh Rathore, PhD,** is a retired professor and chairman, Department of Applied Chemistry, Zakir Husain College of Engineering and Technology, Aligarh Muslim University, Aligarh, India. His research work is in the area of applied/industrial chemistry, with more emphasis on synthetic inorganic ion exchangers, pesticides, and development of analytical techniques. He received MSc, MPhil, and PhD degrees from Aligarh Muslim University in 1967, 1970, and 1971, respectively. He worked as a postdoctoral fellow on biosensors with Professor Marco Mascini in the Second University of Rome, Italy, on a fellowship awarded to him in 1987 by the Third World Academy of Sciences, Trieste, Italy.

Professor Rathore taught several courses of applied chemistry to the students of BTech and MTech. After retirement, he became actively engaged in teaching modern instrumental techniques to MSc students (industrial chemistry), writing a book, and conducting research work in the university.

His research work has been published in the form of 125 papers in journals of international repute. Twenty students have been awarded MPhil and an equal number of PhD degrees under his supervision. Professor Rathore has presented his research work and delivered lectures and invited talks in several national and international conferences held in India and abroad (United States, Italy, Spain, Finland, Hungary, Brazil, etc.). He has attended a dozen summer schools and short-term courses at different institutes, universities, and research laboratories. He is a member of several national and international scientific bodies. He is the editor or referee of some national and international journals.

Professor Rathore has contributed to the following books: *Basic Practical Chemistry* (1982) and *Experiments in Applied Chemistry* (1990), both of which were edited by Ishtiaq Ali and H.S. Rathore and published by AMU, Aligarh; *Handbook of Chromatography: Liquid Chromatography of Polycyclic Aromatic Hydrocarbons* (CRC Press, 1993), edited by Joseph Sherma and Hamir S. Rathore; and *Handbook of Pesticides: Methods of Pesticide Residues Analysis* (2009), *Pesticides: Evaluation of Environmental Pollution* (2012), and *Handbook of Biopesticides* (2014), all edited by Leo M.L. Nollet and H.S. Rathore and published by CRC Press/Taylor & Francis.

He has contributed six chapters in the following handbooks: the *Handbook of Food Analysis* (1996), *Handbook of Water Analysis* (2000), and the revised and enlarged edition of the *Handbook of Food Analysis* (2004). All three books were edited by Leo M.L. Nollet and published by Marcel Dekker.

# Contributors

**Ranjida Ahmed**
Department of Entomology
Tocklai Tea Research Institute
Tea Research Association
Jorhat, Assam, India

**Hanan M. Al-Yousef**
Department of Pharmacognosy
College of Pharmacy
King Saud University
Riyadh, Saudi Arabia

**Carlos Ariel Cardona Alzate**
Instituto de Biotecnología y Agroindustria
Departamento de Ingeniería Química
Universidad Nacional de Colombia
Manizales, Colombia

**N.C. Basantia**
Cavinkare Research Centre
Cavinkare Private Limited
Chennai, Tamilnadu, India

**Irshad Ul Haq Bhat**
Faculty of Earth Science
Universiti Malaysia Kelantan
Kelantan, Malaysia

**Raveed Yousuf Bhat**
Chemistry Department
Aligarh Muslim University
Aligarh, India

**Vasakorn Bullangpoti**
Department of Zoology
Faculty of Science
Kasetsart University
Bangkok, Thailand

**Rosaria Ciriminna**
Istituto per lo Studio dei Materiali
    Nanostrutturati
CNR
Palermo, Italy

**Lêda R.A. Faroni**
Departamento de Engenharia Agrícola
Universidade Federal de Viçosa
Viçosa, Brazil

**Peter Follett**
USDA-ARS
U.S. Pacific Basin Agricultural Research
    Center
Hilo, Hawaii

**R.T. Gahukar**
Arag Biotech Pvt. Ltd.
Nagpur, Maharashtra, India

**A. Onur Girisgin**
Department of Parasitology
Faculty of Veterinary Medicine
Uludag University
Bursa, Turkey

**Christian David Botero Gutiérrez**
Instituto de Biotecnología y Agroindustria
Departamento de Ingeniería Química
Universidad Nacional de Colombia
Manizales, Colombia

**Khalid Haddi**
Entomology Department
Federal University of Viçosa
Viçosa, Brazil

**Gautam Handique**
Department of Entomology
Tocklai Tea Research Institute
Tea Research Association
Jorhat, Assam, India

**Farah Hossain**
Research Laboratories in Sciences Applied
    to Food
Canadian Irradiation Centre
INRS-Institute Armand-Frappier
Québec, Canada

**Efstathia Ioannou**
Department of Pharmacognosy
    and Chemistry of Natural Products
Faculty of Pharmacy
School of Health Sciences
National and Kapodistrian University
    of Athens
Athens, Greece

**Murray B. Isman**
Faculty of Land and Food Systems
University of British Columbia
Vancouver, Canada

**Majid Jamshidian**
Research Laboratories in Sciences Applied
    to Food
Canadian Irradiation Centre
INRS-Institute Armand-Frappier
Québec, Canada

**Cuthbert Katsvanga**
Department of Environmental Science
Faculty of Agriculture and Environmental
    Science
Bindura, Zimbabwe

**Zakia Khanam**
Faculty of Agro Based Industry
Universiti Malaysia Kelantan
Kelantan, Malaysia

**Aikaterini Koutsaviti**
Department of Pharmacognosy
    and Chemistry of Natural Products
Faculty of Pharmacy
School of Health Sciences
National and Kapodistrian University
    of Athens
Athens, Greece

**Monique Lacroix**
Research Laboratories in Sciences Applied
    to Food
Canadian Irradiation Centre
INRS-Institute Armand-Frappier
Québec, Canada

**Alok Lehri**
Council of Scientific and Industrial
    Research
National Botanical Research Institute
Lucknow, India

**Aurelio López-Malo**
Departamento de Ingeniería Química,
    Alimentos y Ambiental
Universidad de las Américas Puebla
Puebla, Mexico

**Ana C. Lorenzo-Leal**
Departamento de Ingeniería Química,
    Alimentos y Ambiental
Universidad de las Américas Puebla
Puebla, Mexico

**Emma Mani-López**
Departamento de Ingeniería Química,
    Alimentos y Ambiental
Universidad de las Américas Puebla
Puebla, Mexico

**Valentina Aristizábal Marulanda**
Instituto de Biotecnología y Agroindustria
Departamento de Ingeniería Química
Universidad Nacional de Colombia
Manizales, Colombia

**Rory Mc Donnell**
Department of Crop and Soil Science
Oregon State University
Corvallis, Oregon

**Francesco Meneguzzo**
Istituto di Biometeorologia
CNR
Firenze, Italy

**Sheetal Mital**
Applied Sciences
Krishna Institute of Engineering &
    Technology
Ghaziabad, Uttar Pradesh, India

**Basil K. Munjanja**
Department of Chemistry
Faculty of Natural and Agricultural
    Sciences
University of Pretoria
Pretoria, South Africa

**N. Muraleedharan**
Department of Entomology
Tocklai Tea Research Institute
Tea Research Association
Jorhat, Assam, India

**Abhishek Niranjan**
Council of Scientific and Industrial
    Research
National Botanical Research Institute
Lucknow, India

**Eugênio E. Oliveira**
Departamento de Entomologia
Universidade Federal de Viçosa
Viçosa, Brazil

**Mario Pagliaro**
Istituto per lo Studio dei Materiali
    Nanostrutturati
CNR
Palermo, Italy

**Enrique Palou**
Departamento de Ingeniería Química,
    Alimentos y Ambiental
Universidad de las Américas Puebla
Puebla, Mexico

**Panos V. Petrakis**
H.A.O. DEMETER
Institute of Mediterranean Forest
    Ecosystems and Forest Products
    Technology
Athens, Greece

**Somnath Roy**
Department of Entomology
Tocklai Tea Research Institute
Tea Research Association
Jorhat, Assam, India

**Vassilios Roussis**
Department of Pharmacognosy
    and Chemistry of Natural Products
Faculty of Pharmacy
School of Health Sciences
National and Kapodistrian University
    of Athens
Athens, Greece

**Stephane Salmieri**
Research Laboratories in Sciences Applied
    to Food
Canadian Irradiation Centre
INRS-Institute Armand-Frappier
Québec, Canada

**Shafiullah**
Chemistry Department
Aligarh Muslim University
Aligarh, India

**Ompal Singh**
Chemical Research Unit
Department of Research in Unani Medicine
Aligarh Muslim University
Aligarh, Uttar Pradesh, India

**Jun-Hyung Tak**
Faculty of Land and Food Systems
University of British Columbia
Vancouver, Canada

**S.K. Tewari**
Council of Scientific and Industrial Research
National Botanical Research Institute
Lucknow, India

**Olga Tzakou**
Department of Pharmacognosy
   and Chemistry of Natural Products
Faculty of Pharmacy
School of Health Sciences
National and Kapodistrian University
   of Athens
Athens, Greece

**Khanh Dang Vu**
Research Laboratories in Sciences Applied
   to Food
Canadian Irradiation Centre
INRS-Institute Armand-Frappier
Québec, Canada

# Section I

# Essential Oils as Green Pesticides

# 1

# Green Pesticides for Organic Farming: Occurrence and Properties of Essential Oils for Use in Pest Control

Hamir Singh Rathore

## CONTENTS

## 1.1 Introduction

Green pesticides, also called ecological pesticides, are pesticides derived from organic sources that are considered environmentally friendly and cause less harm to human and animal health, to habitats, and to the ecosystem. Green revolution is defined as an increase in crop production because of the use of new varieties of seeds, the use of pesticides, and new technologies and improved management.

Organic farming, "originally grown food," is food grown and processed using no synthetic fertilizers or pesticides. Pesticides derived from natural sources (such as biological pesticides) may be used in producing organically grown food. Botanical insecticides are becoming a key element for pest control in organic agriculture and stored products. The use of essential oils or their components adds to this natural concept, owing to their volatility, limited persistence under field conditions, and several of them having exemption from regulatory protocols.

Essential oil–based insecticides started two decades ago but have not reached their full potential.

## 1.2 Definitions

Essential oils are naturally occurring, pleasant-smelling, highly volatile liquids that are widely distributed in several plants. Eucalyptus oil, clove oil, and turpentine oil are a few examples of this class. Essential oils are produced in different parts of plants, such as buds, flower petals, bark, leaves, stems, seeds, roots, and resin or fruit rinds. They are concentrated volatile aromatic compounds that contain different functional groups, such as the alcoholic group in linalool (sandalwood oil and lavender oil), aldehyde group in citral (lemongrass oil), ester group in eucalyptus oil (wintergreen), hydrocarbon group in cymene, and phenolic group in eugenol (bay oil). An individual functional group or a set of functional groups imparts a peculiar, specific scented smell to an essential oil. A plant can be recognized or grouped on the basis of its specific scenting odor. Each and every plant species originates in a certain region of earth, acquiring particular environmental conditions, so the oil extracted from a given species possesses its own characteristic scenting smell.

Essential oils are colorless in pure state but light yellow in crude state. They are soluble in organic solvents in all proportions. They are steam volatile, with decomposition in some cases. They impart stain on paper that disappears upon warming or solvent washing.

Essential oils are frequently referred to as the "life force" of plants. A large amount (tons) of plant material is required to collect just a few hundred kilograms of oil, as each plant contains a low percentage of oil (0.01%–10%). Essential oils possess a wide range of therapeutic constituents, and these are often used for flavor, therapeutic purposes, or odoriferous characteristics in foodstuffs, beverages, medicines, and cosmetics. Pure oil is a complex mixture of certain molecules, and it cannot be duplicated. Recent investigations indicate that some chemical constituents of these oils interfere with the octopaminergic nervous system in insects. As this target site is not shared with mammals, most essential oil chemicals are relatively nontoxic to mammals and fish in toxicological tests, and they fulfill the criteria for "reduced-risk" pesticides. As these oils are used in food and beverages, they are even exempt from pesticide registration. This special regulatory status, combined with the easy availability of essential oils from plant parts, has made it possible to fast-track commercialization of oilbased pesticides. Besides their use against home and garden pests, these green pesticides may also prove effective in agricultural situations, particularly for organic food production.

It is a fact that synthetic chemical pesticides have been very effective, but the continuous resistance development is an issue for many of them. It is likely that resistance may develop more slowly to essential oil–based pesticides owing to the complex mixtures of many constituents of different functional groups. In developing countries that are rich in endemic plant biodiversity, the green pesticides may ultimately have their great impact in coming days in integrated pest management (IPM) programs due to their safety to nontarget organisms and the environment.

## 1.3 Plants and Essential Oils

Essential oils are present in different parts of plants. A brief description of oil-containing plant parts is as follows:

- *Leaves*: Basil, bay leaf, cinnamon, eucalyptus, lemongrass, melaleuca, oregano, patchouli, peppermint, pine, rosemary, spearmint, tea tree, wintergreen, thyme, and so forth
- *Flowers*: Chamomile, clary sage, clove, geranium, hyssop, jasmine, lavender, manuka, marjoram, orange, rose, ylang-ylang, and so forth
- *Peel*: Bergamot, grapefruit, lemon, lime, orange, tangerine, and so forth
- *Seeds*: Almond, anise, celery, cumin, nutmeg, and so forth
- *Wood*: Camphor, cedar, rosewood, sandal, and so forth
- *Berries*: Allspice, juniper, and so forth
- *Bark*: Cassia, cinnamon, and so forth
- *Resins*: Frankincense, myrrh, and so forth
- *Rhizome*: Ginger, and so forth
- *Root*: Valerian, and so forth

Essential oils are significantly different from fatty oils, which are also found in parts of various plants. Essential oils are used by the plants and humans for very similar purposes, such as fighting infection, initiating cellular regeneration, and working as a chemical defense against fungal, viral, and animal foes. They also contain hormone-like compounds. Despite their foliar origins, the chemical structure of essential oils is similar to that of some of the compounds found in blood and tissues, which allows them to be compatible with our own physiology.

## 1.4 Technology of Using Eucalyptus Oil

Eucalyptus oil can be applied externally and internally (inhalation) both ways. Inhalation is more effective than external use, but great care is required during inhalation to avoid harmful effects (skin irritation, etc.) caused by the excess concentration of vapors of the oil. The oil is used in body oils, compresses, cosmetic lotions, baths, hair rinses, perfumes, and room sprays, and for inhalation by steam.

For example, generally, 1%–10% solutions of eucalyptus oil are used in base oils like almond, jojoba, or kernel. A very pleasant method of inhalation may be created by using a diffuser or oil lamp for releasing eucalyptus oil vapors into a given atmosphere.

## 1.5 Aromatherapy

Aromatherapy may be defined as "the treatment of anxiety or minor medical conditions by rubbing pleasant-smelling natural oils into the skin or breathing in their smelling vapors." For example, nose infection due to cold can be cured by smelling jasmine flowers. In other words, aromatic essential oils are used to benefit the body in emotional and physical health and beauty. It has been investigated that our sense of smell plays a significant role in our overall health.

Many essential oils have medicinal properties, and they have been used in medicines since ancient times until today. Essential oils are complex mixtures, so different essential oils have diverse medicinal properties. For example, a great number of essential oils possess antiseptic properties.

### 1.5.1 History of Modern Aromatherapy

The distillation of rose essential oil was carried out first by the Persian philosopher Avicenna (980–1037 AD). He extracted the essence of rose petals through the "enfleurage process." This pioneering work and subsequent use of rose oil in perfume led him eventually to contribute a book on the healing properties of essential oil of rose [1].

In the early twentieth century, a French chemist, Rane-Maurice Gattefosse, began a new chapter in science, which he called "aromatherapy." He was struggling with his studies, and his arm was suffering after a burning accident in his laboratory. After several accidental burnings, he thrust his arm into the nearest available liquid, which happened to be a tub full of lavender oil. The quick healing of his arm by this oil surprised and impressed

Dr. Gattefosse, so he spent his life in researching the value of essential oils [2]. His work made aromatherapy popular and well known in Europe.

### 1.5.2 Mechanisms of Aromatherapy

An essential oil, either inhaled or taken by the olfactory system directly, goes to the limbic system of the brain. The brain responds to the particular scent and then affects our emotions and chemical balance. The essential oils are absorbed by the skin and carried throughout the body parts (internal organs) via the circulatory system. Due to the uniqueness of each person's system, one particular oil sample may not suit all people; that is, the benefits depend on the unique nature of an individual's response to an aromatic stimulus. Thus, by choosing one or more oils, one can experience beneficial effects promoting overall health and even specific targets.

## 1.6 Pharmacological Properties of Essential Oils

Essential oils possess many pharmacological properties. Some properties are discussed briefly in the following sections.

### 1.6.1 Antiseptic

Essential oils are active against a wide range of bacteria and also against antibioresistance strains. They are also known to control fungi and yeasts (*Candida*). Cinnamon, thyme, clover, *Eucalyptus*, culin savory, lavender, citral, geraniol, linalool, and thymol are well-known antiseptics. They are much more potent than phenol.

### 1.6.2 Expectorant and Diuretic

The external application of essential oils like "L'essence de terebenthine" increases microcirculation and provides a slight local anesthetic action. Some essential oils are used in ointments, creams, and gels that are effective in relieving sprains and other articular pains. Eucalyptus or pine oils are administered orally to stimulate ciliated epithelial cells to secrete mucus. They are known to affect the renal system in order to increase vasodilatation and, as a result, bring about a diuretic effect.

### 1.6.3 Spasmolytic and Sedative

The essential oils from the Umbelliferae family (*Mentha* species and *Verbena*) are documented to decrease or eliminate gastrointestinal spasms. They are known to increase secretion of gastric juices. They are reported to be effective against insomnia.

### 1.6.4 Other Related Properties

Some essential oils are known to be anti-inflammatory and cicatrizing and acts as cholagogues.

## 1.7 Pesticidal Properties

A literature survey points out that the plants of the Myrtaceae, Lamiaceae, Asteraceae, Apiaceae, and Rutaceae families are thoroughly investigated for anti-insecticide activities against specific insect orders like Lepidoptera, Coleoptera, Diptera, Isoptera, and Hemiptera [3]. The following essential oils have been reported for their insecticidal activities: essential oils of *Artemisia* species are reported for vapor toxicity and repellent activities against coleopteran beetle, and essential oils of *Cinnamomum camphora, C. cassia,* and *C. zeylanicum* for repellent action against mosquitoes. The toxic effect of essential oil of *Curcuma zedoaria* (with an $LC_{50}$ of 5.44–8.52 µg/mg) was determined against mosquito adults. Both field and laboratory experiments showed that β-ocimene is a good repellent to the leaf cutter ant and *Atta cephalotes*. The aphid can be captured in traps baited with carvone, which occurs in the essential oils of several plants of the Apiaceae. Linalool is reported to be a good repellent of aphids. A monoterpene, β-Thujaplicin is toxic to larvae of the old house borer. Cineole, geraniol, and piperidine extracted from bay leaves are good repellents of cockroaches. Linalool extracted from sour oranges is toxic to adult bean weevils, but it is attractant to male Mediterranean fruit flies.

There is a growing interest in this area. Koul et al. [4] have reported that many plant essential oils show a broad spectrum of activity against pest insect and plant pathogenic fungi, including insecticidal, antifeedant, repellent, oviposition deterrent, growth regulatory, and antivector activities. These oils also have a long tradition of use in the protection of stored products. Some constituents of these oils also interfere with the octopaminergic nervous system in insects. As this target site is not shared with mammals, most essential oil chemical constituents are relatively nontoxic to mammals and fish in toxicological tests, and meet the criteria for reduced-risk pesticides. Some of these oils are widely used as flavoring agents in foodstuffs and beverages and are even exempt from pesticide registration. This special regulatory status, combined with the wide availability of essential oils from the flavor and fragrance industries, has made it possible to fast-track commercialization of essential oil–based pesticides. They may be an integral part of future integrated pest management programs due to their safety to nontarget organisms and the environment.

A team of researchers [5] working on biofertilizers and biopesticides has published a very clear-cut point of opinion. They have reported that the loss during growing crops and postharvest handling, processing, storage, and distribution is 20%–60% of oil, depending on the facilities available, such as technical know-how provided by the government and the skill of users. Three major groups of enemies are fungi, insects, and rodents. Many synthetic insecticides are in use for the protection of stored cereals and pulses. The growing awareness of environment pollution and health hazards due to synthetic pesticides has prompted a search for alternative pesticides. Researchers and users are trying green pesticides and biopesticides for grain storage purposes, that is, for organic farming. One of the alternatives is oils from plant origin that have been found to possess pesticidal properties. They are easy to apply, leave harmless residues, and are safe to natural enemies of pests. However, a serious disadvantage of oil pesticides is they have an adverse effect on the germination power of the treated seeds.

Pugazhvendan et al. [6] have tested the insecticidal and repellent activities of five plant oils, *Citrus autantium, Cinnamomum zeylanicum, Gaultheria fragrantissima, Lavandula officinalis,* and *Ocimum sanctum,* against *Tribolium castaneum* by using a standard protocol *in vitro*. The maximum repellent activity was found to be in tulsi oil (0.5 µg). Wintergreen oil and lavender oil give a good response at a higher concentration (10 µl). The maximum percentages of mortality were found in tulsi oil and wintergreen oil at 76%–92% and 86%, respectively. These results suggest the presence of active principles in the plant oils.

From another part of Asia, Tariq et al. [7] studied the Pakistanian *Acorus calamus* (Araceae), a locally available plant. The common name of *A. calamus* is sweet flag. They analyzed the essential oil obtained from *A. calamus* and found that it is a complex mixture of seven major compounds. This oil prevents cuts and wounds from fungal growth, and cuts and wounds heal rapidly compared with a control. Systematic control of root knot nematode was also recorded in cotton and brinjal plants by using 0.25% solution. The scale insects were also controlled by the same systematic method by using 0.5% solution in the infected cotton plants. Control of mealy bugs on cotton brinjal and *Abutilon indicum* was achieved by a 0.5% solution spray, repeated weekly for 1 month in Sindh and Baluchistan. They recommended this oil as a biopesticide in the agriculture and health sectors.

## 1.8 Chemical Composition of Essential Oils

A pure essential oil is a complex mixture of more than 200 chemical compounds. Normally, it is a mixture of terpenes or phenylpropanic derivatives, in which structural differences between compounds are minimal. The components of an essential oil can mainly be classified into two groups:

1. *Volatile fraction.* Essential oil (90%–95% by weight) consists of hydrocarbons and terpenes and their oxygenated derivatives, such as camphor, along with acids, aliphatic aldehydes, ketones, alcohols, and esters.
2. *Nonvolatile residue.* Essential oil consists of hydrocarbons, fatty acids, carotenoids, sterols, waves, and flavonoids. They comprise at maximum 10% weight of the oil.

### 1.8.1 Volatile Fraction

#### 1.8.1.1 Hydrocarbons

Essential oils consist of chemical compounds containing hydrogen and carbon as their building blocks. The basic hydrocarbon found in plants is isoprene, $CH_2=C(CH_3)CH=CH_2$. It is also called hemiterpene. A fascinating area of research linking organic chemistry and biology is the study of the biogenesis of natural products: the detailed sequence of reactions by which a compound is formed in living systems, plant or animal. All the isoprene units in nature, it appears, originate from the same compound, "isopentenyl" pyrophosphate.

#### 1.8.1.2 Terpenes

When two molecules of an isoprene unit join head to tail, the result is a monoterpene. Similarly, when three units join, it is a sesquiterpene, and when four units join, a diterpene [8–10].

##### 1.8.1.2.1 Monoterpenes

Monoterpenes are naturally occurring compounds consisting of 10 carbon hydrocarbons, the majority of which are unsaturated. Oxygenated derivatives of monoterpenes, such as alcohols, ketones, and carboxylic acids, are known as monoterpenoids. Monoterpenoids are the most representative molecules, constituting 90% of the essential oils, and allow a great variety of structures with diverse functions. The branched-chain 10 carbon hydrocarbons are composed of two isoprene units and are widely distributed in nature. More than 400 naturally occurring monoterpenes have been identified. Monoterpenes are linear derivatives, such as geraniol and citronellol; they can be cyclic molecules, such as menthol (monocyclic), camphor (bicyclic), pinenes, and (α,β)-pinene genera, as well. Thujone (monoterpene) is a toxic agent found in *Artemisia absinthium* (wormwood) from which the liqueur absinthe is made. Boreol and camphor are two common monoterpenes. Borneol, derived from pine oil, is used as a disinfectant and deodorant. Camphor is used as a counterirritant, anesthetic, expectorant, and antipruritic. Camphene and pinene are present in cypress oil. Similarly, camphene, pinene, and thujene are available in black pepper.

### 1.8.1.2.2 Sesquiterpenes

The sesquiterpenes (*sesqui-* = "one and a half") are hydrocarbons of formula $C_{15}H_{24}$, or derivatives of them, which are very common in essential oils of plants. Sesquiterpenes are biogenetically derived from farensyl pyrophosphate and in structure may be linear, monocyclic, or bicyclic. They constitute a very large group of secondary metabolites; some of them are reported to be stress compounds formed as a result of disease or injury. Sesquiterpenes are anti-inflammatory, antiseptic, analgesic, and antiallergic.

### 1.8.1.2.3 Sesquiterpene Lactones

More than 500 compounds of this group are known. They are particularly characteristic of the Compositae but do occur sporadically in other families. They have been found to be of interest from a chemical and chemotaxonomic viewpoint. But they also possess many antitumor, antileukemia, cytotoxic, and antimicrobial activities. They may be responsible for skin allergies in humans, and they may also act as insect-feeding deterrents. Their classification is based on their carboxylic skeletons. For example, guaianolides, pseudoguaianolides, eudesmanolides, eremophilaolides, and xanthanolides can be derived from germacranolides. Lactones, farnesene, and β-caryophyllene are present in chamomile and lavender extracts and in basil and black pepper extracts, respectively.

### 1.8.1.2.4 Diterpenes

Diterpenes are made up of four isoprene units. This molecule is too heavy to steam volatile in steam distillation process (diterpenes are high molecular weight compounds, so they may not be distilled by steam distillation), so it is rarely found in steam-distilled essential oils. Diterpenes occur in all plants and consist of compounds having a $C_{20}$ skeleton. About 2500 diterpenes are known (of 20 major structural types). Well-known diterpene derivatives are plant hormones, gibberellins, and phytol occurring as a side chain on chlorophyll. The biosynthesis occurs in plastids, and interestingly, mixtures of monoterpenes and diterpenes are the major constituents of plant resins. The diterpenes arise from metabolism of geranyl geranyl pyrophosphate (GGPP) in a manner similar to that of monoterpene. Sclareol in clary sage is an example of a diterpene alcohol. Diterpenes are antifungal, expectorant, hormonal balancers and hypotensive. They have limited therapeutic importance, and they are used in certain sedatives (coughs), as well as in antispasmodic antioxiolytics.

## 1.8.1.3 Alcohols

Alcohols exist naturally, as either free compound or part of terpenes. When a hydroxyl group is attached to terpene, the product is an alcohol. Alcohols have a very low or negligible toxic reaction in the body or on the skin. Therefore, they are considered safe to use. Examples of acyclic alcohols are linalool, geraniol, and citronellol. Cyclic alcohols are menthol, isopulegol, and terpineol. Bicyclic alcohols are borneol and verbenol. Other hydroxyl compounds, like phenols, are thymol and carvacrol. Linalool is found in ylang-ylang and lavender. Geraniol is found in geranium and rose. Nerol is found in neroli.

### 1.8.1.4 Aldehydes

Some of the common examples of aldehydes are citral and citronellal. Citral (geranial) is the aldehyde corresponding to geraniol, from which it may be prepared by careful oxidation. Citral imparts to oil of lemon its characteristic odor and is an important constituent of orange, mandarin, and certain kinds of eucalyptus oil. It is very abundant (70%–80%) in lemongrass and lemon balm. Citronellal is found in lemongrass, lemon balm, and eucalyptus. Aldehydes are antifungal, anti-inflammatory, antiseptic, antiviral, bactericidal, disinfectant, and sedative. Essential oils containing aldehydes are effective in treating *Candida* and other fungal infections.

### 1.8.1.5 Acids

Organic acids such as chrysanthemic acid, cinnamic acid, citric acid, and lactic acid in their free state are generally found in very small quantities in essential oils. They possess anti-inflammatory properties. They also act as components or buffer systems to provide a constant pH.

### 1.8.1.6 Esters

Esters such as linalyl acetate, geranyl formate, and methyl salicylate are found in plants. Linalyl acetate and geranyl formate are found in bergamot and geranium, respectively. Methyl salicylate is found in birch and wintergreen, which is toxic within the system. Essential oils containing esters are used for their soothing and balancing effects. Esters are effective antimicrobial agents. They are used medicinally as antifungals and sedatives, with a balancing action on the nervous system. Esters as such are free from precautions.

### 1.8.1.7 Ketones

Ketones such as carvone, menthone, and thujone are found in essential oils. Ketones assist the flow of mucus and ease congestion, so they are used for upper respiratory complaints. Essential oils containing ketones are beneficial for promoting wound healing and encouraging the formation of scar tissue. A few ketones, such as jasmine in jasmine oil, fenchone in fennel oil, carvone in spearmint and dill oil, and menthone in peppermint oil, are nontoxic. But generally, ketones are very toxic. The most toxic ketone is thujone, which is found in mugwort, sage, tansy, thuja, and wormwood oils. Other toxic ketones found in essential oils are pulegone in pennyroyal and pinocamphone in hyssops.

### 1.8.1.8 Lactones

Lactones are cyclic esters, which may be obtained by dehydration of a γ- or δ-hydroxy acid. Treatment of lactone with base rapidly opens the lactone ring to give the open-chain salt. Lactones are anti-inflammatory, antiphlogistic, expectorant, and febrifuge. The essential oils containing lactones are known to be particularly effective for their anti-inflammatory action, possibly by their role in the reduction of prostaglandin synthesis and expectorant actions. Lactones possess stronger expectorant action than ketones.

### 1.8.1.9 Oxides

Oxides such as cineole and camphor are also the constituents of essential oils. 1,8-Cineole, $C_{10}H_{18}O$, boiling point (b.p.) 174.4°C, occurs in eucalyptus oil. There is also 1,4-cineole, which also occurs naturally [11]. Camphor is obtained by distilling the wood and leaves of camphor laurel with steam. Camphor is used as a moth repellant, as a preservative in cosmetics, in medicine, and as a plasticizer in the manufacture of celluloid, smokeless powder, and photographic films. The essential oils containing camphor are useful for medicinal purposes.

## 1.9 Mode of Action of Essential Oils

Essential oils are volatile complex mixtures of several compounds, so many of their chemical constitutes serve as chemical messengers for insects and other animals [12]. Efforts have been made to elucidate the target sites and mode of action for the monoterpenoids, and limited success has been obtained. Most monoterpenes are cytotoxic to plants and animal tissues, causing a drastic reduction in the number of intact mitochondria and Golgi bodies, impairing respiration and photosynthesis and decreasing cell membrane permeability. It is also known that most monoterpenes serve as a signal of relatively short duration, making them especially useful for synomones (and alarm pheromones).

### 1.9.1 Insecticidal Action

Evans has demonstrated that a monoterpenoid (linalool) acts on the nervous system by affecting ion transport and the release of acetylcholine esterase in insects [13]. In insects, octopamine acts as a neurotransmitter, neurohormone, and circulating neurohormone–neuromodulator. When octopamine interacts with at least two classes of receptors, octopamine-I and octopamine-II, it exerts its effects [14]. The interruption in the functioning of octopamine results in a total breakdown of the nervous system in insects. Therefore, the octopaminergic system of insects represents a biorational target for insect control. The lack of octopamine receptors in vertebrates likely accounts for the profound mammalian selectivity of essential oils as insecticides. Many constituents of essential oils have been demonstrated to act on the octopaminergic system of insects [15]. Enan [16] has also shown that eugenol mimics octopamine in increasing intracellular calcium levels in cloned cells from the brain of *Periplaneta americana* and *Drosophila melanogaster*, and this was also found to be mediated via octopamine receptors. It has also been reported that the toxicity of eugenol was increased in mutant *D. melanogaster* that were deficient in octopamine synthesis, suggesting that the toxicity is mediated through the octopaminergic system, although this was not the case for geraniol. It was suggested that these cellular changes induced by eugenol are responsible for its insecticidal properties. Another research group [17] reached a similar conclusion, suggesting possible competitive activation of octopaminergic receptors by essential oil. They recorded significant effects at low concentration in the abdominal epidermal tissue of *Helicoverpa armigera*.

## 1.9.2 Repellent Action

It is known that the repellent molecules interact with the female mosquito olfactory receptors and thereby block the sense of smell. It is also known that the hairs on the mosquito antennae are temperature and moisture sensitive. Little is known about the receptors responsible for the repellent response in cockroaches. Literature [18] shows that oleic acid and linoleic acid are listed in death recognition and death aversion (repellency) in cockroaches, and the term *necromone* has been proposed to describe a compound responsible for this type of behavior. Essential oils seem to have good repellent potential, as they possess a characteristic smell and volatile nature. Unfortunately, no appreciable work has been reported in this direction.

## 1.9.3 Fumigant Action

Many essential oils have been claimed to be used as fumigants [19]. Essential oils obtained from *Artemisia annua*, *Anethum sowu*, *Curcuma longa*, and *Lippia alba*, and isolates such as d-limonene, carvones, and 1,8-ceneole, are well documented as fumigants. It seems that the route of action for these oils is in the vapor phase via the respiratory system. Unfortunately, the exact mode of action is unknown.

## 1.10  Phytotoxicity and Safety

Although commonly considered safe, oils can injure susceptible plants. The effects of plant injury or phytotoxicity may be acute or chronic. The injury may include leaf scorching and browning, reduced flowering, and stunted growth. Phytotoxicity may be associated with plant stress, ambient temperature and humidity, application rate, and nature of formulation. It may vary among plant species and cultivars. However, the plant phytotoxicity of essential oils is lower than that of mineral oils. Many of the essential oil formulations, for example, a natural repellent such as eucalyptus oil reported for use against mosquitoes, can serve as attractants to another blood-feeding pest, namely, midges. For a long time, the oil of the tea tree, *Melaleuca alternifolia*, has been used in Australia as a traditional remedy for insect bite relief. It has now been found to be responsible for contact dermatitis in humans [20]. Oil degradation products such as peroxides, epoxides, and endoperoxides may be responsible for this erratic behavior. Another constituent, d-limonene, of essential oils of lemon, orange, mandarin, and grapefruit, has been recommended as a pesticide for the indoor control of pests. It has also been reported that d-limonene causes dermatitis [21]. An active ingredient of *Artemisia absinthium*, thujone is a potent neurotoxin impacting the γ-amino butyric system [22]. In addition, a major constituent of *Mentha pulegium* oil, pulegone (insecticide), is oxidized upon ingestion by the cytochrome P-450 system into toxic metabolites, including methofuran [23]. These metabolites bind to proteins to cause loss of oxygen of organ function, seizures, acute poisoning, and death. These problems commonly occur whenever pesticides are used. Therefore, the following safety measures may be observed. When pesticidal oils are to be used, it is especially important to read and follow the label instructions and recommendations. Manufacturers are assumed to provide useful information based on extensive testing. In the case of repellent applications, the repellent formulations may be applied at levels lower than those of compounds that have been found to be acutely toxic, thereby lowering the pesticide load on the environment. Oils may separate from the carrier, so agitation is necessary to keep oil in the solution. Oils evaporate quickly and do not generally contaminate the environment, including soil and water, resulting in minimum risk to nontarget organisms and the environment. A proper storage procedure may be followed to preserve essential oil pesticides.

## 1.11 Indoor Reactions

Unsaturated monoterpenes interact with oxidants such as ozone, hydroxyl, and nitrate radicals in common places, and they produce a variety of secondary organic pollutants in the gaseous and particulate phase. In places like modern India (Hindustan), where over-crowded small houses (flats) are fully equipped with electrical and electronic devices and have limited cross-ventilation, the indoor air is ionized. In ionized air, the reactions discussed above are prone to occur. The oxidation products, such as d-limonene, α-pinene, and linalool, including aldehydes, ketones, and organic acids, have been tested in indoor air [24–26]. One of the major pollutants in the above reactions is formaldehyde. The fine particulate particles penetrate into the lower respiratory system, and they also cause skin irritation [27]. Therefore, attention is required, especially when essential oils are exposed to higher temperature during uses and storage.

## 1.12 Synergistic Formulations

A synergist is also called an activator or adjuvant. *Synergetic* or *synergic* means working together. These terms are derived from the Greek word *synergid*, which means cooperation (*syn* = "together," *ergon* = "work"). Synergists are chemicals that have little or no insecticidal activity of their own but, when added to an insecticide, enhance its toxicity manifold. Thus, the synergistic formulation or combination is to reduce the dose of insecticide and thereby reduce the risk of developing resistance and environmental pollution. A broad array of pest-repellent products, including homemade herbal teas, plant extracts, fermentation products (vinegar), and industrial clay and rock powder products (kaolin), are being used in home gardens and organic farming. Now, the use of homemade products is fast declining because of the commercialization of standardized industrial products. The damage to cotton by the bollworm, *Helicoverpa armigera*, can be controlled by formulation of conventional insecticides at 50% of the recommended concentration by combining extracts of three local plants (*Azariracthta indica*, *Khaya senegalensis*, and *Hyptis sauveolens*). This admixture or formulation provides greater efficacy than that of the insecticide alone. It has been reported that low pH and salinity (5%) significantly affect the activity of essential oils such asthyme, anise, and saffron [28]. A proprietary monoterpene mixture containing 0.9% active ingredient has been developed for use against foliar feeding pests [29]. EcoSMART Technologies has developed a distinct combination of different essential oils that appreciably enhances the activity of these oils against insects (EcoPCO).

## 1.13 Structure–Activity Relationships

Recently developed computer-based theoretical modeling methods, such as quantitative structure–property relationship (QSPR) analysis and structure–activity relationship (SAR) analysis, are applied to the prediction and characterization of chemical toxicity of essential oils [30]. Ngoh et al. [31] have evaluated nine essential oil constituents composed of benzene derivatives and terpenes for toxic properties against *P. americana*. They found that benzene derivatives, namely, eugenol, isoeugenol, methyl eugenol, safrole, and isosafrole, are generally more toxic and a better repellent of *P. americana* than the terpenes, such as cineole, limonene, p-cymine, and α-pinene. The distance of the side-chain double bond from the aromatic ring and the substitution of the methoxy group may be responsible for their toxicity and repellency. Similarly, Aggarwal et al. [32] have evaluated L-menthol extracted from *Mentha arvensis* and its seven acyl derivatives for their bioactivities against four stored product insects. They found that the number of methyl groups in the side chain is responsible for the better activity of the derivatives than that of L-menthol. (L-menthol crystals by Sigma-Aldrich are available.)

## 1.14 Essential Oil–Based Commercial Products

In spite of favorable mammalian toxicity of essential oils and their constituents, surprisingly few pest control products based on plant essential oils appear in the market [33–36]. This may be a consequence of regulatory barriers to commercialization (cost of toxicological and environmental evaluation) or the fact that the efficacy of essential oils toward pests and diseases is not apparent or as obvious as that seen with currently available products. Two American companies have recently produced the following products:

- Mycotech Corporation: An aphidicide–miticide–fungicide, emulsifiable concentrate (EC) formulation, which consists of cinnamon oil with 30% cinnamaldehyde for greenhouse and horticultural use and for bush and tree fruits
- EcoSMART Technologies
  - EcoPCO, which contains eugenol and 2-phenyl propionate for controlling crawling and flying insects
  - EcoTrol Plus, an insecticide–miticide, which contains rosemary oil for horticultural crops
  - Sporan, which contains rosemary oil for fungus
  - Matran™, which contains clove oil for weed control

Many commercial products, like Buzz Away®, contain oils of citronella, cedar wood, eucalyptus, and lemongrass. Green Ban® contains oils of citronella, cajuput, lavender, safrole-free sassafras, peppermint, and bergaptene-free bergamot oil. Skin-So-Soft contains some essential oils and stearates. Similarly, many other formulations have been developed by American and European companies on a small scale that contain either garlic oil or mint oil for pest control in the home and garden, or menthol for tracheal mite control in beehives. Italy has developed a formulation, ApilifeVar™, containing menthol and lesser amounts of cineol, menthol, and camphor, to control varroa in honeybees. In Asian countries, several blended formulations containing pleasant-smelling essential oils and chemical pesticides are used. Research work on new formulations and new technologies for storage and usage is being carried out in many laboratories in Canada, France, and Israel. The scarcity of the natural resource, the need for chemical standardization and quality control, and difficulties in registration are barriers to the commercialization of new products.

## 1.15 Prospects of Essential Oils

A literature survey from 2004 to 2014 shows that several symposia [37,38] have been organized and numerous patents [39,40], books [41–43], and reviews [3,4,44,45] have been published on the preparation of admixtures (formulations) and their monitoring for different domestic and agricultural purposes, and on the fabrication of equipment such as sprayers for the application of liquid formulations of essential oil pesticides. The following patents reflect the genuineness of the research work:

- A formulation containing eucalyptus oil, pyrethroid oil, and borax is used for wood preservation and other building materials [46].
- Veneer-faced panels can be protected against insects by deep impregnation of the polymer layer with *Thujopsis dolobrata* essential oil [47].
- The biorational repellent based on nepetalactone and dihydronepetalactone from *Nepeta cataria* is used against cockroaches, mosquitoes, mites, ticks, and other household insects [48].
- Similarly, nootkatone from vetiver oil and its derivatives tetrahydronootkatone and 1,10-dihydronootkatone are used as a repellent against mosquitoes, cockroaches, termites, and ants [49].
- Thyme oil and monoterpenoids like thymol, anethole, eugenol, and citronellal are used against cockroaches and green peach aphids [50].
- Citronellal, citronellol, citronellyl, or a mixture of these is used against the human louse [51].

Recent patents involving essential oils show that the majority of the inventions are focused on household uses. Some patents are in the area of the protection of domestic animals. Many patents are related to the preservation of cloth from destruction by moths and beetles. A good number of patents are related to the protection of agricultural produce, foodstuffs, and building materials.

## 1.16 Global Essential Oil Pesticide Market

Annual sales of essential oils are estimated to be US$700.00 million and a total world production of 45,000 tons. About 90% of this production is focused on mint and citrus plants [45]. The global market for essential oils is increasing day by day. This is due to the fact that essential oils are considered a new class of ecological products for controlling pests. Therefore, the production of safe and scientifically approved herbal products may be given priority. There has been a regular growth of 7%–10% in the production of crop-protecting products during the last 40 years. It comes out to be a total turnover of $25 billion. The growth is partly due to a deep reorganization of the industrial sector because of takeovers and amalgamations of companies and partly due to modifications in the number and nature of commercialized pesticidal molecules. The expensive cost of commercialization licenses and the numerous cases of ecotoxicity or toxicity of pesticides to mammals that occurred in the past are responsible for this change.

One more economic aspect, the continuous and bulk requirement of raw materials, is important to producers. In order to fulfill this requirement, large-scale cultivation is required, which in turn generates good business opportunities and human resource development.

## 1.17 Registration of Green Pesticides

The plant protection process is profit-induced poisoning of the environment. Therefore, policies to reduce the use of plant protection products, especially chemical pesticides, are being developed globally. On the other hand, to feed the ever-growing world population, plant protection and food preservation agents are genuinely needed. In order to meet such aims and objectives, viable low-risk alternative products need to be developed and authorized by their proper registration. At present, the regulatory procedures based on synthetic chemical pesticides have been regarded as a barrier to the commercialization of green pesticides. Steps are being taken to encourage the development of green pesticides by proposing reduced data requirements. The Organisation for Economic Cooperation and Development (OECD), under the Pesticide Programme, has been working on the harmonization of its plant protection product review procedures, sharing the evaluation of plant protection products and proposing policies for the reduction of risks associated with plant protection product use. The OECD has maintained its active role in this area, developing a guidance document for the preparation of a dossier to support pheromones and other semiochemicals [52]. They recognize that semiochemicals act by modifying the behavior of pest species rather than killing them and may be target specific. They are used at very low rates, are nontoxic, and dissipate rapidly. Therefore, they can be regarded as low risk, and the data requirements are not as enormous as for synthetic chemical pesticides.

It has been proposed that semiochemicals may be exploited as pest management agents by synthesizing their derivatives and analogues using simple procedures and environment-friendly, as well as remunerative, approaches [53].

The naturally available terpenoids can be derivatized to molecules possessing pesticidal characteristics by using the structure–activity relationship approach. Monoterpenoids can be selected as lead compounds for the production of new pest management agents because of their bulk availability in plants [54].

## 1.18 Extraction Methods

Many techniques are available for the extraction of essential oils. One technique may produce nicer-smelling oil, while another yields oil with greater aromatherapeutic value. It turns out that essential oil production, like wine making, is an art form as well as a science. Therefore, the procedure applied for oil extraction from plant is important because some processes use solvents that can destroy the therapeutic properties. Some plants, especially flowers, do not lend themselves to steam distillation. Their fragrance and therapeutic essences cannot be completely released by water. For example, jasmine oil and rose oil are often found in "absolute" form. Thus, selection of the method is based on the experience of the user, as well as the application of the product. Therefore, each method has its own merit and importance in the processing of aromatherapy-grade essential oil. A few techniques are described briefly:

- *Maceration*: The plant is soaked in vegetable oil, heated, and strained at a temperature suitable for massage. It gives infused oil rather than an essential oil. It is one of the oldest herbal oil massage treatments.

- *Expression method*: In the expression method, the fruit is halved, placed in water for some time (5–6 hours), and then pressed against a sponge to eject oil.

- *Cold pressing*: This method is used to extract the oil from the rind of citrus, such as orange, lemon, grapefruit, and bergamot, by simply pressing the rind. The rind may be ground or chopped, and then be pressed to obtain a good yield. Thus, the product is a watery essential oil, and the water may be removed as an aqueous layer by keeping the sample for some time. The product obtained by this method possesses bright, fresh, uplifting aromas that smell similar to those of ripe fruit. However, the products have a relatively short life (±6 months).

- *Distillation method*: Essential oils are steam-volatile compounds, so the steam distillation is an affordable, readily available, and commonly used method for their extraction from plant material. To streamline the functioning and cost of the equipment, several modifications are made in the preliminary equipment of steam distillation. Some of the techniques are

  - *Hydrodistillation*: This is a simple method that is still used in places where sophisticated and costly equipment is not available; it is used in third world countries. The weakness of the method is that it can run dry i.e., no proper way to check water level or water presence or be overheated, burning the aromatics and resulting in an essential oil with a burnt smell. Hydrodistillation (HD) is generally used for spice powders and ground wood, roots, or nuts.

  - *Steam distillation*: This is the most commonly used technique for the extraction of essential oils. In this method, the distillation is carried out under atmospheric pressure and then the condensate is fractionated. Generally, plants or flowers are placed on a screen, and then steam is passed through the biomass. The steam is charged with the essence, and then steam passes through the area, where it cools and condenses. The aqueous layer containing essential oil is separated and bottled. As the plants contain essential oil in traces, several hundred pounds may be needed to produce a single ounce.

  - *Turbo distillation*: Another commonly used technique, turbo distillation extraction is used for hard-to-extract or coarse plant material, such as bark, roots, and seeds. In this method, the plant material is soaked in water and then the steam is circulated through the soaked material. The water and steam are continually recycled through the plant material in order to perform rapid and efficient extraction.

- *Cold fat extraction or enfleurage method*: In cold fat extraction method, an extracting mixture that consists of tallow (one part), lard (two parts), and benzoin (0.6%) is spread in thin layers on rectangular wooden frames over which flower petals are distributed and allowed to remain there for about 8–10 weeks. During this period, the fat becomes saturated with oil. The alcoholic solution of the fat extract deposits on cooling, leaving oil in the solution. The solvent is removed by evaporation and the oil is recovered.

- *Extraction with volatile solvent*: The extraction is carried out with volatile solvent, such as petroleum ether, and the solvent is removed by evaporation. The oil is generally obtained in a semisolid form because of accompanying wax. The residue is treated with alcohol and filtered, and the alcohol is removed to obtain the oil. This method is not considered to be the best extraction method due to the fact that traces of solvent may be as impurities, which could cause allergies and affect the immune system.

- *Supercritical carbon dioxide extraction*: The recently developed techniques are carbon dioxide fluid extraction (CFE) and supercritical carbon dioxide fluid extraction (SCFE). Both techniques involve the use of carbon dioxide as the solvent, which carries the essential oil from the raw plant material. The low-pressure carbon dioxide extraction involves chilling carbon dioxide (35°F–55°F), which is pumped through the plant material at about 1000 psi. The carbon dioxide under these conditions condenses to a liquid. SCFE involves carbon dioxide heated to 87°F and pumped through the plant material at around 8000 psi. The carbon dioxide becomes a dense fog or vapor under these conditions. In both processes, when pressure is released, the carbon dioxide escapes as a gas, leaving the essential oil behind. These are ultrasensitive and high-priced methods, as they are working at low temperature and low pressure. However, they give good-quality essential oil, possessing the essence, as well as the therapeutic value, of the original plant.

- *Solvent-free microwave extraction*: Solvent-free microwave extraction (SFME) is a combination of microwave heating and fry distillation, performed at atmospheric pressure without adding any solvent or water [55]. The isolation and concentration of volatile compounds are performed by a single stage. SFME has been compared with a conventional technique, hydrodistillation, for the extraction of oil from three aromatic herbs: basil (*Ocimum baslicum*), garden mint (*Mentha crispa*), and thyme (*Thymus vulgaris*). The essential oils extracted by SFME for 30 minutes were quantitatively (yield) and qualitatively (aromatic profile) similar to those obtained by conventional hydrodistillation for 4.5 hours. The SFME method yields an essential oil with higher amounts of more valuable oxygenated compounds, and allows substantial savings of cost, in terms of time, energy, and plant material. SFME is a green technology and appears as a good alternative for the extraction of essential oils from aromatic plants. However, no comments have been made about the applicability of this technique on the spot, that is, plant-growing agricultural field where sophisticated facilities are not available or have not been installed so far, especially in the third world countries.

## 1.19 Methods of Analysis

Some new techniques have been developed for the estimation of essential oils. Most of the already known methods have been modified for use in essential oil analysis. A brief description of the currently used methods is given.

### 1.19.1 Gas Chromatography

Gas chromatography (GC) with a fused-silica capillary column and flame ionization detector (FID) is the routinely used method for use in essential oil analysis. In third world countries, this instrument in installed in the market of small places (in India, gas chromatography is used to fix the price of the oil; prior to the sale of oil, its chromatograph is seen, i.e., analyze the oil by gas chromatography to fix its price) to evaluate the price of essential oils. GC with capillary columns is quick in producing results. It can easily detect and identify major components of the oil, and gives good authenticity of the oil. Although the technique has no limitations, many minor components of essential oil (<0.01%) do not register on the chromatogram. The packing materials of the column are variable, depending on the polarity of the components to be separated, which varies from sample to sample. The chiral phases (mostly based on cyclodextrin derivatives) provide better resolution of the enantiomers of a volatile complex mixture because of the large changes in solute relative retention time. Gas chromatography with mass spectrometry (GC-MS) gives better resolution than GC alone. The information obtained from GC-MS analysis is sufficient to estimate whether the product is genuine. If the product is adulterated, the nature and level of adulteration is detected. A selective and accurate separation is absolutely necessary to decide the price of oil in industries [56,57].

### 1.19.2 Thin-Layer Chromatography

Thin-layer chromatography (TLC) is a simple and inexpensive tool of analysis. It is used in laboratories where either sophisticated or costly instruments have not been installed or repair facilities are not available. TLC can be installed with a minimum budget and operated by a semiskilled technician. It gives fairly good results for semiquantitative analysis. Now, it has been replaced by high-performance thin-layer chromatography (HPTLC), which gives accurate and reproducible results. TLC can be applied to analyze volatile essential oils after their conversion to nonvolatile substances by derivatization. HPTLC, with all its advantages, may be seen as a complementary technique, but a standard method suitable for all essential oils is still missing. HPTLC is a rapid, reliable, cost-effective, and extremely flexible method of analysis. The results are fast and clearly visible. With the introduction of automated equipment, chromatograms can be well documented, and this method fulfills the good manufacturing practices (GMP) guidelines [58,59].

### 1.19.3 High-Performance Liquid Chromatography

High-performance liquid chromatography (HPLC) is a modern, sophisticated, and ultrasensitive technique, but it is costly. It gives better resolution and is used to identify the minor constituents of essential oils that are not registered by GC-MS. HPLC is one of the most precise techniques for quantitative determination of plant constituents (complicated matrixes). A validated HPLC method for the assay of thymol and carvacrol in *T. vulgaris* oil has been introduced, and concentrations of these two phenolic compounds in the essential oil obtained by HPLC are compared to those obtained by the GC method [60]. The analysis of orange and mandarin essential oils (nonvolatile fraction) was carried out by HPLC in normal and reverse phase modes with ultraviolet (UV) and spectrofluorimetric detection in series. For the identification of chromatographic peaks, a preparative HPLC was used, and the purified fractions were analyzed by GC-MS and by LC-MS. Some flavones were identified in these oils [61].

### 1.19.4 Current Methods

The limitations of GC and GC-MS have pushed chromatographers to dig deep in search of better methods to analyze essential oil volatiles, such as improvement in sample preparation prior to injection methods, such as solid-phase matrix extraction (SPME), headspace GC, or coupling of analytical instruments to increase the separation power of one-dimensional techniques. In the area of high-speed gas chromatography, over the past few years, instruments and methods have been developed to dramatically increase the analysis speed of capillary GC. Efforts have also been made to develop techniques such as multidimensional gas chromatography (MDGC), comprehensive multidimensional chromatography (CMC), and comprehensive two-dimensional superfluid chromatography and gas chromatography (SFCXGC) [62,63].

## 1.20 Advantages of Essential Oil–Based Pesticides

- Broad-spectrum pesticides possess insecticidal, antifeedant, repellent, oviposition deterrent, growth regulatory, and antivector activities.
- They are useful in foodstuffs and stored foods.
- Reduced-risk pesticides are nontoxic to mammals and fish.
- They are widely used as flavoring agents in beverages and foodstuffs.
- Commercialization is possible due to abundant availability.
- Green pesticides are largely used against home and garden pests.
- There is slow pest resistance due to complex mixtures of several compounds.
- There is a unique impact on integrated pest management.
- There is limited persistence and high volatility.
- There is no harm to predators, parasitoids, and pollinators.

## 1.21 Limitations of Essential Oil–Based Pesticides

- Few pest control products are available on the market.
- They have a lower efficacy than chemical pesticides.
- A greater application rate and frequent reapplication are required.

- There is poor specificity due to the presence of several compounds.
- There is a lack of sufficient supply, protection technology, and regulatory approval.
- There is inconsistency in raw material composition obtained from plants grown in different geographical, genetic, climatic, and seasonal areas.
- There is minimum involvement from low-budget companies.
- Oil can have an adverse effect on the germination power of the treated seeds.

## 1.22 Conclusion

In this area, several books have been published [64–78]. Many books have been written on topics such as essential oils and their use as pesticides. However, none of the books do what the *Green Pesticides Handbook* does—covers some pesticidal essential oils that can really protect foodstuffs. Experts from all over the world were invited to contribute chapters in order to make this handbook readable and useful to farmers, industrialists, producers, students, researchers, and teachers.

## References

1. Encyclopaedia Brittanica. Avicenna. http://www.britannica.com/biography/Avicenna.
2. Gattefosse, R.-M. 1993. *Gattefosse's Aromatherapy*. 2nd ed. Ebury Publishing, London.
3. Tripathi, A.K., Upadhyay, S., Bhuyan, M., Bhattacharya, P.R. 2009. A review on prospects of essential oils as biopesticide in insect-pest management. *J. Pharmacog. Phytother.* 1(15): 52–63.
4. Koul, O., Walia, S., Dhaliwal, G.S. 2008. Essential oils as green pesticides: Potential and constraints. *Biopestic. Int.* 4(1): 63–84.
5. Singh, A., Khare, A., Singh, A.P. 2012. Biofertilizers & biopesticides: Use of vegetable oils as biopesticides in grain protection—A review. *J. Biofertil. Biopestic.* 3(1): 1000114.
6. Pugazhvendan, P.R., Ross, P.R., Elumalai, K. 2012. Insecticidal and repellent activities of plants oil against stored grain pest, *Tribolium castaneum* (Herbst) (Coleoptera: Tenebrionidae). *Asian Pac. J. Trop. Dis.* 2(1): S412–S415.
7. Tariq, R.M., Naqvi, N.-H., Choudhary, M.I., Qadri, M.A.H. 2010. Importance and implementation of essential oil of Pakistanian *Acorus calamus* Linn., as a biopesticide. *Pak. J. Bot.* 42(3): 2043–2050.
8. Soni, P.L. 1970. *Textbook of Organic Chemistry*. 6th rev. ed. Sultan Chand & Sons, Publishers, Daryaganj, Delhi.
9. Brewster, R.Q., McEwen, W.E. 1977. *Organic Chemistry*. 3rd ed. Prentice Hall of India, New Delhi.
10. Morris, R.T., Boyd, R.N. 1976. *Organic Chemistry*. 3rd ed. Prentice Hall of India, New Delhi.
11. Finar, I.L. 1989. *Organic Chemistry*, vol. 2, *Stereochemistry and the Chemistry of Natural Products*. 5th ed. Longman Scientific and Technical, Longman Group U.K.
12. Lee, S., Tsao, R., Peterson, C., Coats, J.R. 1997. Insecticidal activity of monoterpenoids to the western corn root worm (Coleoptera: Chrysomelidae), two spotted spider mite (Acari: Tetranychidae) and housefly (Diptera: Muscidae). *J. Econ. Entomol.* 90: 883–892.
13. Evans, P.D. 1980. Biogenic amines in the insect nervous system. *Adv. Insect. Physiol.* 15: 317–473.
14. Evans, P.D. 1981. Multiple receptor types for octopamine in the locust. *J. Physiol. London* 318: 88–122.

15. Enan, E., Beigler, M., Kende, A. 1996. Insecticidal action of terpenes and phenols to cockroaches: Effect on octopamine receptors. Presented at International Symposium on Plant Protection, Ghent, Belgium.

16. Enan, E.E. 2005. Molecular and pharmacological analysis of an octopamine receptor from American cockroaches and fruit fly in response to plant essential oils. *Arch. Insect Biochem. Physiol.* 59: 161–171.

17. Kostyukovsky, M., Rafaeli, A., Gileadi, C., Demchenko, N., Shaaya, E. 2002. Activation of octopaminergic receptors by essential oil constituents isolated from aromatic plants: Possible mode of action against insect pests. *Pest Manag. Sci.* 58: 1101–1106.

18. Rollo, C.D., Borden, J.H., Caey, I.B. 1995. Endogenenously produced repellant from American cockroach (Bilattaria: Blattidae): Function in death recognition. *Environ. Entomol.* 87: 116–124.

19. Sim, M.J., Choi, D.R., Ahn, Y.J. 2006. Vapour phase toxicity of plant essential oils to *Cadra cautella* (Lepidoptera: Pyralidae). *J. Econ. Entomol.* 99: 593–598.

20. Hausen, B.M., Reichling, J., Harkenthal, M. 1999. Degradation products of monoterpenes are the sensitizing agents in tea tree oil. *Am. J. Contact Dermat.* 10: 68–77.

21. Nilsson, U., Magnusson, K., Karlberg, O., Karlberg, A.T. 1999. Are contact allergens stable in patch test preparations? Investigation of the degradation of d-limonene hydroperoxides in petrolatum. *Contact Dermatitis* 40: 1227–1132.

22. Hold, K.M., Sirisoma, M.S., Ikeda, T., Narahashi, T., Casida, E. 2000. Alpha thujone (the active component of absinthe): Gamma-aminobutyric acid type A receptor modulation and metabolic detoxification. *Proc. Natl. Acad. Sci. USA* 97: 3826–3831.

23. Nelson, S.D., McClanahan, R.H., Thomassen, D., Gordon, W.P., Knebal, N. 1992. Investigations of mechanism of reactive metabolite formation from (R)-(+)-pulegone. *Xenobiotica* 22: 1157–1164.

24. Weschler, C.J. 2000. Ozone in indoor environments: Concentration and chemistry. *Indoor Air* 10: 269–288.

25. Shu, Y., Kwok, E.S.C., Tauzon, E.C., Atkinson, R., Arey, J. 1997. Products of the gas phase reactions of linalool with OH radicals, $NO_2$ radicals and $O_3$. *Environ. Sci. Technol.* 31: 896–904.

26. Reissel, A., Harry, C., Aschmann, S.M., Atkin, R., Arey, J. 1999. Formation of acetone from the OH radicals and $O_3$-initiated reactions of a series of monoterpenes. *J. Geophys. Res.* 104: 13869–13879.

27. Wolkoff, P., Clausen, P.A., Wilkenes, C.K., Nielsen, G.D. 2000. Formation of strong airway irritants in terpene/ozone mixture. *Indoor Air* 10: 8–91.

28. Youssef, N.S. 1997. Toxic and synergistic properties of several volatile oils against larvae of the house fly, *Musca domenstica vicina* Maquart (Diptera: Muscidae). *Egyptian German Soc. Zool.* 22: 131–149.

29. Hummelbrunner, L.A., Isman, M.B. 2001. Acute sublethal, antifeedant and synergistic effects of monoterpenoid essential oil compounds on the tobacco cutworm, *Spodoptera litura* (Lep. Noctuidae). *J. Agric. Food Chem.* 49: 715–720.

30. McKinney, J.D., Richard, A., Walter, C., Newman, M.C., Gerberick, F. 2000. The practice of structure activity relationship (SAR) in toxicology. *Toxicol. Sci.* 56(1): 8–17.

31. Ngoh, K., Choo, L.E.W., Pang, F.Y., Huang, Y., Kini, M.R., Mo, S.H. 1998. Insecticidal and repellent properties of nine volatile constituents of essential oils against the American cockroach, *Periplaneta americana* (L). *Pest. Sci.* 54: 261–268.

32. Aggarwal, K.K., Tripathi, A.K., Prajapati, V., Verma, N., Kumar, S. 2001. Toxicity of menthol and its derivatives against four storage insects. *Insect Sci. Applic.* 21: 229–236.

33. Braverman, Y.A., Chizov-Ginsberg Mullens, B.A. 1999. Mosquito repellent attracts *Culicoides imicola* (Diptera: Ceratpogonidae). *J. Med. Entomol.* 36: 113–115.

34. Budhiraja, S.S., Cullium, M.S., Sioutis, S.S., Evagelista, L., Habanota, S.T. 1999. Biological activity of *Melaleuca alternifolia* (tea tree) oil component, terpinene-4-ol, in human myelocutic cell line HL-60. *J. Manipulative Physiol. Ther.* 22: 447–453.

35. Haysen, B.M., Reichling, J., Harkenthal, M. 1999. Degradation products of monoterpenes are the sensitizing agents in tea tree oil. *Am. J. Contact Dermat.* 10: 68–77.

36. Burkhard, P.R., Burkrdt, K., Haenggeli, C.C., Landlis, T. 1999. Plant induced seizures: Reappearance of an old problem. *J. Neurol.* 46: 667–670.
37. Isman, M.B. 2004. Plant essential oils as green pesticides for pest and disease management. In Nelson, W.M. (ed.), *Agricultural Applications in Green Chemistry*, vol. 887, ACS Symposium Series. American Chemical Society, Washington, DC, pp. 41–51.
38. Seigler, D.S. 1983. Role of lipids in plant resistance to insects. In Hedin, P.A. (ed.), *Plant Resistance to Insects*. ACS Symposium Series, vol. 208. American Chemical Society, Washington, DC, pp. 303–328.
39. Miresmalli, S., Ojha, H.D., Drury, J.W. 2013. Apparatus and method for controlled release of botanical fumigant pesticides. WO 2013155438A1, PCT/US2013/036410, October 17.
40. Bessette, S.M., Beigler, M.A. 2005. Synergistic and residual pesticidal compositions containing plant essential oils. U.S. Patent 6,849,614.
41. Villalobos, M.J.P. (ed.). 2014. Virtual special issue: Essential oils as natural pesticides. *Ind. Crops Prod.* 62: 272–279.
42. Soundarajan, R.P. (ed.). 2012. *Agricultural and Biological Sciences: Pesticides—Advances in Chemical and Botanical Pesticides*. Intech Open Sciences. Under CC BY 3.0 License, Institute of Analytical Chemistry, Faculty of Chemical & Food Technology, University of Technology, Bratislava.
43. Oshawa, H., Miyagawa, H., Lee, P.W. 2007. *Pesticide Chemistry: Crop Protection, Environmental Safety*. Wiley-VCH Verlag, Weinhein, Germany.
44. Mohan, M., Haider, S.Z., Andola, H.C., Prohit, V.K. 2011. Essential oils as green pesticides: For sustainable agriculture. *Res. J. Pharm. Biol. Chem. Sci.* 2(4): 100–106.
45. Verlet, N. 1993. Overview of the essential oils economy. *Acta Hortic.* 333: 65–72.
46. Urabe, C. 1992. Insect control in wood. Patent JP 92-308238, 921021.
47. Akita, K. 1991. Insect-repellent and mildewcidal laminate boards. Patent JP 91-242869, 910927.
48. Hallahan, D.L. 2007. Insect repellent compounds. U.S. Patent 20070231357, A1.
49. Zhu, B.C.R., Henerson, G., Laine, R.A. 2005. Dihydronootkatone and tetrahydronootkatone as repellent to arthropods. U.S. Patent 5,897,244.
50. Ninkov, D. 2007. Pesticidal compounds and compositions. U.S. Patent 7,208,519.
51. Ping, L.H. 2007. Pest treatment composition. U.S. Patent 7,282,211.
52. OECD (Organisation for Economic Cooperation and Development). 2002. OECD guidance for industry data submissions for pheromones and other semiochemicals and their active substances: (Dossier guidance for pheromones and other semiochemicals). Part 1. OECD Series on Pesticides, No. 16. OECD, Paris.
53. Kumbhar, P.P., Nikumbh, V.P., Chaudhary, L.S., Chvan, K.M., Bendre, R.S., Patil, V.V., Dewang, P.M. 2000. Pesticide potency of some citronellol derivatives. *Pestol.* 24: 16–18.
54. Kumbhar, P.P., Dewange, P.M. 2001. Monoterpenoids: The natural pest management agents. *Frag. Flavor Assoc. India* 3: 49–56.
55. Lucchesi, M.E., Chemat, F., Smadia, J. 2004. Solvent-free microwave extraction of essential oil from aromatic herbs: Comparison with conventional hydro-distillation. *J. Chromatogr. A* 1043(2): 323–327.
56. Derwich, E., Benziane, Z., Taouil, R. 2010. GC/MS analysis of volatile compounds of the essential oil of the leaves of *Mentha pulegium* growing in Morocco. *Chem. Bull. "POLITEHNICA" Univ. (Timisoara)* 55(69): 2, 103–106.
57. Falodun, A., Siraj, R., Choudhary, M.I. 2009. Research article: GC-MS analysis of insecticidal leaf essential oil of *Pyrenacatha staudtii* Hutch and Dalz (Icacinaceae). *Trop. J. Pharm. Res.* 8(2): 139–143.
58. Gessler, K. 2005. Development of an optimized method for analysis of essential oils. Diploma thesis, Universitat Basel.
59. Shivatare, R.S., Nagore, R.S., Nipanikar, S.U. 2013. 'HPTLC' an important tool in standardization of herbal medical product—A review. *J. Sci. Innov. Res.* 2(6): 1086–1096.
60. Hajimehdipoor, H., Sekarchi, M., Khanavi, M., Adib, N., Amri, M. 2010. A validated high performance liquid chromatography method for analysis of thymol and carvarol in *Thymus vulgaris* L. volatile oil. *Pharmacogn. Mag.* 6(23): 154–158.

61. Buiarelli, F., Cartyoni, G.P., Coccioli, F., Ravazzi, E. 1991. Analysis of orange and mandarin essential oil by HPLC. *Chromatographia* 31(10): 489–492.
62. Makgwane, P.R. 2006. Chapter 3 pdf. upetd.up.ac.za/thesis/available/etd-02222007-184250/-03.
63. Mondello, L., Dugo, P., Bartle, K.D., Dugo, G., Cortroneo, A. 1995. Automated HPLC-HRGC: A powerful method for essential oils analysis. Part V. Identification of terpene hydrocarbons of bergamot, lemon, mandarin, sweet orange, grapefruit, clementine and Mexican lime oils by coupled HPLC-HRGC-MS(ITD). *Flavour Fragr. J.* 10(1): 33–42.
64. Dhaliwal, S.S., Koul, O. 2007. *Biopesticides and Pest Management: Conventional and Biotechnological Approaches.* Kalyani Publishers, New Delhi.
65. Dev, S., Koul, O. 1997. *Insecticides of Natural Origin.* Harwood Academic Publishers, Amsterdam.
66. Koul, O., Dhaliwal, S.S., Arora, J.K. (eds.). 2003. *Biopesticides and Pest Management.* Vol. 2. Campus Books International, New Delhi.
67. Regnault-Roger, C., Philogène, B.J.R., and Vincent, C. (eds.). 2005. *Biopesticides of Plant Origin.* Lavoiseir, Paris.
68. Jacobson, M., Halber, L. 1947. *The Chemistry of Organic Medicinal Plants.* Chapman & Hall, New York.
69. Parmar, B.S., Walia, S. (eds.). 1995. *Pesticides, Crop Protection and Environment.* Oxford and IBH Publishing Co., New Delhi.
70. Koul, O. 2005. *Insect Antifeedants.* CRC Press, Boca Racon, FL.
71. Rai, M., Carpinella, M.C. (eds.). 2006. *Naturally Occurring Bioactive Compounds.* Elsevier, Amsterdam.
72. Bruneton, J. 1999. *Pharmacognosy, Phytochemistry, Medicinal Plants: Essential Oils.* 2nd ed. Lavoisier Publishing, New York.
73. D'Mello, J.P. (ed.). 1997. *Handbook of Plant and Fungal Toxicants.* CRC Press, Boca Raton, FL.
74. Ishaaya, R.N.R., Rami, H. (eds.). 2007. *Insecticides Design Using Advanced Technologies.* Springer-Verlag, Berlin.
75. Ohkawa, H., Miyagawa, H., Lee, P.W. 2007. *Pesticide Chemistry.* Wiley-VCH Verlag, Berlin.
76. Kydonieus, A.F. 1980. *Controlled Release Technologies: Methods, Theory, and Applications.* Vols. 1 and 2. CRC Press, Boca Raton, FL.
77. Shumakov, E.M., Chekmenev, S.Y., Ivanova, T.V. (eds.). 1979. *Biologia Aklutualis Veshchestva Rastenij.* Izd. Kolos, Moscow.
78. Ishaaya, I., Degheele, D. (eds.). 1998. *Insecticides with Novel Modes of Action, Mechanisms and Application.* Springer, Berlin.

# 2

# Commercialization of Insecticides Based on Plant Essential Oils: Past, Present, and Future

Murray B. Isman and Jun-Hyung Tak

## CONTENTS

## 2.1 Introduction

The use of aromatic plants and the oils derived from them dates back at least 6500 years in Egypt, 4500 years in China, and 3000 years in India, and perhaps much further back than those estimates. Most plant essential oils used today on a commodity scale are obtained by hydrodistillation, a process first attributed to the Persian physician Ali-Ibn Sana (also known as Avicenna), who lived 1000 years ago. Essential oils from *Citrus* peels are obtained by cold pressing, and a handful of other essential oils are obtained via extraction with lipophilic solvents.

These oils were first used largely for medicinal and/or ceremonial purposes, but anecdotal records suggest their use in repelling insects dates back hundreds of years. For example, certain aromatic oils were burned during the Black Death (1346–1353) "to ward off evil spirits" (healingscents.net), but may have inadvertently reduced the spread of bubonic plague by repelling or killing fleas that vectored the pathogen. Scientific investigation of the repellent and toxic effects of plant essential oils in insects is far more recent. Jacobson and Haller (1947) reported the insecticidal action of 1,8-cineole from rosemary oil (*Rosmarinus officinalis*) to the house fly *Musca domestica*, while the insecticidal actions of thymol from thyme oil (*Thymus vulgaris*), d-limonene from orange oil (*Citrus × sinensis*), and eugenol from clove oil (*Syzygium aromaticum*) were reported in the 1970s (Sharma and Saxena 1974; Miki 1978; Marcus and Lichtenstein 1979).

However, interest in the bioactivity of plant essential oils to insects within the scientific community has exploded since 2000, with more than 1100 papers published on the subject between 2007 and 2012 according to a recent bibliometric analysis (Isman and Grieneisen 2014). By 2012, the proportion of papers on plant essential oils had risen to almost a quarter of all papers published on botanical insecticides. Unfortunately, the vast majority of these papers reported bioactivity of essential oils derived from a single plant species, with no chemical characterization, and often without a positive control against which bioactivity can be judged. Much of this published work is based on laboratory-scale experiments, and often describes new sources of bioactive oils, indicating a strong skew toward the discovery end of the research and development (R&D) spectrum. It is perhaps not surprising, then, that only a relative handful of plant essential oils have seen successful commercialization and use as pesticides (Isman 2016).

The major uses of plant essential oils are in the fragrance and flavoring industries, with numerous applications in cosmetics, personal health care products, foods and beverages, and household cleaning products. Some are seeing increasing use as alternative medicines, attributable in part to their antibacterial and antifungal properties (Kalemba and Kunicka 2003), and in part due to their use in aromatherapy.

## 2.2 Effects on Insects and Modes of Action

Plant essential oils can have a wide array of both acute and sublethal effects on insects and related arthropods. It is worthwhile to divide effects into those that are primarily behavioral (repellence, attractance, and deterrence; moderated by peripheral chemosensory systems) and those that are primarily physiological (toxicity, inhibition of growth, and developmental impairment). In practice, most essential oils will have both behavioral and physiological effects on a given insect species, sometimes even at one given dose or concentration, and investigators must be careful to design bioassays with very specific end points, to minimize confusion in interpreting their results. For many oils, however, behavioral effects will be observed at lower doses or concentrations than those required to produce acute toxicity or other physiological effects.

These effects can overlap or co-occur for two reasons. First, the majority of chemical constituents of essential oils—monoterpenoids, phenylpropenes, and sesquiterpenes— are relatively volatile, and as such, insects will encounter them in the vapor phase. In this form, they can be detected by chemosensory sensilla—potentiating behaviors—*and* be absorbed through the insect's integument, producing physiological effects. Second, one important mechanism of action is antagonism of the neuromodulator or neurohormone, octopamine, which can influence numerous behavioral and physiological systems in arthropods (Hollingworth and Lund 1982). While there is some compelling evidence indicating that octopamine receptors are important target sites in insects for essential oil constituents (Enan 2001, 2005; Kostyukovsky et al. 2002), there is also evidence that certain constituents (e.g., thymol from *T. vulgaris*) interfere with GABA-gated chloride channels in nerve synapses (Priestley et al. 2003; Tong and Coats 2012). Some constituents have been reported to inhibit acetylcholine esterase *in vitro* (Lopez and Pascual-Villalobos 2015; Yeom et al. 2015), but there are no studies showing a strong correlation between *in vitro* inhibition of acetylcholine esterase by essential oils or their constituents and *in vivo* toxicity in one or more insects. Finally, there is evidence that thymol, through its interaction with tyramine receptors, increases intracellular calcium, an effect correlated with *in vivo* toxicity, at least in *Drosophila melanogaster* (Enan 2005). It is likely that essential oil terpenoids have multiple targets and mechanisms of action in insects, and while it is clear that the insect nervous system is a primary site of action, it may be difficult to ascribe acute toxicity to any one of these specific biochemical targets without much more detailed investigation (Gross et al. 2014a,b).

Many investigations into the repellent activity of plant essential oils to blood-feeding insects have been conducted (principally targeting mosquitoes, but reduviid bugs and ticks have also been subjects), with the goal of producing personal protectants applied either to the skin or to clothing or bed nets (Prajapati et al. 2005; Trongtokit et al. 2005; Nerio et al. 2010). Oil of citronella has been used for many years as a natural alternative to the conventional repellent DEET (*N*,*N*-dimethyl-*m*-toluamide) (Peterson and Coats 2001). However, few other plant oils have seen major commercial success as insect repellents or personal protectants, and therefore the balance of this chapter will focus on the primary use of plant essential oils as insecticides.

This is not to say that behavioral effects—repellence, feeding deterrence, and oviposition deterrence—cannot make important contributions to the overall efficacy of plant essential oil–based insecticides under field or greenhouse conditions. However, the contributions of such effects to efficacy can be difficult to evaluate in the field, unlike the situation in the laboratory where carefully controlled and confined experiments can be conducted. Indeed, there are many demonstrations of deterrent effects from laboratory and greenhouse experiments (e.g., Miresmailli et al. 2006; Isman et al. 2011; Akhtar et al. 2012; Da Camara et al. 2015).

## 2.3 Toxicity of Plant Essential Oils in Insects

The toxicity of plant essential oils in insects has been the subject of numerous investigations, as shown by a literature survey conducted in late 2014 using the Web of Science database. A search for papers on essential oils associated with insect and arthropod species between 2004 and 2014 produced more than 3600 publications, among which 73% reported acute toxicity. With respect to target species, the greatest number of papers dealt with dipterans (mosquitoes and house flies), coleopterans (stored product pests), acarines (spider mites and ticks), and lepidopterans (armyworms, cutworms, and other noctuids) (Tak 2015), as shown in Table 2.1.

Previously, a number of studies in which essential oils were screened against insect pests (Isman and Machial 2006), including eight studies where more than 20 twenty oils were

**TABLE 2.1**

Number of Publications on Essential Oil Research Based on Major Orders of Insects and Other Arthropods (2004–2014)

| Order | Number of Publications[a] | Search Word[b] | Major Target of Research |
|---|---|---|---|
| Diptera | 705 | "Full" + (and diptera*) | Mosquitoes, house fly |
| Coleoptera | 585 | "Full" + (and coleoptera*) | Stored product pests |
| Acari | 420 | "Full" + (and acari*) | Two-spotted spider mites, ticks |
| Lepidoptera | 222 | "Full" + (and lepidoptera*) | Armyworms, cutworms |
| Hemiptera | 167 | "Full" + (and hemiptera* or heteroptera* or homoptera*) | Aphids, stinkbugs |

[a] Search was conducted on December 24, 2014, using the Web of Science database.
[b] "Full" indicates the search query (essential oil or terpene*) and (repell* or insecticid* or larvicid* or acaricid* or deterr* or antifeed* or growth regulat* or oviposit*).

**TABLE 2.2**

Topical Toxicity of Some Common Essential Oils to the Third Instar Cabbage Looper *Trichoplusia ni* and Adult House Fly *Musca domestica*

| Essential Oil | LD$_{50}$ (μg insect$^{-1}$) | |
|---|---|---|
| | *T. ni* | *M. domestica* |
| Clove (*Syzygium aromaticum*) | 47.8[a] | 40[e] |
| Thyme (*Thymus vulgaris*) | 54.1[a,b] | 18[e] |
| Cinnamon (*Cinnamomum zeylandicum*) | 101.5[a] | – |
| Lemongrass (*Cymbopogon citratus*) | 167.5[b] | – |
| Rosemary (*Rosmarinus officinalis*) | 215.8[c] | 28[e] |
| Peppermint (*Mentha piperita*) | 233.8[d] | – |

[a]  From Jiang, Z. et al., *J. Agric. Food Chem.*, 57, 4833–4837, 2009.
[b]  From Tak, J.H. et al., *J. Pest Sci.*, 89, 183–193, 2016.
[c]  From Tak, J.H. et al., *Pest Manag. Sci.*, 72, 474–480, 2016.
[d]  From Tak, unpublished data.
[e]  From Pavela, R., *J. Essential Oil Bearing Plants*, 11, 451–459, 2008.

evaluated, were reviewed. Insects used in these studies included the stored product pests *Acanthoscelides obtectus* (Regnault-Roger et al. 1993; Regnault-Roger and Hamraoui 1994) and *Sitophilus oryzae* (Singh et al. 1989; Shaaya et al. 1991), the greenhouse whitefly *Trialeurodes vaporariorum* (Choi et al. 2003), the codling moth *Cydia pomonella* (Landolt et al. 1999), the house fly *M. domestica* (Singh and Singh 1991), and the aphid *Lipaphis pseudobrassicae* (Sampson et al. 2005). More recent studies on the topical and fumigant toxicities of essential oils to the house fly have been published (Pavela 2008; Palacios et al. 2009; Tarelli et al. 2009). An overall assessment of this literature emphasizes considerable variation in susceptibilities between insect taxa, even at the level of species (Franzios et al. 1997; Shaaya et al. 1997; Kim et al. 2003; Regnault-Roger et al. 2012), although essential oils rich in the phenols thymol and carvacrol (e.g., thyme, oregano, and savory oils) tend to be the most toxic when several oils are compared in a single study. Table 2.2 compares the topical toxicity of several common essential oils to the cabbage looper, an agricultural pest, and the house fly, a public health pest.

It is worth noting that the oils listed in Table 2.2 are considerably less toxic than pyrethrins to the house fly (LD$_{50}$ ~8 μg insect$^{-1}$) (Lee et al. 1997) and far less toxic to the cabbage looper (LD$_{50}$ ~0.1 μg insect$^{-1}$; unpublished data).

## 2.4 Toxicity of Monoterpenoid Constituents of Essential Oils to Insects

Numerous investigators have screened large numbers of monoterpenoid (and some sesquiterpenoid) constituents from plant essential oils for their toxicity to several different insect and mite pests, including the house fly (Lee et al. 1997; Pavela 2008), the maize weevil (Yildirim et al. 2013), the rice weevil (Lee et al. 2001), the two-spotted spider mite (Lee et al. 1997; Choi et al. 2004), and the tobacco cutworm (Hummelbrunner and Isman 2001). However, interspecific comparisons of toxicity can seldom be made owing to differences in methods used to measure toxicity. In particular, fumigant or residual contact toxicity is normally assessed with stored product pests, whereas topical toxicity is often assessed with adult house flies and noctuid larvae. For small, soft-bodied insects (aphids, thrips, and whiteflies) and phytophagous mites, toxicity is often assessed based on residual contact or direct contact (spray) bioassays. Some key monoterpenoid constituents of common plant essential oils with notable toxicity to insects are shown in Figure 2.1.

Almost all essential oil terpenoids and related phenols have some volatility, but differences in volatility between specific compounds can give rise to profound differences in fumigant versus topical toxicity. This is well exemplified by a study of the acute toxicity of six monoterpenoids to the adult house fly (Pavela 2008), in that the three most toxic compounds via fumigation (1,8-cineole, γ-terpinene, and *p*-cymene) were the three least toxic via topical administration, and vice versa for the most toxic compounds topically (thymol, carvicrol, and eugenol). A similar result was obtained when comparing topical and fumigant toxicities of the major constituents of thyme oil (*T. vulgaris*) to the cabbage looper *Trichoplusia ni* (Tak and Isman 2015).

Studies based on topical bioassays consistently point to thymol as the single most active monoterpenoid isolated from common plant essential oils. This result has been obtained using the tobacco cutworm, *Spodotpera litura* (Hummelbrunner and Isman 2001); the cabbage looper, *T. ni* (Tak and Isman, unpublished data) (Table 2.3); the house fly, *M. domestica* (Lee et al. 1997; Pavela 2008); and the German cockroach, *Blattella germanica* (Yeom et al. 2015). As noted above, thymol tends not to be nearly as effective as a fumigant, which is not surprising because, as a solid at room temperature, it has considerably less volatility than most other monoterpenoids that are oils at room temperature.

Apart from the notable toxicity of thymol in insects, there are few consistent trends in toxicity among monoterpenoids from essential oils (Isman and Machial 2006), even from detailed investigations of structure–activity relations (Rice and Coats 1994).

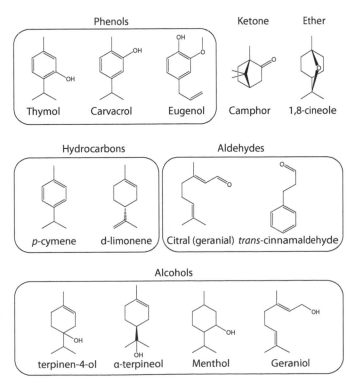

**FIGURE 2.1**
Some key monoterpenoid constituents of common plant essential oils with notable toxicity to insects.

**TABLE 2.3**

Topical Toxicity of Selected Essential Oil Monoterpenoids to the Third Instar Cabbage Looper *Trichoplusia ni* and Adult House Fly *Musca domestica*

| Compound | LD$_{50}$ (µg insect$^{-1}$) | |
| --- | --- | --- |
| | *T. ni*[a] | *M. domestica*[b] |
| Thymol | 25.7 | 29 |
| Terpinen-4-ol | 87.1 | 79 |
| Citral | 97.4 | 54 |
| Cinnamaldehyde | 119.5 | – |
| Geraniol | 151.8 | 73 |
| Menthol | 163.4 | 147 |
| 1,8-Cineole | 278.9 | 281 |
| d-Limonene | 318.6 | 68 |
| Eugenol | 355.6 | 77 |

[a] From Tak and Isman, unpublished data.
[b] From Lee, S. et al., *J. Econ. Entomol.*, 90, 883–892, 1997.

## 2.5 Relationships between Chemical Composition of Essential Oils and Toxicity to Insects

In a number of commodity essential oils, chemical composition can be dominated by one or two monoterpenoid constituents. Examples include eugenol in cloves (*S. aromaticum*), cinnamaldehyde in cinnamon (*Cinnamomum* spp.), d-limonene in *Citrus* species, menthol and menthone in peppermint (*Mentha piperita*), and 1,8-cineole in *Eucalyptus* species (Isman 2016). For these oils, the major constituents are also the most toxic ones to insects, so the relationship between chemical composition and toxicity is easy to confirm. Such a study has confirmed that the major constituents of thyme oil (*T. vulgaris*), thymol, and lemongrass oil (*Cymbopogon citratus*), citral, account for most of the toxicity of those oils to the cabbage looper, *T. ni* (Tak et al. 2016a).

However, for a number of oils that are more chemically complex, the relationship may be far less clear. One particularly well-studied example is that of rosemary oil (*R. officinalis*). In an investigation into the active principles of rosemary oil to the two-spotted spider mite (*Tetranychus urticae*), experimental deletions of individual constituents from artificially reconstituted rosemary oil pointed to 4 of 10 constituents making significant contributions to bioactivity: 1,8-cineole, α-pinene, α-terpineol, and bornyl acetate. Surprisingly, when these four putatively "active" monoterpenes were mixed and tested on adult spider mites, only 25% of the bioactivity of rosemary oil (at an equivalent concentration) was observed (Miresmailli et al. 2006). Mixing these four with the six remaining, putatively "inactive" constituents restored the full bioactivity seen in the natural rosemary oil, strongly suggesting synergisitic action between the active and inactive constituents. Synergy between individual monoterpenes has been observed several times in insects (Hummelbrunner and Isman 2001; Pavela 2008, 2014), but in the case of rosemary oil, the synergy occurs between natural constituents within the same essential oil. Further confirmation of this phenomenon came from a study that attempted to correlate the chemical composition of 10 different commercial rosemary oils with their toxicity to the cabbage looper and the armyworm, *Pseudaletia unipuncta* (Isman et al. 2008). Two minor constituents, d-limonene and α-terpineol, were significantly correlated to toxicity of the oils in *T. ni* (the latter only weakly correlated), whereas none of the nine measured constituents were correlated with toxicity to *P. unipuncta*. An additional study investigated the relative contributions of constituents of an essential oil from *Litsea pungens* to *T. ni*, using the same approach as that taken in the spider mite study with rosemary essential oil (Miresmailli et al. 2006). The two major constituents of *L. pungens* oil, 1,8-cineole and carvone, were also the two most toxic to *T. ni*, both when tested in isolation and from a deletion experiment, as described above. As before, when these two putative active constituents were mixed and tested together, they could only reproduce about 27% of the bioactivity seen with the intact essential oil, suggesting a role for the putatively inactive constituents as synergists (Jiang et al. 2009).

Returning to rosemary oil, a thorough investigation of synergy among constituents using *T. ni* as the test insect ultimately revealed that a binary mixture of the two major constituents, 1,8-cineole and camphor (in their natural proportions in rosemary oil), accounted for at least 90% of the toxicity of the intact oil (Tak and Isman 2015; Tak et al. 2016b). Moreover, enhanced penetration through the insect's integument was elucidated as the mechanism underlying the observed synergy between 1,8-cineole and camphor. What was surprising was that 1,8-cineole was found to facilitate the penetration of camphor, which proved to be the more toxic of the two compounds when injected directly into the insect's hemocoel (Tak and Isman 2015). These observations point to the importance of the penetration of monoterpenes and related constituents of essential oils to their toxicity in insects, and differential penetration and uptake as a potential source of interspecific variation in the toxicity of essential oils in insects. A comparison of selected monoterpenoids based on their topical and injected toxicity to the cabbage looper shows a lack of correlation: some compounds that are quite toxic when injected (e.g., camphor) are relatively nontoxic via topical administration, whereas others that are moderately toxic topically (e.g., citral) are relatively nontoxic when injected (Figure 2.2) (Tak and Isman, unpublished data).

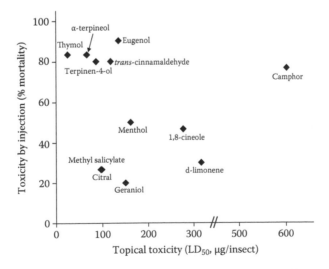

**FIGURE 2.2**
Relationship between toxicity of monoterpenoids *via* topical administration and *via* injection (at 500 μg insect$^{-1}$) in the cabbage looper.

## 2.6 Commercial Insecticides Based on Plant Essential Oils: Current Status

In spite of the voluminous and rapidly growing scientific literature documenting the repellent and insecticidal bioactivities of plant essential oils and/or their constituents to pestiferous insects and mites (Isman and Grieneisen 2014), few insecticides based on essential oils have been commercialized, and even fewer have seen success in the marketplace (Isman 2016). In most jurisdictions, costly and time-consuming regulatory requirements—that emerged and evolved based on synthetic crop protection chemicals—pose a formidable barrier, especially for smaller, specialized producers who often lack the resources needed to secure regulatory approval. A notable exception to this is the United States, where the FIFRA List 25B ("exempt active ingredients") facilitated entry of such products into the marketplace. The list includes almost a dozen commodity essential oils, all of which are widely used in the fragrance and flavoring industries. EcoSMART Technologies took advantage of this opportunity to produce an array of pesticides containing different mixtures of rosemary, peppermint, lemongrass, clove, and cinnamon oils, initially developed for professional (domestic and industrial) pest control (in 1998), but with products introduced later for agriculture, animal health, vector control, and retail consumer markets (Figure 2.3). Their products are currently approved for use in seven other countries, including Mexico, Peru, Uruguay, Singapore, and the United Arab Emirates. Exempt insecticides from other producers in the United States are based on these oils, in addition to some based on cedar and garlic oils.

A relatively important product, in terms of commercialization, is Requiem®, approved for use in the United States by the Environmental Protection Agency (EPA) in 2011 and now marketed by Bayer CropScience. This product, under the name "Terpenoid Blend QRD-460," received tentative approval as an insecticide by the European Commission in late 2015. Rather than a plant essential oil per se, this product is a synthetic blend of three monoterpenoids (sourced independently) that are the major naturally occurring constituents of an essential oil from wormseed, *Dysphania* (*Chenopodium*) *ambrosioides* near *ambrosioides*. More than 9 tonnes (active ingredient) of this product was applied in California alone in 2013 (California Department of Pesticide Regulation [Cal DPR] 2013), making it third only to d-limonene and pyrethrins among botanical insecticides used that year.

At present, the only plant essential oil–based insecticide fully approved for use in the European Union is Prev-AM®, a product developed (and approved for use) in South Africa, and approved for use in France and Belgium as of 2014. This product is based on cold-pressed orange oil, which typically consists predominantly (>95%) of d-limonene. d-Limonene itself is an EPA-registered insecticide in the United States.

In Australia, Eco-Oil® is a miticide–insecticide based on a 2% mixture of eucalyptus and tea tree (*Melaleuca alternifolia*) oils. It is surprising that there are not more products of this type in the country, as the Australia Pesticides and Veterinary Medicines Authority lists 20 common essential oils that fall under Section 14A of the Agvet Code as "active constituents not requiring evaluation" (Australian Pesticides and Veterinary Medicines Authority 2015).

Perhaps even more surprising is the fact that the country producing the greatest volume of published research on botanical insecticides—India—to date has no approved insecticide based on plant essential oils. Brazil, another country with abundant plant diversity and resources and prolific research on botanicals, also lacks any approved insecticides based on essential oils. In contrast, China has several insecticides based either on the

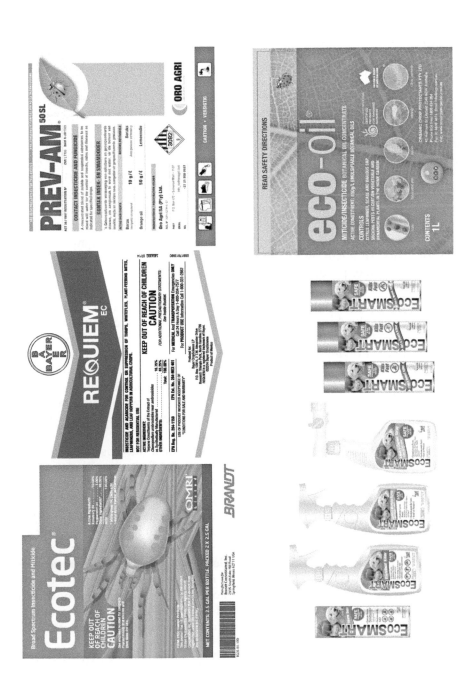

**FIGURE 2.3**
Some successful commercial insecticides based on plant essential oils and/or monoterpenoids as active ingredients.

monoterpenoid camphor (likely from the camphor wood tree, *Cinnamomum camphora*) or eucalyptol (= 1,8-cineole; likely from *Eucalyptus* spp.).

---

## 2.7 Targets, Applications, and Future Prospects

Several emerging factors should combine in the next 10 years to provide excellent opportunities for the introduction and market success of insecticides and other pesticides based on plant essential oils. These include (1) ever-increasing restrictions and limitations on the use of conventional synthetic insecticides, with many current products being eliminated from the pest manager's arsenal; (2) increasing demand for organically produced food, including animal products; and (3) continued introduction and spread of exotic pests facilitated by global commerce and global climate change. For example, recent expansions of the geographic range of several species of mosquitoes that vector human pathogens and concomitant epidemics of mosquito-borne arboviruses (dengue, chikungunya, and zika) have created urgent demand for vector control products that can be used in areas of dense human habitation, without increasing undue risks to exposed human populations.

In this context, plant essential oil–based pesticides benefit from their broad spectrum of activity against pests, their rapid action, their relative safety to nontarget organisms (especially humans), and their lack of environmental persistence. In built landscapes—industrial and professional uses and as consumer products—essential oil pesticides are beginning to be embraced owing to their wide margin of safety to users, particularly in sensitive settings such as schools, hospitals, seniors' homes, and in the hospitality industry (restaurants, hotels, and cruise ships), where the battle against cockroaches, flies, and bed bugs is constantly being fought. They are also making inroads where animal safety is a primary concern, for example, for ecoparasite control on companion animals (dogs and cats), and for fly control in dairy and poultry operations.

In food plant agriculture, they have already had modest success not only in protected cultivation (e.g., greenhouse vegetable production), but also in row crops, in part because of their safety to field workers (no reentry restrictions after application) and in part owing to the growing demand for organic produce free of pesticide residues. In addition to this, their compatibility with conventional insecticides—allowing reductions in volumes used and therefore costs for the latter—either as tank mixes or in rotation, creates additional opportunities in large-scale conventional food production systems. At this point in time, it is unclear to what extent observed efficacy against pests is a direct consequence of contact toxicity, an indirect consequence of repellence or deterrence, or most likely, some additive combination of both effects. Other agricultural operations where essential oil–based pesticides offer promise are in the fumigation of stored grains and legumes and for the disinfestation of fresh produce in shipping.

Finally, many essential oils have efficacy against key vector species (Norris et al. 2015) and might therefore be useful as part of large-scale mosquito abatement programs.

Industry experts have recently predicted strong growth in markets for biopesticides over the next 5 years (Snyder 2015), and manufacturers of plant essential oil–based pesticides should be poised to take advantage of expanding market opportunities to fulfill the promise of these products first predicted more than 20 years ago.

# References

Akhtar, Y., E. Pages, A. Stevens et al. 2012. Effect of chemical complexity of essential oils on feeding deterrence in larvae of the cabbage looper. *Physiological Entomology* 37:81–91.

Australian Pesticides and Veterinary Medicines Authority. 2015. Active constituents not requiring evaluation. Active constituents approved under section 14A of the Agvet Code. http://apvma.gov.au/node/4176 (accessed January 28, 2016).

California Department of Pesticide Regulation. 2013. Summary of Pesticide Use Report Data 2013. http://www.cdpr.ca.gov/docs/pur/pur13rep/chmrpt13.pdf.

Choi, W.I., E.H. Lee, B.B. Choi et al. 2003. Toxicity of plant essential oils to *Trialeurodes vaporariorum* (Homoptera: Aleyrodidae). *Journal of Economic Entomology* 96:1479–1484.

Choi, W.I., S.G. Lee, H.M. Park, and Y.J. Ahn. 2004. Toxicity of plant essential oils to *Tetranychus urticae* (Acari: Tetranychidae) and *Phytoseilus persimilis* (Acari: Phytoseiidae). *Journal of Economic Entomology* 97:553–558.

Da Camara, C.A.G., Y. Akhtar, M.B. Isman et al. 2015. Repellent activity of essential oils from two species of *Citrus* against *Tetranychus urticae* in the laboratory and greenhouse. *Crop Protection* 74:110–115.

Enan, E. 2001. Insecticidal activity of essential oils: Octopaminergic sites of action. *Comparative Biochemistry and Physiology C* 130:325–337.

Enan, E.E. 2005. Molecular and pharmacological analysis of an octopamine receptor from American cockroach and fruit fly in response to plant essential oils. *Archives of Insect Biochemistry and Physiology* 56:161–171.

Franzios, G., M. Mirotsou, E. Hatziapostolou et al. 1997. Insecticidal and genotoxic activities of mint essential oils. *Journal of Agricultural and Food Chemistry* 45:2690–2694.

Gross, A.D., M.J. Kimber, and J.R. Coats. 2014a. G-protein-coupled receptors (GPCRs) as biopesticide targets: A focus on octopamine and tyramine receptors. In *Biopesticides: State of the Art and Future Opportunities*, ed. A.D. Gross, J.R. Coats, S.O. Duke, and J.N. Seiber, 44–56. Washington, DC: American Chemical Society.

Gross, A.D., M.J. Kimber, T.A. Day et al. 2014b. Investigating the effect of plant essential oils against the American cockroach octopamine receptor (Pa oa1) expressed in yeast. In *Biopesticides: State of the Art and Future Opportunities*, ed. A.D. Gross, J.R. Coats, S.O. Duke, and J.N. Seiber, 113–130. Washington, DC: American Chemical Society.

Hollingworth, R.M., and A.E. Lund. 1982. Biological and neurotoxic effects of amidine pesticides. In *Insecticide Mode of Action*, ed. J.R. Coats, 189–227. New York: Academic Press.

Hummelbrunner, L.A., and M.B. Isman. 2001. Acute, sublethal, antifeedant and synergistic effects of monoterpenoid essential oil compounds on the tobacco cutworm, *Spodoptera litura* (Lep., Noctuidae). *Journal of Agricultural and Food Chemistry* 49:715–720.

Isman, M.B. 2016. Pesticides based on plant essential oils: Phytochemical and practical considerations. In *Medicinal and Aromatic Crops: Production, Phytochemistry and Utilization*, ed. V. Jeliazkov, and C.L. Cantrell. ACS Symposium Series 1218, pp. 13–26. Washington, DC: American Chemical Society.

Isman, M.B., and M.L. Grieneisen. 2014. Botanical insecticide research: Many publications, limited useful data. *Trends in Plant Science* 19:140–145.

Isman, M.B., and C.M. Machial. 2006. Pesticides based on plant essential oils: From traditional practice to commercialization. In *Naturally Occurring Bioactive Compounds*, ed. M. Rai and M.C. Carpinella, 29–44. Amsterdam: Elsevier B.V.

Isman, M.B., S. Miresmailli, and C. Machial. 2011. Commercial opportunities for pesticides based on plant essential oils in agriculture, industry and consumer products. *Phytochemistry Reviews* 10:197–204.

Isman, M.B., J.A. Wilson, and R. Bradbury. 2008. Insecticidal activities of commercial rosemary oils (*Rosmarinus officinalis*) against larvae of *Psudaletia unipuncta* and *Trichoplusia ni* in relation to their chemical compositions. *Pharmaceutical Biology* 46:82–87.

Jacobson, M., and H.L. Haller. 1947. The insecticidal component of *Eugenia haitiensis* identified as 1,8-cineole. *Journal of the American Chemical Society* 69:709–710.

Jiang, Z., Y. Akhtar, R. Bradbury et al. 2009. Comparative toxicity of essential oils of *Litsea pungens* and *Litsea cubeba* and blends of their major constituents against the cabbage looper, *Trichoplusia ni. Journal of Agricultural and Food Chemistry* 57:4833–4837.

Kalemba, D., and A. Kunicka. 2003. Antibacterial and antifungal properties of essential oils. *Current Medicinal Chemistry* 10:813–829.

Kim, S.I., J.Y. Roh, D.H. Kim et al. 2003. Insecticidal activities of aromatic plant extracts and essential oils against *Sitophilus oryzae* and *Callosobruchus chinensis. Journal of Stored Products Research* 39:293–303.

Kostyukovsky, M., A. Rafaeli, C. Gileadi et al. 2002. Activation of octopaminergic receptors by essential oil constituents isolated from aromatic plants: Possible mode of action against insect pests. *Pest Management Science* 58:1101–1106.

Lee, B.H., W.S. Choi, S.E. Lee, and B.S. Park. 2001. Fumigant toxicity of essential oils and their constituent compounds towards the rice weevil, *Sitophilus oryzae* (L.). *Crop Protection* 20:317–320.

Lee, S., R. Tsao, C. Peterson, and J.R. Coats. 1997. Insecticidal activity of monoterpenoids to western corn rootworm (Coleoptera: Chrysomelidae), twospotted spider mite (Acari: Tetranychidae), and house fly (Diptera: Muscidae). *Journal of Economic Entomology* 90:883–892.

Lopez, M.D., and M.J. Pascual-Villalobos. 2015. Are monoterpenoids and phenylpropanoids efficient inhibitors of acetylcholinesterase from stored product insect strains? *Flavour and Fragrance Journal* 30:108–112.

Marcus, C., and E.P. Lichtenstein. 1979. Biologically active components of anise: Toxicity and interactions with insecticides in insects. *Journal of Agricultural and Food Chemistry* 27:1217–1223.

Miki, K. 1978. Insecticide against maggots. Japanese Patent 53-066420.

Miresmailli, S., R. Bradbury, and M.B. Isman. 2006. Comparative toxicity of *Rosmarinus officinalis* L. essential oil and blends of its major constituents against *Tetranychus urticae* (Acari: Tetranychidae) on two different host plants. *Pest Management Science* 62:366–371.

Nerio, L.S., J. Olivero-Verbel, and E. Stashenko. 2010. Repellent activity of essential oils: A review. *Bioresource Technology* 101:372–378.

Norris, E.J., A.D. Gross, B.M. Dunphy et al. 2015. Comparison of the insecticidal characteristics of commercially available plant essential oils against *Aedes aegypti* and *Anopheles gambiae* (Diptera: Culicidae). *Journal of Medical Entomology* 52:993–1002.

Palacios, S.M., A. Bertoni, Y. Rossi et al. 2009. Efficacy of essential oils from edible plants as insecticides against the house fly, *Musca domestica* L. *Molecules* 14:1938–1947.

Pavela, R. 2008. Acute and synergistic effects of some monoterpenoid essential oil compounds on the house fly (*Musca domestica* L.). *Journal of Essential Oil Bearing Plants* 11:451–459.

Pavela, R. 2014. Acute, synergisitic and antagonistic effects of some aromatic compounds on the *Spodoptera littoralis* Boisd. (Lep., Noctuidae) larvae. *Industrial Crops and Products* 60:247–258.

Peterson, C., and J.R. Coats. 2001. Insect repellents—Past, present and future. *Pesticide Outlook* 12(4):154–158.

Prajapati, V., A.K. Tripathi, K.K. Aggarwal, and S.P.S. Khanuja. 2005. Insecticidal, repellent and oviposition-deterrent activity of selected essential oils against *Anopheles stephensi*, *Aedes aeypti* and *Culex quinquefasciatus. Bioresource Technology* 96:1749–1757.

Priestley, C.M., E.M. Williamson, K.A. Wafford, and D.B. Sattelle. 2003. Thymol, a constituent of thyme essential oil, is a positive allosteric modulator of human $GABA_A$ receptors and a homo-oligomeric GAMA receptor from *Drosophila melanogaster. British Journal of Pharmacology* 140:1363–1372.

Regnault-Roger, C., and A. Hamraoui. 1994. Reproductive inhibition of *Acanthoscelides obtectus* Say (Coleoptera), bruchid of kidney bean (*Phaseolus vulgaris* L.) by some aromatic essential oils. *Crop Protection* 13:624–628.

Regnault-Roger, C., A. Hamraoui, M. Holeman et al. 1993. Insecticidal effect of essential oils from Mediterranean plants upon *A. obtectus* Say (Coleoptera, Bruchidae), a pest of kidney bean (*Phaseolus vulgaris* L.). *Journal of Chemical Ecology* 19:1231–1242.

Regnault-Roger, C., C. Vincent, and J.T. Arnason. 2012. Essential oils in insect control: Low-risk products in a high-stakes world. *Annual Review of Entomology* 57:405–424.

Rice, P.J., and J.R. Coats. 1994. Insecticidal properties of several monoterpenoids to the house fly (Diptera: Muscidae), red flour beetle (Coleoptera: Tenebrionidae), and southern corn rootworm (Coleoptera: Chrysomelidae). *Journal of Economic Entomology* 87:1172–1179.

Sampson, B.J., N. Tabanca, N. Kirimer et al. 2005. Insecticidal activity of 23 essential oils and their major compounds against adult *Lipaphis pseudobrassicae* (Davis) (Aphididae: Homoptera). *Pest Management Science* 61:1122–1128.

Shaaya, E., M. Kostjukovski, J. Eilberg, and C. Sukprakarn. 1997. Plant oils as fumigants and contact insecticides for the control of stored-product insects. *Journal of Stored Products Research* 33:7–15.

Shaaya, E., U. Ravid, N. Paster et al. 1991. Fumigant toxicity of essential oils against four major stored-product insects. *Journal of Chemical Ecology* 17:499–504.

Sharma, R.N., and K.N. Saxena. 1974. Orientation and developmental inhibition in the house fly by certain terpenoids. *Journal of Medical Entomology* 11:617–621.

Singh, D., M.S. Siddiqui, and S. Sharma. 1989. Reproduction retardant and fumigant properties in essential oils against rice weevil (Coleoptera: Curculionidae) in stored wheat. *Journal of Economic Entomology* 82:727–733.

Singh, D., and A.K. Singh. 1991. Repellent and insecticidal properties of essential oils against housefly, *Musca domestica* L. *Insect Science and Applications* 12:487–491.

Snyder, C. 2015. The Biopesticide Market is Thriving. [Online] Farm Chemicals International. Available at www.agribusinessglobal.com.

Tak, J.H. 2015. A study on insecticidal synergy of plant essential oil constituents against *Trichoplusia ni*. PhD dissertation, University of British Columbia.

Tak, J.H., and M.B. Isman. 2015. Enhanced cuticular penetration as the mechanism of synergy of insecticidal constituents of rosemary essential oil in *Trichoplusia ni*. *Scientific Reports* 5:12690.

Tak, J.H., E. Jovel, and M.B. Isman. 2016a. Contact, fumigant, and cytotoxic activities of thyme and lemongrass essential oils against larvae and an ovarian cell line of the cabbage looper, *Trichoplusia ni*. *Journal of Pest Science* 89:183–193.

Tak, J.H., E. Jovel, and M.B. Isman. 2016b. Comparative and synergistic activity of *Rosmarinus officinalis* L. essential oil constituents against the larvae and an ovarian cell line of the cabbage looper, *Trichoplusia ni* (Lepidoptera: Noctuidae). *Pest Management Science* 72:474–480.

Tarelli, G., E.N. Zerba, and R.A. Alzogaray. 2009. Toxicity to vapor exposure and topical application of essential oils and monoterpenes on *Musca domestica* (Diptera: Muscidae). *Journal of Economic Entomology* 102:1383–1388.

Tong, F., and J.R. Coats. 2012. Quantitative structure-activity relationships of monoterpenoid binding activities to the housefly GABA receptor. *Pest Management Science* 68:1122–1129.

Trongtokit, Y., Y. Rongsriyam, N. Komalamisra et al. 2005. Comparative repellency of 38 essential oils against mosquito bites. *Phytotherapy Research* 19:303–309.

Yeom, H.J., C.S. Jung, J. Kang et al. 2015. Insecticidal and acetylcholine esterase inhibition activity of Asteraceae plant essential oils and their constituents against adults of the German cockroach (*Blattella germanica*). *Journal of Agricultural and Food Chemistry* 63:2241–2248.

Yildirim, E., B. Emsen, and S. Kordali. 2013. Insecticidal effects of monoterpenes on *Sitophilus zeamais* Motschulsky (Coleoptera: Curculionidae). *Journal of Applied Botany and Food Quality* 86:198–204.

# Section II

# Essential and Vegetable Oils

# 3

# Eucalyptus Oil: Extraction, Analysis, and Properties for Use in Pest Control

Hamir Singh Rathore and Leo M.L. Nollet

## CONTENTS

## 3.1 Introduction

*Eucalyptus* is a diverse genus of trees (rarely shrubs). It is native of Australia. There are more than 800 species of *Eucalyptus,* mostly native to Australia. A few native species are found in parts of adjacent New Guinea and Indonesia. A very small number of native species are also found in a distant country, the Philippines. The *Eucalyptus* genus is a part of the myrtle family (Myrtaceae), which is a family of dicotyledon plants, that is, flowering plants. The seed of this family of plants typically contains two embryonic leaves or cotyledons.

Eucalyptus is also known as gum tree [1]. The Myrtaceae is a family of evergreen trees and shrubs. Eucalyptus plants vary in size from shrub to tree. Commonly, most eucalyptus shrubs grow to heights of 6–25 feet, with a spread of between 5 and 10 feet. Eucalyptus shrubs grow upright with an open shape. The foliage is paired, with apposite blue-green leaves. Some species of eucalyptus shrubs, such as *E. globulus,* commonly known as Tasmanian blue gum, are invasive plants that are not ideal for garden landscapes.

The Tasmanian blue gum, southern blue gum, or blue gum (*E. globulus*) typically grows from 30 to 55 m (98–180 feet) tall. The tallest currently known specimen in Tasmania is 90.7 m (297 feet) tall. It has also been reported that even taller trees are known (101 m or 330 feet). The natural distribution of the species includes Tasmania and southern Victoria (particularly the Otway Ranges and southern Gippsland). The species also occurs rarely on King Island and Flinders Island in Bass Strait and on the summit of You Yangs near Geelong. There are naturalized nonnative occurrences in Spain, Portugal, Akmas, and other parts of southern Europe, southern Africa, New Zealand, western United States (California), Hawaii, Micronesia, Caucasus (western Georgia), and Asia (India).

The d'Entrecasteaux expedition made immediate use of the species when they discovered it. Its timber was used to improve their oared boats. The Tasmanian blue gum was proclaimed as the floral emblem of Tasmania on November 27, 1962. The species name is from the Latin *globulus,* "a little button," referring to the shape of the opercula.

Eucalyptus trees are used primarily for their wood, and they also provide other products, including oil, which is used for a variety of industrial and pharmaceutical purposes. Most species of eucalyptus trees produce some essential oil, but only about 20 have commercially viable concentrations of oil in their leaves. Several hundred species of *Eucalyptus* have been found to contain volatile oil. The following species are utilized currently in different countries:

- *E. globulus* Labill (Tasmanian blue gum) in the People's Republic of China, Portugal, Spain, India, Brazil, Chile, Bolivia, Uruguay, and Paraguay
- *E. smithii* R. Baker (gully gum) in South Africa, Swaziland, and Zimbabwe
- *E. polybractea* R. Baker (syn. *E. fruticetorum* F. Muell. ex Miq.) (blumallee) in Australia
- *E. exserta* F. Muell (Queensland peppermint) in the People's Republic of China
- *E. radiate* Sieber ex DC (syn. *E. australiana* and *E. radiata* var. *australiana*) (narrow-leaved peppermint) in South Africa and Australia
- *E. dives* Schauer (cineole variant) (broad-leaved peppermint) in Australia
- *E. camaldulensis* Dehnh (syn. *E. rostrate* Schdl.) (river red gum) in Nepal

Other species, such as *E. cinerea* and *E. cineorifolia,* has also been used for medical purposes, while *E. macarthurtii* has been used in perfumery in the past. *E. citrodora* Hook (lemon-scented gum) and *E. staigeriana* F. Muell. ex Bailey (lemon-scented ironbark) are being used currently in perfumery in the People's Republic of China, Brazil, and India.

The deliberate introduction of eucalyptus to other parts of the word, as an invasive species, has sometimes resulted in unintended consequences, including deadly wild fires, reduced habitat for native plants and animals, and reduction of valuable wetland.

As stated above, several species of *Eucalyptus* are known. A good number of publications are available on each species.

## 3.2 *Eucalyptus melliodora*

*E. melliodora* is a species of trees. There are more than 800 species of eucalypts, and almost all of them are in Australia. Eucalyptus can be found in almost every part of Australia, and the species is adapted to many different habitats. *Eucalyptus* is one of three similar genera that are commonly referred to as "eucalypts," the other being *Corymbia* and *Angophora*. They are also known as gum tree because they exude copious sap from any break in the bark. Eucalypts have many local names, like "gum trees," "melee," "box," "ironbark," "stringybark," and "ash."

### 3.2.1 Flowers and Leaves

Eucalypts have special flowers and fruits that no other trees have. When they flower, a bud cap made of petals grows around the flower until it is ready to open. Then, the bud cap falls off to reveal a flower with no petals.

The woody fruits are called gumnuts. They are roughly cone shaped and open at one end to release the seeds. Almost all eucalypts are evergreen. However, some tropical species lose their leaves at the end of summer. The leaves are covered with oil glands. The shape of the leaves of the older eucalypts differs with time. The leaves of eucalypts of a few years old become longer and spearhead or sickle shaped. Some species keep the round leaf shape throughout their life span. In most cases, their flowering starts when adult leaves begin to appear.

### 3.2.2 Bark

There is a different type of bark on different species of eucalyptus trees. Some trees have smooth bark at the top but rough bark lower down. In smooth-barked trees, most of the bark falls off the tree, leaving a smooth surface that is often colorfully marked. The bark dies every year. With rough-barked trees, the dead bark stays on trees and dies out. Eucalyptus trees are classified in the following groups on the basis of their kind of bark. The classification allows for easy understanding of different species of *Eucalyptus* for better usage.

- *Stringybark*: It consists of strands that can be pulled off in long pieces. It is usually thick with a spongy texture.
- *Ironbark*: It is hard, rough, and deeply furrowed. It is soaked with dried sap exuded by the tree, which gives it a dark red or even black color.

- *Tessellated*: Its bark is broken up to many distinct flakes. These flakes are like cork and can flake off.
- *Box*: It has short fibers.
- *Ribbon*: It has the bark coming off in long thin pieces but still loosely attached in some places; can be long ribbons, firmer strips, or twisted curls.

### 3.2.3 Tall Timer

Currently, specimens of the Australian "mountain ash" are among the tallest trees (92 m) in the world [1] and the tallest of all flowering plants, such as coast redwood, that is, conifers.

### 3.2.4 Frost

A few eucalypts, the hardiest eucalyptus, which are known as snow gums, such as *E. pauciflora*, are capable of withstanding cold and frost (–20°C). Two subspecies (wild varieties) of this tree can survive even colder winters. However, most eucalypts cannot survive frost, or they can only survive light frost (–3°C to –5°C). Several other species, especially from the high plateau and mountains of central Tasmania, have produced extreme cold-hardy forms. The seeds of these hardy strains are planted for ornamental trees in colder parts of the world.

### 3.2.5 Animal Kingdom

It is known that essential oil obtained from eucalyptus leaves is a strong, natural disinfectant, so it is used in some medicines. Eucalyptus oil at high concentration is poisonous, and it is also used as a pesticide. Several marsupials, such as koalas and some possums, are partly resistant to eucalyptus oil. These animals can recognize which plants are safe to eat by their smell. Eucalypts make a lot of nectar, which provides food for many insects.

### 3.2.6 Dangers

All branches of eucalypts drop off as they grow, so eucalypt forests are littered with dead branches. As a result, a large number of tree-felling workers are killed by falling branches. Many deaths are caused by simple camping under eucalypts. The camping is made as trees shed whole and very large branches above the soil to save the water of this soil during droughts. Therefore, the Australian ghost gum eucalyptus is also called "widow maker."

### 3.2.7 Fire

In summer or on hot days, eucalyptus oil (essential oil) vapor rises above the bush to create the well-known distant blue haze of the Australian landscape. The essential oil catches fire very easily, and bush fires can move rapidly through the oil-rich air of the tree crowns. The fallen leaves and branches and dead bark are also flammable. Most species are dependent on trees for spread and regeneration, and eucalypts are well adapted for periodic fires. They regenerate in several ways [2]: (1) by sprouting from underground tubers [3], (2) through hidden buds under their bark, and (3) from seeds sprouting from underground in the ashes after fire has opened them.

### 3.2.8 Fire and the Growth of Eucalypts

As stated above, eucalypts grow back quickly after fire. When the first humans arrived about 50,000 years ago, fires became much more frequent and the fire-loving eucalypts increased many times, for roughly 70% of the Australian forest. On the other hand, valuable timber trees, such as alpine ash and mountain ash, are decreased drastically as they are killed by fire and only grow back from fresh seeds.

## 3.3 Uses and Abuses of Eucalypts

- Eucalypts are important because of their fast growth, especially for providing wood that is a dire need of people.
- They are also planted in parks and gardens to provide timber, firewood, and pulpwood.
- Eucalypts take a very large quantity of water from the soil. Therefore, they are planted in some areas to lower the water table and reduce the salt concentration in the soil.
- Eucalypts have also been used in malaria eradication by draining the soil in places such as Algeria, Sicily [3], and California [4]. Drainage removes swamps, which provide a habitat for mosquito larvae. But such drainage may also destroy harmless habitats.
- Eucalyptus oil is used for many purposes [5,6–9]—for cleaning and deodorizing, as in traces in food supplements such as sweets, cough syrups, and decongestants. It is also used as an insect repellent [5]. The oil also has therapeutic, perfumery, flavoring, antimicrobial, and biopesticide properties [6–9].
- Tasmanian blue gum leaves are used as a therapeutic herbal tea. In India, it is part of green tea [10].
- *E. globulus* bark contains various phenolics, such as guinic, dihydroxyphenylacetic, and caffeic acids; methylellagic acid (MEA)–pentose conjugate, and so forth [11].
- The nectar of eucalyptus produces a high-quality honey that is known for its buttery taste. In fact, eucalyptus flowers in late January, before the flowering of other nut and fruit trees in the United States.
- The ghost gum's leaves were used by Aborigines to catch fish. Aqueous extract of eucalyptus leaves are a mild tranquilizer that stuns fish, making them easy to catch.
- Eucalyptus is also used to fabricate the didgeridoo, a musical wind instrument. It was popularized by Aborigines.
- Eucalypts are grown on plantations in many parts of the world, including the United States, Brazil, Morocco, Portugal, South Africa, Israel, and Spain, to obtain pulpwood for the production of paper and eucalyptus oil.
- Eucalyptus wood for energy is considered to hold great potential as a renewable energy source [12].
- Most species have been found to be invasive (spreading out of the area they are planted in), and they are developing hazards for local flora and fauna (wildlife).

### 3.4 Worldwide Cultivation of Eucalyptus

A botanist, Sir Joseph Banks, was the first to introduce eucalyptus to other parts of the world on the Cook expedition in 1770. It has since been planted in several parts of the world.

#### 3.4.1 Spain

Eucalypts are planted in pulpwood plantations in place of the native oak woodland in Spain. The native woodland was supportive to native animal life, while the eucalypt groves are totally unfavorable to local wildlife. Therefore, there is a decline of wildlife population that is leading to silent forests. Eucalyptus helps some industries, such as saw-milling, pulp, and charcoal producing.

#### 3.4.2 California

In the 1850s, many Australians reached California to participate in the California gold rush. The climate of a great part of California is similar to that of most areas of Australia. So, some people planned to introduce eucalyptus in California. By the early 1900s, eucalyptus was planted in thousands of hectares with the encouragement of the state government, which was hopeful to have a renewable source of timber for construction and furniture purposes. However, it was not possible for two reasons: (1) the trees were cut when they were too long, and (2) the Americans were untrained in processing trees to prevent the wood from twisting and splitting [13]. Mainly the blue gum, a kind of eucalyptus, was found useful in providing windbreaks for highways, orange groves, and other farms in the treeless central part of the state. In many cities and gardens, eucalyptus was admired as ornamental trees and used for shade.

Eucalyptus forests in California have been discarded due to the fact that they drive out the native plants and do not support native animals. They also create fire problems. For example, in 1991 a firestorm destroyed about 3000 homes and killed 25 people in Oakland Hills [14]. As a result, in some parts of California, eucalyptus forests are being replaced by native trees and plants. In order to save the flora and fauna, some eucalypts are being destroyed by insects managed from Australia [15].

#### 3.4.3 Brazil

In 1910, eucalyptus was introduced in Brazil in order to fulfill the requirement of the timber and charcoal industry. Unfortunately, the aftereffects of high consumption of water by eucalyptus were not studied. It causes the soil to dry out, killing many local native plants on which the fauna survive.

#### 3.4.4 India

*E. tereticornis* and *Eucalyptus* hybrid are the two most widely planted eucalypts in India. The species have spread to most parts of the country. Gujarat took the lead in this direction and was followed by Karnataka, Punjab, Haryana, Uttar Pradesh, and other states. Further, farmers went ahead in states such as Gujarat, Haryana, and Punjab in raising irrigated plantations, and the concept of high-density plantations evolved as a suitable alternative to agricultural crops. Unfortunately, this has not had much success. This futile approach was tested because of the versatile nature and amenability of eucalyptus for harvesting in short rotations. Because of the green revolution, India is self-sufficient in food grains, but there is an acute shortage of fuel. In India, despite the unclear situation regarding water requirements, particularly in semiarid areas, eucalyptus has commercial acceptability and is being grown for various purposes, namely, paper and pulp, honey and oil, rural small-scale industries, timber, poles (eucalyptus poles are good for transmission purposes and are also used in the construction of dwelling houses), charcoal, and fuel. Eucalyptus is a versatile, fast-growing, and strongly coppicing tree possessing a wide range of soil and climatic adaptability. Basically, a light demander, the growth of the species is very much reduced under shade. Eucalyptus is known for its drought hardiness. The species is also moderately salt tolerant and relatively fire resistant. Eucalyptus is generally regarded as frost sensitive. There is an impression that eucalypts and wildlife do not go together. While it is true that the natural forest is a better habitat for wildlife, eucalyptus plantation also supports wildlife in places where there is acute deforestation due to a low water table and shortage of rain.

In India, eucalyptus was first planted around 1790 by Tippu Sultan, in his garden on Nandi Hills near Bangalore. The sultan obtained seed from Australia and grew about 16 species of eucalyptus [16]. Subsequently, eucalyptus was planted in Nilgiri Hills, Tamil Nadu, in 1843, and then from 1856, the regular plantation of *E. globulus* was raised to meet the demand for firewood [17]. There were several other attempts to introduce eucalypts in different parts of the country. Today, in fertile land eucalypts have been ignored due to their excessive water absorption tendency and unfriendly nature to native flora and fauna.

About 170 species, varieties, and provenances of eucalypt were tried in India; out of this number, the most sustainable and favorable has been *E. hybrid*, a form of *E. tereticornis* that is known as Mysore gum. In India, it has been found to be fast growing, capable of overtopping weeds, fire hardy, and browse resistant; it coppices well; and it has adaptability to a wide range of edaphoclimatic conditions [11,18]. Other species that are grown on a plantation scale are *E. grandis*, *E. citriodora*, *E. globulus*, and *E. camaldulensis*.

## 3.5 Eucalyptus Oil

### 3.5.1 Methods of Extraction

Many techniques are available for the extraction of essential oil. One technique may produce nicer-smelling oil, while another gives oil with greater aromatherapeutic value. It turns out that essential oil production, like wine making, is an art form as well as a science. Therefore, the procedure applied for oil extraction from plants is important because some processes use solvents that can destroy the therapeutic properties. Some plants, especially flowers, do not lend themselves to steam distillation. Their fragrance and therapeutic essences cannot be completely released by water. For example, jasmine oil and rose oil are often found in "absolute" form. Thus, selection of the method is based on the experience of the user, as well as the application of the product. Therefore, each method has its own merit and importance in the processing of aromatherapy-grade essential oil. For a description of the techniques (maceration, expression method, cold pressing, distillation, hydrodistillation, steam distillation, turbo distillation, extraction with volatile solvent, and supercritical carbon dioxide extraction), the reader is directed to Chapter 1.

## 3.6 Extraction of Eucalyptus Oil

Generally, the fresh or partially dried leaves and young twigs are steam distilled to extract eucalyptus oil. *E. globulus* is the primary source of global eucalyptus oil production. China is the largest commercial producer of this oil [19,20]. The oil yield ranges from 1.0% to 2.4% (fresh weight). It contains cineol as its major part. It is virtually phellandrene-free, a necessary characteristic for internal pharmaceutical use [21]. In 1870, Cloez identified and ascribed the name *eucalyptol*; now it is more often called cineole. It is the dominant portion of *E. globulus* oil [21]. Eucalyptus oil is a complex product. It is made up of many, sometimes hundreds, distinct molecules that come together to form the oil's aroma and therapeutic properties. Most of these molecules are fairly delicate, so their structure may be altered or they may be converted into simple molecules. Therefore, a given method of extraction may give more complete oil, or it may lead to the accumulation of more artifacts than normal. Several papers are published in the literature on methods used for the extraction of eucalyptus oil. A few selected publications are summarized in the following paragraphs.

Fadel et al. have studied [22] the effect of extraction techniques such as hydrodistillation and supercritical fluid extraction (SFE) on the composition of eucalyptus oil. They found that the oil obtained by SFE possesses higher antioxidative activity than the HD extract. Two compounds, p-cymen-7-ol and thymol blue, present in eucalyptus oil are responsible for the antioxidative activity. A newly identified compound, p-cymen-7-ol, in the leaves of eucalyptus species exhibited superior antioxidative activity compared with butylated hydroxyanisol.

The hydrodistillation described in the European Pharmacopoeia [23] has been applied for the isolation of essential oils from eucalyptus leaves collected in Australia and two species, *E. cinerea* and *E. globulus*, obtained from a botanical garden in Jena, Germany.

Robinson [24] has suggested the following procedure for the extraction of eucalyptus oil: Load about 8 tons of fresh, uncompressed eucalyptus leaves in a large vat with a series of steam pipes that run along its bottom. Secure the lid tightly to the vat with locking clamps. Turn the boiler on and begin feeding steam through the pipes at the bottom of the vat. This will cause steam to pass through the leaves, which will vaporize the essential oils in the leaves. This vaporization process requires 3–4 hours and will leave black liquor, which will drain through a hole in the bottom of the vat. Collect the eucalyptus vapor through outlets at the top of the vat. The vapor will collect in a central pipe that is surrounded by pipes that contain cold water. The water circulates through this condenser and cools the vapor back to a liquid state. Drain the condensed distillate into a collection vessel. The distillate contains a water and oil component that will eventually separate. The oil may then be skimmed off the surface. Pour the oil into drums for further refinement. Rectify the unrefined eucalyptus oil. This is a more specific type of distillation that includes the use of chemical reagents that remove the impurities in the eucalyptus oil.

## 3.7 Mechanisms of Steam Distillation of Eucalyptus Oil

It is known that steam distillation may provide a temperature that is substantially below that of the boiling point of the individual constituent. This is a basic requirement when dealing with temperature-sensitive material like essential oil, which is insoluble in water and may decompose at its own boiling point. Eucalyptus oil contains compounds possessing boiling points up to 200°C or higher. In the presence of the aqueous phase (either water or steam), these compounds volatilize at a temperature close to 100°C at atmospheric pressure. Hence, it seems that there is still a need to carry on more research work in order to develop advanced techniques, such as hydrodiffusion, a sort of inverse steam distillation, in which steam is introduced at the top of the organic material–packed chamber, and oil and condensate are obtained from the bottom. The product obtained contains higher ester contents due to less thermally induced hydrolysis.

Therefore, fresh or dried eucalyptus leaves are placed in the plant chamber of the still, and the steam is allowed to pass through the leaves under pressure. The steam treatment softens the cells and allows the oil to escape in vapor form. The steam temperature is regulated to a temperature that is high enough to vaporize the oil, but not too high to burn the leaves or oil. The steam carries the vapors of essential oil through the condenser. Ultimately, the essential oil layer floats on the water surface, which may be separated by decantation or skimmed off.

Thus, a number of factors (including the quality of the leaves, the quality of the equipment, and most important, time, temperature, and pressure) determine the final quality of steam-distilled oil. The color, odor, and therapeutic effects of the oil obtained depend on the technology of the distillation process.

## 3.8 By-Products of Eucalyptus Oil

The water obtained after decantation of oil, a by-product of distillation, is called floral water, distillate, or hydrosol. It possesses many therapeutic properties, so it is valuable in skin care for facial mists and toners. The chemical reactions that occur during steam distillation result in the formation of some artificial chemicals, which are called artifacts. The artifacts have been found to be useful in massage, as they possess mild antiseptic and soothing properties, as well as a pleasing floral aroma.

## 3.9 Physical Properties

Eucalyptus oil has a clear, sharp, pungent, and cooling taste, and a fresh and very distinctive smell (camphoraceous odor). It is a pale yellow liquid and is watery in viscosity. It is insoluble in water, miscible in alcohol having a high concentration or in anhydrous alcohol, and also miscible in oil, fats, paraffins, ether, chloroform, and glacial acetic acid. The boiling point of its major compound, cineole (eucalyptol), is 176°C–177°C.

## 3.10 Methods of Analysis of Eucalyptus Oil

Gas chromatography with fused-silica capillary columns is the routinely used technique in analyzing the essential oils. It can easily analyze major components of essential oil, which can be used to estimate the quality and quantity of the sample. The resolution of gas chromatography can be improved by using the chiral phase (cyclodextrin derivatives) in the column. Several new columns and detectors have been developed to make gas chromatography a commercial technique for the detection and estimation of the quality of eucalyptus oil on the spot; that is, the market value of the oil is decided on the basis of the chromatograph. A few recently reported papers on essential oils are summarized in the following.

Eucalyptus essential oil was extracted from *E. globulus*. It is also known as Tasmanian blue gum, southern blue gum, or blue gum. Gas chromatographic analysis shows [25] that the main components of eucalyptus oil are α-pinene (13.85%), β-pinene (0.78%), sabinene (—), limonene (4.31%), 1,8-cineole (67.67%), p-cymene (13.13), linalool L (—), terpinen-4-ol (—), α-terpineol (45.68%), α-terpinenyleacetate (—), d-carvone (—), α-phellandrene (—), aromadendrene (—), epiglobulol (—), piperitone (—), and globulol (—).

*E. camaldulensis* var. *brevirostris* leaves were extracted by hydrodistillation and supercritical fluid extraction. Both extracts were analyzed by gas chromatography with mass spectrometry (GC-MS), and 90 compounds were identified [22]. In both extracts, the main compounds were found to be β-phellandrene (8.94% and 4.09%), p-cymene (24.01% and 10.61%), cryptone (12.71% and 9.82%), and spathulenol (14.43%, 13.43%, and 13.14%). The yield of the monoterpene hydrocarbons in HD extract (0.288 g/100 g fresh leaves) was slightly higher than that in SFE extract (0.242 g/100 g fresh leaves). The SFE extract possessed a higher concentration of the sesquiterpenes, light oxygenated compounds, and heavy oxygenated compounds than the HD extracts.

A literature survey [26] shows that efforts have been made to analyze *in situ* the main components of eucalyptus oil by means of Raman spectroscopy. The two-dimensional Raman maps obtained by this method demonstrate a unique possibility to study the essential oil distribution in the intact plant tissue. Fourier transform (FT)–Raman and attenuated total reflection infrared (ATR-IR) spectra are recorded for essential oils isolated from several eucalyptus species by hydrodistillation. The density functional theory (DFT) calculations were made in order to interpret the spectra of the essential oils. It has been shown that the main components of essential oils can be recognized by both techniques by comparison of spectra obtained with the spectra of pure terpenoids. It has been found that the vibrational spectroscopic data are very comparable to those obtained by gas chromatography. Both complementary spectroscopic techniques have the potential to replace the existing standard procedures available for quality control purposes. It has also been claimed that both methods can be used in the flavor and fragrance industries, as well as in the pharmaceutical industry, in order to monitor fast quality checks of incoming raw materials and regular control of distillation processes. The *in situ* Raman procedure also provides the possibility of nondestructive analysis of essential oil cells in the intact plant tissue without sample preparation.

Krock et al. [27] have evaluated parallel cryogenic trapping multidimensional gas chromatography coupled with Fourier transform infrared spectrometry and mass spectrometry (MDGC-FT-IR-MS) for the analysis of essential oils. It has been found that MDGC-FT-IR-MS is a good tool to differentiate essential oils, and it can be used as a general method for the analysis of complex mixtures. The data obtained for an authentic sample and for known adulterated eucalyptus Australian oil can be used as the standard for analyzing the unknown samples, as well as for the samples from an unknown source. It has also been claimed that the presence of camphor or a combination of α-thujene, decane, sabinene, β-phellandrene, and γ-terpinene indicates the adulteration of eucalyptus oil.

Maciel et al. estimated the chemical composition of essential oils [28] from three species of plants belonging to the *Eucalyptus* genus [28]. GC and GC-MS were used to analyze *E. staigeriana* (I), *E. citrodora* (II), and *E. globulus* (III). Their percent composition obtained is given in the parentheses in places I, II, and III: α-pinene (3.27, 1.1, 4.15), o-cymene (1.75, —, 2.93), (+) limonene (28.82, —, 8.16), 1,8-cineole (5.39, 0.8, and 83.89), α-terpinolene (9.4, —, —), (–) isopulegol (—, 7.3, —), β-citronellal (0.8, 71.77, —), isopulegol (—, 4.3, —), Z-citral (—, 2.9, —), *trans*-geraniol (10.77, —, —), E-citral (4.2, —, —), methyl gernate (14.16, —, —), and geraniol acetate (3.66, —, —).

## 3.11 Composition of Eucalyptus Oil for Pharmaceutical Use

The properties and uses of the major constituents [25] of pharmaceutical British Pharmacopoeia (BP)-grade eucalyptus oil are discussed in the following. The required percentage of the constituent is given in parentheses.

- *1,8-Cineole (eucalyptol)*, boiling point (b.p.) 174.4°C (at least 70%): This monocyclic monoterpene ether (oxide) is isomeric with α-terpineol, but contains neither a hydroxyl group nor a double bond. The oxygen atom in cineole is inert; for example, it is not attached by sodium or by usual reducing agents. The inertness suggests that the oxygen atom is of the ether type. It is colorless liquid with a light fresh eucalyptus fragrance and a spicy clean cooling taste. It is an antibacterial, cough suppressant, expectorant, nasal decongestant, and respiratory anti-inflammatory. It is also a flavor and fragrance and strong solvent and penetrating oil.
- *α-Pinene* (up to 9%): This bicyclic monoterpene hydrocarbon has a dry woody, resinous-piney odor used in flavors and perfumery. It is mildly antibacterial and anti-inflammatory. It is also a disinfectant and deodorant. It is a strong natural solvent.
- *γ-Limonene* (up to 12%): The monocyclic monoterpene hydrocarbon has the odor of orange–citrus. It is mildly antibacterial. It is a strong solvent used for the removal of oil and grease and in cleaning products, such as hand cleaners. Pure limonene oxidizes readily, and it is a skin and respiratory irritant. It is stable as a constituent of eucalyptus oil.
- *α-Terpineol*: The monocyclic monoterpene alcohol is of lilac odor and strongly antibacterial. It is a common ingredient in perfumes, cosmetics, and flavors.
- *p-Cymene*: The monocyclic monoterpene aromatic hydrocarbon has a strong aromatic hydrocarbon odor. It possesses good antibacterial activity and is used in soaps and preparations to help overcome undesirable odors.
- *Terpinen-4-ol*: It is a tricyclic monoterpene alcohol that is colorless or with a pale yellow color, herbaceous peppery and woody odor.
- *Cuminal aldehyde*: It is a colorless to yellowish oil liquid that possesses a strong persistent odor, acrid burning taste, and antibacterial activities.

- *Globulol*: It is a tricyclic sesquiterpene alcohol with a sweet rose-like odor, and it is antibacterial.
- *p-Isoproplyphenol (australol)*: It is a phenol, and it is a very strong antibacterial.
- *Eudesmol*: It is a bicyclic sesquiterpene alcohol that possesses mild antibacterial activity. It exists in α, β, and γ forms.
- *Aromadendrene*: It is a tricyclic sesquiterpene hydrocarbon of a woody odor, and it possesses antibacterial as well as antifungal activity.

## 3.12 Common Uses of Eucalyptus Oil

- Pharmaceutical-grade eucalyptus oil contains a wide range of natural compounds with a broad spectrum of antibacterial activity. Therefore, it is used in antiseptics, arthritis and sports muscular rubs, liniments, cough mixtures, and many other products.
- Eucalyptus oil for commercial use is a lower-cost alternative to pharmaceutical-grade eucalyptus oil. It is a natural cleaner, freshener, disinfectant, sensitizer, and deodorizer. A water-soluble version is also available.
- Eucalyptus oil for industrial use is an economical blend with a strong eucalyptus fragrance, especially for manufacturers to incorporate into their own formulations.
- Special blends of aqueous solutions of eucalyptus are also available for specific purposes, such as body lotions and massage.
- This oil possesses impressive therapeutic, perfumery properties. *E. globulus* oil has established itself internationally because it is virtually phellandrene-free, a necessary characteristic for internal pharmaceutical use [19]. In 1870, Cloez identified it and gave it the name *EUCALYPTOL*; now, it is commonly known as 1,8-cineole—the major portion of *E. globulus* [21].
- The therapeutic properties of eucalyptus oil are analgesic, antibacterial, anti-inflammatory, antineuralgic, antirheumatic, antiseptic, antispasmodic, antiviral, astringent, balsamic, cicatrisant, decongestant, deodorant, depurative, diuretic, expectorant, febrifuge, hypoglycemic, rubefacient, stimulant, vermifuge, and vulnerary.

## 3.13 Pesticidal Properties and Uses of Eucalyptus Oil

- It is formulated with other ingredients and applied locally to cattle horn stumps following dehorning [29].
- *E. cintrodora* showed the highest mortality of the larval population of the root knot nematode, *Meloidogyne incognita*, at 100 ppm concentration after 48 hours [30].
- As a nematicide, *E. rudis* showed significant activity at all dilutions (1:0, 1:1, and 1:2), while *E. globulus* showed activity at a lower concentration (1:2) for inhibition

in the hatching of eggs of *Meloidogyne javanica in vitro* and *in vivo* and reduction of the number of knots.

- The oral $LD_{50}$ (mg/kg) for rat of 1,8-cineole is 2480, of eugenol for rat 2680, of thymol for mice 1800, and of menthol for rat 3180.

- Eucalyptus oil (3%) can be used to control both hoppers and aphids with the least damage to natural enemies [31].

- Eucalyptus oil possesses antimicrobial and biopesticide properties [21].

## 3.14 Mechanisms of Pesticidal Action of Eucalyptus Oil

Apart from essential oils used mainly in foods, the best-known essential oil worldwide might be eucalyptus oil, produced from the leaves of *E. globulus* [32]. Steam-distilled eucalyptus oil is used throughout Asia, Africa, Central and South America as a primary cleaning and disinfecting agent added to soap mop and countertop cleaning solutions; it also possesses insect and limited vermin control properties. Note that there are hundreds of species of eucalyptus, and perhaps dozens are used to various extents as sources of essential oils. Not only do the products of different species differ greatly in characteristics and effects, but also products from the very same tree can vary grossly.

The oils are generally composed [33,34] of complex mixtures of monoterpenes, biogenetically related phenols, and sesquiterpenes. Examples include 1,8-cineole, the major constituent of eucalyptus oil. They act as fumigant and contact insecticide. Their mode of action is neurological. They interfere with the neuromodulator octopamine [35] and GABA-gated chloride channels [36]. The purified terpenoid constituents of essential oils are moderately toxic to mammals, but with few exceptions, the oils themselves or products based on the oils are mostly nontoxic to mammals, birds, and fish [4]; therefore, they are called "green pesticides." Eucalyptus oils are volatile, so they have limited persistence under field conditions. Therefore, although natural enemies are susceptible via direct contact, predators and parasitoids reinvading a treated crop 1 or more days after treatment are unlikely to be poisoned by residue contact, as often occurs with conventional insecticides.

## 3.15 Technology of Eucalyptus Oil as a Pesticide

In 2013, Mousa et al. [31] reported "the effect of garlic and eucalyptus oils in comparison to organophosphate insecticides against some piercing-sucking faba bean insect pest and natural enemies population." They applied a 3% solution of dimethoate (30%), pestban (48%), garlic *Allium sativa* oil, and eucalyptus oil four times in each spray, and the samples were taken at the first, fifth, seventh, and tenth days after application to determine the reduction in numbers. The population of leafhoppers and plant hoppers was found to be reduced in the following order: garlic oil (68.07%) > dimethoate (67.90%) > pestban (64.02%) > eucalyptus oil (43.27%). Similarly, the efficiency for controlling aphids was found to be as follows: garlic oil (90.96%) > pestban (89.44%) > eucalyptus oil (80.66%) > dimethoate (76.14%). Garlic and eucalyptus oils were found to be superior to organophosphorus insecticides in partly maintaining the natural enemies. This suggests the possible use of eucalyptus oil in crop protection without environmental pollution.

The insecticidal effects [28] of *E. straigeriana*, *E. citrodora*, and *E. globulus* oils on eggs, larvae, and adults of *Lutzomyia longipalpis* have been assessed. In these *in vitro* tests, aqueous solutions of plant oils were used at concentrations of 20, 10, 5, 2.5, and 1.2 mg/ml (*E. staigeriana*) and 40, 20, 10, 5, and 2.5 mg/ml (*E. citrodora* and *E. globulus*). The eggs, larvae, and adults were sprayed with the oils. The hatched larvae were counted for 10 consecutive days and observed until pupation. Insect mortality was observed after 24, 48, and 72 hours. *E. staigeriana* oil was the most effective on all three phases of the insect, followed by *E. citrodara* and *E. globulus* oils. The major constituents of the oils were Z-citral and α-citral (*E. staigeriana*), citronellal (*E. citrodora*), and 1,8-cineole (*E. globulus*). This shows that eucalyptus oils may be used to control *L. longipalpis*—its chemical constituents are already known for their insecticidal activity—and these oils are produced on a commercial scale in Brazil.

## 3.16 Conclusion

Eucalyptus oil may be used as an effective green pesticide for organic food production. It might be a unique partner of future integrated pest management (IPM). The yield of eucalyptus oil ranges from 1.02% to 2.4% (fresh weight of leaves). A literature survey shows that there is a genuine interest in the harvest of eucalypts all over the world. Publications made by different countries in this area are in the following order: China > Portugal = India = South Africa > Australia > Chile = Spain = Brazil = other countries. In Asian countries, especially in India, overhasty plantation of eucalypts results in failure. The data available on the cultivation of eucalypts for wood and pulp are many folds more than that on the cultivation of eucalypts for oil production, while the market value of the latter is many times more than that of the former. This may be due to the fact that the former is easy to adopt and perform, and the latter requires skilled workers. The returns may be maximized on the eucalypts planted for the production of oil by understanding the effect of climatic and edaphic conditions, the selection of species, the application of fertilizers, the design of plantation (spacing, etc.), and the nutrient recycling. This work may be made easier by the collation of published and unpublished data that are at present scattered in the literature.

## References

1. "Eucalyptus L'Hér." 2009. Germplasm Resources Information Network, U.S. Department of Agriculture. https://npgsweb.ars-grin.gov/gringlobal/taxonomygenus.aspx?id=4477.
2. Hickey, J.E., Kostoglou, P., Sargison, G.J. Tasmania's tallest trees. Forestry Tasmania. Retrieved January 27, 2005.
3. Grieve, M. A modern herbal eucalyptus. http://www.botanical.com/botanical/mgmh/e/eucaly14.html. Retrieved January 27, 2005.
4. Santos, R.L. The eucalyptus of California. Alley-Cass Publications, Denair, CA. http://library.csustan.edu/sites/default/files/Bob_Santos-The_Eucalyptus_of_California.pdf. Retrieved January 27, 2005.

5. Fradin, M.S., and Day, J.F. 2002. Comparative efficacy of insect repellents against mosquito bites. *N. Engl. J. Med.* 347:13–18.
6. *Eucalyptus globulus* monograph. Australian Naturopathic Network.
7. Herbal monograph: *Eucalyptus globulus*. Himalaya Healthcare.
8. *Eucalyptus globulus*. Association of Societies for Growing Plants.
9. Yang, Y.-C., Cheoi, H.-Y., Choi, W.-S., Clark, J.M., Ahn, Y.-J. 2004. Ovicidal and adulticidal activity of *Eucalyptus globulus* leaf oil terpenoids against *Pediculuishu manus capitis* (Anoplura: Pediculidae). *J. Agric. Food Chem.* 52(9): 2507–2511.
10. *Eucalyptus globulus* Labill leaf pieces tea.
11. European Pharmacopoeia. 1983. Maisouneuve SA, Sainte Ruffine.
12. Williams, T. 2002. America's largest weeds. *Audubon.* 104: 24–31.
13. Eucalyptus of California. www.library.csustan.edu. Retrieved May 24, 2009.
14. Audubon: Incite. www.audubonmagazine.org. Retrieved May 24, 2009.
15. @UCSD: Tree wars. Alumni.ucsd.edu. Retrieved May 24, 2009.
16. Shyam Sunder, S. 1984. Some aspects of Eucalyptus hybrid. Presented at Workshop on Eucalyptus Plantation, Indian Statistical Institute, Bangalore.
17. Wilson, J. 1973. Rational utilization of montane temperate forests of South India. *Indian Forests* 99(12): 707–715.
18. Kushalappa, K.A. 1985. Productivity and nutrient recycling in Mysore gum plantations near Bangalore. PhD thesis, Mysore University.
19. Boland, D.J., Brophy, J.J., House, A.P.N. (eds.). 1991. *Eucalyptus Leaf Oils—Use Chemistry, Distillation and Marketing*. Inkata Press, Melbourne, Australia.
20. Eucalyptus oil. FAO Corporate Document Repository.
21. Boland, D.J., Brophy, J.J., House, A.P.N. (eds.). 1991. *Eucalyptus Leaf Oils—Use Chemistry, Distillation and Marketing*. Inkata Press, Melbourne, Australia. pp. 3, 78–82.
22. Fadel, H., Marx, F., El-Sawy, A., El-Ghorab, A. 1999. Effect of extraction techniques on the chemical composition and antioxidant activity of *Eucalyptus camaldulensis* var. *brevirostris* leaf oils. *Z. Lebensm. Unters. Forsch. A* 208: 212–216.
23. European Pharmacopoeia. 1983. Maisonneuve SA, Sainte Ruffine.
24. Robinson, A. How to extract eucalyptus oil. eHow. http://www.ehow.com/how_5244668 _extract-eucalyptus-oil.html.
25. Rao, V.P.S., Pandey, D. 2007. Project report on extraction of essential oil and its applications in partial fulfillment of the requirements of Bachelor of Technology (Chemical Engineering). Department of Chemical Engineering, National Institute of Technology, Rourkela, Orissa, India.
26. Baranska, M., Schulz, H., Reitzenstein, S., Uhlemann, U., Strehle, M.A., Krüger, H., Quilitzsch, R., Foley, W., Popp, J. 2005. Vibrational spectroscopic studies to acquire a high quality control method of Eucalyptus essential oils. *Biopolymers* 78(55): 237–248. Article first published online, April 26.
27. Krock, K.A., Ragunathan, N., Wilkins, C.L. 1994. Multidimensional gas chromatography coupled with infrared and mass spectrometry for analysis of eucalyptus essential oils. *Anal. Chem.* 66(4): 425–430.
28. Maciel, M.V., Morais, S.M., Bevilaqua, C.M., Silva, R.A., Barros, R.S., Sousa, R.N., Sousa, L.C., Bitro, E.S., Souza-Neto, M.A. 2010. Chemical composition of *Eucalyptus* spp. essential oils and their insecticidal effects on *Lutzomyia longipalpis*. *Vet. Parasitol.* 167: 1–7.
29. *Farm Chemicals Handbook*. 1980. Meister Publishing, Willoughby, OH, p. D 133.
30. Singh, R. 2008. In *Crop Protection by Botanical Pesticides*, ed. N.A. Shakil. CBS Publishers & Distributors, New Delhi, pp. 93, 97.
31. Mousa K.M., Khodeir, I.A., EI-Dakhakhni, T.N., Youssef, A.F. 2013. Effect of garlic and eucalyptus oils in comparison to organophosphat insecticides against some piercing-sucking faba bean insect pests and natural enemies populations. *Egypt. Acad. J. Biolog. Sci.* 5(2): 21–27.
32. Wikipedia. Essential oil. https://en.wikipedia.org/wiki/Essential_oil.

33. Kaul, O., Walia, S., Dhaliwal, G.S. 2008. Essential oils as green pesticides: Potential and constraints. *Biopestic. Int.* 4(1): 63–84.

34. Kostyukovsky, M., Rafaeli, A., Gileadi, C., Demchenko, N., Shaaya, E. 2002. Activation of octopaminergic receptors by essential oil constituents isolated from aromatic plants: Possible mode of action against insect pests. *Pest Manag. Sci.* 58: 1101–1106.

35. Priestley, C.M., Williamson, E.M., Wafford, K.A., Sattelle, D.B. 2003. Thymol, a constituent of thyme essential oil, a positive allosteric modulator of human GABA receptors and a homo-aligomeric GABA receptor from *Drosophila melanogaster*. *Br. J. Pharmacol.* 140: 1363–1372.

36. Stroh, J., Wan, M.T., Isman, M.B., Moul, D.J. 1998. Evaluation of the acute toxicity to juvenile Pacific coho essential oils, a formulated product, and the carrier. *Bull. Environ. Contam. Toxicol.* 60: 923–930.

# 4

## *Mentha* Oil

Hamir Singh Rathore, Shafiullah, and Raveed Yousuf Bhat

## CONTENTS

## 4.1 Introduction

*Mentha* is an aromatic herb plant having pleasant odor leaves. In Japan, it is known by the name *Pudina*. In India, it is also known by the name *Pudina*. The native place of *Mentha* is Japan. So, *Mentha* is also known as Japanese mint. Peppermint oil is extracted from *M. piperita* of the Labiatae family and is also known as brandy mint and balm mint. This cooling and refreshing essential oil is used in aromatherapy to stimulate the mind, to increase mental agility, and to increase focus, while cooling the skin, reducing redness, and calming irritation and itchiness. It furthermore helps to ease spastic colon, migraine, headaches, sinus, and chest congestion, and boosts the digestive system.

### 4.1.1 Origin of Peppermint Oil

It is a native of the Mediterranean, but is now also cultivated in Italy, the United States, Japan, and Great Britain. It is a perennial herb that grows up to 1 m (3 feet) high and has slightly hairy serrated leaves with pinkish-mauve flowers arranged in a long conical shape.

It has underground runners by which it easily propagates. This herb has many species, and peppermint piperita is a hybrid of watermint (*M. aquatica*) and spearmint (*M. spicata*).

According to Greek mythology, the nymph Mentha was hotly pursued by Pluto, whose jealous wife, Persephone, trod her ferociously into the ground, whereupon Pluto then turned her into an herb, knowing that people would appreciate her for years to come.

It has been cultivated since ancient times in Japan and China. Evidence of use was found in Egypt in a tomb dating back from 1000 BC.

### 4.1.2 Different Species of Mint

There are 25 species of mint, and at least 14 different varieties are growing in South Africa [1]. These include unusual ones such as "Black Peppermint," "Basil Mint," "Mint Julep," and "Slender Mint" (also known as "Aussie Mint"), as well as the better known peppermint, and the fruity ones, such as apple, pineapple, ginger, and eau de cologne mint. There are also several other hybrid species of mint.

Some species [1], with their characteristics and uses, are summarized in Table 4.1.

### 4.1.3 Commercial Cultivars

Commercial cultivars [2,3] may include Dulgo pole, Zefir, Bulgarian population, Clone 11-6-22, Clone 80-121-33, Mitcham Digne 38, Mitcham Ribecourt 19, Todd's#x2019, and Todd's Mitcham, a *Verticillium* wilt-resistant cultivar produced from a breeding and test program of atomic gardening at Brookhaven National Laboratory from the mid-1950s.

A number of cultivars have been selected for garden use: *Mentha × piperita* "Candymint" (stems reddish) [4]; *Mentha × piperita* "Chocolate Mint" [5–7]; *Mentha × piperita* "Citrata" (includes a number of varieties, including eau de cologne mint [8], grapefruit mint, lemon mint [9], and orange mint); *Mentha × piperita* "Crispa" (leaves wrinkled) [10]; *Mentha × piperita* "Lavender Mint" [11]; *Mentha × piperita* "Lime Mint" [12,13]; and *Mentha × piperita* "Variegata" [14].

### 4.1.4 Difference between *Mentha*, Peppermint, and *M. piperita* Essential Oils

*Mentha* oil (*M. arvensis*) is also commonly known as the Japanese mint oil. It has a high menthol content and a very good "cooling" effect associated with it. *M. piperita* is commonly known as "peppermint" in the West. It gives a milder and warm-smelling essential oil. Peppermint oil is obtained from var. *Mentha piperita*, which is a hybrid species developed in India, and it is a crossing between the above two with a higher menthol content than the common *M. piperita*. The reason for this distinction is that most of the peppermint oil being sold in the market is from the hybrid variety of *M. piperita* (with a higher menthol content). Therefore, if somebody wants the warmer-smelling peppermint, one has to go for *M. piperita*.

### 4.1.5 Etymology of *Mint*

Mint descends from the Latin word *mentha*, which is rooted in the Greek word *minthe*, personified in Greek mythology as Minthe, a nymph who was transformed into a mint plant [15].

**TABLE 4.1**

Different Types of Mint and Their Uses

| Name | Characteristics | Used In |
|---|---|---|
| 1. Garden mint (*Mentha spicata*) | Most popular mint; deep green and very aromatic leaves with spearmint flavor | Mint sauce, jellies, cakes, cosmetics, natural insecticides, and medicines |
| 2. Spearmint (*Mentha spicata aquatic*) | Most popular mint; deep green and very aromatic leaves with more spearmint flavor | Mint sauce, jellies, cakes, cosmetics, natural insecticides, and medicines |
| 3. Apple mint (*Mentha suaveolens*) | Tall growing with hairy leaves and mauve flowers | Mint sauce, jellies, cooked vegetables, and salad |
| 4. Basil mint (*Mentha piperita* f. "Citrata") | Small leaves with basil mint aroma | Flavoring melon, tomatoes, and fruit salad |
| 5. Black peppermint (*Mentha × piperita*) | Dark brown, oval, strongly peppermint-scented leaves | Medicines especially for relieving indigestion and chest infections |
| 6. Eau de cologne mint (*Mentha piperita* "Citrata") | Large, round dark green leaves with an orange and purple tinge | Oils and vinegars |
| 7. Chocolate mint (*Mentha piperita* spp.) | Dark green-brown leaves with a chocolate peppermint flavor | Pudding, ice cream, and drinks |
| 8. Ginger mint (*Mentha gracilis*) | Variegated gold and green ginger-scented leaves | Salads, teas, drinks, and floral decorations |
| 9. Mint julep (*Mentha spicata* "Julep") | Sweetly scented leaves and a striking fresh flavor | Beverages, as well as medicinally, and as a natural insect repellent |
| 10. Pineapple mint (*Mentha suaveolens variegata*) | Green and cream variegated leaves with a strong pineapple scent | Salads, fruit salads, and as a garnish |
| 11. Ground cover mints, including slender mint (*Mentha diemenica* "Aussie Mint"), lawn pennyroyal (*Mentha pulegium* var.), and Corsican mint (*Mentha requienii*) | Flat, low-growing ground covers that are ideal between paving, stepping stones, and walkways; they release their scent when trod upon | Foodstuffs |
| 12. Ant and flea repellent mint (penny royal—*Mentha pulegium*) | Upright growing | Repellents |
| 13. Garden mint (*Mentha spicata*) and spearmint (*Mentha spicata aquatic*) | Clusters of mauve flowers; crushed leaves release a strong peppermint fragrance | Repellent to ants and fleas |

Mint leaves, without a qualifier like "peppermint" or "apple mint," generally refer to spearmint leaves.

In Spain and Central and South America, mint is known as *mentha*. In Lusophone countries, especially in Portugal, mint species are popularly known as *hortelã*. In many Indo-Aryan languages, it is called *pudīna*.

The taxonomic family Lamiaceae is known as the mint family. It includes many other aromatic herbs, including most of the more common cooking herbs, such as basil, rosemary, sage, oregano, and catnip. As an English colloquial term, any small mint-flavored confectionery item can be called a mint [16].

In common usage, other plants with fragrant leaves may be called "mint," although they are not in the mint family.

Vietnamese mint, commonly used in Southeast Asian cuisine, is *Persicaria odorata* of the family Polygonaceae, collectively known as smartweeds or pinkweeds.

Mexican mint, marigold, is *Tagetes lucida* of the sunflower family (Asteraceae).

### 4.1.6 Global Scenario

The native place of *Mentha* is Japan. After Japan, its cultivation spread to Argentina, Brazil, and China. In India, it came late, that is, after China. Now India stands in the top position regarding *Mentha* oil production and exports. Other countries, such as Brazil, China, and the United States, are also major producers and suppliers of *Mentha* oil. These countries also produce the by-products obtained in *Mentha* oil processing. However, China is among the biggest importers of *Mentha* oil. It does further refinement of the oil and preparation of its by-products, and then these products are supplied all over the world.

The total *Mentha* oil produced is derived from *M. arvensis* (75%), and it is mainly to produce menthol. Of the total *M. arvensis* oil produced, India contributes 73%, China 18%, and other countries 9%. The consumption *Mentha* produced is 30%–40% in India, followed by China, the United States, and European countries.

Total production of the oil is 75%–80% in India, 18% in China, and the remaining in Brazil and the United States. The total production and consumption of the oil has been found to be steady over a period of time. Recently, the global output was further enhanced. The total output from India alone is estimated to be 35,000 tons in 2009–2010. The export of mint products such as menthol flakes, menthol crystals, and mint oil is increasing day by day. This increase is due to the increasing demand for *Mentha* oil products abroad. The major countries for Indian export are Argentina, Brazil, France, Germany, Japan, the United Kingdom, the United States, and others.

## 4.2 Botany of the *Mentha* Plant

All mints require the same kind of growing conditions: moist and rich soil in a partially shaded spot. The plant dies down in winter but will come up again in spring. Most varieties are rampant growers that will smother other plants around them, and so they should be grown on their own, in pots or in pots that are sunk into the ground. This makes it easier to lift the pots and trim off the runners. Mint will grow from root cuttings, and it is a good idea to make or buy new plants every year, as they will start to lose their vigor.

### 4.2.1 *Mentha piperita* L. Emend. Hudson

*M. piperita* L. Emend. Hudson is currently one of the most economically important aromatic and medicinal crops. It is commonly known as peppermint, brandy mint, candy mint, vilayt ipudina, or paparaminta.

The plant is a strongly scented, perennial, glabrous herb and is 30–90 cm in height. The square stems are usually reddish-purple and smooth. The leaves are short, 2.5–5 cm long, oblong-ovate, and serrate. The flowers are purple-pinkish and appear in the summer months. The sowing of the crop starts in January and continues until March, and it is harvested during May–June. In India, during this period there are sunny days and a hot season, and plants give a good yield of oil. The *Mentha* oil is extracted from the leaves by processing and steam distillation. In India, during May–June dried plant–based fuel (plant material including *Mentha* plant) is also available for steam distillation. Monsoon starts in July, and in the rainy season, the percentage of oils falls and steam distillation becomes costly. The plant has runners above- and belowground, and propagation takes place through these runners. It is originally a native of Asia, Europe, Canada, and the United States, and has been naturalized in several parts of India. It is cultivated in Japan, India, China, Europe, America, Australia, South Africa, Brazil, and some other countries. The leaves and flower tops are collected as soon as flowers begin to open, and are dried as a crude drug for their oil and peppermint [17,18]. The arrival of the oil at the physical market starts in May and extends until November. Generally, two or three cuttings can be done for one crop.

### 4.2.2 *Mentha longifolia* L.

It is commonly known as wild mint (English); kruisement balderjan (African); koena-ya-thaba (southern Sotho); inixina, inzinziniba (Xhosa); and ufuthana lomhlanga (Zulu) [19–22]. Because of its strong mint smell and taste, this herb is grown in kitchen gardens, as well as in pots, where it is used in foodstuffs and as a medicine. As discussed above, there are different mints—different species, many hybrids, and special selections that are grown all over the world. Almost all the species are water lovers and are usually grown in wet and damp places.

#### 4.2.2.1 Description

*M. longifolia*, or wild mint, is a fast-growing perennial herb that has creeps along an underground rootstock. It reaches up to 1.5 m height in favorable climatic conditions; usually it goes to 0.5–1 m, and it remains shorter in dry conditions. It is strongly aromatic; its leaves are formed in pairs opposite each other along the square-shaped stem. The soft, lanceolate leaves (long and narrow with a sharp point) are between 45 and 100 mm long and 7.20 mm wide. The leaves are usually coarsely hairy and the edges sparsely toothed. The color of leaves varies from light and dark green to gray. The small-sized flowers of *M. longifolia* are crowded into spikes at the tip of the stem. The flowers of this wild mint vary in color from white to mauve throughout the summer months.

Two mint species, *M. longifolia* and *M. aquatica* (wild water mint), are indigenous to South Africa. Both are commonly found in marshes and along streams from the cape through Africa and Europe.

*M. longifolia* is identified by its stalkless leaves and white to mauve flowers that are grouped in a long spike. The leaves of *M. aquatica* (aquatica means living in water) are broader and more egg shaped, and its flowerheads are roundish whorls (approximately 25 mm in diameter), with pink or mauve flower clusters formed one above another. There are three subspecies: (1) *wissi* (cape velvet mint), which has long and thin gray-green leaves with an unpleasant aromatic smell; (2) *capensis* (balderjan), which possesses a strong peppermint scent; and (3) *polyadena* (spearmint).

#### 4.2.2.2 Growing *Mentha longifolia*

It is worth finding a place in the garden for the wild mint in order to achieve a strong fragrance and many uses. It is easy to grow, as it is a fast grower with underground runners and heavy feeders, and it is a water lover. To encourage new fresh growth, mint should be cut back often. The mints grow in semishade and full sun. They may be grown in pots but need to be repotted every year or two in new compost-rich soil. It is better to provide small leaks while keeping the mint moist. Many crops of mint may be obtained by cutting back often, and this process encourages new fresh growth of the plant. Mint plants are easy to multiply by division, as the small piece of healthy rootstock quickly grows into a new clump with regular water and compost, and cuttings of young actively growing shoot roots are easily possible throughout the year.

The large mint family, Lamiaceae, with about 250 genera and 6700 species, includes many well-known herbs and garden plants, such as lavender, sage, basil, rosemary, and mint. *Mentha* (Latin for "mint") is a cosmopolitan genus with about 20–30 species that are mainly found in temperate regions. Being a very easy hybridization of mint species, it is quite difficult to identify a particular species of mint. Therefore, the same common name is given to different plants, or the same plant may have different names in different areas and languages. For example, *M. longifolia* is known as horse mint in England, as its leaves are usually unpleasantly scented [23]. *M. spicata* is also known as spearmint; it is not indigenous to South Africa, but it is often found as a garden escape in wet areas. This exotic mint is a very popular herb and has been cultivated in Europe since ancient times. Its origin has been lost, but it has been naturalized throughout the world in different forms [23].

## 4.3 Ecology

In the flowering season, bees and butterflies are attracted to the wild mint. Peppermint typically occurs in moist habitats, including stream sides and drainage ditches. Being a hybrid, it is usually sterile, producing no seeds and reproducing only vegetatively, spreading by its rhizomes. If placed, it can grow anywhere, with a few exceptions [24,25].

Outside of its native range, areas where peppermint was formerly grown for oil often have an abundance of feral plants, and it is considered invasive in Australia, the Galápagos Islands, New Zealand [26], and the United States.

## 4.4 Toxicology

The toxicity studies of the plant have received controversial results. Some authors reported that the plant may induce hepatic diseases (liver disease), while others found that it protects against liver damage that is caused by heavy metals [27,28]. In addition, the toxicities of the plant seem to vary from one cultivar to another [29] and are dose dependent [27,30].

## 4.5 Extraction of Peppermint Oil

Peppermint oil is extracted from the whole plant aboveground just before flowering. The oil is extracted by steam distillation from the fresh or partly dried plant, and the yield is 0.1%–1.0% [31]. Supercritical fluid extraction (SFE) has been performed [32], and the results obtained have been compared with those obtained by hydrodistillation (HD). The extraction was carried out under two conditions: SFE-1 and SFE-2. Higher concentrations of menthone, menthol, 1,8-cineole, and piperitone, and lower concentrations of methyl acetate, α-caryophulene, and α-cadinene were obtained by SFE-1 than by SFE-2. Oxygenated monoterpenes, which are responsible for the peppermint fragrance, come out to 79.2% by SFE-1 and 74.4% by SFE-2. Sesquiterpenes were distilled out only 7.7% by SFE-1 and 11.6% by SFE-2. Hydrodistillation gave a higher concentration of terpene acetate (12.5%), while SFE-1 gave a 12.0% concentration of terpene acetate. A recently developed technique, microwave hydrodiffusion and gravity, has been used for the extraction of peppermint oil [33]. It is much faster than conventional hydrodistillation.

Gill et al. [34] have carried out experiments to study the kinetics of *Mentha* oil extraction from *Mentha* leaves (*Mentha arvensis* L.), and quality analysis was carried out for the oil extracted. The oil was extracted from *Mentha* leaves at three different moisture contents of 74.30% (fresh leaves), 42.30% (shade dried), and 19.35% (sun dried) using the hydrodistillation method. Various physicochemical tests were carried out on the oil extracted. The results revealed that the hydrodistillation process took more time for oil extraction and oil recovery was less. Various physicochemical properties, such as acid value, refractive index, specific gravity, saponification value, and solubility in water, did not show significant variations with respect to oil extracted by differently pretreated *Mentha* leaves.

## 4.6 Consumption and Export

India's domestic consumption of *Mentha* oil is hovering in the range of 8,000–10,000 tonnes, with an annual growth of around 7%–8%, while the export demand ranges from 18,000 to 22,000 tonnes (inclusive of menthol, menthol crystals, and mint oil), growing by 10%–12% annually.

India is the largest exporting country for *Mentha* oil. Mint products, including mint oils, menthol crystals, and menthol powder, are the single largest product group in the export basket, accounting for 22%–27% of spices exported from India. Mint products are exported to the United States and China, and these two countries together account for more than 53% of total mint exports. The other major buyers are Singapore, Germany, the Netherlands, and Japan.

The slowdown in major consuming markets such as the European Union and the United States adversely affected export of value-added spices, including mint products like mint oils, menthol crystals, and menthol powder, from 2007 to 2009 [35].

The export of mint products in 2009–2010 was 19,000 tonnes valued at 1189.72 crore, against 20,500 tonnes valued at 1420.25 crore in 2008–2009. Per the latest release by the Spices Board of India, mint exports in April to August 2010 surged by 2% to 723.95 lakhs, against 595.57 lakhs reported last year in the same period. In 2010, a strong demand was seen in the international market as the export prices of *Mentha* bold were quoting at around \$20–\$20.50 per kilogram, against \$16 per kilogram reported May 2010. Traders are estimating that the total export of 2010–2011 is nearly 22,000 tonnes; in 2009–2010 it was 19,000 tonnes. Export from India has crossed the level of around 10,000 tons per annum, and that is likely to increase in the coming years.

## 4.7 Analysis

Chromatographic profiling of peppermint oil has been made by using gas chromatography with a flame ionization detector [36–39]. Quantification, as well as identification, of *Mentha* oil is difficult due to the fact that its components, like menthol, methone, and methyl acetate, consist of several stereoisomers. For example, menthol has three chiral centers, for a total of eight stereoisomers, which makes the chromatographic separation difficult. This problem was resolved by using gas chromatography–mass spectrometry (GC/MS) [40]. Visible fluorescence analysis of *Mentha* oil has been reported [38,41]. The results obtained show that the same group of organic compounds dominate in the oils of peppermint and spearmint, while different compounds are present in Japanese mint oil. Estimation of menthone, menthofuran, and methyl acetate in peppermint oil has been reported by using capillary gas chromatography equipped with a dual flame ionization detector and dual injector [42]. Gas liquid chromatography has been used for the estimation of pulegone [43], which was found to be unresolved by gas chromatography. In fact, pulegone has a retention time that is very similar to that of menthol and isomenthol. It is either very near to that of menthol, with a consequent overlap, or very similar to those of isomenthol and some sesquiterpene hydrocarbons (cadinen and caryophyllene) [39].

## 4.8 Chemical Composition

The chemical components of peppermint oil are menthol, menthone, 1,8-cineole, methyl acetate, methofuran, isomenthone, limonene, β-pinene, α-pinene, germacrene-d, trans-sabinene hydrate, and pulegone.

The constituents reported in the International Pharmacopoeia [37,44] are limonene (1.0%–5.0%), cineole (3.5%–14.0%), menthone (14.0%–32.0%), menthofuran (1.0%–9.0%), isomenthone (1.5%–10.0%), menthyl acetate (2.8%–10.0%), isopulegone (maximum 0.2%), menthol (30.0%–55.0%), pulegone (maximum 4.0%), and carvone (maximum 1.0%). The ratio of cineole content to limonene content should be a minimum of 2.

The following composition with minor details has also been reported in the literature: the major constituent is volatile oil, of which the principal component is usually (–) menthol, together with menthol stereoisomers such as (+) neomenthol and (+) isomenthol. Other monoterpenes include menthone (10%–40%), methyl acetate (1%–10%), menthofuran (1%–10%), cineol (eucalyptol, 2%–13%), and limonene (0.2%–6%). Monoterpenes like pinene, terpinene, murcene, β-caryophyllene, piperitone, piperitenone, piperitone oxide, pulegone, eugenol, menthone, isomenthone, carvone, cadinene, dipentene, linalool, α-phellendrene, ocimene, sabinene, terpinolene, ϒ-terpinene, fenchrome, p-menthane, and β-thujone are also present in small quantities [45,46].

About 85 constituents of the oil have been identified, and a further 40 are unidentified. Composition of the oil varies from place to place, as it depends on temperature, photoperiod, nutrition, salinity, water stress, plant age, harvesting, and planting time [47]. Flavanoids like luteolin and its 7-glucocide (cynaroside), menthoside, isorhoifolin, and others, including a number of highly oxygenated flavones, have been reported [48,49].

Phenolic acids, including caffeic, chlorogenic, and rosmarinic acids and pseudotannis derived from them, are reported to be present. Triterpenes in microamounts, including squalene, α-amyrin, urosolic acid, sitosterol, and other constituents, azulene, and minerals are also present [50].

Peppermint oil possesses a greater antihydrolytic effect than commercial preservatives, such as butylated hydroxytoluene [51].

Scavroni et al. [52] have evaluated the effects of biosolid levels on the yield and chemical composition of *Mentha piperita* L. essential oil. Mint plants were grown in a greenhouse in pots containing the equivalent of 0, 28, 56, and 112 tonnes/ha biosolids. Three evaluations were made at 90, 110, and 120 days after planting (DAP). The oil was extracted from the dry matter of shoots by hydrodistillation, and composition was determined by GC/MS. Oil production was slightly affected by the biosolid, increasing when plants were grown with 28 tonnes/ha, a condition that did not result in quality improvement. Methyl acetate was the component obtained at the highest percentage in all treatments. At 90 DAP, plants showed a higher percentage of menthol. Under these conditions, plant harvesting is recommended at 90 DAP, a period in which the menthol level was higher.

Verma et al. [53] have evaluated the essential oil content and composition of menthol mint (*Mentha arvensis* L.) and peppermint (*M. piperita* L.) cultivars grown in the Kumaon region at different stages of crop growth. In menthol mint cultivars, that is, Kosi, Saksham, Himalaya, and Kalka, the essential oil content was found to vary from 0.3% to 1.2%, 0.42% to 1.1%, 0.38% to 1.0%, and 0.26% to 1.2% at different days after transplanting (DAT), respectively, while in cultivars Kukrail, CIM-Madhurus, and CIM-Indus of peppermint, it varied from 0.28% to 0.6%, 0.19% to 0.55%, and 0.17% to 0.37%, respectively, at different DAT. The menthol content in the menthol mint cultivars reached higher values at 120 and 150 DAT. In the case of peppermint cultivars, that is, Kukrail, CIM-Madhurus, and CIM-Indus, the menthol content varied from 32.92% to 39.65%, 34.29% to 42.83%, and 22.56% to 32.77%, respectively, during the crop growth. It was concluded that peppermint and menthol mint cultivars should be harvested in June (150 DAT) if the crop is sown in January and July–August (120–150 DAT) if the crop is sown in March, respectively, under the climatic conditions of the Kumaon region of the western Himalayas in Uttarakhand, India.

Peppermint oil has a high concentration of natural pesticides, mainly pulegone (found mainly in *Mentha arvensis* var. *piperascens*, and to a lesser extent [6530 ppm] in *Mentha* × *piperita* subsp. *notho* subsp. *piperita*) [54] and menthone [55].

## 4.9 Physical Properties

*Mentha* or peppermint oil is a colorless or pale clear liquid. It has a pleasant aroma and a pungent taste, followed by a cool aftertaste. It is 95% miscible in methanol, 99.5% in ethanol, 95% in warm ethanol, 95% in diethyl ether, and 95% in dichloromethane. It is practically insoluble in water. The following parameters have been reported in the International Pharmacopoeia [37]: relative density, 0.900–0.916; refractive index, 1.457–1.467; and acid value, maximum 1.4, determined on 5.0 g diluted in 50 ml of the prescribed mixture of solvent.

## 4.10 Pest Control

A renewed interest in the use of essential oils for insect pest control has originated from the need for pesticide products with a less negative environmental and health impact than that of highly effective synthetic pesticides [56–59]. Esmaili et al. [56] have assessed the fumigant toxicity of essential oil from *Mentha pulegium* on the adults of *Callosobruchus maculatus*, *Tribolium castaneum*, *Lasioderma serricorne*, and *Sitophilus oryzae*. An experiment was carried out at 28°C ± 2°C and 65% ± 5% relative humidity (RH) in five concentrations and three replications in 24 hours, under dark conditions. The results demonstrated that mortality increased with the increase in concentration. The results indicated that the $LC_{50}$ values at 24 hours after treatment were 97.3, 165.5, 419.2, and 12.7 μl/L air for adults of *C. maculatus*, *T. castaneum*, *L. serricorne*, and *S. oryzae*, respectively. *S. oryzae* was more susceptible to *M. pulegium* than other pests. It was found that the plant essential oil *M. pulegium* had high potential in controlling different stored pests [60,61].

Khani and Asghari [62] have tested essential oils extracted from the foliage of *M. longifolia* (L.) and *Pulicaria gnaphalodes* Ventenat (Asterales: Asteraceae), and flowers of *Achillea wilhelmsii* C. Koch (Asterales: Asteraceae) in the laboratory for volatile toxicity against two stored product insects, the flour beetle, *T. castaneum*, and the cowpea weevil, *C. maculatus*. The chemical composition of the isolated oils was examined by gas chromatography–mass spectrometry. In *M. longifolia*, the major compounds were piperitenone (43.9%), tripal (14.3%), oxathiane (9.3%), piperitone oxide (5.9%), and d-limonene (4.3%). In *P. gnaphalodes*, the major compounds were chrysanthenyl acetate (22.38%), 2L-4L-dihydroxy eicosane (18.5%), verbenol (16.59%), dehydroaromadendrene (12.54%), β-pinene (6.43%), and 1,8-cineol (5.6%). In *A. wilhelmsii*, the major compounds were 1,8-cineole (13.03%), caranol (8.26%), α-pinene (6%), farnesyl acetate (6%), and p-cymene (6%). *C. maculatus* was more susceptible to the tested plant products than *T. castaneum*. The oils of the three plants displayed the same insecticidal activity against *C. maculatus* based on $LC_{50}$ values (between 1.54 μl/L air in *P. gnaphalodes* and 2.65 μl/L air in *A. wilhelmsii*). While the oils of *A. wilhelmsii* and *M. longifolia* showed the same strong insecticidal activity against *T. castaneum* ($LC_{50}$ = 10.02 and 13.05 μl/L air, respectively), the oil of *P. gnaphalodes* revealed poor activity against the insect ($LC_{50}$ = 297.9 μl/L air). These results suggest that essential oils from the tested plants could be used as potential control agents for stored product insects.

### 4.10.1 Antimicrobial Activity

The peppermint oil exerts antidermatophytic activity against (+) and (–) strains of *Nannizzia fulva* and *N. gypsea* [63]. It also shows antibacterial activity against *Staphylococcus aureus*, *Streptococcus pyogenes*, *Escherichia coli*, *Bacillus subtilis*, and *Proteus vulgaris* [64,65]. It possesses repellent activity against *T. castaneum*, and it is a moderately effective fumigant of both *C. maculatus* and *T. castaneum* [66]. It has a moderate antimyotic property against *Aspergillus fumigatus*, *Candida albicans*, *Geotrichum candidum*, *Rhodotarula rubra* [67], *Phytophthora cinnamini*, *Pyrenochaeta lycopersici*, and *Verticillium dahlia* [68]. Peppermint oil has shown antifungal activity against *Aspergillus niger*, *Alternaria alternata*, and *Fusarium* sp. by the agar well diffusion method [68].

### 4.10.2 Larvicidal and Mosquito Repellent

Oil of *Mentha piperita* L. (peppermint oil), a widely used essential oil, was evaluated for larvicidal activity against different mosquito species [69]—*Aedes aegypti*, *Anopheles stephensi*, and *Culex quinquefasciatus*—by exposing third-instar larvae of mosquitoes in enamel trays (6 feet 4 inches) filled to a depth of 3 inches with water. The oil showed strong repellent action against adult mosquitoes when applied on human skin. The protection obtained against *Anopheles annularis, Anopheles culicifacies*, and *C. quinquefasciatus* was 100%, 92.35%, and 84.5%, respectively. The repellent action of *Mentha* oil was comparable to that of mylol oil consisting of dibutyl and dimethyl phthalates.

## 4.11 Conclusion

*Mentha piperita* oil possesses diversified potential in the areas of food, cosmetics, medicines, and pest control [70]. It has been proven helpful in symptomatic relief of the common cold. It may also decrease symptoms of irritable bowel syndrome and decrease digestive symptoms such as dyspepsia and nausea. It is used topically as an analgesic and to treat headaches. It may also be used as an insecticide, repellent, fumigant, and fungicide in stored foodstuffs. Although *Mentha piperita* oil is on the Food and Drug Administration's (FDA) generally recognized as safe (GRAS) list, more research work is needed, as this herb has a few side effects. It can cause heartburn or perianal irritation and is contraindicated in patients with bile duct obstruction, gallbladder inflammation, and severe liver damage. Caution is required in patients with gastrointestinal (GI) reflux, and its products may not be used directly under the noses of children due to the risk of apnea.

This plant is now well acclimatized and cultivated in different countries, including India. This oil enjoys strong export potential. Its various formulations, for example, Pudin Hara for gastrointestinal disturbances like flatulence and indigestion and Itch Guard for skin disorders, are commercially available.

## References

1. Spenser-Higg, A. Many faces of mint. Home Herb article. Healthy Living Herbs 2010–2011. Available online. http://healthyliving-herbs.co.za. Retrieved January 14, 2015.
2. Stanev, S., Zheljazkov, V.D. 2009. Study on essential oil and free menthol accumulation in 19 cultivars, populations, and clones of peppermint (*Mentha × piperita*). Retrieved June 6, 2009.
3. Jullien, F., Diemer, F., Colson, M., Faure, O. 1998. An optimising protocol for protoplast regeneration of three peppermint cultivars (*Mentha × piperita*). *Plant Cell Tissue Organ Cult.* 54(3): 153–159.
4. Vaughan and Geissler. White Peppermint-Mentha piperita. *The Herbarist.* 1997. Herb Society of America, p. 39. Retrieved July 24, 2013.
5. *Mentha piperita* cv. Chocolate Mint. Mountainvalleygrowers.com. Retrieved July 24, 2013.
6. De Rovira, D. 2008. *Dictionary of Flavors.* Hoboken, NJ: John Wiley & Sons, p. 420.
7. *Mentha × piperita* 'Chocolate Mint': Peppermint. 2007. Hortiplex.gardenweb.com. Retrieved July 24, 2013.
8. *Mentha × piperita* 'Citrata': Eau de cologne mint. Hortiplex.gardenweb.com. 2007. Retrieved July 24, 2013.
9. *Mentha ×* piperita var. citrata: Lemon mint. Hortiplex.gardenweb.com. 2007. Retrieved July 24, 2013.
10. *Mentha ×* piperita 'Crispa': Eau de cologne mint. Hortiplex.gardenweb.com. 2007. Retrieved July 24, 2013.

11. HortiPlex Plant Database: Info, images and links on thousands of plants. Hortiplex.gardenweb.com. Retrieved July 24, 2013.
12. *Harrowsmith Country Life.* 1990. Camden House Pub., Cornell University, Ithaca, New York, p. 48.
13. *Mentha* × piperita 'Lime Mint': Eau de cologne mint. Hortiplex.gardenweb.com. 2007. Retrieved July 24, 2013.
14. *Mentha* × piperita 'Variegata': Variegated mint. Hortiplex.gardenweb.com. 2007. Retrieved July 24, 2013.
15. Quattrocchi, U. 1947. *CRC World Dictionary of Plant Names: Common Names, Scientific Names, Eponyms, Synonyms, and Etymology.* Boca Raton, FL: CRC Press, p. 1658.
16. Davidson, A. 1999. *The Oxford Companion to Food.* Oxford: Oxford University Press, p. 508.
17. Fleming, T. 1998. *PDR for Herbal Medicines.* Montvale, NJ: Medical Economics Company.
18. Sastri, B.N. 1962. *The Wealth of India: A Dictionary of Indian Raw Materials and Industrial Products.* Raw Materials Series, vol. VI. New Delhi: Council of Scientific and Industrial Research, pp. 342–343.
19. Goldblatt, P., Manning, J. 2000a. Wild flowers of the fairest cape. Cape Town: National Botanical Institute.
20. Goldblatt, P., Manning, J. 2000b. Cape plants. A conspectus of the cape flora of South Africa. Cape Town: National Botanical Institute, Cape Town and Missouri Botanical Gardens.
21. Pooley, E. 2003. *Mountain Flowers.* Durban, South Africa: Flora Publications Trust.
22. Van Wyk, B., Gericke, N. 2000. *People's Plants.* Pretoria: Briza Publications.
23. Codd, L.E.W. 1985. *Lamiaceae: Flora of Southern Africa.* Vol. 28, part 4. Pretoria: Botanical Research Institute.
24. Flora of NW Europe: *Mentha* × *piperita.* http://wbd.etibioinformatics.ul/bis/flora.php.
25. *PDR for Herbal Medicines.* 4th ed. New York: Thomson Healthcare, p. 640.
26. Leung, A.Y. 1980. *Encyclopedia of Common Natural Ingredients Used in Food, Drugs and Cosmetics.* New York: John Wiley & Sons, p. 231.
27. Akdogan, M., Ozguner, M., Aydin, G., Gokalp, O. 2004. Investigation of biochemical and histopathological effects of *Mentha piperita* Labiatae and *Mentha spicata* Labiatae on liver tissue in rats. *Hum. Exp. Toxicol.* 23(1): 21–28.
28. Sharma, A., Sharma, M.K., Kumar, M. 2007. Protective effect of *Mentha piperita* against arsenic-induced toxicity in liver of Swiss albino mice. *Basic Clin. Pharmacol. Toxicol.* 100(4): 249–257.
29. Akdogan, M., Kilinç, I., Oncu, M., Karaoz, E., Delibas, N. 2003. Investigation of biochemical and histopathological effects of *Mentha* piperita L. and *Mentha* spicata L. on kidney tissue in rats. *Hum. Exp. Toxicol.* 22(4): 213–219.
30. Akdogan, M., Gultekin, F., Yontem, M. 2004. Effect of *Mentha piperita* (Labiatae) and *Mentha spicata* (Labiatae) on iron absorption in rats. *Toxicol. Ind. Health* 20(6–10): 119–122.
31. http://www.essentialoils.co.za/essential-oils/peppermint.htm. Retrieved on August 25, 2008.
32. Gainar, I., Vilcu, R., Mocan, M. 2002. Supercritical fluid extraction and fractional separation of essential oils. Analele Universitǎ̆Nii din Bucureşti—Chimie, Anul XV (serie nouǎ), vol. II, pp. 79–83.
33. Maryline, A.V., Fernandezb, X., Visinonic, F., Chemata, F. 2008. Microwave hydrodiffusion and gravity, a new technique for extraction of essential oils. *J. Chromatogr. A* 1190: 14–17.
34. Gill, K., Gupta, R., Bhise, S., Bansal, M., Gill, G. 2014. Effect of hydro distillation process on extraction time and oil recovery at various moisture contents from *Mentha* leaves. *Int. J. Eng. Sci.* 4(6): 8–12.
35. Spices Board of India. http://www.indianspices.com/.
36. Indian Pharmacopoeia. 1996. Monograph of peppermint oil.
37. International Pharmacopoeia. Monograph of peppermint oil. http://www.newdruginfo.com/pharmacopeia/bp2003/British%20Pharmacopoeia%20Volume%20I%20and%20II%5CMonographs%20Medicinal%20and%20Pharmaceutical%20substances%5CP%5CPeppermint%20Oil.htm.
38. Rai, A.K., Singh, A.K. 1990. Spectroscopic study of *Mentha* oil. *Spectrochim. Acta A* 46(8): 1269–1272.

39. Shrivastava, A. 2009. A review on peppermint oil. *Asian J. Pharm. Clin. Res.* 2(2): 27–33.
40. Kowalski, J. 2005. Evaluating peppermint oils by chiral GC/Ms. http://www.restek.com/pdfs/adv_2005_04_07.pdf.
41. Katayama, K., Takahashi, O., Matsui, R., Morigaki, S., Aiba, T., Kakemi, M., Koizumi, T. 1992. Effect of 1-menthol on the permeation of indomethacin, mannitol and cortisone through excised hairless mouse skin. *Chem. Pharm. Bull.* 40: 3097–3099.
42. Sang, J.P. 1982. Estimation of menthone, menthofuran, methyl acetate and menthol in peppermint oil by capillary gas chromatography. *J. Chromatogr.* 253: 109–112.
43. Bichi, C., Frattini, C. 1980. Quantitative estimation of minor components in essential oils and determination of pulegone in peppermint oil. *J. Chromatogr.* 190: 471–474.
44. Eccls, R. 1994. Menthol and related cooling compounds. *J. Pharm. Pharmacol.* 46: 618–630.
45. Basla, R.K. 1977. Essential oil of *Mentha piperita* (L.) raised in Kumaon region (India). *Nat. Appl. Sci. Bull.* 29(2): 75.
46. Baslas, R.K., Saxana, S. 1984. Chromatographic analysis of dementholised essential oil of *Mentha piperita*. *Indian J. Phys. Nat. Sci.* 4A: 32.
47. Charles, D.J., Jolly, R.J., Simonj, J.E. 1990. Effect of osmotic stress on the essential oil content and composition of peppermint. *Phytochemistry* 28: 2837–2840.
48. Orani, G.P., Anderson, J.W., Sant's Ambrogio, G., Sant's Ambrogio, F.B. 1991. Upper airway cooling and 1-menthol reduce ventilation in the guinea pig. *J. Appl. Physiol.* 70: 2080–2086.
49. Rastogi, R.P., Mehrota, B.N. 1990. *Compendium of Indian Medicinal Plants.* Vol. 1. Lucknow: Central Drug Research Institute; New Delhi: Publication and Information Directorate, p. 272.
50. Lucida, G.M., Wallace, J.M. 1998. *Herbal Medicines: A Clinician's Guide.* New York: Pharmaceutical Products Press, pp. 85–86.
51. Singh, G., Kapoor, I.P.S., Pandey, S.K. 1998. Studies on essential oils. Part 14. Natural preservatives for butter. *J. Med. Arom. Plant Sci.* 20: 735–739.
52. Scavroni, J., Boaro, C.S.F., Marques, M.O.M., Ferreira, L.C. 2005. Yield and composition of the essential oil of *Mentha piperita* L. (Lamiaceae) grown with biosolid. *Braz. J. Plant Physiol.* 17(4): 345–352.
53. Verma, R.S., Rahman, L., Verma, R.K., Chuhan, A., Yadav, A.J., Singh, A. 2010. Essential oil composition of menthol mint (*Mentha arvensis*) and peppermint (*Mentha piperita*) cultivars at different stages of plant growth from Kumaon region of western Himalaya. *Open Access J. Med. Arom. Plants* 1(1): 13–18.
54. Duke's database. http://www.ars-grin.gov/cgi-bin/duke/highchem.pl.
55. Robert, I.K. 2001. *Handbook of Pesticide Toxicology: Principles.* Amsterdam: Academic Press, p. 823.
56. Esmaili, M., Vojoudi, S., Parsaeyan, E. 2013. Fumigant toxicity of essential oils of *Mentha pulegium* L. on adults of *Callosobruchus maculatus*, *Tribolium castaneum*, *Lasioderma serricorne* and *Sitophilus oryzae* in laboratory conditions. *TJEAS* 3(9): 732–735.
57. Tapondjou, L.A., Adler, C., Bouda, H., Fontem, D.A. 2002. Efficacy of powder and essential oil from *Chenopodium ambrosioides* leaves as post-harvest grain protectants against six-stored product beetles. *J. Stored Prod. Res.* 38(2): 395–402.
58. Lee, B.-H., Annis, P.C., Tumaalii, F., Choi, W.-S. 2004. Fumigant toxicity of essential oils from the Myrtaceae family and 1,8-cineole against 3 major stored-grain insects. *J. Stored Prod. Res.* 40(5): 553–564.
59. Kéita, S.M., Vincent, C., Schmit, J.-P., Arnason, J.T., Bélanger, A. 2001. Efficacy of essential oil of *Ocimum basilicum* L. and *O. gratissimum* L. applied as an insecticidal fumigant and powder to control *Callosobruchus maculatus* (Fab.). *J. Stored Prod. Res.* 37(4): 339–349.
60. Leila, M., Senfi, F., Frouzan, M., Khodabandeh, F. 2014. Fumigant toxicity of *Artemisia annua* L. essential oil against cowpea weevil, *Callosobruchus maculatus* F. (Col.: Bruchidae). *Int. J. Agric. Crop Sci.* 7(11): 805–807.
61. Bittner, M., Casauev, M.E., Arbert, C.C., Aguilera, M.A., Hernanadez, V.J., Becerra, J.V. 2008. Effect of essential oils from five plant species against the granary weevils *Sitophilus zeamais* and *Acanthoscelides obtectus* (Coleoptera). *J. Chil. Chem. Soc.* 53(1): 1.

62. Khani, A., Asghari, J. 2012. Insecticide activity of essential oils of *Mentha longifolia*, *Pulicaria gnaphalodes* and *Achillea wilhelmsii* against two stored product pests, the flour beetle, *Tribolium castaneum*, and the cowpea weevil, *Callosobruchus maculatus*. *J. Insect. Sci.* 12: 73–83.
63. Gautam, M.P., Jain, P.C., Singh, K.V. 1980. The in vivo action of essential oils of different organisms. *Indian Drugs* 17: 269.
64. Kokate, C.K., Varma, K.C. 1970. Antibacterial activity of volatile oil of seven species of *Mentha*. *Indian J. Microbiol.* 10(2): 45.
65. Sawney, S.S., Suri, R.K., Thind, T.S. 1977. Antimicribial efficacy of some essential oils in vitro. *Indian Drugs* 15: 30.
66. Tripathi, A.K., Prajapati, V., Agarwal, K.K., Kumar, S. 2000. Effects of volatile oil constituents of *Mentha* species against the stored grain pests, *Callosobruchus maculatus* and *Tribolium castaneum*. *J. Med. Arom. Plant Sci.* 22: 549–556.
67. Blaszcy, K.T., Krzyonowaska, J., Lamber Zarawska, E. 2000. Screening for antimycotic properties of 56 traditional Chinese drugs. *Phytother Res.* 14: 210–212.
68. Giamperi, L., Fraternale, D., Ricci, D. 2002. The in vitro action of essential oils of different organisms. *Essent. Oil Res.* 14(4): 312–318.
69. Ansaria, M.A., Padma, V., Tandonb, M., Razdana, P.K. 2000. Larvicidal and mosquito repellent action of peppermint (*Mentha piperita*) oil. *Bioresour. Technol.* 71: 267–271.
70. Shah, P.P., Mello, P.M.D. 2004. A review of medicinal uses and pharmacological effects of *Mentha piperita*. *Nat. Prod. Radiance* 3: 214–221.

# 5

# Basil (*Ocimum basilicum* L.) Oil: As a Green Pesticide

**N.C. Basantia**

## CONTENTS

## 5.1 Introduction

Basil (*Ocimum basilicum*) is one of the most popular herbs grown in the world. The name *basil* is thought to be derived from the Greek words *okimon* ("smell") and *basilikon* ("royal or king") (Selvakkumar et al. 2007). It is often referred to as the king of the herbs. Basil originated in Asia and Africa. In Hindu houses, basil is used to protect the family from evil spirits. In early 1600, the English used basil in their food and doorways to ward off uninvited pests, such as flies, as well as evil spirits. Several interesting beliefs are ascribed to the historical use of basil. Some Europeans considered it to be funeral and dreamt of it as unlucky, whereas in Italy, women wear it in their hair and the youth stick a spring of it above ear when they are courting. In India, Hindus believed that a leaf of basil buried with them would serve as their passport to heaven (Sekar et al. 2009). Sweet basil has been grown and sold in New York State since the end of the eighteenth century.

O. *basilicum* is known by different names in different languages around the world, including the Indian subcontinent. In English, it is known as basil or sweet basil, whereas in Hindi and Bengali it is called babui tulsi. The plant is known as badroaj, hebak, or rihan in Arabic, as nasabo or sabje in Gujrati, and as jungle tulsi in Urdu.

Basil oil is the oil obtained from basil (O. *basilicum*) through different conventional (e.g., steam distillation), nonconventional, or novel techniques having major constituents such as linalool, methyl chavicol, methyl cinnamate, eugenol, and cineole. It is reported to possess antimicrobial, bactericidal, and antioxidant activity, and fungicidal, marked insecticidal, lavricidal, and insect repellant properties against mosquitoes (Jayasinghe et al. 2003; Bozin et al. 2006; Kumar et al. 2011; Nour et al. 2012; Shirazi et al. 2014). Due to these activities, the oil is potent for use in perfumery, cosmetics, food, pharmaceutical formulation, and insecticidal and insect repellent formulations.

## 5.2 Botany of the Plant

### 5.2.1 Taxonomic Classification

Kingdom: Plantae
Subkingdom: Tracheobionta
Super division: Spermatophyta
Division: Magnoliophyta
Class: Magnoliophyta
Subclass: Asteriadae
Order: Lamiales
Family: Lamiaceae
Genus: Ocimum
Species: Basilicum
Botanical name: Ocimum basilicum

### 5.2.2 Habitat and Distribution

Basil originated in Asia and Africa. Sweet basil is indigenous to Persia and Sindh and the lower hills of Punjab in India. The plant is widely grown as an ornamental and field crop throughout the greater part of India, Cameroon, Ceylon, and several Mediterranean countries, including Turkey.

### 5.2.3 Botanical Description of the Plant

Sweet basil is an autogamous, aromatic, and herbaceous plant that is annual and perennial. It is an erect branching herb, 0.6–0.9 m high, glabrous, and more or less pubescent. The stems and branches are green and sometimes purplish.

The leaves of *O. basilicum* are simple opposite. They are around 2–2.5 cm long, ovate acute entire or more or less toothed or lobed with a cuneate and entire base. The petiole is 1.3–2.5 cm long. The leaves have numerous dot-like oil glands that secrete strongly scented volatile oil. The whorls are densely racemose, where the terminal raceme is usually much longer than the lateral ones. The bracts are stalked, shorter than the calyx, ovate, and acute. The calyx is 5 mm long, with enlarging fruit, and very shortly pedicelled. Its lower lip with two central teeth is longer than the rounded upper lip. The corolla, 8–13 mm long, is white, pink, or purplish in color, glabrous or variously pubescent. The upper filaments of slightly exerted stamen are toothed at the base. Nutlets are about 2 mm long, ellipsoid, black, and pitted. The sepals of the flower are and remain fused into a two-lipped calyx. The ovary is superior, and there are two carpellaxy, four-locular, and a four-partite fruit of achenes.

## 5.3 Methods of Extraction of Oil

Essential oils are complex mixtures of volatile substances generally present at low concentrations. There are several methods for extracting the essential oil from the natural sources. Different methods can be used for that purpose.

### 5.3.1 Conventional Methods

Distillation and Soxhlet extraction are the conventional methods.

#### 5.3.1.1 Distillation

Distillation is an extracting oil process. It converts volatile liquid (essential oils) into a vapor state and then condenses the vapor into a liquid state. There are different distillation processes, such as water distillation, steam distillation, and hydrodiffusion (Ranjitha and Vijiyalakshmi 2014).

*5.3.1.1.1 Water Distillation*

In this process, the botanic material is completely immersed in water and the still is brought to boil. It is used to protect the oils to a certain degree since the surrounding water acts as a barrier to prevent overheating. When condensed material cools down, the water and essential oil are separated and the oil is decanted to use as an essential oil. Water distillation can also be done at reduced pressure (under vacuum) to reduce the temperature to less than 100°C. This is useful in protecting the botanical material in obtaining the essential oil.

It is a simple and easy-to-operate extraction method of oil from plant species. Due to the use of heat in this method, it may not be used on very fragile plant material. Oil components like esters are sensitive to hydrolysis, while others like cyclic monoterpene hydrocarbons, and aldehydes are susceptible to polymerization. Oxygenated compounds such as phenols have a tendency to dissolve in distilled water, so their complete removal is not possible. As water distillation tends to be a small operation, it takes a long time to accumulate much oil (Ranjitha and Vijiyalakshmi 2014).

### 5.3.1.1.2 Steam Distillation

In the steam distillation method, the botanical material is placed in a still and steam is forced over the material. The hot steam is used to release the aromatic molecules from the plant material. The steam forces the pockets to open, and then the molecules of these volatile oils escape from the plant material and evaporate in the steam. The steam containing the essential oil is passed through a cooling system to condense the steam, which forms a liquid form of essential oil. Then, the water is separated. The steam is produced at greater pressure than atmospheric pressure, and therefore it helps to boil at 100°C without raising the temperature beyond 100°C. This is used to remove the essential oil from the plant material (Masango 2005; Ranjitha and Vijiyalakshmi 2014).

The major advantage of steam distillation is that the temperature never goes above 100°C, so temperature-sensitive compounds can be distilled. A disadvantage is that not many compounds can be steam distilled—only the aromatic ones.

### 5.3.1.1.3 Hydrodiffusion

The hydrodiffusion method is similar to the steam distillation process. The main difference between these two methods is how the steam is introduced into the still. In the case of hydrodiffusion, the steam is fed into the top, onto the botanical material, instead of from bottom, as in normal steam distillation. The steam containing the essential oil is passed through a cooling system to condense steam, which forms a liquid of essential oil, and then water is separated (Ranjitha and Vijiyalakshmi 2014).

The main advantages of this method are that less steam is used, the processing time is shorter, and there is a higher yield.

### 5.3.1.2 Soxhlet Extraction

Soxhlet extraction is a well-established technique that surpasses other conventional extraction techniques in performance, except for limited fields of application, for example, the extraction of thermolabile compounds. Most of the solvent extraction units worldwide are based on Soxhlet principles with recycling of solvents. Basically, the equipment consists of a drug holder extractor, a solvent storage vessel, a reboiler kettle, a condenser, a breather system, and supporting structures like a boiler, a refrigerated chilling unit, and a vacuum unit (William 2007).

This technique is based on the choice of solvent coupled with heat or agitation. In this process, the circulation of solvents causes the displacement of transfer equilibrium by repeatedly bringing fresh solvent into contact with the solid matrix. This method maintains a relatively high extraction temperature and no filtration of extract is required (Shams et al. 2015).

However, the limitation of this technique is that there is a possibility of thermal decomposition of thermolabile targeted compounds because the extraction usually occurs at the boiling point of the solvent for a long time.

Nidia et al. (2013) compared supercritical fluid extraction (SFE) using carbon dioxide with the Soxhlet and hydrodistillation processes for extraction of basil oil and reported a higher yield of oil by Soxhlet extraction.

## 5.3.2 Novel Extraction Methods

### 5.3.2.1 Microwave Extraction Method

Solvent-free microwave extraction (SFME) is used to separate the essential oil from plant material. The method involves placing the sample in a microwave reactor without any addition of organic solvent or water. The internal heating of the water within the sample distends its cells and leads to rupture of the glands and oleiferous receptacles. This process frees essential oil, which is evaporated by the *in situ* water of plant material.

A cooling system outside the microwave oven continuously condenses the vapors, which are collected in specific glassware. The excess of water is refluxed back to the extraction vessel in order to restore the *in situ* water to the sample.

The microwave isolation offers a net advantage in terms of yield and better oil composition. Furthermore, it is environmentally friendly. In this method, low-boiling-point hydrocarbon compounds undergo decomposition (Marie et al. 2004; Ranjitha and Vijiyalakshmi 2014).

### 5.3.2.2 Subcritical Water Extraction

Subcritical water extraction (SWE) is the extraction using hot water under pressure. It has recently emerged as a useful tool to replace traditional extraction methods. Subcritical water extraction is an environmentally clean technique that, in addition, provides higher extraction yields to extract solid samples. Subcritical water extraction is carried out using hot water (from 100°C to 374°C, the latter being the water critical temperature under high pressure [usually up to 10 bars]), enough to maintain water in the liquid state. The most important factor to take into account in this type of extraction procedure is the dielectric constant. This parameter can be modulated easily within a wide range of values by only tuning the extraction temperature.

Water at room temperature is a very polar solvent, with a dielectric constant close to 80. However, this level can be significantly decreased to values close to 27 when water is heated up to 250°C while maintaining its liquid state applying pressure. The dielectric constant value is similar to that of ethanol, and therefore is appropriate for solubilizing less polar compounds.

Basically, the experimental setup needed to use this technique is simple. The instrumentation consists of a water reservoir coupled to a high-pressure pump to introduce the solvent into the system, an oven where the extraction cell is placed and where the extraction takes place, and a restrictor to maintain the pressure along the extraction line. Extracts are collected in the collector vial placed at the end of the extraction system (Shams et al. 2015).

### 5.3.2.3 Supercritical Fluid Extraction

Supercritical fluid extraction is used for the extraction of flavors and fragrances. SFE is a separation technology that uses supercritical fluid as solvent. Every fluid is characterized by a critical point, which is defined in terms of the critical temperature and critical pressure. Fluids cannot be liquefied above the critical temperature regardless of the pressure applied, but may reach the density close to the liquid state. A substance is considered to be a supercritical fluid when it is above its critical temperature and critical pressure.

The main supercritical solvent used is carbon dioxide. Carbon dioxide (critical condition 30.9°C and 73.8 bar) is cheap, environmentally friendly, and generally recognized as safe. Supercritical carbon dioxide is also attractive because of its high diffusivity and easily tunable solvent strength. Another advantage is that carbon dioxide is gaseous at room temperature and ordinary pressure, which makes analyte recovery very simple (Taylor 1996; Ranjitha and Vijiyalakshmi 2014).

### 5.3.2.4 Accelerated Solvent Extraction

Accelerated solvent extraction (ASE) is sometimes called pressurized solvent extraction (PSE). It uses organic solvents at elevated pressure and temperature in order to increase the efficiency of the extraction process. Increased temperature accelerates the extraction kinetics, and elevated pressure keeps the solvent in a liquid state, thus enabling safe and rapid extraction. Furthermore, high pressure forces the solvent into the matrix pores, and hence facilitates the extraction of analyte. High temperature decreases the viscosity of the liquid solvent, allowing a better penetration of the matrix and a weakened solute matrix interaction. Elevated temperature enhances diffusivity of the solvent, resulting in increased extraction speed. The solvent is selected based on the polarity of the analyte and compatibility with postextraction processing equipment. In ASE applications, generally organic solvents are used in conventional techniques, such as methanol. The use of hot water as the extraction solvent under atmospheric or higher pressure is very efficient for extracting phytochemicals.

ASE has been reported to be more efficient than other extraction methods by consuming less solvent and allowing faster extraction. However, since the extraction is performed at elevated temperature, the thermal degradation is a cause of concern, especially for thermo-labile compounds in extracts.

The efficiency of ASE is influenced by factors such as pressure, temperature, static extraction time, flush volume, and vessel void volume (Shams et al. 2015).

### 5.3.2.5 Ultrasound-Assisted Extraction

The mechanical effect of ultrasound accelerates the release of organic compounds within the plant body due to cell wall disruption, mass transfer intensification, and easier access of the solvent to the cell content. Ultrasound-assisted extraction (UAE) is reported to be one of the important techniques for extracting valuable compounds from the vegetable material (Vilkhu et al. 2008). General ultrasonic devices are the ultrasonic cleaning bath and ultrasonic probe system.

The efficiency of UAE depends on various factors, such as the nature of the tissue being extracted, the location of the component to be extracted, the treatment of the tissue prior to extraction, the effect of ultrasonics, the surface mass transfer, and intraparticle diffusion.

UAE can extract analytes under a concentrated form and free from any contaminants or artifacts. It also demonstrates advantages in terms of yield, selectivity, operating time, energy input, and preservation of thermolabile compounds (Shams et al. 2015).

## 5.4 Composition of Oil

Essential oils extracted from the plants are a complex mixture of terpenes, sesquiterpenes, oxygenated derivatives, and other aromatic compounds. These compounds are characteristic for basil aroma, which is a precursor to the presence of 1,8-cineole, methyl cinnamate, methyl chavicol, and linalool. In general, these substances are volatile and are present at low concentrations.

Lawrence (1985) classified basil oil into three large groups, European type, exotic or Reunion type, and African type, according to their chemical composition and geographical origin. He established four essential oil chemotypes: methyl chavicol, linalool, methyl eugenol, and methyl cinnamate.

The major constituents that have been isolated from different *Ocimum* oils include 1,8-cineole, linalool, piene, eugenol, camphor, methyl chavicol, ocimene, terpinene, and limonene. Within the *Ocimum* species, there is a clear variation in their composition in terms of types of constituents. The chemical composition of sweet basil oil has been investigated, and by now, more than 200 chemical components have been reported from many regions of the world (Marwat et al. 2011).

A substantial number of studies conducted on the composition of essential oil of basil revealed a huge diversity in constituents with different chemotypes from many regions of the world. The major constituents that have been isolated from different *O. basilicum* oils include linalool, methyl chavicol, eugenol, methyl cinnamate, 1,8-cineole, and bergamotene (Pandey et al. 2014). In the Czech Republic, Guinea, and Reunion, the major compounds of basil essential oil are linalool and eugenol (Klimankova et al. 2008). Marotti et al. (1996) reported the presence of linalool, methyl chavicol, and eugenol as components of Italian basil oil. Four major compounds reported in basil essential oil from Austria are linalool, methyl chavicol, methyl cinnamate, and α-cadinol. Linalool was reported as the main component in basil oil from Romania (Benedec et al. 2009).

However, the major compound from French basil leaf essential oil is methyl cinnamate (Adam et al. 2009). Regarding Pakistan basil, four major compounds were found: linalool, epi-α-cardinol, α-bergamonotene, and γ-cadinene (Hussain et al. 2008). In the case of Egypt basil oil, linalool, 1,8-cineole, eugenol, and methyl cinnamate are the dominant components (Simon et al. 1990). Concerning basil oil from Turkey, three major compounds, methyl chavicol, limonene, and p-cymene, were cited (Chalchat and Ozcan 2008). Essential oil from Iran and Thailand is rich in methyl chavicol (Bunrathep et al. 2007). Camphor, followed by limonene and β-selenene, was the major compound in *O. basilicum* essential oil from northeast India. Egyptian basil oil is very similar to the European type characterized by linalool and methyl chavicol as major compounds (Simon et al. 1990).

The observed difference in the constituents of basil essential oil between countries may be due to environmental conditions and genetic factors, different chemotypes, and nutritional elements of the plants, as well as other factors.

### 5.4.1 Effect of Extraction Method on Composition and Yield of Basil Oil

Techniques commonly employed for extracting essential oils include hydrodistillation, steam distillation, solvent extraction, and liquid carbon dioxide extraction. The composition of extracted oil may vary from one extraction method to another.

Charles and Simon (1990) compared the influences of three extraction methods on measurements of essential oil content. No significant difference was observed between steam distillation and the hydrodistillation method.

The extraction yields obtained by SFE using $CO_2$ were from 0.719% to 1.483% (w/w), depending on $CO_2$ density. The predominant compounds of the $CO_2$ extracts were linalool (10.14%–16.60%), eugenol (5.91%–9.78%), and $\partial$-cadinene (3.94%–7.21%) depending on SFE condition. The extraction yields using SFE were higher for some compounds (eugenol, α-beragmotene, and $\partial$-cadinene) than those obtained by hydrodistillation (Zekovic et al. 2015).

## 5.5 Methods of Analysis of Basil Oil

The chemical composition of basil oil is analyzed generally using gas chromatographic methods, such as by gas chromatography with a flame ionization detector (GC-FID) or gas chromatography with mass spectrometry (GC-MS).

### 5.5.1 Gas Chromatography with Flame Ionization Detector

Basil oil may be analyzed using GC-FID. Different types of column (nonpolar, polar, and a combination of polar and nonpolar) have been used for the analysis of essential oil. Based on the required resolution and target analyte, different types of column and types and quantities of mobile phase and temperature conditions have been used. Quantification of compound is done based on area normalization or peak area of external standards. Retention indices (RIs) using n-alkanes (octane, nonane, decane, dodecane, octadecane, eicosane, docosane, tetracosane, and hexacosane) were used as the basis.

### 5.5.2 Gas Chromatography with Mass Spectrometry

Today, GC-MS plays an important role in the identification of the chemical composition of basil essential oil. Various researchers worldwide have used this technique and identified a number of analytes, as shown in Table 5.1.

Nevertheless, incomplete identification was achieved by GC-MS due to the complex nature of constituents of essential oil. Comprehensive two-dimensional gas chromatography (GC×GC) is a powerful technique that has been successfully used for separation of the volatile constituents in highly complex samples. This technique is a combination of two columns with different separation mechanisms coupled via a cryogenic modulator interface. Many coeluting components on the first column are separated in the second column. The application of comprehensive two-dimensional gas chromatography coupled to time-of-flight mass spectrometry (GC×GC-TOFMS) has been employed to analyze the aromatic compounds of basil samples (Klimankova et al. 2008). The relative abundance of different constituents allowed differentiation between examined cultivars.

**TABLE 5.1**

Methods Used for Analysis of Basil Oil

| Serial No. | Source of Oil | Method of Extraction | Number of Compounds | Method Used | Reference |
|---|---|---|---|---|---|
| 1 | Leaves of *O. basilicum* | Steam distillation | 4 | GC-MS | Mindaryani and Rahayu (2007) |
| 2 | Leaves and flowers of *O. basilicum* | Supercritical fluid extraction | 20 | GC-MS and GC-FID | Zekovic et al. (2015) |
| 3 | Aerial parts of *O. basilicum* | Distillation | 73 | GC-MS<br>Column: HP-5MS capillary 30 m × 0.25 mm i.d., 0.25 μ thickness<br>Detector: MSD<br>Temp program: 100 @ 2°C/min to 220°C<br>Detector temp: 280°C<br>Injector temp: 250°C | Pripdeevech et al. (2010) |
| 4 | Fresh leaves of *O. basilicum* | Hydrodistillation | 16 | Column: RXi-5MS capillary 30 m × 0.25 mm i.d., 0.25 μ thickness<br>Detector: mass spectroscopy detector (MSD)<br>Temp program: 100 @ 2°C/min to 220°C<br>Detector temp: 280°C<br>Injector temp: 250°C | Astuti et al. (2016) |
| 5 | Fresh aerial parts of *O. basilicum* | Hydrodistillation | 20 | GC-MS<br>Column: Rtx-5 capillary 30 m × 0.25 mm i.d., 0.25 μ thickness<br>Detector: MSD<br>Temp program: 50 @ 5°C/min to 250°C<br>Detector temp: 230°C<br>Injector temp: 240°C | Ismail (2006) |
| 6 | Leaves of *O. basilicum* | Solvent-free microwave extraction and hydrodistillation | 65 | GC-MS and GC-FID<br>Column: HP-1 capillary 50 m × 0.2 mm i.d., 0.25 μ thickness<br>Detector: FID<br>Temp program: 45 @ 2°C/min to 250°C<br>Detector temp: 250°C<br>Injector temp: 250°C | Chenni et al. (2016) |
| 7 | Fresh leaves of *O. basilicum* | Steam distillation | 13 | GC-MS<br>Column: fused silica capillary 30 m × 0.25 mm i.d., 0.25 μ thickness<br>Detector: MSD<br>Temp program: 100 @ 2°C/min to 220°C<br>Detector temp: 280°C<br>Injector temp: 250°C | Pripduvech et al. (2010) |

(*Continued*)

**TABLE 5.1 (CONTINUED)**

Methods Used for Analysis of Basil Oil

| Serial No. | Source of Oil | Method of Extraction | Number of Compounds | Method Used | Reference |
|---|---|---|---|---|---|
| 7 | Fresh leaves of *O. basilicum* | Hydrodistillation | 16 | Column: RXi-5MS capillary 30 m × 0.25 mm i.d., 0.25 μ thickness<br>Detector: MSD<br>Temp program: 70°C for 2 min, from 70°C @ 3°C/min to 230°C, @ 5°C to 240°C<br>Detector temp: 250°C<br>Injector temp: 250°C | Nour et al. (2012) |

*Note:* MSD, mass spectroscopy detector; Temp, temperature.

## 5.6 Physicochemical Properties of Basil Oil

The physicochemical properties include appearance, color, solubility, density, specific gravity, refractive index, and optical rotation (Table 5.2). These may vary from type to type and by method of extraction of basil oil. There are several standards at the national and international level to maintain the quality of the oil.

Chenni et al. (2016) reported that there are no significant differences between the physical constants of essential oil obtained by hydrodistillation and solvent-free microwave extraction, except for differences in color and odor. The color of the essential oil obtained by SFME was lighter and the odor more pleasant.

**TABLE 5.2**

Physicochemical Properties of Basil Oil as per the Food Chemical Codex

| Serial No. | Parameter | Basil Oil, Comoros Type | Basil Oil, European Type | Reference |
|---|---|---|---|---|
| 1 | Description | Light yellow liquid with camphoraceous odor | Pale yellow liquid with a more floral odor | Food Chemical Codex (2014) |
| 2 | Specific gravity at 20°C | 0.952–0.973 | 0.900–0.920 | Food Chemical Codex (2014) |
| 3 | Angular rotation at 20°C | −2° to +2° | −5° to +5° | Food Chemical Codex (2014) |
| 4 | Refractive index at 20°C | 1.512–1.520 | 1.483–1.493 | Food Chemical Codex (2014) |
| 5 | Acid value | 1.0 max | 2.5 max | Food Chemical Codex (2014) |
| 5 | Saponification value | 4–10 | Not specified | Food Chemical Codex (2014) |
| 6 | Ester value of acetylation | 25–45 | 140–180 | Food Chemical Codex (2014) |
| 7 | Solubility in 80% ethanol | Should pass the test | Should pass the test | Food Chemical Codex (2014) |

## 5.7 General Uses of Basil Oil

Sweet basil (*O. basilicum*) essential oil has been used for centuries in perfumery, cosmetics, and medicine, and has been added to food as a part of spices or herbs. The essential oil has been investigated, and now more than 200 chemical components have been reported from many regions of the world. The essential oil or the components have been shown to possess not only a broad range of antimicrobial (antibacterial and antifungal) properties (Table 5.3), but also antiproliferative or anticancer, antioxidant, and antiwormal activities, and it also affects the central nervous system (CNS) activities, as summarized in Table 5.4.

**TABLE 5.3**

Antimicrobial Activities of Various Constituents of Basil Oil (*Ocimum basilicum*)

| Serial No. | Constituents | Biological Activity | Organism | Reference |
|---|---|---|---|---|
| 1 | Essential oil | Antibacterial | *Staphylococcus aureus, Salmonella enteritidis, Escherichia coli* | Marwat et al. (2011) |
| 2 | Essential oil | Antibacterial | *Bacillus cereus* | Budka and Khan (2010) |
| 3 | Essential oil | Antibacterial | *Haemophilus influenza*, pneumococci | Kristinsson et al. (2005) |
| 4 | Essential oil | Antibacterial | *Staphylococcus aureus* | Nguefack et al. (2004) |
| 5 | Linalool, methyl chavicol, methyl cinnamate | Antibacterial | *Staphylococcus aureus, Enterococcus, Pseudomonas* | Opalchenova and Obreshkova (2003) |
| 6 | Essential oil | Antibacterial | *Staphylococcus aureus, Escherichia coli, Bacillus subtilis, Pasteurella multocida* | Suppakul et al. (2003) |
| 7 | Linalool | Antibacterial | *Giardia lamblia* | Almeida et al. (2007) |
| 8 | Rosamarinic acid | Antibacterial | *Pseudomonas aeruginosa* | Bais et al. (2002) |
| 9 | Essential oil | Antibacterial | *Bacillus cereus, Bacillus subtilis, Bacillus megaterium, Escherichia coli, Staphylococcus aureus, Listeria monocytogenes, Shigella boydii, Shigella dysenteriae, Vibrio mimicus, Vibrio parahaemolyticus, Salmonella typhi* | Hossain et al. (2010) |
| 10 | Essential oil | Antifungal activity | *Cadida albicans, Penicillium notatum, Microsporeum gyseum* | Marwat et al. (2011) |
| 11 | Essential oil | Antifungal activity | Yeast and mold | Suppakul et al. (2003) |
| 12 | Essential oil | Antifungal activity | *Aureobasidium pullulans, Debaryomyces hansenii, Penicillium citrinum, Penicillium expansum* | De Martino et al. (2009) |
| 13 | Essential oil | Antifungal activity | *Alternaria alternata, Fusarum solani* var. *coeruleum* | Zhang et al. (2009) |
| 14 | Linalool, methyl chavicol, cineole, eugenol | Antifungal activity | *Fusarium oxysporum* f. sp. *vasinfectum, Rhizopus nigricans* | Reuveni et al. (1984) |
| 15 | Linalool and eugenol | Antifungal activity | *Sclerotonia sclerotiorum, Rhizopus stolonifer* | Edris and Farrag (2003) |

**TABLE 5.4**

Pharmacological Actions of Basil Oil (*Ocimum basilicum*)

| Serial No. | Biological Activity | Constituents | Observation | Potential Use | Reference |
|---|---|---|---|---|---|
| 1 | Antimicrobial and antibiofilm activity | Essential oil | Antibacterial activity against *Streptococcus mutans* and biofilm formation inhibition and biofilm degradation | As active ingredient in antibacterial and antibiofilm product formulation | Astuti et al. (2016) |
| 2 | Antioxidant activity | Essential oil | In hypoxanthine/xanthine oxidase assay, strong antioxidant activity was evidenced | | Jayasinghe et al. (2003) |
| 3 | Anticancer activity | Essential oil | Had an $IC_{50}$ value of 0.0362 mg/ml (12.7 times less potent than 5-Fluorouracil [5-FU]) in P388 cell lines | Potential for cancer treatment | Manosroi et al. (2006) |
| 4 | Antiviral | Apigenin, linalool, and ursolic acid | Broad spectrum of antiviral activity against herpes virus, hepatitis B, and adenoviruses | | Chiang et al. (2005) |
| 5 | Dermatologic effect | Essential oil | Effect on acne vulgaris in humans | | Balambal et al. (1985) |
| 6 | Antigiardial activity | Essential oil | Inhibits proteolytic activity mainly of cysteine proteases | | Almeida et al. (2007) |
| 7 | Antiwormal response | Volatile oil | Two fractions showed antiwormal response under test condition | | Marwat et al. (2011) |
| 8 | Spemicidal effect | Essential oil | Spermicidal action | Spermicidal purpose | |

### 5.7.1 Food

The food industry primarily uses basil oil as flavoring. However, it also represents an interesting source of natural antimicrobials for food preservation. The addition of basil oil results in a better flavor and less microbial contamination. The flavor of sausage was better retained when basil oil was added in the form of microencapsulated oleoresin, which penetrated even in intramuscular fats in sausage (Flint and Seal 1985). A basil flavor has been developed that can be directly sprayed on foods that are particularly suited for the microwave.

Basil essential oil and its principal constituents exhibit antimicrobial activity against a wide range of gram-negative and gram-positive bacteria, yeasts, and molds. According to Budka and Khan (2010), essential oil from basil oil exhibited bactericidal properties against *Bacillus cereus* in rice-based food. The methanolic extract of *O. bacilicum* showed antibacterial activity against *Pseudomonas aeruginosa*, *Shigella* sp., *Listeria monocytogenes*, *Staphylococcus aureus*, and two different strains of *Escherichia coli*. Basil oil had the strongest antimicrobial activity against *Salmonella* sp. (*S. enteritidis* SE3). The oil at a concentration of 50 ppm reduced the number of bacteria in the food from 5 to 2 log cfu/g after storage for 3 days. At a level of 100 ppm in ham, the bacteria count was reduced without affecting the acceptance of consumers.

### 5.7.2 Cosmetics

Basil essential oil is used topically and is massaged into the skin. It enhances the luster of dull-looking skin and hair. As a result, it is frequently used in many skin care supplements that claim to improve the tone of the skin. Basil oil is appreciated by the cosmetic industry because it has antimicrobial and antioxidant properties. Basil oil showed an enhancing activity for accelerating the transdermal delivery of indomethacin (Fang et al. 2004). The mechanism of action is probably due to the increased skin–vehicle partitioning by the oil.

### 5.7.3 Pharmaceutical

The essential oil of *O. basilicum* has been used in folk medicine to treat various diseases. It has been studied in different parts of the world, and it has been reported to exhibit various biological activities, such as antibacterial and antifungal properties (summarized in Table 5.4), antioxidant activities, anticancer activity, antiwormal response, antiviral activity, dermatological effects, spermicidal effect, and effects on the central nervous system. The constituents or parts of basil oil studied for different biological activities and their observation and potential use are described in Table 5.4.

#### 5.7.3.1 Antibacterial and Antifungal Activities

The antibacterial and antifungal activity of basil oil leads to its use in dental care. Due to its antifungal and disinfectant action, it cleans and disinfects the mouth and relieves inflammation of the gums. The antibacterial potency of the essential oil suggests it has potential activity against *Streptococcus mucans*, which is the main cause of dental carries. The oil has been used in a microemulsion mouth wash formula and exhibited antibacterial and antibiofilm activity and was stable for 3 months at accelerated storage conditions. It was suggested as an active ingredient in an antibacterial and antibiofilm formulation (Astuti et al. 2016). As a component of mouth wash, the volatile oil completely inhibited the growth of organisms at a concentration of 0.5% (Ahonkhal et al. 2009). Volatile oil of *O. basilicum* at 0.01% in toothpaste showed antibacterial activities against most resistant organisms.

## 5.8 Pesticidal Uses of Basil Oil

This essential oil also possesses properties to act as an insecticide. Basil oil also has marked repellency and larvicidal properties against mosquitoes.

### 5.8.1 Insecticidal Activity of Basil Oil

Essential oils of plants contain a number of bioactive compounds that may exert regulatory or inhibitory influence on insect life processes, such as growth and development, reproduction, or orientation. Among the compounds present in essential oils are monoterpenes, which are consequently regarded as a candidate for insecticidal activity. These natural compounds have been proposed as lead compounds for the development of safe, effective, and fully biodegradable insecticides. Most of the monoterpenes are cytotoxic to plant and animal tissue, causing a drastic reduction in the number of mitochondria and Golgi bodies, impairing respiration and photosynthesis and decreasing cell membrane permeability (Tripathi et al. 2009). At the same time, they are volatile and may serve as chemical messengers for insects. The doses of essential oils needed to kill insects or pests and their mechanism of action are important for the safety of humans and other vertebrates. Therefore, the target sites and mode of action need to be understood and should be well elucidated. Although little is known about the physiological actions of essential oils on insects, treatment with various essential oils causes symptoms that suggest a neurotoxic mode of action. A monoterpenoid linalool has been demonstrated to act on the nervous system, affecting ion transport and releasing acetylcholine esterase in insects (Re et al. 2000).

Digilio et al. (2008) assayed essential oil extracted from *O. basilicum* for insecticidal activity against the aphid pests *Acyrthosiphon pisum* (Harris) and *Myzus persicae* (Sulzer). Basil essential oil resulted in high mortality (100%) against *A. pisum* and 96.15% against *M. persicae*, even applied at low doses, and activity was dose dependent. The pea aphid *A. pisum* is a phloem-feeding insect that colonizes leguminous crops where it produces direct damage in terms of nutritive subtraction and the injection of toxic saliva, besides being responsible for the transmission of more than 30 viral diseases (Blackman and Eastop 2000). The green peach aphid or peach potato aphid *M. persicae*, a major pest worldwide, has an extreme host range over 40 different plant families, where it not only produces severe direct damage but also is considered the most important aphid virus vector, being able to transmit more than 100 plant viruses. *M. persicae* has resistance to many synthetic insecticides. The insecticidal activity of basil essential oil is assumed to be due to the presence of monoterpenes (Diglio et al. 2008).

Basil oil and three major active constituents (*trans*-anethole, estragole, and linalool) were tested on three tephritid fruit fly species for insecticidal activity. All tested chemicals acted fast and showed a step dose–response relationship (Chang et al. 2009).

### 5.8.2 Larvicidal Activity

To minimize and eradicate the occurrence of mosquito-borne diseases, many steps have been taken to prevent their spread to different extents, for example, mosquito eradication at an early stage, disease prevention via prophylactic drugs and vaccines, and the prevention of mosquito bites using repellants. Out of these, larviciding has the greatest control impact on the mosquito population because the larvae are concentrated, immobile, and accessible. Nour et al. (2012) studied the essential oils from two basil accessions for larvicidal activity against instar *Aedes aegypti*, the major vector for dengue and yellow fever disease larvae. The steam-distilled essential oil showed larvicidal activity against *A. aegypti*. The larval mortality increased as the essential oil concentration increased, and the essential oil from the different accessions had different activities. The highest dose (500 µg/ml) caused 100% mortality after 3 hours' incubation with oil of the methylchavicol accession (MCV), whereas the same dose caused 100% mortality after 6 hours with geranial–geraniol accession (GGV).

### 5.8.3 Basil Oil as a Repellant

How repellents work in different arthropods is unclear, since conflicting evidence exists, such as different insects detecting the same substance by different organs, but there is a difference in sensitivity to the same repellent. Hairs on the mosquito antennae are temperature and moisture sensitive. The repellant molecules interact with the female mosquito olfactory receptors, thereby blocking the sense of smell. Very little is known about the receptors responsible for the repellent responses in cockroaches. Oleic acid and linoleic acid have been indicated in death recognition and death aversion in cockroaches, and the term *necromone* has been proposed to describe the compound responsible for this type of behavior (Rollo et al. 1995).

The hexane extract of *O. basilicum* in melted petroleum jelly has been studied for repellent activities against mosquito *A. aegypti*. At a concentration of 2%, repellency was observed to be 91.45%. Complete protection was obtained at 3% of the oil, but beyond this concentration, there was no significant change in activity observed. Furthermore, mosquito paralysis occurred at this concentration, which is similar to the reported effect of DEET (Kiplang'at and Mwangi 2013). The essential oil from *O. basilicum* at a 2% level showed a significant mortality, repellency, and an antireproductive effect against the rice weevil *Sithophilus oryzae* L. (Popovic et al. 2006).

## 5.9 Advantages of Basil Oil as a Pesticide

The constituents of basil oil have little or no harmful effect on the environment and the nontarget organism. Due to the multiple sites of action through which the constituents can act, the probability of developing a resistant population is very low. These botanical insecticides degrade rapidly in air and moisture, and detoxification enzymes break them readily. Due to rapid breakdown, they have less persistence in the environmental, larvicidal, and adulticidal activities to sublethal effects, including oviposition, deterrence, and repellent actions (Koul et al. 2008).

## 5.10 Constraints of Basil Oil as a Pesticide

Although essential oils as pesticides have a number of advantages over synthetic pesticides, there are some specific constraints. The efficacy of these materials falls short when compared with synthetic pesticides. The commercial application of plant essential oil–based pesticides includes the availability of sufficient quantities of plant material, standardization and refinement of pesticide products, protection of technology (patents), and regulatory approval. In addition, as the chemical profile of plant species can vary naturally depending on geographic, genetic, climatic, or seasonal factors, pesticide manufacturers have to take additional steps to ensure that their product will perform consistently. All this requires substantial costs, and smaller companies are not willing to invest the required funds unless there is high probability of recovering the costs through some form of market exclusivity. Finally, once all these issues are addressed, regulatory approval is required (Mohan et al. 2011).

## 5.11 Basil Oil–Based Insecticides

In spite of considerable research efforts in many laboratories throughout the world and an ever-increasing volume of scientific literature on the pesticidal properties of essential oils and their constituents, surprisingly few pest control products based on basil essential oils have appeared in the marketplace. This may be a consequence of regulatory barriers to commercialization, or the fact that the efficacy of essential oils toward pests and diseases is not apparent or as obvious as that seen with currently available products.

Krishnamurthy et al. (2012) studied a 25% emulsifiable concentrate (EC) formulation of basil and geranium, each at a 1 ml/L dose, that is, 0.0125% active ingredient of essential oil. This formulation at this dose was again found to be effective against chilli thrips. This EC formulation of essential oils evaluated under field conditions and consistently gave good results against thrips in *Capsicum* under polyhouse conditions by basil formulation. Hence, there is great potential in using essential oils and their formulations in the integrated pest management (IPM) of thrips in these crops. The main constraint in using essential oils as such is their cost, as well as the quality. The cost can be reduced by using commercial ready-to-use essential oil formulations; a few are available in the United States.

Environmentally and user-friendly emulsifiable concentrates of basil oil using biodegradable vegetable oil and a suitable emulsifier such as Tween 80 and Tween 60 have been developed. The EC with 5% (w/w) using methyl oleate as a solvent has been found to have antifungal activity, which may be due to the presence of linalool, beta linalool, and methyl chavicol in the basil oil (Thakur et al. 2014).

## 5.12 Conclusion

The constituents of basil oil exhibit various activities, such as antimicrobial, antioxidative, anticarcinogenic, antiviral, and insecticidal activities. Therefore, this oil has great potential not only in food, pharmaceuticals, and cosmetics, but also as insect repellant and insecticide, and in a variety of ways to control a large number of pests. Although the efficacy of these basil oil constituents is comparatively less than that of synthetic pesticides, it is gaining momentum as far as environment pollution and human health are concerned. It is expected that the innovative formulations of pesticides based on this essential oil will find their greatest commercial applications in urban pest control, vector control vis-à-vis human health, and pest control in agriculture, and will help in organic food production systems, where few alternative pesticides are available.

## References

Adam, F., Vahirua-Lechat, I., Deslandes, E., Bessiere, J.M., Menut, C. Aromatic plants of French Polynesia. III. Constituents of the essential oil of leaves of *Ocimum basilicum* L. *J. Essent. Oil Res.* 2009; 21: 237–240.

Ahonkhal, I., Ayinda, B.A., Edogun, O., Uhuwmangho, M.U. Antimicrobial activities of the volatile oils of *Ocimum basilicum* L. and *Ocimum gratissimum L. (Lamiaceae)* against some aerobic dental isolates. *Pak J. Pharm. Sci.* 2009; 22: 405.

Almeida, I., Alviano, D.S., Vieira, D.P., Alves, P.B., Blank, A.F., Lopes, A.H., Alviano, C.S., Mdo, S.R. Antigiardial activity of *Ocimum basilicum* essential oil. *Parasitol. Res.* 2007; 101: 443.

Astuti, P., Saifullah, T.N., Walanjati, M.P., Yosephine, A.D., Ardianti, D. Basil essential oil (*Ocimum basilicum* L.) activities on *Streptococcus mutans* growth, biofilm formation and degradation and its stability in microemulsion mouth wash formula. *Int. J. Pharm. Clin. Res.* 2016; 8(1): 26–32.

Bais, H.P., Walker, T.S., Schweizer, H.P., Vivanco, J.M. Plant physiology. *Biochemistry* 2002; 40: 983.

Balambal, R., Thiruvengadam, K.V., Kameswarant, L., Janaki, V.R., Thambiah, A.S. *Ocimum basilicum* in acne vulgaris: A controlled comparison with a standard regime. *J. Assoc. Physicians India* 1985; 33(8): 507–508.

Blackman, R.L., Eastop, V.F. *Aphids on the World's Crops: An Identification and Information Guide*, 2nd ed. John Wiley & Sons Ltd., U.K., p. 414, 2000.

Benedec, D., Oniga, I., Oprean, R., Tamas, M. Chemical compositions of essential oils of *Ocimum basilicum* L. cultivated in Rumania. *Formacia* 2009; 57: 625–629.

Bozin, B., Mimica-Dukic, N., Simin, N., Anackov, G. Characterisation of the volatile composition of essential oils of some Lamiaceae species and the antimicrobial and antioxidant activities of the entire oils. *J. Agric. Food Chem.* 2006; 54(5): 1822–1828.

Budka, D., Khan, N.A. The effect of *Ocimum basilicum, Thymus vulgaris, Origanum vulgare* essential oils on *Bacillus cereus* in rice-based foods. *Eur. J. Biol. Sci.* 2010; 2: 17–20.

Bunrathep, S., Palanuvej, C., Ruangrungsi, N. Chemical compositions and antioxidative activities of essential oils from four *Ocimum* species endemic to Thailand. *J. Health Res.* 2007; 21: 201–206.

Chalchat, J.C., Ozcan, M.M. Comparative essential oil composition of flowers, leaves and stems of basil (*O. basilicum*) used as herb. *Food Chem.* 2008; 110: 501–503.

Chang, C.L., Cho, K.I., Li, Q.X. Insecticidal activity of basil oil, *trans* anethole, estragole, and linalool to adult fruit flies of *Ceratitis capitata, Bactrocera dorsalis*, and *Bactrocera cucurbitae*. *J. Econ. Entomol.* 2009; 102(1): 203–209.

Charles, D.J., Simon, J. E. Comparision of extraction methods for the rapid determination of essential oil content and composition of basil (*Ocimum* spp.). *J. Am. Soc. Hortic. Sci.*, Alexandria, 1990; 115(3): 458–462.

Chenni, M., El-Abed, D., Rakotomanomana, N., Fernadez, X., Chemat, F. Comparative study of essential oils extracted from Egyptian basil leaves (*Ocimum basilicum* L.) using hydrodistillation and solvent free microwave extraction. *Molecule* 2016; 21: E113.

Chiang, L.C., Ng, L.T., Cheng, P.W., Chiang, W., Lin, C.C. Antiviral activities of extracts and selected pure constituents of *Ocimum basilicum*. *Clin. Exp. Pharmacol. Physiol.* 2005; 32(10): 811–816.

De Martino, L., De Feo, V., Nazzaro, F. Chemical composition and *in vitro* antimicrobial and mutagenic activities of seven Lamiaceae essential Oils. *Molecules* 2009; 14: 4213.

Digilio, M.C., Mancini, E., Voto, E., De Feo, V. Insecticide activity of Mediterranean essential oils. *J. Plant Interact.* 2008; 3(1): 17–23.

Edris, A.E., Farrag, E.S. Antifungal activity of peppermint and sweet basil essential oils and their major aroma constituents on some plant pathogenic fungi from the vapor phase. *Nahrung* 2003; 47: 117.

Fang, J. Y., Leu, Y.L., Hawang, T.L., Chang, H.C. Essential oils from sweet basil (*Ocimum basilicum*) as novel enhancer to accelerate transdermal drug delivery. *Biol. Pharma. Bull.* 2004; 27(11): 1819–1825.

Flint, F., Seal, R. The sausage seasoning scene. *Food Manuf.* 1985; 60: 43–45.

*Food Chemical Codex*. 9th ed. National Academy Press, Washington, DC, 2014.

Hossain, M.A., Kabir, M.J., Salihuddin, S.M., Rahman, S.M., Das, A.K., Singh, S.K., Alam, M.K., Rahman, A. Antibacterial properties of essential oils and methanol extracts of sweet basil *Ocimum basilicum* occuring in Bangladesh. *Pharm. Biol.* 2010; 48: 504–511.

Hussain, A.I., Anwar, F., Sherazi, S.T.H., Przybylski, R. Chemical composition, antioxidant and antimicrobial activities of basil (*O. basilicum*) essential oil depends on seasonal variations. *Food Chem.* 2008; 108: 986–995.

Ismail, M. Central properties and chemical composition of *Ocimum basilicum* essential oil. *Pharm. Biol.* 2006; 44: 619–626.

Jayasinghe, C., Gotoch, N., Aloki, T., Wada, S. Phenolics composition and antioxidant activity of sweet basil (*Ocimum basilicum* L.). *J. Agric. Food Chem.* 2003; 51: 4442–4449.

Kiplang'at, K.P., and Mwangi, R.W. Repellant activities of *Ocimum basilicum, Azadiricta indica* and *Eucalyptus citriodora* extracts on rabbit skin against *Aedes aegypti*. *J. Entomol. Zool. Studies* 2013; 1(5): 84–91.

Koul, O., Wallia, S., Dhaliwal, G.S. Essential oil as green pesticides: Potential and constraints. *Biopestic. Int.* 2008; 4(1): 63–84.

Klimankova, E., Holadova, K., Hajslova, J., Cajka, T., Poustka, J., Koudela, M. Aroma profiles of five basil (*Ocimum basilicum* L.) cultivars grown under conventional and organic conditions. *Food Chem.* 2008; 107: 464–472.

Krishnamurthy, P.N., Saroja, S., Shivaramu, K., Achala Pariporna, K. Bio-efficacy of essential oil formulations of mint, *Basil* and *Geranium* against onion thrips, *Thrips tabaci* Lindeman and chilli thrips, *Scirtothrips dorsalis* Hood under field conditions. Pest Manag. Hort. Ecosyst. 2014; 20(2): 137–140.

Kristinsson, K.G., Magnusdottir, A.B., Peterson, H., Hermansson, A. Effective treatment of experimental acute otitis media by application of volatile fluids into the ear canal. *J. Infect. Dis.* 2005; 191: 1876.

Lawrence, B.M. A review of the world production of essential oil. *Perfum. Flavor.* 1985; 10: 2–16.

Marie. E.L., Farid, C., Jacqueline, S. Solvent free microwave extraction of essential oil from aromatic herbs: Comparision with conventional hydro-distillation. *J. Chromatogr. A* 2004; 1043(2): 323–327.

Marotti, M., Piccaglia, R., Giovanelli, E. Differences in essential oil composition of basil (*Ocimum basilicum* L.) Italian cultivars related to morphological characteristics. *J. Agric. Food Chem.* 1996; 14: 3926.

Marwat, K.S., Rehman, U.F., Khan, S.H., Ghulam, S., Naveed, A., Mustafa, G., Usman, K. Phytochemical constituents and pharmacological activities of sweet basil—*Ocimum basilicum* L. (Lamiaceae). *Asian J. Chem.* 2011; 23(9): 3773–3782.

Masango, P. Cleaner production of essential oils by steam distillation. *J. Cleaner Prod.* 2005; 29(1): 171–176.

Manosroi, J., Dhumtanom, P., Manosroi, A. Antiproliferative activity of essential oil extracted from Thai medicinal plants on KB and P388 cell lines. *Cancer Lett.* 2006; 235(1): 114–120.

Mindaryani, A., Rahayu, S. Essential oil from extraction and steam distillation of *Ocimum basilicum*. In Proceedings of the World Congress on Engineering and Computer Science (WCES 2007) San Francisco, October 24–26, 2007, pp. 1–5.

Mohan, M., Haider, S.Z., Andola, H.C., Purohit, V.K. Essential oils as green pesticides; for sustainable agriculture. *Res. J. Pharm. Biol. Chem. Sci.* 2011; 2(4): 100–106.

Nguefack, J., Budde, B.B., Jakobsen. Five essential oils from aromatic plants of Cameroon: Their antibacterial activity and ability to permeabilize the cytoplasmic membrane of *Listeria innocua* by flow cytometry. *Lett. Appl. Microbiol.* 2004; 39: 395.

Nidia, A.de.B, Rabson, R.R., Andre von, R. de. A., Marisa, F.M. Extraction of basil oil (*Oscimum basilicum* L.) using supercritical fluid, III Iberoamerican Conference on Supercritical Fluids Cartagena de Indias (Columbia). 2013; pp. 1–8.

Nour, A., Abdurahman, N., Yusoff, M., Sandanasamy, J. Bioactive compounds from Basil (*Ocimum basilicum*) essential oils with larvicidal activity against *Aedes aegypti* larvae. In *Third International Conference on Biology, Environment and Chemistry (IPCBEE 2012)*, 2012, vol. 46, pp. 21–24.

Opalchenova, G., Obreshkova, D. Comparative studies on the activity of basil—An essential oil from *Ocimum basilicum* L. against multidrug resistant clinical isolates of the genera *Staphylococcus*, *Enterococcus* and *Pseudomonas* by using different test methods. *J. Microbiol. Methods* 2003; 54: 105–110.

Pandey, A.K., Singh, P., Tripathi, N.N. Chemistry and bioactivities of essential oils of some *Ocimum* species: An overview. *Asian Pac. J. Trop. Biomed.* 2014; 4: 682–694.

Popovic, Z., Kostic, M., Popvic, S., Skorik, S. Bioactivities of essential oils from basil and sage to *Sitophilus oryzae* L. *Biotechnol. Biotechnol. Equip.* 2006; 20: 36–40.

Pripdeevech, P., Chumpolsri, W., Suttiarporn, P., Wongpornchai, S. The chemical composition and antioxidant activities of basil from Thailand using retention indices and comprehensive two-dimensional gas chromatography. *J. Serbian Chem. Soc.* 2010; 75: 1503–1513.

Rahman, A. Antibacterial properties of essential oils and methanol extracts of sweet basil *Ocimum basilicum* occurring in Bangladesh. *Pharma Biol.* 2010; 48: 504.

Ranjitha, J., Vijiyalakshmi, S. Facile methods for the extraction of essential oil from the plant species—A review. *Int. J. Pharm. Sci. Res.* 2014; 5(4): 1107–1115.

Re, L., Barocci, S., Sonnino, S., Menacarelli, A., Vivani, C., Paolucci, G., Scarpantonio, A., Rinaldi, L., Mosca, E. Linalool modifies the nicotinic receptor-ion channel kinetics at the mouse neuromuscular function. *Pharmacol. Res.* 2000; 42: 177–181.

Reuveni, R., Fleischer, A., Putievsky, E. Fungistatic activity of essential oils from *Ocimum basilicum* chemotypes. *J. Phytopathol.* 1984; 110(1): 20–22.

Rollo, C.D., Borden, J.H., Caey, I.B. Endogenously produced repellant from American cockroach (Blattaria: Blattidae) function in death recognition. *Environ. Entomol.* 1995; 24: 116–124.

Sekar, K., Thangaraj, S., Babu, S.S., Harisaranraj, R., Suresh, K. Phytochemical constituent and antioxidant activity of extract from the leaves of *Ocimum basilicum*. *J. Phytol.* 2009; 1: 408.

Selvakkumar, C., Gayathri, B., Vinaykumar, K.S., Lakshmi, B.S., Balakrishnan, A. Potential anti-inflammatory properties of crude alcoholic extract of *Ocimum basilicum* L. in Human peripheral blood mononuclear cells. *J. Health Sci.* 2007; 53(4): 500–505.

Shams, K.A., Abdel-Azim, N.S., Saleh, I.A., Hegazy, M.-E.F., El-Missiry, M.M., Hammouda, F.M. Green technology: Economically and environmentally innovative methods for extraction of medicinal & aromatic plants (MAP) in Egypt. *J. Chem. Pharm. Res.* 2015; 7(5): 1050–1074.

Simon, J.E., Quin J., Murray R.G. Basil: A source of essential oils. In Janik J., Simon J.E. (eds.), *Advances in New Crops*. Timber Press, Portland, OR, 1990, pp. 484–489.

Shirazi, M.T., Ghalami, H., Kavoosi, G., Rowshan, V., Tofsiry, A. Chemical composition, antioxidant, antimicrobial and cytotoxic activities of *Tagets minuta* and *Ocimum basilicum* essential oils. *Food Sci. Nutr.* 2014; 2: 146–155.

Suppakul, P., Miltz, J., Sonneveld, K., Bigger, S.W. Antimicrobial properties of basil and its possible application in food packaging. *J. Agric. Food Chem.* 2003; 51: 3197–3207.

Taylor, L.T. *Supercritical Fluid Extraction*. John Wiley & Sons, New York, 1996.

Thakur, L.K., Roy, S., Prajapati, R., Singh, M.K., Raza, S.K. Development and evaluation of basil oil emulsifiable concentrates. *Afr. J. Sci. Res.* 2014; 3(1): 6–9.

Tripathi, A.K., Upadhyay, S., Bhuiyan, M., Bhattacharya, P.R. A review on prospects of essential oils as biopesticide in insect pest management. *J. Pharmacogn. Phytother.* 2009; 1(5): 53–63.

Vilkhu, K., Mawson, R., Simons, L., Bates, D. Applications and opportunities for ultrasound assisted extraction in the food industry—A review. *Innov. Food Sci. Emerg. Technol.* 2008; 9: 161–169.

William, B.J. The origin of the Soxhlet extractor. *J. Chem. Educ.* 2007; 84(12): 1913.

Zekovic, Z., Filip, S., Vidovik, S., Adamovik, D., Elgndi, A. Basil (*Ocimum basilicum* L.) essential oil and extracts obtained by supercritical fluid extraction. *APTEFF* 2015; 46: 259–269.

Zhang, J.W., Li, S.K., Wu, W.J. The main chemical composition and in vitro antifungal activity of the essential oils of *Ocimum basilicum* Linn. var. *pilosum* (Willd.) Benth. *Molecules* 2009; 14: 273–278.

# 6

# Lemongrass Oil: As a Green Pesticide

**N.C. Basantia**

## CONTENTS

## 6.1  Introduction

Lemongrass (*Cymbopogon* sp.) is an herb that is widely distributed in the tropical and sub-tropical regions of Asia, Africa, Australia, and America. The genus *Cymbopogon* comprises 144 species and is famous for its high content of essential oil; it contains essential oil with a fine lemon flavor. On account of their diverse use in the pharmaceutical, cosmetics, food, flavor, and agriculture industries, the commercial value of some *Cymbopogon* species is further enhanced by their ability to grow in moderate and extremely harsh climatic conditions (Padalia et al. 2011). *Cymbopogon citratus* is ranked as one of the most widely distributed of the genus, being used in every part of the world. The traditional application in different countries shows high applicability as a common tea, medicinal supplement, insect repellant, insecticide, anti-inflammatory, and analgesic, and in flu control.

## 6.2  Botany of the Plant

### 6.2.1  Taxonomic Classification (Shah et al. 2011)

Kingdom: Plantae

Division: Nagmoleophyta

Class: Liliopsida

Order: Poales

Family: Poaceae or Gramineae

Genus: *Cymbopogon*

Species: *citratus, flexiuosus*

Botanical name: *Cymbopogon citratus*

### 6.2.2  Habitat and Distribution

Most lemongrass is native to South Asia and Australia. The crop grows well in both tropical and subtropical climates at an elevation up to 900 m (above sea level). However, the ideal conditions for lemongrass are a warm and humid climate with sufficient sunshine and 250–330 cm of rainfall per annum, evenly distributed over most of the year. A temperature ranging from 20°C to 30°C and sunshine throughout the year is conducive to a high crop yield. Lemongrass can also be grown in semiarid regions receiving low to moderate rainfall. Well-drained sandy loam is most suitable for the growth of the plant. It can be grown on a variety of soils, ranging from loam to poor laterite. Java citronella is mainly produced by Taiwan, Guatemala, Honduras, Brazil, Ceylon, India, Argentina, Equador,

Madagascar, Mexico, and the West Indies. In India, lemongrass is widely cultivated in the states of Kerala, Karnataka, and Tamil Nadu in the southern region; parts of Uttar Pradesh and Uttaranchal in the northern region; and Assam in the northeast region.

### 6.2.3 Botanical Description of the Plant

Lemongrass is a tall, perennial grass, about 1–1.8 m in height, and it throws up dense fascicles of leaves from a short rhizome. The culm is stout and erect.

### 6.2.4 Leaves

The leaves of *Cymbopogon* sp. are long, glaucous, green, and linear, tapering upward. Along the margins, the ligule is very short, the sheaths are terete, and those of the barren shoots are wide and tightly elapsing at the base, and others are narrow and separating.

### 6.2.5 Flower

It is a short-day plant, and it produces profuse flowering. The inflorescence is a long spike about 1 m in length.

## 6.3 Methods of Extraction of Oil

The common methods to extract essential oil from lemongrass are hydrodistillation (HD), steam distillation, and water distillation. Although the process is very simple, it can induce thermal degradation, hydrolysis, and water solubility of some fragrance constituents. In order to overcome these drawbacks, a number of novel methods have been studied from time to time, for example, microwave-assisted hydrodistillation (MAHD), subcritical water extraction (SWE), supercritical fluid extraction (SFE), accelerated solvent extraction (ASE), and ultrasound-assisted extraction (UAE).

### 6.3.1 Conventional Methods

#### 6.3.1.1 Distillation

Distillation is an extracting oil process that converts volatile liquid (essential oils) into a vapor state and then condenses the vapor into a liquid state. There are different categories of distillation processes, such as water distillation, steam distillation, and hydrodiffusion (Ranjitha and Vijiyalakshmi 2014).

##### 6.3.1.1.1 Water Distillation

In this process, the botanic material is completely immersed in water and the still is brought to boil. It is used to protect the oils to a certain degree since the surrounding water acts as a barrier to prevent overheating when condensed material cools down. The water and essential oil are separated, and the oil is decanted to use as an essential oil. Water distillation can also be done at reduced pressure (under vacuum) to reduce the temperature to less than 100°C, which is useful in protecting the botanical material for obtaining the essential oil.

It is a simple and easy-to-operate extraction method of oil from plant species. Due to the use of heat in this method, it may not be used on very fragile plant material. Oil components like esters are sensitive to hydrolysis, while others like cyclic monoterpene hydrocarbons, and aldehydes are susceptible to polymerization. Oxygenated compounds such as phenols have a tendency to dissolve in distilled water, so their complete removal is not possible. As water distillation tends to be a small operation, it takes a long time to accumulate much oil (Ranjitha and Vijiyalakshmi 2014).

### 6.3.1.1.2 Steam Distillation

In the steam distillation method, the botanical material is placed in a still and steam is forced over the material. The hot steam is used to release the aromatic molecules from the plant material. The steam forces the pockets to open, and then the molecules of these volatile oils escape from the plant material and evaporate in the steam. The steam containing the essential oil is passed through a cooling system to condense the steam, which forms a liquid form of essential oil. Finally, the water is separated (Masango 2005; Ranjitha and Vijiyalakshmi 2014).

The major advantage of steam distillation is that the temperature never goes above 100°C, so temperature-sensitive compounds can distilled. A disadvantage is that not many compounds can be steam distilled—only the aromatic ones.

### 6.3.1.1.3 Hydrodiffusion

The hydrodiffusion method is similar to the steam distillation process. The main difference between these two methods is how the steam is introduced into the still. In the case of hydrodiffusion, the steam is fed into the top, onto the botanical material, instead of from bottom, as in normal steam distillation. The steam containing the essential oil is passed through a cooling system to condense steam, which forms a liquid of essential oil, and then water is separated (Ranjitha and Vijiyalakshmi 2014).

The main advantages of this method are that less steam is used, the processing time is shorter, and there is a higher yield.

## 6.3.1.2 Soxhlet Extraction

Soxhlet extraction is a general and well-established technique that surpasses other conventional extraction techniques in performance, except for limited fields of application, for example, the extraction of thermolabile compounds. Most of the solvent extraction units worldwide are based on Soxhlet principles with recycling of solvents. Basically, the equipment consists of a drug holder extractor, a solvent storage vessel, a reboiler kettle, a condenser, a breather system, and supporting structures like a boiler, a refrigerated chilling unit, and a vacuum unit (William 2007).

This technique is based on the choice of solvent coupled with heat or agitation. In this process, the circulation of solvents causes the displacement of transfer equilibrium by repeatedly bringing fresh solvent into contact with the solid matrix. This method maintains a relatively high extraction temperature and no filtration of extract is required (Shams et al. 2015).

However, the limitation of this technique is that there is a possibility of thermal decomposition of thermolabile targeted compounds because the extraction usually occurs at the boiling point of the solvent for a long time.

Nidia et al. (2013) compared supercritical fluid extraction using carbon dioxide with the Soxhlet and hydrodistillation processes for extraction of basil oil and reported a higher yield of oil by Soxhlet extraction. The higher yield may be due to extraction of polar and nonpolar compounds.

### 6.3.2 Novel Extraction Methods

#### 6.3.2.1 Microwave Extraction Method

Solvent-free microwave extraction is used to separate the essential oil from plant material. The method involves placing the sample in a microwave reactor without any addition of organic solvent or water. The internal heating of the water within the sample distends its cells and leads to rupture of the glands and oleiferous receptacles. This process frees essential oil, which is evaporated by the *in situ* water of plant material.

A cooling system outside the microwave oven continuously condenses the vapors, which are collected in specific glassware. The excess of water is refluxed back to the extraction vessel in order to restore the *in situ* water to the sample.

The microwave isolation offers a net advantage in terms of yield and better oil composition. Furthermore, it is environmentally friendly. In this method, low-boiling-point hydrocarbon compounds undergo decomposition (Marie et al. 2004; Ranjitha and Vijiyalakshmi 2014).

#### 6.3.2.2 Subcritical Water Extraction

Subcritical water extraction is the extraction using hot water under pressure. It has recently emerged as a useful tool to replace traditional extraction methods. Subcritical water extraction is an environmentally clean technique that, in addition, provides higher extraction yields to extract solid samples. Subcritical water extraction is carried out using hot water (from 100°C to 374°C, the latter being the water critical temperature under high pressure [usually up to 10 bars]), enough to maintain water in the liquid state. The most important factor to take into account in this type of extraction procedure is the dielectric constant. This parameter can be modulated easily within a wide range of values by only tuning the extraction temperature.

Water at room temperature is very polar solvent, with a dielectric constant close to 80. However, this level can be significantly decreased to values close to 27 when water is heated up to 250°C while maintaining its liquid state applying pressure. The dielectric constant value is similar to that of ethanol, and therefore is appropriate for solubilizing less polar compounds.

Basically, the experimental setup needed to use this technique is simple. The instrumentation consists of a water reservoir coupled to a high-pressure pump to introduce the solvent into the system, an oven where the extraction cell is placed and where the extraction takes place, and a restrictor to maintain the pressure along the extraction line. Extracts are collected in the collector vial placed at the end of the extraction system (Shams et al. 2015).

The use of subcritical water extraction provides a number of advantages over traditional extraction techniques. These are low extraction time, higher-quality extracts, lower cost of the extractant agent, and being an environmentally cleaner technique (Shams et al. 2015).

#### 6.3.2.3 Supercritical Fluid Extraction

Supercritical fluid extraction is used for the extraction of flavors and fragrances. SFE is a separation technology that uses supercritical fluid as solvent. Every fluid is characterized by a critical point, which is defined in terms of the critical temperature and critical pressure. Fluids cannot be liquefied above the critical temperature regardless of the pressure applied, but may reach the density close to the liquid state. A substance is considered to be a supercritical fluid when it is above its critical temperature and critical pressure. Several compounds have been examined as SFE solvents (e.g., hexane, pentane, butane, nitrous oxide, sulfur hexafluoride, and fluorinated hydrocarbons).

The main supercritical solvent used is carbon dioxide. Carbon dioxide (critical condition 30.9°C and 73.8 bar) is cheap, environmentally friendly, and generally recognized as safe. Supercritical carbon dioxide is also attractive because of its high diffusivity and easily tunable solvent strength. Another advantage is that carbon dioxide is gaseous at room temperature and ordinary pressure, which makes analyte recovery very simple and results in a solvent-free analyte (Taylor 1996; Ranjitha and Vijiyalakshmi 2014).

### 6.3.2.4 Accelerated Solvent Extraction

Accelerated solvent extraction is sometimes called pressurized solvent extraction (PSE). It uses organic solvents at elevated pressure and temperature in order to increase the efficiency of the extraction process. Increased temperature accelerates the extraction kinetics, and elevated pressure keeps the solvent in a liquid state, thus enabling safe and rapid extraction. Furthermore, high pressure forces the solvent into the matrix pores, and hence facilitates the extraction of analyte. High temperature decreases the viscosity of the liquid solvent, allowing a better penetration of the matrix and a weakened solute matrix interaction. Elevated temperature enhances diffusivity of the solvent, resulting in increased extraction speed. The solvent is selected based on the polarity of the analyte and compatibility with postextraction processing equipment. In ASE applications, generally organic solvents are used in conventional techniques, such as methanol. The use of hot water as the extraction solvent under atmospheric or higher pressure is very efficient for extracting phytochemicals.

ASE has been reported to be more efficient than other extraction methods by consuming less solvent and allowing faster extraction. However, since the extraction is performed at elevated temperature, the thermal degradation is a cause of concern, especially for thermolabile compounds in extracts.

The efficiency of ASE is influenced by factors such as pressure, temperature, static extraction time, flush volume, and vessel void volume (Shams et al. 2015).

### 6.3.2.5 Ultrasound-Assisted Extraction

The mechanical effect of ultrasound accelerates the release of organic compounds within the plant body due to cell wall disruption, mass transfer intensification, and easier access of the solvent to the cell content. Ultrasound-assisted extraction is reported to be one of the important techniques for extracting valuable compounds from the vegetable material (Vilkhu et al. 2008). General ultrasonic devices are the ultrasonic cleaning bath and ultrasonic probe system.

The efficiency of UAE depends on various factors, such as the nature of the tissue being extracted, the location of the component to be extracted, and pretreatment of the tissue prior to extraction. UAE can extract analytes under a concentrated form, free from any contaminants or artifacts. It has also advantages in terms of yield, selectivity, operating time, energy input, and preservation of thermolabile compounds (Shams et al. 2015).

## 6.4 Composition of Oil

The genus *Cymbopogon* is known to include about 140 species, of which more than 52 have been reported to occur in Africa, 45 in India, 6 each in Australia and South America, 4 in Europe, 2 in North America, and the remaining are distributed in South Asia. There is a considerable variation in the qualitative and quantitative composition of essential oils from different cultivars of *Cymbopogon*. On the basis of chemical similarity, the cultivars of the genus *Cymbopogon* are divided into five chemical variants or groups within two series, that is, Citrati and Rusae.

As explained earlier, lemongrass oil (*Cymbopogon winterianus*) contains a number of fragrant fractions, of which citronellal, geraniol, and citronellol are the major components and are responsible for the real chemistry of this essential oil (Leung 1980; Evans 1989). Citronella oil has two chemotypes: Ceylon type and Java type (Jowitt 1908; Guenther 1950).

- *Ceylon type*: The oil obtained from *Cymbopogon nardus* consists of camphene, dispentene, citronellal, geraniol, granylacetate, nerol, citronellol, thuzylalcohol, borneol, farnesol, linalool, and methyl eugenol. In this type, the content is 18%–20% for geraniol, 9%–11% for limonene, 7%–11% for methyl isoeugenol, 6%–8% for citronellol, and 5%–15% for citronellal.

- *Java type*: This type of oil is obtained from *C. winterianus* Jowitt and consists of limonene, citronellal, citral, geraniol, citronellol, citronellate, eugenol, methyl eugenol, chavicol, sesquicitronellene, elemol, citronellyl oxide, γ- and ∂-cadinene, vanillin, isovaleraldehyde, hexane-2-al, and 3-methyl pentanal. In this oil, the content is 32%–45% for citronellal, 11%–13% for geraniol, 3%–8% for geranyl acetate, and 1%–4% for limonene. The differences in two varieties and the chemical composition of the essential oil have been recorded since early times (Jowitt 1908; Guenther 1950). It was believed that the Java type variety contained around 85% of geraniol. On the other hand, the Ceylon type variety was reported to contain only 55%–65% of geraniol. A geraniol-rich mutant containing as high as 60% of geraniol content has been developed (Ranaweera and Dayananda 1996).

Gas chromatography–mass spectrometry (GC-MS) analysis of citronella oil revealed the presence of many monoterpene hydrocarbons, amounting to more than 20% of the oil in the Ceylon type, but only 3%–4% in the Java type. There is a high proportion of hydrocarbons in the Ceylon type; the most abundant was found to be camphene. The other hydrocarbons present were α- and β-pinene, sabinene, myrcene, car-3-ene, α- and β-phellandrene, α- and β-terpenes, cis/*trans*-ocemene, terpinolene, and p-cymene. The Java-type oil contains more oxy-terpenes than the Ceylon type. There is a great difference in the amount of geraniol in the two oils. However, the Java type contains much more citronellal and citronellol. Another distinguishing feature of the Ceylon type is the presence of methyl eugenol and methyl isoeugenol. Different *Cymbopogon* species contain varying major compounds, such as citral, geraniol, citronellol, piperitone, and elemin. The major components of the *Cymbopogon* species observed are mentioned in Table 6.1.

**TABLE 6.1**

Major Components of *Cymbopogon* Species

| Serial Number | Compound | Molecular Formula | Species | Country/Region | Content % | Reference |
|---|---|---|---|---|---|---|
| 1 | Citronellal | $C_{10}H_{18}O$ | *C. winterianus* | India | 32.7 | Wany et al. (2013) |
| | | | *C. nardus* | Malaysia | 29.6 | Wei et al. (2013) |
| | | | *C. winterianus* | Brazil | 36.19 | Leite et al. (2011) |
| | | | *C. winterianus* | Southeast Brazil | 27.44 | Quintans-Júnior et al. (2008) |
| 2 | Citronellol | $C_{10}H_{20}O$ | *C. winterianus* | India | 15.9 | Wany et al. (2013) |
| | | | *C. winterianus* | Brazil | 11.34 | Leite et al. (2011) |
| | | | *C. winterianus* | Southeast Brazil | 10.45 | Quintans-Júnior et al. (2008) |
| 3 | Geraniol | $C_{10}H_{18}O$ | *C. winterianus* | India | 23.9 | Wany et al. (2013) |
| | | | *C. martini* | India | 84.16 | Dubey et al. (1999) |
| | | | *C. winterianus* | Brazil | 32.82 | Leite et al. (2011) |
| | | | *C. winterianus* | S.E. Brazil | 40.06 | Quintans-Júnior et al. (2008) |
| 4 | Myrcene | $C_{10}H_{16}$ | *C. citratus* | Egypt | 15.69 | Mohamed et al. (2012) |
| | | | *C. citratus* | Zambia | 18.0 | Chisowa et al. (1998) |
| | | | *C. citratus* | Nigeria | 25.3 | Kasali et al. (2001) |
| | | | *C. citratus* | Mali | 9.1 | Sidibé et al. (2001) |
| 5 | Neral | $C_{10}H_{16}O$ | *C. flexuosus* | India | 30.0 | Chowdhury et al. (2010) |
| | | | *C. flexuosus* | Burkina Faso | 34.6 | Bassolé et al. 2011 |
| | | | *C. flexuosus* | Brazil (North) | 30.1 | Andrade et al. (2009) |
| | | | *C. flexuosus* | Egypt | 34.98 | Mohamed et al. (2012) |
| | | | *C. flexuosus* | Zambia | 29.4 | Chisowa et al. (1998) |
| | | | *C. flexuosus* | Kenya | 33.31 | Matasyoh et al. (2011) |
| | | | *C. giganteus* | Benin Republic | 19.93 | Gbenou et al. (2013) |
| | | | *C.giganteus* | Nigeria | 26.5 | Kasali et al. (2001) |
| | | | *C.citratus* | Angola | 28.26 | Soares et al. (2013) |
| | | | *C.citratus* | Malaysia | 50.81 | Ranitha et al. (2014) |
| | | | *C.citratus* | Brazil | 4.53 | Leite et al. (2011) |
| 6 | Geranial | $C_{10}H_{16}O$ | *C. flexuosus* | India (Kumaon) | 33.1 | Chowdhury et al. (2010) |
| | | | | India (Bihar) | 42.4 | Kumar (2013) |
| | | | | Brazil | 50.0 | Andrade et al. (2009) |
| | | | | Egypt | 40.72 | Mohamed et al. (2012) |
| | | | | Zambia | 39.0 | Chisowa et al. (1998) |
| | | | | Kenya | 39.53 | Chisowa et al. (1998) |

*(Continued)*

**TABLE 6.1 (CONTINUED)**

Major Components of *Cymbopogon* Species

| Serial Number | Compound | Molecular Formula | Species | Country/ Region | Content % | Reference |
|---|---|---|---|---|---|---|
| | | | *C. citratus* | Nigeria | 33.7 | Kasali et al. (2001 |
| | | | | Angola | 40.55 | Soares et al. (2013) |
| | | | | Ivory Coast | 34.0 | Sidibé et al. (2001) |
| | | | | Mali | 45.3 | Sidibé et al. (2001) |
| | | | | Iran | 39.16 | Farhang et al. (2012) |
| 7 | Camphene | $C_{10}H_{16}$ | *C. pendulus* | India | 9.1 | Wei et al. (2013) |
| | | | *C. winterianus* | India | 8.0 | Wany et al. (2013) |
| 8 | Lemonene | $C_{10}H_{16}$ | *C. giganteus* | Cameroon | 7.4 | Jirovetz et al. (2007) |
| | | | *C. giganteus* | Burkina Faso | 42.0 | Bassolé et al. (2011) |
| | | | *C. proximus* | Burkina Faso | 3.9 | Menut et al. (2011) |
| 9 | Elemecin | $C_{12}H_{16}O_3$ | *C. pendulus* | India | 53.7 | Shahi et al. (1997) |
| 10 | Linalool | $C_{10}H_{18}O$ | *C. flexuosus* | India | 2.6 | Chowdhury et al. (2010) |
| | | | *C. winterianus* | India | 1.5 | Wany et al. (2013) |
| | | | *C. martin* | India | 2.0 | Dubey et al. (1999) |
| | | | *C. nardus* | Malaysia | 11.0 | Wei et al. (2013) |
| 11 | Pipertone | $C_{10}H_{18}O$ | *C. oliovieri* | Iran | 72.8 | Mahboubi et al. (2012) |
| | | | *C. parkeri* | Iran | 80.8 | Bagheri et al. (2007) |

## 6.4.1 Effect of Method of Extraction on Composition of Lemongrass Oil

It has been proven through a number of studies that the quality of essential oil mainly depends on the procedure used to extract it. In contrast, these common methods can induce thermal degradation, hydrolysis, and water solubilization of some fragrance constituents. Ranitha et al. (2014) evaluated the effect of microwaves on the extraction of essential oil. The concentrations of key compounds found in lemongrass oil were similar for both methods, but the oil composition revealed that a higher amount of oxygenated monoterpenes, such as linalool, geranic acid, and citronellol, are present in essential oil isolated by MAHD. The quality of lemongrass oil obtained by pressurized liquid extract (PLE) was compared with that of conventional extraction methods, hydrodistillation, and the Soxhlet extraction method. PLE gave the significantly highest amount of neral and geraniol, followed by Soxhlet and hydrodistillation. This was due to the ability of the solvent n-hexane to extract almost all nonvolatile and volatile compounds, compared with hydrodistillation, which can only extract the volatile compounds (Nur et al. 2013). Obtained from dried lemongrass stems found to be better quality containing 90% citral in comparision to steam distillation method (Ha et al. 2008).

## 6.5 Methods of Analysis

The classical methods of analysis of the essential oil of *Citronella* were primarily based, on the one hand, on the estimation of total acetylizable material and, on the other hand, on various rough solubility checks, such as Schimmel's test, raised Schimmel's test, and the London solubility test. In addition, refractive index and optical rotation were specified. As the new instrumental methods emerged, the new techniques of the characterization of chemical compounds based on spectroscopic methods resulted in a major surge in natural products research in the early 1960s, followed by the development of chromatographic techniques such as gas–liquid chromatography (GLC), GC-MS, and GC–ion mobility spectrometry (IMS).

The quality of lemongrass oil is determined by its citral content. Various methods have been reported in literature for the estimation of citral in lemongrass oil, and also for the separation of citral from lemongrass oil. The common methods for the estimation and separation of citral are the bisulfite method, neutral sulfite method, hydroxylamine method, and colorimetric methods.

### 6.5.1 Bisulfite Method

The bisulfite method is based on adduct formation. Upon shaking of a measured quantity of oil with a hot aqueous solution of sodium bisulfite, an adduct is formed, which dissolves on heating the solution. The noncitral portion of the oil separates as an oily layer, which can be measured conveniently in the neck of a Cassia flask, and thereby the citral content of the oil can be determined.

### 6.5.2 Neutral Sulfite Method

This is also an adduct formation reaction. In this method, the liberated sodium hydroxide has to be neutralized periodically with acid to permit the reaction to go to completion. However, the solution must not be permitted to turn acidic, as this would result in the formation of the stable dihydrosulfonic compound, from which citral cannot be regenerated. This method has all the disadvantages of the bisulfite method as a method of estimation and also of separation. However, it offers certain advantages over the bisulfite method. By using the indicator (phenolphthalein), the exact end point of the reaction can be determined.

### 6.5.3 Hydroxylamine Method

This method is also used for the estimation of citral in lemongrass oil. It makes use of both hydroxylamine and hydroxylamine hydrochloride. After the reaction of this with the carbonyl group, the mixture is titrated with standard alkali. The hydroxyl amine method also has some defects. All the carbonyl groups present in lemongrass oil will react with hydroxyl amine, and the value obtained will be much higher. However, this method offers some advantages over the adduct formation process. Relatively small amounts of the oil are required for estimation. The reaction of hydroxylamine with aldehyde is rapid, thereby shortening the time required for the estimation. This method proves to be exceptionally applicable to oils that contain large amounts of aldehydes. The solution used for the standard procedure is stable and can be kept for longer periods.

### 6.5.4 Colorimetric Methods

The citral content of lemongrass oil has also been estimated by the coloring agent of Ehrlich Miller. This coloring agent has been found to give better results, and the development of color takes place rapidly and remains quite stable for a long time. The coloring agent is prepared according to Ehrlich Miller and consists of the following solutions:

- 5% p-dimethylaminobenzaldehyde solution in acetic acid
- 10% phosphoric acid solution in acetic acid

One milliliter of each of the above solutions is added to different amounts of citral in acetic acid, whereby a marked color change from blue to pink can be observed. The percentage absorbance and extinction of the colored citral is then measured using a colorimeter, and calibration graphs are plotted. The amount of citral in solutions can be compared with that of known strength, and thus the percentage of citral can be determined. Here, we also need solutions of citral with known strength.

### 6.5.5 Gas–Liquid Chromatography and Gas Chromatography–Mass Spectrometry Methods

The GLC technique depends on the effectiveness of the volatility of the compounds and constituents of the essential oils. However, in most cases a large number of chemical compounds in essential oils were dependent on the extent up to which they could be effectively separated. Thus, GLC offered a method of separation that could be achieved with a very minute quantity of sample. The separated constituents are subjected to the new technique, such as mass spectrometry and nuclear magnetic resonance (NMR) spectroscopy, to determine their chemical structure. The separation of constituents is affected by the type of stationery phase (column) and the temperature condition of separation. Various researchers used various methods for the separation, identification, and quantification of constituents (Table 6.2).

## 6.6 Physicochemical Properties of Lemongrass Oil

The physicochemical properties, such as appearance, color, solubility, specific gravity, refractive index, and angular rotation, have been studied by various researchers. These properties may vary from type to type and by method of extraction. There are several standards framed at the national and international levels to maintain the quality of the oil. The specifications for physicochemical properties per the Food Chemical Codex are shown in Table 6.3.

Hazwan et al. (2014) studied the physical properties of citronella oil from different sources obtained by three different extraction methods, such as ohmic heated hydrodistillation, hydrodistillation, and steam distillation. The refraction index and specific gravity at 20°C for all three different sources of oil were about 1.47 and 0.89, respectively. In addition, the color, which is an important feature to determine the consumer's acceptability of citronella oil products, was also measured. Normally, commercial citronella oil is yellow. In this study, the intensities of yellow and red were measured. The citronella oil extracted by ohmic heated hydrodistillation was in the range of 0.7 R 2.0 Y, whereas extracted citronella oil by steam distillation exhibited color in the range of 0.5 R 3 Y. The characteristics of the color red were possibly due to *trans*-β-caryophyllene and γ-cadinene constituents in the citronella oil (Abena et al. 2007; Harjeet et al. 2011). However, no red color was found

**TABLE 6.2**

Methods Used for Analysis of Lemongrass Oil

| Serial Number | Source of Oil | Method of Extraction | Number of Compounds | Method Used | Reference |
|---|---|---|---|---|---|
| 1 | Aerial parts of *C. citratus* | Hydrodistillation | 18 | GC-FID<br>Column: HP-5 capillary 30 m × 0.32 mm i.d., 0.25 μ thickness<br>Detector: FID<br>Temp program: 50°C for 2 min @ 8°C/min to 240°C<br>Detector temp: 280°C<br>Injector temp: 240°C | Mohamed et al. (2012) |
| 2 | Aerial parts of *C. citratus* | Hydrodistillation | 18 | GC-MS<br>Column: VF-5MS capillary 30 m × 0.25 mm i.d., 0.25 μ thickness<br>Detector: MSD<br>Temp program: 50°C–180°C @ 5°C/min to 250°C<br>Carrier gas: Helium<br>Flow rate: 1 ml/min<br>Split ratio: 1:20<br>Ionization energy: 70 eV | Mohamed et al. (2012) |
| 3 | Aerial parts of *C. nardus* | Distillation | 13 | GC-MS<br>Column: HP-5MS capillary 30 m × 0.25 mm i.d., 0.25 μ thickness<br>Detector: MSD.<br>Temp program: 70°C for 5 min @ 3°C/min to 325°C<br>Carrier gas: Helium<br>Flow rate: 1 ml/min<br>Split ratio: 100:1<br>Ionization EI mode energy: 70 eV<br>Detector temp: 280°C<br>Injector temp: 240°C | Hazwan et al. (2014) |
| 4 | Aerial parts of *C. citratus* | Pressurized liquid extraction | 2 | GC-MSD<br>Column: DB-5 capillary 20 m × 0.188 mm i.d., 0.4 μ thickness<br>Detector: FID<br>Temp program: 100°C for 1 min @ 1°C/min to 120°C<br>Detector temp: 250°C<br>Injector temp: 300°C | Nur Ain et al. (2013) |

*(Continued)*

**TABLE 6.2 (CONTINUED)**

Methods Used for Analysis of Lemongrass Oil

| Serial Number | Source of Oil | Method of Extraction | Number of Compounds | Method Used | Reference |
|---|---|---|---|---|---|
| 5 | Leaves of *C. citratus* | Microwave-assisted hydrodistillation | 7 | GC-MSD<br>Column: HP-5MS capillary 30 m × 0.25 mm i.d., 0.25 μ thickness<br>Detector: MSD<br>Temp program: 50°C for 5 min, then rise @ 3°C/min to 240°C, then @ 5°C/min to 300°C<br>Carrier gas: Helium<br>Flow rate: 1 ml/min<br>Split ratio: 1:10<br>Ionization EI energy: 70 eV | Ranitha et al. (2014) |
| 6 | Citronella oil, Ceylon and Java type | Distillation | 60 | GC-FID<br>Column: 10% Carbowax Chromosorb W (2.7 × 3.2 mm)<br>Detector: FID<br>Temp program: 60°C @ 2°C/min to 220°C<br>Base attenuation: ×16 | Wijesekara (1973) |

*Note:* EI, electron ionization; FID, flame ionization detector; Temp, temperature.

**TABLE 6.3**

Physicochemical Properties of Lemongrass Oil as per Food Chemical Codex

| Serial Number | Parameter | Lemongrass Oil, East Indian Type | Lemongrass Oil, West Indian Type | Reference |
|---|---|---|---|---|
| 1 | Description | Dark yellow to light brown–red liquid with lemon odor | Light yellow to light brown liquid with light lemon odor | Food Chemical Codex (2014) |
| 2 | Specific gravity at 20°C | 0.894–0.904 | 0.869–0.894 | Food Chemical Codex (2014) |
| 3 | Angular rotation at 20°C | −10° to +0° | −10° to +0° | Food Chemical Codex (2014) |
| 4 | Refractive index at 20°C | 1.483–1.489 | 1.483–1.489 | Food Chemical Codex (2014) |
| 7 | Solubility | Soluble in mineral oil, freely soluble in propylene glycol but insoluble in water and glycerine, and dissolves readily in alcohol | Soluble in mineral oil, freely soluble in propylene glycol but insoluble in water and glycerine, and yields cloudy solution with alcohol | Food Chemical Codex (2014) |

when the citronella oil was obtained by the hydrodistillation method. The appearance of red color when using other extraction methods may also be caused by lipid oxidation occurring in the extraction system. Essential oil obtained from *C. citratus* by microwave-assisted hydrodistillation had the following characteristics: refractive index at 20°C, density (g/ml) at 27°C, color parameters are 1.483, 0.873 g/ml, and color as (L* = 97, a* = –2.44, b* = 6.29), respectively (Vazquez-Briones et al. 2015).

## 6.7 General Uses of Lemongrass Oil

Lemongrass oil is one of the 20 most important essential oils that are traded globally (Lawrence 1993). Citronella oil is highly demanded because of its wide usage in perfumes and the soap manufacturing, cosmetics, and flavoring industries, and because it is effective as an insect repellant. This essential oil is characterized by a high content of citral, which is used as a raw material for the production of ionone, vitamin A, and β-carotene. Due to its appealing citric flavor and strong antimicrobial potential, it has been used as flavoring agent in several food products and as a natural preservative for extending the shelf life of food products. In addition to food and cosmetics, this oil is also used in many traditional medicines and has great potential in various pharmaceutical applications. The general uses of lemongrass oil are in food, cosmetics and personal care products, traditional medicine, and pharmaceuticals.

### 6.7.1 Food

A recent consumer trend toward a preference for products with a lower salt and sugar content presents a greater need for efficient food preservatives. However, an increasingly negative consumer perception of synthetic food additives has spurred an interest in finding natural alternatives. Essential oils like lemongrass oil are natural compounds that have shown promising properties, such as antifungal, antibacterial, and antiviral activities. Moreover, essential oils have also been proven to have other diverse beneficial functions, such as antidiabetic, antiradical, and antioxidant effects.

Apart from its appealing flavor, lemongrass (*C. citratus*) essential oil has been shown to have antimicrobial potential. This makes it susceptible for incorporation in food products. The oil of lemongrass could suppress the growth of mesophiles and psychrophiles in fresh-cut apples (Raybaudi-Massilia et al. 2008). Tzortzakis and Economakis (2007) reported that the essential oil of lemongrass oil inhibited the growth of *Botrytis cinerea*. Essential oil lemongrass controls food spoilage and shows antibacterial activity against *Listeria monocytogenes* (Nguefack et al. 2004) and *Staphylococcus aureus* (Baratta et al. 1998). Although this essential oil has shown to be a promising alternative to chemical preservatives against foodborne pathogens, it presents special limitations that preclude its use in food products. Low water solubility, high volatility, strong odor, and toxicological effects at high doses make it difficult for food application. In addition to these drawbacks, the incorporation of oil–based compounds like essential oil in aqueous food products is a big challenge since it shows physical and chemical instability when it is applied in food systems (McClements et al. 2004). Therefore, several studies have shown that the use of nanoemulsion can be a great choice for the application of essential oil in a food matrix. Kim et al. (2013) studied the plum coatings of lemongrass oil incorporating carnauba wax–based nanoemulsion to evaluate antimicrobial properties and physical and chemical changes in plums. The nanoemulsion was able to inhibit the *Salmonella* and *Escherichia coli* population without changing the flavor, fracturability, or glossiness of the product, and it reduces the ethylene production and retards the changes in lightness and the concentration of phenolic compounds.

Essential oils as natural sources of phenolic components attract investigators to evaluate their activity as antioxidants or free radical scavengers. Vazquez-Briones et al. (2015) studied the antioxidant properties of essential oil obtained from *C. citratus* by the microwave-assisted hydrodistillation extraction method. In this study, the phenolic content and antioxidant capacity were reported to the tune of 149.2 ± 6 mg gallic acid equivalent (GAE) per 100 ml of oil and 44.06 ± 0.20 mg Trolox per ml of essential oil, respectively. Different antioxidant capacity values were reported by different researchers (Selim 2011; Mirghani et al. 2012). The difference in values is attributed to factors such as climate, soil composition, season, part of the plant, age, and stage of growing of the plant (Angioni et al. 2006). *C. citratus* could be of great interest in the food industry to be used as a natural additive for flavoring.

### 6.7.2 Cosmetics and Personal Care Products

The redolence of the oil enables its use in soaps, detergents, and the application in the perfumery. Wuthi-udomlert et al. (2011) evaluated the antifungal activity of lemongrass oil against *Malassezia furfur*, an opportunistic yeast associated with dandruff. Two percent lemongrass oil shampoo provided the required qualities necessary for commercial use. After being kept for 6 weeks at 28°C–30°C and 45°C, this formulated shampoo gave minimum fungal concentrations (MFCs) against *M. furfur* of 75 and 18.75 µl/ml, respectively. The 2% concentration of lemongrass oil was selected because of the smell, consistency, and stability of the shampoo.

### 6.7.3 Traditional Medicine

In traditional medicine, the oil has been used as an aromatic tea, vermifuge, diuretic, and antispasmodic. Lemongrass is a folk remedy for coughs, elephantiasis, flu, gingivitis, headache, leprosy, malaria, ophthalmic, pneumonia, and vascular disorders. Studies have shown that lemongrass has antibacterial and antifungal properties. The traditional use includes treatment of fever, intestinal parasites, and digestive and menstrual problems. Mixed with pepper, it is a homeotherapy for menstrual troubles and nausea. Lemongrass is a good cleanser that helps to detoxify the liver, pancreas, kidney, bladder, and digestive tract. It cuts down uric acid, cholesterol, excess fats, and other toxins in the body, while stimulating digestion, blood circulation, and lactation; it also alleviates indigestion and gastroenteritis. It is said that lemongrass also helps improve the skin by reducing acne and pimples and acts as a muscle and tissue toner. Also, it can reduce blood pressure. A recent study by the Food and Nutrition Research Institute of the Department of Science and Technology (DOES) showed that lemongrass can help prevent cancer. It has many uses in aromatherapy (Karkala and Bhushan 2014). It can be used as massage oil for aching joints and muscles. When mental illness has to be treated, citronella can be clarifying and balancing. Combining it with lemon oil can bring an even greater brightening effect to the mind.

### 6.7.4 Pharmaceutical Uses

A vast array of ethnopharmacological applications of lemongrass exist today. Its health restorative capacity may be ascribed to the diverse secondary metabolites it produces. The pharmacological activity of lemongrass oil is summarized in Table 6.4. Batubara et al. (2015) confirmed the ability of β-citronellol, the major component of lemongrass

**TABLE 6.4**

Pharmacological Actions of Lemongrass Oil (*Cymbropogon citratus*)

| Serial Number | Biological Activity | Constituents | Observation | Potential Use | Reference |
|---|---|---|---|---|---|
| 1 | Antimicrobial activity | Ethanolic extracts of the leaves Flavonoids and tannins | Antibacterial property against *Staphylococcus aureus* | Potential antibacterial property against *Staphylococcus aureus* | Danlami et al. (2011) |
| 2 | Antifungal activity | Lemongrass oil and citral | The antifungal activity of lemongrass and citral against *Candida* species | Formulating herbal drugs for oral healthcare | Taweechaisupapong et al. (2012) |
| 3 | Antiprotozoan activity | Citral and major constituents of lemongrass oil | The promasigotes of *Leishmania infantum* undergo programmed cell death upon exposure to citral and constituents of lemongrass oil | Lemongrass may be foreseen as an antiprotozoan drug of the future | Machado et al. (2012) |
| 4 | Antioxidant activity | Phenolic acids present in the plant | Showed the antioxidant profile | | Garg et al. (2012) |
| 5 | Antidiarrheal activity | Citral | Relief in diarrhea | | Tangpu and Yadav (2006) |
| 6 | Anticancerous activity | Emulsion of citral and lemongrass oil | Anticancerous properties on cervical cell lines | The constituents of lemongrass may be used to form potent anticancer drugs in the future | Ghosh (2013) |
| 7 | Antiviral activity | Lemongrass oil and citral | Reduced viral infectivity by coating the viral capsid and preventing it from binding to the host cell | Can be used to sanitize food and surfaces to prevent viral infection | Gilling et al. (2014) |

oil, to bring about a reduction in weight of rats fed a high-fat diet. Inhalation of vapors of β-citronellol enhances the sympathetic nerve activity of the rats, which leads to increased activity in the adipose tissue, resulting in weight loss without affecting the concentration and activity of the liver enzymes. The essential oil of lemongrass is also used to maintain oral health. The antagonistic activity of lemongrass against the planktonic and biofilm forms of *Candida dubliniensis*, a common oral pathogen, has been reported. Therefore, lemongrass may be used in formulating herbal drugs for oral healthcare (Taweechaisupapong et al. 2012). The essential oil of *C. citrus* has been shown to have anti-inflammatory, anticonvulsant, analgesic, and anxiolytic effects (Blanco et al. 2009; Sforcin et al. 2009; Gbenou et al. 2013). It has been reported that the lemongrass is bestowed with hypolipidemic, hypocholesterimic, and hypoglycemic properties. The antagonistic activity of lemongrass toward different pathogenic bacteria, protozoa, and fungi has also been reported. Research on antimicrobial and anti-inflammatory activities, along with GC-MS analysis of lemongrass oil, revealed that the major constituents, like limonene, nerol, gerianal, geraniol, and myrcene, may be responsible for its microbicidal and anti-inflammatory effects. It has been reported that the promastigotes of *Leishmania infantum* undergo programmed cell death upon exposure to citral, the major component of lemongrass oil (Machado et al. 2012).

The combination of silver nanoparticles and the oil has synergistic inhibitory action on the growth of pathogens like *E. coli*, *Staphylococcus*, *Moraxella*, *Enterococcus*, and *Candida* sp. Citronella oil also exhibits antifungal activity against *Aspergillus niger*. The anti-inflammatory effect is through inhibition of production of IL-1β by bioactive compounds of lemongrass oil (citral, neral, and geranial) (Perez et al. 2011). An antiviral effect of lemongrass against an enveloped murine novovirus has been reported (Gilling et al. 2014). The bioactive compounds in citronella oil are studied for their anticancerous properties. An emulsion of citral and lemongrass oil exhibited anticancerous properties on cervical cell lines by reducing cell proliferation and initiating apoptosis (Ghosh 2013). Hence, it is envisaged that the constituents of lemongrass may be used to form potent anticancer drugs in the future.

## 6.8 Insecticidal Activity of Lemongrass Oil

The development of natural products for pest control is increasingly important, since some synthetic pesticides are associated with environmental concerns or are being withdrawn for economic and regulatory reasons. In addition, pesticides sometimes lose their effectiveness due to the difficulty of managing pest resistance, and the search for new synthetic compounds is increasingly time-consuming and expensive. Plant secondary metabolites play an important role in plant–insect interaction, and such compounds may have insecticide, hormonal, or antifeedant activity against insects (Bernays and Chapman 1940). The essential oil compounds and their derivatives are considered to be an alternative means of controlling many harmful insects because these compounds are very specific to harmful insects but do not affect the beneficial insects, and they degrade rapidly unlike synthetic compounds. Essential oils of plants contain a number of bioactive compounds that may exert regulatory or inhibitory influence on insect life processes, such as growth and development, reproduction, and orientation. Recent research has demonstrated their larvicidal and antifeedant activity, capacity to delay development, adult emergence and fertility, deterrent effects on oviposition, and arrestant and repellent action.

### 6.8.1 Insecticidal Activity

Among the compounds present in essential oils, monoterpenes are usually the main component, and are consequently regarded as a candidate for insecticidal activity. These natural compounds have been proposed as lead compounds for the development of safe, effective, and fully biodegradable insecticides. Most of the monoterpenes are cytotoxic to plants and animal tissue, causing a drastic reduction in the number of mitochondria and Golgi bodies, impairing respiration and photosynthesis and decreasing cell membrane permeability (Tripathi et al. 2009). At the same time, they are volatile and may serve as chemical messengers for insects. The doses of essential oils needed to kill insects or pests and their mechanism of action are potentially important for the safety of humans and other vertebrates. Therefore, the target sites and mode of action need to be understood and well elucidated. Although a little is known about the physiological actions of essential oils on insects, treatment with various essential oils and their constituents causes symptoms that suggest a neurotoxic mode of action. A monoterpenoid linalool has been demonstrated to act on the nervous system, affecting ion transport and the release of acetylcholine esterase in insects (Re et al. 2000).

Sudiarta et al. (2013) studied the effect of lemongrass oil extracted from *C. citratus* on *Plutella xylostella*, which causes club root disease in cabbage. The phytotoxicity test was conducted in the field with several concentrations of lemongrass (5%, 2.5%, 1%, 0.5%, 0.25%, and 0.1%). The results showed that a high concentration of lemongrass (10%) was effective as a phytotoxic, with burn symptoms of the cabbage leaf. However, low concentrations (1% and 0.5%) of lemongrass oil can control the population of *P. xylostella* without any phytotoxicity. The essential oils of *Cymbopogon martini* have been studied and found to display high anthelmintic activity against *Caenorhabditis elegans* at an $ED_{50}$ value of 125.4 µg/ml. Essential oils of *C. citratus* in West Africa displayed about a 100% mortality rate against adult *Anopheles gambiae* (Nonviho et al. 2010). The essential oil from *C. winterianus* has caused a dose-dependent mortality of *Culex quinquefasciatus*.

### 6.8.2 Larvicidal Activity

To minimize and eradicate the occurrence of mosquito-borne diseases, many steps have been taken to prevent their spread to different extents, for example, mosquito eradication at an early stage, disease prevention via prophylactic drugs and vaccines, and the prevention of mosquito bites using repellants. Out of these, larviciding has the greatest impact on the mosquito population because the larvae are concentrated, immobile, and accessible. Nazar et al. (2009) studied the larvicidal effect of *C. citratus* essential oil against *Cx. quinquefasciatus* larva and reported an $LC_{50}$ value of 24 mg/L. The essential oil from *C. citratus* had a larvicidal activity against *Aedes aegypti*, causing 100% mortality at a concentration of 100 ppm (Cavalcanti et al. 2004).

### 6.8.3 Lemongrass Oil as a Repellant

Very little is known about the receptors responsible for the repellent responses in cockroaches. Oleic acid and linoleic acid have been indicated in death recognition and death aversion in cockroaches, and the term *necromone* has been proposed to describe the compound responsible for this type of behavior (Rollo et al. 1995).

*Citronella* oil has repellency activity against *Ae. aegypti* mosquitoes. The extracted oil was microencapsulated (1.5% gelatin and 1.5% arabic gum by a complex coacervation method). This citronella oil was treated on cotton fabrics using gelatin and gum acacia microcapsules by the pad dry method, in which 15%, 30%, and 50% repellency effects were studied. The 50% concentrated repellents gave the best mosquito repellency. However, the microencapsulated oil gave a better repellant effect for a longer time (Murugan et al. 2012).

## 6.9 Advantages of Lemongrass Oil as a Pesticide

The constituents of lemongrass oil are selective and have little or no harmful effect on the environment and the nontarget organism. Due to the multiple sites of action through which the constituents can act, the probability of developing a resistant population is very low. These botanical insecticides degrade rapidly in air and moisture, and detoxification enzymes break them readily. Due to rapid breakdown, they are less persistent in the environment (Koul et al. 2008).

## 6.10 Constraints of Lemongrass Oil as a Pesticide

The efficacy of these materials falls short when compared with synthetic pesticides. Essential oils also require somewhat greater application rates (as high as 1% active ingredients) and may require frequent reapplication when used outdoors. The commercial application of plant essential oil–based pesticides has challenges, like having sufficient quantities of plant material, the standardization and refinement of pesticide products, the protection of technology, and regulatory approval. In addition, as the chemical profile of plant species can vary naturally, depending on geographic, genetic, and climatic annual or seasonal factors, pesticide manufacturers have to take additional steps to ensure that their product will perform consistently. All this requires substantial costs, and smaller companies are not willing to invest the required funds unless there is a high probability of recovering the costs through some form of market exclusivity. Finally, once all these issues are addressed, regulatory approval is required (Mohan et al. 2011).

## 6.11 Lemongrass Oil–Based Insecticides

The commercial plant product based on lemongrass oil, available as the trade name Green Match EX™, contains essential oils of lemongrass from *C. nardus*, *C. citratus*, and *Cymbopogon flexiosus* containing citronellal and citral as the main bioactive compounds. This formulation is used as an insecticide (Fischer et al. 2013).

## 6.12 Conclusion

Lemongrass oil (*Cymbopogon* sp.) consists of a diverse array of bioactive compounds and exhibits a wide range of activities, such as antimicrobial, antioxidative, anticarcinogenic, antiviral, and insecticidal activities. Therefore, this oil has great potential not only in food, pharmaceuticals, and cosmetics, but also as an insect repellant. It is expected that the innovative formulation of pesticides based on this essential oil will find their greatest commercial application in urban pest control, vector control vis-à-vis human health, and pest control in agriculture, and will help in organic food production systems, where a few alternative pesticides are available.

## References

Abena, A.A., Gbenoub, J.D., Yayib, E., Moudachiroub, M., Ongokac, R.P., Oumbac, J.M., Siloud, T. Comparative chemical and analgesic properties of essential oils of *Cymbopogon nardus* (L.) Rendle of Benin and Congo. *Afr. J. Tradit. Complement. Altern. Med.* 2007; 4(2): 267–272.

Andrade, E.H., Zoghbi, M.D., Lima, M.D. Chemical composition of the essential oils of *Cymbopogon citratus* (DC.) Stapf cultivated in north of Brazil. *J. Essent. Oil Bear. Plants* 2009; 12: 41–45.

Angioni, A., Barra, A., Coroneo, V., Dessi, S., Cabras, P. Chemical composition, seasonal variability, and antifungal activity of *Lavandula stoechas* L. ssp. *stoechas* essential oils from stem/leaves and flowers. *J. Agric. Food Chem.* 2006; 54: 4364–4370.

Bagheri, R., Mohamadi, S., Abkar, A., Fazlollahi, A. Essential oil components of *Cymbopogon parkeri* STAPF from Iran. *Pak. J. Biol. Sci.* 2007; 10: 3485–3486.

Baratta, M.T., Dorman, H.J.D., Deans, S.G., Figueiredo, A.C., Barroso, J.G., Ruberto, G. Antimicrobial and antioxidant properties of some commercial essential oils. *Flavour Fragr. J.* 1998; 13: 235–244.

Batubara, I., Suparato, I.H., Sadiah, S., Matusuoka, R., Mitsunaga, T. Effect of inhaled citronella oil and related compounds on rat body weight and brown adipose tissue sympathetic nerve. *Nutrients* 2015; 7: 1859–1870.

Bernays, E.A., Chapman, R.F. *Host Plant Selection by Phytophageous Insects.* Chapman & Hall, New York, 1994.

Blanco, M.M., Coasta, C.A.R.A., Freire, A.O., Santosh, J.G., Costa, M. Neurobehavioral effect of essential oil of *Cymbopogon citratus* in mice. *Phytomedicine* 2009; 16: 265–270.

Baratta, M.T., Dorman, H.J.D, Deans, S.G., Figueiredo, A.C., Barroso, J.G. and Ruberto, G. Antimicrobial and antioxidant properties of some commercial essential oils. *Flavour Fragr J.* 1998; 13: 235–244.

Bassolé, I.H., Lamien-Meda, A., Bayala, B., Obame, L.C., Ilboudo, A.J., Franz, C., Novak, J., Nebié, R.C., Dicko, M.H. Chemical composition and antimicrobial activity of *Cymbopogon citratus* and *Cymbopogon giganteus* essential oils alone and in combination. *Phytomedicine* 2011; 18: 1070–1074.

Cavalcanti, E.S., Morais, S.M., Lima, M.A., Santana, E.W. Larvicidal activity of essential oils from Brazilian plants against *Aedes aegypti* L. *Mem. Inst. Oswaldo Cruz* 2004; 99: 541–544.

Chisowa, E.H., Hall, D.R., Farman, D.I. Volatile constituents of the essential oil of *Cymbopogon citratus* Stapf grown in Zambia. *Flavour Fragr. J.* 1998; 13: 29–30.

Chowdhury, S.R., Tandon, P.K., Chowdhury, A.R. Chemical composition of the essential oil of *Cymbopogon flexuosus* (Steud) Wats. growing in Kumaon region. *J. Essent. Oil Bear. Plants* 2010; 13: 588–593.

Danlami, U., Rebeca, A., Machan, D.B., Asuquo, T.S. Comparative study on the antimicrobial activities of the ethanolic extracts of lemon grass and polyalthia longifolia. *J. Appl. Pharm. Sci.* 2011; 1(9): 174–176.

Dubey, V.S., Mallavarapu, G.R., Luthra, R. Changes in the essential oil content and its composition during palmarosa (*Cymbopogon martini* (Roxb.) Wats. var. motia) inflorescence development. *Flavour Fragr. J.* 1999; 15: 309–314.

Evans, W.C. *Trease Evans' Pharmacognosy*, 13th ed. Bailliere Tindall, London, 1989.

Farhang, V., Amini, J., Javadi, T., Nazemi, J., Ebadollahi, A. Chemical composition and antifungal activity of essential oil of *Cymbopogon citratus* (DC.) Stapf. against three *Phytophthora* species. *Greener J. Biol. Sci.* 2012; 3: 292–298.

Fischer, D., Imholt, C., Pelz, H.J., Wink, M., Prokopc, A., Jacoba, J. The repelling effect of plant secondary metabolites on water voles, *Arvicola amphibious*. *Pest Manag. Sci.* 2013; 69: 437–443.

*Food Chemical Codex*. 9th ed. National Academy Press, Washington, DC, 2014.

Gbenou, J.D., Ahounou, J.F., Akakpo, H.B., Laleye, A., Yayi, E., Gbaguidi, F., Baba-Moussa, L. et al. Phytochemical composition of *Cymbopogon citratus* and *Eucalyptus citriodora* essential oils and their anti-inflammatory and analgesic properties on Wistar rats. *Mol. Biol. Rep.* 2013; 40: 1127–1134.

Guenther, E. *The Essential Oils 4*. Van Nostrand, New York, 1950.

Garg, D., Muley, A., Khare, N., Marar, T. Comparative analysis of phytochemical profile and antioxidant activity of some Indian culinary herbs. *Res. J. Pharm. Biol. Chem. Sci.* 2012; 3(3): 845–854.

Ghosh, K. Anticancer effect of lemongrass oil and citral on cervical cancer cell lines. *Pharmacogn. Commun.* 2013; 3: 41–48.

Gilling, D.H., Kitajima, M., Torrey, J.R., Bright, K.R. Mechanism of antiviral action of plant antimicrobials against murine norovirus. *Appl. Environ. Microbiol.* 2014; 80: 4898–4910.

Ha, H.K.P., Maridable, J., Gaspillo, P.D., Kawasaki, J. Essential oil from lemongrass extracted by supercritical carbon dioxide and steam distillation. *Philippine Agric. Sci.* 2008; 91(1): 36–41.

Harjeet, S., Gupta, V.K., Rao, M.M., Sannd, R., Mangal, A.K. Evaluation of essential oil composition of *Cymbopogon* spp. *Int. J. Pharma Recent Res.* 2011; 3(1): 40–43.

Hazwan, M.H., Man, H.C., Abidin, Z.Z., Jamaludin, H. Comparison of citronella oil extraction methods from *Cymbopogon nardus* by ohmic-heated hydrodistillation, hydrodistillation and steam distillation. *Bioresources* 2014; 9(1): 256–272.

Jensen, W.B. The origin of Soxhelt Extractor. *J. Chem. Educ.* 2007; 84(12): 1913.

Jirovetz, L., Buchbauer, G., Eller, G., Ngassoum, M.B., Maponmetsem, P.M. Composition and antimicrobial activity of *Cymbopogon giganteus* (Hochst.) Chiov. essential flower, leaf and stem oils from Cameroon. *J. Essent. Oil Res.* 2007; 19: 485–489.

Jowitt, I.F. Annals of the Royal Botanical Gardens, Peradeniya 4: 185. In Gildemiester, A., Hoffman, A. (eds.), *Volatile Oils*. 2nd ed. Longmans, London, 1908.

Karkala, M., Bhushan, B. Review on pharmacological activity of *Cymbopogon citratus*. *Int. J. Herbal Med.* 2014; 1(6): 5–7.

Kasali, A.A., Oyedeji, A.O., Ashilokun, A.O. Volatile leaf oil constituents of *Cymbopogon citratus* (DC.) Stapf. *Flavour Fragr. J.* 2001; 16: 377–378.

Kim, I.H., Lee, H., Kim, J.E. Plum coatings of lemongrass oil–incorporating carnauba wax–based nanoemulsion. *J. Food Sci.* 2013; 78: E1551–E1559.

Koul, O., Walia, S., Dhaliwal, G.S. Essential oils as green pesticides: Potential and constraints, *Biopestic. Int.* 2008; 4: 63–84.

Kumar, B.S. Essential oil of *Cymbopogon citratus* against diabetes: Validation by *in vivo* experiments and computational studies. *J. Bioanal. Biomed.* 2013; 5: 194–203.

Lawrence, B.M. A planning scheme to evaluate new aromatic plants for the flavor and fragrance industries. In Janick, J., Simon, J.E. (eds.), *New Crops*. Wiley, New York, 1993, pp. 620–627.

Leite, B.L.S., Souza, T.T., Antoniolli, A.R., Guimarães, A.G., Siqueira, R.S., Quintans, J.S.S., Bonjardim, L.R. et al. Volatile constituents and behavioral change induced by *Cymbopogon winterianus* leaf essential oil in rodents. *Afr. J. Biotechnol.* 2011; 10: 8312–8319.

Leung, A.Y. *Encyclopedia of Common Natural Ingredients Used in Food, Drugs, and Cosmetics.* John Wiley & Sons, New York, 1980.

Machado, M., Pires, P., Dinis, A.M., Santos-Rosa, M., Alves, V., Salgueiro, L., Caveleiro, C., Sousa, M.C. Monoterpenic aldehydes as potential anti-*Leishmania* agents: Activity of *Cymbopogon citratus* and citral on *L. infantum, L. tropica* and *L. major. Exp. Parasitol.* 2012; 130: 223–231.

Mahboubi, M., Kazempour, N. Biochemical activities of Iranian *Cymbopogon olivieri* (Boiss) Bor. essential oil. *Indian J. Pharm. Sci.* 2012; 74: 356–360.

Marie, E.L., Farid, C., Jacqueline, S. Solvent free microwave extraction of essential oil from aromatic herbs: Comparision with conventional hydrodistillation. *J. Chromatogr. A.* 2004; 1043(2): 323–327.

Masango, P. Cleaner production of essential oils by steam distillation. *J. Cleaner Prod.* 2005; 29(1): 171–176.

McClements, D.J. *Food Emulsions: Principles, Practices and Techniques.* CRC Press, Boca Raton, FL, 2004.

Matasyoh, J.C., Wagara, I.N., Nakavuma, J.L., Kiburai, A.M. Chemical composition of *Cymbopogon citratus* essential oil and its effect on mycotoxigenic *Aspergillus* species. *Afr. J. Food Sci.* 2011; 5: 138–142.

Menut, C., Bessiére, J.M., Samaté, D., Djibo, A.K. Aromatic plants of tropical west Africa. XI. Chemical composition, antioxidant and antiradical properties of the essential oils of three *Cymbopogon* species from Burkina Faso. *J. Essent. Oil Res.* 2011; 12: 37–41.

Mirghani, M.E.S., Liyana, Y., Parveen, J. Bioactivity analysis of lemongrass (*Cymbopogon citratus*) essential oil. *Int. Food Res. J.* 2012; 19(2): 569–575.

Mohamed, H.R., Sallam, Y.I., el-Leithy A.S., Aly, S.E. Lemongrass (*Cymbopogon citratus*) essential oil as affected by drying methods. *Ann. Agric. Sci.* 2012; 57: 113–116.

Mohan, M., Haider, S.Z., Andola, H.C., Purohit, V.K. Essential oils as green pesticides; for sustainable agriculture. *Res. J. Pharm. Biol. Chem. Sci.* 2011; 2(4): 100–106.

Murugan, V.K., Masthan, M.K., Vediappan, V.K. Potential and controlled repellent activity of microencapsulated citronella oil treated textile cotton fabrics against *Aedes aegypti. Hitek. J. Bio. Sci. Bioengg.* 2012; 1(1): 1–8.

Nazar, S., Ravikumar, S., Prakash, W.G., Syed, A.M., Suganthi, P. Screening of Indian coastal plant extracts for larvicidal activity of *Culex quinquefasciatus. Indian J. Sci. Technol.* 2009; 2: 24–27.

Nidia, A. de. B., Rabson, R.R., Andre von, R. de. A., Marisa, F.M. Extraction of basil oil (*Oscimum basilicum* L.) using supercritical fluid. III iberoamerican Conference on Supercritical Fluids Cartegena de Indias (Columbia). 2013; pp. 1–8.

Nguefack, J., Leth, V., Amvam Zollo, P.H., Mathur, S.B. Evaluation of essential oils from aromatic plants of Cameroon for controlling food spoilage and mycotoxin producing fungi. *Int. J. Food Microbiol.* 2004; 94: 329–334.

Nonviho, G., Wotto, V.D., Noudogbessi, J., Avlessi, F., Akogbeto, M., Sohounhloué, D.C. Original research paper insecticidal activities of essential oils extracted from three species of *Poaceae* on *Anopheles Gambiae* Spp., major vector of malaria. *Sci. Study Res.* 2010; 11: 411–420.

Nur Ain, A.H., Zaibunnisa, A.H., Halimahton Zahrah, M.S., and Norashikin, S. An experimental design approach for the extraction of lemongrass (*Cymbopogoan citratus*) oleoresin using pressurised liquid extraction. *Int. Food Res. J.* 2013; 20: 451–455.

Padalia, R.C., Verma, R.S., Chanotiya, C.S., Yadav, A. Chemical fingerprinting of the fragrant volatiles of nineteen Indian cultivars of *Cymbopogon* Sprenz. (Poaceae). *Rec. Nat Prod.* 2011; 5(4): 290–299.

Perez, G.S., Zavala, S.M., Arias, G.L., Ramos, L.M. Anti-inflammatory activity of some essential oils. *J. Essential Oil Res.* 2011; 23: 38–44.

Quintans-Júnior, L.J., Souza, T.T., Leite, B.S., Lessa, M.N., Bonjardim, L.R., Santos, M.R., Alves, P.B., Blank, A.F., Antoniolli, A.R. Phythochemical screening and anticonvulsant activity of *Cymbopogon winterianus* Jowitt (Poaceae) leaf essential oil in rodents. *Phytomedicine* 2008; 15: 619–624.

Ranaweera, S.S., Dayananda, K.R. Mosquito-larvicidal activity of Ceylon citronella (*Cymbopogon nardus* [L.] Rendle) oil fractions. *J. Nat. Sci. Council* 1996; 24: 247–252.

Ranitha, M., Abdurahman, H.N., Ziad, A.S., Azhari, H.N., Thana Raj, S. A comparative study of lemongrass (*Cymbopogon citratus*) essential oil extracted by microwave-assisted hydro-distillation and conventional hydrodistillation method. *Int. J. Chem. Eng. Appl.* 2014; 5(2): 104–108.

Ranitha, M., Nour, A.H., Sulaiman, A.Z., Nour, A.H., Thani, R.S. A comparative study of lemongrass (*Cymbopogon citratus*) essential oil extracted by microwave-assisted hydrodistillation (MAHD) and conventional hydrodistillation (HD) method. *Int. J. Chem. Eng. Appl.* 2014; 5: 104–108.

Ranjitha, J., Vijiyalakshmi, S. Facile methods for the extraction of essential oil from the plant species—A review. *Int. J. Pharm. Sci. Res.* 2014; 5(4): 1107–1115.

Raybaudi-Massilia, R.M., Rojas-Grau, M.A. Mosqueda-Melgar, J., Martin Belloso, O. Comparative study on essential oils incorporated in to an alginate based edible coating to assure the safety and quality of fresh cut Fuji apples. *J. Food Prot.* 2008; 71: 1150–1161.

Re, L., Barocci, S., Sonnino, S., Menacarelli, A., Vivani, C., Paolucci, G., Scarpantonio, A., Rinaldi, L., Mosca, E. Linalool modifies the nicotinic receptor-ion channel kinetics at the mouse neuro muscular function. *Pharmacol. Res.* 2000; 42: 177–181.

Rollo, C.D., Borden, J.H., Caey, I.B. Endogenously produced repellant from American cockroach (Blattaria: Blattidae) function in death recognition. *Environ. Entomol.* 1995; 24: 116–124.

Selim, S.A. Chemical composition, antioxidant and antimicrobial activity of the essential oil and methanol extract of the Egyptian lemongrass *Cymbopogon proximus stapf. Grasas Aceites* 2011; 62(1): 55–61.

Sforcin, J.M., Amaral, J.T., Fernades, A., Sousa, J.P.B., Bastos, J.K. Lemongrass effects on IL-1β and IL-6 production by macrophages. *Nat. Prod. Res.* 2009; 23: 1511–1519.

Shah, G., Shri, R., Panchal, V., Sharma, N., Singh, B., Mann, A.S. Scientific basis for the therapeutic use of *Cymbopogon citratus* stapf (lemongrass). *J. Adv. Pharm. Technol. Res.* 2011; 2: 3–8.

Shahi, A.K., Sharma, S.N., Tava, A. Composition of *Cymbopogon pendulus* (Nees ex Steud) Wats, an elemicin-rich oil grass grown in Jammu region of India. *J. Essent. Oil Res.* 1997; 9: 561–563.

Shams, K.A., Abdel-Azim, N.S., Saleh, I.A., Hegazy, M.-E.F., El-Missiry, M.M., Hammouda, F.M. Green technology: Economically and environmentally innovative methods for extraction of medicinal & aromatic plants (MAP) in Egypt. *J. Chem. Pharm. Res.* 2015; 7(5): 1050–1074.

Sidibé, L., Chalchat, J.-C., Garry, R.-P., Lacombe, L., Harama, M. Aromatic plants of Mali (IV): Chemical composition of essential oils of *Cymbopogon citratus* (DC.) Stapf and *C. giganteus* (Hochst.) Chiov. *J. Essent. Oil Res.* 2001; 13: 110–112.

Soares, M.O., Vinha, A.F., Barreira, S.V., Coutinho, F., Aires-Goncalves, S., Oliveira, M.B., Pires, P.C., Castro, A. *Cymbopogon citratus* EO antimicrobial activity against multi-drug resistant Gram-positive strains and non-*albicans-Candida* species. *FORMATEX* 2013; 1081–1086.

Sudiarta, P., Sumiartha, K., Antara, N.S. Utilization of essential oil of lemongrass (*Cymbopogon citratus*) as a bio-pesticide to control *Plutella xylostella* (Lepidoptera: Plutellidae). *E-Jurnal Agroeteknologi Tropika* 2013; 2(1): 1–5.

Tangpu, V., Yadav, A.K. Antidiarrhoeal activity of *Cymbopogon citratus* and its main constituent, citral. *Pharmacologyonline* 2006; 2: 290–298.

Taweechaisupapong, S., Ngaonee, P., Patsuk, P., Pitiphat, W., Khunkitti, W. Antibiofilm activity and post antifungal effect of lemongrass oil on clinical *Candida dubliniensis* isolate. *South Afr. J. Bot.* 2012; 78: 37–43.

Taylor, L.T. *Supercritical Fluid Extraction.* John Wiley & Sons, New York, 1996.

Tripathi, A.K., Upadhyay, S., Bhuiyan, M., Bhattacharya, P.R. A review on prospects of essential oils as biopesticide in insect pest management. *J. Pharmacogn. Phytother.* 2009; 1(5): 53–63.

Tzortzakis, N.G., Economakis, C.D. Antifungal activity of lemongrass (*Cymbopogon citratus* L.) essential oil against key post harvest pathogens. *Innov. Food Sci. Emerg. Technol.* 2007; 8: 253–258.

Vazquez-Briones, M.C., Herandez, L.R., Guerrero-Beltran, J.A. Physicochemical and antioxidant properties of *Cymbopogon citratus* essential oil. *J. Food Res.* 2015; 4(3): 36–45.

Vilkhu, K., Mawson, R., Simons, L., Bates, D. Application and opportunities for ultrasound assisted extraction in food industry: A review. *Innov. Food Sci. Emerg. Technol.* 2008; 9: 161–169.

Wany, A., Jha, S., Nigam, V.K., Pandey, D.V. Chemical analysis and therapeutic uses of citronella oil from *Cymbopogon winterianus*: A short review. *Int. J. Adv. Res.* 2013; 1: 504–521.

Wei, L.S., Wee, W. Chemical composition and antimicrobial activity of *Cymbopogon nardus* citronella essential oil against systemic bacteria of aquatic animals. *Iran. J. Microbiol.* 2013; 5: 147–152.

Wijesekara, R.O.B. The chemical composition and analysis of Citronella oil. *J. Natl. Sci. Counc. Srilanka.* 1973; 1: 67–81.

William B.J. *Journal of Chemical Education.* 2007; 84(12): 1913.

Wuthi-udomlert, M., Chotipatoomwan, P., Panyadee, S., Gritsanapan, W. Inhibitory effect of formulated lemongrass shampoo on *Malassezia furfur*: A yeast associated with dandruff. *Southeast Asian J. Trop. Med. Public Health* 2011; 42(2): 363–369.

# 7

# Cinnamon Oil

Khalid Haddi, Lêda R.A. Faroni, and Eugênio E. Oliveira

## CONTENTS

## 7.1 Introduction

Cinnamon is a common spice that has been used for several centuries by different cultures around the world. It is obtained from different parts of a tropical evergreen tree belonging to the genus *Cinnamomum*. Various reports have dealt with the numerous properties of cinnamon and its major components not only for human health but also for agriculture applications. In this chapter, important aspects of trees from the *Cinnamomum* genus, and their products, such as botany, pharmacology, toxicology, and some end uses, with a special focus on the pesticidal potential for agriculture and indoor uses, are covered.

## 7.2 Botany of the Plant

The genus *Cinnamomum* (Lauraceae) includes more than 250 aromatic evergreen trees and shrubs of up to 10–20 m, primarily distributed in Southeast Asia, China, and Australia (Barceloux 2009). Investigations conducted at the beginning of the 1980s have shown that this genus has a center of diversity in south India (Ravindran et al. 2003). However, although formerly thought to be a purely Asiatic genus, *Cinnamomum* has been enriched with species such as *Phoebe*, transferred from neotropical genera based on studies and investigations carried out by taxonomists such as Kostermans. Kostermans has also defined the key characteristics for the *Cinnamomum* species identification (Kostermans 1980, 1983), leading the genus to include not only the Asiatic species but also New World ones. A very detailed botanical characterization of different species of the genus *Cinnamomum* can be found in the monograph on cinnamon and cassia written by Ravindran et al. (2003).

There are mainly four types of cinnamon:

1. True cinnamon, *Cinnamomum verum* J. Presl, also called Ceylon cinnamon, *Cinnamomum zeylanicum*, or Mexican cinnamon, *Laurus cinnamomum* L. Moderately sized (10–15 m) evergreen trees with smooth and brown branches when young. The leaves are opposite or subopposite, leathery, ovate or elliptic to broadly ovate, triplinerved with the three main nerves prominent on both surfaces. Young leaves are reddish and later turn dark green. The bark is smooth, light pinkish brown, and up to 10 mm thick. Small, pale yellow, campanulate flowers, arranged in cymes, are borne in axillary or terminal panicles. The flowering time is from October to February. The fruit is a fleshy, ellipsoid to oblong-ovoid drupe, which contains one seed and turns dark purple or black when ripe between May and June.

2. Cassia cinnamon, *Cinnamomum aromaticum* Nees, or Chinese cinnamon, *Cinnamomum cassia* J. Presl. This evergreen tree grows up to 18–20 m high. Young branches are smooth and brown, and the bark is gray to brown colored and is 13–15 mm thick when mature. The leaves are simple, opposite to subopposite, oblong lanceolate or oblanceolate, with three prominent veins. The leaves are glabrous above and with microscopic hairs below. These leaves are reddish when young and dark green when mature. The small, white flowers are borne in axillary or terminal panicles with characteristics similar to those of *C. verum*. The flowering takes place from

October to December. The fruit is a green, one-seeded, fleshy, globose drupe and turns pink-violet when mature. This fruit is similar in size to a small olive.

3. Vietnamese cinnamon, *Cinnamomum loureiroi*. Although considered for a long time as a different species, the Vietnamese cinnamon seems to be a *C. cassia*. The difference seen in the final product is a result of different harvesting processes between Vietnam and China. The confusion is believed to derive from the fact that the original *C. loureiroi*, described by Loureiro and on which he based his study, is a very rare species or even could be mislaid or lost (Dao 2003).

4. Indonesian cinnamon, *Cinnamomum burmannii*. It is a small evergreen tree, up to 15 m tall. Bark smooth, grayish brown, 2–3 mm. Leaves are opposite or subopposite, elliptical-ovate to oblong and triplinerved. They are pale red and finely hairy when young. Older leaves are glabrous, glossy green above and glaucous pruinose below. Inflorescences axillary or subterminal, slender, paniculate-cymose. Flowers are green to dark red. The fruit is ellipsoid or oblanceoloid with a pointed lip.

5. Other species include Indian cassia, *Cinnamomum tamala*, and camphor, *Cinnamomum camphora*. Indian cassia is a moderate-sized (around 8 m) evergreen tree with four morphotypes. Leaves are alternate, subopposite or opposite, glabrous, three-nerved from the base and are pink when young. Morphotypes are differentiated according to the morphology of the leave. The flowering starts from May. The fruit is slender, ellipsoid, acutish cup obconical, and fleshy, and fruits ripen between June and July. *C. camphora* is a small to medium-sized tree with small triplinerved leaves that are glabrous on both surfaces or sparsely puberulent beneath only when young. Flowers are small, yellowish-white, and similar to those of *C. verum*. They appear in April to May. The fruit is a small, purplish-black, ovate or subglobose drupe and ripens in August to November.

## 7.3 Methods of Extraction of Oil

Traditionally, the main products of the *Cinnamomum* genus are formed by its leaves and the dried, inner bark extracted from shoots, traded as quills, quillings, ships, and powder, and extensively used in flavoring of various dishes and processed food. But recently, interest in the value-added products, such as bark oil and leaf oil, extracted mainly from *C. verum* cinnamon bark, has been consistently growing. These oils are used in food, pharmaceutical, and perfume industries. The high value of these oils is a result of time- and effort-consuming extraction processes. The high price of these oils depends to a large extent not only on the quality of raw material used but also on the extraction process, methods, and final use of the oil.

Like most other essential oils, cinnamon oils can be extracted using a large array of techniques (Ravindran et al. 2003; Wang and Weller 2006; Tongnuanchan and Benjakul 2014; El Asbahani et al. 2015). Such techniques and methods can be classified into two broad classes: conventional and advanced methods (De Castro and García-Ayuso 1998; Huie 2002; Doughari 2012; El Asbahani et al. 2015).

The conventional methods include distillation and organic solvent extraction, while the advanced methods include a number of innovative techniques, such as supercritical fluid extraction (SFE), subcritical extraction liquids, ultrasound-assisted extraction, microwave-assisted extraction, solvent-free microwave extraction, microwave hydrodiffusion and gravity, microwave steam distillation (MSD), and microwave steam diffusion (Doughari 2012; Dima and Dima 2015; El Asbahani et al. 2015). A subclass of the innovative methods mostly used at laboratory and microsampling analysis scales includes Clevenger distillation, microdistillation, and headspace solid-phase microextraction (Dima and Dima 2015).

### 7.3.1 Conventional Methods for Oil Extraction

Distillation is one of the oldest, simplest, and most widespread methods of extracting cinnamon essential oils, especially at commercial levels (Meyer-Warnod 1984; Ravindran et al. 2003; Wong et al. 2014; Dima and Dima 2015). In the cinnamon bark or leaf hydrodistillation process, water vapors are used as solvent driving, at boiling temperature, the cinnamon essential oil molecules (codistillation). The extraction device is simple and includes a heating source surmounted by a copper or steel tank, partially filled with water, where bark or leaves are added. The distillate produced from the tank passes through a precooling system to a condenser, consisting of copper tubing immersed in a large water tank, and a decanter to allow condensation and separation of essential oil and water. In some cases, there are various tanks connected to the same condenser (Ravindran et al. 2003). When the raw material is not immersed in water but maintained at a certain distance above the water surface using a grid or perforated support in a way that allows the vapor circulation from the bottom upward and across the raw material, the process is termed vapor hydrodistillation. The last variant of the distillation process is called steam distillation, and it uses two separate tanks for vapor generation and essential oil extraction. The steam produced in the first tank is introduced into the lower part of the second tank (extractor) and allowed to pass through the raw material.

Although widely used, especially in small-scale extraction units, the distillation method suffers from some drawbacks: prolonged extraction time (3–6 hours); degradation of some temperature-sensitive molecules; simultaneous extraction of other components, such as plant pigments; and environmental negative impacts (El Asbahani et al. 2015).

Together with steam distillation, solvent extraction has been widely used for the extraction of essential oils from various plant parts, including cinnamon bark and leaves. This technique uses either pure organic solvents or mixtures of them. Different solvents, including hexane, petroleum ether, methanol, propanol, methylene chloride, and ethanol, can be used for extraction (Areias et al. 2000; Pizzale et al. 2002; Kosar et al. 2003; Tongnuanchan and Benjakul 2014), but acetone is the most commonly used one. Basically, in this method, the solvent is mixed with the fine grounded plant material, heated to extract the essential oil, and then filtrated. Subsequently, the filtrate is concentrated by solvent evaporation. The resulting concentrate is a resin or a combination of wax, fragrance, and essential oil (concrete), from which the absolute essential oil is obtained using an alcohol-based distillation (Tongnuanchan and Benjakul 2014). At an industrial level, the Soxhlet extraction method is the most used among the solvent-based extraction methods. In conventional Soxhlet, the sample is placed in a "thimble" made of strong filter paper, which is placed in a thimble holder in the chamber of the Soxhlet apparatus, and gradually filled with condensated fresh solvent from a distillation heated flask. When the liquid reaches the overflow level, a siphon aspirates the solute of the thimble holder into the distillation flask in a continuous process until complete extraction is achieved (De Castro and García-Ayuso 1998; Ravindran et al. 2003).

### 7.3.2 Advanced Methods for Oil Extraction

Supercritical fluid extraction is one of the innovative techniques used for cinnamon oil and other essential oil extraction (Aghel et al. 2004; Khajeh et al. 2004; Braga et al. 2005; Carvalho et al. 2005; Moura et al. 2005; Fornari et al. 2012). It is using the supercritical state of a solvent fluid, usually $CO_2$, achieved when the temperature and the pressure of the solvent are raised above its critical value (31°C; 74 bar) (Wang and Weller 2006). Fluids reaching their supercritical state have both gas and liquid characteristics and present a density similar to that of liquids, low viscosity, and a high diffusion coefficient. Moreover, fluids like $CO_2$ gas are cheap and available at high purity, nontoxic and nonflammable, easily manipulated, and adjusted by varying the pressure and temperature, which is non-aggressive for thermosensitive molecules. The technique has great versatility, and the end product is virtually free from any solvent traces. In this process, the temperature and pressure of $CO_2$ are adjusted to reach the supercritical state; after that, the solvent is allowed to pass through finely ground raw plant material. The fluid and the dissolved compounds are transported to one or more separators for a depression step, where the $CO_2$ is gradually decompressed, resulting in lower solubility of the solute and leading to the separation of the solute from the solvent. Once the material is separated, the gas is compressed and recycled back to be reused again in the extraction process (Fornari et al. 2012). When water is used as the fluid, the technique is called superheated water extraction.

Although the supercritical extraction technique is considered a good tool to overcome the disadvantages of the conventional methods, it is not practiced for commercial cinnamon or cassia oil production because this technology is very costly and has little final product quality enhancement compared with solvent-extracted product (Doughari 2012; Ravindran et al. 2003).

Two other relatively newer methods used to extract cinnamon oil mainly for small-scale and laboratory analyses are ultrasound-assisted extraction and microwave-assisted extraction (Gallo et al. 2010; Gursale et al. 2010; Dvorackova et al. 2015; Sowbhagya 2016). The ultrasound-assisted extraction uses ultrasonic waves with frequency higher than 20 kHz to induce mechanical vibration leading to the destruction of cell and storage gland walls of plant material immersed in water or solvent and the release of cell contents, including essential oils (Wang and Weller 2006; El Asbahani et al. 2015). The microwave-assisted extraction uses electromagnetic radiations (frequency higher than 0.3 GHz) that can interact with cellular water and create heat, leading to cell disruption and facilitating the release of cell contents.

### 7.3.3 Conventional versus Advanced Methods for Oil Extraction

A continuous search for economically and ecologically sound extraction technologies as an alternative to conventional extraction methods has been growing over recent years. These novel techniques aim to overcome the drawbacks of conventional ones with respect to extraction time, solvent consumption, extraction yields and purity of the essential oil, reproducibility, and energy consumption and operating costs.

Various studies have compared two or more methods for cinnamon oil extraction considering several of the above-cited criteria. Gallo et al. (2010) analyzed the total polyphenols of different extracts of *C. zeylanicum*, and other species, using microwave-assisted and ultrasound extraction methods. Considering factors such as the extraction time and the solvent wastage, the results suggested that the microwave-assisted method was more effective than the ultrasound extraction method, as higher recoveries for *C. zeylanicum* were obtained. The microwave-assisted method was considered effective in extracting antioxidant components from cinnamon. Using the ethanol extraction technique, Yang et al. (2012) obtained a higher yield than with the supercritical $CO_2$ extraction when studying the antioxidant activity of various parts of *C. cassia*, while Golmohammad et al. (2012) compared two aqueous solutions of *C. zeylanicum* bark obtained by superheated water extraction and distillation methods, followed by a solid-phase extraction method. Golmohammad et al.'s results showed that the distillation yielded a higher quantity of essential oil, but the superheated water extraction improved the purity of the oil extracted. Another recommendation of Golmohammad et al. is the use of the solid-phase extraction method as an alternative to liquid–liquid extraction for its simplicity and low cost. More recently, Wong et al. (2014) compared steam distillation with Soxhlet methods and concluded that although the quantity of oil extracted was higher with the Soxhlet method, steam distillation was the most suitable method for extracting cinnamaldehyde, as it uses a lower temperature. Dvorackova et al. (2015) concluded that the best option to extract phenolic compounds from cinnamon was the classical solvent-based extraction method, and they considered the extraction methods based on sonication and shaking to be inappropriate when they studied phenolic compounds from *C. cassia* using four different extraction methods: classical solvent, ultrasonication, maceration, and shaking.

## 7.4 Methods of Analysis of Oil

The important demand for essential oils by the flavor, cosmetic, health, and phytomedicine industries is leading to large quantities of cinnamon extracts, and particularly cinnamon essential oil, being produced and traded worldwide. Thus, analysis of the physical and chemical compositions of such extracts and oils is becoming a pertinent issue to ensure quality, consumer safety, and fair trade (Figueiredo et al. 1997, 2008; Do et al. 2015). It is also well known that the chemical composition of essential oils, including cinnamon oils, depends on many factors, such as growing conditions, harvest periods and techniques, drying processes, and extraction and isolation methods used. Such issues were reported earlier, in the 1970s and 1980s, as problems faced by the food industry to distinguish between commercially available cinnamons and cassia (Lawrence 1967; Archer 1988). Moreover, cases of falsification and fraud have been reported (Kubeczka 2002; Price and Price 2007; Do et al. 2015). Thus, for both consumers and chemical companies, it is necessary to determine a profile (e.g., physical, organoleptic, or chemical characteristics) of the constituents of essential oils (Dima and Dima 2015; Do et al. 2015).

Traditionally, essential oil analysis was performed to investigate their quality aspects, focusing on their purity and identity. However, with the improvements in instrumental analytical chemistry, the characterization of essential oils has allowed the scanning of a greater number of molecular constituents of essential oils. Most of the data available on cinnamon's physical and chemical composition were determined earlier by conventional methods (reviewed in Wijesekera 1977; Senanayake et al. 1978; Senanayake and Wijesekera 2003). These data have been fine-tuned with more recent and innovative methods that include gas chromatography (GC), chiral GC, isotope-ratio mass spectrometry, high-performance liquid chromatography (HPLC), high-performance thin-layer chromatography (HPTLC) analysis, vibrational spectroscopy (infrared [IR], Fourier transform infrared [FTIR], and near infrared [NIR]), and their coupled and multidimensional chromatography variants (Jayaprakasha et al. 2002, 2003; El-Baroty et al. 2010; Jayawardena and Smith 2010; Jayaprakasha and Rao 2011; Golmohammad et al. 2012; Kamaliroosta et al. 2012; Khoddami et al. 2013; Li et al. 2013a,b; Wong et al. 2014; Dvorackova et al. 2015). Very detailed descriptions of all these methods can be found in Kubeczka (2002) and Zellner et al. (2009).

The early analytical techniques were used for physical measurements, such as relative density, optical activity, and refractive index, or melting, congealing, and boiling point determinations of cinnamon essential oil. But when combined with modern analytical techniques, such as column-based liquid chromatography and mass spectrometry, all these techniques resulted in the identification of essential compounds present in very small quantities (Ravindran et al. 2003). Chromatography is a separation procedure based on the relative affinities of the compounds to be separated toward stationary and mobile phases. The mixture of compounds to be separated is subjected to flow by mobile liquid through the stable stationary phase. Compounds with higher affinity to the stationary phase travel slower and for a shorter distance, while compounds with lower affinity travel faster and longer. The separated compounds are further identified by other techniques, like ultraviolet (UV)–visible, infrared, nuclear magnetic resonance (NMR), and mass spectroscopy. Chromatography can be planar with the stationary phase consisting of a plane surface, like in thin-layer chromatography (TLC) and paper chromatography (PC), or columnar with the stationary phase lying in the walls of a capillary tube, while the mobile phase is flushed through the column like in column chromatography, gas chromatography, and high-pressure liquid chromatography. In general, the volatile fraction of an essential oil is analyzed by GC, while the nonvolatile by liquid chromatography (LC).

Finally, organoleptic properties and nutritive and mineral values of cinnamon essential oil can be assessed with different methods. They include olfactive and sensory analyses by specialized individuals, with high risks of inconsistency deriving from variability between individuals and official methods of analysis based on standard procedures like those of the Association of Official Analytical Chemists (AOAC 2003) used by Gul and Safdar (2009) to determine the nutritive and mineral compositions of cinnamon.

## 7.5 Composition of Cinnamon Oil

The chemical composition of cinnamon oils varies depending on several factors that include the part of the plant used, age of trees, growing season and location, and extraction methods (Kaul et al. 2003; Rajeswara et al. 2007; Barceloux 2009; Wang et al. 2009; Paranagama et al. 2010; Geng et al. 2011; Li et al. 2013a,b; Pandey et al. 2014; Wong et al. 2014; Chakraborty et al. 2015).

One of the first detailed studies of cinnamon oil composition was carried out by Senanayake et al. (1978). Different parts of the cinnamon plant have different primary constituents: cinnamaldehyde is majorly found in bark oil, eugenol in leaf oil, and camphor in root-bark oil (Wijesekera 1977). In the European Pharmacopoeia (2008), and according to a summary report on the essential oil of cinnamon bark by the Committee for Veterinary Medicinal Products, the cinnamon bark essential oil mainly contains cinnamaldehyde (55%–76%), eugenol (5%–18%), and saffrole (up to 2%). Cinnamon bark contains up to 4% of essential oil, consisting primarily of cinnamaldehyde (60%–75%), eugenol (1%–10%), cinnamyl acetate (1%–5%) (WHO 1999), β-caryophyllene (1%–4%), linalool (1%–3%), and 1.8-cineole (1%–2%) (ESCOP 2003). Wang et al. (2009) reported that the main constituents found in the leaves of *C. zeylanicum* are eugenol (79.75%), *trans*-cinnamaldehyde (16.25%), and linalool (0.14%).

Leela (2008) summarized the results of several authors and reached 124 and 66 different volatiles that could be found in different parts of *C. verum* and *C. cassia* plants, respectively. At least 94 volatile components have been found in cinnamon bark (Gong et al. 2004). A total of 26 compounds have been characterized from the *C. zeylanicum* flower oil, with (E)-cinnamyl acetate, *trans*-alpha-bergamotene, and caryophyllene oxide being the major compounds (Jayaprakasha et al. 2002, 2003; Jayaprakasha and Rao 2011). The oil of *C. zeylanicum* buds contains 34 compounds consisting of terpene hydrocarbons and oxygenated terpenoids, with alpha-bergamotene and alpha-copaene found to be the major compounds (Jayaprakasha et al. 2002). The volatile oil from *C. zeylanicum* fruit stalks contains more than 27 compounds, including (E)-cinnamyl acetate and (E)-caryophyllene as major compounds (Jayaprakasha et al. 2003). Geng et al. (2011) found that the majority of compounds in different parts of *C. cassia* oil belonged to the sesquiterpene hydrocarbon and oxygenated sesquiterpene fractions, with *trans*-cinnamaldehyde (33.95%–76.4%), cinnamyl alcohol acetate (0.09%–49.63%), 2-methoxycinnamaldehyde (0.09%–6.69%), and copaene (1.09%–14.3%) as major compounds.

The main components of the essential oil obtained from the bark of *C. zeylanicum* are eugenol, cinnamaldehyde, and linalool (Kubeczka 2002; Kubeczka and Formáček 2002), while *C. cassia* bark contains cinnamaldehyde, cinnamic acid, cinnamyl alcohol, and coumarin (Ranasinghe et al. 2013). Other *Cinnamomum* species were found to have lower contents of cinnamaldehyde (He et al. 2005).

Some constituents frequently encountered in cinnamon bark oil include eugenol, eugenol acetate, cinnamyl acetate, cinnamyl alcohol, methyl eugenol, benzaldehyde, cuminaldehyde, benzyl benzoate, linalool, monoterpene hydrocarbons (e.g., pinene, phellandrene, and cymene), carophyllene, and safrole. Cinnamon leaf oil also contains many of the major constituents present in cinnamon bark oil (e.g., cinnamaldehyde, cinnamyl acetate, eugenol acetate, and benzaldehyde), as well as other minor compounds, like humulene, isocaryophyllene, alpha-ylangene, coniferaldehyde, methyl cinnamate, and ethyl cinnamate (Leung and Foster 1996).

Other minor constituents also reported to be found in cinnamon essential oil include oligopolymeric procyanidins, cinnamic acid, phenolic acids, pentacyclic diterpenes, cinnzeylanol and its acetyl derivative cinnzeylanine, and the sugars mannitol, L-arabino-D-xylanose, L-arabinose, D-xylose, and α-D-glucose, as well as mucilage polysaccharides (ESCOP 2003). Several nonvolatile compounds (e.g., cinncassiols, cinnzeylanol, cinnzeylanin, anhydrocinnzeylanol, anhydrocinnzeylanin, several benzyl isoquinoline alkaloids, flavanol glucosides, coumarin, b-sitosterol, cinnamic acid, protocatechuic acid, vanillic acid, and syringic acid) have been also reported to be found in cinnamon essential oils (Leela 2008).

## 7.6 Physical and Chemical Properties of Oil

Essential oils are oily aromatic liquids that are soluble only in organic solvents. They are immiscible with water due to their hydrophobic nature and lower density compared with that of water. In the ISO standards lists (ISO 2003), cinnamon oil is described as the essential oil obtained by steam distillation of the leaves of *Cinnamomum zeylanicum* (Lauraceae), growing mainly in Sri Lanka. It is a clear, mobile liquid with a light to dark amber color. It is characterized by a spice-like odor reminiscent of eugenol. At 20°C, its relative density is between 1.037 and 1.053; its refractive index is between 1.527 and 1.540; and its optical rotation ranges from −2.5° to + 2°. It shall not be necessary to use more than 2 volumes of ethanol, 70% (volume fraction), to obtain a clear solution with 1 volume of essential oil. The phenol content of cinnamon leaves oil should be between 75% and 85%.

Sri Lanka is the major world exporter of cinnamon essential oil. The Sri Lanka Standards Institution specifies that the cinnamon leaves' oil refractive index has to be between 1.530 and 1.540, its specific gravity has to be between 1.034 and 1.050, its solubility should be 1.5 volumes of 70% (v/v) ethanol at 28°C, and it should contain no less than 75% of total phenols. For the cinnamon bark oil, the values are as follows: refractive index between 1.555 and 1.580, specific gravity between 1.010 and 1.030, solubility of 1.5 volumes of 70% (v/v) ethanol at 28°C, and containing no more than 18% of total phenols for the superior special and average grade. Its content in cinnamic aldehyde was also specified for superior grade (not less than 60% m/m), special grade (55%–60%), average grade (45%–55%), and ordinary grade (30%–45%).

Leela (2008), based on the works of Baslas and Baslas (1970), reported some of the physicochemical properties of *C. verum* leaf's oil. The specific gravity ranges from 1.044 to 1.062, the refractive index is between 1.522 and 1.530, its optical rotation is 3.60, and its eugenol content is estimated at 65%–87.2%. It shall not be necessary to use more than 2 volumes of ethanol, 70% (volume fraction), to obtain a clear solution with 1 volume of essential oil. All these data were obtained at 30°C.

For the Open Chemistry Database (NCBI 2016), the physicochemical characteristics of a product named cinnamon oil are listed under the reference PubChem CID 6850781. This reference lists the synonyms of cinnamon oils as cassia oil, Chinese cinnamon, cassia bark oil, and cinnamon bark oil. The cinnamon oil is described there as a slightly water-soluble liquid that darkens and thickens on exposure to air with a molecular formula of $C_{19}H_{22}O_2$ and a molecular weight of 282.37678 g/mol. When heated to decomposition, it emits acrid smoke and irritating fumes. Its density at 20°C is 1.052–1.070 for cassia oil, 1.037–1.053 for cinnamon leaf oil, and 1.010–1.030 for cinnamon bark oil (at 25°C). The optical rotation at 20°C varies from −2.5 to +2 for cinnamon leaf oil and from −2 to 0 for cinnamon bark oil. The index of refraction at 25°C ranges from 1.6 to 1.5910 for cassia oil, 1.5730 to 1.5910 for cinnamon bark oil, and 1.53 to 1.54 for cinnamon leaf oil (20°C). The solubility at 20°C is 1 volume in 3 volumes of 70% ethanol for cassia oil, 1 volume in 2 volumes of 70% ethanol for cinnamon leaf oil, and 1 volume in at least 3 volumes of 70% ethanol for cinnamon bark oil.

The Open Chemistry Database also gives a detailed physicochemical description of the major constituents of the cinnamon oils. A summary of the physicochemical properties of cinnamaldehyde, eugenol, linalool, coumarin, and camphor are given in Table 7.1.

## 7.7 General Uses of Oil

### 7.7.1 Usage in the Ancient Periods

Cinnamon and cassia have been used since ancient times as flavoring and medicinal ingredients. Cinnamon is mentioned in the Bible as a component of the oil used by Moses

**TABLE 7.1**

Physicochemical Properties of Major Constituents of the Cinnamon Oils

| Compound | Molecular Formula | Molecular Weight (g/mol) | Boiling Point (°C at 760 mmHg) | Melting Point (°C) | Density (g/cm³) | Refractive Index (at 20°C) | Description |
|---|---|---|---|---|---|---|---|
| Cinnamaldehyde | $C_9H_8O$ | 132.15922 | 253 | −7.5 | 1.048–1.052 (25°C) | 1.618–1.623 | Clear yellow oily liquid with an odor of cinnamon and a sweet taste<br>Solubility in water: 1420 mg/L (25°C)<br>Dissolves in 1:2.5 (v/v) of 70% alcohol |
| Eugenol | $C_{10}H_{12}O_2$ | 164.20108 | 225 | −9.2 to −9.1 | 1.0652 (20°C) | 1.5405 | Clear colorless pale yellow or amber-colored liquid; odor of cloves; spicy pungent taste; darkens and thickens on exposure to air<br>Solubility in water: 2460 mg/L (25°C)<br>1 ml dissolves in 2 ml 70% alcohol |
| Linalool | $C_{10}H_{18}O$ | 154.24932 | 198 | – | 0.858–0.868 (25°C) | 1.4627 | Colorless liquid; odor similar to that of bergamot oil and French lavender with a good stability<br>Solubility in water: 1600 mg/L (25°C)<br>Soluble in alcohol, ether, fixed oils, propylene glycol; insoluble in glycerin |
| Coumarin | $C_9H_6O_2$ | 146.14274 | 301.71 | 71 | 0.935 (20°C) | – | Colorless crystals, flakes, or colorless to white powder with a pleasant fragrant vanilla odor and a bitter aromatic burning taste<br>Solubility in water: 1.900 mg/L (20°C)<br>Soluble in ethanol; very soluble in ether and chloroform |
| Camphor | $C_{10}H_{16}O$ | 152.23344 | 209 | 174–179 | 0.992 (25°C) | 1.5452 | Colorless or white crystals with a penetrating, aromatic odor<br>Solubility in water: 1600 mg/L (25°C)<br>1 g dissolves in about in 1 ml alcohol, 1 ml ether, 0.5 ml chloroform |

for the purpose of anointment (to make a person holy). It has also been used as an ingredient in many ancient Indian medicinal preparations. In Egypt, cassia, along with other exotic herbs, was used not only in daily life, like cooking and bathing, by the privileged classes, but also as the botanical ingredients in mummification rituals. Cinnamon was sometimes exchanged in the barter system with other goods under the Roman Empire extension.

### 7.7.2 Food Uses

Cinnamon is used as a spice, a condiment, and flavoring material principally in cookery; chocolate preparation, especially in Mexico; many dessert recipes, such as apple pie, doughnuts, and cinnamon buns, as well as spicy candies; coffee; tea; hot cocoa; and liqueurs. In the Middle East, in Turkish and Persian cuisine, cinnamon is often used in chicken and lamb meat dishes and in a variety of thick soups, drinks, and sweets.

Cinnamon is an excellent spice used with meat and poultry in Indian and Moroccan dishes. It is an essential part of the curry pastes used across Asia. It is also used, along with other spices, in pickles, sauces, soups, confectionaries, and canned fruits. Cinnamon is a popular flavoring in numerous alcoholic beverages, such as "cinnamon liqueur," which is popular in Europe (Willard 2013). Krishnamoorthy and Rema (2003) reported that cinnamon bark oil is used frequently in the food, pharmaceutical, and perfume industries. It has largely replaced cinnamon powder in the processing industry, since it can be measured accurately according to well-established replacement ratios for ground spice using oils and oleoresins, such as the ones elaborated by Tainter and Grenis (1993).

### 7.7.3 Medicinal Uses

Cinnamon and cassia are believed to have a broad spectrum of medicinal and pharmacological applications. In folk medicine, cinnamon is used for the treatment of impotence, frigidity, dyspnea, eye inflammations, leukorrhea, vaginitis, rheumatism, and neuralgia, as well as wounds and toothaches (WHO 1999). In African and Chinese pharmacopoeias and traditional systems of medicine, cinnamon is indicated for the treatment of dyspeptic conditions, including mild spastic conditions of the gastrointestinal tract, fullness and flatulence, and loss of appetite. Cinnamon is also known to be a carminative, expectorant, and antidiarrheal, and to be useful for bronchitis, itching, and urinary disease (Leela 2008). Cassia is traditionally used for digestive problems, such as flatulence, colic, dyspepsia, diarrhea, and nausea, as well as colds, influenza, fevers, arthritis, and rheumatism (Barceloux 2009). Recent pharmacological studies have shown that besides its role as a spice, cinnamon can be used as a hypoglycemic and cholesterol-lowering (Khan et al. 2003), wound pro-healing (Kamath et al. 2003), and anti-inflammatory compound (Chao et al. 2005). It is a risk-reducing agent for colon cancer (Wondrak et al. 2010) and can prevent bleeding due to its anticoagulant properties (Husain and Ali 2013). Several studies have reported the anti-inflammatory activity of cinnamon and its essential oils (Sosa et al. 2002; Li et al. 2003; Matu and Staden 2003; Chao et al. 2005; Tung et al. 2008, 2010). Cinnamaldehyde, the major compound of cinnamon, exhibited anti-inflammatory activity by inhibiting the activation of the nuclear factor kappa-light-chain enhancer of activated B cells (Reddy et al. 2004; Lee and Balick 2005). Other constituents of cinnamon oil belonging to the flavonoid group have been demonstrated to possess anti-inflammatory activities (Kim et al. 2004; Stoner and Wang 2013).

### 7.7.3.1 Antioxidant Properties

Cinnamon bark has been shown to contain very high concentrations of antioxidants (Dragland et al. 2003). The higher antioxidant activities of cinnamon, compared with those of other spices, have been previously described (Murcia et al. 2004; Shan et al. 2005). Considerable antioxidant activities of various extracts of cinnamon have been reported by Mancini-Filho et al. (1998), and their inhibition of fatty acid oxidation and lipid peroxidation was demonstrated *in vitro* (Shobana and Naidu 2000). Singh et al. (2007) evaluated the antioxidant potential of cinnamon volatile oils and oleoresins of leaf and bark and their major components by comparing their lipid inhibitory activities with selected antioxidant activities and concluded that the volatile oils and oleoresins of cinnamon leaf and bark have good antioxidant properties. Yang et al. (2012) compared the antioxidant activities of extracts of various parts of *C. cassia* (barks, buds, and leaves) obtained by supercritical carbon dioxide extraction and ethanol extraction and showed that the extracts of *Cinnamon* barks exhibited higher antioxidant activity than other parts of cinnamon. Moreover, Yang et al. (2012) also demonstrated that ethanol is the best solvent to obtain the main antioxidant constituents.

Etheric, methanolic, and aqueous cinnamon extracts also inhibited oxidative processes *in vitro* (Dhuley 1999; Mathew and Abraham 2006; Lin et al. 2007). Several *in vitro* studies have demonstrated the antioxidant effects of the essential oil obtained from the bark of *C. zeylanicum* and its main components (Lee et al. 2002; Jayaprakasha et al. 2003; Chericoni et al. 2005; Lee and Balick 2005). Studying the free radical–scavenging activities of various medicinal plants, Okawa et al. (2001) concluded that different flavonoids extracted from cinnamon have good antioxidant properties. The study of Lee et al. (2002) showed that cinnamaldehyde and other compounds of cinnamon have inhibitory activities against the production of nitric oxide. *In vivo* study carried out by Lin et al. (2003) showed that the ethanolic extract of *C. cassia* exhibited significant antioxidant activity compared with the natural antioxidant α-tocopherol. El-Baroty et al. (2010) found that cinnamon essential oil exhibited appreciable *in vitro* antioxidant activity. Volatile oils of *C. zeylanicum* showed significant antioxidant activities, as reported by Jayaprakasha and Rao (2011).

### 7.7.3.2 Hypoglycemic Properties

Pharmacological studies, on human and animals both *in vitro* and *in vivo*, have recently been trying to show that cinnamon may play a possible role in improving glucose and insulin metabolism (Imparl-Radosevich et al. 1998; Onderoglu et al. 1999; Broadhurst et al. 2000; Kar et al. 2003; Khan et al. 2003), and that cinnamaldehyde may have a potential role as an antidiabetic agent with contrasting results.

In his analysis of randomized controlled trials including more than 500 patients, Allen et al. (2013) established that when taken in a dose ranging from 0.12 to 6.0 g/day for approximately 4 months, cinnamon contributed to a statistically significant decrease in the levels of fasting plasma glucose, coupled with an improvement in the lipid profile. The same conclusions were reached by Alanazi and Khan (2015) in their meta-analysis of 16 randomized control trials with 638 patients, where they concluded that the consumption of cinnamon is associated with a statistically significant decrease in the levels of fasting plasma glucose, total cholesterol, and triglyceride. However, the high degree of heterogeneity in the studies analyzed may limit the ability to apply these results to patient care.

Regarding the cinnamon compound mechanism of action in diabetes, Sheng et al. (2008) explained that its role in insulin resistance derived from the increased expression of peroxisome proliferator-activated receptors (PPARs) α and γ. Moreover, the effect of cinnamon on PPAR γ was found to be analogous to that of the thiazolidinediones in type 2 diabetes (Rafehi et al. 2012). A different role played by *C. cassia* in mitigation of insulin resistance was advanced by Jitomir and Willoughby (2009) and consisted of the enhancement of expression of insulin-sensitive glucose transporters by acting on the phosphorylation of signaling proteins. Finally, cinnamon has also been reported to have an insulin mimetic and insulin-sensitizing action (Howard and White 2013). A good review of potential mechanisms of action may be found in Medagama (2015).

### 7.7.3.3 Other Medicinal Uses

Cinnamon is also frequently used as flavor in chewing gums due to its effects and ability to remove bad breath, and it has been traditionally used as tooth powder and for dental problems such as toothaches and bad breath (Aneja et al. 2009; Jakhetia et al. 2010; Gupta et al. 2012). The active component cinnamaldehyde is said to be cardioprotective (Song et al. 2013) and has a vasorelaxative effect (Alvarez-Collazo et al. 2014). A systemic review of previous studies has suggested that cinnamon-supplemented diets can result in a significant fall in blood pressure (Wainstein et al. 2011; Akilen et al. 2013).

Much research has been done to see the effect of cinnamon on melanoma cells, and the results of a study suggested that *C. cassia* can inhibit the survival, viability, and proliferation of tumor cells *in vitro* without having a significant effect on the normal cells (Han et al. 2004). *C. cassia* bark extracts also effectively inhibited the virus-induced cytopathogenicity in MT-4 cells infected with HIV (Premanathan et al. 2000). Cinnamon, cinnamon extracts and essential oils, and constituents of cinnamon, such as monoterpenoids and cinnamaldehyde, have all been reported to exhibit anticancer, antitumor, antiproliferative, and antimutagenic effects (Shaughnessy et al. 2006; Bhattacharjee et al. 2007; King et al. 2007; Wu and Ng 2007; Duessel et al. 2008; Dong et al. 2009; Lin et al. 2009; Sharififar et al. 2009).

## 7.8 Pesticidal Uses of Cinnamon Essential Oil

### 7.8.1 Antibacterial Properties

Cinnamon oils have been widely studied for their antimicrobial effects on various bacteria (Hili et al. 1997; Chao et al. 2000; Matan et al. 2006; Shan et al. 2007; Singh et al. 2007; Abdollahzadeh et al. 2014). Cinnamon oils and extracts, as well as their major components cinnamaldehyde and eugenol, have been found to exhibit antimicrobial effects on both gram-positive and gram-negative bacteria such as *Salmonella enterica*, *Escherichia coli* (Friedman et al. 2004), and *Listeria monocytogenes* (Yuste and Fung 2002). El-Baroty et al. (2010) found that cinnamon essential oil exhibited a strong antibacterial activity against *Bacillus subtilis*, *Bacillus cereus*, *Staphylococcus aureus*, *Streptococcus faecalis*, and *Micrococcus luteus* and gram-negative bacteria *Alcaligenes faecalis*, *Enterobacter cloacae*, *Pseudomonas aeruginosa*, *Klebsiella pneumoniae*, and *Serratia marcescens* (Chao et al. 2000). Cinnamon bark oil and its major components showed antibacterial effects on the major respiratory and gastrointestinal tract pathogens *Haemophilus influenzae*, *Streptococcus pneumoniae*, *Streptococcus pyogenes*, and *S. aureus* (Inouye et al. 2001a,b). Furthermore, when incorporated in biofilms, cinnamaldehyde has also been reported to have negative effects on *E. coli* and *Pseudomonas* spp. (Niu and Gilbert 2004), *Burkholderia* spp. (Brackman et al. 2009), uropathogenic *E. coli* (Amalaradjou et al. 2010), *Vibrio* spp. (Brackman et al. 2011), methicillin-resistant *S. aureus* and *Staphylococcus epidermidis* (Jia et al. 2011; Kavanaugh and Ribbeck 2012), *Candida* spp. (Khan and Ahmad 2012), *Listeria* spp. (Upadhyay et al. 2013), *Salmonella* spp. (Zhang et al. 2014), *S. pyogenes* (Shafreen et al. 2014), and *P. aeruginosa* (Kim et al. 2015). Ouattara et al. (2000) reported that incorporating cinnamaldehyde into chitosan film reduced the growth of *Lactobacillus sakei*, *Serratia liquefaciens*, and Enterobacteriaceae on the surface of meat products.

Antibacterial activities of cinnamon bark oil and cinnamaldehyde have been attributed to considerable alterations in the structure of cell envelopes (Di Pasqua et al. 2007). The membrane permeability may be affected by an inhibition of energy generation, probably due to the inhibition of glucose uptake or utilization of glucose (Gill and Holley 2004). The cinnamaldehyde of the biofilm is partially caused by the downregulation of quorum-sensing systems (Kim et al. 2015).

The antibacterial actions of natural extracts of cinnamon have been suggested to be a relevant tool in the control of pathogens of aquatic animals. In their investigations, Yeh et al. (2009) demonstrated that shrimp treated with natural extracts of cinnamon exhibited an enhanced disease resistance to *Vibrio alginolyticus*.

## 7.8.2 Antifungal Properties

Based on several *in vivo* and *in vitro* studies, cinnamon essential oils and its major components have been found to exhibit significant inhibitory effects against several fungi, including *Coriolus versicolor*, *Laetiporus sulphureus*, *Eurotium* spp., *Aspergillus* spp., and *Penicillium* (Chipley and Uraih 1980; Cao 1993; Mastura et al. 1999; Guynot et al. 2003; Simić et al. 2004; Cheng et al. 2006). *trans*-Cinnamaldehyde, a component in the oil of *C. zeylanicum*, was the most active compound against 17 micromycetes (Simić et al. 2004). The essential oils of several *Cinnamomum* species showed anticandidal and antidermatophytic activity *in vitro* (Lima et al. 1993; Mastura et al. 1999). Quale et al. (1996) reported that the use of *C. zeylanicum* allowed overcoming the resistance to fluconazole in *Candida* isolates. Singh et al. (2007), using several methods to study the antifungal efficacy of cinnamon essential oil and its oleoresin, reported that the volatiles elicited from the essential oils extracted from cinnamon leaves were found to be 100% antifungal against *Aspergillus niger*, *Aspergillus terreus*, *Fusarium moniliforme*, *Fusarium graminearum*, *Penicillium citrinum*, and *Penicillium viridicatum*, but not against *A. ochraceus* and *A. terreus*. The leaf oleoresin showed complete mycelial zone inhibition for *P. citrinum*, and volatiles elicited from the essential oils extracted from cinnamon barks showed complete inhibition against fungi such as *F. graminearum*, *F. moniliforme*, *P. citrinum*, *P. viridicatum*, and *A. terreus* (Singh et al. 2007). Moreover, Singh et al. (2007) also suggested that among cinnamon oil constituents, cinnamaldehyde possessed the best antifungal activity. El-Baroty et al. (2010) found that cinnamon essential oil has a strong antifungal activity against four fungal strains: *A. niger*, *Penicillium notatum*, *Mucora heimalis*, and *Fusarim oxysporum*. Cinnamon oils and extracts showed good antifungal activities against important plant diseases. Wilson et al. (1997) found that among 49 essential oils tested, *C. zeylanicum* demonstrated a great antifungal activity against *Botrytis cinerea*, while Montes-Belmont and Carvajal (1998) reported that *A. flavus* was totally inhibited with *C. zeylanicum*. In other studies, *C. zeylanicum* was fungicidal against pathogens isolated from banana, including *Colletotrichum musae*, *Lasiodiplodia thebromae*, and *Fusarium proliferatum* (Ranasinghe et al. 2002); exerted antifungal activity toward *Oidium murrayae* (Chu et al. 2006); and inhibited conidial germination of *Colletotrichum gloesporioides* (Barrera-Necha et al. 2008). In *in vitro* experiments, it was found to have a good mycelial inhibition of the corn rot *F. oxysporum* f.sp. *gladioli* (Barrera-Necha et al. 2009), to be highly effective against the growth of *Rhizoctonia solani* (Nguyen et al. 2009), and to have an excellent antifungal activity against early blight of tomato *Alternaria solani* (Yeole et al. 2014). The investigations of Wang et al. (2014) showed that cinnamon microemulsions had high *in vivo* control activity against gray mold of pears *Botritys cinerea*.

### 7.8.3 Insecticidal Properties

#### 7.8.3.1 Against Vector of Human Diseases

As many other essential oils, cinnamon essential oils offer great potential in medical entomology, especially against mosquitoes, which represent one of the most relevant vectors of human diseases. They have been shown to be effective larvicides against mosquitoes (Cheng et al. 2004, 2009; Chang et al. 2006). Larvicidal tests demonstrated that the components of leaf essential oils, such as cinnamaldehyde–cinnamyl acetate and cinnamyl alcohol, had an excellent inhibitory effect against the fourth-instar larvae of the yellow fever mosquito *Aedes aegypti* (Cheng et al. 2004; Chang et al. 2006). Results of mosquito larvicidal assays also showed that the most effective constituents in leaf essential oils were cinnamaldehyde, eugenol, anethole, and cinnamyl acetate. Cinnamon has also shown excellent repellency in tests conducted on blood-starved females of *Ae. aegypti* mosquitoes (Chang et al. 2006). Reviewing the literature of mosquito larvae control using botanical larvicides, Pavela (2015) concluded that from 122 initially studied plant species, 3 *Cinnamomum* species were among the 7 most significant botanical larvicides that may be considered suitable sources for substances to control mosquito larvae.

#### 7.8.3.2 Against Agricultural Insect Pests

Cinnamon oils and its components, such as cinnamaldehyde, are well-known insecticidal compounds that have been studied against a variety of other insects (Huang and Ho 1998; Lee et al. 2001; Chang and Cheng 2002; Lee et al. 2008). The antitermitic activities of the essential oils from the leaves of *C. osmophloeum* and its chemical ingredients against the Formosan subterranean termite *Coptotermes formosanus* were investigated by direct contact application (Chang and Cheng 2002). Results have demonstrated that the indigenous cinnamon leaf essential oil has a good effective antitermitic activity, and that cinnamaldehyde, eugenol, and α-terpineol extracted from indigenous cinnamon leaf essential oil are responsible for the high antitermitic effectiveness. Cheng et al. (2008) described that the leaf essential oil of *C. osmophloeum* exhibits effective toxicity in both open and closed exposure against red imported fire ants, with *trans*-cinnamaldehyde as the major component in the essential oil playing the key role in controlling the red imported fire ant *Solenopsis invicta*. Park et al. (2000), using a fumigation test, found that the *Cinnamomum* bark–derived compounds were much more effective against larvae of the oak nut weevil *Mechoris ursulus* in closed cups than in open ones, indicating that the insecticidal activity of tested compounds was attributable to fumigant action. Park et al. (2000) concluded that the *Cinnamomum* bark–derived materials could be useful as a preventive agent against damage caused by *M. ursulus*. Ovicidal activity of *C. zeylanicum* oil was reported for the rice moth *Corcyra cephalonica* (Bhargava and Meena 2001). Passino et al. (1999) reported insecticidal activity of *C. zeylanicum* oil against the Mediterranean fruit fly *Ceratitis capitata*.

### 7.8.3.3 Against Stored Product Pests

In stored product pests, the susceptibility of the rice weevil *Sitophilus oryzae* to fumigant actions of cinnamon oil was investigated, along with other essential oils, and resulted in 100% mortality within 1 day of treatment in closed containers (Kim et al. 2003). In a different study, Paranagama et al. (2003) concluded that *C. zeylanicum* leaf essential oils can be used as stored paddy rice protectant, as it kept the samples studied free of *S. oryzae* and the angoumois grain moth *Sitotroga cerealella*, two major pests of stored grains, without altering the quality of the stored rice.

C. *cassia* and its major constituent, cinnamaldehyde, exhibited fumigant toxicity and residual effects against *S. oryzae* (Lee et al. 2008), and *C. cassia* vapor caused the highest mortality of various life stages of the red flour beetle *Tribolium castaneum* (Mondal and Khalequzzaman 2009). The insecticidal effects of cinnamon oil and other essential oils were evaluated by Karahroudi et al. (2010) for three of the most important stored product pests of the Indian mealmoth *Plodia interpunctella*, the confused flour beetle *Tribolium confusum*, and the pulse beetle *Callosobruchus chinensis*, and the results indicated that cinnamon essential oil has good fumigant activity against the three tested species.

In their study, Jumbo et al. (2014) evaluated the insecticidal (e.g., lethal toxicities, disturbances on reproductive traits, and persistence of action) and repellent activities of cinnamon, *C. zeylanicum*, and clove, *Syzygium aromaticum*, essential oils on the bean weevil *Acanthoscelides obtectus* in a nonfumigant manner. Jumbo et al. (2014) concluded that cinnamon not only has a good insecticidal activity but also significantly reduced the bean weight losses caused by *A. obtectus*. Similar results were reported by a study on *C. maculatus* and *S. oryzae* (Brari and Thakur 2015), where the essential oil of *C. zeylanicum* and its two components (cinnamaldehyde and linalool) were found to exhibit contact and fumigant toxicity against the adults of both insect species tested.

### 7.8.3.4 Against Medical–Veterinary Insect Pests

In veterinary area, the application of camphor and cinnamon oils to water buffalo, at concentrations similar to the ones used in the laboratory studies, resulted in a large decline in numbers of the ungulate lice *Haematopinus tuberculatus* up to 6 days after application (Khater et al. 2009). The same study reported a decrease in a number of three fly species (*Stomoxys calcitrans*, *M. domestica*, and *Hippobosca equina*) on treated cattle. Results also indicated that the essential oils from cinnamon and its most predominant compound had high ovicidal activity against various harmful flies (Shen et al. 2007).

In veterinary use, although anthelmintic effects of Ceylon cinnamon were reported as early as in the 1950s by Cavier (1950), it is only recently that Williams et al. (2015) showed for the first time that cinnamon bark has anthelmintic potential *in vitro* using swine nematode *Ascaris suum*, and this derives both from its proanthocyanidin tannins and most notably from *trans*-cinnamaldehyde. However, their *in vivo* experiments with pigs and poultry made them reach a conclusion that for the potential of *trans*-cinnamaldehyde to be used as an anthelmintic against intestinal helminthes, appropriate formulations to stabilize and protect the compound will likely be necessary.

### 7.8.4 Acaricidal Effects

Cinnamon displayed acaricidal activity against the poultry red mite *Dermanyssus gallinae* (Kim et al. 2004). In investigations conducted by Bahadon and Azarhoosh (2013), it was demonstrated that plant preparations from *C. cinnamon* and other aromatic plants can be used for controlling *D. gallinae*. On the basis of $LC_{50}$ values, essential oil extracted from cinnamon leaves was one of the most active oils among 24 Tai herbal oils tested against house dust mites *Dermatophagoides pteronyssinus* (Veeraphant et al. 2011), which was attributed to be due to its eugenol content (Veeraphant et al. 2011). By evaluating the acaricidal and repellent effects of cinnamon essential oil on the house dust mite, *Dermatophagoides farina* and *D. pteronyssinus*, Oh (2011) demonstrated that cinnamon bark essential oil was a very effective acaricide at the concentration of 0.125 µl; it had good repellent effect when used at the concentration of 0.094 µl.

The results of dose–mortality experiments, carried out by Shen et al. (2012) to test the effects of *trans*-cinnamaldehyde (a component of cinnamon essential oil) on the common worldwide parasite of rabbits, the rabbit ear mite *Psoroptes cuniculi*, indicated that this compound had a good killing activity against *P. cuniculi* adults, and that *trans*-cinnamaldehyde can be considered as a promising agent for mite control. Similar results were already reported by Fichi et al. (2007), as he showed cinnamon leaf to have high levels of acaricidal efficacy against *P. cuniculi* in rabbits at concentrations of 2.5%.

However, the mechanism of cinnamon oils' and extracts' acaricidal activity is not yet well understood. Ellse and Wall (2014) suggest that the acaricidal efficacy of essential oils may be linked to the vapor pressure to which the mites are exposed, as it affects the concentration of volatile. Vapor assays conducted by Na et al. (2011) using 34 compounds extracted from *Cassia* spp. and *Cinnamomum* spp. showed that α-methyl-E-cinnamaldehyde and E-cinnamaldehyde had acaricidal efficacy comparable to that of the chemical acaricide dichlorvos.

### 7.8.5 Nematicidal Effects

Park et al. (2005) reported nematicidal activity of plant essential oils, including *C. verum* and its components, against the pine wood nematode *Bursaphelenchus xylophilus*. Analyzing the activity of 88 commercial essential oils against mixed-stage of *B. xylophilus*, Kong et al. (2006) identified highly active *C. zeylanicum* bark essential oils showing nematicidal activity that proved to be higher than those obtained for some commercial synthetic nematicides, such as fenitrothion.

### 7.8.6 Repellency Effects

In investigations conducted by Prajapati et al. (2005), the essential oil of *C. zeylanicum* proved to be oviposition deterrent and repellent against three mosquito species tested (*Anopheles stephensi*, *Ae. aegypti*, and *Culex quinquefasciatus*). Yang et al. (2004) investigated the repellent activity of methanol extracts and steam distillate from 23 aromatic medicinal plant species against female blood-starved *Ae. aegypti*, and they found that at a dose of 0.1 mg/cm², the repellency of extracts of *C. cassia* bark and *C. camphora* steam distillate was comparable to that of deet. The duration of the effectiveness for extracts from *C. cassia* bark was comparable to that of deet.

Laboratory studies suggested that cinnamon may be useful as an insect repellant (Cloyd et al. 2009). Hori (2003) described *C. cassia* as having repellent activity against the cigarette beetle *Lasioderma serricorne*, while in the study of Jumbo et al. (2014), cinnamon oil exhibited repellent actions against *A. obtectus*, in accordance with other investigations, such as the one carried out by Liu et al. (2006), and where they described a good repellent activity exhibited by essential oil from the seeds of *C. camphora* against storage pests *S. oryzae* and

*Bruchus rugimanus.* In addition, cinnamon essential oils displayed repellant action against the red bud borer *Resseliella oculiperda* (Van tol et al. 2007).

Hanifah et al. (2012) reported that *C. zeylanicum* showed the highest repellency rate compared with the other plants extracts in a study carried out to evaluate the repellency of six plant extracts against the larval stage of *Leptotrombidium deliense*, the mite vector of scrub typhus.

### 7.8.7 Herbicide Effects

In a laboratory and greenhouse experiments with essential oils from different plants (including cinnamon), Tworkoski (2002) tried to determine the herbicidal effect of plant-derived oils and identify the active ingredient with herbicide activity. Essential oils in aqueous concentrations from 5% to 10% (v/v) added of two adjuvants (nonionic surfactant and paraffinic oil blend at 0.2% [v/v]) were applied to shoots of the common lambsquarters *Chenopodium album*, the common ragweed *Ambrosia artemisiifolia*, and johnsongrass *Sorghum halepense* in the greenhouse; shoot death occurred within 1 hour to 1 day after application. Essential oil (1%, v/v) from cinnamon was one of the most phytotoxic and caused electrolyte leakages, resulting in cell death. Eugenol (one of the major components of cinnamon) was confirmed to be the active ingredient in the essential oil of cinnamon.

Campiglia et al. (2007) evaluated and compared the inhibition effect exerted by the essential oils of cinnamon, peppermint *Mentha × piperita* L., and lavender *Lavandula* spp. on seed germination of some of the most common weed species of the Mediterranean environment, like pigweed *Amarantus retroflexus* L., wild mustard *Sinapis arvensis* L., and ryegrass *Lolium* spp. Their results highlighted a control in the weed germination, with cinnamon oil exhibiting the highest inhibition effect compared with lavender and peppermint ones. The dicotyledonous species have been more susceptible to the cinnamon oil inhibition of seed germination than the monocotyledonous ones.

Under controlled and semicontrolled conditions (laboratory and greenhouse), Cavalieri and Caporali (2010) studied the allelopathic effects of essential oil extracted from *C. zeylanicum* on the seed germination of seven Mediterranean weed species (i.e., redroot pigweed *A. retroflexus* L., black nightshade *Solanum nigrum* L., common purslane *Portulaca oleracea* L., common lambsquarters *C. album* L., wild mustard *S. arvensis* L., ryegrass *Lolium* spp. and common vetch *Vicia sativa* L.). Cinnamon oil showed drastic inhibitory effects, and in a semicontrolled condition, the 345.6 mg/L concentration of cinnamon essential oil totally inhibited the seed germination of *A. retroflexus* L. (Cavalieri and Caporali 2010).

## 7.9 Advantages as a Pesticide

The use of essential oils as pesticides has drawn a large and continuous interest, exemplified by the very high number of studies dealing with extraction, chemistry, toxicology, and uses of essential oils (Ravindran et al. 2003; Bakkali et al. 2008; Barceloux 2009; El-Baroty et al. 2010; Doughari 2012; Rao and Gan 2014; Tongnuanchan and Benjakul 2014; Chakraborty et al. 2015; Dima and Dima 2015; El Asbahani et al. 2015; Medagama 2015). Cinnamon essential oils are no exception to this rule. This wide interest found its basis in the problems faced with chemical pesticides, such as risk to human health and environment, pest resurgence, secondary pest problems, and the development of resistance, as well as the concerns expressed by consumers and pressure groups about the safety of pesticide residues in food. The essential oils, including cinnamon oils, are considered natural, and their use as spices and in food industry is often regarded as sufficient evidence of their safety. Pesticide products containing certain of these essential oils are exempt from toxicity data requirements by the U.S. Environmental Protection Agency (USEPA).

Essential oils and their main constituents are generally regarded as safe products owing to this positive perception to their very low mammalian toxicity. The estimated oral intake $LD_{50}$ for rat is 1.160 mg/kg (body weight) for cinnamaldehyde, and it is 500 mg/kg for eugenol (Shivanandappa and Rajashekar 2014) and 2790 mg/kg for linalool (OECD 2002), representing very low mammalian toxicity values compared with other insecticides. In the fact sheet of cinnamaldehyde issued by the USEPA, it is written, "Cinnamaldehyde is Generally Recognized As Safe (GRAS) by the Flavoring Extract Manufacturers' Association and is approved for food use by the Food and Drug Administration. Cinnamon oil, which contains 70% to 90% cinnamaldehyde, is also classified as GRAS, and, like cinnamaldehyde, is used in the food and flavoring industry" (USEPA 2015).

This mammalian selectivity was partially attributed to the mode of action of various components of essential oils like eugenol (Enan et al. 1998). In fact, Enan et al. (1998) demonstrated that a number of essential oil compounds act on the octopaminergic system of insects. For instance, eugenol was found to mimic octopamine in increasing intracellular calcium levels in cloned cells from the brain of *Periplaneta americana* and *Drosophila melanogaster* (Enan 2005). Altering the functioning of octopamine by a compound like eugenol results in total interruption of nervous system functioning in insects, which makes the octopaminergic system of insects a sound and rational target for insect control, as these receptors are not found in vertebrates.

Essential oils such as cinnamon oils are a mixture of many biosynthetically diverse compounds and analogs. This characteristic diversity is related not only to the age and developmental stage of the part source of the oil, but also to the growing conditions (Regnault-Roger et al. 2012). In the case of cinnamon oil, the cinnamaldehyde is acting on the energy production system, possibly interfering with glucose uptake or utilization, and besides eugenol acting on the octopaminergic system, linalool, a frequently reported monoterpenoid in cinnamon oil, has been demonstrated to act on the nervous system, affecting ion transport and the release of acetylcholine esterase in insects (Re et al. 2000). Finally, synergistic effects between the components of essential oils have been reported by previous studies (Berenbaum 1985; Miresmailli et al. 2006; Joffe et al. 2012; Koul et al. 2013; Faraone et al. 2015; Omolo et al. 2005). This indicates that the effect of the major components needs synergism from secondary constituents in the essential oils. This mixture of compounds with various sites of action and synergized effects between the essential oil constituents may be behind the improved efficacy of essential oils as insecticides and is surely playing a crucial role as a barrier for resistance development.

The essential oils and their components are generally considered safe for the environment, as they are majorly nonpersistent (Isman 2000). The major components of cinnamon oil are also nonpersistent, and hence have little impact on the environment when used as pesticides. The USEPA reports that because cinnamaldehyde is not soluble in water and rapidly degraded in the soil, it is not expected to pose any hazard to nontarget organisms (USEPA 2015). Eugenol is anticipated to be short-lived in the environment, and is rapidly dissipated and degraded via volatilization and atmospheric decomposition (Marin Municipal Water District 2008). Most linalool, both natural and synthetic, is released to the atmosphere, where it is rapidly degraded abiotically with a typical half-life below 30 minutes (OECD 2002). In water, linalool is readily biodegraded under both aerobic and anaerobic conditions; the same is predicted for soil and sediment. This nonpersistent characteristic is linked to the susceptibility of the essential oil and its compounds to temperature and UV light degradation, resulting in short residual activity with shorter restriction intervals for the treated areas (Miresmailli and Isman 2006).

Historically, aromatic plants and plant extracts were widely used for insect control in traditional agricultural systems in many developing countries. Using techniques as easy and affordable as steam distillation, plants products such as essential oils may be affordable for small farmers under the conditions of improving safety use and knowledge about both the accurate compositions and the pesticidal activities.

## 7.10  Limitations as a Pesticide

Essential oils such as cinnamon oil have been intensively studied for their pesticidal activities and have been described as a sustainable, effective, and affordable alternative to chemical insecticides. However, such oils still face various challenges regarding their use as pesticides. Various reports have described human toxicity cases of cinnamon and cinnamon oil involving local irritation and allergic reactions (Barceloux 2009). Because of its skin-sensitizing property, the use of bark oil in perfume and cosmetic industry is very limited (Ravindran et al. 2003), and occupational allergic contact dermatites, although rare (Kanerva et al. 1996), have been reported among workers with cinnamon (Kanerva et al. 1996).

The cinnamon essential oil biological activities are linked mainly to their major components: cinnamaldehyde (for bark oil) and eugenol (for leaf oil). Although synergism among major and minor components is well known, each of the dominant constituents of the oils is acting with a single mode of action, making them in this regard similar to conventional synthetic insecticides (Copping and Menn 2000). In fact, Correa et al. (2015) evaluated the toxicity (including the effects on the population growth rates) of cinnamon essential oils, as well as its effects on the behavioral (locomotory) and respiratory rates of four Brazilian populations of *Sitophilus zeamais* with distinct susceptibilities to traditional insecticides (phosphine and pyrethroids) and concluded that although cinnamon essential oil has the potential to control *S. zeamais* populations, insects from the studied populations that are resistant to traditional insecticides (e.g., pyrethroids and phosphine) might share some physiological and behavioral mechanisms to mitigate the actions of such essential oils. Moreover, Haddi et al. (2015), in a recent study, reported that the sublethal exposure to cinnamon essential oil of a population of *S. zeamais* susceptible to conventional insecticides resulted in stimulatory responses in the median survival time and the number of larvae per grain. The sublethal exposure elicited behavioral and physiological mechanisms that these insects normally use to overcome the actions of insecticides. Such findings showed that replacing synthetic insecticides with botanical insecticides like cinnamon essential oils to control insect pests still needs further investigation and scrutiny. More studies are also needed on the potential effects of the use of cinnamon oils on the nontarget and beneficial insects (Isman 2000), especially because a study demonstrated negative effects of the alcohol extract of *C. camphora* on two aphid parasitoids, *Aphidius gifuensis* and *Diaeretiella rapae* (Zhou and Liang 2003).

The nonpersistence of essential oils, first looked at as an environmental advantage, may turn out to be an issue, as higher quantities will be needed to reach the same levels of control achieved with conventional insecticides, which may lead to higher residues in the environment (Copping and Menn 2000). Pest control in foodstuffs may face a problem of odor acceptability by consumers if cinnamon essential oils are applied as contact and fumigant, due to its strong flavored smell.

The availability of the raw material is another concern for the use of cinnamon oil as a pesticide to a large extent (Isman 2000). Furthermore, the wide use of these oils will depend deeply on the possibilities of growers to provide the essential oil industry with standardized material. Variability and inconsistencies in composition will influence the extraction efficiency and may jeopardize the investors' interest in such a risky sector.

## TABLE 7.2

List of Commercially Available Cinnamon-Based Pesticides

| Products (Trade Name) | Uses and Activity Description | Composition |
| --- | --- | --- |
| Snail & Slug Away | Kills snails and slugs and their eggs | Active ingredients: Cinnamon oil<br>Other ingredients: Soap bark, water, soybean oil, sunflower oil |
| Weed Zap | Contact, nonselective, broad-spectrum, foliar applied herbicide<br>Controls both annual and perennial broadleaf and grassy weeds; does not translocate | Clove oil: 45%<br>Cinnamon oil: 45%<br>Other ingredients (lactose and water): 10% |
| Weed-A-Tak | Kills broadleaf weeds, grasses, vines, and brush | Citric acid: 4.0%<br>Cinnamon oil: 1.0%<br>Clove leaf oil: 1.0%<br>Other ingredients: 94% (lecithin, water) |
| Cinnacure | Fungicide, insecticide, miticide | 30% cinnamaldehyde |
| Cinnamite | Fungicide, insecticide, miticide | 30% cinnamaldehyde |
| Valoram II | Kills and repels numerous insect pests | Clove/cinnamon/mint oils |
| Armorex II | Kills and repels numerous insect pests | Clove/cinnamon/mint oils |
| Kinnamon 70 | Fungicide, insecticide, accaricide | Cinnamon extract 50% w/w |
| Citrokinnamon 50–30 | Fungicide, insecticide, accaricide | Cinnamon extract 50% w/w, citrus extract 30% w/w |
| Cinnamon extract JBQ | Combats powdery mildew and spider mite pests<br>Preventive and curative | Cinnamon extract (70%) and conditioners |
| Flower farm pesticide | Natural insecticide/miticide/fungicide | Cinnamon oil, cottonseed oil, rosemary oil |
| Final Stop® Pest Control Killer Spray | Controls: Ants, cockroaches, spiders, fleas, wasps, stink bugs, moths, silverfish, mosquitoes, centipedes, earwigs, gnats, chiggers, ticks, pillbugs, crickets, and other nasty creepy-crawly insects | Active ingredients: Cinnamon oil, rosemary oil, sesame oil, peppermint oil, thyme oil, garlic extract<br>Inert ingredients: Beeswax, calcium carbonate, carrageenan, cellulose, citric acid, glycerin, kaolin, lecithin, mustard powder, sodium bicarbonate, sodium chloride, soybean oil, wintergreen oil, distilled water |
| Spider Killer | Against spider mites, thrips, nematodes, cochineal, aphids, and other insect pathogens | Rich in cinnamon (*Cinnamomum zeylanicum*) |

## 7.11 Essential Oil–Based Pesticides

The review of the available experimental data (Section 7.8) supports the hypothesis that cinnamon oil is an effective natural pesticide and repellant against insects. Cinnamon essential oils and/or their constituents have shown a broad spectrum of insecticidal, miticidal, nematicidal, fungicidal, and bactericidal activity, as well as having a good repellency potential. Moreover, registration specifications indicate that they are user and environment safe. Nevertheless, few cinnamon-based pesticides are available in the market for wide use (see Table 7.2). The majority of the pesticides with composition containing cinnamon or its constituents are targeting mainly small-scale uses, such as home garden and indoor applications. The only major exception is Cinnacure, which is recommended for large-scale uses, as both an insecticide and a fungicide.

## 7.12 Conclusions

Although cinnamon tree products have played key roles in nutrition, medicine, and religion for centuries, their use as green pesticides is still limited. The chemistry of the genus *Cinnamomum* is interesting, but efforts in the research and development of these tree spices and their products have been restricted mainly to the volatile oil and its constituents. Various studies have dealt with the potential benefits of *Cinnamomum* for human health with controversial results. The pesticidal activities of cinnamon species have received little attention, and more research and scientific investigations are needed to unleash the huge potential of a tree qualified often as the "spice of life."

## References

Abdollahzadeh, E., M. Rezaei, and H. Hosseini. 2014. Antibacterial activity of plant essential oils and extracts: The role of thyme essential oil, nisin, and their combination to control *Listeria monocytogenes* inoculated in minced fish meat. *Food Control* 35 (1):177–183.

Aghel, N., Y. Yamini, A. Hadjiakhoondi, and S.M. Pourmortazavi. 2004. Supercritical carbon dioxide extraction of *Mentha pulegium* L. essential oil. *Talanta* 62 (2):407–411.

Akilen, R., Z. Pimlott, A. Tsiami, and N. Robinson. 2013. Effect of short-term administration of cinnamon on blood pressure in patients with prediabetes and type 2 diabetes. *Nutrition* 29 (10):1192–1196.

Alanazi, A.S., and M.U. Khan. 2015. Cinnamon use in type 2 diabetes: An updated meta-analysis. *World Journal of Pharmacy and Pharmaceutical Sciences* 4 (5):1838–1852.

Allen, R.W., E. Schwartzman, W.L. Baker, C.I. Coleman, and O.J. Phung. 2013. Cinnamon use in type 2 diabetes: An updated systematic review and meta-analysis. *Annals of Family Medicine* 11 (5):452–459.

Alvarez-Collazo, J., L. Alonso-Carbajo, A.I. López-Medina, Y.A. Alpizar, S. Tajada, B. Nilius, T. Voets, J.R. López-López, K. Talavera, and M.T. Pérez-García. 2014. Cinnamaldehyde inhibits L-type calcium channels in mouse ventricular cardiomyocytes and vascular smooth muscle cells. *Pflügers Archiv: European Journal of Physiology* 466 (11):2089–2099.

Amalaradjou, M.A.R., A. Narayanan, S.A. Baskaran, and K. Venkitanarayanan. 2010. Antibiofilm effect of trans-cinnamaldehyde on uropathogenic *Escherichia coli*. *Journal of Urology* 184 (1):358–363.

Aneja, K.R., R. Joshi, and C. Sharma. 2009. www.jpronline.info. *Journal of Pharmacy Research* 2 (9):1387–1390.

AOAC. 2003. *Official Methods of Analysis of the Association of Official's Analytical Chemists*, 17th ed. Association of Official Analytical Chemists, Arlington, Virginia.

Archer, A.W. 1988. Determination of cinnamaldehyde, coumarin and cinnamyl alcohol in cinnamon and cassia by high-performance liquid chromatography. *Journal of Chromatography A* 447:272–276.

Areias, F.M., P. Valentão, P.B. Andrade, M.M. Moreira, J. Amaral, and R.M. Seabra. 2000. HPLC/DAD analysis of phenolic compounds from lavender and its application to quality control. *Journal of Liquid Chromatography and Related Technologies* 23:2563–2572.

Bahadon, S.R., and F. Azarhoosh. 2013. Study on acaricidal activity of cinnamon, mint and eucalyptus extracts in control of poultry red mite (*Dermanyssus gallinae*). *Journal of Veterinary Research* 68 (3):203–208.

Bakkali, F., S. Averbeck, D. Averbeck, and M. Idaomar. 2008. Biological effects of essential oils— A review. *Food and Chemical Toxicology* 46 (2):446–475.

Barceloux, D.G. 2009. Cinnamon (cinnamomum species). *Disease-a-Month* 55 (6):327–335.

Barrera-Necha, L.L., S. Bautista-Baños, H.E. Flores-Moctezuma, and A. Rojas-Estudillo. 2008. Efficacy of essential oils on the conidial germination, growth of *Colletotrichum gloeosporioides* (Penz.) Penz. and Sacc and control of postharvest diseases in papaya (*Carica papaya* L.). *Plant Pathology Journal* 7 (2):174–178.

Barrera-Necha, L.L., C. Garduno-Pizana, and L.J. Garcia-Barrera. 2009. In vitro antifungal activity of essential oils and their compounds on mycelial growth of *Fusarium oxysporum* f. sp. gladioli (Massey) Snyder and Hansen. *Plant Pathology Journal* 8 (1):17–21.

Baslas, R.K., and K.K. Baslas. 1970. Chemistry of Indian essential oils. Part VIII. *Flavour Industry* 1:473–474.

Berenbaum, M. 1985. Brementown revisited: Interactions among allelochemicals in plants. In *Chemically Mediated Interactions between Plants and Other Organisms*, eds. G.A. Cooper-Driver, T. Swain, and E.E. Conn, 139–169. Berlin: Springer.

Bhargava, M.C., and B.L. Meena. 2001. Effect of some spice oils on the eggs of *Corcyra cephalonica* Stainton. *Insect Environment* 7 (1):43–44.

Bhattacharjee, S., T. Rana, and A. Sengupta. 2007. Inhibition of lipid peroxidation and enhancement of GST activity by cardamom and cinnamon during chemically induced colon carcinogenesis in Swiss albino mice. *Asian Pacific Journal of Cancer Prevention* 8 (4):578–582.

Brackman, G., S. Celen, U. Hillaert, S. Van Calenbergh, P. Cos, L. Maes, H.J. Nelis, and T. Coenye. 2011. Structure-activity relationship of cinnamaldehyde analogs as inhibitors of AI-2 based quorum sensing and their effect on virulence of *Vibrio* spp. *PLoS One* 6 (1):e16084.

Brackman, G., U. Hillaert, S. Van Calenbergh, H.J. Nelis, and T. Coenye. 2009. Use of quorum sensing inhibitors to interfere with biofilm formation and development in *Burkholderia multivorans* and *Burkholderia cenocepacia*. *Research in Microbiology* 160 (2):144–151.

Braga, M.E.M., P.A.D. Ehlert, L.C. Ming, M. Angela, and A. Meireles. 2005. Supercritical fluid extraction from *Lippia alba*: Global yields, kinetic data, and extract chemical composition. *Journal of Supercritical Fluids* 34 (2):149–156.

Brari, J., and D.R. Thakur. 2015. Insecticidal efficacy of essential oil from *Cinnamomum zeylanicum* Blume and its two major constituents against *Callosobruchus maculatus* (F.) and *Sitophilus oryzae* (L.). *Journal of Agricultural Technology* 11 (6):1323–1336.

Broadhurst, C.L., M.M. Polansky, and R.A. Anderson. 2000. Insulin-like biological activity of culinary and medicinal plant aqueous extracts in vitro. *Journal of Agricultural and Food Chemistry* 48 (3):849–852.

Campiglia, E., R. Mancinelli, A. Cavalieri, and F. Caporali. 2007. Use of essential oils of cinnamon, lavender and peppermint for weed control. *Italian Journal of Agronomy* 2 (2):171–178.

Cao, G.Y. 1993. Prevention and treatment of oral candidiasis with cortex cinnamon solution. *Zhonghua Hu Li Za Zhi* 28 (12):711.

Carvalho, R.N., L.S. Moura, P.T.V. Rosa, and M.A.A. Meireles. 2005. Supercritical fluid extraction from rosemary (*Rosmarinus officinalis*): Kinetic data, extract's global yield, composition, and antioxidant activity. *Journal of Supercritical Fluids* 35 (3):197–204.

Cavalieri, A., and F. Caporali. 2010. Effects of essential oils of cinnamon, lavender and peppermint on germination of Mediterranean weeds. *Allelopathy Journal* 25 (2):441–452.

Cavier, R. 1950. Anthelmintic properties of the essences of Ceylon cinnamon and of clove. *Therapie* 5 (3):140.

Chakraborty, A., V. Sankaran, M. Ramar, and D.R. Chellappan. 2015. Chemical analysis of leaf essential oil of *Cinnamomum verum* from Palni hills, Tamil Nadu. *Journal of Chemical Pharmaceutical Sciences* 8 (3):476–479.

Chang, K.-S., J.-H. Tak, S.-I. Kim, W.-J. Lee, and Y.-J. Ahn. 2006. Repellency of *Cinnamomum cassia* bark compounds and cream containing cassia oil to *Aedes aegypti* (Diptera: Culicidae) under laboratory and indoor conditions. *Pest Management Science* 62 (11):1032–1038.

Chang, S.-T., and S.-S. Cheng. 2002. Antitermitic activity of leaf essential oils and components from *Cinnamomum osmophleum*. *Journal of Agricultural and Food Chemistry* 50 (6):1389–1392.

Chao, L.K., K.-F. Hua, H.-Y. Hsu, S.-S. Cheng, J.-Y. Liu, and S.-T. Chang. 2005. Study on the antiinflammatory activity of essential oil from leaves of *Cinnamomum osmophloeum*. *Journal of Agricultural and Food Chemistry* 53 (18):7274–7278.

Chao, S.C., D.G. Young, and C.J. Oberg. 2000. Screening for inhibitory activity of essential oils on selected bacteria, fungi and viruses. *Journal of Essential Oil Research* 12 (5):639–649.

Cheng, S.-S., J.-Y. Liu, Y.-R. Hsui, and S.-T. Chang. 2006. Chemical polymorphism and antifungal activity of essential oils from leaves of different provenances of indigenous cinnamon (*Cinnamomum osmophloeum*). *Bioresource Technology* 97 (2):306–312.

Cheng, S.-S., J.-Y. Liu, C.-G. Huang, Y.-R. Hsui, W.-J. Chen, and S.-T. Chang. 2009. Insecticidal activities of leaf essential oils from *Cinnamomum osmophloeum* against three mosquito species. *Bioresource Technology* 100 (1):457–464.

Cheng, S.-S., J.-Y. Liu, C.-Y. Lin, Y.-R. Hsui, M.-C. Lu, W.-J. Wu, and S.-T. Chang. 2008. Terminating red imported fire ants using *Cinnamomum osmophloeum* leaf essential oil. *Bioresource Technology* 99 (4):889–893.

Cheng, S.-S., J.-Y. Liu, K.-H. Tsai, W.-J. Chen, and S.-T. Chang. 2004. Chemical composition and mosquito larvicidal activity of essential oils from leaves of different *Cinnamomum osmophloeum* provenances. *Journal of Agricultural and Food Chemistry* 52 (14):4395–4400.

Chericoni, S., J.M. Prieto, P. Iacopini, P. Cioni, and I. Morelli. 2005. In vitro activity of the essential oil of *Cinnamomum zeylanicum* and eugenol in peroxynitrite-induced oxidative processes. *Journal of Agricultural and Food Chemistry* 53 (12):4762–4765.

Chipley, J.R., and N. Uraih. 1980. Inhibition of *Aspergillus* growth and aflatoxin release by derivatives of benzoic acid. *Applied and Environmental Microbiology* 40 (2):352–357.

Chu, Y.L., W.C. Ho, and W.H. Ko. 2006. Effect of Chinese herb extracts on spore germination of *Oidium murrayae* and nature of inhibitory substance from Chinese rhubarb. *Plant Disease* 90 (7):858–861.

Cloyd, R.A., C.L. Galle, S.R. Keith, N.A. Kalscheur, and K.E. Kemp. 2009. Effect of commercially available plant-derived essential oil products on arthropod pests. *Journal of Economic Entomology* 102 (4):1567–1579.

Copping, L.G., and J.J. Menn. 2000. Biopesticides: A review of their action, applications and efficacy. *Pest Management Science* 56 (8):651–676.

Correa, Y.D.C.G., L.R.A. Faroni, K. Haddi, E.E. Oliveira, and E.J.G. Pereira. 2015. Locomotory and physiological responses induced by clove and cinnamon essential oils in the maize weevil *Sitophilus zeamais*. *Pesticide Biochemistry and Physiology* 125:31–37.

Dao, N.K. 2003. Chinese cassia. In *Cinnamon and Cassia: The Genus Cinnamomum*, ed. P.N. Ravindran, K. Nirmal-Babu, and M. Shylaja, 156. Boca Raton, FL: CRC Press.

De Castro, M.D.L., and L.E. Garcia-Ayuso. 1998. Soxhlet extraction of solid materials: An outdated technique with a promising innovative future. *Analytica Chimica Acta* 369 (1):1–10.

Dhuley, J.N. 1999. Anti-oxidant effects of cinnamon (*Cinnamomum verum*) bark and greater carda-mom (*Amomum subulatum*) seeds in rats fed high fat diet. *Indian Journal of Experimental Biology* 37:238–242.

Dima, C., and S. Dima. 2015. Essential oils in foods: Extraction, stabilization, and toxicity. *Current Opinion in Food Science* 5:29–35.

Di Pasqua, R., G. Betts, N. Hoskins, M. Edwards, D. Ercolini, and G. Mauriello. 2007. Membrane tox-icity of antimicrobial compounds from essential oils. *Journal of Agricultural and Food Chemistry* 55 (12):4863–4870.

Do, T.K.T., F. Hadji-Minaglou, S. Antoniotti, and X. Fernandez. 2015. Authenticity of essential oils. *TrAC Trends in Analytical Chemistry* 66:146–157.

Dong, L., H. Schill, R.L. Grange, A. Porzelle, J.P. Johns, P.G. Parsons, V.A. Gordon, P.W. Reddell, and C.M. Williams. 2009. Anticancer agents from the Australian tropical rainforest: Spiroacetals EBC-23, 24, 25, 72, 73, 75 and 76. *Chemistry—A European Journal* 15 (42):11307–11318.

Doughari, J.H. 2012. *Phytochemicals: Extraction Methods, Basic Structures and Mode of Action as Potential Chemotherapeutic Agents*. Rijeka, Croatia: INTECH Open Access Publisher.

Dragland, S., H. Senoo, K. Wake, K. Holte, and R. Blomhoff. 2003. Several culinary and medicinal herbs are important sources of dietary antioxidants. *Journal of Nutrition* 133 (5):1286–1290.

Duessel, S., R.M. Heuertz, and U.R. Ezekiel. 2008. Growth inhibition of human colon cancer cells by plant compounds. *Clinical Laboratory Science* 21 (3):151.

Dvorackova, E., M. Snoblova, L. Chromcova, and P. Hrdlicka. 2015. Effects of extraction methods on the phenolic compounds contents and antioxidant capacities of cinnamon extracts. *Food Science and Biotechnology* 24 (4):1201–1207.

El Asbahani, A., K. Miladi, W. Badri, M. Sala, E.H. Aït Addi, H. Casabianca, A. El Mousadik, D. Hartmann, A. Jilale, and F.N.R. Renaud. 2015. Essential oils: From extraction to encapsulation. *International Journal of Pharmaceutics* 483 (1):220–243.

El-Baroty, G.S., H.H. Abd El-Baky, R.S. Farag, and M.A. Saleh. 2010. Characterization of antioxi-dant and antimicrobial compounds of cinnamon and ginger essential oils. *African Journal of Biochemistry Research* 4 (6):167–174.

Ellse, L., and R. Wall. 2014. The use of essential oils in veterinary ectoparasite control: A review. *Medical and Veterinary Entomology* 28 (3):233–243.

Enan, E., M. Beigler, and A. Kende. 1998. Insecticidal action of terpenes and phenols to cockroaches: Effect on octopamine receptors. Presented at Proceedings of the International Symposium on Plant Protection. Gent, Belgium.

Enan, E.E. 2005. Molecular and pharmacological analysis of an octopamine receptor from American cockroach and fruit fly in response to plant essential oils. *Archives of Insect Biochemistry and Physiology* 59 (3):161–171.

ESCOP (European Scientific Cooperative on Phytotherapy). 2003. *ESCOP Monographs*. 2nd ed. World Press. Amsterdam, the Netherlands.

European Pharmacopoeia, 6th ed.; Council of Europe: Strasbourg, France, 2008.

Faraone, N., N.K. Hillier, and G.C. Cutler. 2015. Plant essential oils synergize and antagonize toxicity of different conventional insecticides against *Myzus persicae* (Hemiptera: Aphididae). *PloS One* 10 (5):e0127774.

Fichi, G., G. Flamini, L.J. Zaralli, and S. Perrucci. 2007. Efficacy of an essential oil of *Cinnamomum zeylanicum* against *Psoroptes cuniculi*. *Phytomedicine* 14 (2):227–231.

Figueiredo, A.C., J.G. Barroso, L.G. Pedro, and J.J.C. Scheffer. 2008. Factors affecting secondary metabolite production in plants: Volatile components and essential oils. *Flavour and Fragrance Journal* 23 (4):213–226.

Figueiredo, A.C., J.G. Barroso, L.G. Pedro, S.S. Fontinha, A. Looman, and J.C. Scheffer. 1997. Essential oils: Basic and applied research. In *Proceedings of the 27th International Symposium on Essential Oils*, eds. Ch. Franz, Á. Máthé, and G. Buchbauer, 95–107. Allured, Carol Stream, IL.

Fornari, T., G. Vicente, E. Vázquez, M.R. García-Risco, and G. Reglero. 2012. Isolation of essential oil from different plants and herbs by supercritical fluid extraction. *Journal of Chromatography A* 1250:34–48.

Friedman, M., P.R. Henika, C.E. Levin, and R.E. Mandrell. 2004. Antibacterial activities of plant essential oils and their components against *Escherichia coli* O157:H7 and *Salmonella enterica* in apple juice. *Journal of Agricultural and Food Chemistry* 52 (19):6042–6048.

Gallo, M., R. Ferracane, G. Graziani, A. Ritieni, and V. Fogliano. 2010. Microwave assisted extraction of phenolic compounds from four different spices. *Molecules* 15 (9):6365–6374.

Geng, S., Z. Cui, X. Huang, Y. Chen, D. Xu, and P. Xiong. 2011. Variations in essential oil yield and composition during *Cinnamomum cassia* bark growth. *Industrial Crops and Products* 33 (1):248–252.

Gill, A.O., and R.A. Holley. 2004. Mechanisms of bactericidal action of cinnamaldehyde against *Listeria monocytogenes* and of eugenol against *L. monocytogenes* and *Lactobacillus sakei*. *Applied and Environmental Microbiology* 70 (10):5750–5755.

Golmohammad, F., M.H. Eikani, and H.M. Maymandi. 2012. Cinnamon bark volatile oils separation and determination using solid-phase extraction and gas chromatography. *Procedia Engineering* 42:247–260.

Gong, F., Y.-Z. Liang, and Y.-S. Fung. 2004. Analysis of volatile components from *Cortex cinnamomi* with hyphenated chromatography and chemometric resolution. *Journal of Pharmaceutical and Biomedical Analysis* 34 (5):1029–1047.

Gul, S., and M. Safdar. 2009. Proximate composition and mineral analysis of cinnamon. *Pakistan Journal of Nutrition* 8 (9):1456–1460.

Gupta, C., A. Kumari, and A.P. Garg. 2012. Comparative study of cinnamon oil and clove oil in some oral microbiota. *Acta Bio Medica Atenei Parmensis* 82 (3):197–199.

Gursale, A., V. Dighe, and G. Parekh. 2010. Simultaneous quantitative determination of cinnamaldehyde and methyl eugenol from stem bark of *Cinnamomum zeylanicum* Blume using RP-HPLC. *Journal of Chromatographic Science* 48 (1):59–62.

Guynot, M.E., A.J. Ramos, L. Seto, P. Purroy, V. Sanchis, and S. Marin. 2003. Antifungal activity of volatile compounds generated by essential oils against fungi commonly causing deterioration of bakery products. *Journal of Applied Microbiology* 94 (5):893–899.

Haddi, K., E.E. Oliveira, L.R.A. Faroni, D.C. Guedes, and N.N.S. Miranda. 2015. Sublethal exposure to clove and cinnamon essential oils induces hormetic-like responses and disturbs behavioral and respiratory responses in *Sitophilus zeamais* (Coleoptera: Curculionidae). *Journal of Economic Entomology* 108 (6):2815–2822.

Han, D.C., M.-Y. Lee, K.D. Shin, S.B. Jeon, J.M. Kim, K.-H. Son, H.-C. Kim, H.-M. Kim, and B.-M. Kwon. 2004. 2′-Benzoyloxycinnamaldehyde induces apoptosis in human carcinoma via reactive oxygen species. *Journal of Biological Chemistry* 279 (8):6911–6920.

Hanifah, A.L., H.T. Ming, V.V. Narainasamy, and A.T. Yusoff. 2012. Laboratory evaluation of six crude plant extracts as repellents against larval *Leptotrombidium deliense* (Acari: Trombiculidae). *Asian Pacific Journal of Tropical Biomedicine* 2 (1):S257–S259.

He, Z.-D., C.-F. Qiao, Q.-B. Han, C.-L. Cheng, H.-X. Xu, R.-W. Jiang, P.P.-H. But, and P.-C. Shaw. 2005. Authentication and quantitative analysis on the chemical profile of cassia bark (*Cortex cinnamomi*) by high-pressure liquid chromatography. *Journal of Agricultural and Food Chemistry* 53 (7):2424–2428.

Hili, P., C.S. Evans, and R.G. Veness. 1997. Antimicrobial action of essential oils: The effect of dimethylsulphoxide on the activity of cinnamon oil. *Letters in Applied Microbiology* 24 (4):269–275.

Hori, M. 2003. Repellency of essential oils against the cigarette beetle, *Lasioderma serricorne* (Fabricius) (Coleoptera: Anobiidae). *Applied Entomology and Zoology* 38 (4):467–473.

Howard, M.E., and N.D. White. 2013. Potential benefits of cinnamon in type 2 diabetes. *American Journal of Lifestyle Medicine* 7 (1):23–26.

Huang, Y., and S.H. Ho. 1998. Toxicity and antifeedant activities of cinnamaldehyde against the grain storage insects, *Tribolium castaneum* (Herbst) and *Sitophilus zeamais* Motsch. *Journal of Stored Products Research* 34 (1):11–17.

Huie, C.W. 2002. A review of modern sample-preparation techniques for the extraction and analysis of medicinal plants. *Analytical and Bioanalytical Chemistry* 373 (1–2):23–30.

Husain, S.S., and M. Ali. 2013. Analysis of volatile oil of the stem bark of *Cinnamomum zeylanicum* and its antimicrobial activity. *International Journal of Research in Pharmacy & Science* 3 (4):40–49.

Imparl-Radosevich, J., S. Deas, M.M. Polansky, D.A. Baedke, T.S. Ingebritsen, R.A. Anderson, and D.J. Graves. 1998. Regulation of PTP-1 and insulin receptor kinase by fractions from cinnamon: Implications for cinnamon regulation of insulin signalling. *Hormone Research in Paediatrics* 50 (3):177–182.

Inouye, S., T. Takizawa, and H. Yamaguchi. 2001a. Antibacterial activity of essential oils and their major constituents against respiratory tract pathogens by gaseous contact. *Journal of Antimicrobial Chemotherapy* 47 (5):565–573.

Inouye, S., H. Yamaguchi, and T. Takizawa. 2001b. Screening of the antibacterial effects of a variety of essential oils on respiratory tract pathogens, using a modified dilution assay method. *Journal of Infection and Chemotherapy* 7 (4):251–254.

Isman, M.B. 2000. Plant essential oils for pest and disease management. *Crop Protection* 19 (8):603–608.

ISO (International Organization for Standardization). 2003. Oil of cinnamon leaf, Sri Lanka type (*Cinnamomum zeylanicum* Blume). ISO 3524:2003(E). Geneva: ISO.

Jakhetia, V., R. Patel, P. Khatri, N. Pahuja, S. Garg, A. Pandey, and S. Sharma. 2010. Cinnamon: A pharmacological review. *Journal of Advanced Scientific Research* 1 (2):19–23.

Jayaprakasha, G.K., and L.J.M. Rao. 2011. Chemistry, biogenesis, and biological activities of *Cinnamomum zeylanicum*. *Critical Reviews in Food Science and Nutrition* 51 (6):547–562.

Jayaprakasha, G.K., L.J. Rao, and K.K. Sakariah. 2002. Chemical composition of volatile oil from *Cinnamomum zeylanicum* buds. *Zeitschrift für Naturforschung C* 57 (11–12):990–993.

Jayaprakasha, G.K., L.J.M. Rao, and K.K. Sakariah. 2003. Volatile constituents from *Cinnamomum zeylanicum* fruit stalks and their antioxidant activities. *Journal of Agricultural and Food Chemistry* 51 (15):4344–4348.

Jayawardena, B., and R.M. Smith. 2010. Superheated water extraction of essential oils from *Cinnamomum zeylanicum* (L.). *Phytochemical Analysis* 21 (5):470–472.

Jia, P., Y.J. Xue, X.J. Duan, and S.H. Shao. 2011. Effect of cinnamaldehyde on biofilm formation and sarA expression by methicillin-resistant *Staphylococcus aureus*. *Letters in Applied Microbiology* 53 (4):409–416.

Jitomir, J., and D.S. Willoughby. 2009. *Cassia cinnamon* for the attenuation of glucose intolerance and insulin resistance resulting from sleep loss. *Journal of Medicinal Food* 12 (3):467–472.

Joffe, T., R.V. Gunning, G.R. Allen, M. Kristensen, S. Alptekin, L.M. Field, and G.D. Moores. 2012. Investigating the potential of selected natural compounds to increase the potency of pyrethrum against houseflies *Musca domestica* (Diptera: Muscidae). *Pest Management Science* 68 (2):178–184.

Jumbo, L.O.V., L.R.A. Faroni, E.E. Oliveira, M.A. Pimentel, and G.N. Silva. 2014. Potential use of clove and cinnamon essential oils to control the bean weevil, *Acanthoscelides obtectus* Say, in small storage units. *Industrial Crops and Products* 56:27–34.

Kamaliroosta, L., M. Gharachorloo, Z. Kamaliroosta, and K.H. Alimohammad Zadeh. 2012. Extraction of cinnamon essential oil and identification of its chemical compounds. *Journal of Medicinal Plant Research* 6 (4):609–614.

Kamath, J.V., A.C. Rana, and A.R. Chowdhury. 2003. Pro-healing effect of *Cinnamomum zeylanicum* bark. *Phytotherapy Research* 17 (8):970–972.

Kanerva, L., T. Estlander, and R. Jolanki. 1996. Occupational allergic contact dermatitis from spices. *Contact Dermatitis* 35 (3):157–162.

Kar, A., B.K. Choudhary, and N.G. Bandyopadhyay. 2003. Comparative evaluation of hypoglycaemic activity of some Indian medicinal plants in alloxan diabetic rats. *Journal of Ethnopharmacology* 84 (1):105–108.

Karahroudi, Z.R., F. Seifi, and A. Rahbarpour. 2010. Study on the fumigant insecticide effect of essential oil of five medicinal plants on three stored product pests. *Plant Protection Journal* 2 (3):197–207.

Kaul, P.N., A.K. Bhattacharya, B.R.R. Rao, K.V. Syamasundar, and S. Ramesh. 2003. Volatile constituents of essential oils isolated from different parts of cinnamon (*Cinnamomum zeylanicum* Blume). *Journal of the Science of Food and Agriculture* 83 (1):53–55.

Kavanaugh, N.L., and K. Ribbeck. 2012. Selected antimicrobial essential oils eradicate *Pseudomonas* spp. and *Staphylococcus aureus* biofilms. *Applied and Environmental Microbiology* 78 (11):4057–4061.

Khajeh, M., Y. Yamini, Fa. Sefidkon, and N. Bahramifar. 2004. Comparison of essential oil composition of *Carum copticum* obtained by supercritical carbon dioxide extraction and hydrodistillation methods. *Food Chemistry* 86 (4):587–591.

Khan, A., M. Safdar, M.M.A. Khan, K.N. Khattak, and R.A. Anderson. 2003. Cinnamon improves glucose and lipids of people with type 2 diabetes. *Diabetes Care* 26 (12):3215–3218.

Khan, M.S.A., and I. Ahmad. 2012. Antibiofilm activity of certain phytocompounds and their synergy with fluconazole against *Candida albicans* biofilms. *Journal of Antimicrobial Chemotherapy* 67 (3):618–621.

Khater, H.F., M.Y. Ramadan, and R.S. El-Madawy. 2009. Lousicidal, ovicidal and repellent efficacy of some essential oils against lice and flies infesting water buffaloes in Egypt. *Veterinary Parasitology* 164 (2):257–266.

Khoddami, A., M.A. Wilkes, and T.H. Roberts. 2013. Techniques for analysis of plant phenolic compounds. *Molecules* 18 (2):2328–2375.

Kim, S.-I., J.-Y. Roh, D.-H. Kim, H.-S. Lee, and Y.-J. Ahn. 2003. Insecticidal activities of aromatic plant extracts and essential oils against *Sitophilus oryzae* and *Callosobruchus chinensis*. *Journal of Stored Products Research* 39 (3):293–303.

Kim, S.-I., J.-H. Yi, J.-H. Tak, and Y.-J. Ahn. 2004. Acaricidal activity of plant essential oils against *Dermanyssus gallinae* (Acari: Dermanyssidae). *Veterinary Parasitology* 120 (4):297–304.

Kim, Y.-G., J.-H. Lee, S.-I. Kim, K.-H. Baek, and J. Lee. 2015. Cinnamon bark oil and its components inhibit biofilm formation and toxin production. *International Journal of Food Microbiology* 195:30–39.

King, A.A., D.T. Shaughnessy, K. Mure, J. Leszczynska, W.O. Ward, D.M. Umbach, Z. Xu, D. Ducharme, J.A. Taylor, and D.M. DeMarini. 2007. Antimutagenicity of cinnamaldehyde and vanillin in human cells: Global gene expression and possible role of DNA damage and repair. *Mutation Research/Fundamental and Molecular Mechanisms of Mutagenesis* 616 (1):60–69.

Kong, J.-O., S.-M. Lee, Y.-S. Moon, S.-G. Lee, and Y.-J. Ahn. 2006. Nematicidal activity of plant essential oils against *Bursaphelenchus xylophilus* (Nematoda: Aphelenchoididae). *Journal of Asia-Pacific Entomology* 9 (2):173–178.

Kosar, M., H.J.D. Dorman, O. Bachmayer, K.H.C. Baser, and R. Hiltunen. 2003. An improved on-line HPLC-DPPH method for the screening of free radical scavenging compounds in water extracts of Lamiaceae plants. *Chemistry of Natural Compounds* 39 (2):161–166.

Kostermans, A.J.G.H. 1980. A note on two species of *Cinnamomum* (Lauraceae) described in *Hortus Indicus Malabaricus*. In *Botany and History of Hortus Malabaricus*, ed. K. Manilal, 163–167. New Delhi: Oxford & IBH Publ. Co.

Kostermans, A.J.G.H. 1983. The South Indian species of *Cinnamomum* Schaeffer (Lauraceae). *Nelumbo* 25 (1–4):90–133.

Koul, O., R. Singh, B. Kaur, and D. Kanda. 2013. Comparative study on the behavioral response and acute toxicity of some essential oil compounds and their binary mixtures to larvae of *Helicoverpa armigera*, *Spodoptera litura* and *Chilo partellus*. *Industrial Crops and Products* 49:428–436.

Krishnamoorthy, B., and J. Rema. 2003. End uses of cinnamon and cassia. In *Cinnamon and Cassia: The Genus Cinnamomum*, ed. P.N. Ravindran, K. Nirmal-Babu, and M. Shylaja, 311. Boca Raton, FL: CRC Press.

Kubeczka, K.H. 2002. *Essential Oil Analysis*. New York: Wiley.

Kubeczka, K.-H., and V. Formáček. 2002. *Essential Oils Analysis by Capillary Gas Chromatography and Carbon-13 NMR Spectroscopy*. Hoboken, NJ: John Wiley & Sons.

Lawrence, B.M. 1967. A review of some of the commercial aspects of cinnamon. *Perfumer and Essential Oil Review* 58:236–241.

Lee, B.-H., W.-S. Choi, S.-E. Lee, and B.-S. Park. 2001. Fumigant toxicity of essential oils and their constituent compounds towards the rice weevil, *Sitophilus oryzae* (L.). *Crop Protection* 20 (4):317–320.

Lee, E.-J., J.-R. Kim, D.-R. Choi, and Y.-J. Ahn. 2008. Toxicity of cassia and cinnamon oil compounds and cinnamaldehyde-related compounds to *Sitophilus oryzae* (Coleoptera: Curculionidae). *Journal of Economic Entomology* 101 (6):1960–1966.

Lee, H.-S., B.-S. Kim, and M.-K. Kim. 2002. Suppression effect of *Cinnamomum cassia* bark-derived component on nitric oxide synthase. *Journal of Agricultural and Food Chemistry* 50 (26):7700–7703.

Lee, R., and M.J. Balick. 2005. Sweet wood—Cinnamon and its importance as a spice and medicine. *Explore: The Journal of Science and Healing* 1 (1):61–64.

Leela, N.K. 2008. Cinnamon and cassia. In *Chemistry of Spices*, eds. V.A. Parthasarathy, B. Chempakam, T.J. Zachariah, 124. Vol. 3. Wallingford, Oxfordshire, UK: CAB International.

Leung, A.Y., and S. Foster. 1996. *Encyclopedia of Common Natural Ingredients Used in Food, Drugs and Cosmetics*. 2nd ed. John Wiley & Sons, New York.

Li, R.W., G.D. Lin, S.P. Myers, and D.N. Leach. 2003. Anti-inflammatory activity of Chinese medicinal vine plants. *Journal of Ethnopharmacology* 85 (1):61–67.

Li, Y.-Q., D.-X. Kong, R.-S. Huang, H.-L. Liang, C.-G. Xu, and H. Wu. 2013a. Variations in essential oil yields and compositions of *Cinnamomum cassia* leaves at different developmental stages. *Industrial Crops and Products* 47:92–101.

Li, Y.-Q., D.-X. Kong, and H. Wu. 2013b. Analysis and evaluation of essential oil components of cinnamon barks using GC–MS and FTIR spectroscopy. *Industrial Crops and Products* 41:269–278.

Lima, E.O., O.F. Gompertz, A.M. Giesbrecht, and M.Q. Paulo. 1993. In vitro antifungal activity of essential oils obtained from officinal plants against dermatophytes. *Mycoses* 36 (9–10): 333–336.

Lin, C.-C., S.-J. Wu, C.-H. Chang, and L.-T. Ng. 2003. Antioxidant activity of *Cinnamomum cassia*. *Phytotherapy Research* 17 (7):726–730.

Lin, C.-W., C.-W. Yu, S.-C. Wu, and K.-H. Yih. 2009. DPPH free-radical scavenging activity, total phenolic contents and chemical composition analysis of forty-two kinds of essential oils. *Journal of Food & Drug Analysis* 17 (5):386–395.

Lin, K.-H., S.-Y. Yeh, M.-Y. Lin, M.-C. Shih, and S.-Y. Hwang. 2007. Major chemotypes and antioxidative activity of the leaf essential oils of *Cinnamomum osmophloeum* Kaneh. from a clonal orchard. *Food Chemistry* 105 (1):133–139.

Liu, C.H., A.K. Mishra, R.X. Tan, C. Tang, H. Yang, and Y.F. Shen. 2006. Repellent and insecticidal activities of essential oils from *Artemisia princeps* and *Cinnamomum camphora* and their effect on seed germination of wheat and broad bean. *Bioresource Technology* 97 (15):1969–1973.

Mancini-Filho, J., A. Van-Koiij, D.A. Mancini, F.F. Cozzolino, and R.P. Torres. 1998. Antioxidant activity of cinnamon (*Cinnamomum Zeylanicum*, Breyne) extracts. *Bollettino Chimico Farmaceutico* 137 (11):443–447.

Marin Municipal Water District. 2008. Vegetation management plan—Herbicide risk assessment. In *Clove Oil*, 25–27. Available at http://www.marinwater.org/documentcenter/view/253.

Mastura, M., M.A. Nor Azah, S. Khozirah, R. Mawardi, and A. Abd Manaf. 1999. Anticandidal and antidermatophytic activity of *Cinnamomum* species essential oils. *Cytobios* 98 (387):17–23.

Matan, N., H. Rimkeeree, A.J. Mawson, P. Chompreeda, V. Haruthaithanasan, and M. Parker. 2006. Antimicrobial activity of cinnamon and clove oils under modified atmosphere conditions. *International Journal of Food Microbiology* 107 (2):180–185.

Mathew, S., and T.E. Abraham. 2006. In vitro antioxidant activity and scavenging effects of *Cinnamomum verum* leaf extract assayed by different methodologies. *Food and Chemical Toxicology* 44 (2):198–206.

Matu, E.N., and J.V. Staden. 2003. Antibacterial and anti-inflammatory activities of some plants used for medicinal purposes in Kenya. *Journal of Ethnopharmacology* 87 (1):35–41.

Medagama, A.B. 2015. The glycaemic outcomes of cinnamon, a review of the experimental evidence and clinical trials. *Nutrition Journal* 14 (1):1.

Meyer-Warnod, B. 1984. Natural essential oils: Extraction processes and application to some major oils. *Perfumer & Flavorist* 9 (2):93–104.

Miresmailli, S., R. Bradbury, and M.B. Isman. 2006. Comparative toxicity of *Rosmarinus officinalis* L. essential oil and blends of its major constituents against *Tetranychus urticae* Koch (Acari: Tetranychidae) on two different host plants. *Pest Management Science* 62 (4):366–371.

Miresmailli, S., and M.B. Isman. 2006. Efficacy and persistence of rosemary oil as an acaricide against twospotted spider mite (Acari: Tetranychidae) on greenhouse tomato. *Journal of Economic Entomology* 99 (6):2015–2023.

Mondal, M., and M. Khalequzzaman. 2009. Ovicidal activity of essential oils against red flour beetle, *Tribolium castaneum* (Coleoptera: Tenebrionidae). *Journal of Bio-Science* 17:57–62.

Montes-Belmont, R., and M. Carvajal. 1998. Control of *Aspergillus flavus* in maize with plant essential oils and their components. *Journal of Food Protection* 61 (5):616–619.

Moura, L.S., R.N. Carvalho Jr., M.B. Stefanini, L.C. Ming, M. Angela, and A. Meireles. 2005. Supercritical fluid extraction from fennel (*Foeniculum vulgare*): Global yield, composition and kinetic data. *Journal of Supercritical Fluids* 35 (3):212–219.

Murcia, M.A., I. Egea, F. Romojaro, P. Parras, A.M. Jiménez, and M. Martínez-Tomé. 2004. Antioxidant evaluation in dessert spices compared with common food additives. Influence of irradiation procedure. *Journal of Agricultural and Food Chemistry* 52 (7):1872–1881.

Na, Y.E., S.-I. Kim, H.-S. Bang, B.-S. Kim, and Y.-J. Ahn. 2011. Fumigant toxicity of cassia and cinnamon oils and cinnamaldehyde and structurally related compounds to *Dermanyssus gallinae* (Acari: Dermanyssidae). *Veterinary Parasitology* 178 (3):324–329.

NCBI (National Center for Biotechnology Information). 2016. Cinnamon oil PubChem CID 6850781.

Nguyen, V.-N., D.-J. Seo, R.-D. Park, and W.-J. Jung. 2009. Antimycotic activities of cinnamon-derived compounds against *Rhizoctonia solani* in vitro. *Biocontrol* 54 (5):697–707.

Niu, C., and E.S. Gilbert. 2004. Colorimetric method for identifying plant essential oil components that affect biofilm formation and structure. *Applied and Environmental Microbiology* 70 (12):6951–6956.

OECD (Organisation for Economic Cooperation and Development). 2002. SIDS initial assessment report for SIAM 14. Paris: OECD.

Oh, M.J. 2011. The acaricidal and repellent effect of cinnamon essential oil against house dust mite. *World Academy of Science, Engineering and Technology* 60:710–714.

Okawa, M., J. Kinjo, T. Nohara, and M. Ono. 2001. DPPH (1,1-diphenyl-2-picrylhydrazyl) radical scavenging activity of flavonoids obtained from some medicinal plants. *Biological and Pharmaceutical Bulletin* 24 (10):1202–1205.

Omolo, M.O., D. Okinyo, I.O. Ndiege, W. Lwande, and A. Hassanali. 2005. Fumigant toxicity of the essential oils of some African plants against *Anopheles gambiae* sensu stricto. *Phytomedicine* 12 (3):241–246.

Onderoglu, S., S. Sozer, K. Erbil, R. Ortac, and F. Lermioglu. 1999. The evaluation of long-term effects of cinnamon bark and olive leaf on toxicity induced by streptozotocin administration to rats. *Journal of Pharmacy and Pharmacology* 51 (11):1305–1312.

Ouattara, B., R.E. Simard, G. Piette, A. Begin, and R.A. Holley. 2000. Diffusion of acetic and propionic acids from chitosan-based antimicrobial packaging films. *Journal of Food Science* 65 (5):768–773.

Pandey, S., R. Pandey, and R. Singh. 2014. Phytochemical screening of selected medicinal plant *Cinnamon Zeylanicum* bark extract, area of research; Uttarakhand, India. *International Journal of Scientific and Research Publications* 4.

Paranagama, P.A., T. Abeysekera, L. Nugaliyadde, and K. Abeywickrama. 2003. Effect of the essential oils of *Cymbopogon citratus*, *C. nardus* and *Cinnamomum zeylanicum* on pest incidence and grain quality of rough rice (paddy) stored in an enclosed seed box. *Food, Agriculture & Environment* 1 (2):134–136.

Paranagama, P.A., S. Wimalasena, G.S. Jayatilake, A.L. Jayawardena, U.M. Senanayake, and A.M. Mubarak. 2010. A comparison of essential oil constituents of bark, leaf, root and fruit of cinnamon (*Cinnamomum zeylanicum* Blum) grown in Sri Lanka. *Journal of the National Science Foundation of Sri Lanka* 29 (3–4).

Park, I.-K., H.-S. Lee, S.-G. Lee, J.-D. Park, and Y.-J. Ahn. 2000. Insecticidal and fumigant activities of *Cinnamomum cassia* bark-derived materials against *Mechoris ursulus* (Coleoptera: Attelabidae). *Journal of Agricultural and Food Chemistry* 48 (6):2528–2531.

Park, I.-K., J.-Y. Park, K.-H. Kim, K.-S. Choi, I.-H. Choi, C.-S. Kim, and S.-C. Shin. 2005. Nematicidal activity of plant essential oils and components from garlic (*Allium sativum*) and cinnamon (*Cinnamomum verum*) oils against the pine wood nematode (*Bursaphelenchus xylophilus*). *Nematology* 7 (5):767–774.

Passino, G.S., E. Bazzoni, L. Moretti, and R. Prota. 1999. Effects of essential oil formulations on *Ceratitis capitata* Wied. (Dipt., Tephritidae) adult flies. *Journal of Applied Entomology* 123 (3):145–149.

Pavela, R. 2015. Essential oils for the development of eco-friendly mosquito larvicides: A review. *Industrial Crops and Products* 76:174–187.

Pizzale, L., R. Bortolomeazzi, S. Vichi, E. Überegger, and L.S. Conte. 2002. Antioxidant activity of sage (*Salvia officinalis* and *S. fruticosa*) and oregano (*Origanum onites* and *O. indercedens*) extracts related to their phenolic compound content. *Journal of the Science of Food and Agriculture* 82 (14):1645–1651.

Prajapati, V., A.K. Tripathi, K.K. Aggarwal, and S.P.S. Khanuja. 2005. Insecticidal, repellent and oviposition-deterrent activity of selected essential oils against *Anopheles stephensi*, *Aedes aegypti* and *Culex quinquefasciatus*. *Bioresource Technology* 96 (16):1749–1757.

Premanathan, M., S. Rajendran, T. Ramanathan, and K. Kathiresan. 2000. A survey of some Indian medicinal plants for anti-human immunodeficiency virus (HIV) activity. *Indian Journal of Medical Research* 112:73.

Price, S., and L. Price. 2007. *Aromatherapy for Health Professionals*. Amsterdam: Elsevier Health Sciences.

Quale, J.M., D. Landman, M.M. Zaman, S. Burney, and S.S. Sathe. 1996. In vitro activity of *Cinnamomum zeylanicum* against azole resistant and sensitive *Candida* species and a pilot study of cinnamon for oral candidiasis. *American Journal of Chinese Medicine* 24 (2):103–109.

Rafehi, H., K. Ververis, and T.C. Karagiannis. 2012. Controversies surrounding the clinical potential of cinnamon for the management of diabetes. *Diabetes, Obesity and Metabolism* 14 (6):493–499.

Rajeswara Rao, B.R., D.K. Rajput, and A.K. Bhattacharya. 2007. Essential oil composition of petiole of *Cinnamomum verum* Bercht. & Presl. *Journal of Spices and Aromatic Crops* 16 (1):38–41.

Ranasinghe, L., B. Jayawardena, and K. Abeywickrama. 2002. Fungicidal activity of essential oils of *Cinnamomum zeylanicum* (L.) and *Syzygium aromaticum* (L.) Merr et LM Perry against crown rot and anthracnose pathogens isolated from banana. *Letters in Applied Microbiology* 35 (3):208–211.

Ranasinghe, P., S. Pigera, G.A. Sirimal Premakumara, P. Galappaththy, G.R. Constantine, and P. Katulanda. 2013. Medicinal properties of 'true' cinnamon (*Cinnamomum zeylanicum*): A systematic review. *BMC Complementary and Alternative Medicine* 13 (1):1.

Rao, P.V., and S.H. Gan. 2014. Cinnamon: A multifaceted medicinal plant. *Evidence-Based Complementary and Alternative Medicine* 2014:642942.

Ravindran, P.N., K. Nirmal-Babu, and M. Shylaja. 2003. *Cinnamon and Cassia: The Genus Cinnamomum*. Boca Raton, FL: CRC Press.

Re, L., S. Barocci, S. Sonnino, A. Mencarelli, C. Vivani, G. Paolucci, A. Scarpantonio, L. Rinaldi, and E. Mosca. 2000. Linalool modifies the nicotinic receptor–ion channel kinetics at the mouse neuromuscular junction. *Pharmacological Research* 42 (2):177–182.

Reddy, A.M., J.H. Seo, S.Y. Ryu, Y.S. Kim, K.R. Min, and Y. Kim. 2004. Cinnamaldehyde and 2-methoxy-cinnamaldehyde as NF-kappaB inhibitors from *Cinnamomum cassia*. *Planta Medica* 70 (9):823–827.

Regnault-Roger, C., C. Vincent, and J.T. Arnason. 2012. Essential oils in insect control: Low-risk products in a high-stakes world. *Annual Review of Entomology* 57:405–424.

Senanayake, U.M., T.H. Lee, and R.B.H. Wills. 1978. Volatile constituents of cinnamon (*Cinnamomum zeylanicum*) oils. *Journal of Agricultural and Food Chemistry* 26 (4):822–824.

Senanayake, U.M., and R.O.B. Wijesekera. 2003. Chemistry of cinnamon and cassia. In *Cinnamon and Cassia: The Genus Cinnamomum*, ed. P.N. Ravindran, K. Nirmal-Babu, and M. Shylaja, 80. Boca Raton, FL: CRC Press.

Shafreen, B., R. Mohmed, C. Selvaraj, S. Kumar Singh, and S. Karutha Pandian. 2014. In silico and in vitro studies of cinnamaldehyde and their derivatives against LuxS in *Streptococcus pyogenes*: Effects on biofilm and virulence genes. *Journal of Molecular Recognition* 27 (2):106–116.

Shan, B., Y.-Z. Cai, J.D. Brooks, and H. Corke. 2007. Antibacterial properties and major bioactive components of cinnamon stick (*Cinnamomum burmannii*): Activity against foodborne pathogenic bacteria. *Journal of Agricultural and Food Chemistry* 55 (14):5484–5490.

Shan, B., Y.Z. Cai, M. Sun, and H. Corke. 2005. Antioxidant capacity of 26 spice extracts and characterization of their phenolic constituents. *Journal of Agricultural and Food Chemistry* 53 (20):7749–7759.

Sharififar, F., M.H. Moshafi, G. Dehghan-Nudehe, A. Ameri, F. Alishahi, and A. Pourhemati. 2009. Bioassay screening of the essential oil and various extracts from 4 spices medicinal plants. *Pakistan Journal of Pharmaceutical Sciences* 22:317–322.

Shaughnessy, D.T., R.M. Schaaper, D.M. Umbach, and D.M. DeMarini. 2006. Inhibition of spontaneous mutagenesis by vanillin and cinnamaldehyde in *Escherichia coli*: Dependence on recombinational repair. *Mutation Research/Fundamental and Molecular Mechanisms of Mutagenesis* 602 (1):54–64.

Shen, F., M. Xing, L. Liu, X. Tang, W. Wang, X. Wang, X. Wu, X. Wang, X. Wang, and G. Wang. 2012. Efficacy of trans-cinnamaldehyde against *Psoroptes cuniculi* in vitro. *Parasitology Research* 110 (4):1321–1326.

Shen, L.R., H.Y. Li, Y.G. Zhou, S. Gu, and Y.G. Lou. 2007. Ovicidal activity of nine essential oils against *Chrysomya megacephara* in bacon and kipper. *Ying Yong Sheng Tai Xue Bao* 18 (10):2343–2346.

Sheng, X., Y. Zhang, Z. Gong, C. Huang, and Y.Q. Zang. 2008. Improved insulin resistance and lipid metabolism by cinnamon extract through activation of peroxisome proliferator-activated receptors. *PPAR Research* 2008:581348.

Shivanandappa, T., and Y. Rajashekar. 2014. Mode of action of plant-derived natural insecticides. In *Advances in Plant Biopesticides*, ed. D. Singh, 323–345. Berlin: Springer.

Shobana, S., and K.A. Naidu. 2000. Antioxidant activity of selected Indian spices. *Prostaglandins, Leukotrienes and Essential Fatty Acids* 62 (2):107–110.

Simić, A., M.D. Soković, M. Ristić, S. Grujić-Jovanović, J. Vukojević, and P.D. Marin. 2004. The chemical composition of some Lauraceae essential oils and their antifungal activities. *Phytotherapy Research* 18 (9):713–717.

Singh, G., S. Maurya, and C.A.N. Catalan. 2007. A comparison of chemical, antioxidant and antimicrobial studies of cinnamon leaf and bark volatile oils, oleoresins and their constituents. *Food and Chemical Toxicology* 45 (9):1650–1661.

Song, F., H. Li, J. Sun, and S. Wang. 2013. Protective effects of cinnamic acid and cinnamic aldehyde on isoproterenol-induced acute myocardial ischemia in rats. *Journal of Ethnopharmacology* 150 (1):125–130.

Sosa, S., M.J. Balick, R. Arvigo, R.G. Esposito, C. Pizza, G. Altinier, and A. Tubaro. 2002. Screening of the topical anti-inflammatory activity of some Central American plants. *Journal of Ethnopharmacology* 81 (2):211–215.

Sowbhagya, H.B. 2016. Microwave impact on the flavour compounds of cinnamon bark (*Cinnamomum Cassia*) volatile oil and polyphenol extraction. *Current Microwave Chemistry* 2.

Stoner, G., and L.-S. Wang. 2013. Natural products as anti-inflammatory agents. In *Obesity, Inflammation and Cancer*, eds. A.J. Dannenberg and N.A. Berger, 341–361. Berlin: Springer.

Tainter, D.R., and A.T. Grenis. 1993. *Spices and Seasonings: A Food Technology Handbook*. New York: VCH Publishers.

Tongnuanchan, P., and S. Benjakul. 2014. Essential oils: Extraction, bioactivities, and their uses for food preservation. *Journal of Food Science* 79 (7):R1231–R1249.

Tung, Y.-T., M.-T. Chua, S.-Y. Wang, and S.-T. Chang. 2008. Anti-inflammation activities of essential oil and its constituents from indigenous cinnamon (*Cinnamomum osmophloeum*) twigs. *Bioresource Technology* 99 (9):3908–3913.

Tung, Y.-T., P.-L. Yen, C.-Y. Lin, and S.-T. Chang. 2010. Anti-inflammatory activities of essential oils and their constituents from different provenances of indigenous cinnamon (*Cinnamomum osmophloeum*) leaves. *Pharmaceutical Biology* 48 (10):1130–1136.

Tworkoski, T. 2002. Herbicide effects of essential oils. *Weed Science* 50 (4):425–431.

Upadhyay, A., I. Upadhyaya, A. Kollanoor-Johny, and K. Venkitanarayanan. 2013. Antibiofilm effect of plant derived antimicrobials on *Listeria monocytogenes*. *Food Microbiology* 36 (1):79–89.

USEPA (U.S. Environmental Protection Agency). 2015. Pesticides: Regulating pesticides, cinnamaldehyde (040506) fact sheet. Washington, DC: Office of Pesticide Programs, USEPA. Available at http://epa.gov/pesticides/biopesticies/ingredients/factsheets/factsheet_040506.htm.

Van tol, R.W.H.M., H.J. Swarts, A. van der Linden, and J.H. Visser. 2007. Repellence of the red bud borer *Resseliella oculiperda* from grafted apple trees by impregnation of rubber budding strips with essential oils. *Pest Management Science* 63 (5):483–490.

Veeraphant, C., V. Mahakittikun, and N. Soonthornchareonnon. 2011. Acaricidal effects of Thai herbal essential oils against *Dermatophagoides pteronyssinus*. *Mahidol University Journal of Pharmaceutical Sciences* 38:1–12.

Wainstein, J., N. Stern, S. Heller, and M. Boaz. 2011. Dietary cinnamon supplementation and changes in systolic blood pressure in subjects with type 2 diabetes. *Journal of Medicinal Food* 14 (12):1505–1510.

Wang, L., and C.L. Weller. 2006. Recent advances in extraction of nutraceuticals from plants. *Trends in Food Science & Technology* 17 (6):300–312.

Wang, R., R. Wang, and B. Yang. 2009. Extraction of essential oils from five cinnamon leaves and identification of their volatile compound compositions. *Innovative Food Science & Emerging Technologies* 10 (2):289–292.

Wang, Y., R. Zhao, L. Yu, Y. Zhang, Y. He, and J. Yao. 2014. Evaluation of cinnamon essential oil microemulsion and its vapor phase for controlling postharvest gray mold of pears (*Pyrus pyrifolia*). *Journal of the Science of Food and Agriculture* 94 (5):1000–1004.

WHO (World Health Organization). 1999. *WHO Monographs on Selected Medicinal Plants*. Vol. 2. Geneva: WHO.

Wijesekera, R.O. 1977. Historical overview of the cinnamon industry. *CRC Critical Reviews in Food Science and Nutrition* 10 (1):1–30.

Willard, H. 2013. 11 cinnamon-flavored liquors for the holidays. Available at http://www.thedaily meal.com/11-cinnamon-flavored-liquors-holidays/121613.

Williams, A.R., A. Ramsay, T.V.A. Hansen, H.M. Ropiak, H. Mejer, P. Nejsum, I. Mueller-Harvey, and S.M. Thamsborg. 2015. Anthelmintic activity of trans-cinnamaldehyde and A- and B-type proanthocyanidins derived from cinnamon (*Cinnamomum verum*). *Scientific Reports* 5:14791.

Wilson, C.L., J.M. Solar, A. El Ghaouth, and M.E. Wisniewski. 1997. Rapid evaluation of plant extracts and essential oils for antifungal activity against *Botrytis cinerea*. *Plant Disease* 81 (2):204–210.

Wondrak, G.T., N.F. Villeneuve, S.D. Lamore, A.S. Bause, T. Jiang, and D.D. Zhang. 2010. The cinnamon-derived dietary factor cinnamic aldehyde activates the Nrf2-dependent antioxidant response in human epithelial colon cells. *Molecules* 15 (5):3338–3355.

Wong, Y.C., M.Y. Ahmad-Mudzaqqir, and W.A. Wan-Nurdiyana. 2014. Extraction of essential oil from cinnamon (*Cinnamomum zeylanicum*). *Oriental Journal of Chemistry* 30 (1):37–47.

Wu, S.-J., and L.-T. Ng. 2007. MAPK inhibitors and pifithrin-alpha block cinnamaldehyde-induced apoptosis in human PLC/PRF/5 cells. *Food and Chemical Toxicology* 45 (12):2446–2453.

Yang, C.-H., R.-X. Li, and L.-Y. Chuang. 2012. Antioxidant activity of various parts of *Cinnamomum cassia* extracted with different extraction methods. *Molecules* 17 (6):7294–7304.

Yang, Y.-C., E.H. Lee, H.S. Lee, D.K. Lee, and Y.J. Ahn. 2004. Repellency of aromatic medicinal plant extracts and a steam distillate to *Aedes aegypti*. *Journal of the American Mosquito Control Association* 20 (2):146–149.

Yeh, R.-Y., Y.-L. Shiu, S.-C. Shei, S.-C. Cheng, S.-Y. Huang, J.-C. Lin, and C.-H. Liu. 2009. Evaluation of the antibacterial activity of leaf and twig extracts of stout camphor tree, *Cinnamomum kanehirae*, and the effects on immunity and disease resistance of white shrimp, *Litopenaeus vannamei*. *Fish & Shellfish Immunology* 27 (1):26–32.

Yeole, G.J., N.P. Teli, H.M. Kotkar, and P.S. Mendki. 2014. *Cinnamomum zeylanicum* extracts and their formulations control early blight of tomato. *Journal of Biopesticides* 7 (2):110.

Yuste, J., and D.Y.C. Fung. 2002. Inactivation of *Listeria monocytogenes* Scott A 49594 in apple juice supplemented with cinnamon. *Journal of Food Protection* 65 (10):1663–1666.

Zellner, B.A., P. Dugo, G. Dugo, and L. Mondello. 2009. Analysis of essential oils. In *Handbook of Essential Oils: Science, Technology, and Applications*, eds. K. Hüsnü, C. Baser, and G. Buchbauer, 151. Boca Raton, FL: CRC Press.

Zhang, H., W. Zhou, W. Zhang, A. Yang, Y. Liu, Y. Jiang, S. Huang, and J. Su. 2014. Inhibitory effects of citral, cinnamaldehyde, and tea polyphenols on mixed biofilm formation by foodborne *Staphylococcus aureus* and *Salmonella enteritidis*. *Journal of Food Protection* 77 (6):927–933.

Zhou, Q., and G. Liang. 2003. Effect of plant alcohol extracts on vegetable aphids and their parasitoids. *Ying Yong Sheng Tai Xue Bao* 14 (2):249–252.

# 8

## Citronella Oil

**Valentina Aristizábal Marulanda, Christian David Botero Gutiérrez, and Carlos Ariel Cardona Alzate**

## CONTENTS

## 8.1 Citronella: Overview

*Cymbopogon* (Poaceae) is a genus of aromatic plants that includes about 140 species distributed across all continents. A tentative distribution of these plants is mentioned in [1]. The vast majority of these aromatic plants are in Africa, with 52 reported species, followed by 45 in India. Australia and South America have six species each, Europe hosts four species, and North America two. The remaining species are located in South Asia. Genus *Cymbopogon* dates to 1815, when Sprengel named it for the first time. Years later, a group of taxonomists concluded that this plant was a subgenus of *Andropogon*, a convention that was accepted for a few years. In 1906, Stapf stated that *Cymbopogon* should again be considered a genus, as it is known today [1].

The great importance of this genus of aromatic plants lies in its commercial power, because the essential oils extracted from the plants are used in various chemical syntheses. Industries such as the perfume, pesticide, and pharmaceutical industries have the greatest demand for essential oils, especially the genus *Cymbopogon*. Some of the active compounds with higher concentrations in the genus *Cymbopogon* are citronellol, geraniol, citronellal, and linalool. These compounds are used for the production of fragrances and as food additives [1].

According to taxonomy, species of *Cymbopogon* are classified into three groups: "Schoenathi," "Rusae," and "Citrati." The difference between these groups lies in the physiological characteristics of their leaves. For "Schoenathi," the leaves are laminations; in "Rusae," the leaves are heart shaped (subcordate); and "Citrati" has broad leaves at the bottom, but with a sharp point (lanceolate). A clear classification of *Cymbopogon* has been very difficult to carry out due to the large number of species and varieties of this genus and the constant appearance of transitional forms that are generated by hybridization [1]. Table 8.1 shows some of the species and varieties of *Cymbopogon*.

Experimentally, the *Cymbopogon* species can be classified according to their morphology and chemotypical characteristics, although in many cases the distinction is confusing [2].

In this genus, *C. martinii* var. Sofia and *C. martinii* var. Motia present similar morphological characteristics, not distinguishable to the eye, although they have different chemical composition. On the other hand, *C. flexuosus* and *C. citratus* have similar chemotypical features but different morphology [1]. The different chemical compositions that exist between some species and even with plants of the same species may be the work of human intervention, geography, and weather conditions where the plant is [1]. Hybridization can also play an important role in the chemical composition of essential oils, since it can develop plants with morphological and chemotypical intermediate characteristics that cannot be defined taxonomically [1]. At the time, studies performed to discern the extent of genetic

**TABLE 8.1**

Some Species and Varieties of *Cymbopogon*

| | |
|---|---|
| *C. ambiguus* A. Camus. | *C. annamensis* A. Camus. |
| *C. arabicus* Nees ex Steud. | *C. arundinaceus* Schult. |
| *C. bassacensis* A. Camus. | *C. bhutanicus* Noltie |
| *C. bombycinus* var. townsvillensis Domin. | *C. bracteatus* Hitchcock |
| *C. calcicola* C.E. Hubb. | *C. chevalieri* A. Camus |
| *C. citratus* Stapf | *C. citriodorus* Link |
| *C. densiflorus* Stapf | *C. divaricatus* Stapf |
| *C. elegans* Spreng. | *C. excavatus* Stapf |
| *C. flexuosus* Stapf | *C. floccosus* Stapf |
| *C. giganteus* Chiov. | *C. gratus* Domin. |
| *C. iwarancusa* Schult. | *C. khasianus* (Hackel) Stapf ex Bor |
| *C. lividus* (Thwaites) Willis | *C. martinii* Stapf |
| *C. nardus* (L.) Rendle | *C. nyassae* Pilg. |
| *C. polyneuros* Stapf | *C. princeps* Stapf |
| *C. winterianus* Jowitt | *C. virgatus* Stapf ex Bor |

variation between species are presented, aiming at a more accurate classification of the species in this genus [2].

The aim of this chapter is to show the potential of citronella oil given its ovicidal, antimicrobial, fungicidal, and insecticidal ability, as an alternative to synthetic products of this type. This work also presents the chemical composition of citronella oil according to different authors, characteristics, uses, and methods of extraction. Finally, a technoeconomic and environmental analysis of citronella oil extraction is performed with the most relevant extraction technologies, such as solvents, supercritical fluids, and steam distillation.

## 8.2  Citronella Oils: *Cymbopogon winterianus* (Java) and *Cymbopogon nardus* (Ceylon)

The *Cymbopogon* genus belongs to the family Poaceae (Gramineae), and some species produce essential oils that are a valuable source for the flavor industry, as the case of *C. winterianus* and *C. nardus* [3,4]. The famous citronella oil is derived from *C. nardus* and *C. winterianus*. *C. nardus* (L.) Rendle is also called Ceylon citronella oil or Lanabatu oil, and *C. winterianus* Jowitt is also called Java citronella. Between these oils, there are some differences, such as morphology by length, shape of its leaves, and the chemical composition of its essential oil. The characteristics, uses, and production of citronella essential oil are shown in the following subsections.

### 8.2.1  Characteristics

Java citronella oil is grown in tropical and subtropical countries of Asia, America, and Africa, among which Haiti, Honduras, Taiwan, Guatemala, China, and Brazil are the most recognized [4]. Citronella Java is an aromatic herbal plant native of southeastern Asia [3] that is of great importance for both the industry and the Aborigine cultures of the countries where it grows. In India, one of the countries with higher commerce of citronella oil, this plant was introduced in 1959 [5]. In America, the plant was incorporated in Brazil, around the eighteenth century [3]. Citronella Java (*C. winterianus*) comes from citronella Ceylon (*C. nardus*) [5]. At the beginning of the last century, the latter was the largest-producing species of citronella oil, but because citronella Java has higher yields of essential oil, it gradually replaced citronella Ceylon in the market. Currently, Brazil is the largest producer of citronella oil in the Americas [6]. Figure 8.1 shows a picture of citronella Java (*C. winterianus*) [7].

Citronella Java is known for its long leaves, which can grow to be more than 100 cm long and 4 cm wide. Leaves have the highest concentration of oil, which is rich in citronellol, geraniol, and citronellal, ranging from 1% to 1.2% of these compounds [8]. Ceylon-type oil contains phenolic derivatives (methyl eugenol and methyl isoeugenol)—the most significant difference between it and Java-type oil. Both types contain similar amounts of geraniol, although the Java type presents a high quantity of citronellol and citronellal. Also, it contains tricyclene, eugenol, and 1-borneol and small amounts of monoterpene hydrocarbons, such as limonene [4]. Citronella oil comprises about 30–40 compounds extracted from the stem and leaves, among which aromatic compounds like monoterpenes, which are approximately 80%, and sesquiterpenes can be highlighted [9]. Table 8.2 shows the composition of the essential citronella oil Java and the compounds with a higher percentage content (citronellol, citronellal, and geraniol) [10]. These active compounds are highly sought in the chemical industry because they are raw materials for various chemical syntheses [5].

Citronella oil production is affected by factors such as weather conditions and geographical location of plants [8]. Citronella Java generates good yields in warm and high-humidity locations. It is calculated that annual rain precipitations of 200–250 cm are necessary in the crop area [8]. Citronella Java, as well as almost all the plants in the *Cymbopogon* genus, is tolerant to alkalinity and soil salinity [8], but this aromatic plant is very susceptible to flooding in the sown area, so it is necessary to take account of the weather conditions [9]. The necessary conditions for the plant to provide the most oil yield have been studied, and it has been shown that fertile sandy soils are ideal for growing *C. winterianus* [4]. The high concentration of nitrogen in the soil and the plant can generate a high amount of biomass but a low yield of essential oil [4]. Similarly, agricultural factors such as amount of light,

**FIGURE 8.1**
Citronella Java (*C. winterianus*). (From Singh, N. K. et al., *Parasitol. Res.*, 113(1), 341–350, 2014.)

**TABLE 8.2**

Composition of the Citronella Oil Java (*C. Winterianus*)

| Compounds | % |
|---|---|
| Limonene | 1.58 |
| (Z)-Ocimene | 0.19 |
| (E)-Ocimene | 0.09 |
| Bergamal | 0.10 |
| NI | 0.25 |
| Linalool | 0.88 |
| Isopulegol | 0.60 |
| Citronellal | 27.44 |
| *iso*-Isopulegol | 0.20 |
| NI | 0.09 |
| NI | 0.16 |
| *n*-Decanal | 0.41 |
| NI | 0.07 |
| Citronellol | 10.45 |
| Neral(-citral) | 6.02 |
| Geraniol | 40.06 |
| Geranial(-citral) | 8.05 |
| Citronellyl acetate | 0.79 |
| Geranyl acetate | 1.77 |
| Caryophyllene | 0.55 |
| Cardinene | 0.23 |
| Total identified | 99.41 |

*Source:* Souza, T. et al., *Phytomedicine*, 15(8), 619–624, 2008.
*Note:* NI, not identified.

humidity, and time of seeding alter oil composition significantly [4]. The supply of light provided by the environment for the plant is a priority for its growth, and low temperatures and moderate relative humidity generate high yields of oil [4].

C. *winterianus* oil is sold at a higher cost than that of *C. nardus* because geraniol and citronellal are precursors of various chemical synthesis, so consumers prefer high concentrations of these compounds to be fractionated and used in different industrial applications [5].

Due to its benefit, citronella Java has been moved to different countries, and because of problems such as low germination percentage, different researchers have modified its DNA, seeking to reduce these problems and get more oil yield from the plant [11]. Essential oils obtained from genetically modified plants may vary the concentration of some compounds compared with Java oil from unmodified plants, generating advantages for crops that are modified, because they theoretically have higher yields, improved adaptability to the environment, and increased composition of value-added compounds in the produced essential oil.

## 8.2.2  Uses

Citronella essential oils are used in Chinese and Brazilian medicine. Chinese medicine implements this oil to provide massages and treat rheumatoid problems. It is also used as a painkiller and treatment for women with menstrual problems [12]. Citronella oil has been used for the treatment of parasites so that they can be expelled from the body, to treat muscle

**TABLE 8.3**

Chemical and Physical Characteristics of Citronella Oil

| Guideline 151B-17 | Characteristics/Description |
|---|---|
| Color | Light yellow/yellowish-brown |
| Physical state | Liquid |
| Odor | Sweet–floral/grassy/camphoraceous |
| Melting point | Not applicable |
| Boiling point | 170°C |
| Density | 0.891–0.901 (at 25°C) |
| Solubility | Very soluble in water at 20°C |
| Vapor pressure (major components) | Camphene 3.0/limonene 1.4/geraniol 0.02 Citronellal 0.23/citronellol 0.015 |
| Flammability | Flash point at 170°C (TCC)[a] |
| Storage stability | Stable under normal conditions |
| Viscosity | Not known |
| Miscibility | Not to be diluted with petroleum solvents |
| Corrosion characteristic | Noncorrosive |
| Octanol–water partition coefficient | Very large, because of high solubility in octanol |

[a] U.S. Environmental Protection Agency, "Re-registration eligibility decision oil of citronella list C case 3105."

spasms, and as a diuretic in people with liquid retention [13]. Similarly, Brazilian medicine has employed it as an anxiolytic and analgesic, but its greatest application is as an anticonvulsant [10]. The fact that some people do not respond to conventional treatments with drugs obtained by chemical synthesis, and knowing that, in many cases, these drugs cause serious side effects, prompted natural medicine doctors to test the infusion of the leaves of citronella Java for the treatment of epileptic patients [10]. This treatment resulted in eyelid ptosis, ataxia, analgesic sedation, and reduction of motor activity. With this purpose, the depressant action of the active compounds of the essential oil on the central nervous system of the individual was demonstrated, by improving chronic conditions of the disease in the patient [10]. Because of its high amount of monoterpenes, citronella essential oil has antitumor activity, which can be used for the treatment of cancerous tumors in their stages of initiation and promotion or progression. It can also be used in patients with advanced stages of this disease. Citronella oil has geraniol among its main components, which is highly used in the food industry. Although this substance is needed in low amounts (45 ppm) [12], it is used as a flavoring component of black currant, melon, red apple, orange, lemon, pineapple, watermelon, and blueberry [9]. Geraniol has a smell of roses and is colorless; on the other hand, citronellal also has a pleasant smell, although it is less refined than the aroma of geraniol. Therefore, geraniol has gained a strategic position in the synthesis of fragrances for the perfume industry in the world [9]. For its sweet aromas, it is also used in soap, candles, and even the cosmetics and pharmaceuticals industry [10]. Citronella oil has a repellence property, so it is used in the pesticide industry, but this will be shown in more detail in the next section. Table 8.3 shows the chemical and physical characteristics of citronella oil.

## 8.2.3 Market

Information on citronella oil, such as its production, cultivated area, price, and market, is not easily available. Indonesia and China are the major producers of citronella oil, with more

**TABLE 8.4**

U.S. Imports and Prices of Citronella Oil, 2013

| Country | Tonnes | USD/kg |
|---|---|---|
| Indonesia | 197.2 | 18 |
| China | 63.8 | 15 |
| France | 15.3 | 21 |
| Sri Lanka | 11.6 | 15 |
| Spain | 3.8 | 30 |
| India | 3 | 12 |
| United Kingdom | 1.2 | 28 |
| Taiwan | 3.1 | 10 |
| Total | 298.9 | 18.5 |

*Source:* Adapted from Market Insider, Essential oils and oleoresins, 2014.

than 40% of the world's production. Other producers are Taiwan, Guatemala, Sri Lanka, Argentina, Honduras, Ecuador, Jamaica, Madagascar, Mexico, and Brazil. The leading exporters of citronella oil are Indonesia and China, followed by Sri Lanka, India, Taiwan, and European countries such as France and the United Kingdom [4]. The United States is the world's largest importer of citronella oil, and Table 8.4 indicates the reported values of the imports and prices in 2013. Other countries, such as France, the United Kingdom, Germany, the Netherlands, Japan, and Hong Kong, are also importers. European countries prefer Java-type oil due to the presence of the world-famous perfumery industry [4,14].

## 8.3 Citronella Oil Used as a Pesticide

Rural areas are in constant search for solutions to control pests in crops and animals, becoming a firsthand necessity. Thus, synthetic pesticide consumption has increased in recent years to avoid the loss of products; however, excessive use generates resistance to the plague. About 2.5 million tonnes of pesticides are used worldwide on crops each year, causing great damage to soils. These damages are associated with high toxicity and low biodegradability [15]. Also, synthetic pesticides and plagues can be the cause of environmental problems and the deterioration of human health [13,16]. These reasons have forced the pesticides industry to look for environmentally friendly alternatives, and that is how "green pesticides" were born [15,17]. The concept refers to all types of plant materials that contribute to reducing pests and increasing the sustainability of crops. They are safe for application and consumption. Substances such as plant extracts, toxins of organic origin, hormones, and pheromones are considered green pesticides, and their properties are comparable to, or even better than, those of synthetics [17]. For example, essential oil repellents tend to have an effective short life that depends on volatility [17].

Essential oils are a complex mixture of volatile organic compounds and by-products of plant metabolism (secondary metabolites) that have strong aromatic compounds, and odor and flavor characteristics. Essential oils are contained in the cavities of the plant cell wall (leaves, roots, fruits, stem, etc.) in the form of liquid droplets, and the presence of these compounds in the plants allows their use as defense materials, and in attracting or repelling insects.

For example, the U.S. Environmental Protection Agency (USEPA) has registered citronella, lemon and eucalyptus oils as insect repellent ingredients for application on the skin. These oils have the characteristics of relatively low toxicity and comparable efficacy [17]. Citronella (*C. nardus*) essential oil has been used for more than 50 years as an insect repellent. The larvicidal activity of citronella oil has been mainly attributed to the high content of bioactive compounds such as citronellal, citronellol, and geraniol [15]. Some applications of citronella oil as a pesticide are shown in the following sections.

### 8.3.1 Ovicidal Potential

There has been research on citronella oil (*C. nardus*) applied for the biological control of mosquitoes, such as *Aedes aegypti* Linn, *Culex quinquefasciatus* Say, and *Helicoverpa armigera* Hubner [18–20]. According to Warikoo et al. [18], 1% of pure oil diluted in water has the capacity to avoid 100% the deposition of eggs in places where it has been applied, and the appearance of larvae is zero. Mosquitoes can perceive different chemical signals through the sensory receptors present on their antennas and select or reject specific oviposition sites. The full deterrent that applying citronella oil against mosquitoes offers is a major improvement over new prospects for pest and vector management. In this sense, citronella oil interrupts the cycle life, since it inhibits mosquito breeding and promotes the control of pests. According to Ramar et al. [19], citronella oil was tested for ovicidal activity against *C. quinquefasciatus* Say (commonly known as the southern house mosquito) at two different concentrations, 12.5 and 200 ppm, with the oil registering 30% and 83.75% of ovicidal activity, respectively. The dose-dependent oviposition activity of citronella oil against *C. quinquefasciatus* is 7.7% and 79.6% at 12.5 and 200 ppm, respectively. Setiawati et al. [20] reported that a 4000 ppm concentration of citronella oil reduced egg laying of *H. armigera* by 53%–66% on chili peppers.

### 8.3.2 Repellent Potential

*Cymbopogon* has been traditionally used in tropical regions as a repellent of mosquitoes. Extracts and essential oils of these plants have presented a good repellent effect against different kinds of arthropods [17]. *C. winterianus* oil, mixed with 5% vanillin, gave 100% protection for 6 h against *Ae. aegypti*, *C. quinquefasciatus*, and *Anopheles dirus*, compared with the results observed for 25% DEET (*N,N*-diethyl-3-methylbenzanmide) [21]. *C. citratus* has been formulated successfully in a liquid paraffin solution [22].

Citronella oil is marketed in different concentrations and presentations to be used as a natural repellent. Oil has been researched as an excellent repellent of mosquitoes through impregnation on skin, candles, or incense, producing an insecticidal effect against some arthropods [23–25]. The mechanism of repellency for citronella oil against arthropods is not clearly known [23]. Some authors report that citronella oil is harmless, but others have published that it can cause allergic reactions on the skin (irritancy) and has problems due to its rapid volatility [25,26]. Some examples of the repellent effect of citronella oil are

- *Triatoma rubida* (Uhler), *Triatoma protracta* (Uhler), and *Triatoma recurva* (Stal) are hematophagous insects that produce severe allergic reactions and possess the potential to transmit the blood parasite *Trypanosoma cruzi* [23]. Zamora et al. [23] evaluated the main components of citronella oil (geraniol, citronellol, limonene, and citronellal) and found them to be an excellent deterrent of the feeding of *T. rubida* on a restrained mouse. They determined that all components have some inhibition of feeding (from a mild inhibition in the case of limonene to a considerable inhibition in the case of geraniol and citronellol).

- As a volatile repellent, the effect is achieved through lighting a candle when an uncontrolled release of oil is given in the environment, generating concentration gradients. This method is not the most suitable, because in the majority of cases, it does not reach the concentration in which the oil produces repellency. Otherwise, if the candle releases high amounts of oil, the repellency effect lasts a short time [16].

The storage of agricultural products is an important stage in their production and commercialization. Insects and pests changing the properties of products, such as nutritional value, weight, quality, and hygiene, can affect this stage. Thus, the products lose economic value in the market. Citronella oil is presented as an alternative to end this type of problem. For example, the oil has repellent action against *Callosobruchus maculatus*, considered the most important plague for the cowpea. At a concentration of 622 ppm of *C. winterianus*, a 100% reduction of *C. maculatus* in cowpea grains is achieved [27]. However, the addition of botanical oils can change the organoleptic properties of products. At this point, the quality and toxicity of essential oil play an important role.

### 8.3.3 Insecticidal Activity

In tropical countries, the livestock sector faces problems related to the proliferation of ticks in the animals, causing weight loss and thus economic losses. The use of synthetic insecticides has caused some species of ticks to generate immunity to the conventional chemicals and also the contamination of meat and dairy products, affecting human health [28]. Citronella oil has insecticidal effects for some species of ticks [7,28]. According to Singh et al. [7], the aqueous and ethanolic extracts of leaves of *C. winterianus* are assessed for their acaricidal activity against the larvae of deltamethrin (synthetic acaricidal)-resistant *Hyalomma anatolicum*. As a result, this work obtained that "the ethanolic extracts produced a concentration dependent increase in larval tick mortality, whereas the aqueous extracts exhibited a much lower mortality. The highest mortality ($93.7 \pm 0.66\%$) was observed at the 5.0% concentration of ethanolic extract of leaves of *C. winterianus*" [28]. Also, Singh et al. [28] reported that "the acaricidal activity of aqueous and ethanolic extracts of leaves of *C. winterianus* against the SP resistant engorged females of *Rhipicephalus* (*Boophilus*) *microplus* is evaluated. A high activity was found with the ethanolic extract of leaves of *C. winterianus* with $LC_{50}$ (95%CL) values of 0.46% (0.35– 0.59%). The results of this study indicate that the extract can be used for the control of SP resistant ticks" [7]. Torres et al. (2012) show a work where "the influence of *C. winterianus* fractionation on acaricidal activity against the cattle tick *R. (B.) microplus* is studied. The oil is fractionated by vacuum distillation yielding fractions. The obtained results indicate that fractions 4 (100°C–125°C) and 5 (>125°C) of the *C. winterianus* essential oil are the most active, showing $LC_{50}$ values of 1.20 and 1.34 µl/ml, respectively. The $LC_{50}$ of the total oil is 3.30 µl/ml, while the effect of fractions 1–3 is less pronounced, with $LC_{50}$ values of 4.37, 4.24 and 3.49 µl/ml, respectively" [29].

### 8.3.4 Antifungal Activity

The necessity to avoid the fungal contamination of industrial and agricultural products has generated the search for alternatives to conventional processes such as temperature control, ultraviolet irradiation, and dehumidification. These alternatives are focused to develop nontoxic, environmentally friendly, and natural fungicides. For example, the antifungal activity of citronella oil on *Aspergillus niger* conidia is determined, and the experimental results indicate that "the citronella oil has strong antifungal activity: 0.125 (v/v) and 0.25% (v/v) citronella oil inhibited the growth of $5 \times 10^5$ spore/ml conidia separately for 7 and 28 days, while 0.5% (v/v) citronella oil could completely kill the conidia of $5 \times 10^5$ spore/ml. Moreover, the fungicidal kinetic curves revealed that more than 90% of the conidia (initial concentration is $5 \times 10^5$ spore/ml) was killed in all the treatments with 0.125 to 2% citronella oil after 24 h" [30]. Billerbeck et al. [31]) reported that essential oil of *C. nardus* at a concentration of 400 mg/L caused growth inhibition of 80% after 4 days of incubation of *A. niger*. Chen et al. [32] studied the antifungal activity of citronella oil against postharvest *Alternaria alternata* in cherry tomato. "The results indicate that citronella oil possesses strong antifungal activity against *A. alternata* in vitro and in vivo. For in vivo culture, the most effective dosage of the oil was 1.5 μl/ml, with 52% reduction, and the oil had no negative effect on fruit quality. Citronella oil could be a promising natural product for use as an anti-*A. alternata* agent to control black rot in cherry tomato" [32].

### 8.3.5 Antimicrobial Activity

Some works report studies where the antimicrobial activity of citronella oil is analyzed. Thus, Oussalah et al. [33] indicated that *C. nardus* and *C. winterianus* (herb grass) showed a high antimicrobial activity at concentrations of 4 and >8 mg/ml respectively, against *Pseudomonas putida* CRDAV 372 isolated from fresh beef [33]. According to Wei and Wee [34], "the antimicrobial activity values of the *C. nardus* ranged from 0.244 μg/ml to 0.977 μg/ml when tested against *Edwardsiella* spp., *Vibrio* spp., *Aeromonas* spp., *Escherichia coli*, *Salmonella* spp., *Flavobacterium* spp., *Pseudomonas* spp. and *Streptococcus* spp. isolated from internal organs of aquatic animals" [34]. Luangnarumitchai et al. [35] presents for "*C. nardus* a value of antibacterial activity of approximately 18 mm, represented as inhibition zone, against strains of *Propionibacterium acnes* that plays an important role in the pathogenesis of acne inflammation" [35].

### 8.4 Citronella Oil Composition

The experimental analysis of citronella oil is generally carried out by methods such as gas chromatography (GC) coupled to mass spectrometry (MS) [3,6,9,10,12,27,34,36], gas–liquid chromatography (GLC) [1], and gas chromatography [37]. Some authors report the chemical composition of citronella oil as indicated in Table 8.5.

### 8.5 Toxicity

#### 8.5.1 Mammalian Toxicity

Table 8.6 shows the results for a toxicological analysis of Ceylon- and Java-type oils applied to mammals, supported in the Registration Eligibility Decision (RED) test [38]. According

**TABLE 8.5**

Chemical Composition of Citronella Oil

| Component (%) | Java [39] | Java [3] | Java [12] | Ceylon [12] | Ceylon [34] | Java [6] | Java [27] | Java [10] | Java [9] | Java [37] | Java [36] |
|---|---|---|---|---|---|---|---|---|---|---|---|
| Limonene | 8.02 | 2.2 | 1.3 | 9.7 | 2.7 | 3.0 | 3.90 | 1.58 | 3.41 | 2.6 | 2.15 |
| Citronellyl acetate | 10.22 | 2.5 | 3.0 | 1.9 | – | 3.5 | 2.51 | 0.79 | 4.58 | 4.5 | 4.41 |
| Citronellal | 11.05 | 26.5 | 32.7 | 5.2 | 29.6 | 36.1 | 35.47 | 27.44 | 40.23 | 28.8 | 35.28 |
| Citronellol | 14.61 | 7.3 | 15.9 | 8.4 | 4.8 | – | 10.94 | 10.45 | 13.39 | 9.4 | 10.93 |
| Geraniol | 15.40 | 16.2 | 23.9 | 18.0 | – | 19.9 | 21.83 | 40.06 | 17.70 | 17.6 | 21.99 |
| Linalool | 18.55 | 0.7 | 1.5 | 1.2 | – | 0.1 | 1.15 | 0.88 | 0.97 | 0.7 | 1.61 |
| Geranial | 19.48 | 0.7 | – | – | – | 0.6 | 0.50 | 8.05 | 1.13 | – | 1.22 |
| Isopulegol | 24.13 | – | – | – | – | 0.1 | 1.22 | 0.60 | – | – | – |
| Neryl acetate | 24.90 | 0.1 | – | 0.3 | – | – | 0.03 | – | – | – | – |
| Myrcene | – | 3.3 | – | – | – | 0.1 | 0.07 | – | – | – | – |
| (E)-β-Ocimene | – | 0.7 | – | – | – | – | – | 0.09 | – | – | – |
| allo-Ocimene | – | 0.2 | – | – | – | – | – | – | – | – | – |
| (E)-Isocitral | – | 0.2 | – | – | – | – | – | – | – | – | – |
| Nerol | – | 0.4 | 7.7 | 0.9 | – | 0.3 | – | – | – | 0.3 | – |
| Neral | – | 0.5 | – | – | – | 0.4 | 0.33 | 6.02 | – | – | – |
| Geranyl acetate | – | 3.4 | – | – | – | 3.8 | – | 1.77 | 4.67 | 6.3 | 4.52 |
| β-Elemene | – | 4.4 | – | – | 3.3 | 1.6 | 1.67 | – | 2.71 | – | 2.83 |
| β-Ylangene | – | 0.3 | – | – | – | – | – | – | – | – | – |
| β-Gurjunene | – | 0.2 | – | – | – | – | – | – | – | – | – |
| Aromadendrene | – | 0.1 | – | – | – | – | – | – | – | – | – |
| Neryl Propanoate | – | 0.1 | – | – | – | – | – | – | – | – | – |
| α-Humulene | – | 0.1 | – | – | – | 0.1 | 0.11 | – | – | – | – |
| cis-Cadina 1,(6),4-diene | – | 0.1 | – | – | – | – | 0.05 | – | – | – | – |
| cis-Muurola-4,(14),5-diene | – | 0.1 | – | – | – | – | – | – | – | – | – |
| γ-Muurolene | – | 0.1 | – | – | – | – | 0.14 | – | – | – | – |
| Germacrene D | – | 1.1 | – | – | 2.3 | 2.6 | 1.93 | – | – | – | – |
| trans-Muurola-4,(14),5-diene | – | 0.1 | – | – | – | – | 0.05 | – | – | – | – |
| Viridiflorene | – | 0.1 | – | – | – | – | – | – | – | – | – |

(Continued)

**TABLE 8.5 (CONTINUED)**

Chemical Composition of Citronella Oil

| Component (%) | Java [39] | Java [3] | Java [12] | Ceylon [12] | Ceylon [34] | Java [6] | Java [27] | Java [10] | Java [9] | Java [37] | Java [36] |
|---|---|---|---|---|---|---|---|---|---|---|---|
| γ-Cardinene | — | 0.4 | — | — | — | — | — | — | — | — | — |
| α-Muurolene | — | 0.4 | — | — | — | 0.4 | 0.45 | — | — | — | — |
| δ-Cadinene | — | 2.5 | — | — | 1.8 | 1.9 | — | — | 2.87 | — | 2.80 |
| Zonarene | — | 0.1 | — | — | — | — | — | — | — | — | — |
| α-Cadinene | — | 0.1 | — | — | — | — | 0.08 | — | — | — | — |
| Elemol | — | 14.5 | 6.0 | 1.7 | — | 5.8 | 3.73 | — | 4.77 | 12.3 | 4.62 |
| 10-epi-γ-Eudesmol | — | 0.1 | — | — | — | 0.2 | 0.07 | — | — | — | — |
| 1-epi-Cubenol | — | 0.1 | — | — | — | — | 0.05 | — | — | — | — |
| γ-Eudesmol | — | 0.8 | — | — | — | — | 0.58 | — | — | — | — |
| epi-α-Cadinol | — | 0.5 | — | — | — | — | — | — | — | — | — |
| epi-α-Muurolol | — | 0.7 | — | — | — | 1.0 | 0.86 | — | — | — | — |
| β-Eudesmol | — | 0.2 | — | — | — | — | 0.33 | — | — | — | — |
| α-Eudesmol | — | 0.2 | — | — | — | 1.6 | — | — | — | — | — |
| α-Cadinol | — | 2.7 | — | — | — | — | 1.61 | — | — | — | — |
| (2E,6Z)-Farnesol | — | 0.2 | — | — | — | — | — | — | — | — | — |
| Methyl heptenone | — | — | Traces | 0.2 | — | — | 0.06 | — | — | — | — |
| Bourbonene | — | — | Traces | 1.0 | — | — | — | — | — | — | — |
| Linalyl acetate | — | — | 2.0 | 0.8 | — | — | — | — | — | 0.3 | — |
| β-Caryophyllene | — | — | 2.1 | 3.2 | — | 0.1 | 0.11 | — | — | 1.0 | — |
| Geranyl formate | — | — | 2.5 | 4.2 | — | — | — | — | — | 1.5 | — |
| Citronellol butyrate | — | — | Traces | Traces | — | — | — | — | — | — | — |
| Methyl eugenol | — | — | Traces | 1.7 | — | — | — | — | — | — | — |
| Methyl isoeugenol | — | — | 2.3 | 7.2 | — | — | — | — | — | — | — |
| Farnesol | — | — | 0.6 | Traces | — | 0.2 | — | — | — | — | — |
| Tricyclene | — | — | — | 1.6 | — | — | — | — | — | — | — |
| α-Pinene | — | — | — | 2.6 | — | — | 0.01 | — | — | — | — |
| Camphene | — | — | — | 8.0 | — | — | — | — | — | — | — |
| β-Pinene | — | — | — | Traces | — | — | — | — | — | — | — |

*(Continued)*

**TABLE 8.5 (CONTINUED)**

Chemical Composition of Citronella Oil

| Component (%) | Java [39] | Java [3] | Java [12] | Ceylon [12] | Ceylon [34] | Java [6] | Java [27] | Java [10] | Java [9] | Java [37] | Java [36] |
|---|---|---|---|---|---|---|---|---|---|---|---|
| Sabinene | — | — | — | Traces | — | — | — | — | — | — | — |
| Car-3-ene | — | — | — | Traces | — | — | — | — | — | — | — |
| α-Phellandrene | — | — | — | 0.8 | — | — | 0.02 | — | — | — | — |
| cis-Ocimene | — | — | — | 1.4 | — | — | — | — | — | — | — |
| trans-Ocimene | — | — | — | 1.8 | — | — | — | — | — | — | — |
| p-Cymene | — | — | — | Traces | — | — | — | — | — | — | — |
| Terpinolene | — | — | — | 0.7 | — | 0.1 | 0.07 | — | — | — | — |
| 1-Hexanol | — | — | — | 0.1 | — | — | — | — | — | — | — |
| Camphor | — | — | — | 0.5 | — | — | — | — | — | — | — |
| α-Terpineol | — | — | — | Traces | — | 0.1 | 0.06 | — | — | — | — |
| 4-Terpineol | — | — | — | Traces | — | 0.1 | 0.06 | — | — | — | — |
| Menthol | — | — | — | Traces | — | — | 0.21 | — | — | — | — |
| 1-Borneol | — | — | Traces | 6.6 | — | — | — | — | — | — | — |
| Geranyl butyrate | — | — | — | 1.5 | — | — | — | — | — | — | — |
| Nerolidol | — | — | — | 0.3 | — | — | — | — | — | — | — |
| 2,6-Octadienal, 3,7-dimethyl-, (E)- | — | — | — | — | 11 | — | — | — | — | — | — |
| cis-2,6-Dimethyl-2,6-octadiene | — | — | — | — | 6.9 | — | — | — | — | — | — |
| Propanoic acid, 2-methyl-, 3,7-dimethyl-2,6-octadienyl ester, (E)- | — | — | — | — | 6.9 | — | — | — | — | — | — |
| Caryophyllene | — | — | — | — | 6.5 | — | 0.55 | — | — | — | — |
| Phenol, 2-methoxy-3-(2-propenyl)- | — | — | — | — | 4.5 | — | — | — | — | — | — |
| 2,6-Octadien-1-ol, 3,7-dimethyl-, (E)- | — | — | — | — | 2.4 | — | — | — | — | — | — |
| 2,6-Octadiene, 2,6-dimethyl- | — | — | — | — | 1.6 | — | — | — | — | — | — |
| Eugenol | — | — | — | — | 1.5 | — | 0.82 | — | — | — | — |
| 3,7-Cyclodecadiene-1-methanol, α,α,4,8-tetramethyl-, [s-(z,z)] | — | — | — | — | 1.3 | — | — | — | — | — | — |
| Cyclohexane, 1-ethenyl-1-methyl-2,4-bis(1-methylethenyl)-,[1S-(1α,2α,4α)]- | — | — | — | — | 1.3 | — | — | — | — | — | — |

*(Continued)*

**TABLE 8.5 (CONTINUED)**

Chemical Composition of Citronella Oil

| Component (%) | Java [39] | Java [3] | Java [12] | Ceylon [12] | Ceylon [34] | Java [6] | Java [27] | Java [10] | Java [9] | Java [37] | Java [36] |
|---|---|---|---|---|---|---|---|---|---|---|---|
| Cyclohexanemethanol, 4-ethenyl-α,α,4-trimethyl-3-(1-methylethenyl)-,[1R-(1α,3α,4α)]- | – | – | – | – | 1.3 | – | – | – | – | – | – |
| 2,6-Octadien-1-ol,3,7-dimethyl-,acetate,(E)- | – | – | – | – | 1.2 | – | – | – | – | – | – |
| α-Caryophyllene | – | – | – | – | 0.3 | – | – | – | – | – | – |
| Naphthalene, 1,2,4α,5,6,8α-hexahydro-4,7-dimethyl-1-(1-methylethyl)-,(1α,4αα,8αα)- | – | – | – | – | 1.1 | – | – | – | – | – | – |
| Naphthalene, 1,2,3,4,4α,5,6,8α-octahydro-7-methyl-4-methylene-1-(1-methylethyl)-, (1α, 4αα, 8aa)- | – | – | – | – | 0.6 | – | – | – | – | – | – |
| 2-Furanmethanol,5-ethenyltetrahydro-α,α-5-trimethyl-, cis- | – | – | – | – | 0.2 | – | – | – | – | – | – |
| β-Phellandrene | – | – | – | – | – | 0.1 | – | – | – | – | – |
| Decanal | – | – | – | – | – | 0.1 | 0.08 | – | – | – | – |
| β-Citronellol | – | – | – | – | – | 9.9 | – | – | – | – | – |
| β-Bourbonene | – | – | – | – | – | 0.2 | – | – | – | – | – |
| γ-Cadinene | – | – | – | – | – | 0.4 | 0.36 | 0.23 | – | – | – |
| Germacrene D-4-ol | – | – | – | – | – | 1.7 | 0.45 | – | – | – | 1.48 |
| O-Cymene | – | – | – | – | – | – | 0.02 | – | – | – | – |
| (Z)-β-Ocymene | – | – | – | – | – | – | – | – | – | – | – |
| Bergamal | – | – | – | – | – | – | 0.05 | 0.10 | – | – | – |
| γ-Terpinene | – | – | – | – | – | – | 0.02 | – | – | – | – |
| (Z)-Rose oxide | – | – | – | – | – | – | 0.03 | – | – | – | – |
| (E)-Rose oxide | – | – | – | – | – | – | 0.01 | – | – | – | – |
| Menthone (iso) | – | – | – | – | – | – | 0.03 | – | – | – | – |
| Isopelugol (neoiso) | – | – | – | – | – | – | 0.08 | – | – | – | – |
| Methyl chavicol | – | – | – | – | – | – | 0.04 | – | – | – | – |
| (E)-Anethole | – | – | – | – | – | – | 0.72 | – | – | – | – |
| Thymol | – | – | – | – | – | – | 0.03 | – | – | – | – |

*(Continued)*

**TABLE 8.5 (CONTINUED)**

Chemical Composition of Citronella Oil

| Component (%) | Java [39] | Java [3] | Java [12] | Ceylon [12] | Ceylon [34] | Java [6] | Java [27] | Java [10] | Java [9] | Java [37] | Java [36] |
|---|---|---|---|---|---|---|---|---|---|---|---|
| α-Copaene | – | – | – | – | – | – | 0.02 | – | – | – | – |
| Geranil acetate | – | – | – | – | – | – | 3.15 | – | – | – | – |
| β-Copaene | – | – | – | – | – | – | 0.03 | – | – | – | – |
| Muurola-3,5-diene | – | – | – | – | – | – | 0.03 | – | – | – | – |
| Dauca-5,8-diene | – | – | – | – | – | – | 0.03 | – | – | – | – |
| β-Selinene | – | – | – | – | – | – | 0.06 | – | – | – | – |
| Cubebol | – | – | – | – | – | – | 0.13 | – | – | – | – |
| Germacrene-A | – | – | – | – | – | – | 0.38 | – | 1.75 | – | – |
| Cadina-1,4-diene | – | – | – | – | – | – | 0.05 | – | – | – | – |
| Eudesmol (5-epi-7-epi-α) | – | – | – | – | – | – | 0.03 | – | – | – | – |
| α-Muurulol | – | – | – | – | – | – | 0.15 | – | – | – | – |
| Bulnesol | – | – | – | – | – | – | 0.16 | – | – | – | – |
| -(Z)-Ocimene | – | – | – | – | – | – | – | 0.19 | – | – | – |
| iso-Isopulegol | – | – | – | – | – | – | – | 0.20 | – | – | – |
| n-Decanal | – | – | – | – | – | – | – | 0.41 | – | – | – |
| 1,6-Germacradien-ol | – | – | – | – | – | – | – | – | 1.8 | – | – |
| Cedrenus | – | – | – | – | – | – | – | – | – | – | 0.76 |

**TABLE 8.6**

Report of Acute Mammalian Toxicity for Citronella Oil

| Guideline | Test Material | Results | Toxicity Category |
|---|---|---|---|
| 152B-10: Acute oral tox. (rat) | Citronella oil 100% (Ceylon) | $LD_{50} > 5000$ mg/kg | IV |
| | Citronella oil 100% (Java) | $LD_{50} > 4380$ mg/kg | III |
| 152B-11: Acute dermal tox. (rabbit) | Citronella oil 100% (Ceylon) | $LD_{50} > 2000$ mg/kg | III |
| | Citronella oil 100% (Java) | $LD_{50} > 2000$ mg/kg | III |
| 152B-12: Acute inhalation (rat) | Citronella oil 100% (Ceylon) | $LC_{50} > 5000$ mg/kg | IV |
| | Citronella oil 100% (Java) | 4 h exposure $LC_{50} >$ 3.1 mg/L | IV |
| 152B-13: Primary eye irritation (rabbit) | Citronella oil 100% (Ceylon) | Irritation cleared in 72 h | III |
| | Citronella oil 100% (Java) | Irritation cleared within 7 days | III |
| 152B-14: Primary dermal irritation (rabbit) | Citronella oil 100% (Ceylon) | Irritation present at 21 days | II |
| | Citronella oil 100% (Java) | All irritation resolved by 48 h Citronella mild irritant | III |
| 152B-15: Dermal sensitization (guinea pig) | Citronella oil 100% (Ceylon) | Sensitizer (Buehler test) | Not applicable |
| | Citronella oil 100% (Java) | Nonsensitizer (Buehler test) | Not applicable |
| 152-16: Hypersensitivity | All products | All incidents must be reported to the agency | |

*Source:* Adapted from USEPA, Registration eligibility decision—Oil of citronella, USEPA, Washington, DC, 1992; USEPA, R.E.D. FACTS oil of citronella, USEPA, Washington, DC, 1997.

*Note:* Categories: I, very highly or highly toxic; II, moderately toxic; III, slightly toxic; IV, practically nontoxic. $LC_{50}$, median lethal concentration. A statistically derived concentration of a substance that can be expected to cause death in 50% of test animals. It is usually expressed as the weight of substance per weight or volume of water, air, or feed (e.g., mg/L, mg/kg, or ppm). $LD_{50}$, median lethal dose. A statistically derived single dose that can be expected to cause death in 50% of the test animals when administered by the indicated route (oral, dermal, or inhalation). It is expressed as a weight of substance per unit weight of animal (e.g., mg/kg).

to the USEPA, the Ceylon-type oil is most appropriate to be qualified in toxicity category III due to dermal irritation, and therefore the products that contain citronella oil must be labeled with precautions [26]. For Java-type oil, there is no presence of dermal irritation in the animals when the test is carried out, as shown in Table 8.5. In the primary eye irritation for both oils, the results are similar (toxicity category III—all irritation cleared within 7 days). As part of the RED test, the oils are reevaluated for the studies of dermal sensitization. The Ceylon-type oil is a sensitizer and Java-type oil is a non-sensitizer. Thus, the USEPA requires additional precautions for Ceylon-type oil about dermal sensitization [26].

## 8.5.2 Ecological Toxicity

Table 8.7 shows the ecological toxicity data of citronella oil to perform an assessment of the environmental effects for the use of oil. Applications such as lotion, candle, and spray

**TABLE 8.7**

Report of Ecological Toxicity for Citronella Oil

| Guideline | Study | Results |
|-----------|-------|---------|
| 154B-6 | Avian acute oral (bobwhite quail) | $LC_{50} > 2250$ mg/kg; practically nontoxic; NOEL 1350 mg/kg |
| 154B-7 | Avian subacute dietary | Waived because of low avian acute toxicity and no mortality observed at upper test limits |
| 154B-8 | Fish toxicity (rainbow trout) | $LC_{50} > 17.3$ mg/L (based on nominal concentration); slightly toxic; minimal exposure to aquatic sources |
| 154B-9 | Invertebrate toxicity (*Daphnia magna*) | $EC_{50} > 24.6$ mg/L (based on nominal concentration); slightly toxic; minimal exposure to aquatic invertebrate species |
| 154B-10 | Nontarget plants | Waived because exposure to nontarget plants will be minimal |
| 154B-11 | Nontarget insects | Waived because exposure to nontarget insects will be minimal |

*Source:* Adapted from USEPA, Registration eligibility decision—Oil of citronella, USEPA, Washington, DC, 1992.
*Note:* NOEL, no observed effect level.

do not represent dangerous exposure situations for avian, aquatic, and nontarget species. For ornamental and dump uses, the citronella oil presents a major exposure potential. Table 8.7 indicates that for avian, aquatic, or insect species, adverse effects are not likely. The USEPA reported that the effectiveness of citronella oil diminishes over time as insect repellent; the reasonable effectiveness lasts for 1–2 h [26].

## 8.6 Extraction of Citronella Oil

Essential oils can be extracted from different parts of a plant, such as peels, flowers, seeds, leaves, and bark. Different extraction methods are used according to botanical use and the state and form of material. The quality of an essential oil depends on the extraction method. The extraction can be carried out by various methods, such as extraction with solvent and distillation; for example, the steam distillation technique has been a method widely used for industrial-scale production. The use of an inappropriate procedure can damage or alter the action of the chemical composition of essential oil, affecting its bio-activity and natural characteristics. Also, the oil can present physical changes, such as discoloration, loss of odor and flavor, and increase of viscosity [40]. These changes are due to loss of components or the presence of solvent residues, and they can affect the economic value of the product in the market.

A brief description of extraction methods is presented in the next sections.

### 8.6.1 Solvent Extraction

#### 8.6.1.1 Conventional Solvents

This method is used for extraction of delicate components that are not tolerant to heat. Various solvents can be used, depending on the plant material and interesting compounds. Ethanol, hexane, water, methanol, and acetone, among others, can be used as extraction solvents [41]. To achieve a good efficiency, the extraction technique should consider parameters such as particle size, solvent-to-feed ratio, solvent characteristics,

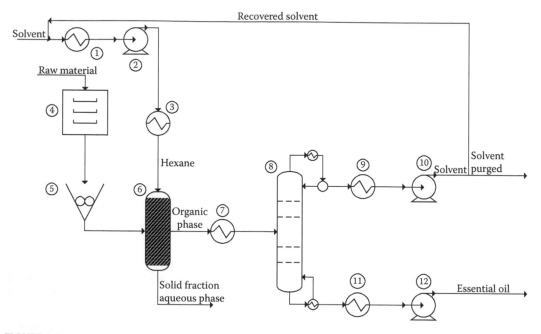

**FIGURE 8.2**
Flowsheet of solvent extraction. 1, heat exchanger; 2, pump; 3, heat exchanger; 4, dryer; 5, grinder; 6, contact container; 7, cooler; 8, recovery column; 9 and 11, heat exchanger; 10 and 12; pump.

and temperature. The method consists of mixing the material with solvent, heating up to extract the essential oil, and filtering to separate the solid from the liquid phase. The liquid contains the organic phase and essential oil. This is concentrated, and the solvent is recovered by evaporation. The obtained resin (waxy mass) is diluted in alcohol (ethanol) to purify the essential oil and remove the unwanted material by distillation at low temperature to avoid the thermal degradation of oil. Figure 8.2 shows the process scheme of solvent extraction (SE). This technique has the disadvantages of requiring a long time in comparison with other methods, the presence of solvent traces in oil that can be toxic, the cost of solvents, and that it might be unfriendly to the environment [40–43].

### 8.6.1.2 Supercritical Carbon Dioxide

Supercritical fluid extraction (SFE) is presented as a novel, environmentally benign, and green technology to obtain natural extracts. This technique is carried out at mild temperatures in the absence of air, avoiding thermal and oxidative degradation of thermolabile components, and therefore, it can be successfully used for the recovery of volatile compounds. The technique consists in taking the $CO_2$ to supercritical conditions and contacting it with the plant material. A stream rich in essential oil mixed with supercritical $CO_2$ is obtained. Then, it is slowly depressurized to prevent loss of product of interest. Figure 8.3 shows the process scheme of SFE. SFE is presented as an alternative to conventional techniques to correct some problems linked to low efficiency, high processing time, and energy consumption. For example, supercritical fluid extraction is more economically viable than steam distillation due to the lower yield and the higher energy consumption that

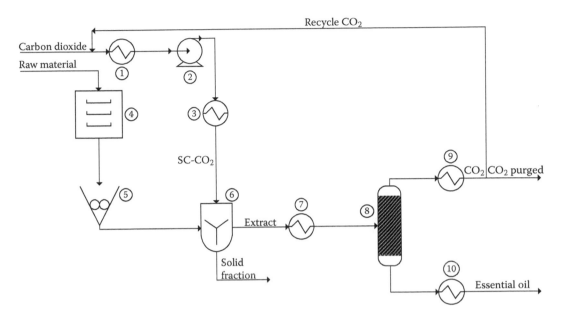

**FIGURE 8.3**
Flowsheet of supercritical fluid extraction. 1, heat exchanger; 2, pump; 3, heat exchanger; 4, dryer; 5, grinder; 6, extraction chamber; 7, heat exchanger; 8, container; 9 and 10, heat exchanger.

the second technique presents [44]. Hydrodistillation (HD) cannot recover some organic compounds that SFE can obtain [40].

Carbon dioxide is the most commonly used supercritical fluid for extraction because it has nontoxic and nonflammable characteristics, and it is an inexpensive solvent [44–46]. $CO_2$ is converted into liquid under high-pressure conditions, and thus generates a safe medium to extract aromatic components from plant material. In the final product, there are no residues of solvent because at ambient temperature and atmospheric pressure, it is gas and evaporates. The extraction efficiency depends on the solubility between interesting products and supercritical fluid. Therefore, the extraction efficiency is low for polar compounds because $CO_2$ is a nonpolar solvent and cannot be solubilized. The solubility of natural compounds in supercritical $CO_2$ is improved through the variation of temperature and pressure of extraction and the addition of polar cosolvents. In supercritical $CO_2$ extraction, ethanol is commonly used as a cosolvent due to its properties of high miscibility in $CO_2$ and low toxicity [4,40,44–47].

### 8.6.1.3 Subcritical Water

This method of extraction is also known as pressurized hot water or superheated water (subcritical water extraction [SWE]), and it is a technique based on using water as an extracting agent at certain temperature and pressure conditions (100°C–374°C, high pressure) to maintain it in the liquid state. Figure 8.4 indicates the flowsheet of subcritical water extraction. SWE is presented as a powerful alternative of extraction by features such as rapid extraction, use of low temperatures, low cost in terms of energy and material, and favorable environmental impact. Subcritical water extraction is a method of high efficiency because, compared with hydrodistillation, it consumes less time and obtains high-quality

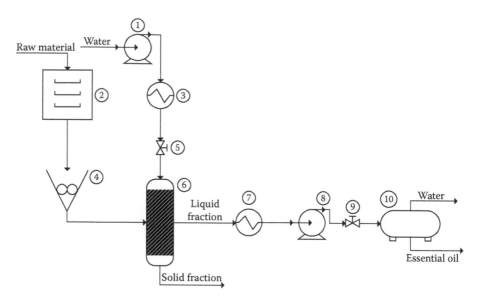

**FIGURE 8.4**
Flowsheet of subcritical water extraction. 1, pump; 2, dryer; 3, heat exchanger; 4, grinder; 5, valve; 6, contact container; 7, heat exchanger; 8, pump; 9, valve; 10, decanter.

and terpene-free essential oils. The work at low temperatures avoids the degradation or loss of volatile and thermolabile compounds [40,48,49].

### 8.6.2 Solvent-Free Microwaves

The conventional methods to extract essential oil have some failures, such as low extraction efficiency, loss of some volatiles, long extraction times, and thermal degradation of compounds. These factors have led to consideration of the use of other techniques, such as solvent-free microwave extraction (SFME), which is based on the combination of microwave heating and dry distillation, and is carried out at atmospheric pressure. The method consists in putting the vegetal material in a microwave reactor. If the plant material is fresh, do not add any water or solvent. Otherwise, the sample must be rehydrated by soaking in water for some time, and then the water excess is drained off. The internal heating of the *in situ* water within the material distends the plant cells and leads to rupture of the glands, releasing the essential oil that is evaporated by the *in situ* water of the material. The distillate is condensed with a cooling system outside the microwave oven and is collected in a receiving vessel. Figure 8.5 indicates the flowsheet of solvent-free microwave extraction. SFME has advantages involving shorter time and higher yields and selectivity, and it is environmentally friendly [40,50–52].

### 8.6.3 Distillation

#### 8.6.3.1 Steam Distillation

The essential oil extraction by steam distillation is the most widely used method for commercial-scale production, although in some cases, the high costs of installation and use of steam can be limiting. The steam distillation method has advantages over other

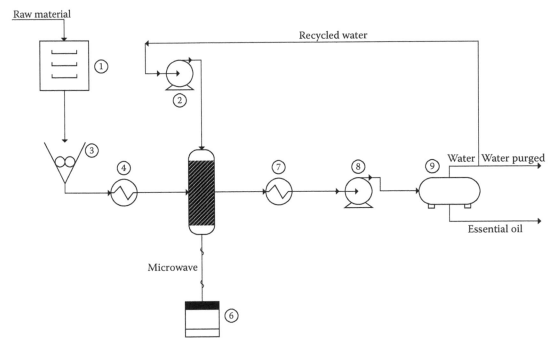

**FIGURE 8.5**
Flowsheet of solvent-free microwave extraction. 1, dryer; 2, pump; 3, grinder; 4, heat exchanger; 5 and 6, microwave reactor; 7, heat exchanger; 8, pump; 9, decanter.

methods; for example, it is relatively fast to operate at a basic level, and the properties of oil are not altered if there is good temperature control. The proportion of essential oils extracted by steam distillation is 93%, and the remaining 7% is extracted by other methods [40,53]. Steam distillation is a special type of separation process that considers the sensitivity of materials to the temperature and their low water solubility (e.g., oil and hydrocarbons). Figure 8.6 shows the process diagram for steam distillation extraction. The method consists in drying the material and reducing particle size to promote high contact between the raw material and steam. Then, the plant sample gets in contact with boiling water or steam, where the hot solvent breaks the cell structure of material, and the aromatic compounds and essential oils are released. The heat supply must be sufficient to break the plant material and vaporize the oil present, but not so high that it can destroy the plant or burn the oil. After the extraction process, steam containing essential oil is rapidly cooled to form two liquid fractions, rich in oil and rich in water, which are separated in a decanter [40,43,53,54]. Masango [53] reported a method to reduce the loss of compounds in wastewater and increase oil yield. "The system is composed of a packed bed of material that is located above the steam source and only steam passes through it and the boiling water is not mixed with material. Therefore, the process requires the minimum amount of steam and the amount of water in the distillate is reduced" [53].

### 8.6.3.2 Hydrodiffusion

The hydrodiffusion method consists in putting the steam in contact with plant material, as described in the steam distillation technique, but the difference is that the steam

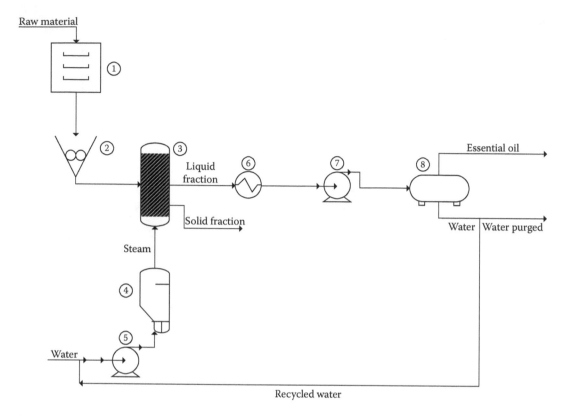

**FIGURE 8.6**

Flowsheet of steam distillation extraction. 1, dryer; 2, grinder; 3, contact container; 4, heat exchanger; 5, pump; 6, cooler; 7, pump; 8, decanter.

inlet is at the top of the container. The technique can operate to low pressure or vacuum, and Figure 8.7 shows the process scheme. This method is used when the material is not degraded at boiling temperature and has been dried and crushed [55]. The hydrodiffusion method has advantages over steam distillation due to a shorter processing time and less steam used, with a higher oil yield [40].

### 8.6.3.3 Hydrodistillation

Also called water distillation, hydrodistillation consists in the complete immersion of the material in water, followed by boiling and condensation in an aqueous fraction of the steam and essential oil vapor. Figure 8.8 shows the flowsheet of the hydrodistillation process. Direct contact, steam jackets, coils, and electric resistances can carry out the heat transfer to a material–water mixture. The material is constantly stirred to avoid deposits on the bottom of the container, and it has a thermal degradation. The material size is considered, in order to generate good contact between material and water and obtain essential oil of quality.

The extraction of essential oils considered in the hydrodistillation technique has become a standard method that is often used to isolate non-water-soluble natural products with a high boiling point [40]. This method has advantages, such as protection against overheating of the extracted oil by the barrier exerted by the surrounding water, that the required

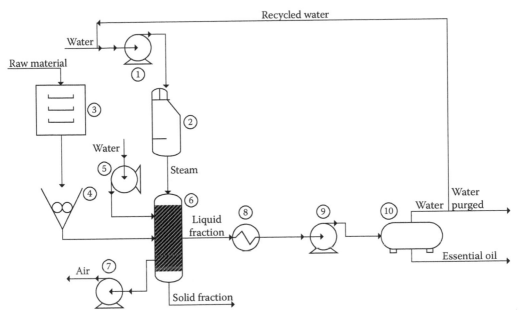

**FIGURE 8.7**
Flowsheet of hydrodiffusion method. 1, pump; 2, heat exchanger; 3, dryer; 4, grinder; 5, pump; 6, contact container; 7, vacuum pump; 8, cooler; 9, pump; 10, decanter.

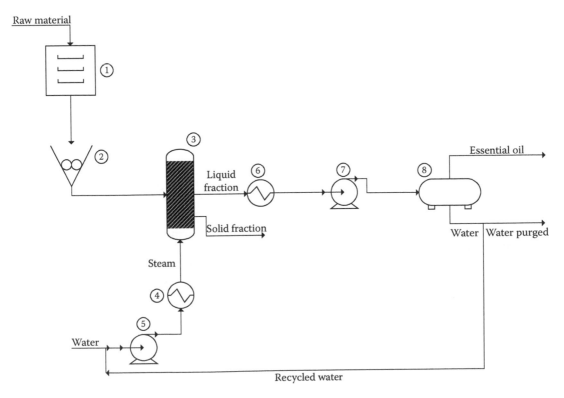

**FIGURE 8.8**
Flowsheet of hydrodistillation process. 1, dryer; 2, grinder; 3, contact container; 4, pump; 5 and 6, heat exchanger; 7, pump; 8, decanter.

material can be distilled below 100°C, and a moderate demand of capital costs. Some problems that the technique presents are the formation of hydrolyzed compounds, the polymerization of aldehydes, and the extraction time [40,45,56].

## 8.7 Analysis of the Citronella Oil Production

The objective of this section is to analyze as an example the feasibility of three extraction technologies of citronella oil cropped in Colombia from technical, economic, and environmental perspectives.

### 8.7.1 Methodology

#### 8.7.1.1 Raw Material

Citronella (*C. winterianus*) is obtained from a farm placed in Chinchiná town, located in the department of Caldas (west center of Colombia), with an average temperature of 21°C and an altitude of 1368 m above sea level.

#### 8.7.1.2 Experimental Extraction of Citronella Essential Oil

An experimental analysis is proposed to determine the chemical characterization of the essential oil and to use this information as a starting point to feed a simulation procedure.

The steps considered in the experimental analysis are (1) drying at 30°C until reaching 10% of moisture, (2) grinding for particle size reduction below 500 µm, (3) extraction by water distillation (2 h), (4) condensation and separation, and (5) storage and sample preparation.

#### 8.7.1.3 Analysis of Essential Oil

Ten microliters of essential oil obtained from citronella is diluted in 1.5 ml of chromatographic-grade hexane and added to a test tube. This mixture is analyzed by mass spectra (Electron Impact [IE] I, 70 eV) obtained with a gas chromatograph (Agilent Technologies 6850 Series II) equipped with a mass-selective detector (MSD 5975B). The injector temperature is operated at 260°C. The chromatographic separation is performed using a HP-INNOWAX capillary column (30 m length, 0.25 mm internal diameter). Helium (99.99%) is used as the carrier gas (split ratio 30:1). The GC oven temperature is programmed from 50°C for 2 min, temperature ramp of 7°C/min to 85°C for 5 min, temperature ramp of 10°C/min to 130°C for 5 min, temperature ramp of 10°C/min to 200°C for 3 min, temperature ramp of 10°C/min to 250°C for 3 min.

#### 8.7.1.4 Process Description

The technologies included in the technoeconomic and environmental assessment of the extraction of essential oil from *Citronella* are supercritical fluid extraction with carbon dioxide, solvent extraction with n-hexane, and extraction by hydrodistillation. The description of each technology is presented in the following sections.

### 8.7.1.5 Supercritical Fluid Extraction

The process to extract essential oil using supercritical fluids begins with the reception of the raw material. First, the material is cut to an average particle size of 5 cm, and subsequently dried at 30°C in order to reach a moisture content of 10% in weight. After this, the solid is grinded to a particle size below 500 µm in order to expose the oily fraction, allowing a proper extraction condition. Subsequently, the reduced size solid is directed to the extraction vessel to pack it. Once the last procedure is done, the carbon dioxide must be prepared to the supercritical conditions, which begins with a cooling process to approximately –30°C in order to avoid pump cavitation in the compression stage. The cooling fluid used for this stage is an ethylene glycol–water mixture (70:30). After this, the $CO_2$ is compressed to 200 bars, and its temperature is increased to 35°C, reaching the supercritical conditions [57]. Later, the supercritical fluid is passed through the extraction vessel previously packed with the solid material. It is very important to consider that temperature and pressure should be kept constant during the extraction stage, because a little perturbation on these variables will notoriously affect the carbon dioxide density and viscosity, and therefore the extraction yield. After the extraction process, two streams are obtained: one including the exhausted solid and another with the essential oil diluted in supercritical $CO_2$. Considering that the extract is obtained at high pressure, it is necessary to depressurize the stream and separate the $CO_2$ from the essential oil. The depressurization is done in two stages, reducing the pressure to 50 bars and gasifying the carbon dioxide at 25°C and 35°C, respectively. The ratio of grams of essential oil per kilogram of $CO_2$ is 29.03. Besides, up to 97.95% of $CO_2$ can be recovered and recycled. However, it is first cooled to –30°C and then pressurized to 200 bars. The power required to compress the carbon dioxide is 21.71 MJ/t of $CO_2$.

### 8.7.1.6 Solvent Extraction

In the case of extraction using solvents, the process begins with the reception of the raw material. First, the material is cut to an average particle size of 5 cm, and subsequently dried at 30°C, in order to reach a moisture content of 10% in weight. After this, the solids are ground to a particle size below 500 µm in order to expose the oily fraction, allowing a proper extraction condition, and then the solid is sent to the extraction column. It is very important to take into account that important conditions should be met to conduct a proper extraction process, such as particle size, solvent-to-feed ratio, and temperature. The solvent-to-feed ratio (dry basis) was kept at 4:1, while the extraction temperature was kept at 50°C; therefore, the extracting solvent is prepared to the extraction conditions. Hexane was selected as a solvent because of its selectivity to extract organic compounds [58]. On the other hand, after the extraction process, two streams are obtained: one stream including the solid and the aqueous phase, and a second stream containing the organic phase, which is rich in hexane and essential oil. After the essential oil extraction, the solvent is recovered in a vacuum distillation column with a solvent-rich top stream and a bottom stream rich in essential oil.

### 8.7.1.7 Hydrodistillation

The process includes the reception, material cutting to an average particle size of 5 cm, drying at 30°C in order to reach a moisture content of 10% in weight, and grinding to a particle size below 500 µm in order to expose the oily fraction, allowing a proper extraction condition to pack the solid in the extraction column. After this, steam is generated in a boiler to pass subsequently through the packed column. The water-to-feed ratio is 5:1 (dry basis). In the case of water distillation, the energy required to generate steam uses low-pressure steam (3 bars) and cools water for the other units. After the extraction process, steam containing essential oil is rapidly cooled to form two liquid fractions: an oily rich one and a water-rich one, which are separated in a decanter.

### 8.7.1.8 Simulation Procedure

Simulations of the extraction of essential oil from citronella are carried out using Aspen Plus v8.0. The simulations considered a plant capacity to process 200 kg/h of fresh feedstock. The physical property data for components missing in the Aspen Plus databases, and required in the simulations, are estimated by correlating data available from the National Institute of Standards and Technology (NIST) to the Aspen properties, and missing properties are estimated using the method reported by Marrero and Gani [59], which is suitable for complex molecules. Additional data are obtained from the work of Wooley and Putsche [60] (i.e., hemicellulose and lignin). The Unifac–Dortmund thermodynamic model is used to calculate the activity coefficients in the liquid phase, and the Hayden–O'Connell equation of state is used to model the vapor phase.

### 8.7.1.9 Technoeconomic Assessment

In the economic assessment, the capital and operating costs are calculated using the software Aspen Economic Analyzer (Aspen Technologies, Inc.). However, specific parameters regarding some Colombian conditions, such as raw material costs, income tax (33%), labor salaries, and interest rate (16.02%), among others, are incorporated in order to calculate the production costs per unit of essential oil at the Colombian conditions. The above-mentioned software estimates the capital costs of process units, as well as the operating costs, among other valuable data. This software uses the design information provided by Aspen Plus and data introduced by the user for specific conditions, for instance, project location. Equipment calculations are performed following the Aspen Economic Analyzer v8.0 user guide. Utilities, civil works, pipelines, person-hours, and many different parameters are estimated using the same software. Table 8.8 shows prices used in the economic evaluation.

**TABLE 8.8**

Prices/Costs Used in the Economic Assessment

| Item | Value | Unit |
| --- | --- | --- |
| Citronella[a] | 18 | USD/ton |
| Citronella oil[b] | 10 | USD/kg |
| $CO_2$[b] | 1.55 | USD/kg |
| Hexane[b] | 0.31 | USD/L |
| Water[c] | 1.252 | USD/m³ |
| Electricity[c] | 0.1 | USD/kWh |
| Operator[c] | 2.14 | USD/h |
| Supervisor[c] | 4.29 | USD/h |

[a] Price due to transport charges and average market. Average traveled distance, 100 km; type of truck, 10 t truck; diesel price, 4.11 USD/gal.
[b] Prices based on ICIS pricing indicatives [61].
[c] Colombian national average.

### 8.7.1.10 Environmental Assessment

The waste reduction algorithm (WAR), developed by the National Risk Management Research Laboratory from the USEPA, was used as the method for the calculation of the potential environmental impact (PEI). This method proposes to add a conservation reaction over the PEI, based on the impact of input and output flow rates from the process. The PEI for a given mass or energy quantity could be defined as the effect that those (energy and mass) would have on the environment if they were arbitrarily discharged. The environmental impact is a quantity that cannot be directly measured. However, it can be calculated from different measurable indicators [62,63]. The WAR includes eight categories: human toxicity by ingestion (HTPI), human toxicity by dermal exposition or inhalation (HTPE), terrestrial toxicity potential (TTP), aquatic toxicity potential (ATP), global warming potential (GWP), ozone depletion potential (ODP), photochemical oxidation potential (PCOP), and acidification potential (AP). The weighted sum of all impacts ends in the final impact per mass of products. It is very important to clarify that this environmental assessment only corresponds to the possible impact generated in the processing stage.

### 8.7.2 Results and Discussion

#### 8.7.2.1 Experimental Analysis

In the experimental extraction, a yield of 8.27 g of essential oil per kilogram of citronella is obtained. The relative density of the citronella extract is 0.898 at 25°C, the refraction index is 1.466 at 20°C, and the solubility in ethanol at 80% and 20°C is 1:2 (v/v). The main compounds identified in the citronella essential oil are (mass percentage): citronellal (54.16%), d-limonene (6.49%), citronellyl acetate (3.29%), β-bourbonene (5.60%), β-cubebene (4.54%), β-elemene (2.41%), α-bergamotene (2.19%), β-caryophyllene (2.34%), β-gurjunene (2.85%) and α-selinene (2.02%).

#### 8.7.2.2 Process Simulation

The analysis of process simulation for the extraction of essential oil from citronella is focused on the processing yield, which is defined as kilogram of essential oil per tonne of fresh raw material. In this sense, the extraction yield depends on the technology. The moisture content of citronella is around 60%–70%. Yields for citronella correspond to 10.16, 9.97, and 8.75 kg/t for SFE, SE, and HD, respectively. SFE technology led to a high processing yield, which can affect the energy consumption and total production cost. However, there is not a big difference between SFE and SE. SFE yield is only 1.92% higher than SE, while SFE is 13.88% higher than HD. The yields obtained in simulations are in concordance with typical values reported in literature [57,64,65].

The energy consumption corresponds to 280.33, 244.76, and 303.76 MJ/tonne of fresh citronella, for SFE, SE, and HD, respectively. The technology with the highest energy consumption is HD, because of the energy required to produce the steam required for the extraction. In the case of SFE, the required energy is mainly due to the compression and depressurization process for adapting and separating the carbon dioxide, respectively. In the case of SE, most of the required energy is due to the recovery process of the solvent. It can be inferred that each process has a specific unit that consumes most of the energy.

### 8.7.2.3 Economic Evaluation

The economic evaluation is presented for each technology in Table 8.9. The production costs obtained for SFE, SE, and HD are 7.93, 8.77, and 8.19 USD/kg, respectively. SE technology presents the higher cost due to the energy demand to heat and recover the solvent. In all extraction methods, raw material costs and utilities contribute to the majority of the total costs. Considering the production cost, it is possible to calculate the economic margin for each technology. Thus, values of 20.7%, 12.3%, and 18.1% are obtained for SFE, SE, and HD, respectively. At this point, it is also very important to note that all technologies can be feasible. In this sense, this could serve as a basis to draw recommendations on the selection of technologies to extract essential oils. Furthermore, this selection should consider different aspects, such as the environmental impact. This is significantly relevant since all technologies should be seen not only from the economics perspective, but also from the environmental one. The environmental assessment is presented next.

### 8.7.2.4 Environmental Evaluation

The environmental assessment is based on the criteria of the impacts named in the methodology, which includes the analysis for the different technologies. The results of the potential environmental impact per kilogram of product (essential oil) are presented in Figure 8.9. The results show that the friendliest configuration is SFE, followed by HD. The analysis is very similar to the one presented in the economic evaluation. In this sense, it can be inferred that energy consumption affects the environmental impact, as it directly affects the economics. On the other hand, it is shown that the SE technology shows the highest impact. This is due to the solvent selected for the extraction, because the hexane purge affects impacts such as ATP and PCOP in a most dramatic way compared with technologies using $CO_2$ and water. The toxicity of hexane and its harmful effects significantly

**TABLE 8.9**

Citronella Essential Oil Production Using SFE, SE, and HD as Technologies

| Item | Cost (USD/kg) and Share (%) | SFE | SE | HD |
|---|---|---|---|---|
| Raw materials | Cost | 2.54 | 2.23 | 2.06 |
| | Share | 32.09 | 25.54 | 25.08 |
| Operating labor | Cost | 0.95 | 1.27 | 1.10 |
| | Share | 12.02 | 14.50 | 13.44 |
| Utilities | Cost | 2.60 | 2.72 | 3.30 |
| | Share | 32.89 | 31.10 | 40.25 |
| Operating charges, plant overhead, maintenance | Cost | 0.65 | 0.95 | 0.64 |
| | Share | 8.26 | 10.84 | 7.85 |
| General and administrative cost | Cost | 0.31 | 0.37 | 0.34 |
| | Share | 3.87 | 4.25 | 4.15 |
| Depreciation of capital | Cost | 0.86 | 1.20 | 0.76 |
| | Share | 10.87 | 13.77 | 9.23 |
| Total | Cost | 7.91 | 8.74 | 8.20 |
| | Share | 100.00 | 100.00 | 100.00 |
| Total production cost (USD/kg) | | 7.93 | 8.77 | 8.19 |

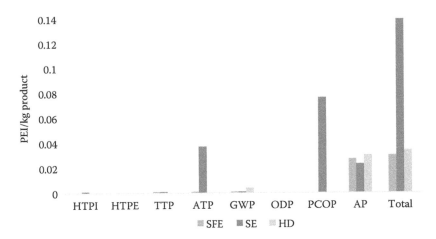

**FIGURE 8.9**
Potential environmental impact of extraction technologies.

increase the impact if it is not properly handled at the processing level. Therefore, it can be inferred that the potentials affected by energy are the GWP and the AP. At this point, it is very important to take into account that total environmental impact is the result of the weighted sum of all categories, and as mentioned in methodology, WAR is used to compare different process configurations.

SFE can be considered a promising technology due to its similarity to HD. Another important aspect to consider in the selection of technologies is the possible impact of traces on the final product. In the case of SFE, the essential oil can be completely separated from the $CO_2$ at normal conditions. The HD leads to a very small amount of water in the final product because of the insolubility of the oil in an aqueous phase. In addition, the water is not considered a substance with high environmental potential impact at normal conditions. Against this, a technology using solvents would lead to traces in the final composition of the essential oil, which can certainly affect the quality of product. In this way, greener solvents should be proposed, or newer technologies following the green engineering and green chemistry rules [66,67], for instance, the use of ionic liquids as green solvents [68].

## 8.8 Conclusions

Citronella oil is considered a good option as an alternative to synthetic pesticides. The chemical composition of oil gives representative characteristics like ovicidal and repellent potential and insecticidal, antifungal, and antimicrobial activity.

When the production of citronella oil is evaluated, the three technologies taken into account from the technoeconomic perspective are feasible. Based on a potential environmental perspective, the most harmful technology is solvent extraction using hexane because of its high toxicity. From both perspectives, the most promising technology is the extraction by supercritical fluids.

## References

1. S. P. S. Khanuja, A. K. Shasany, A. Pawar, R. K. Lal, M. P. Darokar, A. A. Naqvi, S. Rajkumar, V. Sundaresan, N. Lal, and S. Kumar, Essential oil constituents and RAPD markers to establish species relationship in *Cymbopogon* Spreng. (Poaceae), *Biochemical Systematics and Ecology*, vol. 33, no. 2, pp. 171–186, 2005.

2. S. Adhikari, S. Saha, T. K. Bandyopadhyay, and P. Ghosh, Efficiency of ISSR marker for characterization of *Cymbopogon* germplasms and their suitability in molecular barcoding, *Plant Systematics and Evolution*, vol. 301, no. 1, pp. 439–450, 2014.

3. K. Antonio, F. Rodrigues, C. N. Dias, F. Maria, M. Amaral, D. F. C. Moraes, V. E. M. Filho, E. H. A. Andrade, and J. G. S. Maia, Molluscicidal and larvicidal activities and essential oil composition of *Cymbopogon winterianus*, *Pharmaceutical Biology*, vol. 51, no. 10, pp. 1–5, 2013.

4. A. Akhila, ed., *Essential Oil-Bearing Grasses: The Genus* Cymbopogon, CRC Press/Taylor & Francis Goup, 2010, p. 264.

5. A. K. Shasany, R. K. Lal, N. K. Patra, M. P. Darokar, A. Garg, S. Kumar, and S. P. S. Khanuja, Phenotypic and RAPD diversity among *Cymbopogon winterianus* Jowitt accessions in relation to *Cymbopogon nardus* Rendle, *Genetic Resources and Crop Evolution*, vol. 47, pp. 553–559, 2000.

6. D. Lorenzo, E. Dellacassa, A. C. Santos, C. Frizzo, N. Paroul, P. Moyna, L. Mondello, G. Dugo, D. Farmaco-chimico, and V. S. S. Annunziata, Composition and stereoanalysis of *Cymbopogon winterianus* Jowitt oil from southern Brazil, *Flavour and Fragrance Journal*, vol. 15, no. 3, pp. 177–181, 1999.

7. N. K. Singh, Jyoti, B. Vemu, A. Nandi, H. Singh, R. Kumar, and V. K. Dumka, Acaricidal activity of *Cymbopogon winterianus*, *Vitex negundo* and *Withania somnifera* against synthetic pyrethroid resistant *Rhipicephalus (Boophilus) microplus*, *Parasitology Research*, vol. 113, no. 1, pp. 341–350, 2014.

8. S. Fatima, A. H. A. Farooqi, and S. Sharma, Physiological and metabolic responses of different genotypes of *Cymbopogon martinii* and *C. winterianus* to water stress, *Plant Growth Regulation*, vol. 37, pp. 143–150, 2002.

9. S. C. Beneti, E. Rosset, M. L. Corazza, C. D. Frizzo, M. Di, and J. V. Oliveira, Fractionation of citronella (*Cymbopogon winterianus*) essential oil and concentrated orange oil phase by batch vacuum distillation, *Journal of Food Engineering*, vol. 102, no. 4, pp. 348–354, 2011.

10. T. T. Souza, B. S. Leite, N. M. N. Lessa, and L. R. Bonjardim, Phythochemical screening and anticonvulsant activity of *Cymbopogon winterianus* Jowitt (Poaceae) leaf essential oil in rodents, *Phytomedicine*, vol. 15, no. 8, pp. 619–624, 2008.

11. T. Dey, S. Saha, and P. D. Ghosh, Somaclonal variation among somatic embryo derived plants—Evaluation of agronomically important somaclones and detection of genetic changes by RAPD in *Cymbopogon winterianus*, *South African Journal of Botany*, vol. 96, pp. 112–121, 2015.

12. A. Wany, S. Jha, V. K. Nigam, and D. M. Pandey, Chemical analysis and therapeutic uses of citronella oil from *Cymbopogon winterianus*: A short review, *International Journal of Advanced Research*, vol. 1, no. 2320, pp. 504–521, 2013.

13. K. Devi, B. Dehury, M. Phukon, M. K. Modi, and P. Sen, Novel insights into structure-function mechanism and tissue-specific expression profiling of full-length dxr gene from *Cymbopogon winterianus*, *FEBS Open Bio*, vol. 5, pp. 325–334, 2015.

14. Market Insider, Essential oils and oleoresins, 2014.

15. O. Koul, S. Walia, and G. S. Dhaliwal, Essential oils as green pesticides: Potential and constraints, *Biopecticides International*, vol. 4, pp. 63–84, 2008.

16. B. Solomon, F. F. Sahle, T. Gebre-Mariam, K. Asres, and R. H. H. Neubert, Microencapsulation of citronella oil for mosquito-repellent application: Formulation and in vitro permeation studies, *European Journal of Pharmaceutics and Biopharmaceutics*, vol. 80, no. 1, pp. 61–66, 2012.

17. L. S. Nerio, J. Olivero-Verbel, and E. Stashenko, Repellent activity of essential oils: A review, *Bioresource Technology*, vol. 101, no. 1, pp. 372–378, 2010.

18. R. Warikoo, N. Wahab, and S. Kumar, Oviposition-altering and ovicidal potentials of five essential oils against female adults of the dengue vector, *Aedes aegypti* L., *Parasitology Research*, vol. 109, no. 4, pp. 1125–1131, 2011.

19. M. Ramar, S. Ignacimuthu, and M. G. Paulraj, Ovicidal and oviposition response activities of plant volatile oils against *Culex quinquefasciatus* Say, *Journal of Entomology and Zoology Studies*, vol. 2, no. 4, pp. 82–86, 2014.

20. W. Setiawati, R. Murtiningsih, and A. Hasyim, Laboratory and field evaluation of essential oils from *Cymbopogon nardus* as oviposition deterrent and ovicidal activities against *Helicoverpa armigera* Hubner on chili pepper, *Indonesian Journal of Agricultutal Science*, vol. 12, pp. 9–16, 2011.

21. A. Tawatsin, S. D. Wratten, R. R. Scott, U. Thavara, and Y. Techadamrongsin, Repellency of volatile oils from plants against three mosquito vectors, *Journal Vector Ecology*, vol. 26, pp. 76–82, 2001.

22. A. O. Oyedele, A. A. Gbolade, M. B. Sosan, F. B. Adewoyin, and O. L. Soyelu, Formulation of an effective mosquito-repellent topical product from lemongrass oil, *Phytomedicine*, vol. 9, pp. 259–262, 2002.

23. D. Zamora, S. A. Klotz, E. A. Meister, and J. O. Schmidt, Repellency of the components of the essential oil, citronella, to *Triatoma rubida, Triatoma protracta,* and *Triatoma recurva* (Hemiptera: Reduviidae: Triatominae), *Journal of Medical Entomology*, vol. 52, no. 4, pp. 719–721, 2015.

24. G. C. Müller, A. Junnila, V. D. Kravchenko, E. Edita, J. Butler, and Y. Schlein, Indoor protection against mosquito and sand fly bites: A comparison between citronella, linalool, and geraniol candles, *American Mosquito Control Association*, vol. 24, pp. 150–153, 2008.

25. N. P. Yadav, V. K. Rai, N. Mishra, P. Sinha, D. U. Bawankule, A. Pal, A. K. Tripathi, and C. S. Chanotiya, A novel approach for development and characterization of effective mosquito repellent cream formulation containing citronella oil, *BioMed Research Internation*, vol. 2014, pp. 1–12, 2014.

26. USEPA (U.S. Environmental Protection Agency), Registration eligibility decision—Oil of citronella, USEPA, Washington, DC, 1992.

27. N. M. S. Gusmão, J. V. de Oliveira, D. M. do A. F. Navarro, K. A. Dutra, W. A. da Silva, and M. J. A. Wanderley, Contact and fumigant toxicity and repellency of *Eucalyptus citriodora* Hook., *Eucalyptus staigeriana* F., *Cymbopogon winterianus* Jowitt and *Foeniculum vulgare* Mill. essential oils in the management of *Callosobruchus maculatus* (FABR.) (Coleoptera: Chrysomeli), *Journal of Stored Products Research*, vol. 54, pp. 41–47, 2013.

28. N. K. Singh, Jyoti, B. Vemu, A. Nandi, H. Singh, R. Kumar, and V. K. Dumka, Laboratory assessment of acaricidal activity of *Cymbopogon winterianus, Vitex negundo* and *Withania somnifera* extracts against deltamethrin resistant *Hyalomma anatolicum, Experimental & Applied Acarology*, vol. 63, no. 3, pp. 423–430, 2014.

29. F. C. Torres, A. M. Lucas, A. Lucia, S. Ribeiro, R. Martins, G. Von Poser, M. S. Guala, H. V. Elder, and E. Cassel, Influence of essential oil fractionation by vacuum distillation on acaricidal activity against the cattle tick, *Brazilian Archives of Biology and Technology*, vol. 55, pp. 613–621, 2012.

30. W.-R. Li, Q.-S. Shi, Y.-S. Ouyang, Y.-B. Chen, and S.-S. Duan, Antifungal effects of citronella oil against *Aspergillus niger* ATCC 16404, *Applied Microbiology and Biotechnology*, vol. 97, no. 16, pp. 7483–7492, 2013.

31. V. G. De Billerbeck, C. G. Roques, J. Bessière, J. Fonvieille, and R. Dargent, Effects of *Cymbopogon nardus* (L.) W. Watson essential oil on the growth and morphogenesis of *Aspergillus niger*, *Canadian Journal of Microbiology*, vol. 17, pp. 9–17, 2001.

32. Q. Chen, S. Xu, T. Wu, J. Guo, S. Sha, X. Zheng, and T. Yu, Effect of citronella essential oil on the inhibition of postharvest *Alternaria alternata* in cherry tomato, *Journal of the Science of Food and Agriculture*, vol. 94, no. 12, pp. 2441–2447, 2014.

33. M. Oussalah, S. Caillet, L. Saucier, and M. Lacroix, Antimicrobial effects of selected plant essential oils on the growth of a *Pseudomonas putida* strain isolated from meat, *Meat Science*, vol. 73, no. 2, pp. 236–244, 2006.

34. L. S. Wei and W. Wee, Chemical composition and antimicrobial activity of *Cymbopogon nardus* citronella essential oil against systemic bacteria of aquatic animals, *Iranian Journal of Microbiology*, vol. 5, no. 2, pp. 147–152, 2013.

35. S. Luangnarumitchai, S. Lamlertthon, and W. Tiyaboonchai, Antimicrobial activity of essential oils against five strains of *Propionibacterium acnes*, *Journal of Pharmaceutical Sciences*, vol. 34, pp. 60–64, 2007.

36. N. Paroul, L. P. Grzegozeski, V. Chiaradia, and H. Treichel, Solvent-free production of bioflavors by enzymatic esterification of citronella (*Cymbopogon winterianus*) essential oil, *Applied Biochemestry and Biotechnology*, vol. 166, no. 1, pp. 13–21, 2012.

37. A. A. Naqvi, Ł. S. Mandal, A. Chattopadhyay, and A. Prasad, Salt effect on the quality and recovery of essential oil of citronella (*Cymbopogon winterianus* Jowitt), *Flavour and Fragance Journal*, vol. 17, no. 2, pp. 109–110, 2002.

38. USEPA (U.S. Environmental Protection Agency), R.E.D. FACTS oil of citronella, USEPA, Washington, DC, 1997.

39. M. R. Silva, R. M. Ximenes, J. G. M. da Costa, L. K. Leal, A. A. de Lopes, and G. S. D. B. Viana, Comparative anticonvulsant activities of the essential oils (EOs) from *Cymbopogon winterianus* Jowitt and *Cymbopogon citratus* (DC) Stapf. in mice, *Naunyn-Schmiedeberg's Archives of Pharmacology*, vol. 381, no. 5, pp. 415–426, 2010.

40. P. Tongnuanchan and S. Benjakul, Essential oils: Extraction, bioactivities, and their uses for food preservation, *Journal of Food Science*, vol. 79, no. 7, pp. R1231–R1249, 2014.

41. M. Koşar, H. J. D. Dorman, and R. Hiltunen, Effect of an acid treatment on the phytochemical and antioxidant characteristics of extracts from selected Lamiaceae species, *Food Chemistry*, vol. 91, no. 3, pp. 525–533, 2005.

42. X.-M. Li, S.-L. Tian, Z.-C. Pang, J.-Y. Shi, Z.-S. Feng, and Y.-M. Zhang, Extraction of *Cuminum cyminum* essential oil by combination technology of organic solvent with low boiling point and steam distillation, *Food Chemistry*, vol. 115, no. 3, pp. 1114–1119, 2009.

43. J. Moncada, J. A. Tamayo, and C. A. Cardona, Techno-economic and environmental assessment of essential oil extraction from citronella (*Cymbopogon winteriana*) and lemongrass (*Cymbopogon citrus*): A Colombian case to evaluate different extraction technologies, *Industrial Crops and Products*, vol. 54, pp. 175–184, 2014.

44. C. G. Pereira and M. A. A. Meireles, Supercritical fluid extraction of bioactive compounds: Fundamentals, applications and economic perspectives, *Food and Bioprocess Technology*, vol. 3, no. 3, pp. 340–372, 2009.

45. P. Costa, J. M. Loureiro, M. A. Teixeira, and A. E. Rodrigues, Extraction of aromatic volatiles by hydrodistillation and supercritical fluid extraction with $CO_2$ from *Helichrysum italicum* subsp. *picardii* growing in Portugal, *Industrial Crops and Products*, vol. 77, pp. 680–683, 2015.

46. L. T. Danh, P. Truong, R. Mammucari, and N. Foster, Extraction of vetiver essential oil by ethanol-modified supercritical carbon dioxide, *Chemical Engineering Journal*, vol. 165, no. 1, pp. 26–34, 2010.

47. E. Reverchon, Supercritical fluid extraction and fractionation related products, *Journal of Supercritical Fluids*, vol. 10, pp. 1–37, 1997.

48. M. H. Eikani, F. Golmohammad, and S. Rowshanzamir, Subcritical water extraction of essential oils from coriander seeds (*Coriandrum sativum* L.), *Journal of Food Engineering*, vol. 80, no. 2, pp. 735–740, 2007.

49. M. Z. Ozel, F. Gogus, and A. C. Lewis, Subcritical water extraction of essential oils from *Thymbra spicata*, *Food Chemistry*, vol. 82, no. 3, pp. 381–386, 2003.

50. B. Bayramoglu, S. Sahin, and G. Sumnu, Solvent-free microwave extraction of essential oil from oregano, *Journal of Food Engineering*, vol. 88, no. 4, pp. 535–540, 2008.

51. M. E. Lucchesi, F. Chemat, and J. Smadja, Solvent-free microwave extraction of essential oil from aromatic herbs: Comparison with conventional hydro-distillation, *Journal of Chromatography A*, vol. 1043, no. 2, pp. 323–327, 2004.

52. M. E. Lucchesi, J. Smadja, S. Bradshaw, W. Louw, and F. Chemat, Solvent free microwave extraction of *Elletaria cardamomum* L.: A multivariate study of a new technique for the extraction of essential oil, *Journal of Food Engineering*, vol. 79, no. 3, pp. 1079–1086, 2007.

53. P. Masango, Cleaner production of essential oils by steam distillation, *Journal of Cleaner Production*, vol. 13, no. 8, pp. 833–839, 2005.

54. H. C. Baser and G. Buchbauer, *Handbook of Essential Oils: Science, Technology, and Applications*, CRC Press/Taylor & Francis Goup, Boca Raton, FL, 2010, p. 994.

55. M. A. Vian, X. Fernandez, F. Visinoni, and F. Chemat, Microwave hydrodiffusion and gravity, a new technique for extraction of essential oils, *Journal of Chromatography*, vol. 1190, no. 1–2, pp. 14–17, 2008.

56. O. O. Okoh, A. P. Sadimenko, and A. J. Afolayan, Comparative evaluation of the antibacterial activities of the essential oils of *Rosmarinus officinalis* L. obtained by hydrodistillation and solvent free microwave extraction methods, *Food Chemistry*, vol. 120, no. 1, pp. 308–312, 2010.

57. M. Khajeh, M. G. Moghaddam, and M. Shakeri, Application of artificial neural network in predicting the extraction yield of essential oils of *Diplotaenia cachrydifolia* by supercritical fluid extraction, *Journal of Supercritical Fluids*, vol. 69, no. 2010, pp. 91–96, 2012.

58. S. M. Fakhr Hoseini, T. Tavakkoli, and M. S. Hatamipour, Extraction of aromatic hydrocarbons from lube oil using n-hexane as a co-solvent, *Separation and Purification Technology*, vol. 66, no. 1, pp. 167–170, 2009.

59. J. Marrero and R. Gani, Group-contribution based estimation of pure component properties, *Fluid Phase Equilibria*, vol. 183, pp. 183–208, 2001.

60. R. J. Wooley and V. Putsche, Development of an ASPEN PLUS physical property database for biofuels components, 1996.

61. ICIS chemical pricing. Price reports for chemicals, 2010. Available at http://www.icis.com/chemicals/channel-info-chemicals-a-z/.

62. D. Young, R. Scharp, and H. Cabezas, The waste reduction (WAR) algorithm: Environmental impacts, energy consumption, and engineering economics, *Waste Management*, vol. 20, pp. 605–615, 2000.

63. D. Y. C. Cardona and V. Marulanda, Analysis of the environmental impact of butylacetate process through the WAR algorithm, *Chemical Engineering Science*, vol. 59, pp. 5839–5845, 2004.

64. J. B. Cannon, C. L. Cantrell, T. Astatkie, and V. D. Zheljazkov, Modification of yield and composition of essential oils by distillation time, *Industrial Crops and Products*, vol. 41, pp. 214–220, 2013.

65. X. Zhang, H. Gao, L. Zhang, D. Liu, and X. Ye, Extraction of essential oil from discarded tobacco leaves by solvent extraction and steam distillation, and identification of its chemical composition, *Industrial Crops and Products*, vol. 39, pp. 162–169, 2012.

66. M. M. Kirchhoff, Origins, current status, and future challenges of green, *Accounts of Chemical Research*, vol. 35, no. 9, pp. 686–694, 2002.

67. P. Anastas and N. Eghbali, Green chemistry: Principles and practice, *Chemical Society Reviews*, vol. 39, no. 1, pp. 301–312, 2010.

68. Q. Yang, H. Xing, B. Su, K. Yu, Z. Bao, Y. Yang, and Q. Ren, Improved separation efficiency using ionic liquid–cosolvent mixtures as the extractant in liquid–liquid extraction: A multiple adjustment and synergistic effect, *Chemical Engineering Journal*, vol. 181–182, pp. 334–342, 2012.

# 9

# Marigold Oil

**Cuthbert Katsvanga**

## CONTENTS

## 9.1 Botany of *Tagetes minuta*

Mexican marigold (*Tagetes minuta*) (Figures 9.1 through 9.4), which is synonymous with *Tagetes glandulifera* or *Tagetes gladulosa*, belongs to the Asterales order and Asteraceae family. The Asteraceae (sunflower) family is the largest family of vascular plants, with more than 23,000 species, which are rich in secondary metabolites and essential oils. The genus contains 27 annual and 29 perennial species (Soule, 1993) from subtropical and tropical America and only 1 from tropical Africa. The genus *Tagetes* consists of 40 species that are endemic from Arizona to Argentina and is well known in the province of Chaco (Argentine) as "chinchilla" (Parodi, 1959). The genus name refers to the Latin name for marigold, *Tages*, an Etruscan god associated with agriculture; *minuta* means "very small" and probably refers to the very small individual flowers within the inflorescences. Mexican marigold is also known as khakhi or stink weed, Omosumo (Kisii), Abuba (Luo), Mũbangi (Kikuyu), Ngwekwe (Kibukusu), and Chemiasoriet (Kalenjin) (Ofori et al., 2013).

It is native to the South American (Argentina, Chile, Bolivia, Peru, and Paraguay) mountainous and grassland regions, but has naturalized itself around the globe: Africa, South Europe, South Asia, and Australia. *T. minuta* has deliberately been spread across the tropics, subtropics, and some temperate countries as an ornamental, medicinal, or perfume species. It is cultivated in several countries for a number of purposes, with a global demand of the essential oil for all applications estimated to be more than 12 tonnes. The current largest commercial producers of tagetes oil are France, South Africa, and North

**FIGURE 9.1**
Young Mexican marigold before flowering.

**FIGURE 9.2**
Flowering Mexican marigold.

America. European production of the oil is negligible due to high cultivation and labor costs for picking the delicate flowers. As such, demand is satisfied through imports. South Africa, India, Zimbabwe, Egypt, France, and Argentina produce significant amounts of *T. minuta* oil. However, there is a significant decline in supplies from Egypt, since many farmers have switched to more reliable crops, as political instability has affected tagetes oil exports. In South Africa, Robertet, the French fragrances and flavorings manufacturer, has the greatest *T. minuta* essential oil market share. In 2003, South Africa produced approximately 6500 tonnes, whereas Zimbabwe and India produced approximately 2 and 4 tonnes, respectively. Since South Africa supplies all its oils to France, France is Europe's major importer. Other European importers have indicated great interest in *T. minuta* oil but are affected by limited supplies.

**FIGURE 9.3**
Mexican marigold showing root system.

**FIGURE 9.4**
Fields infested with dry Mexican marigold.

Mexican marigold is an erect annual plant characterized by a strong scent. It naturally grows from spring until the beginning of winter, completing its life cycle within this period. Depending on geographical location, flowering and fruiting vary with the rainy season. *T. minuta* is capable of fast growth, producing about 29,000 seeds per plant, capacitating it to seasonally invade various types of habitats. Flowering physiology is initiated by a 13 h day length (Luciani-Gresta, 1975). However, in less sunny temperate regions, it may not flower, and in soil, pH values range from 4.3 to 6.6 (Holm et al., 1997). It prefers high nutrient and high soil moisture environments. It is tolerant to low rainfall regimes. The stems are characteristically erect, woody, and grooved or ridged, turning from green to brown at maturity. *T. minuta* takes 105–120 days to mature (Aranha et al., 1982), growing up to 1–2 m with stalked leaves, opposite, light green or slightly glossy green, 7–15 cm long, pinnately dissected into four to six pairs of pinnae. The leaflets are lanceolate with finely serrated margins. Inflorescences are terminal with numerous tubelike clustered panicles of 20 to 80 capitula of dull yellow florets. The fruits are black, narrow ellipsoid, and hard-seed achenes. They are approximately 7 mm long, pilose and pappus of one or two setae to 3 mm long and three or four scales to 1 mm long with ciliate apex. According to Martinez-Ghersa et al. (2000), *T. minuta* seed is relatively small and without a dormancy stage, and 7–8 months' longevity. It has an aerial seed bank, which compensates for the lack of a persistent soil seed bank. The plant is propagated by seed, which germinates within 48 h at temperatures between 20°C and 30°C (optimal 25°C). Of great importance is the under-surface of the leaves, which bear sunken punctuate, multicellular oil glands, orangish in color, with licorice-like aroma when ruptured. Such glands are also available on stems and involucre bracts (Prakaso et al., 1999). Fresh seed may have as high as 95% germination. The taproots are typically short and tapering, with fibrous lateral roots, which form mycorrhizal associations.

In terms of its genetics, there is no record of its chromosome number yet its flowers are hermaphrodite and insect pollinated. Galicia-Fuentes (1995) distinguished four phenotypes of *T. minuta* on the basis of the morphological traits of capitula, whereas Gil et al. (2000) identified three hemotypes, which vary in biomass and inflorescence and leaf essential oil yield.

## 9.2 Methods of Oil Extraction

*T. minuta* essential oil is extracted from nonbloomed plant parts, such as leaves, flowers, and stalks, when the seeds are at the forming stage through steam or vacuum distillation. Steam distillation was a popular special laboratory purification process for high-temperature-sensitive organic compounds. Although the process is fairly outdated, it is still functional in numerous laboratory and industrial sectors. The more recent practice, vacuum distillation, involves the separation of aromatic compounds whereby the pressure over the liquid mixture to be distilled is lowered to below its vapor pressure, resulting in the evaporation of the most volatile liquid(s). The process can run in the presence or absence of heat in the mixture since it is premised on the principle that boiling occurs when the vapor pressure of a liquid surpasses the ambient pressure. The composition of *T. minuta* oil is primarily related to the part of the plant that is distilled. However, the extraction method also affects the composition of the oil.

### 9.2.1 Composition of a Good-Quality *T. minuta* Oil

If you distill the flowers, as opposed to stems or leaves, to comply with buyer specifications, optimize the time of harvesting and minimize the time between harvesting and distillation. Use the extraction method (pressure, time, or temperature), which is most appropriate for production of the essential oil with a composition according to buyer preferences. Prevent contamination by foreign materials (e.g., dust) by keeping facilities and equipment clean.

The collected *T. minuta* vegetable matter is air-dried or oven-dried at 40°C for approximately 72 h, after which it is ground and subjected to steam or vacuum distillation to separate the essential oil. In the process, the herbage is deposited on a perforated plate, which serves as support to homogenize steam flow. The plate is then placed on an extractor and a pipe attached to transfer the vapors to a condenser. The steam carries the volatile aromatic compounds, that is, the essential oil, present in the plant material within the vapor phase. A flask positioned at the end of the condenser separates *T. minuta* essential oil from water. The remaining water from the essential oil isolation is extracted by filtration using anhydrous sodium sulfate and stored in tightly closed dark vials at −4°C, and as such, it is ready for analysis.

## 9.3 Methods of Oil Analysis

The essential oil extract is qualitatively and quantitatively analyzed by gas chromatography–mass spectrometry (GC-MS) process. A gas chromatography–mass spectrometer equipped with a capillary column, a split injector, an automatic injection system, and a selective mass detector is used to perform the test at 250°C. Oven temperature is programmed from 50°C to 210°C at 10°C/min with helium or nitrogen as carrier gases, with a split ratio of 1:50. A gas flow of 0.7 ml/min at a constant speed of 30 cm/s and an interface of 250°C is maintained. The injector temperature of 200°C is sustained at an injection volume of 1 ml. The sample is prepared in $CHCl_3$. The peak area percentages are calculated without correction factors or internal standards. The peaks are identified by contrasting their mass spectra to Wiley's Flavors and Fragrances of Natural and Synthetic Compounds (FFNSC) mass spectral information from an analysis of the fragmentation pattern obtained for each component and on a comparison of their retention indexes (RIs) with the GC-MS solution program.

## 9.4 Oil Composition

So far, 27 compounds have been identified, which constitute approximately 75% of the total oil composition. Several researchers have shown that *T. minuta* L. oil has (Z)-β-ocimene, dihydrotagetone, (Z)- and (E)-tagetone, (Z)- and (E)-tagetenone, (Z)- and (E)-ocimenone as the principal constituents. The constituent percentage of oil *cis*-β-ocimene is 40%–55%, while *cis*-tagetenone and *trans*-tagetenone have 10%–30%, dihydrotagetone 5%–20%, *cis*-tagetone and *trans*-tagetone 5%–20%, and limonene 3%–9%. Table 9.1 shows the chemical composition of the essential oil of *T. minuta*.

A number of *T. minuta* studies indicate variations in essential oil composition, depending on growing conditions and harvesting location (Zygadlo et al., 1993), the growth stage (Moghaddam et al., 2007), plant part (Weaver et al., 1994), and the dissimilar chemotypes (Gil et al., 2000). A chemotype is largely an epigenetic (with some genetic) variation of a plant caused by the effects of sunlight, soil, temperature, and weather conditions (Mohamed et al., 2002). Botanically, the plants are identical, but their chemical compositions are different. Different chemotypes of the same essential oil sometimes have different effects when used in aromatherapy. Users of essential oils can use chemotypes to create interesting variations in personal fragrances, cosmetics, and aromatherapy (Lawrence, 1993).

**TABLE 9.1**

*Tagetes minuta* Chemical Composition

| No. | Compounds |
|-----|-----------|
| 1 | α-Pinene |
| 2 | Sabinene |
| 3 | Limonene |
| 4 | Dihydrotagetone |
| 5 | 1-Methyl-4-isopropenylbenzene |
| 6 | α-Terpinolene |
| 7 | *cis*-Epoxy ocimene |
| 8 | *trans*-Epoxy ocimene |
| 9 | Z-Tagetone |
| 10 | E-Tagetone |
| 11 | p-Cymene-8-ol |
| 12 | Linalyl propionate |
| 13 | Verbenone |
| 14 | Z-Ocimenone |
| 15 | Piperitone |
| 16 | Piperitenone |
| 17 | Piperitone oxide |
| 18 | β-Caryophyllene |
| 19 | Spathulenol |
| 20 | Caryophyllene oxide |
| 21 | Heptadecane |
| 22 | Neophytadiene |
| 23 | Octadecane |
| 24 | Nonadecane |
| 25 | Heneicosane |
| 26 | Docosane |
| 27 | Tricosane |

Biological and environmental factors affect the content and composition of *T. minuta* oil. It therefore follows that different countries have significant differences in *T. minuta* essential oil composition (Lawrence, 2000). In Argentina, GC and GC-MS analyses of the *T. minuta* essential oil showed that dihydrotagetone, α-phellandrene, limonene, *o*-cymene, β-ocimene, *trans*-tagetone, and tagetenone were the major constituents (Gil et al., 2000), while in Egypt, the main constituents were monoterpenes, of which *trans*- and *cis*-tagetone were present at 52.3% and 64.2%, respectively (Mohamed et al., 2002). In South Africa, β-ocimene (32.0%) and dihydrotagetone (16.4%) were the main components (Mohamed et al., 2004), and in southwest Iran, the oil was particularly rich in limonene (13.0%), piperitenone (12.2%), α-terpinolene (11.0%), piperitone (6%), *cis*-tagetone (5.7%), (Z)-ocimenone (5.1%), and eucarvone (4.8%) (Mohamad et al., 2010).

As such, some expected effectiveness may not be experienced in some regions due to the presence or absence of certain compounds. Table 9.2 shows the varying composition of *T. minuta* essential oil at different stages of the plant life cycle. The oils were extracted through steam distillation, independently, from nonbloomed plant leaves, bloomed plant leaves, and flowers. An analysis of the relative composition of flower essential oil showed that the major components are β-ocimene and tagetenone, at the expense of dihydrotagetone

**TABLE 9.2**

Variations in Essential Oil Composition in *Tagetes minuta* Plant Parts at Different Growth Stages

| | | Component Essential Oil Percentage (%) | | | | | |
|---|---|---|---|---|---|---|---|
| Peak No. | Compound | Leaves of Nonbloomed Plant in March | Leaves of Bloomed Plant in April–May | Leaves of Bloomed/ Fructified Plant in June | Flower in April | Flower in May | Flower and Seed in June |
| 1 | β-Phelandrene | 2.6 | 2.1 | 2.2 | .8 | 1.2 | 1.4 |
| 2 | Limonene | 12.9 | 13.1 | 11.8 | 4.4 | 6.4 | 8.6 |
| 3 | β-Ocimene | 11.1 | 17.6 | 15.4 | 47.4 | 38.2 | 31.2 |
| 4 | Dihydrotagetone | 47.0 | 28.1 | 38.3 | 1.5 | 6.6 | 11.7 |
| 5 | Tagetone | 15.9 | 18.2 | 12.3 | 3.7 | 8.9 | 11.8 |
| 6 | Tagetenone | 8.8 | 18.1 | 11.5 | 36.0 | 34.0 | 28.7 |

(major component of leaves from nonbloomed plant). This might explain the increased biocidal activity of flower oil, since its effectiveness is usually associated with the presence of tagetenone.

## 9.5 Physical and Chemical Properties of Oil

*T. minuta* essential oil is characterized by a sweet, fruity, and wild citrus–like aroma. The oil color is reddish-amber. It has a medium viscosity and thickens with increased aerial exposure. *T. minuta* is rich in many secondary compounds, including acyclic, monocyclic, and bicyclic monoterpenes; sesquiterpenes; flavonoids; thiophenes; and aromatics. The physical properties of *T. minuta* essential oil are indicated in Table 9.3.

## 9.6 General Uses of Oil

Various applications of *T. minuta* essential oil are in the cosmetic industry, as ingredients of fragrances, decorative cosmetics, fine fragrances, and flavoring; in the food industry, as aromas and flavors; in the pharmaceutical industry, as active components of medicines and as antibacterials or antimicrobials; and in aromatherapy. *T. minuta* essential oil is a major flavor component in food products such as cola and alcoholic beverages, frozen dairy desserts, candy, baked goods, gelatins, puddings, condiments, and relishes. The essential oil has broad applications as flavor and perfume (Vasudevan et al., 1997). *Tagetes* species were originally used as a source of essential oil for flavoring in food industries. The powders and extracts of tagetes are rich in the orange-yellow carotenoid and are used as a colorant in foods such as pasta, vegetable oil, margarine, mayonnaises, salad dressing, baked goods, confectionery, dairy products, ice cream, yogurt, citrus juice, mustard, and poultry.

   *T. minuta* is also extensively used medicinally as a condiment and herbal tea in a wide variety of fields in its native region, and as a popular traditional folk remedy or in complementary and medical therapy. It has several medical benefits, such as a remedy for

**TABLE 9.3**

Physical Properties of *T. minuta* Essential Oil

| Property | Description |
| --- | --- |
| Color | Yellow to orange |
| Physical state | Liquid |
| Odor | Bright, diffusive, fruity-floral |
| Stability to normal and elevated temperatures, metals and metal ions | Stable under normal and elevated temperatures for 14 days; no evidence of instability under contact with metals and metal ions for 14 days at ambient and elevated temperatures; some color and physical state changes to the metal ions were observed during the study |
| Flammability | Flashpoint = 63°C<br>43°C |
| pH | 3.79 (neat oil)<br>4.43 (1% dilution) |
| Ultraviolet (UV)/visible light absorption | UV profiles are similar under all pH conditions; no significant molar absorbance ($\varepsilon$) at the wavelength range of 290–750 nm<br>Neutral pH: $\varepsilon$ = 44.52, 45.98 at wavelength 237 nm<br>Basic pH: $\varepsilon$ = 40.13, 44.80 at wavelength 237 nm<br>Acidic pH: $\varepsilon$ = 45.38, 47.81 at wavelength 237 nm |
| Viscosity | 1.5–4.05 mPa at shear rates of 5–200 rpm at 20°C |
| Boiling point/range | 190°C |
| Density | Specific gravity = 0.870 at 20°C |
| Dissociation constant in water | Does not dissociate in water |
| Partition coefficient (n-octanol/water) | Values estimated using KOWWIN version 1.68 (U.S. Environmental Protection Agency) were submitted for the major components of tagetes oil, as the oil itself is not amenable to testing.<br>Limonene: Log Kow = 4.83<br>Ocimene: Log Kow = 4.80<br>Dihydrotagetone: Log Kow = 2.92<br>Linalool: Log Kow = 3.38 |
| Water solubility | Insoluble |
| Vapor pressure | Approximately 59% volatilization (measured via weight loss) in 48 h in laboratory study |

colds, respiratory inflammations, and stomach problems; antispasmodic; antiparasitic; antiseptic; insecticide; and sedative. Mexican marigold oil has antibiotic, antiseptic, antimicrobial, antiphlogistic, antispasmodic, cytophylactic, emollient, disinfectant, sedative, fungicide, insecticide, and hypertensive properties.

It is also a good insect repellant and antiparasitic. In humans, the essential oil has been observed to have medicinal properties that effectively treat fungal infections such as athlete's foot.

The oil hampers bacterial, fungal, and parasite growth and development. Tagetes oil is multipurpose and is used to treat sores, ulcers, and gangrene. Thyophenes contained in the essential oil have proven to have an antiviral effect. It can be used as a repellent for several insects, such as bees, lice, mosquitoes, and bed bugs. The oil relieves coughs, cramps, convulsions, and diarrhea. It is also a natural relaxant and can put under sedation excretory, nervous, and neurotic system inflammation and irritation. The essential oil compounds are used variably. Limonene is used as fragrance additive by the perfumery industry and as a component of artificial essential oils, while piperitenone is used in the manufacture of fragrances and flavor concentrates of all types. Piperitenone is an essential intermediate for the production of menthol for cosmetic and pharmaceutical purposes.

α-Terpinolene is characterized by a floral, sweet, and pine-like aroma, whereas piperitone is used as the major raw material for the production of synthetic menthol and thymol.

## 9.7 Pesticidal Uses of Oil (Including Public Health)

*T. minuta* as a plant is a very effective biopesticide (Katsvanga and Chigwaza, 2004) with root secretions that kill subsurface and surface soil pathogens (Nada, 2008). The plant suppresses free-living nematodes and has been used as an intercrop in rotation to protect crops (Kimpinski and Arsenault, 1994). Floral, foliar, and root extracts have insecticidal activity against adult *Coleoptera* and mosquito larvae and adults (Keita et al., 2000; Sarin, 2004). Its root extracts can reduce populations of weed species *Agropyron repens* and *Convolvulus arvensis* (Nada, 2008). In organic farming, it is grown as an intercrop due to its insecticidal and nematocidal activities between lettuce, cabbage, and tomatoes (Krueger et al., 2007). The essential oil is applied on coffee to prevent coffee berry disease and protects peas from blight, mildew, and other fungal diseases because it produces a substance called α-terthienyl, which helps in the reduction of disease-promoting organisms such as fungi, bacteria, insects, and some viruses (Soule, 1993). It also repels aphids, termites, blowflies, caterpillars, ants, maggots, flies, and moths. The secondary compounds in *Tagetes* are effective deterrents of numerous organisms, including fungi (Chan et al., 1975), bacteria (Grover and Rao, 1978), round worms (Loewe, 1974), trematodes (Graham et al., 1980), nematodes (Grainge and Ahmed, 1988), and other numerous insect pests (Jacobson, 1990). *T. minuta* also exhibits insecticidal activity against stored product pests (Sarin, 2004) and mosquitoes (Seyoum et al., 2002). It has also been reported that control of root knot nematodes in rose flower beds with *T. minuta* extract is as effective as furadan, nemacure, and temik insecticides. Rose plants with feeder roots destroyed by root knot nematodes showed significant recovery 1 month after treatment of the soil with an extract from *T. minuta* (Goswani and Vijayalakshmi, 1986). The essential oil has also been used to control various tick (*Rhipicephalus microplus*, *Rhipicephalus sanguineus*, and *Dermacentor nitens*) species on livestock. High concentrations of Mexican marigold powder are effective against maize weevils (*Sitophilus zeamais*) and Mexican bean weevils (*Zabrotes subfasciatus*) when mixed with grain. The liquid plant extract is effective against aphid (*Brevicoryne brassicae*) and red spider mites (*Tetranychus urticae*) in a variety of vegetables (*Brassica capitate* and *Brassica napus*). Decoctions have been used to control intestinal parasites in livestock.

## 9.8 Advantages of Marigold Oil as a Pesticide

Botanical pesticides have a long history and track record of application in pest control and have functioned comparably to synthetic pesticides. However, botanical pesticides have had a reduced market compared with synthetics. Environmental activism has demanded a favorable environment for the preference of botanical pesticides. The advantage of *T. minuta* is its application in organic agriculture, where synthetics are excluded because the industry regulates and restricts synthetic pesticides. It has demonstrated contact, fumigant, and antifeedant effects over a range of pests. Second, public resistance to genetically modified organisms (GMOs) favors the use of biopesticides as alternative measures. The other advantage of marigold oil–based pesticides is its rapid degradation and lack of persistence and bioaccumulation within the environment, unlike synthetic pesticides. Essential oil constituents are nonpersistent in water and soils. Studies have shown tagetes pesticides degrading in the environment within hours or days, which has given them a long history of safe use, although this may not be conclusive for new products. Marigold essential oil pesticides are known for their diversity and redundancy of phytochemicals. Redundancy, which is the presence of numerous analogs of one compound, increases the efficacy of extractives through analog synergism and prevents the development of pesticide resistance over several generations. As a biopesticide, *T. minuta* has the advantage of application to a broader array of insect and mite pests and pathogenic fungi, due to demonstrable contact and fumigant toxicity.

Another advantage as a crop protectant is its favorable mammalian toxicity, which ranges from slightly toxic to a few that may be of acute toxicity. For this, some tagetes essential oil is used as culinary herbs and spices. The essential oil poses little or no threat to human health and the environment. Another advantage is that *T. minuta* essential oil is attractive to manufacturers and consumers as safer and less expensive than synthetics, and normally is exempt from normal registration requirements in a number of countries, for example, the United States (U.S. Environmental Protection Agency). Yet another advantage is that one pesticide can be effective on a broad host of pests. Intercropping of *T. minuta* with vegetable species where it functions as a suppressant, repellant antibacterial, and antibiotic is also possible, even without spraying the vegetables. It can be cultivated in a multicrop system as a cover crop and can be rotated or grown as an intercrop with other plants. Synthetics are toxic on nontarget microorganisms, such as beneficial microorganisms. Marigold essential oil can reduce nematode populations if fumigated to greater soil depths.

## 9.9 Limitations of Marigold as a Pesticide

Large-scale commercialization of the essential oil is hampered by the scarcity of the natural resource, the need for chemical standardization and quality control, and difficulties in registration. Economic supply of *T. minuta* essential oil limits the broader application of the pesticide, which reduces the market size of the plant product. The other challenge is associated with quality control and lack of stability and competition with other biopesticides. Tagetes essential oil is reliably efficacious as a direct spray on pests. One disadvantage is its ineffectiveness in eradicating pests, but simply reducing populations, and another is the increased application frequency. Intercropping takes up more space, thus resulting in reduced crop yields.

## 9.10 Essential Oil–Based Insecticides

Essential oil constituents in pesticides are primarily lipophilic compounds that act as toxins, antifeedants, and deterrents to a variety of pests. Although many monoterpenoids have insecticidal properties, the degree of toxicity of different compounds to one pest differs considerably. In spite of the considerable research on biopesticides, only a few essential oil product-based pesticides are in the market. This could be a result of the fact that the efficacy of essential oils toward pests and diseases is not as apparent and obvious as that of synthetics. In the United States, essential oil–based pesticides have been exempted from rigorous registration if they are in use in the food industry, and this opportunity has increased the production of essential oils. Companies like Mycotech and ecoSMART Technologies have had great interest in developing the products. The known codes of *T. minuta* oil are Chemical Abstracts Service (CAS): 91770-75-1 (replacing 8016-84-0) and Harmonised System (HS).

## 9.11 Conclusions

*T. minuta* essential oil has demonstrated efficacy against a broad range of pests and diseases, and as such, it may be applied as a fumigant, granular formulation, or direct spray with a range of effects, such as lethal toxicity to repellence and oviposition deterrent in insects. In a way, it shows that essential oil–based pesticides can be used in a variety of ways to manage pests and diseases. However, their efficacy falls short compared with synthetics, although there are cases when they indicate some equivalence. Essential oils require high application frequencies and rates when applied outdoors. Sufficient quantities of plant material for large commercial production are a further constraint. Geographical variation also poses a constraint, since the constituents also vary based on that. Regulatory systems in various countries need to be adjusted in order to relax approval. However, essential oils by their volatile nature do not affect pollinators and the environment, and enhance insect activity in the environment. They also can play a role in integrated pest management in developing counties where most of the plants are endemic.

As *T. minuta* oil is a relatively expensive essential oil, it is mainly used in high-value products. *T. minuta* oil with a high tagetone content (20%) fetches the highest prices. Nonetheless, interest in organic essential oils is increasing. As the yield of extraction of *T. minuta* oil amounts to only 0.1%–0.4%, it is a relatively expensive product. The yield per hectare is around 25 tonnes of raw plant material and between 12.5 and 17.5 kg of *T. minuta* oil. In early 2013, Freight on Board (FOB) prices for high-quality oil increased to around €150–190/kg, due to decreased availability. In early 2014, average import prices for Indian tagetes oil amounted to only €80/kg. Prices for low-quality tagetes oil can be as low as €70/kg (CBI Market Information Database, www.cbi.eu). With the very high competition for *T. minuta* in the industry, the likelihood of it being used as a pesticide is low because of competition with the food and beverage industries.

## References

Aranha, C., Bacchi, O., Leitão Filho, H.F. 1982. *Plantas Invasoras de Culturas*, vol. 2, *Campinas*. Sao Paulo: Instituto Campineiro de Ensino Agricola.

Chan, G., Towers, H., Mitchell, J. 1975. Ultraviolet-mediated antibiotic activity of thiophene compounds of *Tagetes*. *Phytochemistry* 14: 2295–2296.

Galicia-Fuentes, S.S. 1995. Floral diversity in different species of marigold. *Revista Fitotecnia Mexicana* 18(1): 43–53.

Gil, A., Ghersa, C.M., Leicach, S. 2000. Essential oil yield and composition of *Tagetes minuta* accessions from Argentina. *Biochemical Systematics and Ecology* 28: 261–274.

Goswani, B., Vijayalakshmi, K. 1986. Nematicidal properties of some indigenous plant material against root-knot nematodes, *Meloidogyne incognita* on tomato. *Indian Journal of Nematology* 16: 65–68.

Graham, K., Graham, A., and Towers, H. 1980. Cercaricidal activity of phenylheptatriyne and alpha-terthienyl, naturally occurring compounds in species of the Asteraceae. *Canada Journal of Zoology* 58: 1955–1958.

Grainge, M., Ahmed, S. 1988. *Handbook of Plants with Pest-Control Properties*. New York: Wiley, pp. 67–77.

Grover, G., and Rao, J. 1978. In vitro antimicrobial studies of the essential oil of *Tagetes erecta*. *Perfumery and Flavoury* 3: 23.

Holm, L.G., Doll, J., Holm, E., Pancho, J.V., Herberger, J.P. 1997. *World Weeds: Natural Histories and Distribution*. New York: John Wiley & Sons.

Jacobson, M. 1990. *Glossary of Plant-Derived Insect Deterrents*. Boca Raton, FL: CRC Press.

Katsvanga, C.A.T., and Chigwaza, S. 2004. Effectiveness of natural herbs, fever tea (*Lippia javanica*) and Mexican marigold (*Tagetes minuta*) as substitutes to synthetic pesticides in controlling aphid species (*Brevicoryne brassica*) on cabbage (*Brassica capitata*). *Tropical and Subtropical Agroecosystems* 4: 101–106.

Keita, S., Vincent, C., Schmit, J., Ramaswamy, S., and Belanger, A. 2000. Effects of various essential oils on *Callosobruchus maculates* (F) (Coleopteran: Bruchidae). *Journal of Stored Products Research* 13: 595–598.

Kimpinski, B., and Arsenault, W. 1994. Nematodes in ryegrass, marigold, mustard, red clover and soybean. *Forage Notes* 37: 52–53.

Krueger, R., Dover, K., McSorley, R., Wang, K. 2007. *Marigolds* (Tagetes *spp.) for Nematode Management*. Vol. 1. Institute of Food and Agriculture Sciences, University of Florida, Gainesville, FL.

Lawrence, B. 2000. Progress in essential oils, tagetes oils. *Perfumery and Flavourist* 10: 73–82.

Lawrence, B.M. 1993. A planning scheme to evaluate new aromatic plants for the flavor and fragrance industries. In *New Crops*, ed. Janic, J., Simon, J. New York: Wiley, pp. 620–627.

Loewe, H. 1974. Recent advances in the medicinal chemistry of anthelminthics. *International Journal of Medicinal Chemistry* 4: 271–301.

Luciani-Gresta, F. 1975. The photoperiodic behaviour of *Tagetes minuta* of Sicily. *Bulletin Societie Botanique France* 122: 363–366.

Martinez-Ghersa, M.A., Ghersa, C.M., Benech-Arnold, R.L., Donough, R.M., Sanchez, R.A. 2000. Adaptive traits regulating dormancy and germination of invasive species. *Plant Species Biology* 15(2): 127–137.

Moghaddam, M., Omidbiagi, R., Sefidkon, F. 2007. Chemical composition of the essential oil of *Tagetes minuta* L. *Journal of Essential Oil Research* 19(1): 3–4.

Mohamad, H.M., Armstrong, K.L., Burge, J.R., Kinnamon, K.E. 2010. Chemical characterization of volatile components of *Tagetes minuta* L. cultivated in southwest of Iran by nanoscale injection. *Digest Journal of Nanomaterial and Biostructures* 5(1): 101–106.

Mohamed, M.A.H., Harris, P.J.C., Henderson, J., Senatore, F. 2002. Effects of drought stress on the yield and composition of volatile oils of drought-tolerant and non-drought-tolerant clones of *Tagetes minuta*. *Planta Medica* 68: 472–484.

Nada, H. 2008. Wild marigold—*Tagetes minuta* L. New weed on the Island of Hvar and new contribution to the knowledge of its distribution in Dalmatia (Croatia). *Agriculturae Conspectus Scientificus* 71: 23–26.

Ofori, D.A., Anjarwalla, P., Mwaura, L., Jamnadass, R., Stevenson, P.C., Smith, P. 2013. Pesticidal plant leaflet. *Tagetes minuta* L. http://projects.nri.org/adappt/docs/Tagetes_factsheet.pdf.

Parodi, L.R. 1959. *Enciclopedia Argentina de Agricultura y Jardineria Tomo 11*. Buenos Aires: Editorial Acme S.A.C.I.

Prakaso, R., Syamasundar, K., Gopinath, C., and Ramesh, S. 1999. Agronomical and chemical studies on *Tagetes minuta* grown in a red soil of a semiarid tropical region in India. *Journal of Essential Oil Research* 11: 259–261.

Sarin, R. 2004. Insecticidal activity of callus culture of *Tagetes erecta*. *Fitoterapia* 75: 62–64.

Senatore, F., Napolitano, F., Mohamed, M.A.H., Harris, P.C.J., Mnkeni, P.N.S., Henderson, J. 2004. Antibacterial activity of *Tagetes minuta* L. (Asteraceae) essential oil with different chemical composition. *Flavor and Fragrance Journal* 19: 574–578.

Seyoum, A., Palsson, K., Kung'a, S., Kabiru, E., Lwande, W., Killeen, G., Hassanali, A., Knols, B. 2002. Traditional use of mosquito repellent plants in western Kenya and their evaluation in semi-field experimental huts against *Anopheles gambiae*. Ethnobotinical studies and application by thermal expulsion and direct burning. *Transactions of the Royal Society of Tropical Medicine and Hygiene* 96: 223–231.

Soule, J. 1993. *Tagetes minuta*: A potential new herb from America. In *New Crops*, ed. Janic, J., Simon, J. New York: Wiley, pp. 649–654

Vasudevan, P., Kashyap, S., Sharma, S. 1997. Tagetes: A multipurpose plant. *Bioresource Technology* 62: 29–35.

Weaver, D.K., Wells, C.D., Dunkel, F.V., Bertsch, W., Sing, S.E., Sriharan, S. 1994. Insecticidal activity of floral, foliar, and root extracts of *Tagetes minuta* (Asterales: Asteraceae) against adult Mexican bean weevils (Coleoptera: Bruchidae). *Journal of Economic Entomology* 87(6): 1718–1725.

Zygadlo, J.A., Lamarque, A., Maestri, D.M., Guzman, C.A., Grosso, N.R. 1993. Composition of the inflorescence oils of some *Tagetes* species from Argentina. *Journal of Essential Oil Research* 5: 679–681.

# 10

## Clove Oil

**A. Onur Girisgin**

## CONTENTS

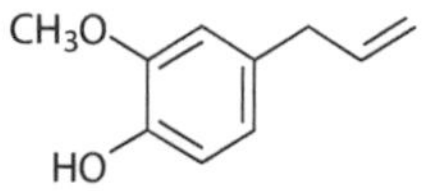

**FIGURE 10.1**
Chemical structure of eugenol.

## 10.1 Introduction

The family Myrtaceae comprises at least 132 genera and 5671 species distributed in tropical and subtropical regions worldwide, with centers of diversity in Australia, Southeast Asia, and South America, but with only poor representation in Africa. *Syzygium aromaticum* (L.) Merrill & Perry (with synonyms of *Eugenia caryophyllus* Bullock and S.G. Harrison, *Caryophyllus aromaticus* L., *Eugenia aromatica* [L.] [Baill.], and *Eugenia caryophyllata* [Thunb.] [Spreng.]), commonly known as clove, is a medium-sized tree from the Myrtaceae family. Its essential oil is isolated from dried flower buds, which are the source of its strongly smelling oil, popularly known as clove oil. The oil extracted from the stem and leaf is used in preparing high-grade eugenol and vanillin. The essential oil is widely used and well known for its medicinal properties. Traditional uses of clove oil include use in dental care as an antiseptic and analgesic. Previous studies have reported antifungal, anticarcinogenic, antiallergic, antimutagenic, antioxidant, and insecticidal properties. The chief constituent of clove oil is 70%–90% eugenol ($C_{10}H_{14}O_2$) (Figure 10.1), followed by β-caryophyllene and, in lesser amounts, α-humulene, caryophyllene oxide, and eugenyl acetate, although in different concentrations (Pandey and Chadha, 1993; reviewed by Razafimamonjison et al., 2013).

## 10.2 Botany of the Plant

The clove tree is a perennial tropical plant that grows to a height ranging from 10 to 20 m, having large oval leaves and crimson flowers in numerous groups of terminal clusters. The flowers are hermaphrodite, borne at the terminal end in small bunches, and produce a two-celled capsule (bud) containing the well-known hot aromatic fruit. The buds have a slightly cylindrical base and are surrounded by the plump ball-like unopened corolla, which is surrounded by four-toothed calyx. Production of flower buds, which is the commercialized part of this tree, starts after 4 years of plantation. Flower buds are collected in the maturation phase before flowering. The leaves are lanceolate on the branched tree and possess plenty of oil glands on the lower surface. The branches are semierect, the limbs are more often brittle with a grayish bark, and the head is bushy and dense (Board, 2010; reviewed by Cortes-Rojas et al., 2014).

## 10.3 Methods of Extraction of Oil

### 10.3.1 Conventional Extraction

Essential oils are defined as products extracted from natural plants by physical means only, such as hydrodistillation, steam distillation, dry distillation, cold press, headspace analysis, liquid $CO_2$ extraction, and solvent extraction. The composition of the extracted oil may vary from one extraction method to another. Headspace analysis offers a potentially rapid method to extract essential oils and requires very little plant material, but complete recovery occurs only for highly volatile materials. Steam distillation and solvent extraction as conventionally applied result in severe losses of volatile materials because the liquid in which the oil is collected must be subsequently removed by evaporation. Solvent extraction results in the recovery of nonvolatile compounds. While liquid $CO_2$ extraction can give reliable and efficient recovery with little or no decompositional changes induced by the extraction process, this process is expensive and used only on a limited basis in commerce. Moreover, similar results can be obtained by hydro- or steam distillation (reviewed by Charles and Simon, 1990).

These processes are not expensive, but they can induce thermal degradation, hydrolysis, and water solubilization of some fragrance constituents. Extracts obtained by solvents contain residues that pollute the foods and fragrances to which they are added. These disadvantages have attracted recent research attention and stimulated the intensification, optimization, and improvement of existing and novel "green" extraction techniques. All these techniques are appropriately applied with a careful consideration of plant organs and the quality of the final products (reviewed by Li et al., 2014).

These conventional extraction techniques could typically extract essential oils from plants ranging from 0.005% to 10%, which are influenced by the distillation duration, temperature, operating pressure, and most importantly, type and quality of raw plant materials.

There are some extraction methods to obtain clove oil, like the other essential oils. The most common ones are steam distillation and supercritical fluid extraction (SFE) methods, which are official approved methods for the isolation of essential oils from plant materials (Guan et al., 2007; reviewed by Li et al., 2014).

In steam distillation, the plant materials charged in the alembic are subjected to the steam without maceration in water. The injected steam passes through the plants from the base of the alembic to the top. The vapor laden with essential oils flows through a "swan neck" column and is then condensed before decantation and collection in a Florentine flask. Essential oils that are lighter or heavier than water form two immiscible phases and can be easily separated. The principle of this technique is that the combined vapor pressure equals the ambient pressure at about 100°C, so that the volatile components with boiling points ranging from 150°C to 300°C can be evaporated at a temperature close to that of water. Furthermore, this technique can also be carried out under pressure, depending on the essential oil's extraction difficulty.

Supercritical fluid extraction has been established as an environmentally benign technique for separating essential oils from the vegetable. Moreover, extracts obtained by SFE can retain the organoleptic characteristics of the starting spice material (Guan et al., 2007).

From these two methods, there are other methods, like hydrodiffusion, hydrodistillation, and Soxhlet extraction (with the solvents of methanol, acetone, and chloroform) (reviewed by Li et al., 2014).

Some studies on the extraction of essential oil of cloves have concluded a yield of 8.6% on a steam distillation process of clove (*S. aromaticum*) for 8 h with the physical properties of essential oils that meet SNI 06-4267-1996. In the supercritical $CO_2$ extraction to get the yield 19.6% and 58.77% of eugenol, concentration at a pressure of 10 MPa and 50°C for 2 h are used. Extraction of essential oil clove buds by the steam distillation process for 8–10 h results in a yield of 10.1%; hydrodistillation for 4–6 h, 11.5%; and Soxhlet extraction for 6 h, 41.85%. On the extraction of essential oil of cloves with *E. caryophyllata*, using the hydrodistillation method for 150 min results in a yield of 5.06%; microwave-assisted extraction (MAE) for 10–30 min, 7.42%; and microwave steam distillation (MSD) for 10–30 min, 16.25% (reviewed by Daryono, 2015).

### 10.3.2 Green Extraction

The principles of green extraction can be generalized as the discovery and design of extraction processes that could reduce the energy consumption, allow the use of alternative solvents and renewable and innovative plant resources to eliminate petroleum-based solvents, and ensure safe and high-quality extracts or products. Turbo distillation, ultrasound or microwave-assisted extraction, instantaneous controlled pressure drop technology and household espresso machine method are used (reviewed by Li et al., 2014; Just et al., 2016).

## 10.4 Methods of Analysis of Oil

The analysis of clove oil is commonly performed with gas chromatography (GC) and gas chromatography–mass spectrometry (GC-MS) equipment. GC is a common type of chromatography used in analytical chemistry for separating and analyzing compounds that can be vaporized without decomposition. Typical uses of GC include testing the purity of a particular substance, or separating the different components of a mixture (the relative amounts of such components can also be determined). In some situations, GC may help in identifying a compound. In preparative chromatography, GC can be used to prepare pure compounds from a mixture (Donald et al., 2006).

GC-MS is an analytical method that combines the features of gas chromatography and mass spectrometry to identify different substances within a test sample. Applications of GC-MS include drug detection, fire investigation, environmental analysis, explosives investigation, and identification of unknown samples. GC-MS can identify trace elements in materials that were previously thought to have disintegrated beyond identification (Donald et al., 2006).

Clove oil can be analyzed by GC-MS practically, and its components can be documented with the help of a device library.

## 10.5 Composition of Oil

The chemical composition of clove oil and the relative concentration of bioactive components may vary, depending on a range of factors, including the extraction method used, geographical location where the plant was cultivated, time of harvest, storage method, and part of the plant used. Clove represents one of the major vegetal sources of phenolic compounds as flavonoids, hidroxibenzoic acids, hidroxicinamic acids, and hidroxiphenyl propens (Table 10.1). Eugenol is the main bioactive compound of clove, which is found in concentrations ranging from 9,381.70 to 14,650.00 mg/100 g of fresh plant material. With regard to the phenolic acids, gallic acid is the compound found in higher concentrations

**TABLE 10.1**

Composition and Percentage of Clove Essential Oil (*Eugenia caryophyllata*)
Obtained with GS-MS Analysis

| No. | Compound[a,b] | Kovats Index[c] (HP-20M) | Percentage (%) |
|-----|---------------|--------------------------|----------------|
| 1 | Eugenol | 2151 | 88.58535 |
| 2 | Eugenyl acetate | 2263 | 5.62086 |
| 3 | β-Caryophyllene | 1595 | 1.38830 |
| 4 | 2-Heptanone | 1172 | 0.93232 |
| 5 | Ethyl hexanoate | 1232 | 0.66098 |
| 6 | Humulenol | 2265 | 0.27527 |
| 7 | α-Humulene | 1668 | 0.19985 |
| 8 | Calacorene | 1918 | 0.11437 |
| 9 | Calamenene | 1828 | 0.10538 |
| 10 | 2-Heptanol | 1304 | tr |
| 11 | Menthyl octanoate | 1384 | tr |
| 12 | 2-Nonanone | 1392 | tr |
| 13 | Ethyl octanoate | 1429 | tr |
| 14 | α-Cubebene | 1459 | tr |
| 15 | Copaene | 1491 | tr |
| 16 | 2-Nonanol | 1499 | tr |
| 17 | Linalool | 1548 | tr |
| 18 | 2-Undecanone | 1588 | tr |
| 19 | Menthyl benzoate | 1619 | tr |
| 20 | Ethyl benzoate | 1647 | tr |
| 21 | Menthyl chavicol | 1669 | tr |
| 22 | α-Amorphene | 1675 | tr |
| 23 | α-Terpinyl acetate | 1695 | tr |
| 24 | α-Muurolene | 1711 | tr |
| 25 | Benzyl acetate | 1714 | tr |
| 26 | Carvone | 1731 | tr |
| 27 | γ-Cadinene | 1756 | tr |
| 28 | 2-Phenyiethyl acetate | 1826 | tr |
| 29 | (E)-Anethole | 1827 | tr |
| 30 | Benzyl alcohol | 1861 | tr |
| 31 | Caryophyllene oxide | 1976 | tr |
| 32 | Menthyl eugenol | 1985 | tr |
| 33 | Humulene oxide | 1986 | tr |
| 34 | Cinnamic aldehyde | 2018 | tr |
| 35 | Ethyl cinnamate | 2072 | tr |
| 36 | Benzyl tiglate | 2103 | tr |
| Total identified | | | 98.2769 |

*Source:* Chaieb, K. et al., *Phytother. Res.*, 21, 501–506, 2007.

*Note:* tr, trace (<0.1%).

[a] Order of elution on HP-20M capillary.

[b] Identified by comparison of the mass spectral and Kovats index data.

[c] Kovats indices on HP-20M column.

(783.50 mg/100 g fresh weight). However, other gallic acid derivatives, such as hydroliz-able tannins, are present in higher concentrations (2375.8 mg/100 g). Other phenolic acids found in clove are the caffeic, ferulic, elagic, and salicylic acids. Flavonoids such as kaemp-ferol, quercetin, and its derivate (glycosilated) are also found in clove in lower concentra-tions. Concentrations up to 18% of essential oil can be found in the clove flower buds. Roughly 89% of the clove essential oil is eugenol (Figure 10.1), and 5%–15% is eugenol ace-tate and β-caryophyllene. Another important compound found in the essential oil of clove in concentrations up to 2.1% is α-humulen. Other volatile compounds present in lower concentrations in clove essential oil are β-pinene, β-selinene, limonene, farnesol, benzalde-hyde, 2-heptanone, ethyl hexanoate, humulenol, α-humulene, calacorene, and calamenene. The number of components of clove oil can reach up to 36 (reviewed by Pasay et al., 2010; reviewed by Cortes-Rojas et al., 2014; Girisgin et al., 2014).

## 10.6 Physical and Chemical Properties of Oil

Physical and chemical properties of clove oil may vary, depending on the extraction method used, geographical location where the plant was cultivated, time of harvest, stor-age method, and part of the clove (bud, leaf, or stem) used. A comparative list of physical and chemical properties of clove oil is presented in Table 10.2. Also, analytical data and standards of clove are presented (Table 10.3).

## 10.7 General Uses of Oil

### 10.7.1 Antioxidant Capacity

Eugenol, the major constituent of clove oil, has many antioxidant properties. The anti-oxidant activity may occur via various mechanisms, such as scavenging the radicals and chelating metal ions. Eugenol reportedly participates in photochemical reactions and dis-plays strong antioxidant activity and photocytotoxicity (reviewed by Chaieb et al., 2007).

**TABLE 10.2**

Physical and Chemical Properties of Clove Oil

|  | Clove Bud Oil | Clove Leaf Oil | Clove Stem Oil |
|---|---|---|---|
| Specific gravity at 25°C | 1.038–1.5250 | 1.036–1.046 | 1.048–1.056 |
| Refractive index at 20°C | 1.5270–1.5350 | 1.5231–1.5350 | 1.5340–1.5380 |
| Optical rotation at 20°C | −1030' to 00 | −20 to 00 | −1°5' to 0° |
| Solubility | 1:2 in 70% EtOH | 1:2 in 70% EtOH | 1:2 in 70% EtOH |
| Phenol content (as eugenol) | Not less than 85% by volume | Not less than 84% and not more than 88% by volume | Not less than 98% and not more than 95% by volume |

*Source:* Panda, H., *Essential Oils Handbook*, National Institute of Industrial Research, New Delhi, 2003, p. 219.

**TABLE 10.3**

Analytical Data and Standards of Clove Oil

|  | Range (%) | BPC Standards (%) | U.S. Standards (%) | Indian Standards (%) |
|---|---|---|---|---|
| Moisture | 5–11.0 | – | – | Max 12 |
| Ash | 4.5–7 | 7.0 (max) | 7.0 (max) | Max 7 |
| Water soluble ash | 2.7–4.2 | – | – |  |
| Acid insoluble ash | 0–0.3 | 1.0 (max) | 0.5 (max) | Max 0.5 |
| Volatile oil | 14.5–20 | 15 (min): Whole 12 (min): Powder | 15 (min): For volume of ethanol extract | Min 15 |
| Fixed oil | 6.2–10.1 | – | – | – |
| Alcohol extract | 13.5–15.5 | – | – |  |
| Crude fiber | 6–10 | – | 10 (max) |  |
| Nitrogen | 0.9–1.2 | – | – |  |
| Foreign organic matter (e.g., fruits) | – | 1.0 (max) |  |  |
| Stalks for stems | – | 5.0 (max) | 5 (max) |  |
| Quercitannic acid | – | – | 12 (min) |  |

*Source:* Singhal, R., Kulkarni, P.R., and Rege, D.V., *Handbook of Indices of Food Quality and Authenticity*, Woodhead Publishing, Cambridge, UK: 1997, p. 402.

*Note:* BPC, British Pharmacopoeia.

Jirovetz et al. (2006) found that the antioxidant action of 0.005% clove oil was identical to that of standard butylated hydroxytoluene at a concentration of 0.01%. It manifests considerable chelating potential against $Fe^{+3}$, resulting in the prevention of the initiation of hydroxyl radicals (Jirovetz et al., 2006). Clove oil thus shows powerful antioxidant activity; moreover, it can be used as an easily accessible source of natural antioxidants and in pharmaceutical applications (reviewed by Chaieb et al., 2007).

Shan et al. (2005) conducted a study on the main phenolic compounds of 26 spices, which were identified and quantified by high-performance liquid chromatography, followed by *in vitro* antioxidant activity analysis by the ABTS method. Results showed a high correlation between the polyphenol content and the antioxidant activity. Clove (buds) was the spice presenting higher antioxidant activity and polyphenol content tetraethylammonium chloride (mmol of Trolox/100 g dried weight) and gallic acid (equivalents/100 g of dried weight), respectively. The major types of phenolic compounds found were phenolic acids (gallic acid), flavonol glucosides, phenolic volatile oils (eugenol and acetyl eugenol), and tannins (reviewed by Cortes-Rojas et al., 2014).

Antioxidants are important compounds for the treatment of memory deficits caused by oxidative stress. Pretreatment with clove essential oil decreases the oxidative stress assessed by malondialdehyde and reduced glutathione levels in the brain of mice. This study concluded that clove oil could revert memory and learning deficits caused by scopolamine in the short and the long term as a result of the reduction in the oxidative stress (Halder et al., 2011). Memory and learning improvements of clove oil were observed in scopolamine-treated mice at doses of 0.025, 0.05, and 0.1 ml/kg when compared with a saline solution control group in an elevated plus maze test. These works prove the benefits of the employment of clove as a rich source of antioxidants for the treatment of memory deficits caused by oxidative stress (reviewed by Cortes-Rojas et al., 2014).

## 10.7.2 Anesthetic Activity

Eugenol is used in a wide range of applications, such as a local anesthetic in dentistry and as an ingredient in dental cement for temporary fillings. It is relatively user-friendly and can be used in lower concentrations than other local anesthetics, and it is rapidly metabolized and excreted, thus requiring no withdrawal period. It has been shown to be effective in anaesthetizing fish such as rainbow trout, *Oncorhynchus mykiss*, and channel catfish, *Ictalurus punctatus*. Eugenol at 65 mg/L was shown to safely and effectively induce all stages of anaesthesia in juvenile and subadult tambaqui fish within the desired time. Further research needs to focus on assessing its efficacy in other tropical species, as well as investigating its lethal dosage (reviewed by Chaieb et al., 2007).

## 10.7.3 Antimicrobial Activity

The antibacterial activity of different clove oil extracts has been demonstrated against pathogenic bacteria, including *Campylobacter jejuni*, *Klebsiella pneumoniae*, *Salmonella enteritidis*, *Escherichia coli*, and *Staphylococcus aureus* (reviewed by Chaieb et al., 2007; Hemalatha et al., 2015). A study reported that the growth rates of *Listeria monocytogenes* strains observed at 15°C and 5°C were significantly reduced by treatment with 1% and 2% clove oil (Mytle et al., 2006), and furthermore, Ogunwande et al. (2005) found that the essential oil of the fruit exhibited strong antibacterial activity against *S. aureus*, while the leaf oil strongly inhibited the growth of *Bacillus cereus*, with a minimum inhibitory concentration (MIC) of 39 µg/ml. The oil was also active against 26 strains of *Staphylococcus epidermidis* isolated from dialysis fluids, 3 human pathogenic Gram-positive cocci, 2 Gram-negative bacilli, and 1 Gram-positive bacillus (diameter of inhibition zone, 11–15 mm). In contrast, the oil was ineffective against *P. aeruginosa* ATCC 27853 (diameter of inhibition zone, 9 mm). These results are in agreement with those of another study reporting that clove essential oil exhibited antibacterial activity against a large number of methicillin-resistant *S. epidermidis* and *S. aureus*. However, the oil also appears to be effective against both Gram-positive and Gram-negative microorganisms, contrasting the results found in some other studies (reviewed by Chaieb et al., 2007).

Sofia et al. (2007) tested the antimicrobial activity of different Indian spice plants, such as mint, cinnamon, mustard, ginger, garlic, and clove. The only sample that showed a complete bactericidal effect against all the food-borne pathogens tested, *E. coli*, *S. aureus*, and *B. cereus*, was the aqueous extract of clove at 3%. At a concentration of 1%, clove extract also showed good inhibitory action (reviewed by Cortes-Rojas et al., 2014).

In addition to the wide spectrum of activity of eugenol against bacteria, a study showed that eugenol and cinnamaldehyde at 2 µg/ml inhibited the growth of 31 strains of *Helicobacter pylori*, after 9 and 12 h of incubation, respectively, being more potent than amoxicillin and without developing resistance. The activity and stability of those compounds was checked at low pH values since *H. pylori* resides in the stomach (Ali et al. 2005; reviewed by Cortes-Rojas et al., 2014).

## 10.7.4 Anticancer–Antitumor Activity

Clove essential oil has been reported to show anticarcinogenic and antimutagenic potential. Volatile oils display cytotoxic action toward the human tumor cell lines PC-3 and Hep G2, and in a recent study, eugenol was shown to induce apoptosis of human cancer cells (Namiki, 1994), with the major antimutagenic compound being identified as dehydrodieugenol. More recently, the antimutagenic activity of cinnamaldehyde was reported in human-derived hepatoma cells, where it suppressed the frequency of micronuclei induced by various heterocyclic amines (reviewed by Chaieb et al., 2007).

Eugenol isolated from clove oil was investigated using human promyelocytic leukemia cells (HL-60) and might be a potent agent in cancer therapy. After treatment with eugenol, the HL-60 cells showed hallmarks of apoptosis, such as DNA fragmentation and formation of DNA ladders in agarose gel electrophoresis. Apoptotic cell death was induced via generation of ROS, inducing a mitochondrial permeability transition, reducing the antiapoptotic protein bcl-2 level, and inducing cytochrome C release to the cytosol (reviewed by Buchbauer, 2010).

### 10.7.5 Fungicidal Activity

The phenolic component of eugenol is known to possess fungicidal characteristics, including activity against fungi isolated from onychomycosis (Gayoso et al., 2005). The main antifungal action appears to be exerted on the cellular membrane (Cox et al., 2001). Eugenol has shown antifungal activity against *Candida albicans* and *Trichophyton mentagrophytes* (Tampieri et al., 2005). Núñez et al. (2001) demonstrated that the mixture of clove oleoresin with concentrated sugar solution produced a strong fungicidal effect by reducing fungi inoculum size. Scanning electron microscope (SEM) micrographs showed significant morphological damage, with cellular deformity to *Saccharomyces cerevisiae* cells by clove oil (Chami et al., 2005). The fungicidal activity of clove essential oil has also been reported on several food-borne fungal species, and it was observed in a study that the essential oil of clove even inhibited the growth of *Aspergillus niger* (Pawar and Thaker, 2006; reviewed by Chaieb et al., 2007).

### 10.7.6 Antiviral Activity

In general, viruses are highly sensitive to the components of essential oils. Satisfactory results have been yielded in some studies on the *in vitro* antiviral activity of clove oil. For example, Hussein et al. (2000) found that *S. aromaticum* extract was highly active at inhibiting replication of the hepatitis C virus (≥90% inhibition at 100 µg/ml). Kurokawa et al. (1998) isolated and identified an anti–herpes simplex virus (HSV) compound, eugeniin, from the extracts of *S. aromaticum*, which showed specificity in inhibiting HSV-1 DNA polymerase activity (reviewed by Chaieb et al., 2007). The antiviral activity of eugeniin, isolated from *S. aromaticum* and *Geum japonicum*, was also tested against herpes virus strains, being effective at 5 µg/ml, and it was deducted that one of the major targets of eugeniin is viral DNA synthesis by inhibition of the viral DNA polymerase (reviewed by Cortes-Rojas et al., 2014).

### 10.7.7 Antinociceptive

The employment of clove as analgesic has been reported since the thirteenth century, for toothache and joint pain and as an antispasmodic, with eugenol being the main compound responsible for this activity. The mechanism involved has been attributed to the activation of calcium and chloride channels in ganglionar cells. The voltage-dependent effects of eugenol in sodium and calcium channels and in receptors expressed in the trigeminal ganglia also contributed to the analgesic effect of clove. Other results show that the analgesic effect of clove is due to its action as a capsaicin agonist. The peripheral antinociceptive activity of eugenol was reported by Daniel et al. (2009), showing significant activity at doses of 50, 75, and 100 mg/kg (reviewed by Cortes-Rojas et al., 2014).

## 10.8 Pesticidal Uses of Oil (Including Public Health)

### 10.8.1 Advantages as a Pesticide

Environmental problems caused by overuse of pesticides have been the matter of concern for both scientists and the public in recent years. It has been estimated that about 2.5 million tons of pesticides are used on crops each year, and the worldwide damage caused by pesticides reaches $100 billion annually. The reasons for this are (1) the high toxicity and nonbiodegradable properties of pesticides and (2) the residues in soil, water resources, and crops that affect public health. Thus, one needs to search the new highly selective and biodegradable pesticides to solve the problem of long-term toxicity to mammals, and on the other hand, one must study the environmentally friendly pesticides and develop techniques that can be used to reduce pesticide use while maintaining crop yields. Natural products are an excellent alternative to synthetic pesticides as a means to reduce negative impacts to human health and the environment. The move toward green chemistry processes and the continuing need for developing new crop protection tools with novel modes of action make discovery and commercialization of natural products as green pesticides an attractive and profitable pursuit that is commanding attention. The concept of green pesticides refers to all types of nature-oriented and beneficial pest control materials that can contribute to reduce the pest population and increase food production. They are safe and ecofriendly. They are more compatible with the environmental components than synthetic pesticides (reviewed by Koul et al., 2008).

Clove oil is one of the most effective pesticides compared with other essential oils, is liquid at room temperature, and is easily transformed from a liquid to a gaseous state at room or slightly higher temperature without undergoing decomposition. Low concentrations can be effective on pests and used as a pesticide. Thus, production of a pesticide from clove oil or any of its components may not be expensive.

### 10.8.2 Limitations as a Pesticide

The toxicity of essential oils is relatively well studied experimentally and clinically because of their use in human and veterinary medicine. The mammalian toxicity of essential oils is generally low. Most essential oils, including clove, have an oral $LD_{50}$ value ranging from 1500 to 5000 mg/kg in rats (reviewed by Regnault-Roger, 2013). Clove oil is generally recognized as safe substance when consumed in concentrations lower than 1500 mg/kg. The World Health Organization (WHO) established that the daily acceptable quantity of clove is 2.5 mg/kg of weight in humans. It can cause dermal toxicity on mammals as an irritant. Eugenol is easily absorbed when administrated by the oral route, rapidly reaching plasma and blood, with mean half-lives of 14.0 and 18.3 h, respectively. A cumulative effect has been hypothesized and associated with relieve of neuropathic pain after repeated daily administrations (reviewed by Cortes-Rojas et al., 2014).

Because risk includes both hazard and exposure, the use of clove oil requires that the applicators carefully follow the labeling recommendations. Essential oils are most often delivered by spraying or fogging that may induce a dermal or respiratory exposure. The need for suitable equipment for handling essential oils products and the treated plant must be observed to avoid accident or chronic intoxication. The essential oil risk, however minimal it may be, must not be ignored simply because essential oils are natural products (reviewed by Regnault-Roger, 2013).

### 10.8.3 Essential Oil–Based Insecticides

There are several studies on the effect of clove oil or its components on several pests and parasites. Clove oil, like some other essential oils, not only acts as a poison or neurotrope on the nervous system of insects, but also can intervene in cellular breathing either by inhibiting cellular oxidation through transfer interruption in the respiratory chain or by asphyxiation through the formation of an impermeable film insulating insects from the air. Moreover, clove oil may also have an inhibitory power to enzyme activity and insect growth with respect to adults, larvae, and eggs (Li et al., 2014).

Clove oil or its component eugenol has been tested as a pesticide on ticks, mites, mosquitoes, cockroaches, and grain pests. Satisfactory results have been obtained from studies on both adults and larvae. Examples presented here are classified according to pest type.

### 10.8.4 Ticks

Usage of environmentally friendly pesticides on animal ectoparasites has been a matter of concern to both scientists and the public in recent years. An acaricide should not leave residue in the meat and milk of both farm animals and poultry flocks. The acaricide or repellent potential of clove oil has been tested on ticks.

Clove leaf oil concentrations of 1%, 5%, 10%, and 15% have been evaluated *in vitro* for acaricidal activity against adults of *Argas* spp. soft ticks collected from poultry flocks. A 15% concentration of clove oil killed 80% of ticks in 24 h, with an $LC_{50} = 3.15\%$ and $LT_{50} = 47.9$ (Hasson and Al-Zubaidi, 2015).

### 10.8.5 Mites

Since *in vitro* and *in vivo* conventional acaricide resistance from human scabies mites has been reported, natural product extracts and compounds have become potential sources of alternative acaricides.

Clove oil and its components have been tested on permethrin-sensitive *Sarcoptes scabei* var. *suis* and permethrin-resistant *S. scabei* var. *canis*, separately at concentrations of 1.56%, 3.12%, 6.25%, 12.5%, and 25%. At all concentrations tested (1.56%–25%), contact with clove oil resulted in 100% mortality of permethrin-sensitive mites after 0.25 h. Permethrin-resistant mites died at the same time but required higher concentrations (≥6.25%) of clove oil. The acaricidal effect was faster in sensitive mites than in resistant mites. The acaricidal activities of eugenol, a major component of clove oil, and its minor component acetyleugenol, as well as the related analogues isoeugenol and methyleugenol, have been tested in contact bioassays against the two scabies mite populations. Mortality in the two mite populations at different concentrations of the compound have been plotted as dose–response curves, and $EC_{50}$ values were derived after an hour of observation. For both mite populations, there was no significant difference between the activity observed for the positive control acaricide (benzyl benzoate) and the test compounds eugenol, acetyleugenol, and isoeugenol ($p > 0.11$). In contrast, methyleugenol had no acaricidal effect in sensitive mites after the first hour of observation. This compound only displayed activity in resistant mites after 24 h at the highest concentration tested (100 mM) (Pasay et al., 2010).

Three components (terpenes) found in essential oils (eugenol, geraniol, and citral) have been tested against the poultry red mite *Dermanyssus gallinae*. All provided 100% mortality in toxicity tests when undiluted. Even at 1% of this dose, eugenol was 20% effective against experimental pest populations, although the remaining terpenes have shown to be largely ineffective at this concentration. $LC_{50}$ values confirmed the superior efficacy of eugenol (10%) over both citral (30%) and geraniol (40%) (Sparagano et al., 2013).

The eradication of house dust mites, *Dermatophagoides pteronyssinus*, by direct contact using the essential clove oil has been evaluated. For this, synthetic fibers have been immersed in 2% clove oil for 30 min, dried in a hot-air oven at 60°C for 2 h, after which 0.5 g of dust mites was exposed to these coated fibers placed in the Siriraj chamber. Ten mites were placed in the chamber and 10 µl of clove oil was pipetted or sprayed onto them. These latter two procedures were each carried out for three consecutive days at 0, 1, 3, and 6 months. The effectiveness of pipetting and spraying was 99% and 81%, respectively, while the placebo mortality was <5% (Mahakittikun et al., 2013).

*Varroa destructor* (Acari: Varroidae) is accepted as the most dangerous agent of honeybees (*Apis mellifera*) worldwide. Essential oil–based home-made or commercial acaricides are widely used to struggle against varroosis. The biological activity of the clove oil applied to *V. destructor* and *A. mellifera* has been evaluated in two laboratory tests. Mite lethality has been estimated using a complete exposure method test with the oil at different concentrations, and a systemic administration method of oil at different concentrations diluted in syrup has been placed in feeders for bees. The $LC_{50}$ for the complete exposure method at 24 h was 0.59 ml/dish. The $LC_{50}$ estimated at 48 h showed a slight decrease compared with that recorded at 24 h. Ratio selection ($LC_{50}$ of *A. mellifera*/$LC_{50}$ of *V. destructor*) for the complete exposure method was 26.46 and 13.35 for 24 and 48 h, respectively. Regarding the systemic administration method, mite's $LC_{50}$ at 24 h was 12,300 ppm. *S. aromaticum* oil was found to be an attractant for *V. destructor* at 4.8% (w/w) concentration. The results have shown that oil toxicity against *V. destructor* differed depending on its administration. Nevertheless, the ratio selection calculated by this oil is expected to enable its application under field conditions with a good safety margin. This oil could also be used in combination with other oils in integrated pest management strategies in bee colonies (Maggi et al., 2010).

### 10.8.6 Mosquitoes

Recently, outdoor pest control has changed to using nonresidual and environmentally friendly alternative pesticides. Clove oil has been evaluated to determine mortality rates, morphological aberrations, and persistence when used against third and fourth larval instars of *Aedes aegypti* and *Anopheles dirus*. The oil has been evaluated at 1%, 5%, and 10% concentrations in mixtures with soybean oil. Persistence of higher concentrations has been measured over a period of 10 days. For *Ae. aegypti*, clove oil has caused various morphological aberrations, to include deformed larvae, incomplete eclosion, white pupae, deformed pupae, dead normal pupae, and incomplete pupal eclosion. All these aberrations have led to larval mortality. In *Ae. aegypti* larvae, there were no significant differences in mortality at days 1, 5, and 10 or between third and fourth larval instar exposure. In *A. dirus*, morphological aberrations were rare and *S. aromaticum* oil was effective in causing mortality among all larval stages. Oil was effective at producing mortality on days 1, 5, and 10, and had slightly increased $LT_{50}$ rates from day 1 to day 10 (Soonwera and Phasomkusolsil, 2016).

Another work evaluated the larvicidal activity of aqueous and methanolic extracts from clove, *E. caryophyllata* Thunberg (Myrtaceae), and its chemical component, eugenol, against malaria and dengue mosquito vectors. Bioassays have been carried out with these extracts and eugenol on *Anopheles darlingi* Root, 1926 and *Ae. aegypti* Linnaeus, 1762 (Diptera, Culicidae) third-instar larvae. The median lethal concentration values obtained with aqueous extract against *Aa. aegypti* ($LC_{50}$ = 6.4 mg/ml) were higher than those observed against *A. darlingi* ($LC_{50}$ = 99 mg/ml). Eugenol exhibited an $LC_{50}$ value of 3.6 mg/ml against *Ae. aegypti* larvae (Medeiros et al., 2013). These findings show eugenol's potential as a larvicide against malaria and dengue vectors.

Clove oil is also used as a mosquito repellent by itself or in combination with some other essential oils, like citronella, hairy basil, catnip, and vetiver (reviewed by Tisgratog et al., 2016).

### 10.8.7 Grain Pests

Grain pests cause severe postharvest losses in all types of grains and legumes. The control of these insects by conventional insecticides may increase the risks associated with pests and hazards to human health and environmental contamination. Thus, protecting grains with alternative chemical control options is needed. Clove oil can be a good alternative instead of the conventional insecticides.

Nonfumigant applications of clove and cinnamon essential oils have been tested for insecticidal and repellent activities on the bean weevil, *Acanthoscelides obtectus*. Clove oil, which includes 92.94% eugenol, showed a toxicity of $LD_{50}$ = 43.6 µl/kg beans, steadily decreased the growth rate of *A. obtectus* in a dose-dependent manner, and lost its insecticidal activity over time. Additionally, the clove oil delayed bean weevil emergence, whereas cinnamon oil repelled the bean weevil (Jumbo et al., 2014).

Similarly, clove oil has been tested at the doses of 35, 17.9, 8.9, 3.6, 1.8, 0.4, and 0.2 µl/g for maize weevil *Sitophilus zeamais* and bean weevil *A. obtectus* under laboratory conditions. The clove essential oil has caused a mortality of 100% for both species 48 h after treatment with concentrations of 17.9 and 35 µl/g. The $LC_{50}$ for *A. obtectus* was 9.45 µl/g, against 10.15 µl/g for *S. zeamais* (Jairoce et al., 2016).

A grain pest population that is resistant to conventional insecticides might have distinct but possibly overlapping mechanisms to mitigate the actions of essential oils and conventional insecticides. This situation has been represented for clove oil on *S. zeamais* (Correa et al., 2015).

Another test was conducted on two-spotted spider mite, *Tetranychus urticae*, one of the most serious pests of crops, with cumin, clove, and spearmint oils. The total preadult developmental time was significantly shorter when treated with *Eugenia caryophyllata* oil than with control. The mean total fecundity (eggs/female) ranged from 31.08 for those treated with clove oil to 64.44 for the control. The highest ovicidal activity was recorded for cumin oil ($LC_{50}$ = 7.65 µl/L air), followed by clove ($LC_{50}$ = 8.73 µl/L air) and spearmint ($LC_{50}$ = 9.01 µl/L air). According to repellency tests, by increasing the concentration of oils, the repellency effects were increased. The most potent repellency effect was recorded for clove, followed by spearmint and cumin oils (Kheradmand et al., 2015).

### 10.8.8 Cockroaches

Indoor use of conventional pesticides always has a risk of contamination of foods and hypersensitive allergic reactions in humans and animals. So, use of natural pesticides or repellents has become more of an issue.

Evaluation of the repellency and fumigant toxicity of clove (*S. aromaticum*) and sesame (*Sesamum indicum*) oils against the American cockroach (*Periplaneta americana*) has shown complete repellency (100%) against first-instar nymphs at a concentration of 2% for clove oil and 6% for sesame oil. The same result has been obtained against fourth-instar nymphs at a concentration of 10% of sesame oil after 48 h. Clove oil completely repelled all fourth-instar nymphs after 24 h at a concentration of 8%. For the adult stage, the greatest repellency percentages have been recorded by clove oil (90.00 ± 5.77%) and sesame oil (83.33 ± 3.33%) after 48 h at a concentration of 10% (Omara et al., 2013). Clove oil provided highly fumigant toxicity against nymphs and adults of *P. americana* after 24 and 48 h, respectively. Complete mortality (100%) has been recorded at a concentration of 7.5 µl/L of air for first-instar nymphs, 10 µl for fourth-instar nymphs, and 17.5 µl for adults after 48 h of fumigation (Omara et al. 2013).

Surface contact toxicities of clove and rosemary, *Rosmarinus officinalis* L., oils have been investigated against the American cockroach, *P. americana* (L.), in the laboratory. Both clove and rosemary oils have shown variable mortality percentages according to concentration, exposure time, and stage of the insect. Clove oil induced highly significant contact toxic effects against *P. americana* nymphs and adults after 4, 24, and 48 h exposure. First- and fourth-instar nymphs were more sensitive to clove oil ($LC_{50}$ values of 0.0001 and 0.0077 µl/ cm², respectively) than rosemary oil ($LC_{50}$ values of 1.92 and 2.25 µl/cm², respectively) after 24 h. Regarding adults, the effect of both oils was very weak after 48 h exposure. Clove oil had lower $LT_{50}$ values than rosemary oil. For clove oil, the lowest $LT_{50}$ values ranged from 14.10 to 4.85 h in the case of the first instar at a low concentration level of 0.0005–0.0020 µl/ cm², and 20.00–5.00 h for adults at a high concentration level of 2.0–3.0 µl/cm². The $LT_{50}$ was high for the fourth-instar nymphs and ranged from 43.67 to 26.68 h at moderate concentrations, 0.002–0.010 µl/cm², respectively (Sharawi et al., 2013).

### 10.8.9 Head Lice

Head lice, *Pediculus humanus capitis* De Geer, are an obligate ectoparasite of humans worldwide that causes pediculosis capitis, with high prevalence in children. *P. h. capitis* has been treated by methods that include the physical removal of lice, various domestic treatments, and conventional insecticides. Head lice resistance to conventional insecticides has increased, and new alternative topical therapies for head lice infestations are needed, especially those containing plant-derivative active ingredients. Moreover, absorption of synthetic chemicals from the skin of the head has the potential hazard of chronic diseases.

Two herbal shampoos based on *Alpinia galanga* and *S. aromaticum* against head lice have been tested on children, and they have been compared with malathion shampoo (1% w/v malathion) and baby shampoo in order to assess their *in vitro* efficacy. The results have revealed that all herbal shampoo at 6 µl/cm² is a more effective pediculicide than malathion shampoo and baby shampoo at 3 µl/cm². The highest pediculicidal activity was shown by *A. galanga* shampoo at a dose of 6 µl/cm², with 100% mortality at 3 min, an $LT_{50}$ value of 0.4 min, and an $LC_{50}$ value of 1.8 µl/cm². At a dose of 6 µl/cm², *S. aromaticum* shampoo showed 100% mortality at 10 min, an $LT_{50}$ value of 0.9 min, and an $LC_{50}$ value of 2.3 μl/cm². However, malathion shampoo showed low toxicity, with 34.5% mortality at 10 min and an $LT_{50}$ value of 35.9 min, and baby shampoo was nontoxic to all head lice during the observation periods. All data in this study showed that two herbal shampoos based on *A. galanga* and *S. aromaticum* have high potential to alternative pediculicides for head lice treatments of children (Soonwera, 2015).

Another study was carried out to determine the pediculocidal activity using hexane flower bud extract of *S. aromaticum* against *P. h. capitis*, examining direct contact and fumigant toxicity (closed- and open-container methods) bioassay. The filter paper contact bioassay study showed pronounced pediculicidal activity in the flower bud hexane extract of *S. aromaticum*. The toxic effect was determined every 5 min in an 80 min treatment. The results showed percent mortalities of 40, 82, and 100 at 5, 10, and 20 min, and the $LT_{50}$ value was 5.83 (0.5 mg/cm²); 28, 82, and 100 at 5, 10, and 30 min ($LT_{50}$ = 6.54; 0.25 mg/cm²); and 13, 22, 42, 80, and 100 at 5, 10, 20, 40, and 80 min ($LT_{50}$ = 18.68; 0.125 mg/cm²), respectively. The vapor phase toxicity was tested at 0.25 mg/cm². There was a significant difference in the pediculicidal activity of *S. aromaticum* extract against *P. h. capitis* between closed- and open-container methods. The mortality was more effective in the closed containers than in the open ones, indicating that the effect of hexane extract was largely a result of action in the vapor phase exhibiting fumigant toxicity (Bagavan et al., 2011).

### 10.8.10 Other Pests

Eugenol, eugenol acetate, and β-caryophyllene were effective in repelling red imported fire ants *Solenopsis invicta* (Hymenoptera: Formicidae), with eugenol being the fastest-acting compound. Clove oil was also an effective spatial repellent for pestiferous social wasps *Vespula pensylvanica* and paper wasps, mainly *Polistes dominulus* (reviewed by Cortes-Rojas et al., 2014).

---

## 10.9 Conclusions

The development of essential oils as pesticides and biocides to control insects is an alternative or complementary approach to synthetic insecticides. They are environmentally friendly products; that is, they have natural origin and are biodegradable, and they have diverse physiological targets within insects that may delay the evolution of insect resistance. Thus, they are especially suited to organic farming, as well as to integrated pest management.

The essential oils have an attractive potential to substitute for synthetic insecticides because of their diversity and efficiency.

Clove oil is a good candidate for a safer control agent that may provide good antipest activity due to its low toxicity to mammals and easy biodegradability.

These features indicate that pesticides based on clove essential oil could be used in a variety of ways to control a large number of pests. However, more research is needed on the tested oils, such as the acaricidal effects of each major compound, and their modes of action and efficacies in field conditions.

# References

Ali, S.M., Khan, A.A., Ahmed, I. et al. 2005. Antimicrobial activities of eugenol and cinnamaldehyde against the human gastric pathogen *Helicobacter pylori*. *Ann Clin Microbiol Antimicrob*, 4: 20.

Bagavan, A., Rahuman, A.A., Kamaraj, C. et al. 2011. Contact and fumigant toxicity of hexane flower bud extract of *Syzygium aromaticum* and its compounds against *Pediculus humanus capitis* (Phthiraptera: Pediculidae). *Parasitol Res*, 109(5): 1329–1340.

Board, N. 2010. *Handbook on Spices*. Reprinted ed. Delhi: Asia Pacific Business Press.

Buchbauer, G. 2010. Biological activities of essential oils. In Başer, K.H.C., Buchbauer, G. (eds.), *Handbook of Essential Oils Science, Technology and Application*. Boca Raton, FL: CRC Press, p. 235.

Chaieb, K., Hajlaoui, H., Zmantar, T. et al. 2007. The chemical composition and biological activity of clove essential oil, *Eugenia caryophyllata* (*Syzigium aromaticum* L. Myrtaceae): A short review. *Phytother Res*, 21: 501–506.

Chami, F., Chami, N., Bennis, S., Bouchikhi, T., Remmal, A. 2005. Oregano and clove essential oils induce surface alteration of *Saccharomyces cerevisiae*. *Phytother Res*, 19: 405–408.

Charles, D.J., Simon, J.E. 1990. Comparison of extraction methods for the rapid determination of essential oil content and composition of basil. *J Am Soc Hortic Sci*, 115(3): 458–462.

Correa, Y.D.C.G., Faroni, L.R., Haddi, K., Oliveira, E.E., Pereira, E.J.G. 2015. Locomotory and physiological responses induced by clove and cinnamon essential oils in the maize weevil *Sitophilus zeamais*. *Pestic Biochem Phys*, 125: 31–37.

Cortes-Rojas, D.F., de Souza, C.R.F., Oliveira, W.P. 2014. Clove (*Syzygium aromaticum*): A precious spice. *Asian Pac J Trop Biomed*, 4(2): 90–96.

Cox, S.D., Mann, C.M., Markham, J.L. 2001. Interactions between components of the essential oil of *Melaleuca alternifolia*. *J Appl Microbiol*, 91: 492–497.

Daniel, A.N., Sartoretto, S.M., Schimidt, G., Caparroz-Assef, S.M., Bersani-Amado, C.A., Cuman, R.K. 2009. Anti-inflamatory and antinociceptive activities of eugenol essential oil in experimental animal models. *Rev Bras Farmacogn*, 19(1B): 212–217.

Daryono, E.D. 2015. Reactive extraction process in isolation of eugenol of clove essential oil (*Syzigium aromaticum*) based on temperature and time process. *Int J Chemtech Res*, 8(11): 564–569.

Donald, P.L., Lampman G.M., Kritz, G.S., Engel, R.G. 2006. *Introduction to Organic Laboratory Techniques*. 4th ed. Thomson Brooks/Cole-Cengage Learning, Hampshire, UK, pp. 797–817.

Gayoso, C.W., Lima, E.O., Olivera, V.T. et al. 2005. Sensitivity of fungi isolated from onychomycosis to *Eugenia caryophyllata* essential oil and eugenol. *Fitoterapia*, 76: 247–249.

Girisgin AO, Barel S, Zilberman Barzilai D, Girisgin O. 2014. Determining the stability of clove oil (eugenol) for use as an acaricide in beeswax. *Israel J Vet Med*, 69(4): 197–201.

Guan, W., Li, S., Yan, R., Tang, S., Quan, C. 2007. Comparison of essential oils of clove buds extracted with supercritical carbon dioxide and other three traditional extraction methods. *Food Chem*, 101(4): 1558–1564.

Halder, S., Mehta, A.K., Kar, R., Mustafa, M., Mediratta, P.K., Sharma, K.K. 2011. Clove oil reverses learning and memory deficits in scopolamine treated mice. *Planta Med*, 77(8): 830–834.

Hasson, R.H., Al-Zubaidi, H.H. 2015. The acaricidal effect of clove leaf plant extract *Eugenia caryophyllus* against *Argas* spp. soft ticks in Iraq. *Int J Adv Res*, 3(9): 257–262.

Hemalatha, R., Nivetha, P., Mohanapriya, C., Sharmila, G., Muthukumaran, C., Gopinath, M. 2015. Phytochemical composition, GC-MS analysis, in vitro antioxidant and antibacterial potential of clove flower bud (*Eugenia caryophyllus*) methanolic extract. *J Food Sci Technol*, 1–10.

Hussein, G., Miyashiro, H., Nakamura, N., Hattori, M., Kakiuchi, N., Shimotohno, K. 2000. Inhibitory effects of Sudanese medical plant extracts on hepatitis C virus (HCV) protease. *Phytother Res*, 14: 510–516.

Jairoce, C.F., Teixeira, C.M., Nunes, C.F., Nunes, A.M., Pereira, C.M., Garcia, F.R. 2016. Insecticide activity of clove essential oil on bean weevil and maize weevil. *Rev Bras Eng Agríc Ambient*, 20(1): 72–77.

Jirovetz, L., Buchbauer, G., Stoilova, I., Stoyanova, A., Krastanov, A., Schmidt, E. 2006. Chemical composition and antioxidant properties of clove leaf essential oil. *J Agric Food Chem*, 54: 6303–6307.

Jumbo, L.O.V., Faroni, L.R., Oliveira, E.E., Pimentel, M.A., Silva, G.N. 2014. Potential use of clove and cinnamon essential oils to control the bean weevil, *Acanthoscelides obtectus* Say, in small storage units. *Ind Crops Prod*, 56: 27–34.

Just, J., Bunton, G.L., Deans, B.J., Murray, N.L., Bissember, A.C., Smith, J.A. 2016. Extraction of eugenol from cloves using an unmodified household espresso machine: An alternative to traditional steam-distillation. *J Chem Educ*, 93(1): 213–216.

Kheradmand, K., Beynaghi, S., Asgari, S., Garjan, S. 2015. Toxicity and repellency effects of three plant essential oils against two-spotted spider mite, *Tetranychus urticae* (Acari: Tetranychidae). *J Agric Sci Technol*, 17: 1223–1232.

Koul, O., Walia, S., Dhaliwal, S. 2008. Essential oils as green pesticides: Potential and constraints. *Biopestic Int*, 4(1): 63–84.

Kurokawa, M., Hozumi, T., Basnet, P. et al. 1998. Purification and characterization of eugeniin as an anti-herpesvirus compound from *Geum japonicum* and *Syzygium aromaticum*. *J Pharmacol Exp Ther*, 284: 728–735.

Li, Y., Fabiano-Tixier A.S., Chemat, F. 2014. *Essential Oils as Reagents in Green Chemistry*. Berlin: Springer.

Maggi, M.D., Ruffnengo, S.R., Gende, L.B., Sarlo, E.G., Eguaras, M.J., Bailac, P.N., Ponzi, M.I. 2010. Laboratory evaluations of *Syzygium aromaticum* (L.) Merr. et Perry essential oil against *Varroa destructor*. *J Essential Oil Res*, 22(2): 119–122.

Mahakittikun, V., Soonthornchareonnon, N., Foongladda, S., Boitano, J.J., Wangapai, T., Ninsanit, P. 2013. A preliminary study of the acaricidal activity of clove oil, *Eugenia caryophyllus*. *Asian Pac J Allergy Immunol*, 32(1): 46–52.

Medeiros, E.D.S., Rodrigues, I.B., Litaiff-Abreu, E., Pinto, A.C.D.S., Tadei, W.P. 2013. Larvicidal activity of clove (*Eugenia caryophyllata*) extracts and eugenol against *Aedes aegypti* and *Anopheles darlingi*. *Afr J Biotechnol*, 12(8): 836–840.

Mytle, N., Anderson, G.L., Doyle, M.P., Smith, M.A. 2006. Antimicrobial activity of clove (*Syzgium aromaticum*) oil in inhibiting *Listeria monocytogenes* on chicken frankfurters. *Food Control*, 17: 102–107.

Namiki, M. 1994. Antimutagen and anticarcinogen research in Japan. In Huang, M.T., Osawa, T., Ho, C.T., Rosen, R.T. (eds.), *Food Phytochemicals for Cancer Prevention III*. Vol. 546. Maple Press: York, PA, pp. 64–81.

Núñez, L., D'Aquino, M., Chirife, J. 2001. Antifungal properties of clove oil (*Eugenia caryophyllata*) in sugar solution. *Braz J Microbiol*, 32: 123–126.

Ogunwande, I.A., Olawore, N.O., Ekundayo, O., Walker, T.M., Schmidt, J.M., Setzer, W.N. 2005. Studies on the essential oils composition, antibacterial and cytotoxicity of *Eugenia uniflora* L. *Int J Aromather*, 15: 147–152.

Omara, S.M., Al-Ghamdi, K.M., Mahmoud, M.A., Sharawi, S.E. 2015. Repellency and fumigant toxicity of clove and sesame oils against American cockroach (*Periplaneta americana* (L.). *Afr J Biotechnol*, 12(9): 963–970.

Panda, H. 2003. *Essential Oils Handbook*. New Delhi: National Institute of Industrial Research, p. 219.

Pandey, S.N., Chadha, A. 1993. *A Text Book of Botany: Plant Anatomy and Economic Botany*. Vol. 3. Vikas Publishing House, Uttar Pradesh, India, p. 485.

Pasay, C., Mounsey, K., Stevenson, G. et al. 2010. Acaricidal activity of eugenol based compounds against scabies mite. *PloS One*, 5(8): e12079.

Pawar, V.C., Thaker, V.S. 2006. In vitro efficacy of 75 essential oils against *Aspergillus niger*. *Mycoses*, 49: 316–323.

Razafimamonjison, G., Jahiel, M., Duclos, T., Ramanoelina, P., Fawbush, F., Danthu, P. 2013. Bud, leaf and stem essential oil composition of clove (*Syzygium aromaticum* L.) from Indonesia, Madagascar and Zanzibar. *Nat Prod Commun*, 8: 1–7.

Regnault-Roger, C. 2013. Essential oils in insect control. In Ramawat, K.G., Merillon, J.M. (Eds.), *Natural Products*. Berlin: Springer, pp. 4087–4107.

Shan, B., Cai, Y.Z., Sun, M., Corke, H. 2005. Antioxidant capacity of 26 spice extracts and character-ization of their phenolic constituents. *J Agric Food Chem*, 53(20): 7749–7759.

Sharawi, S.E., Abd-Alla, S.M., Omara, S.M., Al-Ghamdi, K.M. 2013. Surface contact toxicity of clove and rosemary oils against American cockroach, *Periplaneta americana* (L.). *Afr Entomol*, 21(2): 324–332.

Singhal, R., Kulkarni, P.R., Rege, D.V. 1997. *Handbook of Indices of Food Quality and Authenticity.* Woodhead Publishing, Cambridge, UK, p. 402.

Sofia P.K., Prasad R., Vijay V.K., Srivastava A.K. 2007. Evaluation of antibacterial activity of Indian spices against common food-borne pathogens. *Int J Food Sci Technol*, 42(8): 910–915.

Soonwera, M. 2015. Herbal pediculicides base on *Alpinia galanga* (L.) Willd (Zingiberaceae) and *Syzygium aromaticum* (L.) Merrill & Perry (Myrtaceae) against head louse (*Pediculus humanus capitis* De Geer; Pediculidae). *Int J Agric Technol*, 11(7): 1503–1513.

Soonwera, M., Phasomkusolsil, S. 2016. Effect of *Cymbopogon citratus* (lemongrass) and *Syzygium aromaticum* (clove) oils on the morphology and mortality of *Aedes aegypti* and *Anopheles dirus* larvae. *Parasitol Res*, 1–13.

Sparagano, O., Khallaayoune, K., Duvallet, G., Nayak, S., George, D. 2013. Comparing terpenes from plant essential oils as pesticides for the poultry red mite (*Dermanyssus gallinae*). *Transbound Emerg Dis*, 60(s2): 150–153.

Tampieri, M.P., Galuppi, R., Macchioni, F. et al. 2005. The inhibition of *Candida albicans* by selected essential oils and their major components. *Mycopathologia*, 159: 339–345.

Tisgratog, R., Sanguanpong, U., Grieco, J.P., Ngoen-Kluan, R., Chareonviriyaphap, T. 2016. Plants traditionally used as mosquito repellents and the implication for their use in vector control. *Acta Trop*, 157: 136–144.

# 11

## Sandalwood Oil (*Santalum album* L.): Source of a Botanical Pesticide—Present Status and Potential Prospects

Somnath Roy, Gautam Handique, Ranjida Ahmed, and N. Muraleedharan

### CONTENTS

## 11.1 Introduction

Plant essential oils are naturally occurring volatile components primarily derived from the seeds, stems, bark, and roots of many plants that have been in use for thousands of years in various civilizations for medicinal and health purposes (Jones, 1996). The versatile role of these essential oils ranges from personal beauty care and household cleaning products, aromatherapy, and natural drugs, which comes from the antimicrobial, anti-inflammatory, and antioxidant properties present in them (Anonymous, 2016). These oils are usually present in glandular hairs or secretory cavities of the plant cell wall and as fluid droplets in the bark, flowers, leaves, stems, roots, and/or fruits of diverse plants. The aromatic properties of essential oils execute various functions for the plants, which include attracting or deterring insects, protecting themselves from heat or cold, and utilizing some chemical constituents such as monoterpenes, sesquiterpenes, and phenolics (Bhalla et al., 2013) of the oil as defense materials (Koul et al., 2008).

**TABLE 11.1**

Geographical Distribution and General Uses of *Santalum* L. (Santalaceae)

| Serial No. | Scientific Name | Geographical Distribution | General Uses | Reference |
|---|---|---|---|---|
| 1 | *S. album* L. | India, United States, Japan, Germany, Indonesia, France | Flavor component; fragrance; anti-inflammatory, antipyretic, antiulcerogenic, antifungal, antiviral, antibacterial, antioxidant, antispasmodic (muscle relaxant), astringent, diuretic, and anticancerous activities; neuroleptic | Rani et al., 2013; Burdock and Carabin, 2008; Okugawa et al., 1995; Koch et al., 2008; McKinnel, 1990; Baldovini et al., 2010 |
| 2 | *S. freycinetianum* Gaudich | Sri Lanka, Hawaiian Islands | Perfume, medicinal | Subasinghe et al., 2013; Merlin et al., 2006 |
| 3 | *S. haleakalae* Hillebr | Sri Lanka | Perfume, medicinal | Subasinghe et al., 2013 |
| 4 | *S. ellipticum* Gaudich | Sri Lanka, Hawaiian Islands | Perfume, medicinal | Subasinghe et al., 2013; Merlin et al., 2006 |
| 5 | *S. peniculum* Hook. & Arn. | Sri Lanka | Perfume, medicinal | Subasinghe et al., 2013 |
| 6 | *S. pyrularium* A. Gray. | Sri Lanka | Perfume, medicinal | Subasinghe et al., 2013 |
| 7 | *S. involutum* St. John | Sri Lanka | Perfume, medicinal | Subasinghe et al., 2013 |
| 8 | *S. boniness* (Nakai) Tuyama | Sri Lanka | Perfume, medicinal | Subasinghe et al., 2013 |
| 9 | *S. insulare* Bertero ex A.DC. | Sri Lanka, France | Perfume, medicinal | Subasinghe et al., 2013; Baldovini et al., 2010 |
| 10 | *S. austrocaledonicum* Viell. | Sri Lanka, New Caledonia, Vanuatu, France, Fiji | Perfume, medicinal | Subasinghe et al., 2013; McKinnel, 1990; Baldovini et al., 2010; Thomson et al., 2011 |
| 11 | *S. minutum* N. Halle | Fiji | Perfume, medicinal | Thomson et al., 2011 |
| 12 | *S. pilosulum* N. Halle | Fiji | Perfume, medicinal | Thomson et al., 2011 |
| 13 | *S. yasi* Seem. | Sri Lanka, Fiji | Perfume, medicinal | Subasinghe et al., 2013; McKinnel, 1990; Thomson et al., 2011; Huish et al., 2015 |
| 14 | *S. macgregorii* F. Muell | Sri Lanka, New Guinea | Perfume, medicinal | Subasinghe et al., 2013; McKinnel, 1990 |
| 15 | *S. obtusifolium* R. Br. | Sri Lanka | Perfume, medicinal | Subasinghe et al., 2013 |
| 16 | *S. lanceolatum* R. Br. | Australia, Sri Lanka | Perfume, medicinal | McKinnel, 1990; Subasinghe et al., 2013 |
| 17 | *S. fernanderzianum* F. Phil. | Sri Lanka | Perfume, medicinal | Subasinghe et al., 2013 |
| 18 | *S. salicifolium* Meurisse | Sri Lanka | Perfume, medicinal | Subasinghe et al., 2013 |
| 19 | *S. spicatum* (R. Br.) A.DC. | Australia, Sri Lanka, France | Perfume, medicinal | McKinnel, 1990; Subasinghe et al., 2013; Baldovini et al., 2010 |
| 20 | *S. acuminatum* (R. Br.) A.DC. | Australia | Perfume, medicinal | McKinnel, 1990 |
| 21 | *S. murrayanum* (T. Mitch) C.A. Gardner | Australia | Perfume, medicinal | McKinnel, 1990 |

Among the commercially viable essential oils, sandalwood oil is largely used for a variety of purposes. The oil is derived from the genus *Santalum* of the Santalaceae family, which is hemiparasitic in character that includes 15 extant species with approximately 14 varieties, and 1 extinct species, spread all throughout India, Australia, and the Pacific Islands (Table 11.1) to as far as the Juan Fernandez Islands, 600 miles off the coast of Chile, to the northwest of the Bonin Islands, 600 miles south of Honshu province, Japan (van Balgooy, 1960, 1971; van Balgooy et al., 1996).

## 11.2 Botany of the Plant

*Santalum album* (East Indian sandalwood or sandal) is a small evergreen plant that can grow up to a height of 4 m in Australia, while in India, it can grow up to 20 m with girth diameter of around 2.4 m and lean, flabby branchlets. The bark may be dark brown, reddish, dark gray, or nearly black; tight, smooth in young trees; and rough with deep vertical cracks in older trees, having a reddish color from within. Leaves are thin, usually opposite, ovate or ovate elliptical, 3–8 to 3–5 cm, glabrous and shining green above, glaucous and slightly paler beneath; tip rounded or pointed; stalk grooved, 5–15 cm long; venation noticeably reticulate. Flowers are small; purplish-brown, straw colored, reddish, green, or violet; about 4–6 mm long; up to six small terminal or axillary clusters, unscented in axillary or terminal; paniculate cymes. Fruit is a globose, fleshy drupe; red, purple to black when ripe; about 1 cm in diameter, with hard ribbed endocarp and crowned with a scar; almost stalkless; smooth; single seeded (Sindhu et al., 2010). The generic name is derived from the Greek *santalon*, meaning "sandalwood," and the species name from the Latin *albus*, meaning "white," alluding to the bark.

S. *album* forms haustoria root connections with the roots of a wide range of species, including annual and perennial crops (Tennakoon et al., 1997; Tennakoon and Cameron, 2006), and is renowned for its fragrant heartwood (Shea et al., 1997). In general, S. *album* is a small to medium-sized tree that grows to 8–12 m in height and 2.5 m in girth. The tree has many slender drooping branches, and the bark is smooth gray-brown, thinly furrowed, semicoarse. The bark and sapwood are odorless, and the roots and heartwood contain the essential oil (Kirtikar and Basu, 1993). The flowers are small with numerous short stalks (Grieve, 1971). The form of the crown is flat or a dome. It has an inflorescence terminal or axillary panicle or raceme. Flowers are unisexual or hermaphrodite, small, white or yellowish, 4(-5)-merous; the perianth tube is campanulate. Flowering starts at 3–4 year-old plants (Prasetyaningtyas, 2007). Benencia and Courreges (1999) report that the heartwood is yellowish brown and strongly scented. Leaves are 3.8–6.3 by 1.6–3.2 cm, elliptic lanceolate, and subacute glabrous, and the entire thin base is acute. Petioles are 1–1.3 cm long slender flowers, brownish purple induorous, in terminal and auxiliary paniculate cymes shorter than leaves. They also report that perianth has a campanulated limb of four, valvate triangular segments with four stamens, exerted, alternating with four rounded obtuse scales (Vasundhara et al., 2015).

## 11.3 Methods of Oil Extraction

The sandalwood tree's precious oil is located within the tree's heartwood, and the older the tree, the higher the proportion of its heartwood. As the oil is held tightly within the wood, a distillation process is required, in which the wood is first ground to a powder form. Distillation methods vary, ranging from today's steam distillation to the more traditional hydrodistillation (water), also including $CO_2$ extraction, absolute extraction, and an array of new technologies (Montoro et al., 2015). The sandalwood industry, however, lacks a universally accepted method to determine the heartwood oil content (Hettiarachchi et al., 2010).

### 11.3.1 Steam Distillation

Steam distillation is a process in which steam heated at extremely high temperatures (usually around 140°F–212°F) is passed through the powdered wood. The steam then carries the essential oil that is locked within the cellular structure of the wood. The mixture of steam and oil then flows through a condenser and cools, yielding two separate layers of oil and water. The sandalwood essential oil, which separates from the hydrosol (floral water) and ascends to the top, is collected. The distillation process for sandalwood oil takes 14–36 hours, longer than for many other essential oils. Despite the fact that this method requires a longer process than other distillation methods (Kusuma and Mahfud, 2016b), it is known to produce superior quality oil, yielding up to 84.32% santalol. Steam-distilled sandalwood oil is light pale yellow in color (Nautiyal, 2011).

### 11.3.2 Hydrodistillation

Hydrodistillation (water distillation) is the traditional method of extracting sandalwood essential oil. Hydrodistillation is a widely used oil extraction process to quantify the oil yield and quality (Hettiarachchi et al., 2010). This process produces good-quality oil. Instead of steam passing through the powdered wood, in hydrodistillation the powdered wood is immersed in water to soak. The water is then boiled, often heated in an open fire, and the vapor it carries is collected from the top of the hydrodistiller. The oil is then separated from the hydrosol. Some disadvantages of this method include the fact that it requires heating a large quantity of water, which increases the cost and time needed for sandalwood distillation. Also, the temperature of the boiling water is difficult to control, which causes the rate of distillation to vary and often causes the oil to be "burned," lessening its quality.

### 11.3.3 $CO_2$ Extraction

$CO_2$ extraction, also known as supercritical $CO_2$ extraction or supercritical fluid $CO_2$ extraction, is the latest technique of getting essential oil from pure plant parts (Brunner, 2005). $CO_2$ extracts are similar to essential oils and can be used in aromatherapy or perfumery. For this process, $CO_2$ is used as the solvent in place of steam or water. The $CO_2$ extraction method involves impelling pressurized carbon dioxide into a chamber filled with the powdered heartwood of sandalwood. When $CO_2$ is subjected to high pressure, it attains liquid properties while retaining its gaseous state. Because of these liquid properties of the gas, $CO_2$ acts as a solvent, releasing the sandalwood oil from the wood. While steam distillation requires temperatures of 140°F–212°F, $CO_2$ extraction only requires temperatures of around 95°F–100°F. Oil yielded in $CO_2$ extraction contains about 82.5% santalol, while steam-distilled oil contains about 84.32% santalol, and hydrodistilled oil contains about 52.59% santalol.

### 11.3.4 Absolute (Solvent) Extraction

Absolute (solvent) extraction is a method of extracting essential oils that are most often used by the perfume industry, due to the strong aroma of the oil extracted by this process. Solvent extraction uses solvents, such as ethanol, methanol, hexane, or petroleum ether to extract the essential oil from the sandalwood plant. The solvent additionally pulls out the chlorophyll and other plant tissues, which results in the formation of a deeply colored and viscous extract of the solvent. This first product is termed as a concrete and contains the concentrated extract of waxes and/or fats, as well as the odoriferous materials from the plant. The concrete is then mixed with alcohol, which helps in extracting the major aromatic components of the material. The final product is known as an absolute. This distillation method is, however, not applied to produce therapeutic oils for aromatherapy, such as sandalwood essential oil, as chemicals such as hexane, acetone, and di-methylene-chloride are used in the process. Absolute extraction is rarely used to extract sandalwood. In a study by Nautiyal (2010), the oil produced through solvent extraction methods was more in quantity than that produced through hydrodistillation and steam distillation, with the highest yield in ethyl alcohol extraction (84% santalol), but three of the four solvent-extracted sandalwood oils were recorded as "less pleasant," indicating the generally inferior note of oil derived from the other methods.

### 11.3.5 Phytonics Process

The phytonics process is one of the most recently developed technologies for essential oil extraction using nonchlorofluorocarbons (non-CFCs). It is also called florasol extraction, and the oils are referred to as phytols. Advanced Phytonics Limited, Manchester, United Kingdom, developed and patented this method. A recent statement on essential oil extraction processes reports that the oil mostly extracted by this process is biological or phyto-pharmacological aromatic components of essential oils that can be used directly without further physical or chemical treatments (Handa, 2008). The phytonics process involves the solvent hydrofluorocarbon-134a, which has a boiling point of 25°C (Swapna et al., 2015). Additionally, this solvent is neither toxic nor flammable, and has no negative impact on the ozone layer. By most standards, this substance makes for a "poor solvent," as it does not mix with many other chemicals, such as mineral oils or triglycerides. However, since this solvent also does not dissolve plant wastes, it is perfect for use in the extraction of essential oils. Other advantages of this extraction technology are numerous; for example, unlike other methods, such as steam or water distillation, which employ high temperatures, the phytonics process is "cool and gentle," and the products are not exposed to excess, possibly harmful temperatures. Because the phytonics process also uses minimal electricity and does not release any harmful emissions into the atmosphere, it is thought to be much less threatening to the environment (Handa, 2008).

## 11.4 Methods of Analysis of Oil

Santalol in sandalwood oil is conventionally analyzed by gas chromatography (GC) (Howes et al., 2004), while those in heartwoods of field-grown trees are analyzed by gas chromatography with mass spectrometry (GC-MS)/MS (Jones et al., 2006). Fourier transform infrared (FTIR) qualitative analysis is also found to be quite interpretative in establishing the sandalwood oil measurement (Nautiyal, 2011). More recently, Misra and Dey (2013) evaluated the oil in high-performance thin-layer chromatography (HPTLC)–derived spectral scans, which could be helpful in the identification of sesquiterpenoids present in the oil.

Panto et al. (2015) used the hyphenation of liquid and gas chromatography techniques for collection of the important sesquiterpene alcohols of sandalwood oil, as reported by the international regulations, and found the following components: (Z)-α-santalol (44.6%), (Z)-α-*trans*-bergamotol (5.9%), (Z)-β-santalol (19.2%), epi-(Z)-β-santalol (3.2%), α-bisabolol (0.7%), (Z)-lanceol (1.4%), and (Z)-nuciferol (4.0%). Furthermore, Nautiyal (2011) used hydrodistillation, steam distillation, solvent extraction, and subcritical carbon dioxide for analysis of oil components and reported that subcritical carbon dioxide yielded oil determined to contain 2.56% santalene and 82.5% santalol, benzene extracted oil contained 9.49% santalene and 42.99% santalol, diethyl ether extracted oil contained 2.04% santalene and 72.19% santalol, ethyl alcohol extracted oil contained 1.56% santalene and 83.56% santalol, hydrodistilled oil contained 3.43% santalene and 52.59% santalol, hydrodistilled (alkaline) oil contained 7.26% santalene and 56.79% santalol, and steam-distilled oil contained 2.57% of santalene and 84.32% santalol. Howes et al. (2004) suggested a specification of >43% Z-α-santalol and >18% Z-β-santalol by GC-MS for sandalwood oil.

## 11.5 Composition of Oil

Sandalwood oil consists almost exclusively of closely related sesquiterpenoids (Burdock and Carabin, 2008). Four sesquiterpenols, α-, β-, and epi-β-santalol and α-exo-bergamotol, make up approximately 90% of the oil of *S. album* (Diaz-Chavez et al., 2013). These compounds are the hydroxylated analogues of α-, β-, and epi-β-santalene and α-exo-bergamotene. The oil contains other minor components, including sesquiterpene hydrocarbons (~6%) (Burdock, 2002). The sesquiterpene hydrocarbons are mostly α- and β-santalenes and epi-β-santalene, with small amounts of α- and β-curcumenes and possibly β-farnesene and dendrolasin (Burdock and Carabin, 2008). The volatile composition of *S. album* essential oil contains >90% sesquiterpene alcohols. Current International Organization for Standardization (ISO) standards for (Z)-α-santalol must fall within the 41%–55% range, while (Z)-β-santalol must fall within the 16%–24% range of alcohol content. Different authors reported that sandalwood also contains (Z)-α-*trans*-bergamotol, epi-β-santalol, α-bisabolol, (E,E)-farnesol, (Z)-lanceol, (Z)-nuciferol (Howes et al., 2004; Panto et al., 2015), α-bisabalol, *cis*-bergamatol, γ-curcumen-12-ol (Subasinghe et al., 2013), hydrocarbons (santene and nor-tricyclo-ekasantalene), alcohols (santenol and teresantalol), aldehydes (nor-tricyclo-kasantalal), and acids (α- and β-santalic acids and teresantalic acids) (Sindhu et al., 2010). The alcohols α-santalol and β-santalol are mainly responsible for the odor, with α-santalol comprising approximately 7%–60% of the total santalol and β-santalol comprising approximately 7%–33% of the total santalol, depending on the sourced species (Lawrence, 1991). The other constituents reported include dihydro-β-agarofuran, santene, teresantol, borneol, teresantalic acid, tricycloekasantalal, santalone, and santanol (Khan and Abourashed, 2010). The other minor chemical components reported in sandalwood oil are phenols, lactones, and terpenes. Hentriacontan-16-one is reported in one *S. album* oil sample (Anonymous, 1999).

Two minor components, cyclosantalal (0.21%–2.26%) and isocyclo-santalal (0.11%–1.47%), as new sesquiterpene aldehydes were reported. Also, a new acid ketosantalic (as methyl ester) and gamma-L-glutamyl-S-(*trans*-1-propenyl)-L-cysteine sulfoxide, an interesting natural sulfoxide diastereoisomer, have been isolated from sandal (Sindhu et al., 2010). Zhang et al. (2012) detected a total of 66 volatile compounds from the pericarp of sandalwood. The most prominent compounds were found to be oleic and palmitic acids, which represented about 40%–70% of the oil. Many fragrant constituents and biologically active components, such as α- and β-santalol, cedrol, esters, aldehydes, phytosterols, and squalene, are also present in the pericarp oils.

## 11.6 Physical and Chemical Properties of Oil

The volatile oil derived from the roots and heartwood of *S. album* L. is a somewhat viscous, yellowish liquid of a peculiarly sweet and very lasting odor (Benencia and Courreges, 1999). Vasundhara et al. (2015) reported the physical and chemical properties of *S. album* and found that the refractive index of sandalwood seed oil was 1.4898, and specific gravity was 0.9406. Sandalwood seed oil has a high level of freshness, as it has very low acid value (0.748), and the ester value was 157.108. The specific gravity at 25°C was 0.94061. Further, they reported that the saponification value was 157.856. In addition, Sindhu et al. (2010) reported that volatile oil extracted from *S. album* L. derived from the roots and heartwood is a colorless to yellowish, viscous (reference index 1.499–1.506, specific gravity 0.962–0.985, and rotation 19°–20°) liquid with a peculiar heavy sweet odor, the chief constituent of the oil being santalol.

## 11.7 General Uses of Oil

Sandalwood oil is used as a flavor component in many food products (Rasor and Dunca, 2014), including alcoholic and nonalcoholic beverages, frozen dairy desserts, candy, baked goods, and gelatin and puddings, at use levels generally below 0.001% (10 ppm), except in hard candy. The highest maximum use level for sandalwood oil in food products is approximately 90 ppm. Sandalwood oil is generally used as a natural flavoring substance (Hall and Oser, 1965) or in conjunction with other flavor (Burdock, 2002). Current interests in sandalwood oil are growing in the aromatherapy, cosmetics, and food industries due to its sedative action and fragrance (Burdock and Carabin, 2008). The Flavor and Extract Manufacturers Association (FEMA) has approved sandalwood oil as a "generally recognized as safe" (GRAS) flavoring ingredient for use in food.

Sandalwood oil is mainly used in the perfume industry. The oil is an excellent base and fixative for other high-grade perfumes. Most top-grade perfumes have sandalwood oil as their base, which, in itself, is an excellent, mild, long-lasting, and sweet perfume, yet the industry finds that it can blend very well with other perfumes and does not impart its fragrance when used as a base. It can also fix better perfumes, which are volatile, for longer hours. From perfumery to joss sticks, there are several hundred products that use sandalwood oil. It is also used in the soap industry (Ral, 1990).

Essential oils obtained from seeds, stem bark, and roots of many plants have been widely used in traditional medicine. Sandalwood oil is used medicinally for common colds, bronchitis, fever, infection of the urinary tract, inflammation of the mouth and pharynx, liver and gallbladder complaints, and other maladies (Kusuma and Mahfud, 2016a). In addition, the main α-santalol component is found to prevent the development of skin tumors in mice and reduce the likelihood of actinic keratosis and skin cancer (Dickinson et al., 2014). Sandalwood is mainly used as a coolant, and also has a sedative effect and astringent activity, making it useful as a disinfectant in the genitourinary and bronchial tracts; it is also a diuretic, expectorant, and stimulant. The same is also used as a tonic for the heart, stomach, and liver; as an antipoison; for fever and memory improvement; and as a blood purifier (Sindhu et al., 2010). Satou et al. (2014) reported that fragranced sandalwood oil has an anxiolytic activity on mice. Sandalwood oil is used as an antiseptic, anti-inflammatory, antiphlogistic, antispasmodic, astringent, cicatrisant, carminative, diuretic, disinfectant, emollient, expectorant, hypotensive, memory booster, sedative, and tonic. Sandalwood oil is an active substance of agreeable odor employed in the treatment of subacute and chronic infections of mucous tissues, particularly gonorrhea after the active symptoms have been mitigated. Chronic bronchitis with fetid expectoration, chronic mucous diarrhea, chronic inflammation of the bladder, and pyelitis are also said to be benefited by it. It occasionally disturbs the gastrointestinal tract, and like copaiba, which it was introduced to supersede, it will cause cutaneous eruptions. The dose ranges from 5 to 20 drops, in capsules or emulsion (Nautiyal, 2011). Sandalwood oil is also used to treat gastric ulcers (Ahmed et al., 2013).

## 11.8 Pesticidal Uses of Sandalwood Oil

In general, plant essential oils could be good resources in developing insect pest control agents (Batish et al., 2008; Ebadollahi, 2013), as they are reported to have many bioactivities, including insecticidal and repellant activities, and as oviposition deterrent against flies (Kim et al., 2016). Besides other uses, sandalwood oil can also be used as an insecticide. Roh et al. (2011) reported that sandalwood oil and its constituents have considerable acaricidal properties. Sandalwood oil was significantly active against adults of *Tetranychus urticae*, with more than 80% mortality in adults with oviposition-deterring effects of up to 89.3%. A mixture of α- and β-santalol, which are two of the main components of the oil, also demonstrated higher mortality (85.5%) and oviposition-deterrent effects (94.7% reduction in the number of eggs) than the control. From the experiment, it was conclusive that sandalwood oil could likely be used for the sustainable management of *T. urticae* on various greenhouse vegetables and food crops. Similarly, Roh et al. (2015) reported that both sandalwood oil and santalol had significant insecticidal properties (up to 98.8%) against *Aphis gossypii*, which were comparable to that of imidacloprid, a neonicotinoid insecticide. Further, the control efficacies of sandalwood oil (94.0%), α-santalol (84.2%), β-santalol (90.6%), and a mixture of α- and β-santalols (88.7%) against *A. gossypii* infesting hot peppers were also comparable to each other in greenhouse bioassays. Amer and Mehlhorn (2006) reported that sandalwood oil at 50 ppm concentration inflicted 83.3% mortality within 1 hour of application against third-instar larvae of *Aedes aegypti*, and mortality increased to 100% after 12 hours of treatment. Similar trends, in terms of median lethal dose, were also observed in the case of *Anopheles stephensi* and *Culex quinquefasciatus*, ranging from 100 to 50 ppm, respectively, after 1 hour of treatment.

## 11.9 Advantages as a Pesticide

Natural pesticides based on plant essential oils may represent alternative crop protectants. Many plant essential oils show a broad spectrum of activity against pest insects and plant pathogenic fungi, including insecticidal, antifeedant, repellent, oviposition-deterrent, growth regulatory, and antivector activities (Koul et al., 2008). Plant-derived oils and powders have recently been evaluated and shown to be effective against a number of insect pests (Butler and Henneberry, 1990; Koul et al., 2008; Dhaliwal et al., 2015; Sola et al., 2014). The aromatic characteristics of essential oils provide various functions for the plants, including (1) attracting or repelling insects, (2) protecting themselves from heat or cold, and (3) utilizing chemical constituents in the oil as defense materials (Koul et al., 2008). They are nonenvironmental pollutants, unlike other synthetic pesticides (Isman, 2000). Essential oils are totally nontoxic to other mammals, birds, bees, and even beneficial insects (Anonymous, 2007). One of the most important advantages of essential oils is that they do not leave any residue on the plants (Lokanadhan et al., 2012). Plant-based oils as a natural pesticide are of immense significance in view of the environmental and toxicological implications of the indiscriminate use of synthetic pesticides and overcoming or reducing the problem of increasing pest resistance. It is also obvious that resistance will develop more slowly to essential oil–based pesticides owing to the complex mixtures of constituents that characterize many of these oils (Koul et al., 2008).

## 11.10 Limitations as a Pesticide

The characteristic features of sandalwood oil indicate that it has potential to be used as a pesticide in a variety of ways to control a large number of pests, some of which remain to be explored. However, the rationality of use of sandalwood oil as a pesticide suffers from a few constraints, primarily the high price of the oil and unavailability of sufficient quantities of plant material. Further, the efficacy of the oil may fall short when compared with synthetic pesticides, although there are specific pest contexts where control equivalent to that with conventional products has been observed. Also, the oil may require somewhat greater application rates and frequent reapplication when used in fields. In general, additional challenges to the commercial application of plant essential oil–based pesticides include standardization and refinement of pesticide products, protection of technology (patents), and regulatory approval (Isman, 2005). In addition, as the chemical profile of plant species can vary naturally, depending on geographic, genetic, climatic, annual, or seasonal factors, pesticide manufacturers must take additional steps to ensure that their products will perform consistently (Koul et al., 2008). Then, data on their bioefficacy, product characterization, efficacy, safety, toxicology, and label claim will have to be generated, before applying for registration. Finally, once all these issues are addressed, regulatory approval is required. Accordingly, regulatory approval continues to be a barrier to commercialization and will likely continue to be a barrier until regulatory systems are adjusted to better accommodate these products (Isman and Machial, 2006).

## 11.11 Conclusion

In the future, the use of pesticides will be more firmly regulated because of well-documented environmental risks in the use of synthetic chemicals (Ongley, 1996; Koul et al., 2008). This may lead to a growing demand for biological plant protection agents, including plant-based oils. Sandalwood oil has a wide range of applications, ranging from fragrance to medicinal and pesticidal uses. These features indicate that pesticides based on sandalwood oil could be used in a variety of ways to control a large number of key pests. Since several plant-based essential oils have already been researched and reported successfully for their anti-insect properties, it is high time to refocus the attention of researchers toward the development and application of known plant oils to speed up the scaling-up process, enhancing the availability of eco-friendly pest management techniques. A simplified process facilitating the registration of the essential oils having pesticidal properties of these plant-based formulations would ensure future development, improvement, and applications of these potential biopesticides that form a part of the rich floral diversity. In order to derive and implement a comprehensive management strategy, a paradigm shift in the mindset of pesticide users is needed. The mission must facilitate the successful development, commercialization, and adoption of biopesticides in public–private partnership (PPP) mode (Roy et al., 2016). It is expected that such collective efforts would lead to the successful implementation of plant-based oils in the integrated approach in pest management.

## References

Ahmed, N., M. S. Ali Khan, A. M. Mat Jais et al. 2013. Anti-ulcer activity of sandalwood (*Santalum album* L.) stem hydroalcoholic extract in three gastric-ulceration models of Wistar rats. *Boletín Latinoamericano y del Caribe de Plantas Medicinales y Aromáticas* 12: 81–91.

Amer, A. and H. Mehlhorn. 2006. Larvicidal effects of various essential oils against *Aedes, Anopheles*, and *Culex* larvae (Diptera, Culicidae). *Parasitology Research* 99: 466–472.

Anonymous. 1999. Hentriacontane-1-ol. In *Phytochemical Dictionary: A Handbook of Bioactive Compounds from Plants*, 2nd ed., ed. J. B. Harborne, H. Baxter and G. P. G. Moss, 168–169. Boca Raton, FL: Taylor & Francis, Philadelphia, p. 53.

Anonymous. 2007. Neem benefits: Discover all the benefits of neem oil, leaf, and neem trees. Discover NEEM. http://www.discoverneem.com/neem-benefits.html.

Anonymous. 2016. 101 Essential oil uses & benefits. Dr. Axe Food Is Medicine. https://draxe.com/essential-oil-uses-benefits/.

Baldovini, N., C. Delasalle and D. Jaulain. 2010. Phytochemistry of the heartwood from fragrant *Santalum* species: A review. *Flavor and Fragrance Journal* 26: 7–26.

Batish, D. R., H. P. Singh, R. K. Kohli and S. Kaur. 2008. Eucalyptus essential oil as a natural pesticide. *Forest Ecology and Management* 12: 2166–2174.

Benencia, F. and M. C. Courreges. 1999. Antiviral activity of sandalwood oil against herpes simplex viruses 1 & 2. *Phytomedicine* 6: 119–123.

Bhalla, Y., V. K. Gupta and V. Jaitak. 2013. Anticancer activity of essential oils: A review. *Journal of the Science of Food and Agriculture* 93: 3643–3653.

Brunner, G. 2005. Supercritical fluids: Technology and application to food processing. *Journal of Food Engineering* 67: 21–33.

Burdock, G. A. 2002. *Fenaroli's Handbook of Flavor Ingredients*. 4th ed. CRC Press, Boca Raton, FL, pp. 1684–1685.

Burdock, G. A. and I. G. Carabin. 2008. Safety assessment of sandalwood oil (*Santalum album* L.). *Food and Chemical Toxicology* 46: 421–432.

Butler, Jr., G. D. and T. J. Henneberry. 1990. Pest control on vegetables and cotton with household cooking oils and liquid detergents. *Southwest Entomology* 15: 123–131.

Dhaliwal, G. S., G. A. Burdock, J. Vikas and M. Bharathi. 2015. Crop losses due to insect pests: Global and Indian scenario. *Indian Journal of Entomology* 77: 165–168.

Diaz-Chavez, M. L., J. Moniodis, L. L. Madilao et al. 2013. Biosynthesis of sandalwood oil: *Santalum album* CYP76F cytochromes P450 produce santalols and bergamotol. *PLoS One* 8: e75053. http://journals.plos.org/plosone/article?id=10.1371/journal.pone.0075053.

Dickinson, S. E., E. R. Olson, C. Levenson, J. Janda, J. J. Rusche and D. S. Alberts. 2014. A novel chemopreventive mechanism for a traditional medicine: East Indian sandalwood oil induces autophagy and cell death in proliferating keratinocytes. *Archives of Biochemistry and Biophysics* 558: 143–152.

Ebadollahi, A. 2013. Plant essential oils from Apiaceae family as alternatives to conventional insecticides. *Ecologia Balkanica* 5: 149–172.

Grieve, M. 1971. A modern herbal: The medicinal culinary, cosmetic and economic properties. In *Cultivation and Folk-lore of Herbs, Grasses, Fungi, Shrubs & Trees with All Their Modern Scientific Uses*. Vol. II. United States: Courier Corporation.

Hall, R. L. and B. L. Oser. 1965. Recent progress in the consideration of flavoring ingredients under the food additives amendment. *Food Technology* 19: 151–197.

Handa, S. S. 2008. An overview of extraction techniques for medicinal and aromatic plants. In *Extraction Technologies for Medicinal and Aromatic Plants*, ed. S. S. Handa, S. P. S. Khanuja, G. Longo and D. D. Rakesh, 21–54. Trieste, Italy: International Centre for Science and High Technology, pp. 21–54.

Hettiarachchi, D. S., M. Gamage and U. Subasinghe. 2010. Oil content analysis of sandalwood: A novel approach for core sample analysis. *Sandalwood Research Newsletter* 25: 1–4.

Howes, M. J. R., M. S. J. Simmonds and G. C. Kite. 2004. Evaluation of the quality of sandalwood essential oils by gas chromatography–mass spectrometry. *Journal of Chromatography A* 1028: 307–312.

Huish, R. D., T. Fakaosi, H. Likiafu, J. Mateboto and K. H. Huish. 2015. Distribution, population structure, and management of a rare sandalwood (*Santalum yasi*, Santalaceae) in Fiji and Tonga. *Pacific Conservation Biology* 21: 27–37.

Isman, M. B. 2000. Plant essential oils for pest and disease management. *Crop Protection* 19: 603–608.

Isman, M. B. 2005. Problems and opportunities for the commercialization of botanical insecticides. In *Biopesticides of Plant Origin*, ed. C. Regnault-Roger, B. J. R. Philogene and C. Vincent, 283–291. Paris: Lavoisier.

Isman, M. B. and C. M. Machial. 2006. Pesticides based on plant essential oils: From traditional practice to commercialization. In *Naturally Occurring Bioactive Compounds*, ed. M. Rai and M. C. Carpinella, 29–44. The Netherlands: Elsevier Science.

Jones, C. G., E. L. Ghisalberti, J. A. Plummer and E. L. Barbour. 2006. Quantitative co-occurrence of sesquiterpenes: A tool for elucidating their biosynthesis in Indian sandalwood, *Santalum album*. *Phytochemistry* 67: 2463–2468.

Jones, F. A. 1996. Herbs—Useful plants: Their role in history and today. *European Journal of Gastroenterology and Hepatology* 8: 1227–1231.

Khan, I. A. and E. A. Abourashed. 2010. *Leung's Encyclopedia of Common Natural Ingredients*. New York: John Wiley & Sons Inc.

Kim, S., H. Lee, M. Jang, C. Jung and I. Park. 2016. Fumigant toxicity of Lamiaceae plant essential oils and blends of their constituents against adult rice weevil *Sitophilus oryzae*. *Molecules* 21: 1–10.

Kirtikar, K. R. and B. D. Basu. 1993. Santalum album. In *Indian Medicinal Plants*. Vol. 3, 2nd ed. Allahabad, India: L. M. Basu, pp. 2184–2188.

Koch, C., J. Reichling, J. Schneele and P. Schnitzler. 2008. Inhibitory effect of essential oils against herpes simplex virus type 2. *Phytomedicine* 15: 71–78.

Koul, O., S. Walia and G. S. Dhaliwal. 2008. Essential oils as green pesticides: Potential and constraints. *Biopesticide International* 4: 63–84.

Kusuma, H. S. and M. Mahfud. 2016a. Preliminary study: Kinetics of oil extraction from sandalwood (*Santalum album*) by microwave-assisted hydro-distillation. Paper presented at the International Conference on Innovation in Engineering and Vocational Education, Bandung, Indonesia.

Kusuma, H. S. and M. Mahfud. 2016b. The extraction of essential oil from sandalwood (*Santalum album*) by microwave air-hydrodistillation method. *Journal of Materials and Environmental Science* 7: 1597–1606.

Lawrence, B. M. 1991. Recent progress in essential oils. *Perfumer & Flavorist* 16: 49–58.

Lokanadhan S., P. Muthukrishna and S. Jeyarama. 2012. Neem products and their agricultural application. *Journal of Biopesticides* (Supplementary): 72–76.

McKinnell, F. H. 1990. Status of management and silviculture research on sandalwood in western Australia and Indonesia. https://www.fs.fed.us/psw/publications/documents/psw_gtr122/psw_gtr122_mckinnell.pdf (accessed September 22, 2016).

Merlin, M. D., L. Thomson and C. R. Elevitch. 2006. *Santalum ellipticum, S. freycinetianum, S. haleakalae,* and *S. paniculatum* (Hawaiian sandalwood) Santalaceae (sandalwood family). http://www.doc-development-durable.org/file/Arbres-Bois-de-Rapport-Reforestation/FICHES_ARBRES/Arbres-non-classes/Santalum-Haw-sandalwood.pdf (accessed September 22, 2016).

Misra, B. B. and S. Dey. 2013. Developmental variations in sesquiterpenoid biosynthesis in East Indian sandalwood tree (*Santalum album* L.). *Trees* 27: 1071–1086.

Montoro, P., M. Masullo, S. Piacente and C. Pizza. 2015. Extraction, sample preparation and analytical methods for quality issues of essential oils. In *Aromatherapy: Basic Mechanisms and Evidence Based Clinical Use,* ed. G. Bagetta, M. Cosentino and T. Sakurada, 105–149. Boca Raton: CRC Press/ Taylor & Francis.

Nautiyal, O. H. 2010. Subcritical carbon dioxide and conventional extraction techniques of sandalwood oil: An industry project. *Sandalwood Research Newsletter* 25: 5–7.

Nautiyal, O. H. 2011. Analytical and Fourier transform infrared spectroscopy evaluation of sandalwood oil extracted with various process techniques. *Journal of Natural Products* 4: 150–157.

Okugawa, H., R. Ueda, K. Matsumoto, K. Kowanishi and A. Kato. 1995. Effect of α-santalol and β-santalol from sandalwood on the central nervous system in mice. *Phytomedicine* 2: 119–126.

Ongley, E. D. 1996. Control of water pollution from agriculture- FAO irrigation and drainage paper 55. http://www.fao.org/docrep/w2598e/w2598e00.htm#Contents (accessed September 22, 2016).

Panto, S., D. Sciarrone, M. Maimone et al. 2015. Performance evaluation of a versatile multidimensional chromatographic preparative system based on three-dimensional gas chromatography and liquid chromatography-two-dimensional gas chromatography for the collection of volatile constituents. *Journal of Chromatography A* 1417: 96–103.

Prasetyaningtyas, M. 2007. *Santalum album* L. Seed leaflet 116. http://curis.ku.dk/ws/files/20495554/santalum_album.pdf (accessed October 15, 2016).

Ral, S. N. 1990. Status and cultivation of sandalwood in India. https://www.fs.fed.us/psw/publications/documents/psw_gtr122/psw_gtr122_rai.pdf (accessed October 15, 2016).

Rani, A., P. Ravikumar, M. D. Reddy and A. Kush. 2013. Molecular regulation of santalol biosynthesis in *Santalum album* L. *Gene* 527: 642–648.

Rasor, A. S. and S. E. Dunca 2014. Fats and oils—Plant based. In *Food Processing: Principles and Applications,* ed. S. Clark, S. Jung and B. Lamsal, 457–480. New York: John Wiley & Sons Ltd.

Roh, H. S., J. Kim, E. S. Shin, D. W. Lee, H. Y. Choo and C. G. Park. 2015. Bioactivity of sandalwood oil (*Santalum austrocaledonicum*) and its main components against the cotton aphid, *Aphis gossypii*. *Journal of Pest Science* 88: 621–627.

Roh, H. S., E. G. Lim and J. Kim. 2011. Acaricidal and oviposition deterring effects of santalol identified in sandalwood oil against two-spotted spider mite, *Tetranychus urticae* Koch (Acari: Tetranychidae). *Journal of Pest Science* 84: 495–501.

Roy, S., G. Handique, N. Muraleedharan, K. Dashora, S. M. Roy, A. Mukhopadhyay and A. Babu. 2016. Use of plant extracts for tea pest management in India. *Applied Microbiology and Biotechnology* 100: 4831–4844.

Satou, T., M. Miyagawa, H. Seimiya, H. Yamada, T. Hasegawa, and K. Koike. 2014. Prolonged anxiolytic-like activity of sandalwood (*Santalum album* L.) oil in stress-loaded mice. *Flavour and Fragrance Journal* 29: 35–38.

Shea, S. R., A. M. Radomiljac, J. Brand, and P. Jones. 1997. An overview of sandalwood and the development of sandal in farm forestry in western Australia. Paper presented at the International Seminar on Sandal and Its Products, Canberra, Australia.

Sindhu, R. K., A. K. Upma, and S. Arora. 2010. *Santalum album* Linn: A review on morphology, phytochemistry and pharmacological aspects. *International Journal of Pharm Tech Research* 2: 914–919.

Sola, P., B. M. Mvumi, J. O. Ogendo, O. Mponda, J. F. Kamanula, S. P. Nyirenda, S. R. Belmain, and P. C. Stevenson. 2014. Botanical pesticide production, trade and regulatory mechanisms in sub-Saharan Africa: Making a case for plant-based pesticidal products. *Food Security* 6: 369–384.

Subasinghe, U., M. Gamage, and D. S. Hettiarachchi. 2013. Essential oil content and composition of Indian sandalwood (*Santalum album* L.) in Sri Lanka. *Journal of Forestry Research* 24: 127–130.

Swapna, G., T. Jyothirmai, V. Lavanya, S. Swapnakumari, and P. A. Sri Lakshmi. 2015. Extraction, and characterization of bioactive compounds from plant extracts: A review. *European Journal of Pharmaceutical Science and Research* 2: 1–6.

Tennakoon, K. U. and D. D. Cameron. 2006. The anatomy of *Santalum album* (sandalwood) haustoria. *Canadian Journal of Botany* 84: 1608–1616.

Tennakoon, K. U., J. S. Pate, and D. Arthur. 1997. Ecophysiological aspects of the woody root hemi-parasite *Santalum acuminatum* (R. Br.) A. DC and its common hosts in south western Australia. *Annals of Botany* 80: 245–256.

Thomson, L. A. J., J. Doran, D. Harbaugh, and M. D. Merlin. 2011. Farm and forestry production and marketing profile for sandalwood (*Santalum* species). http://pacificschoolserver.org/content /_public/Local%20Topics/Pacific%20Islands/Agriculture%20for%20Islands/Specialty%20 crops/Sandalwood.pdf (accessed October 15, 2016).

van Balgooy, M. M. J. 1960. Preliminary plant-geographic analysis of the Pacific. *Blumea* 10: 385–430.

van Balgooy, M. M. J. 1971. Plant genera of the Pacific as based on a census of Phanaerogam genera. *Blumea* 6: 1–222.

van Balgooy, M. M. J., P. H. Hovenkamp and P. C. van Welzen. 1996. Phytogeography of the Pacific: Floristic and historical distribution patterns in plants. In *The Origin and Evolution of Pacific Island Biotas, New Guinea to Eastern Polynesia: Patterns and Processes*, ed. A. Keast and S. E. Miller, 191–213. Amsterdam: SPB Academic Publishing.

Vasundhara, M., B. S. Thara, B. Radhika, A. Jayaram, and R. Priyanka. 2015. Assessment of Indian sandalwood (*Santalum album* L.) seeds for seed oil production and fatty acid methyl esters. *World Journal of Pharmaceutical Research* 4: 1416–1425.

Zhang, X. H., J. A. da Silva, Y. X. Jia, J. T. Zhao, and G. H. Ma. 2012. Chemical composition of volatile oils from the pericarps of Indian sandalwood (*Santalum album*) by different extraction methods. *Natural Product Communications* 7: 93–96.

# 12

# Lavender Oil

Irshad Ul Haq Bhat and Zakia Khanam

## CONTENTS

## 12.1 Botany

Lavender belongs to the family Labiatae (Lamiaceae) and subfamily Nepetoideae (synonyms *Lavandula angustifolia*, *Lavandula vera*, *Lavandula latifola*, *Lavandula officinalis*, *Lavandula spica*, and *Lavandula delphinensis*). Lavender is a shrubby plant with aromatic flowers and narrow leaves. The plant grows from 1 to 3 feet in height. The arrangement of leaves is opposite, sessile, with bluntly quadrangular branches, finely pubescent, possessing stellate hairs (Upson, 2002). The flowers consisting of cymes are present in the form of inflorescence at the end of branches either with opposite decussate or an alternate spiral arrangement. The cymes are found in different sizes, shapes, and nervation characteristics of different species. The plant flowers possess either a single flower or three to nine flowers per cyme, without or with bracteoles, respectively. The calyx and corolla vary in shape, size, and color (Upson, 2002).

## 12.2 Methods of Extraction of Lavender Oil

The lavender oil from *L. angustifolia* can be extracted by different methods, although some are not practiced currently. Researchers have reported various methods for extraction of lavender oil, for example, steam distillation, solvent extraction, microwave extraction, and supercritical fluid extraction (SFE). The varieties of methods are employed in order to maximize the yield and reduce the time of extraction.

### 12.2.1 Steam Distillation

Zheljazkov et al. (2013) carried out steam distillation of dried lavender flowers at different time intervals. A total of 250 g of lavender flowers was steam distilled in a 2 L pear-shaped flask by a procedure similar to that described in their earlier reports (Zheljazkov et al., 2010a,b, 2011; Cannon et al., 2013). The water-filled flask attached to a bioflask containing plant sample was heated to yield oil by a condensation process into a well-designed funnel and resembled the commercial-type distillation apparatus. Lesage-Meessen et al. (2015) reported steam distillation of lavender oil by traditional and green ground distillation methods. The latter method was prioritized to increase the productivity at pilot scale in the field, as this method combines mechanical procedures like harvesting, cutting, grinding of lavender stems, and steam distillation. Al-Younis et al. (2015) isolated the lavender oil by steam distilling the plant samples at their full flowering stage. Steam distillation was used by Jianu et al. (2013) in accordance with the method published earlier, and oil was isolated by decantation, followed by drying on anhydrous sodium sulfate. Ben Salah et al. (2009) steam distilled *L. angustifolia* and obtained a liquid with a camphoraceous smell. Bilke and Mosandl (2002) obtained lavender oil by using a glass steam extractor fitted with a filter at the bottom, along with a steam-heated jacket. The process was carried out for half an hour.

### 12.2.2 Solvent Extraction

Wells and Lis-Balchin (2002) stated that oil from *L. angustifolia* P. Miller can be obtained by absolute ethanol after evaporation under reduced pressure. Besides, it can be extracted from water by benzene or petroleum ether, followed by extraction with ethanol. Lavender oil can be extracted by using different solvents. The oil obtained in such a process is high in concentration. Jablonský et al. (2016) used butanol and dichloromethane, polar and non-polar solvents, respectively, for isolation of lavender oil. Similarly, dichloromethane was used to isolate the lavender oil by the solvent extraction method, as reported by Méndez-Tovar et al. (2015).

### 12.2.3 Microwave Extraction

Microwave-assisted extraction (MAE) is the superlative method for extracting superior-quality essential oils, and is considered a sustainable future extraction process at the industrial scale. Compared with steam distillation method, it is more efficient and faster, with environmentally friendly advantages. In the future, microwave-assisted extraction will probably be the leading technique for the production of essential oil. Zhi-ling et al. (2011) reported the microwave process, in addition to the steam distillation process, for the extraction of lavender oil. Their observation revealed the abundance of active ingredients and a relatively high amount of major components of essential oil. MAE, in conjunction with steam distillation, has been exploited by various researchers to a greater extent (Sahraoui et al., 2008; Chen et al., 2011; Périno-Issartier et al., 2013). The only aim of such a combination is to achieve lavender oil of high quality in short time period.

### 12.2.4 Supercritical Fluid Extraction

There is a significant interest in replacing the steam distillation and solvent extraction processes usually used to obtain lavender oil by supercritical fluid extraction. SFE of the lavender oil is a promising option at the industrial scale for extraction of lavender oil using supercritical fluid processing. SFE is advantageous over solvent extraction because solvent extraction can extract tannins, chlorophyll, and minerals also. Oszagyh et al. (1996) reported that notable quantities of essential oil components were extracted by using SFE with carbon dioxide under different extraction conditions. The SFE also enhanced the yield of essential oil components compared with steam distillation, which was attributed to the separation of compounds of higher molecular weight present in lavender samples during the SFE process. The lavender flowers were subjected to SFE by AdaFoglu et al. (1994) for extraction of lavender oil. The different parameters, like pressure, temperature, carbon dioxide flow rate, and particle size, were varied at a fixed SFE extraction time period. The results were supported by second-order central composite design, which revealed that the obtained results were adequate. Reverchon (1997) stated the importance of SFE extraction of lavender oil. SFE avoids the incomplete hydrolysis of lavender oil components as reported in the steam distillation process. The extraction of lavender oil by SFE has been carried out by different researchers, as stated by Danh et al. (2012), Da Porto et al. (2009), AdaFoglu et al. (1994), and Razazadeh et al. (2008). The utilization of SFE at lower temperatures has a tendency to retain thermally sensitive labile components of lavender extract. In addition, the use of $CO_2$ is attributed to its nonharmful and easily available properties.

## 12.3 Methods of Analysis of Oil

Gas chromatography–mass spectrometry (GC-MS), a qualitative analysis, has been considered a benchmark technique for the analysis of lavender oil. GC and GC-MS have been comprehensively studied for the analysis of essential oil. Shellie et al. (2002) reported GC-MS analysis of lavender oil and concluded that essential oils from various lavender cultivators can be differentiated. An et al. (2001) stated that the GC-MS analysis technique is a realistic method for lavender oil analysis. Verma et al. (2010) analyzed lavender oil by GC-MS and found 37 constituents. Furthermore, based on the analysis data provided, a good opportunity for local farmers can be provided to compete in the international market.

Samfira et al. (2015) analyzed lavender oil by the Fourier transform infrared (FTIR) spectrophotometer technique. This technique is focused on the presence of hydrogen bonds, which are difficult to analyze by other optical methods. The FTIR can be considered a fast and safe identification method of lavender oil. Lafhal et al. (2015) analyzed lavender oil components by using a Raman spectroscopic technique, which they attributed to sensitivity to carbon–carbon double bonds and low affinity toward water.

## 12.4 Composition of Lavender Oil

The composition of lavender oil varies from species to species, as well as from the place of origin. Danh et al. (2012) extracted and compared the compounds of lavender oil extracted by different extraction methods and corroborated their results with earlier reports and found the results to be in conformity with those of other researchers. Compounds of lavender oil found are camphene, 1-octen-3-ol, myrcene, 1,4-cineole, limonene, 1,8-cineole, (Z)-β-ocimene, (E)-β-ocimene, α-terpinene, cis-linalool oxide, terpinolene, linalool, octen-1-ol-acetate, camphor, borneol, lavandulol, terpinen-4-ol, cryptone, p-cymen-8-ol, α-terpineol, hexenyl butanoate, hexyl isobutanoate, isobornyl formate, cumin aldehyde, hexyl-2-methyl butyrate, hexyl isovalerate, linalyl acetate, bornyl acetate, lavandulyl acetate, hexyl tiglate daucene, β-bourbonene, neryl acetate, α-cis-bergamotene, (E)-caryophyllene, α-santalene, α-*trans*-bergamotene, β-farnesene, germacrene, β-bisabolene, γ-cadinene, lavandulyl isovalerate caryophyllene oxide, epi-α-cadinol, and α-bisabolol (Danh et al., 2012). In addition, hydrocarbons, alcohols, carbonyl, esters, ketones, aldehydes, and ethers were reported.

## 12.5 Physical and Chemical Properties of Lavender Oil

The physical and chemical properties of lavender oil are given in Table 12.1.

## 12.6 General Uses of Lavender Oil

The lavender oil from different lavender species has been extensively used for raw ingredients production of perfumes, fragrance material, and toiletries. Lavender oil has been utilized as a condiment (nonalcoholic beverages, ice creams, candies, and chewing gums) (Areias et al., 2000). The aromatizing property of lavender oil has earned a good place in phytotherapy; in addition, lavender oil has been used to treat flatulent dyspepsia. The diuretic and spasmolytic uses of lavender oil are well known (Areias et al., 2000). The oil is known to possess sedative, carminative, antidepressive, anti-inflammatory, antiviral, and antibacterial properties (Hass, 2001; Cavanagh, 2005; Verma et al., 2010). Lavender oil

**TABLE 12.1**

Physical and Chemical Properties of Lavender Oil

| Color | Pale yellow to yellow green |
|---|---|
| Odor | Floral, fresh, sweet, herbaceous and slightly fruity, slightly camphorous |
| Taste | Herbaceous, woody, undertone of mint, earthy, smokey, apple-like, and green are descriptive of the taste of lavender; lavender presents a floral, pungent aroma and flavor |
| Boiling point | 204°C |
| Density | 0.879 g/ml at 25°C |
| Solubility | Insoluble in water; soluble in ethanol |

is acknowledged in the cosmetic industries and has gained impetus in food processing, not only as a flavoring agent but also to protect the food. Food can be protected from oxidative rancidity, loss of labile compounds, and the formation of off-flavors by application of lavender oil (Gulcin, 2004; Topal et al., 2008; Hui et al., 2010). Thus, in general lavender oil has been explored in the cosmetic, pharmaceutical, and food industries because of its distinctive chemical composition, which confers both aromatic and biological activities. Therefore, from field to industry, lavender oil has sufficiently contributed to generating income for common masses, as well as stakeholders, by its diverse uses.

## 12.7  Pesticide Uses of Essential Oil

The harm to the environment caused by excessive use of pesticides is an issue of concern for researchers in the public as well as the private domain. The annual consumption of pesticides in crops alone is 2.5 million tons, causing about a hundred-billion-dollars of damage throughout the world (Koul et al., 2008). This huge damage caused to surroundings is the result of the creation of pesticide residues in soil, the hydrosphere, and ultimately, edible plants, as well as by the nonbiodegradable nature of pesticides. Thus, the need at this time is to look for a change in paradigm, an alternative solution to produce pesticides that are environmentally friendly and nontoxic to the health of both humans and animals. The exploitation of naturally occurring compounds or natural products or green pesticides can support such a paradigm. The essential oils also can be a leading source of pesticides because of their volatility and narrow persistence under field conditions.

### 12.7.1  Lavender Oil as a Herbicide

Crop production is severely hindered in terms of quality and quantity due to the invasion of weeds, leading to heavy losses for crop growers. The annual loss due to weed infestation worldwide is 12%, as stated by Anaya (1999). Excessive use of synthetic herbicides has destroyed the healthy soil, as well as groundwater, leading to toxic accumulation of herbicides in microorganisms and humans. Furthermore, the increased resistance offered by weeds to different herbicides has forced researchers to find compounds with better herbicidal activity (Grosso et al., 2010). Essential oils can be an alternative for such challenges, because of their volatility and having less or no toxicity (Anaya, 1999; Heisey and Heisey, 2003; Singh et al., 2003, 2005). The essential oil components can effectively alter the biochemical processes in weeds (Weston and Duke, 2003; Azirak and Karaman, 2007). Grosso et al. (2010) reported the herbicidal activity of lavender oil. Volatile oils from lavender can be alternatives to the synthetic herbicides as less harmful to crop species. Zanellato et al. (2009) investigated the potential genotoxic effect of lavender oil on *Vicia faba* root meristems by genotoxicity tests and hypothesized that lavender oil can be a potential alternative bioherbicide. The herbicidal activity against weeds (common cocklebur, *Xanthium strumarium* L.; sterile wild oat, *Avena sterilis* L., and short spiked canarygrass, *Phalaris brachystachys* L.) was studied by Uremis et al. (2009). The essential oil of *L. angustifolia* can be used as an alternative to herbicides to subdue the germination of *X. strumarium*, *A. sterilis*, and *P. brachystachys* seeds.

## 12.7.2 Lavender Oil as a Fungicide

There is an immense need to find out how to safeguard crop harvests from fungal invasion by substances purely of natural origin, so that an alternative to synthetic fungicides can be generated. A potential alternative is the use of essential oils. Essential oils from plants have been arguably accepted as potential candidates of a new source of fungicide to control pathogenic fungi. Císarová et al. (2016) recently reported that lavender oil exhibited antifungal activity against fungi black *Aspergillus niger* and supported the argument that lavender oil can be used as antifungal agents. Tabassum and Vidyasagar (2013) elaborated the role of essential oils in treating various diseases. In their extensive review, it has been mentioned that the main components, linalool and linalyl acetate, were investigated for their antifungal activity against 50 clinical isolates of *Candida albicans* (D'Auria et al., 2005). Al-Naser and Al-Abrass (2014) studied the effect of lavender oil against *Fusarium solani, F. oxysporum*, and *A. niger* as a developing step to the use of lavender oil as an alternative to synthetic antifungal agents. Furthermore, this study can be further explored to investigate the phytotoxicity of lavender oil on plants and the sensory quality of treated fruits and vegetables. The effect of lavender essential oil was also examined on the growth and sporulation of *Aspergillus terreus, F. oxysporum, Penicillium expansum*, and *Verticillium dahlia* by Kadoglidou et al. (2011). The study revealed that different effects were caused by the lavender oil components on fungal sporulation than on fungal growth. In order to develop an alternative to synthetic fungicides, Soylu et al. (2010) used the control of fungal pathogen *Botrytis cinerea*, the gray mold disease of tomato, by various essential oils, including lavender oil. The antifungal activity of lavender oil reported by some researchers is given in Table 12.2.

## 12.7.3 Lavender Oil as a Bactericide

Bacterial infections have led to severe health and life-threatening diseases. The resistance offered by microorganisms to a variety of drugs has justified research for developing new effective drugs from natural resources, particularly of plant origin. Lavender oil, a secondary plant metabolite, has been attributed this role because of its antibacterial activities (Cavanagh and Wilkinson, 2002; Ait Said et al., 2015). Ait Said et al. (2015) analyzed the antibacterial activity of lavender oil against 11 pathogenic bacteria and concluded that lavender oil can be a potential antibacterial agent. The use of lavender oil against oral bacteria was carried out by Thosar et al. (2013). These authors stated that it can provide a useful intracanal antiseptic solution against oral pathogens. Study of the antibacterial activity against *Shigella flexneri, Staphylococcus aureus, E. coli, Salmonella typhimurium*, and *Streptococcus pyogenes* by

**TABLE 12.2**

Fungicidal Activity of Lavender Oil

| Fungi | Reference |
| --- | --- |
| *Botrytis cinerea* and *Penicillium expansum* | Lopez-Reyes et al. (2010) |
| Dermatophytes and *Dematiaceous fungi* | Tullio et al. (2007) |
| *Phytophthora infestans* | Olanya and Larkin (2006) |
| *Rhizoctonia solani, Fusarium oxysporum, Aspergillus flavus* | Angioni et al. (2006) |
| *Aspergillus niger, Aspergillus fumigatus, Aspergillus repens* (teleomorph *Eurotium repens*), *Cladosporium herbarum, Penicillium frequentans, Trichoderma viride, Chaetomium globosum, Paecilomyces variotii*, and *Stachybotrys atra* | Rakotonirainy and Lavédrine (2005) |

**TABLE 12.3**

Bactericidal Activity of Lavender Oil

| Bacteria | Reference |
| --- | --- |
| *Bacillus subtilis, Staphylococcus aureus, Acinetobacter baumanii, Escherichia coli, Klebsiella pneumoniae,* and *Pseudomonas aeruginosa* | Voravuthikunchai et al. (2012) |
| *Pseudomonas* strain | Végh et al. (2012) |
| *Acidovorax avenae* subsp. *citrulli* (Aac) | Mengulluoglu and Soylu (2012) |
| *Escherichia coli* ATCC 25922, *Pseudomonas aeruginosa* ATCC 43895, *Salmonella enteretidis* ATCC 9027, *Bacillus cereus* ATCC 8739, *Staphylococcus aureus* ATCC 25923 | Stanković et al. (2011) |
| *Propionibacterium acnes* | Zu et al. (2010) |
| *Shigella flexneri, Listeria monocytogenes* | Bakkali et al. (2008) |
| *Escherichia coli* Easter | Preuss et al. (2005) |
| *Porphyromonas gingivalis, Actinobacillus actinomycetemcomitans, Fusobacterium nucleatum, Streptococcus mutans,* and *Streptococcus sobrinus* | Takarada et al. (2004) |
| *Pseudomonas syringae, Erwinia herbicola* | Karamanoli et al. (2000) |

lavender oil was conducted by Jianu et al. (2013). They revealed that the antibacterial properties of lavender oil can be contributed to the major and minor constituents (caryophyllene β-phellandrene, eucalyptol, camphor, and eucalyptol). Lavender oil has been applied against bacteria that affects plants, for example, *Ralstonia solanacearum*, a bacterial species causing bacterial wilt in solanaceous species. Thus, Hosseinzadeh et al. (2013) revealed that lavender oil along with other oils can be promising in exhibiting bactericidal activity against *R. solanacearum*. Similarly, ample work has been reported by different researchers analyzing the antibacterial potential of lavender oil. This potential is summarized in Table 12.3.

## 12.8 Limitations as a Pesticide

The usage of lavender oil for pest management has a promising future. However, some limitations need to be addressed. The limited efficacy of lavender oil–based pesticides compared with synthetic pesticides is one of the major concerns for its pilot-scale application. The outdoor application needs a high concentration of active ingredients, which in turn needs frequent reapplications. The standardization of lavender oil in terms of dosage and quantity production is also a limiting factor for the exploitation of lavender oil as a pesticide on a commercial scale. Moreover, the appropriate technology, depending on the extraction procedures, geographical locations, and genotypic characteristics, is another factor that hinders the utilization of lavender oil as pesticide.

## 12.9 Lavender Oil–Based Insecticide

Various plant essential oils are potentially repellent to arthropods (Nerio et al., 2010). Researchers have studied the efficiency of essential oils for the control of stored product and agricultural insect pests (Odalo et al., 2005; George et al., 2009; Yang et al., 2009; Zapata and Smagghe, 2010). Yoon et al. (2011) carried out a repellency study on *Lycorma delicatula* fourth-instar nymphs using an olfactometer. Linalool, linalyl acetate, terpinen-4-ol, and caryophyllene oxide, components found in various percentages in lavender oil, were studied, and among these, linalool showed promising repellency to *L. delicatula* genders. Mahmoud and Croteau (2002) stated that essential oils having commercial value are used in cooking, medicine, and agriculture; also, plants use them for allelopathy, pollinator attraction, and plant defense. In the agriculture sector, the primary function of essential oil is the attraction of insects for pollination, as well as for a defense mechanism. In addition, some act as insect pheromones, for example, linalool (Mahmoud and Croteau, 2002). Many researchers pointed out that a variety of insect species can be controlled by the repelling activity of essential oils (Mant et al., 2005; Regnault-Roger et al., 2012; Moretti et al., 2015; Peixoto et al., 2015; Pinto et al., 2015; Rossi and Palacios, 2015; Tabata et al., 2015; Yeom et al., 2015). Based on these findings, Erland et al. (2015) reported the insecticidal activity of commercially available lavender oil for utilization in sustainable future eco-friendly insecticidal management. Among three terpenes, 1,8-cineole, 3-carene, and linalool, found in lavender oil, linalool exhibited effective insecticidal activity against spotted wing drosophila (SWD). This was the first kind of study where lavender oil was used against SWD. Papachristos and Stamopoulos (2004) stated that most studies target postembryonic stages, and limited studies have been reported to evaluate the toxicity of essential oils against the eggs of stored product insects. Thus, *Acanthoscelides obtectus* (Say) egg hatchability and subsequent postembryonic mortality were investigated under the influence of lavender oil vapors. Application of these oils, however, for stored beans needs thorough investigation. The approach toward integrated pest management makes lavender oil an important source of insect repellent. Benelli et al. (2012) reported that lavender oil exhibited significant toxic effects against medfly adults and hypothesized that in pest control programs, this initiative can reduce the control costs. Some of the recent insecticidal activity of lavender during the last decade is summarized in Table 12.4.

**TABLE 12.4**

Insecticidal Activity of Lavender Oil

| Insect | Reference |
| --- | --- |
| *Bovicola ocellatus* | Ellse et al. (2016) |
| *Myzus persicae* (*Sulzer*) (Hemiptera: Aphididae) | Faraone et al. (2015) |
| *Meligethes* spp. (Coleoptera: Nitidulidae) | Dorn et al. (2014) |
| Horn flies | Lachance and Grange (2014) |
| Housefly | Sinthusiri and Soonwera (2014) |
| Potato tuber moth | Rafiee-Dastjerdi (2014) |
| *Bactrocera oleae* | Canale et al. (2013) |
| *Bovicola ocellatus* | Ellse et al. (2013) |
| Nasal botfly, *Cephalopina titillator* | Khater et al. (2013) |
| Housefly, *Musca domestica* L. | Sinthusiri and Soonwera (2013) |
| *Bovicola* (*Werneckiella*) *ocellatus* | Talbert and Wall (2012) |
| *Diaphorina citri* Kuwayama (Hemiptera: Psyllidae) | Mann et al. (2012) |
| *Aedes albopictus* (Diptera: Culicidae) | Conti et al. (2010) |
| *Tetranychus cinnabarinus* Boisd. (Acarina: Tetranychidae) | Sertkaya et al. (2010) |

## 12.10 Conclusions

Lavender oil can be obtained by conventional and modern technologies. The alteration in method with variation in parameters has led to different yields of lavender oil. The obtained lavender oil can be analyzed by various analytical and optical techniques. Lavender oil at the industrial scale has primarily been used in cosmetics and perfumery products. However, the growing concern for the environment in the search of biodegradable pesticides has revealed lavender oil as a pesticide of choice, owing to its herbicidal, fungicidal, bactericidal, and insecticidal properties. Researchers have presented data signaling possible pilot-scale use of lavender oil as a pesticide. But, the challenges in outdoor application are worrisome, which needs immediate thought. This problem can be resolved by designing a dose-dependent policy, which can further enhance the use of lavender oil as a pesticide at the commercial scale.

## References

AdaFoglu, N., DinGer, S., Bolat, E. (1994) Supercritical-fluid extraction of essential oil from Turkish lavender flowers. *Journal of Supercritical Fluids*, 7, 93–99.

Ait Said, L., Zahlane, K., Ghalbane, I., El Messoussi, S., Romane, A., Cavaleiro, C., Salgueiro, L. (2015) Chemical composition and antibacterial activity of *Lavandula coronopifolia* essential oil against antibiotic-resistant bacteria. *Natural Product Research*, 29 (6), 582–585.

Al-Naser, Z., Al-Abrass, N. (2014) Chemical composition and fungitoxic activities of *Lavandula officinalis* L. oil and comparison with synthetic fungicide on the growth of some fungi *in vitro*. *International Journal of ChemTech Research*, 6, 4918–4926.

Al-Younis, F., Al-Naser, Z., Al-Hakim, W. (2015) Chemical composition of *Lavandula angustifolia* Miller and *Rosmarinus officinalis* L. Essential oils and fumigant toxicity against larvae of *Ephestia kuehniella* Zeller. *International Journal of ChemTech Research*, 8, 1382–1390.

An, M., Haig, T., Hatfielda, P. (2001) On-site field sampling and analysis of fragrance from living lavender (*Lavandula angustifolia* L.) flowers by solid-phase microextraction coupled to gas chromatography and ion-trap mass spectrometry. *Journal of Chromatography A*, 917 (2001), 245–250.

Anaya, A.L. (1999) Allelopathy as a tool in the management of biotic resources in agroecosystems. *Critical Reviews in Plant Sciences*, 18, 697–739.

Angioni, A., Barra, A., Coroneo, V., Dessi, S., Cabras, P. (2006) Chemical composition, seasonal variability, and antifungal activity of *Lavandula stoechas* L. ssp. *stoechas* essential oils from stem/leaves and flowers. *Journal of Agricultural and Food Chemistry*, 54 (12), 4364–4370.

Areiasa, F.M., Valentãoa, P., Andradea, P.B., Moreiraa, M.M., Amarala, J., Seabrab, R.M. (2000) HPLC/DAD analysis of phenolic compounds from lavender and its application to quality control. *Journal of Liquid Chromatography & Related Technologies*, 23, 2563–2572.

Azirak, S., Karaman, S. (2007) Allelopathic effect of some essential oils and components on germination of weed species. *Acta Agriculturae Scandinavica, Section B*, 58, 88–92.

Bakkali, F., Averbeck, S., Averbeck, D., Idaomar, M. (2008) Biological effects of essential oils—A review. *Food and Chemical Toxicology*, 46 (2), 446–475.

Benelli, G., Flamini, G., Canale, A., Cioni, P.L., Conti, B. (2012) Toxicity of some essential oil formulations against the Mediterranean fruit fly *Ceratitis capitata* (Wiedemann) (Diptera Tephritidae), *Crop Protection* 42, 223–229.

Ben Salah, M., Abderraba, M., Tarhouni, M.R., Abdelmelek, H. (2009) Effects of ultraviolet radiation on the kinetics of in vitro percutaneous absorption of lavender oil. *International Journal of Pharmaceutics*, 382 (1–2), 33–38.

Bilke, S., Mosandl, A. (2002) Authenticity assessment of lavender oils using GC-P-IRMS: 2H/1H isotope ratios of linalool and linalyl acetate. *European Food Research and Technology*, 214 (6), 532–535.

Canale, A., Benelli, G., Conti, B., Lenzi, G., Flamini, G., Francini, A., Cioni, P.L. (2013) Ingestion toxicity of three Lamiaceae essential oils incorporated in protein baits against the olive fruit fly, *Bactrocera oleae* (Rossi) (Diptera Tephritidae). *Natural Product Research*, 27 (22), 2091–2099.

Cannon, J.B., Cantrell, C.L., Astatkie, T., Zheljazkov, V.D. (2013) Modification of yield and composition of essential oils by distillation time. *Industrial Crops and Products*, 41, 214– 220.

Cavanagh, H.M., Wilkinson, J.M. (2002) Biological activities of lavender essential oil. *Phytotherapy Research*, 16, 301–308.

Cavanagh, H.M.A. (2005) Lavender essential oil: A review. *Australian Infection Control*, 10, 35–37.

Chen, Z.-L., Chao, J.-P., Cao, H.-L, Bi, W.-T., Cui, H.-Y., Li, M.-L. (2011) Research on the extraction of plant volatile oils. *Procedia Environmental Sciences*, 8, 426–432.

Císarová, M., Tančinová, D., Medo, J. (2016) Antifungal activity of lemon, eucalyptus, thyme, oregano, sage and lavender essential oils against *Aspergillus niger* and *Aspergillus tubingensis* isolated from grapes, *Potravinarstvo*, 10 (1), 83–88.

Conti, B., Canale, A., Bertoli, A., Gozzini, F., Pistelli, L. (2010) Essential oil composition and larvicidal activity of six Mediterranean aromatic plants against the mosquito *Aedes albopictus* (Diptera: Culicidae). *Parasitology Research*, 107 (6), 1455–1461.

D'Auria, F.D., Tecca, M., Strippoli, V., Salvatore, G., Battinelli, L., Mazzanti, G. (2005) Antifungal activity of *Lavandula angustifolia* essential oil against *Candida albicans* yeast and mycelial form. *Medical Mycology*, 43 (5), 391–396.

Danh, L.T., Triet, N.D.A., Han, L.T.N., Zhao, J., Mammucari, R., Foster, N. (2012) Antioxidant activity, yield and chemical composition of lavender essential oil extracted by supercritical $CO_2$. *Journal of Supercritical Fluids*, 70, 27–34.

Da Porto, C., Decorti, D., Kikic, I. (2009) Flavour compounds of *Lavandula angustifolia* L. to use in food manufacturing: Comparison of three different extraction methods. *Food Chemistry*, 112, 1072–1078.

Dorn, B., Jossi, W., Humphrys, C., Hiltbrunner, J. (2014) Screening of natural products in the laboratory and the field for control of pollen beetles. *Journal of Applied Entomology*, 138, 109–119.

Ellse, L., Burden, F.A., Wall, R. (2013) Control of the chewing louse *Bovicola* (*Werneckiella*) *ocellatus* in donkeys, using essential oils. *Medical and Veterinary Entomology*, 27, 408–413.

Ellse, L., Sands, B., Burden, F.A., Wall, R. (2016) Essential oils in the management of the donkey louse, *Bovicola ocellatus*. *Equine Veterinary Journal*, 48, 285–289.

Erland, L.A.E., Rheault, M.R., Mahmoud, S.S. (2015) Insecticidal and oviposition deterrent effects of essential oils and their constituents against the invasive pest *Drosophila suzukii* (*Matsumura*) (Diptera: Drosophilidae). *Crop Protection*, 78, 20–26.

Faraone, N., Hillier, N.K., Cutler, G.C. (2015) Plant essential oils synergize and antagonize toxicity of different conventional insecticides against *Myzus persicae* (Hemiptera: Aphididae). *PLoS One*, 10 (5), e0127774.

George, D.R., Sparagano, O.A.E., Port, G., Okello, E., Shiel, R.S., Guy, J.H. (2009) Repellence of plant essential oils to *Dermanyssus gallinae* and toxicity to the non target invertebrate *Tenebrio molitor*. *Veterinary Parasitology*, 162, 129–134.

Grosso, C., Coelho, J.A., Urieta, J.S., Palavra, A.M.F., Barroso, J.G. (2010) Herbicidal activity of volatiles from coriander, winter savory, cotton lavender, and thyme isolated by hydrodistillation and supercritical fluid extraction. *Journal of Agricultural and Food Chemistry*, 58, 11007–11013.

Gulcin, W. (2004) Comparison of antioxidant activity of clove (*Eugenia caryophylata* Thunb) buds and lavender (*Lavandula stoechas* L.). *Food Chemistry*, 87, 393–400.

Hass, C.D. (2001) *Lavender—The Most Essential Oil*. Pennon Publishing Pty, Limited, Essendon, VIC, Australia. ISBN10:1877029033 ISBN 13: 9781877029035.

Heisey, R.M., Heisey, T.K. (2003) Herbicidal effects under field conditions of *Ailanthus altissima* bark extract, which contains ailanthone. *Plant Soil*, 256, 85–99.

Hosseinzadeh, S., Shams-Bakhsh, M., Hosseinzadeh, E. (2013) Effects of sub-bactericidal concentration of plant essential oils on pathogenicity factors of *Ralstonia solanacearum*. *Archives of Phytopathology and Plant Protection*, 46, 643–655.

Hui, L., He, L., Huan, L., Xiaolan, L., Aiguo, Z. (2010) Chemical composition of lavender essential oil and its antioxidant activity and inhibition against rhinitis-related bacteria. *African Journal of Microbiology Research*, 4, 309–313.

Jablonský, M., Ramajová, H., Ház, A., Sládková, A., Škulcová, A., Cížová, K. (2016) Comparison of different methods for extraction from lavender: Yield and chemical composition. *Key Engineering Materials*, 688, 31–37.

Jianu, C., Pop, G., Gruia, A.T., Horhat, F.G. (2013) Chemical composition and antimicrobial activity of essential oils of lavender (*Lavandula angustifolia*) and lavandin (*Lavandula × intermedia*) grown in western Romania. *International Journal of Agriculture and Biology*, 15, 772–776.

Kadoglidou, K., Lagopodi, A., Karamanoli, K., Vokou, D., Bardas, G.A., Menexes, G., Constantinidou, H.-I.A. (2011) Inhibitory and stimulatory effects of essential oils and individual monoterpenoids on growth and sporulation of four soil-borne fungal isolates of *Aspergillus terreus*, *Fusarium oxysporum*, *Penicillium expansum*, and *Verticillium dahlia*. *European Journal of Plant Pathology*, 130, 297–309.

Karamanoli, K., Vokou, D., Menkissoglu, U., Constantinidou, H.-I. (2000) Bacterial colonization of phyllosphere of Mediterranean aromatic plants. *Journal of Chemical Ecology*, 26 (9) 2035–2048.

Khater, H.F., Ramadan, M.Y., Mageid, A.D.A. (2013) In vitro control of the camel nasal botfly, *Cephalopina titillator*, with doramectin, lavender, camphor, and onion oils. *Parasitology Research*, 112, 2503–2510.

Koul, O., Walia, S., Dhaliwal, G.S. (2008) Essential oils as green pesticides: Potential and constraints. *Biopesticides International*, 4 (1), 63–84.

Lachance, S., Grange, G. (2014) Repellent effectiveness of seven plant essential oils, sunflower oil and natural insecticides against horn flies on pastured dairy cows and heifers. *Medical and Veterinary Entomology*, 28, 193–200.

Lafhal, S., Vanloot, P., Bombarda, I., Valls, R., Kister, J., Dupuy, N. (2015) Raman spectroscopy for identification and quantification analysis of essential oil varieties: A multivariate approach applied to lavender and lavandin essential oil. *Journal of Raman Spectroscopy*, 46, 577–585.

Lesage-Meessen, L., Bou, M., Sigoillot, J.-C., Faulds, C.B., Lomascolo, A. (2015) Essential oils and distilled straws of lavender and lavandin: A review of current use and potential application in white biotechnology. *Applied Microbiology and Biotechnology*, 99 (8), 3375–3385.

Lopez-Reyes J.G., Spadaro, D., Gullinoa, M.L., Garibaldia, A. (2010) Efficacy of plant essential oils on postharvest control of rot caused by fungi on four cultivars of apples in vivo. *Flavour and Fragrance Journal*, 25, 171–177.

Mahmoud, S.S., Croteau, R.B. (2002) Strategies for transgenic manipulation of monoterpene biosynthesis in plants. *Trends in Plant Sciences*, 7, 366–373.

Mann, R.S., Tiwari, S., Smoot, J.M., Rouseff, R.L., Stelinski, L.L. (2012) Repellency and toxicity of plant-based essential oils and their constituents against *Diaphorina citri* Kuwayama (Hemiptera: Psyllidae). *Journal of Applied Entomology*, 136 (1–2), 87–96.

Mant, J., Brändli, C., Vereecken, N.J., Schulz, C.M., Francke, W., Schiestl, F.P. (2005) Cuticular hydrocarbons as sex pheromone of the bee *Colletes cunicularius* and the key to its mimicry by the sexually deceptive orchid, *Ophrys exaltata*. *Journal of Chemical Ecology*, 31, 1765–1787.

Méndez-Tovar I., Sponza, S., Asensio-S-Manzanera, C., Schmiderer, C., Novak, J. (2015) Volatile fraction differences for Lamiaceae species using different extraction methodologies. *Journal of Essential Oil Research*, 27 (6), 497–505.

Mengulluoglu, M., Soylu, S. (2012) Antibacterial activities of essential oils extracted from medicinal plants against seed-borne bacterial disease agent, *Acidovorax avenae* subsp. *citrulli*. *Research on Crops*, 13 (2), 641–646.

Moretti, A.N., Zerba, E.N., Alzogaray, R.A. (2015) Lethal and sublethal effects of eucalyptol on *Triatoma infestans* and *Rhodnius prolixus*, vectors of Chagas disease. *Entomologia Experimentalis et Applicata*, 154, 62–70.

Nerio, L.S., Verbel, J.O., Stashenko, E. (2010) Repellent activity of essential oils: A review. *Bioresource Technology*, 101, 372–378.

Odalo, J.O., Omolo, M.O., Malebo, H., Angira, J., Njeru, P.M., Ndiege, I.O., Hassanali, A. (2005) Repellency of essential oils of some plants from the Kenyan coast against *Anopheles gambiae*. *Acta Tropica*, 95, 210–218.

Olanya, O.M., Larkin, R.P. (2006) Efficacy of essential oils and biopesticides on *Phytophthora infestans* suppression in laboratory and growth chamber studies. *Biocontrol Science and Technology*, 16 (9), 901–917.

Oszagyh, M., Simandi, B., Sawinsky, J., Kery, A., Lemberkovics, E. (1996) Supercritical fluid extraction of volatile compounds from lavandin and thyme. *Flavour and Fragrance Journal*, 11, 157–165.

Papachristos, D.P., Stamopoulos, D.C. (2004) Fumigant toxicity of three essential oils on the eggs of *Acanthoscelides obtectus* (Say) (Coleoptera: Bruchidae). *Journal of Stored Products Research*, 40, 517–525.

Peixoto, M.G., Bacci, L., Blank, A.F., Araújo, A.P.A., Alves, P.B., Silva, J.H.S., Santos, A.A., Oliveira, A.P., da Costa, A.S., de Fatima Arrigoni-Blank, M. (2015) Toxicity and repellency of essential oils of *Lippia alba* chemotypes and their major monoterpenes against stored grain insects. *Industrial Crops and Products*, 71, 31–36.

Périno-Issartier, S., Ginies, C., Cravotto, G., Chemat, F. (2013) A comparison of essential oils obtained from lavandin via different extraction processes: Ultrasound, microwave, turbohydrodistillation, steam and hydrodistillation. *Journal of Chromatography A*, 1305, 41–47.

Pinto, Z.T., Sanchez, F.F., Santos, A.R.D., Amaral, A.C.F., Ferreira, J.L.P., Escalona-Arranz, J.C., Queiroz, M.M. (2015) Chemical composition and insecticidal activity of *Cymbopogon citratus* essential oil from Cuba and Brazil against housefly. *Revista Brasileira de Parasitologia Veterinária*, 24, 36–44.

Preuss, H.G., Echard, B., Enig, M., Brook, I., Elliott, T.B. (2005) Minimum inhibitory concentrations of herbal essential oils and monolaurin for gram-positive and gram-negative bacteria. *Molecular and Cellular Biochemistry*, 272, 29–34.

Rafiee-Dastjerdi, H., Khorrami, F., Hassanpour, M. (2014) The toxicity of some medicinal plant extracts to the potato tuber moth, *Phthorimaea operculella* (Lepidoptera: Gelechiidae). *Archives of Phytopathology and Plant Protection*, 47, 1827–1831.

Rakotonirainy, M.S., Lavédrine, B. (2005) Screening for antifungal activity of essential oils and related compounds to control the biocontamination in libraries and archives storage areas. *International Biodeterioration and Biodegradation*, 55, 141–147.

Razazadeh, S.H., Baha-Aldini, B.Z.B.F., Vatanara, A., Behbahani, B., Rouho-lamini, N.A., Maleky-Doozzadeh, M., Yarigar-Ravesh, M., Pirali, H.M. (2008) Comparison of super critical fluid extraction and hydrodistillation methods on lavender's essential oil composition and yield. *Journal of Medicinal Plants*, 7, 63–68.

Regnault-Roger, C., Vincent, C., Arnason, J.T. (2012) Essential oils in insect control: Low-risk products in a high-stakes world. *Annual Review of Entomology*, 57, 405–424.

Reverchon, E. (1997) Supercritical fluid extraction and fractionation of essential oils and related products. *Journal of Supercritical Fluids*, 10, 1–37.

Rossi, Y.E., Palacios, S.M. (2015) Insecticidal toxicity of *Eucalyptus cinerea* essential oil and 1,8-cineole against *Musca domestica* and possible uses according to the metabolic response of flies. *Industrial Crops and Products*, 63, 133–137.

Sahraoui, N., Vian, M.A., Bornard, I., Boutekedjiret, C., Chemat, F. (2008) Improved microwave steam distillation apparatus for isolation of essential oils. Comparison with conventional steam distillation. *Journal of Chromatography A*, 1210, 229–233.

Samfira, I., Rodino, S., Petrache, P., Cristina, R.T., Butu, M., Butnariu, M. (2015) Characterization and identity confirmation of essential oils by mid infrared absorption spectrophotometry. *Digest Journal of Nanomaterials and Biostructures*, 10, 557–566.

Sertkaya, E., Kaya, K., Soylu, S. (2010) Acaricidal activities of the essential oils from several medicinal plants against the carmine spider mite (*Tetranychus cinnabarinus* Boisd) (Acarina: Tetranychidae). *Industrial Crops and Products*, 31, 107–112.

Shellie, R., Mondello, L., Marriott, P., Dugo, G. (2002) Characterisation of lavender essential oils by using gas chromatography–mass spectrometry with correlation of linear retention indices and comparison with comprehensive two-dimensional gas chromatography. *Journal of Chromatography A*, 970, 225–234

Singh, H.P., Batish, D.R., Kohli, R.K. (2003) Allelopathic interactions and allelochemicals: New possibilities for sustainable weed management. *Critical Reviews in Plant Sciences*, 22, 239–311.

Singh, H.P., Batish, D.R., Setia, N., Kohli, R.K. (2005) Herbicidal activity of volatile oils from *Eucalyptus citriodora* against *Parthenium hysterophorus*. *Annals of Applied Biology*, 146, 89–94.

Sinthusiri, J., Soonwera, M. (2013) Efficacy of herbal essential oils as insecticides against the housefly, *Musca domestica* L. *Southeast Asian Journal of Tropical Medicine and Public Health*, 44 (2), 188.

Sinthusiri, J., Soonwera, M. (2014) Oviposition deterrent and ovicidal activities of seven herbal essential oils against female adults of housefly, *Musca domestica* L. *Parasitology Research*, 113, 3015–3022.

Soylu, E.M., Kurt, Ş., Soylu, S. (2010) In vitro and in vivo antifungal activities of the essential oils of various plants against tomato grey mould disease agent *Botrytis cinerea*. *International Journal of Food Microbiology*, 143, 183–189.

Stanković, N.S., Čomić, L.R., Kocić, B.D., Nikolić, D.M., Mihajilov-Krstev, T.M., Ilić, B.S., Miladinović, D.L. (2011) Antibacterial activity chemical composition relationship of the essential oils from cultivated plants from Serbia [in Serbian]. *Hemijska Industrija*, 65 (5), 583–589.

Tabassum, N., Vidyasagar, G.M. (2013) Antifungal investigations on plant essential oils. A review. *International Journal of Pharmacy and Pharmaceutical Sciences*, 5, 19–28.

Tabata, J., Teshiba, M., Shimizu, N., Sugie, H. (2015) Mealybug mating disruption by a sex pheromone derived from lavender essential oil. *Journal of Essential Oil Research*, 27, 232–237.

Takarada, K., Kimizuka, R., Takahashi, N., Honma, K., Okuda, K., Kato, T. (2004) A comparison of the antibacterial efficacies of essential oils against oral pathogens. *Oral Microbiology and Immunology*, 19 (1), 61–64.

Talbert, R., Wall, R. (2012) Toxicity of essential and non-essential oils against the chewing louse, *Bovicola* (*Werneckiella*) *ocellatus*. *Research in Veterinary Science*, 93 (2), 831–835.

Thosar, N., Basak, S., Bahadure, R.N., Rajurkar, M. (2013) Antimicrobial efficacy of five essential oils against oral pathogens: An in vitro study. *European Journal of Dentistry*, 7 (5 Suppl.), S71–S77.

Topal, U., Sasaki, M., Goto, M., Otles, S. (2008) Chemical compositions and antioxidant properties of essential oils from nine species of Turkish plants obtained by supercritical carbon dioxide extraction and steam distillation. *International Journal of Food Sciences and Nutrition*, 59, 619–634.

Tullio, V., Nostro, A., Mandras, N., Dugo, P., Banche, G., Cannatelli, M.A., Cuffini, A.M., Alonzo, V., Carlone, N.A. (2007) Antifungal activity of essential oils against filamentous fungi determined by broth microdilution and vapour contact methods. *Journal of Applied Microbiology*, 102, 1544–1550.

Upson, T. (2002) The taxonomy of the genus *Lavandula* L. In M. Lis-Balchin (ed.), *Lavender: The Genus Lavandula: Medicinal and Aromatic Plants—Industrial Profiles*. Taylor & Francis, London, 2–34.

Uremis, I., Arslan, M., Sangun, M.K. (2009) Herbicidal activity of essential oils on the germination of some problem weeds. *Asian Journal of Chemistry*, 21 (4), 3199–3210.

Végh, A., Bencsik, T., Molnár, P., Böszörményi, A., Lemberkovics, É., Kovács, K., Kocsis, B., Horváth, G. (2012) Composition and antipseudomonal effect of essential oils isolated from different lavender species. *Natural Product Communications*, 7, 1393–1396.

Verma, R.S., Rahman, L.U., Chanotiya, C.S., Verma, R.K., Chauhan, A., Yadav, A., Singh, A., Yadav, A.K. (2010) Essential oil composition of *Lavandula angustifolia* Mill. cultivated in the mid hills of Uttarakhand. *India Journal of the Serbian Chemical Society*, 75, 343–348.

Voravuthikunchai, S.P., Minbutra, S., Goodla, L., Jefferies, J., Voravuthikunchai, S. (2012) Mixtures of essential oils in an air conditioning prototype to reduce the prevalence of airborne pathogenic bacteria. *Journal of Essential Oil-Bearing Plants*, 15, 739–749.

Wells, R., Lis-Balchin, M. (2002) Perfumery uses of lavender and lavandin oils. In M. Lis-Balchin (ed.), *Lavender: The Genus Lavandula: Medicinal and Aromatic Plants—Industrial Profiles*. Taylor & Francis, London, 194–199.

Weston, L.A., Duke, S.O. (2003) Weed and crop allelopathy. *Critical Reviews in Plant Sciences*, 22, 367–389.

Yang, J.O., Park, J.H., Son, B.K., Moon, S.R., Kang, S.H., Yoon, C., Kim, G.H. (2009) Repellency and electrophysiological response of caraway and clove bud oils against bean bug *Riptortus clavatus*. *Journal of the Korean Society for Applied Biological Chemistry*, 52, 668–674.

Yeom, H.-J., Jung, C.-S., Kang, J., Kim, J., Lee, J.-H., Kim, D.-S., Kim, H.-S., Park, P.-S., Kang, K.-S., Park, I.-K. (2015) Insecticidal and acetylcholine esterase inhibition activity of Asteraceae plant essential oils and their constituents against adults of the German cockroach (*Blattella germanica*). *Journal of Agricultural and Food Chemistry*, 63, 2241–2248.

Yoon, C., Moon, S.-R., Jeong, J.-W., Shin, Y.-H., Cho, S.-R., Ahn, K.-S., Yang, J.-O., Kim, G.-H. (2011) Repellency of lavender oil and linalool against spot clothing wax cicada, *Lycorma delicatula* (Hemiptera: Fulgoridae) and their electrophysiological responses *Journal of Asia-Pacific Entomology*, 14, 411–416.

Zanellato, M., Masciarelli, E., Casorri, L., Boccia, P., Sturchio, E., Pezzella, M., Cavalieri, A., Fabio, C. (2009) The essential oils in agriculture as an alternative strategy to herbicides: A case study. *International Journal of Environment and Health*, 3, 198–213.

Zapata, N., Smagghe, G. (2010) Repellency and toxicity of essential oils from the leaves and bark of *Laurelia sempervirens* and *Drimys winteri* against *Tribolium castaneum*. *Industrial Crops and Products*, 32, 405–410.

Zheljazkov, V.D., Cantrell, C.L., Astatkie, T., Cannon, J.B. (2011) Lemongrass productivity, oil content, and composition as a function of nitrogen sulfur, and harvest time. *Agronomy Journal*, 103, 805–881.

Zheljazkov, V.D., Cantrell, C.L., Astatkie, T., Ebelhar, M.W. (2010b) Peppermint productivity and oil composition as a function of nitrogen growth stage, and harvest time. *Agronomy Journal*, 102, 124–128.

Zheljazkov, V.D., Cantrell, C.L., Astatkie, T., Hristov, A. (2010a) Yield, content, and composition of peppermint and spearmints as a function of harvesting time and drying. *Journal of Agricultural and Food Chemistry*, 11400–11407.

Zheljazkov, V.D., Cantrell, C.L., Astatkie, T., Jeliazkova, E. (2013) Distillation time effect on lavender essential oil yield and composition. *Journal of Oleo Science*, 62 (4), 195–199.

Zhi-ling, C., Jian-ping, J., Hai-yan, C., Hui-lin, C., Mo-lin, H., Wei-tao, B. (2011) Research on the extraction of plant volatile oils. *Procedia Environmental Sciences*, 8, 426–432.

Zu, Y., Yu, H.C., Liang, L., Fu, Y., Efferth, T.D., Liu, X., Wu, N. (2010) Activities of ten essential oils towards *Propionibacterium acnes* and PC-3, A-549 and MCF-7 cancer cells. *Molecules*, 15 (5), 3200–3210.

# 13

## Pine Needle Oil

Aikaterini Koutsaviti, Efstathia Ioannou, Olga Tzakou,
Panos V. Petrakis, and Vassilios Roussis

### CONTENTS

## 13.1 Introduction

Pines, usually tall and stout trees and less often shrubs, have evergreen foliage leaves in the shape of needles and contain resin in their bark. Pines are monoicus woody plants growing naturally or being naturalized in both hemispheres, mainly distributed over the northern hemisphere, but also occurring in subtropical and tropical regions of Central America and Asia, dominating forests or coexisting with other conifers (Farjon 1984; Gaussen et al. 1993). Pines have received much attention due to their ecological importance as a major component of many temperate forests and their economic significance as a source of timber, pulp and paper, nuts, seeds, resin, construction materials, and other products (Richardson and Rundel 1998).

## 13.2 Botany of the Genus *Pinus*

Pinaceae are among the largest families of conifers (order Coniferales) (Farjon 2005). Among the 225 species belonging to this order, approximately half (ca. 110), considered true pines, belong to the genus *Pinus*, the most widespread tree genus in the northern hemisphere (Price et al. 2000). It is commonly accepted to divide the genus into two subgenera, namely, *Strobus* and *Pinus* (Farjon 2005; Gernandt et al. 2005). Subgenus *Strobus*, also known as *Haploxylon*, includes the Asiatic and North American pines, while subgenus *Pinus*, also known as *Diploxylon*, occupies Middle Europe and the Mediterranean area. Haploxyls, or soft pines, have only one fibrovascular bundle with usually five needles per bundle. In contrast, diploxyls, or hard pines, which could be considered genetically more developed (Norin 1972), have two fibrovascular bundles with two to five needles per bundle. Subgenus *Pinus* is comprised of two sections (*Pinus* and *Trifoliae*) and five subsections (*Australes*, *Contortae*, *Pinaster*, *Pinus*, and *Ponderosae*), whereas two sections (*Parrya* and *Quinquefoliae*) and six subsections (*Balfourianae*, *Cembroides*, *Gerardianae*, *Krempfianae*, *Nelsoniae*, and *Strobus*) are recognized in subgenus *Strobus* (Figure 13.1) (Farjon 2005; Gernandt et al. 2005).

According to Mirov (1967), the genus originated in the late Paleozoic to early Mesozoic period in a circumpolar continent called Beringia. Since then, the genus spread southward toward the equator until the Pleistocene period. The improvement of fossil dating methods and the augmentation of the fossil records position the origin of pines in the early–middle Mesozoic, as Mirov suggested (Millar 2000). In the Eocene, the genus was restricted in refugia that functioned as secondary centers of pine diversity. The extensive speciation in many subsections and the many varieties of the 47 Central American pine taxa are consequences of the evolution of the genus during the Eocene (Millar 1993). Pleistocene climatic fluctuations with glacial–interglacial periods exerted an extinction pressure on pine populations, especially in western Europe. However, in some taxonomic subsections significant speciation occurred, while in others, a significant divergence in several characters, such as the constitution of needle essential oil, appeared (Petrakis et al. 2001). This can explain the great variety observed in the composition of pine needle essential oil, which can only be compared with that of the leaf essential oil of the speciose genus *Eucalyptus*.

## 13.3 Methods of Extraction and Analysis of Pine Needle Oil

Production of volatiles and flavor components from various plants' biomass is one of the oldest chemical operations developed by humankind. The required quality of the final pine needle essential oil dictates the selection of the most appropriate method, often used with various adaptations according to the available infrastructure of the industrial facility. Steam distillation and hydrodistillation are used as the simplest and, in most cases, cost-effective techniques to yield water-immiscible volatile oils. More sophisticated but considerably more expensive methods, such as supercritical fluid extraction, are receiving increasing interest as consumers' concerns about the chemical integrity of natural ingredients is rising. Carbon dioxide is the most frequently used extraction gas because of the low temperatures required to reach the hypercritical state and the short duration of the

Subgenus *Pinus* (Diploxylon or hard pines)
    Section *Pinus*
        Subsection *Pinaster*
            *P. brutia, P. canariensis,*
            *P. halepensis, P. heldreichii,*
            *P. pinaster, P. pinea,*
            *P. roxburghii*
        Subsection *Pinus*
            *P. densata, P. densiflora,*
            *P. hwangshanensis, P. kesiya,*
            *P. luchuensis, P. massoniana,*
            *P. merkusii, P. mugo, P. nigra,*
            *P. resinosa, P. sylvestris,*
            *P. tabuliformis, P. taiwanensis,*
            *P. thunbergii, P. tropicalis,*
            *P. uncinata, P. yunnanensis*
    Section *Trifoliae*
        Subsection *Australes*
            *P. attenuata, P. caribaea,*
            *P. cubensis, P. echinata,*
            *P. elliotti, P. glabra, P. greggii,*
            *P. herrerae, P. jaliscana,*
            *P. lawsonii, P. leiophylla,*
            *P. lumholtzii, P. muricata,*
            *P. occidentalis, P. oocarpa,*
            *P. palustris, P. patula,*
            *P. praetermissa, P. pringlei,*
            *P. pungens, P. radiata, P. rigida,*
            *P. serotina, P. taeda,*
            *P. tecunumanii, P. teocote*
        Subsection *Contortae*
            *P. banksiana, P. clausa,*
            *P. contorta, P. virginiana*
        Subsection *Ponderosae*
            *P. cooperi, P. coulteri, P. donnell-*
            *smithii, P. devoniana,*
            *P. douglasiana, P. durangensis,*
            *P. engelmannii, P. hartwegii,*
            *P. jeffreyi, P. maximinoi,*
            *P. montezumae, P. nubicola,*
            *P. ponderosa, P. pseudostrobus,*
            *P. sabineana, P. torreyana,*
            *P. washoensis*

Subgenus *Strobus* (Haploxylon or soft pines)
    Section *Parrya*
        Subsection *Balfourianae*
            *P. aristata, P. balfouriana,*
            *P. longaeva*
        Subsection *Cembroides*
            *P. cembroides, P. culminicola,*
            *P. discolor, P. edulis, P. johannis,*
            *P. maximartinezii,*
            *P. monophylla, P. pinceana*
            *P. quadrifolia, P. remota,*
            *P. rzedowskii*
        Subsection *Nelsoniae*
            *P. nelsonii*
    Section *Quinquefoliae*
        Subsection *Gerardianae*
            *P. bungeana, P. gerardiana,*
            *P. squamata*
        Subsection *Krempfianae*
            *P. krempfii*
        Subsection *Strobus*
            *P. albicaulis, P. armandii,*
            *P. ayacahuite, P. bhutanica,*
            *P. cembra, P. chiapensis,*
            *P. dabeshanensis, P. dalatensis,*
            *P. fenzeliana, P. flexilis,*
            *P. koraiensis, P. lambertiana,*
            *P. monticola, P. morrisonicola,*
            *P. parviflora, P. peuce, P. pumila,*
            *P. sibirica, P. strobiformis,*
            *P. strobus, P. wallichiana,*
            *P. wangii*

**FIGURE 13.1**
Classification of the genus *Pinus*. (From Gernandt, D.S. et al., *Taxon*, 54, 29–42, 2005.)

extraction process. Carbon dioxide is nontoxic, colorless, odorless, and inert, and therefore does not interact chemically with the ingredients of the extracted biomass.

The chemical composition of the pine needle essential oils is routinely analyzed by gas chromatography using a flame ionization detector (GC-FID) and gas chromatography–mass spectrometry (GC-MS), which are fast and reliable techniques for the identification of volatile constituents. The FID is used mainly for quantitative (using internal standards) analyses, whereas the MS detector, providing mass spectra for comparison with spectra libraries, along with comparison of retention times of the eluted volatiles, is preferred for the qualitative chemical profiling of the oils.

## 13.4  Physical and Chemical Properties and Composition of Pine Needle Oil

The techniques commonly applied for the assessment of the physicochemical properties of pine needle oil include the measurement of specific gravity, the determination of refractive index, and the measurement of optical rotation. In all cases, these vary depending on the chemical composition of the needle oil. When measurements are performed at different temperatures, conversion factors should be applied for the normalization of values before comparison with literature data. The purity of the essential oil can be determined by measurement of its solubility in aqueous saturated sodium chloride solution.

The main chemical constituents of pine needle oils belong to the class of monoterpenes, sesquiterpenes, and diterpenes, which can be in the form of hydrocarbons or oxygenated derivatives. Among pine species, both qualitative and quantitative differences are usually observed. The most abundant constituents of pine needle oils are in most cases α- and β-pinene, owing their name to their initial source of isolation.

The chemical compositions of the different pine needle oils reported in the literature are compiled below, according to their botanical classification. The major constituents and their relative abundance as a result of the origin of the plant source are summarized.

### 13.4.1  Subgenus *Pinus*, Section *Pinus*, Subsection *Pinaster*

*Pinus brutia* Tenore (eastern Mediterranean or Calabrian pine)

Plant source: Greece (Roussis et al. 1995; Ioannou et al. 2014; Koutsaviti et al. 2015), Morocco (Lahlou 2003; Hmamouchi et al. 2001)

Key constituents: β-Pinene (31.2%–33.6%), α-pinene (14.8%–20.6%), β-caryophyllene (4.85%–14.5%), germacrene D (0.0%–17.0%), α-terpinyl acetate (0.0%–5.3%)

*Pinus canariensis* C.Sm. (Canary Island pine)

Plant source: Algeria (Dob et al. 2005), Greece (Roussis et al. 1995; Ioannou et al. 2014; Koutsaviti et al. 2015), Morocco (Hmamouchi et al. 2001)

Key constituents: δ-Cadinene (1.3%–5.6%), α-pinene (1.0%–15.0%), limonene (0.5%–7.9%), germacrene D (0.0%–62.5%), β-caryophyllene (0.05%–16.8%), β-selinene (0.0%–63.7%), myrcene (0.0%–8.8%)

*Pinus halepensis* Mill. (Aleppo pine)

Plant source: Algeria (Dob et al. 2005, 2007; Abi-Ayad et al. 2011; Djerrad et al. 2015), Greece (Roussis et al. 1995; Ioannou et al. 2014; Koutsaviti et al. 2015), Italy (Macchioni et al. 2003), Morocco (Hmamouchi et al. 2001; Lahlou 2003), Tunisia (Amri et al. 2013, 2014; Hamrouni et al. 2014), Turkey (Ustun et al. 2012)

Key constituents: α-Humulene (0.2%–10.5%), caryophyllene oxide (0.0%–48.5%), β-pinene (0.0%–46.8%), *cis*-caryophyllene (0.0%–40.3%), longifolene (0.0%–33.9%), β-caryophyllene (0.0%–28.5%), myrcene (0.0%–27.9%), α-pinene (0.0%–23.3%), terpinolene (0.0%–11.0%), sabinene (0.0%–9.4%), germacrene D (0.0%–8.8%), aromadendrene (0.0%–8.5%), thunbergol (0.0%–8.3%), cembrene (0.0%–7.6%), *cis*-muurrola-4(14),5-diene (0.0%–7.4%)

*Pinus heldreichii* Christ (Bosnian pine)

Plant source: Greece (Petrakis et al. 2001; Ioannou et al. 2014), Italy (Bonesi et al. 2010), Montenegro (Nikolić et al. 2007), former Yugoslav Republic of Macedonia (FYROM) (Nikolic et al. 2015), Serbia (Simić et al. 1996; Nikolić et al. 2007; Bojović et al. 2012)

Key constituents: α-Pinene (8.5%–22.2%), limonene (7.8%–31.9%), β-caryophyllene (4.5%–15.1%), germacrene D (0.7%–45.4%), δ-cadinene (0.6%–7.6%), δ-3-carene (0.0%–18.6%), β-pinene (0.0%–8.4%), aristolene (0.0%–6.0%), terpinolene (0.0%–5.9%)

*Pinus pinaster* Aiton (maritime or cluster pine)

Plant source: Algeria (Dob et al. 2005), France (Pauly et al. 1973; Ottavioli et al. 2008), Greece (Petrakis et al. 2001; Koutsaviti et al. 2015), Italy (Macchioni et al. 2003), Morocco (Hmamouchi et al. 2001; Lahlou 2003), Spain (Garrido et al. 1988; Ioannou et al. 2014), Turkey (Ustun et al. 2012)

Key constituents: β-Pinene (0.3%–42.4%), myrcene (0.1%–7.6%), limonene (0.1%–4.6%), α-pinene (0.0%–29.4%), β-caryophyllene (0.0%–22.2%), γ-muurolene (0.0%–6.8%)

*Pinus pinea* L. (umbrella pine)

Plant source: Algeria (Dob et al. 2005), Greece (Ioannou et al. 2014), Italy (Macchioni et al. 2002), Morocco (Lahlou 2003), Tunisia (Nasri et al. 2011), Turkey (Ustun et al. 2012)

Key constituents: α-Copaene (0.1%–58.8%), β-pinene (0.0%–42.4%), α-pinene (0.0%–37.0%), limonene (0.0%–58.9%), abienol (0.0%–12.3%), germacrene D (0.0%–10.4%), farnesyl acetate (0.0%–9.0%), sylvestrene (0.0%–7.9%), β-phellandrene (0.0%–6.7%)

*Pinus roxburghii* Sarg. (chir pine)

Plant source: Australia (Ioannou et al. 2014), Egypt (Islam 2006; Zafar et al. 2010; Salem et al. 2014), Nepal (Satyal et al. 2013)

Key constituents: δ-3-Carene (2.3%–33.4%), β-caryophyllene (1.07%–31.7%), α-pinene (0.4%–39.0%), terpinen-4-ol (0.0%–30.1%), α-humulene (0.0%–7.3%)

## 13.4.2 Subgenus *Pinus*, Section *Pinus*, Subsection *Pinus*

*Pinus densiflora* Siebold & Zucc. (Japanese red pine)

Plant source: Japan (Yatagai and Sato 1986), South Korea (Hong et al. 2004; Lee et al. 2009; Park and Lee 2010; Jeon and Lee 2012; Kim et al. 2013), Spain (Ioannou et al. 2014)

Key constituents: β-Pinene (5.0%–12.0%), α-pinene (1.8%–25.3%), myrcene (1.1%–12.2%), sabinene (0.0%–16.6%), camphene (0.0%–22.4%), β-phellandrene (0.0%–21.0%), limonene (0.0%–20.2%), α-thujene (0.0%–19.3%), α-humulene (0.0%–18.1%), *trans*-cadina-1(6),4-diene (0.0%–12.7%), germacrene D (0.0%–11.3%), α-fenchyl acetate (0.0%–10.3%), bornyl acetate (0.0%–9.8%), β-caryophyllene (0.0%–8.7%)

*Pinus massoniana* Lamb. (masson pine)

Plant source: Australia (Ioannou et al. 2014), China (Yatagai and Hong 2007)

Key constituents: α-Pinene (26.9%–45.5%), β-pinene (13.7%–16.5%), β-phellandrene (1.4%–6.8%), β-caryophyllene (0.7%–11.6%), germacrene D (0.0%–20.7%), bornyl acetate (0.0%–7.8%)

*Pinus mugo* Turra (dwarf mountain pine)

Plant source: England (Ioannou et al. 2014), Poland (Celiński et al. 2015), Republic of Kosovo (Hajdari et al. 2015), Scotland (Tsitsimpikou et al. 2001), Serbia (Stevanovic et al. 2005)

Key constituents: α-Pinene (13.7%–33.3%), δ-3-carene (9.9%–27.8%), β-caryophyllene (2.4%–8.9%), myrcene (1.4%–6.9%), β-phellandrene (1.2%–7.9%), germacrene D (0.7%–7.1%), bornyl acetate (0.0%–11.5%), β-pinene (0.0%–7.8%), terpinolene (0.0%–7.3%), germacrene D-4-ol (0.0%–6.1%)

*Pinus mugo* **var.** *prostrata*

Plant source: England (Ioannou et al. 2014)

Key constituents: Bornyl acetate (14.1%), α-pinene (12.9%), camphene (6.5%), δ-3-carene (6.4%), germacrene D-4-ol (6.0%), bicyclogermacrene (5.7%)

*Pinus mugo* **var.** *pumilio* Zenari

Plant source: England (Ioannou et al. 2014)

Key constituents: α-Pinene (14.1%), δ-3-carene (12.0%), germacrene (8.0%), bornyl acetate (7.6%), β-caryophyllene (6.8%), camphene (6.3%), sabinene (4.8%), germacrene D-4-ol (4.1%)

*Pinus nigra* J.F. Arnold (black pine)

Plant source: Greece (Roussis et al. 1995; Koutsaviti et al. 2015), Italy (Macchioni et al. 2003), Turkey (Sezik et al. 2010; Ustun et al. 2012)

Key constituents: Germacrene D (6.5%–21.1%), β-caryophyllene (4.9%–21.5%), α-pinene (4.6%–49.3%), β-pinene (2.3%–21.3%)

*Pinus nigra* **var.** *calabrica* C.K. Schneid.

Plant source: Italy (Bonesi et al. 2010)

Key constituents: α-Pinene (24.6%), β-pinene (10.9%), γ-cadinene (9.9%), β-phellandrene (6.3%), manoyl oxide (6.2%)

*Pinus nigra* **subsp.** *caramanica* Businský (Crimean pine)

Plant source: England (Ioannou et al. 2014)

Key constituents: β-Pinene (20.7%), germacrene D (20.0%), α-pinene (18.0%), β-caryophyllene (7.8%)

*Pinus nigra* **subsp.** *laricio* Maire (Corsican pine)

Plant source: England (Ioannou et al. 2014), Tunisia (Amri et al. 2014)

Key constituents: Germacrene D (12.7%–16.9%), β-caryophyllene (8.9%–13.9%), limonene (2.6%–9.7%), α-pinene (2.1%–18.0%), δ-3-carene (0.3%–16.1%)

*Pinus nigra* **subsp.** *nigra* (Austrian or black pine)

Plant source: Italy (Bonesi et al. 2010), Spain (Ioannou et al. 2014)

Key constituents: α-Pinene (19.1%–25.3%), sabinene (12.1%–12.8%), limonene (3.1%–22.6%), β-caryophyllene (3.0%–16.1%), germacrene D (0.5%–32.1%)

*Pinus nigra* subsp. *salzmannii* Franco (Salzmann pine)

Plant source: England (Ioannou et al. 2014)

Key constituents: Germacrene D (18.5%), β-caryophyllene (17.8%), limonene (12.5%), α-pinene (12.2%)

*Pinus resinosa* Sol. ex Aiton (Norway or red pine)

Plant source: Poland (Krauze-Baranowska et al. 2002)

Key constituents: β-Pinene (42.4%), α-pinene (3.3%), myrcene (14.5%), germacrene D (4.9%)

*Pinus sylvestris* L. (Scots pine)

Plant source: England (Ioannou et al. 2014), Estonia (Orav et al. 1996), Finland (Lawrence and Reynolds 2003), France (Lawrence and Reynolds 1991), Germany (Kubeczka and Schulze 1987), Lithuania (Venskutonis et al. 2000; Judzentiene and Kupcinskiene 2008), Nigeria (Fayemiwo et al. 2014), Russia (Góra and Lis 2000), Scotland (Tsitsimpikou et al. 2001), Slovakia (Berta et al. 1997), Turkey (Ustun et al. 2006)

Key constituents: α-Pinene (0.0%–69.1%), δ-3-carene (0.0%–43.4%), manoyl oxide (0.0%–30.2%), α-terpineol (0.0%–27.2%), terpinen-4-ol (0.0%–21.8%), β-terpineol (0.0%–14.1%), β-pinene (0.0%–18.4%), camphene (0.0%–16.8%), caryophyllene oxide (0.0%–12.7%), δ-cadinene (0.0%–11.6%), germacrene D (0.0%–10.3%), β-caryophyllene (0.0%–7.0%), borneol (0.0%–6.7%), limonene (0.0%–5.2%)

*Pinus sylvestris* subsp. *scotica* E.F. Warb.

Plant source: England (Ioannou et al. 2014)

Key constituents: Isoabienol (25.9%), β-pinene (10.7%), germacrene D-4-ol (10.0%), α-pinene (9.5%), δ-3-carene (5.7%), germacrene D (5.1%), α-cadinol (4.1%)

*Pinus tabuliformis* Carrière (Chinese red pine)

Plant source: China (Chen et al. 2006; Zhang and Wang 2010), Spain (Ioannou et al. 2014)

Key constituents: α-Pinene (8.6%–78.5%), β-caryophyllene (0.0%–23.6%), β-pinene (0.0%–16.6%), germacrene D (0.0%–14.5%), caryophyllene oxide (0.0%–7.4%)

*Pinus taiwanensis* Hayata (Taiwan red or Formosa pine)

Plant source: England (Ioannou et al. 2014)

Key constituents: β-Phellandrene (15.6%), β-caryophyllene (14.9%), α-pinene (12.1%), β-pinene (11.5%), myrcene (7.4%), terpinolene (5.6%), germacrene D (5.2%), bornyl acetate (4.2%)

*Pinus thunbergii* Parl. (Japanese black pine)

Plant source: Italy (Ioannou et al. 2014), Japan (Kusumoto and Shibutani 2015), South Korea (Park and Lee 2011; Jeon and Lee 2012; Kim et al. 2013)

Key constituents: β-Pinene (4.0%–32.7%), 2*H*-benzocyclohepten-2-one (0.0%–34.3%), α-humulene (0.0%–19.6%), terpinolene (0.5%–19.3%), α-pinene (0.7%–18.4%), germacrene D (0.0%–18.7%), δ-3-carene (0.0%–16.8%), β-phellandrene (0.0%–14.5%), sabinene (0.0%–10.2%), β-caryophyllene (0.0%–9.4%), γ-terpinene (0.0%–6.3%), limonene (0.0%–5.9%), myrcene (0.0%–5.4%), terpinen-4-ol (0.0%–5.4%), δ-cadinene (0.0%–4%)

*Pinus uncinata* Ramond ex DC (mountain pine)

Plant source: Poland (Bonikowski et al. 2015)

Key constituents: Bornyl acetate (30.2%), α-pinene (12.2%), β-caryophyllene (6.8%), limonene (4.9%)

*Pinus yunnanensis* **Franch. var.** *tenuifolia* W.C. Cheng & Y.W. Law (Yunnan pine)

Plant source: China (Tian et al. 2012)

Key constituents: α-Pinene (22.5%), β-caryophyllene (16.6%), 1,2,4a,5,8,8a-hexahydro-4,7-dimethyl-1-(1-methylethyl)-naphthalene (5.1%), β-cubebene (4.8%), elixene (4.4%)

### 13.4.3 Subgenus *Pinus*, Section *Trifoliae*, Subsection *Australes*

*Pinus attenuata* Lemmon (knobcone pine)

Plant source: Greece (Petrakis et al. 2001; Ioannou et al. 2014)

Key constituents: α-Pinene (9.6%–38.1%), germacrene D (4.5%–29.0%), β-caryophyllene (3.5%–7.9%), limonene (1.4%–5.0), β-pinene (1.0%–19.1%)

*Pinus caribaea* Morelet (Caribbean pine)

Plant source: Bangladesh (Chowdhury et al. 2008), Nigeria (Sonibare and Olakunle 2008)

Key constituents: β-Caryophyllene (10.2%–23.8%), α-pinene (5.0%–6.6%), germacrene D (2.4%–8.4%), caryophyllene oxide (0.2%–21.8%), limonene (0.0%–48.84%), α-caryophyllene (0.0%–6.5%)

*Pinus elliottii* Engelm. (slash pine)

Plant source: Australia (Ioannou et al. 2014), Mozambique (Pagula and Baeckström 2006)

Key constituents: α-Pinene (10.6%–43.0%), β-pinene (12.9%–27.1%), germacrene D (0.0%–24.5%), β-caryophyllene (0.0%–6.6%)

*Pinus muricata* D. Don (bishop pine)

Plant source: Scotland (Ioannou et al. 2014)

Key constituents: Germacrene D (41.5%), α-pinene (17.3%), *trans*-β-ocimene (5.4%), δ-3-carene (5.3%), β-pinene (4.8%)

*Pinus palustris* Mill. (longleaf or Florida pine)

Plant source: Japan (Yatagai and Sato 1986; Kurose et al. 2007)

Key constituents: β-Pinene (25.2%–32.3%), α-pinene (8.7%–11.6%), α-terpineol (8.0%–27.3%), δ-cadinene (5.0%–8.3%), β-caryophyllene (3.8%–6.7%)

*Pinus patula* Schiede ex Schltdl. & Cham. (Mexican weeping pine)

Plant source: Italy (Ioannou et al. 2014), Tunisia (Amri et al. 2011)

Key constituents: α-Pinene (18.5%–35.2%), β-phellandrene (14.7%–19.5%), germacrene D (1.6%–21.3%), β-caryophyllene (0.0%–11.8%)

*Pinus radiata* D. Don (Monterey pine)

Plant source: Greece (Petrakis et al. 2001; Ioannou et al. 2014), commercially available (Sacchetti et al. 2005)

Key constituents: β-Pinene (26.4%–38.7%), α-pinene (18.9%–21.9%), β-phellandrene (1.6%–12.6%), sabinene (0.0%–7.3%), iso-sylvestrene (0.0%–8.4%), germacrene D (0.0%–6.4%)

*Pinus rigida* Mill. (pitch pine)

Plant source: Germany (Ioannou et al. 2014), Japan (Kurose et al. 2007), South Korea (Jeon and Lee 2012; Kim et al. 2013)

Key constituents: β-Pinene (7.4%–23.0%), α-pinene (4.7%–14.1%), α-cadinol (2.6%–8.4%), 2H-benzocyclohepten-2-one (0.0%–24.5%), germacrene D (0.0%–15.5%), bicyclogermacrene (0.0%–14.1%), β-caryophyllene (0.0%–13.6%), ethyl laurate (0.0%–9.6%), caryophyllene oxide (0.0%–7.7%), dodecanoic acid (0.0%–7.1%), *trans*-ocimene (0.0%–5.2%)

*Pinus taeda* L. (lobolly pine)

Plant source: Mozambique (Pagula and Baeckström 2006)

Key constituents: α-Pinene (62.3%), β-pinene (7.1%), tricyclene (3.8%), β-phellandrene (3.7%)

*Pinus teocote* Schiede ex Schltdl. & Cham. (Mexican small cone or Aztec pine)

Plant source: England (Ioannou et al. 2014)

Key constituents: α-Pinene (33.3%), germacrene D (27.6%), β-caryophyllene (9.8%), β-pinene (7.8%), δ-3-carene (4.4%)

## 13.4.4 Subgenus *Pinus*, Section *Trifoliae*, Subsection *Contortae*

*Pinus banksiana* Lamb. (Jack pine)

Plant source: Finland (Ioannou et al. 2014)

Key constituents: Bornyl acetate (15.7%), germacrene D (14.7%), α-pinene (8.2%), β-pinene (7.8%), myrcene (6.3%), β-selinene (5.5%), δ-cadinene (4.6%), α-cadinol (4.4%)

*Pinus contorta* Douglas ex Loudon **var. *contorta*** (shore or beach pine)

Plant source: Scotland (Ioannou et al. 2014)

Key constituents: β-Phellandrene (19.9%), α-cadinol (8.9%), pimarinal (8.9%), δ-3-carene (7.4%), manool (7.4%), α-pinene (5.9%), β-pinene (4.7%), methyl dehydroabietate (4.7%), terpinen-4-ol (5.2%)

*Pinus contorta* **var. *latifolia*** Engelm. (Rocky Mountain lodgepole pine)

Plant source: United States (Pauly and Von Rudloff 1971), Scotland (Ioannou et al. 2014)

Key constituents: β-Pinene (30.5%–32.8%), β-phellandrene (26%–34.3%), α-pinene (5.6%–6.5%), *cis*-β-ocimene (2.5%–4.9%), δ-cadinene (1.8%–8.4%)

*Pinus contorta* var. *murrayana* (Balf.) S. Watson (Sierra Nevada lodgepole pine)

Plant source: Germany (Ioannou et al. 2014)

Key constituents: β-Phellandrene (47.0%), α-pinene (4.8%), *cis*-β-ocimene (4.6%), α-cadinol (3.3%), β-pinene (3.0%), δ-cadinene (2.9%)

### 13.4.5 Subgenus *Pinus*, Section *Trifoliae*, Subsection *Ponderosae*

*Pinus coulteri* D. Don (Coulter or bigcone pine)

Plant source: England (Ioannou et al. 2014)

Key constituents: 4-*epi*-Isocembrol (17.7%), α-pinene (13.6%), germacrene D (8.8%), β-phellandrene (6.2%), α-cadinol (4.7%)

*Pinus jeffreyi* A. Murray bis (Jeffrey pine)

Plant source: Germany (Ioannou et al. 2014), United States (Adams and Wright 2012)

Key constituents: α-Pinene (20.9%–29.8%), β-pinene (4.7%–6.7%), β-phellandrene (3.4%–4.6%), δ-cadinene (2.4%–4.0%), germacrene D (1.0%–11.5%), thunbergol (0.9%–9.2%), α-cadinol (0.7%–3.4%), limonene (0.0%–5.0%)

*Pinus ponderosa* Douglas ex C. Lawson (ponderosa or western yellow pine)

Plant source: Germany (Ioannou et al. 2014), Poland (Krauze-Baranowska et al. 2002), United States (Kurose et al. 2007)

Key constituents: β-Pinene (38.2%–45.7%), α-pinene (10.2%–22.5%), δ-3-carene (8.2%–12.0%), estragole (2.6%–10.5%), myrcene (1.4%–4.6%), β-phellandrene (0.8%–4.4%), α-terpinyl acetate (0.0%–7.5%)

*Pinus sabineana* Douglas ex D. Don (foothill or digger pine)

Plant source: Germany (Ioannou et al. 2014), United States (Adams and Wright 2012)

Key constituents: α-Pinene (39.1%–61.6%), β-pinene (3.3%–4.5%), β-phellandrene (2.0%–10.4%), *cis*-β-ocimene (2.0%–4.6%), methyl chavicol (1.4%–4.5%), limonene (0.7%–10.5%)

*Pinus torreyana* Parry ex Carrière (Torrey pine)

Plant source: United States (Ioannou et al. 2014)

Key constituents: 4-*epi*-Isocembrol (55.7%), cembrene (12.7%), limonene (8.6%), thunbergol (7.7%)

### 13.4.6 Subgenus *Strobus*, Section *Parrya*, Subsection *Balfourianae*

*Pinus aristata* Engelm. (Colorado bristlecone pine)

Plant source: England (Ioannou et al. 2014)

Key constituents: δ-3-Carene (38.4%), β-phellandrene (12.7%), thymol methyl ether (11.4%), terpinolene (8.6%), α-pinene (6.6%), β-pinene (6.4%)

*Pinus longaeva* D.K. Bailey (Great Basin bristlecone pine)

Plant source: Scotland (Tsitsimpikou et al. 2001)

Key constituents: β-Phellandrene (32.5%), β-pinene (17.7%), α-pinene (17.1%), *cis*-γ-bisabolene (6.3%)

### 13.4.7 Subgenus *Strobus*, Section *Parrya*, Subsection *Cembroides*

*Pinus cembroides* Zucc. (Mexican pinyon)

Plant source: Scotland (Ioannou et al. 2014)

Key constituents: α-Pinene (30.9%), β-caryophyllene (19.2%), germacrene D (9.4%), β-pinene (5.6%), myrcene (5.0%), α-humulene (3.2%)

*Pinus culminicola* Andresen & Beaman (Potosi pinyon)

Plant source: Scotland (Ioannou et al. 2014)

Key constituents: α-Pinene (33.6%), β-pinene (20.2%), β-phellandrene (16.9%), germacrene D (7.9%)

*Pinus monophylla* Torr. & Frém. (single-leaf pinyon)

Plant source: England (Ioannou et al. 2014)

Key constituents: β-Pinene (27.2%), α-pinene (18.7%), δ-cadinene (7.2%), limonene (6.8%), germacrene D (4.9%), myrcene (4.6%), β-phellandrene (4.6%)

### 13.4.8 Subgenus *Strobus*, Section *Quinquefoliae*, Subsection *Gerardianae*

*Pinus bungeana* Zucc. ex Endl. (lacebark pine)

Plant source: China (Chen et al. 2006), England (Ioannou et al. 2014), South Korea (Jeon and Lee 2012)

Key constituents: α-Pinene (37.7%–40.9%), β-pinene (5.6%–54.9%), limonene (2.0%–11.3%), camphene (0.6%–21.3%), germacrene D (0.0%–11.2%), β-caryophyllene (0.0%–27.2%), γ-muurolene (0.0%–10.0%)

*Pinus gerardiana* Wall. ex D. Don (chilgoza or Gerard's pine)

Plant source: England (Ioannou et al. 2014)

Key constituents: β-Pinene (39.1%), α-pinene (26.4%), myrcene (5.7%), β-phellandrene (5.3%)

### 13.4.9 Subgenus *Strobus*, Section *Quinquefoliae*, Subsection *Strobus*

*Pinus armandii* Franch. (Chinese white or Armand's pine)

Plant source: China (Chen et al. 2006; Domrachev et al. 2012), Scotland (Tsitsimpikou et al. 2001; Ioannou et al. 2014)

Key constituents: α-Pinene (0.0%–74.2%), γ-muurolene (0.0%–40.7%), β-pinene (0.0%–39.3%), β-caryophyllene (0.0%–36.3%), germacrene D (0.0%–19.0%), α-humulene (0.0%–5.8%)

*Pinus cembra* L. (Swiss stone or Arolla pine)

Plant source: Romania (Apetrei et al. 2013), Russia (Domrachev et al. 2012), Scotland (Ioannou et al. 2014)

Key constituents: α-Pinene (21.1%–69.1%), β-phellandrene (5.3%–13.5%), bicyclogermacrene (2.4%–4.7%), β-pinene (0.9%–4.6%), δ-cadinene (0.3%–5.2%), germacrene D (0.0%–15.7%), germacrene D-4-ol (0.0%–9.4%), methyl daniellate (0.0%–4.1%)

*Pinus flexilis* E. James (limber or Rocky Mountain white pine)

Plant source: England (Ioannou et al. 2014)

Key constituents: α-Pinene (24.5%), germacrene D (12.2%), β-pinene (8.6%), camphene (9.0%), bornyl acetate (3.8%), α-cadinol (3.4%)

*Pinus koraiensis* Siebold & Zucc. (Korean stone pine)

Plant source: Japan (Kurose et al. 2007), Russia (Domrachev et al. 2012), Scotland (Ioannou et al. 2014), South Korea (Jeon and Lee 2012; Kim et al. 2012)

Key constituents: α-Pinene (12.8%–22.3%), bornyl acetate (4.7%–16.3%), camphene (3.6%–21.1%), β-pinene (1.0%–5.0%), limonene (3.5%–20.0%), β-copaene (0.2%–6.7%), δ-3-carene (0.0%–15.3%), germacrene D (0.0%–18.9%), δ-cadinene (0.0%–12.1%), β-caryophyllene (0.0%–8.5%), myrcene (0.0%–7.3%), terpinolene (0.0%–6.7%), γ-cadinene (0.0%–6.7%)

*Pinus monticola* Douglas ex D. Don (western white pine)

Plant source: Scotland (Ioannou et al. 2014)

Key constituents: β-Elemene (15.0%), α-pinene (14.9%), β-pinene (14.2%), germacrene D-4-ol (7.2%), germacrene D (6.8%), β-phellandrene (6.0%)

*Pinus parviflora* Siebold & Zucc. (Japanese white pine)

Plant source: China (Chen et al. 2015), Japan (Kurose et al. 2007), Russia (Domrachev et al. 2012), Scotland (Ioannou et al. 2014), South Korea (Jeon and Lee 2012)

Key constituents: Germacrene D (6.7%–24.2%), α-pinene (5.6%–21.1%), bornyl acetate (3.0%–19.7%), β-pinene (1.4%–16.2%), camphene (0.9%–8.1%), β-caryophyllene (0.4%–11.5%), β-phellandrene (0.0%–31.9%), limonene (0.0%–14.8%), γ-cadinene (0.0%–6.3%), myrcene (0.0%–5.8%)

*Pinus peuce* Griseb. (Macedonian or Balkan white pine)

Plant source: Bosnia and FYROM (Nikolić et al. 2014), Finland (Ioannou et al. 2014), Greece (Koukos et al. 2000; Petrakis et al. 2001), Montenegro and Serbia (Nikolić et al. 2008)

Key constituents: α-Pinene (21.8%–45.6%), β-pinene (4.6%–22.0%), bornyl acetate (4.5%–9.8%), β-caryophyllene (2.9%–5.5%), camphene (2.5%–10.7%), germacrene D (0.0%–18.8%), citronellol (0.0%–13.4%), β-phellandrene (0.0%–6.8%)

*Pinus pumila* (Pall.) Regel (dwarf stone pine)

Plant source: Germany (Ioannou et al. 2014), Japan (Yatagai and Sato 1986; Kurose et al. 2007), Russia (Domrachev et al. 2012), Scotland (Tsitsimpikou et al. 2001)

Key constituents: α-Pinene (5.7%–17.0%), δ-3-carene (4.7%–20.8%), α-cadinol (2.2%–7.3%), terpinolene (1.5%–37.7%), germacrene D (0.1%–7.6%), limonene (0.0%–9.1%), γ-cadinene (0.0%–8.7%), β-copaene (0.0%–7.5%), germacrene D-4-ol (0.0%–6.8%)

*Pinus sibirica* Mayr (Siberian cedar pine)

Plant source: Mongolia (Shatar and Adams 1996), Russia (Domrachev et al. 2012)

Key constituents: α-Pinene (57.2%–88.2%), limonene (2.1%–12.2%), β-pinene (2.1%–7.7%), β-phellandrene (1.4%–19.7%)

*Pinus strobiformis* Engelm. (Mexican white or southwestern white pine)

Plant source: Finland (Ioannou et al. 2014)

Key constituents: Germacrene D (25.5%), β-pinene (12.5%), α-cadinol (8.1%), α-pinene (8.0%), δ-cadinene (6.4%)

*Pinus strobus* L. (eastern white pine)

Plant source: Canada (Von Rudloff 1985), England (Ioannou et al. 2014), Germany (Kubeczka and Schultze 1987), Greece (Koutsaviti et al. 2015), Japan (Yatagai and Sato 1986), Poland (Krauze-Baranowska et al. 2002), South Korea (Jeon and Lee 2012)

Key constituents: α-Pinene (14.7%–57.8%), β-pinene (5.8%–35.5%), myrcene (2.4%–19.5%), germacrene D (2.3%–19.6%), camphene (1.7%–4.6%), β-caryophyllene (1.1%–8.2%)

*Pinus wallichiana* A.B. Jacks. (Himalayan blue pine)

Plant source: England (Ioannou et al. 2014), India (Dar et al. 2012)

Key constituents: β-Pinene (18.1%–46.8%), α-pinene (13.8%–25.2%), β-caryophyllene (1.8%–7.2%), germacrene D (0.0%–10.3%), germacrene D-4-ol (0.0%–6.7%)

## 13.5 General Uses of Pine Needle Oil

Traditionally, several preparations of pines have been used for the treatment of different ailments, such as ptilosis, dermatitis, and toothache (Berendes 1902). Several studies on the pharmacological activities of pine species have shown that turpentine, its extracts, and compounds isolated thereof possess antioxidant, antiviral, analgesic, anti-inflammatory, cytotoxic, and/or antimicrobial activities (Ioannou et al. 2014).

Pine essential oils have exhibited significant biological activities, such as antifungal, acaricidal, and antiplatelet activities (Macchioni et al. 2002; Tognolini et al. 2006; Kolayli et al. 2009) and are commonly used in cosmetic industries due to their fragrant odor (Ekundayo 1988). Their essential oils have also been proven to hold an important role in the defense system of conifers against numerous herbivorous insects and pathogens (Gijzen et al. 1993).

There is an ancient belief that there is a relation between the pine needle oil and good health. This belief is based on its strong disinfectant and therapeutic properties (Hammer et al. 1999; Hong et al. 2004; Koul et al. 2008; Group 2016). In some cases, the therapeutic effect of pine oil was confused with the pesticidal and repellent properties that it exhibits against some common nuisance or microorganism-carrying insects (Süntar et al. 2012). It is known that people in early societies and sometimes in rural societies of developing countries used plant-derived materials to prevent bites from blood-feeding arthropods. Throughout history, humans have tried to avoid the biting of blood-feeding arthropods by using a variety of methods, such as burning plants to produce smoke and drive away pests from indoor or outdoor spaces and applying plant oils or elemental sulfur directly on the skin or clothes (Moore and Debboun 2007). Boas and other social anthropologists (Boas 1911; Group 2016) have reported that Native Americans used fresh-cut pine needles for bedding in order to prevent bites of bedbugs, lice, and mosquitoes (Moore and Debboun 2007; Rogers 2012). The use of conifer needles was a common practice, and the selection of the source species was based on the availability of the trees.

## 13.6 Pesticide Uses of Pine Needle Oil

All insect pests of pine trees locate their plant host by sensing the volatiles emitted by the needles. An important insect pest is the pine scale *Marchalina hellenica* (Hemiptera, Margarodidae), which, when it colonizes the pines, produces sugary excretions that are used as nutrients by bees. This parasitization from the pine scale is detrimental for the physiology of the infested pines (Petrakis et al. 2010) and causes severe reduction of the pine insect diversity (Petrakis et al. 2011). The biodiversity in pine forests is naturally very low, in terms not only of insects but also of understory plants, fungi, and soil microbes. This is due to the acidity of the forests' floor soil and the terpene content contributed mainly from the fallen needles. Since any treatment of the pine scale parasite based on a nonsweeping killing method would affect many species, and among them the natural enemies of the parasite, an ecologically sound chemical formulation was developed (Petrakis et al. 2005). This method modifies the behavior of pine pests by causing "feeding disruption," and therefore the pine bark is not recognized as an acceptable feed. In a similar manner, pine needle oil can be used when the feeding substrate is not a pine tree. Traditional pesticides and treatment of in-house plants (planted in corridors, balconies, and yards) are usually safeguarded by spraying the foliage of the plants and the places that could harbor pests with pine needle oil–water mixtures in emulsified formulations. Pine needle oils are preferred in many cases because of the wide availability of pine trees and the pleasant pine odor. Usually, these formulations are enriched with additional essential oils, such as citronella, verbena, lemongrass, geranium, basil, lavender, thyme, and peppermint, to simultaneously repel mosquitoes (Coats et al. 1991; Koul et al. 2008).

A parasite affected by pine needle oil is the nematode *Bursaphelenchus xylophilus* (Nematoda, Parasitaphelenchidae), which causes the pine wilt disease. The pine wilt nematode can only tolerate up to a certain concentration of terpenoids (Yamada 2008; Nunes da Silva et al. 2015). Needle removal and treatment of the trees with inhibitors of photosynthesis reduced the terpene content and enabled *B. xylophilus* and the related species *Bursaphelenchus mucronatus* to reproduce on *P. thunbergii* and kill its seedlings. However, these findings are still at the research level, and terpenoids produced from resistant pine needles have not so far been used in agricultural-scale applications.

The only case in which the pine needle essential oil has been commercialized is for the production of the attractive blend for *Monochamus galloprovincialis*, the insect vector of *B. xylophilus* (Akbulut and Stamps 2012). For the improvement of attractive blends, pine host tree kairomones have been added to the aggregation pheromones of the insect in order to increase the attraction rate and subsequently reduce the pest populations.

Macchioni and colleagues (2002) evaluated the acaricidal activity of needle essential oils obtained from four common pine species, *P. pinea*, *P. halepensis*, *P. pinaster*, and *P. nigra*, against the stored food mite *Tyrophagus putrescentiae*. All oils were proven effective, with the oil of *P. pinea* and its major constituents 1,8-cineole and limonene exhibiting the highest levels of activity.

The use of pine needle essential oils for the control of mosquitoes and other insects of concern in public health is a promising but not yet commercialized method (Shaalan et al. 2005). Virus-bearing mosquitoes invaded Europe and North America in the last 50 years, probably as a result of their physiological adaptations, such as the light-induced diapause of *Aedes albopictus* (Focks et al. 1994) and the survival of the eggs of *Aedes aegypti* that are laid in densely populated areas that lack reliable water supplies, waste management, and sanitation, and the unpredictable oviposition grounds (Honorio et al. 2009). Recent studies report the potential of pine needle essential oils to kill or repel mosquitoes and the ability of steam-distilled needle oils from Mediterranean pines to kill larvae and repel adults of the invasive species *Ae. albopictus* (Koutsaviti et al. 2015). This action is intensified against other mosquito species when extracts of pine needles that, besides terpenes, contain large amounts of lignin are used (Kanis et al. 2009).

## 13.7 Advantages of Pine Needle Oil as a Pesticide

With the possible exception of *Dendroctonus micans* (Coleoptera, Scolytidae), the colonization of pine hosts results in the death of the tree if the population density of the beetles exceeds a certain level. At this point, the resistance of the host tree plays an important role by keeping the population densities of the insect at low levels (Lieutier et al. 2016). There are generally two lines of defense, the constitutive and the induced defenses, depending on the manifested alterations of their activity before and after the inflicted damage by the attacking insect. Among these, only the constitutive defenses, and specifically only their chemical components, have been used for the control of pests. The resistance of the insects to constitutive chemical defenses has been observed only in evolutionary time. The classic resistance that has been documented in ecological time cannot be developed when pine needle oil and other essential oils are used. According to the Insecticide Resistance Action Committee (IRAC), the definition of resistance of an insect to a particular insecticide is the observable and "heritable change in the sensitivity of a pest population" (Tabashnik et al. 2013; Sparks and Nauen 2015). On an evolutionary time scale, the insects detoxify the chemical defenses of the host plants by involving adaptations that counteract the toxic effects of the plants' secondary metabolites. However, in a microevolutionary time scale, various pests can evolve biochemical mechanisms to develop resistance (Whalon et al. 2008; Osakabe and Goka 2009). This is the main advantage of pine needle essential oil, which in general comprises blends of numerous terpenoids prohibiting the development of such resistance.

Sometimes resistance to pesticides is achieved when pesticide-polluted runoff water is used as a breeding site (Dabiré et al. 2012). Such an incident was observed in Burkina Faso, West Africa, when control of the malaria vector *Anopheles gambiae* was attempted. According to Mamai and colleagues (2016), this type of pollution, in addition to the pesticides used for the control of vector populations, dramatically increased the resistance of pests that use such polluted biotopes as breeding substrates. The biological basis of such an increase is the intense selection of genes responsible for resistance, and its molecular basis has been extensively investigated. In such cases, the pine needle essential oils can be used as an alternative pesticide or insect repellent, although the $LC_{50}$ values are in general high.

The overuse of synthetic pesticides, as well as their lack of biodegradability in the soil and water horizon, has been reported as the main cause of nontarget toxicity in croplands (Koul et al. 2008), while the environmental effect of the involved pesticides becomes even more significant in the presence of additional pollutants. As a consequence, public health is affected, resulting in a rising of public awareness regarding the synthetic pesticides. Pine needle oil and essential oils from other plants are slowly proving their efficacy as good alternatives to synthetic pesticides. In addition to other oils, pine needle oil possesses several advantages. The needle biomass from which the essential oils are obtained is available in huge amounts throughout the northern hemisphere (Mirov 1967) and in several parts of the southern hemisphere due to the invasive ability of some *Pinus* species (Richardson and Rundel 1998). As a result, the pine needle essential oils that have a pleasant aroma reminiscent of cleanness are economically affordable, even in societies with financial impediments.

In indoor environments, there is a need to replace synthetic pesticides because of their high toxicity to humans and livestock. For the control of the house dust mites (*Dermatophagoides farinae* in America and *Dermatophagoides pteronyssinus* in Europe), the needle hydrodistillate of *P. densiflora* has been used due to its contact and fumigant toxicity. The suppression effect of *P. densiflora* needle oil on the American house dust mite populations has been investigated by Lee et al. (2013). It was found that the hydrodistillate was very toxic to the mites, with levels of toxicity higher than those of DEET or dibutyl phthalate, which are two routinely used formulations for the treatment of house dust mites. Furthermore, the *P. densiflora* needle essential oil conferred 95% mortality in comparison with that of permethrin (0%) in vapor phase mortality tests. It is speculated that this type of biological activity can be extended to all species within the family Pyroglyphidae (Acarina), thus acting as a fumigant with contact acaricidal activity.

Another sector in which pine needle oil can be used is the organic arboriculture (e.g., pistachio trees) where farmers frequently spray or brush the trunks of the trees in order to drive away insect colonizers. The bark beetle colonizers of pistachio trees bear a fungus that ferments the wood. Application of pine needle essential oils decreases the living spores and/or mycelia of the fungus, and as a result, the bark beetle is unable to successfully colonize since the number of spores is not adequate (i.e., large enough) for the establishment of the fungus (Kirisits 2004). The same holds true for the number of insects necessary for successful colonization (Lieutier 2004). Thus, in such a case treatment with pine needle essential oils results in a type of acquired immunity.

## 13.8 Limitations of Pine Needle Oil as a Pesticide

The use of pine needle essential oils against human and agricultural pests and parasites has certain limitations relating to the toxicity of the oils and their inefficiency against a number of pest organisms. Other limitations of pine needle oil stem from the fact that the terpenoid constituents can cause endocrine disruption in several juvenile insects, nematodes, annelids, and mollusks (Oehlmann and Schulte-Oehlmann 2003; Oetken et al. 2004). This is associated with the lack of knowledge of the dose-dependent effect that renders them environmentally hazardous (Schulte-Oehlmann et al. 2006). In 2016, the EPA issued a warning (GCID 474858 in EPA ACToR) for the animal toxicity of the dwarf pine needle oil, quoting an $LD_{50}$ value of 6880 mg/kg for rats.

In several cases, sickness was observed, mainly in children of third world countries, who had received a decoction of pine needles or 2–5 ml of pine needle oil as traditional medicine for antihelmintic therapy (Tisserand 2007; Süntar et al. 2012). These effects, though, may be due to the low quality of commercially available essential oils, which might also contain toxic impurities.

Because of their biocide action, genotoxicity has been evidenced for several essential oils, even those that are used as food preservatives. The pine needle oil of *P. sylvestris* has been studied for its genotoxicity, expressed in terms of chromosome aberrations, and exchange of sister chromatids in *in vitro* cultures of human lymphocytes and in *in vivo* somatic mutations and recombination tests in *Drosophila melanogaster* (Lazutka et al. 2001). It was found that pine needle oil caused chromosome aberrations and sister chromatid exchange in a dose-dependent manner, while it was proven cytotoxic to human lymphocytes.

## 13.9 Essential Oil–Based Insecticides

It is surprising that the extensive scientific literature and the associated research in many laboratories throughout the world, in combination with the public awareness of the adverse effects of synthetic insecticides and the increasing resistance of agricultural and public health pests, have resulted in only a small number of commercially available essential oil–based insecticides for professionals and laypersons (Moretti et al. 2002; Kaushik et al. 2005; Koul et al. 2008; Baser and Franz 2015; Buchbauer and Hemetsberger 2015). The few essential oil–based commercial products do not incorporate pine needle oil, as it was judged of inferior efficiency in comparison with that of other oils.

The scarcity of essential oil commercial formulations is due mainly to their cost in contrast to synthetic products, which are extremely inexpensive. In addition to the production cost of essential oils, the expenses for the legal approval of a new formulation in Europe are proportional to the number of active constituents, which, in the case of essential oils, are numerous. The U.S. authorities, in contrast to the European ones, have simplified the registration process of insecticides when they are based on essential oils that are already used in food, feeds, beverages, or pharmaceutical drugs, resulting in a considerably larger number of essential oil–based insecticides. In Europe, the complexity of the legislation for plant protection products and the difficulty imposed for the registration of new products, along with the requirement to provide data in a separate dossier "for each active substance, safener and synergist contained in the plant protection product" (European Community Regulation 2009, Article 3, paragraph 3a–g), impose significant difficulties to the development and commercialization of new essential oil–based insecticides.

The European Chemicals Agency (ECHA) allows the registration of chemicals that are not defined compounds (e.g., essential oils) (Demyttenaere 2015) and are summed up under an abbreviation in the lists of the European Inventory of Existing Commercial Chemical

Substances (EINECS) published in the *Official Journal of the Commission* 146A/15.6.1990 and the European List of Notified Chemical Substances (ELINCS) in support of Directive 92/32/EEC. In this context, pine needle oil is not listed in the European registration list but is included in the Chemical Abstracts Service (CAS) numbering system as 8023-99-2 (pine oil) and 8016-46-4 (pine needle oil) and synonyms. In the CAS system, only the species *P. sylvestris* is explicitly mentioned as the source of the oil. In the European Union (EU) registration list, the entry under the name "Pine, *Pinus sylvestris*, ext." is described as "extractives and their physically modified derivatives such as tinctures, concretes, absolutes, essential oils, oleoresins, terpenes, terpene-free fractions, distillates, residues, etc., obtained from *Pinus sylvestris*, Pinaceae." Pine needle essential oil is not referred to anywhere in the European legislation and can be used only under the umbrella of "Pine, *Pinus sylvestris*, ext."

In the case that essential oil–based insecticides are used as plant protection products, each member state in the European Union may use different pine species as a source of raw materials. In effect, each member state could have a different array of terpenoids in the final blend of an essential oil–based insecticide. This, however, would substantially increase the final registration costs of the commercial product proportionally to the number of its active constituents.

All the above impose serious obstacles for the commercialization of related products in European countries.

Some products use essential oils alone or in combination with insecticides for the treatment of fleas and mites on dogs and cats (Baser and Franz 2015). Usually, a quantity of pine needle essential oil is mixed with cedarwood oil and emulsified in water to yield a formulation that is used to brush or soak the pets' fur. The adult fleas and mites, as well as their eggs, are gathered on the brush or comb and are rinsed off with water. For the impregnation of pet tick repellent collars, oils such as citronella, thyme, geraniol, cinnamon oil, or garlic can be additionally used.

In many rural areas in which *Cedrus atlantica* trees are not accessible, pine needle essential oils and resins from *P. halepensis*, *P. brutia*, *P. pinaster*, and *P. pinea* are used as molluscicides, even though *C. atlantica* oil was shown to be more effective than pine oil. The molluscicidal activity was attributed to α- and β-pinene and myrcene, which are among the major constituents of pine needle essential oils. Such products are commonly produced and traded by small pet shop firms.

## 13.10 Conclusions

Pine needle essential oils possess a wide array of biological properties that exhibit significant variation, depending on their chemical composition, which in turn is related to the investigated species and the origin of the plant source. Regardless of the numerous investigations revealing their potential as active constituents for several applications, their incorporation in commercial products is so far limited in traditionally prepared formulations or locally distributed products, mainly due to the cumbersome European legislation regarding the use of essential oils. Public awareness on the environmental and health problems arising from the use of synthetic pesticides will hopefully force the reformation of the legislation framework in the near future, allowing more frequent use of essential oil–based insecticides.

# References

Abi-Ayad, M., Abi-Ayad, F.Z., Lazzouni, H.A., Rebiahi, S.A., Ziani-Cherif, C. and Bessiere, J.M. 2011. Chemical composition and antifungal activity of Aleppo pine essential oil. *J Med Plants Res* 5:5433–5436.

Adams, R.P. and Wright, J.W. 2012. Alkanes and terpenes in wood and leaves of *Pinus jeffreyi* and *P. sabiniana*. *J Essent Oil Res* 24:435–440.

Akbulut, S. and Stamps, W.T. 2012. Insect vectors of the pinewood nematode: A review of the biology and ecology of *Monochamus* species. *Forest Pathol* 42:89–99.

Amri, I., Hamrouni, L., Hanana, M., Gargouri, S., Fezzani, T. and Jamoussi, B. 2013. Chemical composition, physico-chemical properties, antifungal and herbicidal activities of *Pinus halepensis* Miller essential oils. *Biol Agric Hortic* 29:91–106.

Amri, I., Hanana, M., Jamoussi, B. and Hamrouni, L. 2014. Essential oils of *Pinus nigra* J.F. Arnold subsp. *laricio* Maire: Chemical composition and study of their herbicidal potential. *Arabian J Chem*. Available at http://dx.doi.org/10.1016/j.arabjc.2014.05.026.

Amri, I., Lamia, H., Gargouri, S. et al. 2011. Chemical composition and biological activities of essential oils of *Pinus patula*. *Nat Prod Commun* 6:1–6.

Apetrei, C.L., Spac, A., Brebu, M., Tuchilus, C. and Miron, A. 2013. Composition and antioxidant and antimicrobial activities of the essential oils of a full-grown *Pinus cembra* L. tree from the Calimani Mountains (Romania). *J Serb Chem Soc* 78:27–37.

Baser, K.H.C. and Franz, C. 2015. Essential oils used in veterinary medicine. In *Handbook of Essential Oils: Science, Technology, and Applications*, ed. K.H.C. Baser and G. Buchbauer, 655–669. Boca Raton, FL: CRC Press/Taylor & Francis.

Berendes, J. 1902. *Des Pedanios Dioskurides aus Anazarbos, Arzneimittellehre in fünf Büchern*. Stuttgart: F. Enke.

Berta, F., Spuka, J. and Chladna, A. 1997. The composition of terpenes in needles of *Pinus sylvestris* in a relatively clear and in a city environment. *Biologia* 52:71–78.

Boas, F. 1911. *The Mind of Primitive Man*. New York: MacMillan.

Bojović, S., Nikolić, B., Ristić, M., Orlović, S., Veselinović, M., Rakonjac, L. and Dražic, D. 2012. Variability in chemical composition and abundance of the rare tertiary relict *Pinus heldreichii* in Serbia. *Chem Biodivers* 8:1754–1765.

Bonesi, M., Menichini, F., Tundis, R. et al. 2010. Acetylcholinesterase and butyrylcholinesterase inhibitory activity of *Pinus* species essential oils and their constituents. *J Enzyme Inhib Med Chem* 25:622–628.

Bonikowski, R., Celiński, K., Wojnicka-Półtorak, A. and Maliński, T. 2015. Composition of essential oils isolated from the needles of *Pinus uncinata* and *P. uliginosa* grown in Poland. *Nat Prod Commun* 10:371–373.

Buchbauer, G. and Hemetsberger, S. 2015. Use of essential oils in agriculture. In *Handbook of Essential Oils: Science, Technology, and Applications*, ed. K.H.C. Baser and G. Buchbauer, 670–706. Boca Raton, FL: CRC Press/Taylor & Francis.

Celiński, K., Bonikowski, R., Wojnicka-Półtorak, A., Chudzińska, E. and Maliński, T. 2015. Volatiles as chemosystematic markers for distinguishing closely related species within the *Pinus mugo* complex. *Chem Biodivers* 12:1208–1213.

Chen, J., Gao, Y., Jin, Y., Li, S. and Zhang, Y. 2015. Chemical composition, antibacterial and antioxidant activities of the essential oil from needles of *Pinus parviflora* Siebold & Zucc. *J Essent Oil Bearing Plants* 18:1187–1196.

Chen, H., Tang, M., Gao, J., Chen, X. and Li, Z. 2006. Changes in the composition of volatile monoterpenes and sesquiterpenes of *Pinus armandi*, *P. tabulaeformis* and *P. bungeana* in northwest China. *Chem Nat Compd* 42:534–538.

Chowdhury, J.U., Bhuiyan, M.N.I. and Nandi, N.C. 2008. Essential oil constituents of needles, inflorescences and resins of *Pinus caribaea* Morelet growing in Banglandesh. *Bangladesh J Bot* 37:211–212.

Coats, J.R., Karr, L.L. and Drewes, C.D. 1991. Toxicity and neurotoxic effects of monoterpenoids in insects and earthworms. In *Naturally Occurring Pest Bioregulator*, ed. P.A. Hedin, 306–316. Washington, DC: ACS Symposium Series 449, American Chemical Society.

Dabiré, K.R., Diabaté, A., Namountougou, M. et al. 2012. Trends in insecticide resistance in natural populations of malaria vectors in Burkina Faso, West Africa: 10 years' surveys. In *Insecticides—Pest Engineering*, ed. F. Perveen, 479–502. Rijeka, Croatia: InTech Publications.

Dar, Y.M., Shah, W.A., Mubashir, S. and Rather, M.A. 2012. Chromatographic analysis, anti-proliferative and radical scavenging activity of *Pinus wallichiana* essential oil growing in high altitude areas of Kashmir, India. *Phytomedicine* 19:1228–1233.

Demyttenaere, J.C.R. 2015. Recent EU legislation on flavors and fragrances and its impact on essential oils. In *Handbook of Essential Oils: Science, Technology, and Applications*, ed. K.H.C. Baser and D. Buchbauer, 1053–1070. Boca Raton, FL: CRC Press/Taylor & Francis.

Djerrad, Z., Kadika, L. and Djouahrib, A. 2015. Chemical variability and antioxidant activities among *Pinus halepensis* Mill. essential oils provenances, depending on geographic variation and environmental conditions. *Ind Crop Prod* 74:440–449.

Dob, T., Berramdane, T. and Chelghoum, C. 2007. Essential oil composition of *Pinus halepensis* Mill. from three different regions of Algeria. *J Essent Oil Res* 19:40–43.

Dob, T., Berramdane, T., Dahmane, D. and Chelghoum, C. 2005. Chemical composition of the needles oil of *Pinus canariensis* from Algeria. *Chem Nat Compd* 41:165–167.

Domrachev, D.V., Karpova, E.V., Goroshkevich, S.N. and Tkachev, A.V. 2012. Comparative analysis of volatiles from needles of five-needle pines of northern and eastern Eurasia. *Russ J Bioorg Chem* 38:780–789.

Ekundayo, O. 1988. Volatile constituents of *Pinus* needle oils. *Flavour Fragr J* 3:1–11.

EPA (Environmental Protection Agency). 2016. ACToR: Aggregated computational toxicology resource. Chemical summary: Null (8000-26-8). New York: EPA.

European Community Regulation. 2009. Regulation (EC) No 1107/2009 of the European Parliament and of the council concerning the placing of plant protection products on the market and repealing Council Directives 79/117/EEC and 91/414/EEC. *Official Journal of the European Union* L 309/1:1–50.

Farjon, A. 1984. *Pines: Drawings and descriptions of the genus Pinus*. Leiden: E.J. Brill and W. Backhuys.

Farjon, A. 2005. *Pines: Drawings and description of the genus Pinus*. 3rd ed. Leiden: Brill.

Fayemiwo, K.A., Adeleke, M.A., Okoro, O.P., Awojide, S.H. and Awoniyi, I.O. 2014. Larvicidal efficacies and chemical composition of essential oils of *Pinus sylvestris* and *Syzygium aromaticum* against mosquitoes. *Asian Pac J Trop Biomed* 4:30–34.

Focks, D.A., Linda, S.B., Craig, G.B., Hawley, W.A. and Pumpuni, C.B. 1994. *Aedes albopictus* (Diptera: Culicidae): A statistical model of the role of temperature, photoperiod, and geography in the induction of egg diapause. *J Med Entomol* 31:278–286.

Garrido, D.M., Garcia Martin, D. and Garcia Vallejo, M.C. 1988. The essential oil of needles from Spanish *Pinus pinaster* Ait. *Dev Food Sci* 18:211–229.

Gaussen, H., Heywood, V.H. and Chater, A.O. 1993. *Pinus* L. In *Flora Europaea*, ed. T.G. Tutin, N.A. Burges, A.O. Chater, J.R. Edmondson, V.H. Heywood, D.M. Moore, D.H. Valentine, S.M. Walters, and D.A. Webb, 40–44. 2nd ed., vol. 1. Cambridge: Cambridge University Press.

Gernandt, D.S., Geada López, G., Ortiz García, S. and Liston, A. 2005. Phylogeny and classification of *Pinus*. *Taxon* 54:29–42.

Gijzen, M., Lewinsohn, E., Savage, T.J. and Croteau, R.B. 1993. Conifer monoterpenes: Biochemistry and bark beetle chemical ecology. In *Bioactive Volatile Compounds from Plants*, ed. R. Teranishi, R.G. Buttery, and H. Sugisawa, 8–22. ACS Symposium Series, vol. 525. Washington, DC: American Chemical Society.

Góra, J. and Lis, A. 2000. Olejki sosnowe. *Aromaterapia* 2:5–12.

Group, E.F. 2016. The health benefits of pine oil. Available at http://www.globalhealingcenter.com/natural-health/pine-oil/.

Hajdari, A., Mustafa, B., Ahmeti, G. et al. 2015. Essential oil composition variability among natural populations of *Pinus mugo* Turra in Kosovo. *SpringerPlus* 4:828.

Hammer, K.A., Carson, C.F. and Riley, T.V. 1999. Antimicrobial activity of essential oils and other plant extracts. *J Appl Microbiol* 86:985–990.

Hamrouni, L., Hanana, M., Amri, I., Romane, A.E., Gargouri, S. and Jamoussi, B. 2014. Allelopathic effects of essential oils of *Pinus halepensis* Miller: Chemical composition and study of their antifungal and herbicidal activities. *Arch Phytopathol Pfl* 48:145–158.

Hmamouchi, H., Hamarnouchi, J., Zouhdi, M. and Bessiere, J.M. 2001. Chemical and antimicrobial properties of essential oils of five Moroccan Pinaceae. *J Essent Oil Res* 13:298–302.

Hong, E.J., Na, K.J., Choi, I.G., Choi, K.C. and Jeung, E.B. 2004. Antibacterial and antifungal effects of essential oils from coniferous trees. *Biol Pharm Bull* 27:863–866.

Honorio, N.A., Codeco, C.T., Alves, F.C., Magalhaes, M.A. and Lourenco-De-Oliveira, R. 2009. Temporal distribution of *Aedes aegypti* in different districts of Rio de Janeiro, Brazil, measured by two types of traps. *J Med Entomol* 46:1001–1014.

Ioannou, E., Koutsaviti, A., Tzakou, O. and Roussis, V. 2014. The genus *Pinus*: A comparative study on the needle essential oil composition of 46 pine species. *Phytochem Rev* 13:741–768.

Islam, W.T. 2006. Volatile oils from needles and cones of Egyptian chir pine (*Pinus roxburghii* Sarg.). *Bull Fac Pharm Cairo Univ* 44:77–83.

Jeon, J.-H. and Lee, H.-S. 2012. Volatile components of essential oils extracted from *Pinus* species. *J Essent Oil Bearing Plants* 15:750–754.

Judzentiene, A. and Kupcinskiene, E. 2008. Chemical composition on essential oils from needles of *Pinus sylvestris* L. grown in northern Lithuania. *J Essent Oil Res* 20:26–29.

Kanis, L.A., Antonio, R.D., Antunes, E.P., Prophiro, J.S. and Silva, O.S.D. 2009. Larvicidal effect of dried leaf extracts from *Pinus caribaea* against *Aedes aegypti* (Linnaeus, 1762) (Diptera: Culicidae). *Rev Soc Bras Med Trop* 42:373–376.

Kaushik, P., Satya, S. and Naik, S.N. 2005. Essential oil entrapment in alginate gel to control *Callosobruchus maculates*—Preliminary investigation. Preliminary Publication P-65. Delhi: CRDT Indian Institute of Technology.

Kim, H., Lee, B. and Yun, K.W. 2013. Comparison of chemical composition and antimicrobial activity of essential oils from three *Pinus* species. *Ind Crop Prod* 44:323–329.

Kim, J.-H., Lee, H.-J., Jeong, S.-J., Lee, M.-H. and Kim, S.-H. 2012. Essential oil of *Pinus koraiensis* leaves exerts antihyperlipidemic effects via up-regulation of low density lipoprotein receptor and inhibition of acyl co-enzyme A: Cholesterol acyltransferase. *Phytother Res* 26:1314–1319.

Kirisits, T. 2004. Fungal associates of European bark beetles with special emphasis on the ophiostomatoid fungi. In *Bark and Wood Boring Insects in Living Trees in Europe, a Synthesis*, ed. K.R. Day, A. Battisti, J.-G. Grıgoire, and H.F. Evans, 181–236. Dordrecht: Springer.

Kolayli, S., Ocak, M., Aliyazicioglu, R. and Karaoglu, S. 2009. Chemical analysis and biological activities of essential oils from trunk-barks of 8 trees. *Asian J Chem* 21:2684–2694.

Koukos, P.K., Papadopoulou, K.I., Patiaka, D.Th. and Papagiannopoulos, A.D. 2000. Chemical composition of essential oils from needles and twigs of Balkan pine (*Pinus peuce* Grisebach) grown in northern Greece. *J Agric Food Chem* 48:1266–1268.

Koul, O., Walia, S. and Dhaliwal, G.S. 2008. Essential oils as green pesticides: Potential and constraints. *Biopestic Int* 4:63–84.

Koutsaviti, A., Giatropoulos, A., Pitarokili, D., Papachristos, D., Michaelakis, A. and Tzakou O. 2015. Greek *Pinus* essential oils: Larvicidal activity and repellency against *Aedes albopictus* (Diptera: Culicidae). *Parasitol Res* 114:583–592.

Krauze-Baranowska, M., Mardarowicz, Wiwart, M., Pobłocka, L. and Dynowskad, M. 2002. Antifungal activity of the essential oils from some species of the genus *Pinus*. *Z Naturforsch* 57c:478–482.

Kubeczka, A.-H. and Schulze, W. 1987. Biology and chemistry of conifer oils. *Flavour Fragr J* 2:137–148.

Kurose, K., Okamura, D. and Yatagai, M. 2007. Composition of the essential oils from the leaves of nine *Pinus* species and the cones of three of *Pinus* species. *Flavour Fragr J* 22:10–20.

Kusumoto, N. and Shibutani, S. 2015. Evaporation of volatiles from essential oils of Japanese conifers enhances antifungal activity. *J Essent Oil Res* 27:380–394.

Lahlou, M. 2003. Composition and molluscicidal properties of essential oils of five Moroccan Pinaceae. *Pharm Biol* 41:207–210.

Lawrence, B.M. and Reynolds, R.J. 1991. Progress in essential oils. *Perfum Flav* 16:59–67.

Lawrence, B.M. and Reynolds, R.J. 2003. Progress in essential oils. *Perfum Flav* 28:70–86.

Lazutka, J.R., Mierauskiene, J., Slapsyte, G. and Dedonyte, V. 2001. Genotoxicity of dill (*Anethum graveolens* L.), peppermint (*Mentha × piperita* L.) and pine (*Pinus sylvestris* L.) essential oils in human lyphocytes and *Drosophila melanogaster*. *Food Chem Toxicol* 39:485–492.

Lee, J.-H., Kim, J.-R., Koh, Y.R. and Ahn, Y.-J. 2013. Contact and fumigant toxicity of *Pinus densiflora* needle hydrodistillate constituents and related compounds and efficacy of spray formulations containing the oil to *Dermatophagoides farinae*. *Pest Manag Sci* 69:696–702.

Lee, J.H., Lee, B.K., Kim, J.H., Lee, S.H. and Hong, S.K. 2009. Comparison of chemical compositions and antimicrobial activities of essential oils from three conifer trees; *Pinus densiflora, Cryptomeria japonica*, and *Chamaecyparis obtusa*. *J Microbiol Biotechnol* 19:391–396.

Lieutier, F. 2004. Host resistance to bark beetles and its variations. In *Bark and Wood Boring Insects in Living Trees in Europe, a Synthesis*, ed. K.R. Day, A. Battisti, J.-G. Gregoire, and H.F. Evans, 133–180. Dordrecht: Springer.

Lieutier, F., Mendel, Z. and Faccoli, M. 2016. Bark beetles of Mediterranean conifers. In *Insects and Diseases of Mediterranean Forest Systems*, ed. T.D. Paine and F. Lieutier, 105–197. Cham, Switzerland: Springer.

Macchioni, F., Cioni, P.L., Flamini, G., Morelli, I., Maccioni, S. and Ansaldi, M. 2003. Chemical composition of essential oils from needles, branches and cones of *Pinus pinea, P. halepensis, P. pinaster* and *P. nigra* from central Italy. *Flavour Fragr J* 18:139–143.

Macchioni, F., Cioni, P.L., Flamini, G., Morelli, I., Perrucci, S., Franceschi, A., Macchioni, G. and Ceccarini, L. 2002. Acaricidal activity of pine essential oils and their main components against *Tyrophagus putrescentiae*, a stored food mite. *J Agric Food Chem* 50:4586–4588.

Mamai, W., Simard, F., Couret, D., Ouedraogo, G.A., Renault, D., Dabiré, K.R. and Mouline, K. 2016. Monitoring dry season persistence of *Anopheles gambiae* s.l. populations in a contained semi-field system in southwestern Burkina Faso, West Africa. *J Mel Entomol* 53:130–138.

Millar, C.I. 1993. Impact of the Eocene on the evolution of *Pinus* L. *Ann Missuri Bot Gard* 80:471–498.

Millar, C.I. 2000. Early evolution of pines. In *Ecology and Biogeography of Pinus*, ed. D.M. Richardson, 69–91. Cambridge: Cambridge University Press.

Mirov, N.T. 1967. The genus *Pinus*. New York: Ronald Press.

Moore, S.J. and Debboun, M. 2007. The history of insect repellents. In *Insect Repellents: Principles, Methods, and Use*, ed. M. Debboun, S.P. Frances, and D. Strickman, 3–43. Boca Raton, FL: CRC Press/Taylor & Francis.

Moretti, M.D.L., Sanna-Passino, G., Demontis, S. and Bazzoni, E. 2002. Essential oil formulations useful as a new tool for insect pest control. *AAPS PharmSciTech* 3(2):13. Available at http://www.aapspharmscitech.org.

Nasri, N., Tlili, N., Triki, S., Elfalleh, W., Cheraif, I. and Khaldi, A. 2011. Volatile constituents of *Pinus pinea* L. needles. *J Essent Oil Res* 23:15–19.

Nikolić, B., Ristić, M., Bojović, S., Krivošej, Z., Matevski, V. and Marin, P.D. 2015. Population variability of essential oils of *Pinus heldreichii* from the Scardo-Pindic mountains Ošljak and Galičica. *Chem Biodivers* 12:295–308.

Nikolić, B., Ristić, M., Bojović, S. and Marin, P.D. 2007. Variability of the needle essential oils of *Pinus heldreichii* from different populations in Montenegro and Serbia. *Chem Biodivers* 4:905–916.

Nikolić, B., Ristić, M., Bojović, S. and Marin, P.D. 2008. Variability of the needle essential oils of *Pinus peuce* from different populations in Montenegro and Serbia. *Chem Biodivers* 5:1377–1388.

Nikolić, B., Ristić, M., Bojović, S., Matevski, V., Krivošej, Z. and Marin, P.D. 2014. Essential-oil composition of the needles collected from natural populations of Macedonian pine (*Pinus peuce* Griseb.) from the Scardo-Pindic mountain system. *Chem Biodivers* 11:934–948.

Norin, T. 1972. Some aspects of the chemistry of the order of Pinales. *Phytochemistry* 11:1231–1242.

Nunes da Silva, M., Solla, A., Sampedro, L., Zas, R. and Vasconcelos, M.W. 2015. Susceptibility to the pinewood nematode (PWN) of four pine species involved in potential range expansion across Europe. *Tree Physiol* 35:987–999.

Oehlmann, J. and Schulte-Oehlmann, U. 2003. Endocrine disruption in invertebrates. *Pure Appl Chem* 75:2207–2218.

Oetken, M., Bachmann, J., Schulte-Oehlmann, U. and Oehlmann, J. 2004. Evidence for endocrine disruption in invertebrates. *Int Rev Cytol* 296:1–44.

Orav, A., Kailas, T. and Liiv, M. 1996. Analysis of terpenoic composition of conifer needle oil by steam distillation/extraction, gas chromatography and gas chromatography-mass spectrometry. *Chromatographia* 43:215–219.

Osakabe, R.U. and Goka, K. 2009. Evolutionary aspects of acaricide-resistance development in spider mites Masahiro (Mh.). *Psyche* 947439.

Ottavioli, J., Bighelli, A. and Casanova, J. 2008. Diterpene-rich needle oil of *Pinus pinaster* Ait. from Corsica. *Flavour Fragr J* 23:121–125.

Pagula, F.P. and Baeckström, P. 2006. Studies on essential oil-bearing plants from Mozambique. Part II. Volatile leaf oil of needles of *Pinus elliottii* Engelm. and *Pinus taeda* L. *J Essent Oil Res* 18:32–34.

Park, J.-S. and Lee, G.-H. 2011. Volatile compounds and antimicrobial and antioxidant activities of the essential oils of the needles of *Pinus densiflora* and *Pinus thunbergii. J Sci Food Agric* 91:703–709.

Pauly, G., Gleizes, M. and Bernard-Dagan, C. 1973. Identification des constituants de l'essence des aiguilles de *Pinus pinaster. Phytochemistry* 12:1395–1398.

Pauly, G. and Von Rudloff, E. 1971. Chemosystematic studies in the genus *Pinus*. Leaf oil of *Pinus contorta* var *latifolia. Can J Bot* 49:1201–1210.

Petrakis, P.V., Roussis, V., Vagias, C. and Tsoukatou, M. 2005. A system for the control of the insect pest of pine *Marchalina hellenica* from the volatile constituents of the plants *Thymus capitatus, Pistacia lentiscus* and *Helichrysum stoechas*. Patent OBI 20050100267. Athens: Organization for the Industrial Property.

Petrakis, P.V., Roussis, V., Vagias, C. and Tsoukatou, M. 2010. The interaction of pine scale with pines in Attica, Greece. *Eur J Forest Res* 129:1047–1056.

Petrakis, P.V., Spanos, K. and Feest, A. 2011. Insect biodiversity reduction of pinewoods in southern Greece caused by the pine scale (*Marchalina hellenica*). *Forest Res* 20:27–41.

Petrakis, P.V., Tsitsimpikou, C., Tzakou, O., Couladis, M., Vagias, C. and Roussis, V. 2001. Needle volatiles from five *Pinus* species growing in Greece. *Flavour Frag J* 16:249–252.

Price, R.A., Liston, A. and Strauss, S.H. 2000. Phylogeny and systematics of *Pinus*. In *Ecology and Biogeography of Pinus*, ed. D.M. Richardson, 49–68. Cambridge: Cambridge University Press.

Richardson, D.M. and Rundel, P.W. 1998. Ecology and biogeography of *Pinus*: An introduction. In *Ecology and Biogeography of Pinus*, ed. D.M. Richardson, 3–46. Cambridge: Cambridge University Press.

Rogers, E.L. 2012. *Longleaf Pine Tree Usage by American Indian Tribes of Louisiana: Pine Cones, Needles and Logs*. Alexandria, LA: U.S. Department of Agriculture, Natural Resources Conservation Service.

Roussis, V., Petrakis, P., Ortiz, A. and Mazomenos, B. 1995. Volatile constituents of five *Pinus* species grown in Greece. *Phytochemistry* 39:357–361.

Sacchetti, G., Maietti, S., Muzzoli, M. et al. 2005. Comparative evaluation of 11 essential oils of different origin as functional antioxidants, antiradicals and antimicrobials in foods. *Food Chem* 91:621–632.

Salem, M.Z.M., Ali, H.M. and Basalah, M.O. 2014. Essential oil from wood, bark and needles of *Pinus roxbutghii* Sarg. from Alexandria, Egypt: Antibacterial and antioxidant activities. *Bioresources* 9:7454–7466.

Satyal, P., Paudel, P., Raut, J., Deo, A., Dosoky, N.S. and Setzer, W.N. 2013. Volatile constituents of *Pinus roxburghii* from Nepal. *Pharmacognosy Res* 5:43–48.

Schulte-Oehlmann, U., Albanis, T., Allera, A. et al. 2006. COMPRENDO: Focus and approach, a monograph. *Environ Health Perspect* 114:98–100.

Sezik, E., Üstün, O., Demirci, B. and Baser, K.H.C. 2010. Composition of the essential oils of *Pinus nigra* Arnold from Turkey. *Turk J Chem* 34:313–325.

Shaalan, E., Canyon, A.-S.D., Younes, M.W.F., Abdel-Wahab, H. and Mansour, A.-H. 2005. A review of botanical phytochemicals with mosquitocidal potential. *Environ Int* 31:1149–1166.

Shatar, S. and Adams, R.P. 1996. Analyses of the leaf and resin essential oils of *Pinus sibirica* (Rupr.) Mayr from Mongolia. *J Essent Oil Res* 8:549–552.

Simić, N., Palić, N., Andelković, S., Vajs, V. and Milosavljević, S. 1996. Essential oil of *Pinus heldreichii* needles. *J Essent Oil Res* 8:1–5.

Sonibare, O.O. and Olakunle, K. 2008. Chemical composition and antibacterial activity of the essential oil of *Pinus caribaea* from Nigeria. *Afr J Biotechnol* 7:2462–2464.

Sparks, T.C. and Nauen, R. 2015. IRAC: Mode of action classification and insecticide resistance management. *Pestic Biochem Phys* 121:122–128.

Stevanovic, T., Garneau, F.-X., Jean, F.-I., Gagnon, H., Vilotic, D., Petrovic, S., Ruzic, N. and Pichette, A. 2005. The essential oil composition of *Pinus mugo* Turra from Serbia. *Flavour Fragr J* 20:96–97.

Süntar, I., Tumen, I., Ustüna, O., Keleş, H. and Eakol, E.K. 2012. Appraisal on the wound healing and anti-inflammatory activities of the essential oils obtained from the cones and needles of *Pinus* species by in vivo and in vitro experimental models. *J Ethnopharmacol* 139:533–540.

Tabashnik, B.E., Brévault, T. and Carrière, T.Y. 2013. Insect resistance to Bt crops: Lessons from the first billion acres. *Nat Biotechnol* 31:510–521.

Tian, Y., Li, Z., Wu, S. and Huang, Q. 2012. Antimicrobial activity of essential oil from *Pinus yunnanensis* Franch. var *tenuifolia* growing in China. *Adv Mat Res* 430–432:438–442.

Tisserand, R. 2007. *Challenges Facing Essential Oil Therapy: Proof of Safety*. Denver: Alliance of International Aromatherapists Board.

Tognolini, M., Barocelli, E., Ballabeni, V. et al. 2006. Comparative screening of plant essential oils: Phenylpropanoid moiety as basic core for antiplatelet activity. *Life Sci* 78:1419–1432.

Tsitsimpikou, C., Petrakis, P.V., Ortiz, A., Harvala, C. and Roussis, V. 2001. Volatile needle terpenoids of six *Pinus* species. *J Essent Oil Res* 13:174–178.

Ustun, O., Senol, F.S., Kurkcuoglu, M., Orhan, I.E., Kartal, M. and Baser, K.H.C. 2012. Investigation on chemical composition, anticholinesterase and antioxidant activities of extracts and essential oils of Turkish *Pinus* species and pycnogenol. *Ind Crop Prod* 38:115–123.

Ustun, O., Sezik, E., Kurkcuoglu, M. and Baser, K.H.C. 2006. Study of the essential oil composition of *Pinus sylvestris* from Turkey. *Chem Nat Compd* 42:26–31.

Venskutonis, P.R., Vyskupaityte, K. and Plausinaitis, R. 2000. Composition of essential oils of *Pinus sylvestris* L. from different locations of Lithuania. *J Essent Oil Res* 12:559–565.

Von Rudloff, E. 1985. The leaf oil terpene composition of eastern white pine, *Pinus strobus* L. *Flavour Fragr J* 1:33–35.

Whalon, M., Mota-Sanchez, D. and Hollingworth, R.M. 2008. *Global Pesticide Resistance in Arthropods*. Wallingford, UK: CABI International.

Yamada, T. 2008. Biochemical responses in pine trees affected by pine wilt disease. In *Pine Wilt Disease*, ed. B.G. Zhao, K. Futai, J.R. Sutherland, and Y. Takeuchi, 223–234. Tokyo: Springer.

Yatagai, M. and Hong, Y. 1997. Chemical composition of the essential oil of *Pinus massoniana* Lamb. *J Essent Oil Res* 9:487–489.

Yatagai, M. and Sato, T. 1986. Terpenes of leaf oils from conifers. *Biochem Syst Ecol* 14:469–478.

Zafar, I., Fatima, A., Khan, S.J., Rehman, Z. and Mehmud, S. 2010. GC–MS studies of essential oil from needles of *Pinus roxburghii* and their antimicrobial activity. *Electron J Environ Agric Food Chem* 9:468–473.

Zhang, Y. and Wang, Z. 2010. Comparative analysis of essential oil components of two *Pinus* species from Taibai Mountain in China. *Nat Prod Commun* 5:1295–1298.

# 14

## Turpentine or Pine Oil

Emma Mani-López, Ana C. Lorenzo-Leal, Enrique Palou, and Aurelio López-Malo

### CONTENTS

## 14.1 Botany of the Plant

Pines make up the largest genus of conifers, which includes several *Pinus* species that belong to the Pinaceae botanical family (Barceloux 2008). They are the dominant tree in the vast coniferous forest of the northern hemisphere. These trees have also been extensively planted in the southern hemisphere, but only the Merkus pine (*Pinus merkusii*) occurs there naturally, and its distribution extends poorly south of the equator. Pine trees are some of the world's oldest-known living organisms; some bristlecone pines, native to the White Mountains of eastern central California and the Snake range of the central Nevada–Utah border in the United States, are still standing after 4600 years (Bidlack and Jansky 2011).

A physical description of pine trees is distinctive bundles of long, narrow needles and large, woody cones with tough scales and branches that commonly grow in rings around the main trunk. Pine leaves are needle-like and arranged in clusters (fascicles) of one to eight leaves each, and if the leaves are held together, form a cylindrical rod (Barceloux 2008). The wood is composed of only tracheid cells. Annual rings, spring wood, and summer wood are all visible because large-diameter tracheid is produced in the spring, followed by narrow-diameter tracheid in the summer. Among the tracheid and the rays runs the thick, sticky, antiseptic, and aromatic resin (preventing the development of fungi and insect attacks) in the resin canals. Conspicuous resin canals develop in the mesophyll and consist of tubes lined with black color (Bidlack and Jansky 2011) (Figure 14.1).

## 14.2 Methods of Oil Extraction

There are many extraction methods that have been used to obtain essential oils from vegetable and wood materials, such as steam distillation, Soxhlet extraction, microwave-assisted extraction, pressurized liquid extraction, high-pressure solvent extraction, and hot-water-based extraction, among others (Liazid et al. 2010; Seabra et al. 2012; Chupin et al. 2013; Tongnuanchan and Benjakul 2014). The choice of extraction process, solvents, and operational settings is based on the extract's required quality, as well as on other specifications, like extraction yield and the presence of undesired compounds (Seabra et al. 2012).

The basic process of steam distillation consists in generating steam in a separate vessel, passing the steam through the sample, volatilizing and removing the oils, cooling and condensing into liquids (the steam and oil vapors), and decanting the immiscible oils from water (Kelkar et al. 2006). Turpentine oil is often obtained by distillation with a Clevenger-type extractor for different periods of time, and the oleoresin is then separated by density difference (Tümen and Reunanen 2010; Ulukanli et al. 2014; Missio et al. 2016).

The term *pine oil* or *essential oil of turpentine* comes from the terpenic oil, obtained by various distillation methods of the resinous exudates (turpentine) or wood of coniferous trees. This semifluid substance consists of resins dissolved in volatile oil; the mixture can be separated by different distillation techniques into a volatile portion named oil of turpentine and a nonvolatile portion called rosin. It is also called spirits of turpentine, pine tree terpenic, pine oleoresin, gum turpentine, terpene oil, or turpentine from Bordeaux (Mercier et al. 2009). This oil can be obtained by tapping living species of the genus *Pinus*; the main species for turpentine are listed in Table 14.1.

**FIGURE 14.1**
*Pinus ayacahuite* var. *veitchii* Shaw (Mexican white pine) tree and its main parts: (a) needles, (b) cones, and (c) resinous material.

**TABLE 14.1**

Main Sources of Turpentine from *Pinus* Species

| Common Name | Scientific Name |
| --- | --- |
| Calabrian pine | *P. brutia* |
| Slash pine | *P. elliottii, P. caribaea* |
| Aleppo pine | *P. halepensis* |
| Masson pine or Chinese red pine | *P. massoniana* |
| Merkus pine | *P. merkusii* |
| Maritime pine | *P. pinaster* |
| Monterey pine | *P. radiata* |
| Scots pine | *P. sylvestris* |
| Sumatra pine | *P. merkusii* |
| Longleaf pine | *P. palustris* |
| Taeda pine | *P. taeda* |
| Ponderosa pine | *P. ponderosa* |
| Chir pine | *P. roxburghii* |

Turpentine oil is often classified as sulfate turpentine, wood turpentine, gum turpentine, or crude turpentine, according to the route by how it is produced. The first is widely used in the chemical industry, and it is obtained as a by-product of the kraft or sulfate in the pulping processes during the manufacture of kraft paper. Wood turpentine is obtained from steam distillation of dead aged pines, gum turpentine is obtained via tapping of living pine trees, and crude turpentine is extracted from living pine by tapping, and it usually contains approximately 65% gum rosin and 18% gum turpentine (Pandey et al. 2015; Encyclopedia Britannica 2016) (Figure 14.2).

In spite of using the same extraction method, different yields can be obtained from distillation processes due to factors such as *Pinus* species, location of the plant, and season. For example, using hydrodistillation, Tümen and Reunanen (2010) extracted turpentine oil from three different resins (A, B, and C) of *Pinus sylvestris* from Denizli and obtained volumetric oil yields of 3.96% for resin A (from Acipayam), 4.36% for B (from Çal), and 5.60% for C (from Çameli), showing differences between the locations of the plant. Using the same method but different times of extraction, Ulukanli et al. (2014) obtained essential oil from *Pinus brutia* or *Pinus pinea* with yields of 7.6% (v/w) or 6.4% (v/w), respectively, which were higher than the yields for *P. sylvestris*.

Besides the method of hydrodistillation, the National Institute of Environmental Health Sciences (2002) describes two extraction techniques for crude sulfate turpentine; the first method consists in heating (150°C–180°C) the wood in an aqueous digestion substance (NaOH, $Na_2S$, $Na_2CO_3$ with $Na_2SO_4$, $Na_2SO_3$, and $Na_2S_2O_3$) in vessels with a pressure of 7–13 bars for 1–5 h. The crude sulfate turpentine is then condensed from the vapors of the wood digestion. Sulfur compounds such as methanethiol and dimethyl sulfide are oxidized with sodium hypochlorite solution (60°C) to less volatile sulfonic acids, sulfoxides, or sulfones. Another technique consists of impregnating wood chips with a pulping solution—in acid bisulfite pulping ($Ca(HSO_3)_2$ and $SO_2$ [pH 1.5–2]), in bisulfite pulping ($Mg(HSO_3)_2$ [pH 4]), and in a neutral sulfite process ($Na_2SO_3$, $NaHCO_3$, $(NH_4)_2SO_3$, and $NH_3$ [pH 7–9])—at increasing temperatures, alternating between reduced and increased pressure. In this method, chips are digested at 125°C–160°C, 7–13 bars, for about 6–10 h. After releasing the waste gas, the crude oil (floating on pulping liquor) has to be neutralized with NaOH or lime, and distilled to colorless sulfite turpentine. Fermont (1981) described a method to extract turpentine oil

**FIGURE 14.2**
Typical pine oil.

with supercritical fluids at temperatures above critical and pressures of about 100 psi above critical; this method is described in great detail in U.S. Patent US 4308200 A.

## 14.3 Methods of Oil Analysis

### 14.3.1 Chemical Analysis

Separation techniques, such as chromatography, are among the most suitable analytical techniques for essential oil chemical characterization; they can be performed by gas chromatography (GC) or high-performance liquid chromatography (HPLC) (Jagodzińska et al. 2011). GC is a method that achieves separation of mixtures by partitioning components between a stationary and a mobile gas phase. Using a flame ionization detector (FID), it is possible to determine the presence of each compound, and its identification is carried out by comparing the experimental information with databases. A mass spectrometry (MS) detector coupled with GC can also be used to determine chemical structures. On the other hand, HPLC also separates constituent chemical compounds in a mix solution by means of a chromatographic column; the separation is done in the analytical column, and it depends on the interaction of components of the sample and the mobile and stationary phases. The HPLC technique is often coupled to MS, ultraviolet–visible (UV-Vis), or a diode array detector (DAD) (Stolerman 2010; Jagodzińska et al. 2011).

Among the advantages of GC are the commercial availability of gas chromatographs, the low amount of reagents needed, the adequacy of both solvents and solutes for the technique, and the fact that the solute does not need to be of high purity. However, GC could cause structural alterations of thermally labile compounds, due to high injector temperatures or catalytically active surfaces. An alternative method to prevent this problem may be HLPC because of its versatility, sensitivity, and selectivity (Díaz et al. 2004; Turek and Stintzing 2011).

Most published studies conducted to characterize turpentine oil have used GC, and according to different authors, preferred conditions of GC for analyzing turpentine oil are as follows: helium or nitrogen carrier gas, 0.8–30 ml/min flow rate, injection volume of 0.0003–0.5 µl, and initial column temperature of 50°C–80°C for 0.5–2.0 min, increasing to a maximum of 120°C–290°C (rate of 4°C/min) (Pakdel et al. 2001; Papajannopoulos et al. 2001; Díaz et al. 2004; Wiyono et al. 2006a; Tümen and Reunanen 2010). The identification of compounds is usually based on FID and MS, referring to the NIST98 and WILEY275 mass spectral libraries (Pakdel et al. 2001; Tümen and Reunanen 2010; Ulukanli et al. 2014).

Further, Turek and Stintzing (2011) applied HPLC to characterize the essential oil of turpentine; they used purified water as mobile phase A and acetonitrile as mobile phase B at a flow rate of 0.21 ml/min, starting isocratically with 30% B for 2 min, followed by a linear gradient to 100% B in 40 min. The injection volume of each sample was 10 µl. Moreover, UV-Vis spectra were recorded (190–720 nm) in order to obtain three-dimensional spectral information for each reference compound and essential oil sample.

### 14.3.2 Physical Analysis

Measurements of physical properties, such as viscosity and density, provide insight into the molecular arrangement in liquids and help us to understand some of the thermodynamic properties of liquid mixtures. The viscosity of turpentine oil can be determined by using a viscosity measuring unit equipped with Ubbelhode capillary viscometers. Cell constant $k =$ 0.003146, 0.004993, and 0.009937, with times varying from 180 to 400 s. The viscosimeter has to be calibrated with distilled water. The kinematic viscosity ($v$) is obtained from Equation 14.1:

$$v = k(t - \theta)$$

(14.1)

where $k$ is the capillary constant, $t$ (s) is the flowing time of the sample, and $\theta$ is the Hagenbach correction. Then, the absolute viscosity ($\eta$) is calculated from $\eta = \rho v$, where $\rho$ refers to density (Francesconi and Castellari 2001).

On the other hand, the density of the essential oil is determined by a density meter equipped with a measuring cell. The oscillator period ($\tau$) in the vibrating U-tube is converted to density by the following equation:

$$\tau = A(\tau^2 - B)$$

(14.2)

where $A$ and $B$ are the apparatus constants that are determined by literature density data of dry air and water. Finally, the refractive index of essential oil is usually measured with an Abbe refractometer, keeping the temperature constant during the measurements (Francesconi and Castellari 2001).

## 14.4 Composition of Turpentine Oil

The composition of turpentine oil directly depends on a variety of factors, including extraction and refining methods, plant part, species, geographical location, and season. The major constituents of turpentine are terpene compounds, such as α-pinene, β-pinene, and Δ-3-carene, shown in Figure 14.3.

*Terpene* is a term used to describe a mixture of isomeric plant hydrocarbons of the molecular formula $C_{10}H_{16}$, and its name is derived from the word *turpentine* (Rodrigues-Corrêa et al. 2013). The minor compounds of turpentine oil, such as m-cymenene, d-camphene, α-fenchol, camphor, pinocarvone, borneol, p-mentha-1,5-dien-8-ol, terpinen-4-ol, α-terpinol, myrtenal, myrtenol, verbone, and carvone, seem to be biosynthesized only under particular environmental conditions (Barceloux 2008; Rodrigues-Corrêa et al. 2012).

As mentioned before, the content of the components in turpentine oil can change due to the pine species. The percentages of the main components of turpentine oil from different *Pinus* species are presented in Table 14.2, where it can be seen that there is an important

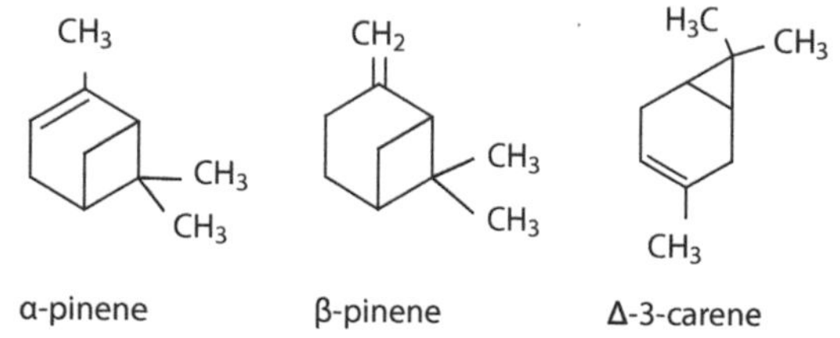

**FIGURE 14.3**
Structures of the major constituents of turpentine oil.

**TABLE 14.2**

Contents of the Principal Components of Turpentine Oil from Different Pine Species and Its Characterization Method

| Pine Species | α-Pinene (%) | β-Pinene (%) | Δ-3-Carene (%) | Other Components Reported | Characterization Method | Reference |
|---|---|---|---|---|---|---|
| P. merkusii | 82–86 | 2.2–2.4 | 8–12 | d-Camphene, sabinene, myrcene, d-limonene | GC-FID | Wiyono et al. 2006b |
| P. pinea | 21.4 | 9.7 | 0.05 | Camphene, β-phellandrene, β-myrcene, α-phellandrene, d-limonene | GC-FID | Ulukanli et al. 2014 |
| P. brutia | 25.4 | 9.69 | 0.16 | | | |
| P. merkusii from Jantho[a] | 73.3 | 1.7 | 19.3 | Camphene, β-myrcene, α-thujune, sabinene, limonene, α-tepinolene | GC-MS | Sukarno et al. 2015 |
| P. merkusii from Takengon[a] | 73.9 | 1.5 | 19.1 | | | |
| P. merkusii from Blangkejeren[a] | 81.7 | 1.8 | 11.9 | | | |
| P. merkusii from Java[a] | 87.2 | 2.2 | 7.3 | | | |
| P. sylvestris from Acipayam[b] | 43.6 | 17.1 | 7.7 | Camphene, 3,7,7-trimethylcyclohepta-1,3,5-triene, limonene, o-cymene, trans-pinocarveol | GC-FID | Tümen and Reunanen 2010 |
| P. sylvestris from Çal[b] | 34.9 | 15.3 | 6.5 | | | |
| P. sylvestris from Çamlibel[b] | 42.8 | 29.4 | 9.7 | | | |
| P. halapensis | 23.2 | 0.3 | 0.3 | Limonene, terpinolene, sesquiterpenes | GC-MS | Papajannopoulos et al. 2001 |
| P. brutia | 8.5 | 2.9 | 3.4 | | | |

Note: GC-FID, gas chromatography–flame ionization detector; GC-MS, gas chromatography–mass spectrometry.
[a] Subpopulations from Indonesia.
[b] Locations from Denizli, Turkey.

difference among the contents of α-pinene, β-pinene, and Δ-3-carene for selected species of pine tree (*P. merkusii, P. pinea, P. brutia, P. sylvestris,* and *Pinus halepensis*). Furthermore, the composition could change because of the location of the plant, as can be seen from data reported for *P. merkusii* and *P. sylvestris*. It is important to mention that the method to determine the composition of the oil is different in some of the cases presented in Table 14.2.

Regarding the location of pine trees, data shown by the National Institute of Environmental Health Sciences (2002) demonstrate that turpentine oil from Greece, Mexico, China, and Portugal has up to 70% of α-pinene and less than 20% of β-pinene in its composition, while turpentine oil from New Zealand, the United States, and Canada has more than 20% of β-pinene and less than 60% of α-pinene.

## 14.5 Physical and Chemical Properties of Turpentine Oil

Turpentine is a colorless liquid with the following physical properties: boiling point of 154°C–170°C, melting point of –60°C to –50°C, density of 0.854–0.868 g/cm³ (20°C), relative vapor density of 4.6–4.8 g/cm³, flash point of 30–46°C, insolubility in water, and solubility in benzene, chloroform, ether, carbon, disulfide, petroleum ether, and oils (in this order) (National Institute of Environmental Health Sciences 2002; National Institute for Occupational Safety and Health 2015; Encyclopedia Britannica 2016).

Panda (2011) mentions that American turpentine oils, which consist essentially of dextro-pinene, have a specific gravity of 0.864–0.866 and a boiling point of 156°C–160°C, are completely distilled at 170°C, and are clear liquids with a peculiar and characteristic odor. French spirits of turpentine have a more uniform behavior than the American oil and are colorless, with a boiling point of 161°C and a density of 0.864–0.880 g/cm³ (at 16°C). Russian spirits of turpentine are very similar to the American ones, but their density varies from 0.864 to 0.870 g/cm³ and they begin to boil at 155°C (Leyva et al. 2009; Panda 2011).

## 14.6 General Uses of Pine Oil

### 14.6.1 Typical Uses

Turpentine oil is used in a wide range of products; typical uses of turpentine include insect repellant, cleaners, fragrances and flavoring compounds, pesticides, pharmaceutical products, and as raw product for the isolation of chemicals (like camphor, citral, citronellal, isobornyl acetate, linalool, menthol, and pine oil), as well as a solvent for varnish, paint, and polish products (Díaz et al. 2004; Barceloux 2008; Mercier et al. 2009; Rodrigues-Corrêa et al. 2012).

### 14.6.2 Traditional Medicine and Aromatherapy

In ancient times, doctors like Hippocrates, Dioscordie, and Galien used terpenic oil to cure lung diseases and biliary lithiasis. In France, the oil was recommended against blennorrhoea and cystitis, and it was also used against neuralgias, rheumatism, sciatica, nephritis, drip, constipation, and mercury salivation. In Germany, Slovenia, and Poland, drugs that were used for renal and hepatic diseases contained α- and β-pinenes (Mercier et al. 2009).

Some of the current uses in medicine of this essential oil are as an antiparasitic, analgesic, revulsive, balsamic, active against bronchial and pulmonary secretions, antispasmodic, disinfectant (external use), active against genital-urinary tract infections, hemostatic, diuretic, and antidote for poisoning caused by phosphorus (Dorow et al. 1987; National Institute of Environmental Health Sciences 2002; Mercier et al. 2009; Tümen and Reunanen 2010).

Specifically, the essential oil obtained from *Pinus roxburghii* is claimed to have different medicinal uses, for example, as a hemostatic, stimulant, antihelmitic, digestive, liver tonic, diaphoretic, or diuretic. The effects of this essential oil could be due to the presence of its main constituents (α-pinene, β-pinene, and Δ-3-carene), resin acids, and camphene and polymeric terpenes (Kaushik et al. 2012). Concerning aromatherapy, turpentine oil extracted from *P. sylvestris* has been used in treatments for bronchitis, asthma, arthritis, inflammations, and asthenia (Sanz Bascuñana 2015; Can Başer and Buchbauer 2016).

### 14.6.3 Disinfectants (Household Uses)

Oil of turpentine has been utilized as a disinfectant since 1876 in Germany, because the terpene alcohol terpineol has antimicrobial activity and, with pinenes, shares the property of modifying the action of phenols in solubilized disinfectant formulations, although its antimicrobial activity is not the same for every microbial species (Ascenzi 1996; Mercier et al. 2009; Fraise et al. 2012). Recently, formulations of pine oil cleaning products contain smaller concentrations of pine oil and higher concentrations of constituents in order to reduce the viscosity (and aspiration potential) of the product (Rodrigues-Corrêa et al. 2012).

Terpineol is an important terpene alcohol; it is one of the three isomeric alcohols having the basic molecular formula $C_{10}H_7OH$, and is a colorless liquid that tends to darken during storage. It is used in manufacturing disinfectants, as well as in the perfumery industry (Fraise et al. 2012).

Cleaning products are important with a wide range of utilities in assisting in dirt and grime removal from surfaces. Their composition should be clear and phase stable at normal use and storage temperatures. Some of the patents for cleaning products containing pine oil are listed in Table 14.3. Most of these claim to have an invention related to improved blooming-type cleaning compositions and concentrates, especially useful for cleaning hard surfaces and/or disinfecting applications, with the main difference among them being their composition (specially with regard to pine oil). Some of the patents shown in this table (US 5135743 A, US 5183655 A, US 5189987 A, and US 5016568 A) have a controlling effect on animal litter (for domestic animals) as well.

### 14.6.4 Paints and Coatings

Turpentine is a constituent of paints, coatings, and paint thinners and solvents because the terpenes present in turpentine oil are capable of accelerating the drying of oils and other film formers. Turpentine has been used as dilute printer's ink for a long time, but its use has diminished recently because there are less expensive solvents made from petroleum. Nevertheless, turpentine is still used as a diluent for various products, such as black shoe polishes, furniture, the ceramic coating industry, and pottery (National Institute of Environmental Health Sciences 2002; Ortuño Sánchez 2006; Mercier et al. 2009).

**TABLE 14.3**

U.S. Patents of Cleaning Products That Contain Pine Oil

| Name of Patent | Patent No. | Composition Claimed for Terpene or Pine Oil | Reference |
|---|---|---|---|
| Biphenyl-based solvents in blooming-type hard-surface cleaners | US 6100231 A | 1%–10% by weight (wt) of pine oil constituent | Cheung and Smialowicz 2000a |
| Biphenyl-based solvents in blooming-type germicidal hard-surface cleaners | US 6075002 A | 1%–10% wt of pine oil constituent | Cheung and Smialowicz 2000b |
| Blooming-type disinfecting cleaning composition | US 6110295 A, US 6066606 A | 0.1%–10% wt of terpene (for both patents) | Lu and Kloeppel 2000a,b |
| Blooming pine oil–containing compositions | US 5985819 A | 6%–15% wt of pine oil preparation | Lu and Smialowicz 1999 |
| Cleaning composition containing pine oil extenders | US 6010998 A | 5%–20% wt of pine oil concentrate | Merchant et al. 2000 |
| Combined odor-controlling animal litter | US 5135743 A, US 5183655 A | 0.003%–50% wt of pine oil (for both patents) | Stanislowski et al. 1992a,b |
| Germicidal pine oil cleaning compositions | US 5629280 A | 0.01%–3% wt of pine oil | Richter and Taraschi 1997 |
| Odor-controlling animal litter with pine oil | US 5189987 A, US 5016568 A | 0.001%–50% of pine oil (for both patents) | Stanislowski and England 1991, 1993 |
| Pine oil cleaning composition | US 6465411 B2 | 5%–20% wt of pine oil | Manske and McPherson 2002 |
| Pine oil hard-surface cleaning compositions | US 5591708 A, US 5728672 A | 0.1%–4% wt of a pine oil preparation (for both patents) | Richter 1997, 1998 |
| Pine oil–containing hard-surface cleaning composition | US 5308531 A | 1%–5% wt of pine oil | Urfer and Lazarowitz 1994 |
| Stable liquid cleaners containing pine oil | US 5962394 A | 5%–25% wt of pine oil | Gryj et al. 1999 |

## 14.6.5 Other Uses

Aside from the uses of turpentine oil as a mixture, raw products isolated from this oil, like pinenes and other mono- and sesquiterpenes present in the oleoresin, are used as food additives (tricyclene, myrcene, 3-carene, α-terpinene, α-phellandrene, limonene, myrtenol, aromadendrene, bisabolene, β-caryophyllene, germacrene, and camphor), polymers (myrcene), explosives (camphene), cosmetics (borneol, limonene, and terpineol), household cleaners (borneol and limonene), antioxidants (α-terpinene), pharmaceuticals (isolongifolence), insect attractants (camphor), disinfectants (*p*-cymene and terpineol), solvents (*p*-cymene and limonene), and textile auxiliaries (terpineol) (National Institute of Environmental Health Sciences 2002; Rodrigues-Corrêa et al. 2012). According to Mercier et al. (2009), the principal uses of α-pinene are as a liphophilic, bactericidal, fungicidal, insecticidal, pesticidal, anticarcinogenic, diuretic, antioxidant, immune-stimulant, anti-inflammatory, anticonvulsive, sedative, antistress, or hypoglycemic agent, while β-pinene is commonly used as a lipophilic, bactericidal, fungicidal, insecticidal (acting against osteoclasts), pesticidal, antioxidant, or sedative agent.

## 14.7 Pesticide Uses of Pine Oil (Including Public Health)

Pests include weeds, insects, rodents, fungi, bacteria, and other organisms. Pests, especially weeds, pathogens, and insects, are implicated in losses in agriculture and cause public health problems. Crop production can be reduced up to 20%–25% by pests. Weeds alone are responsible for nearly a 34% reduction of crop yields (Oerke 2006). Mosquitoes are the main virus vectors of malaria and encephalitis; meanwhile, bacteria cause infections at home. Current pest management is carried out with synthetic pesticides; however, their environmental impact, pest resistance, and need for natural alternatives favor the application of essential oils (like pine oil) as pesticides.

### 14.7.1 Herbicides

Few scientific reports are available for pine oil as an herbicide. In contrast, herbicidal commercial products based on pine oil are common, especially in organic products; however, their weed control is limited, their cost is high, and their efficiency is low. Some improvements have been suggested in order to enhance pine oil's effect on weeds. Giepen et al. (2014) included 8% sodium chloride in the formulation of a pine oil product (8% pine oil + 2% emulsifier) and observed that product cost decreased, while it maintained its weed control at lower pine oil concentrations, whereas commercial products suggest or contain 13.6%–20% pine oil as an active ingredient. Other strategies focus on modified weed seed environmental conditions before the germination process begins, when pine oil is used as weed seed control. McLaren et al. (2012) evaluated pine oil's (5%) and sugar's (0.31 kg/m$^2$) effect on the germination of seeds of *Nassella trichotoma* (a common weed in Australia); treatment reduced 98%–100% of germination, and results were similar to those obtained with flupropanate (a typical preemergent herbicide).

### 14.7.2 Insecticides

An ideal insecticide should be effective, ecologically sound, sustainable, and cost-effective and exhibit low mammalian toxicity. In addition, larvicides should not significantly change the characteristics of environmental water (WHOPES 2011); however, consumers can only choose a few of these features in most commercial products.

Insect pests cause huge losses if there is no adequate control in the crop, postharvest handling, and/or storage of selected products and commodities. Utilization of insecticides decreased 21.7% in the United States between 2000 and 2007, from 100,243 to 78,471 tons of active ingredients for product production (FAOSTAT 2015). It has been suggested that the total utilization of insecticides is greater than that reported by the Food and Agriculture Organization Corporate Statistical Database (FAOSTAT), since many other applications are not usually registered. Traditional insecticides are effective; however, health effects and environmental impact have motivated the development of alternative natural-based insecticides. Pine oil has evidenced interesting applications for insect control.

#### 14.7.2.1 Mosquitoes

Mosquitoes are considered the main vectors of viral diseases, and some of humanity's deadliest illnesses. *Anopheles* mosquitoes are responsible for the transmission of malaria, filariasis, and encephalitis. *Culex* mosquitoes carry encephalitis, filariasis, and the West Nile virus, while *Aedes* mosquitoes transmit yellow fever, dengue, and encephalitis.

Commercial pine oil of *Pinus longifolia* has shown larvicidal activity against *Anopheles stephensi*, *Culex quinquefasciatus*, and *Aedes aegypti*, with a lethal concentration for 50% of the population ($LC_{50}$) of 112.6, 85.7, and 82.1 µl/L, respectively. Required doses of pine oil seem high; however, potential applications could focus on small breeding places, such as domestic and peridomestic containers and desert coolers, where water is stagnant (Ansari et al. 2005). On the other hand, Lucia et al. (2007) reported concentrations of 14.7 µl/L of commercial pine oil for an $LC_{50}$ against *Ae. aegypti* larvae after 24 h; in addition, α- and β-pinene from the pine oil showed the same larvicidal effect at the same concentration of pine oil (14.7 µl/L). These results evidence that larvicidal effects are due to pinenes of the oil.

Larvicidal effects of pine oil on *Ae. aegypti* at 21.4 mg/L (lethal dose [$LD_{50}$]) had been reported for oil from *Pinus caribaea*; this essential oil also had ovicidal effects by reducing the eclosion of the eggs (51.2%) at 66 mg/L (Leyva et al. 2009). Control at different stages of the life cycle of yellow fever mosquito (*Ae. aegypti*) was studied for modified pine oil (*P. caribaea* and *Pinus tropicalis*). Pine oil was treated with photochemical isomerization, and its major components α-pinene and β-pinene were converted to limonene derivatives with oxygen functions in order to obtain pulegone and verbenone. The modified oil killed 90.3% of larvae and also affected pupa and semiemerged adult mosquitoes; thus, total mortality was 94.14%. Besides, emerged adults were also reduced (35%) by this modified pine oil (Leyva et al. 2010).

### 14.7.2.2 Crop Pests

Rice weevil (*Sitophilus oryzae* L.), adzuki bean weevil (*Callosobruchus chinensis* L.), and rice moth (*Corcyra cephalonica* S.) are susceptible to commercial pine oil. The mortality was 80% for the rice weevil, 92% for the bean weevil, and 90% for the rice moth when they were exposed to 130 µg/cm² of pine oil, and the $LC_{50}$ was 77, 75, and 64 µg/cm² of pine oil, respectively; tests were carried out by direct contact on adult insects. A repellent effect was also observed during the study; insects moved toward the lid and preferred to stay away from the discs containing oil. This research also considered a fumigant effect; thus, pine oil was also evaluated on the effect of the vapor phase on the insects. Pine oil was more effective in the vapor phase, where insect mortality was 84% for the rice weevil, 98% for the bean weevil, and 94% for the rice moth after 72 h of exposure at 130 µg/cm². The amount of pine oil was lower than that for liquid application, achieving an $LC_{50}$ of 48, 33, and 34 µg/cm² for the rice weevil, bean weevil, and rice moth, respectively (Usha 2012).

Pine oil reduced oviposition of female onion maggots (*Delia antiqua* Meigen) by 72% at a release of 200 µg/day. In addition, the main individual components of the commercial pine oil (Norpine 65™), ρ-cymene, limonene, and 3-carene, and their mixture were evaluated for deterred oviposition. The mixture was the most effective (prevented 87%, 320 µg/24 h), followed by 3-carene (320 µg/24 h), which deterred 73%; meanwhile, limonene (320 µg/24 h) and ρ-cymene (220 µg/24 h) obtained 65% and 56% deterrence of oviposition (Ntiamoah et al. 1996). A few studies have evaluated the insecticidal effects of pine oil on crop pests, and the reported results are encouraging; however, more research is necessary in order to obtain potential applications.

### 14.7.2.3 Forest Pests

Trees can be damaged by moths, caterpillars, beetles, termites, and northern walking stick. Big populations of these insects can cause complete loss of forests. *Thaumetopoea pityocampa*

Schiff larvae (a moth pine defoliator) can be eliminated with pine oil of *P. brutia* Ten, due to its camphene component, a typical insecticidal compound, and limonene, which is a repellent to this moth (Kanat and Alma 2003). The times required to kill 50 larvae (100% of the population) when 2 ml of pine oil at 25% or 100% was poured on larvae were 0.79 and 0.25 min, respectively (Kanat and Alma 2003); these authors studied nine essential oils, and pine oil was the most effective. Terpenes from *Pinus pinaster* exhibited a toxic effect on three species of termites, *Reticulitermes santonensis*, *Reticulitermes* (*lucifugus*) *grassei*, and *Reticulitermes* (*lucifugus*) *banyulensis*; geranyl-linalool was the most toxic terpene, with an $LD_{50}$ of 2.74, 4.48, and 10.64 μg/mg of worker termites for *R. grassei*, *R. santonensis*, and *R. banyulensis*, respectively. Further, α-pinene was less effective, with an $LD_{50}$ between 14.23 and 20.63 for the three species of worker termites (Nagnan and Clement 1990). Thus, pine oil or its components can be utilized as an insecticide against termites based on these reports.

### 14.7.3 Mite Control

Pine oil of branches and its main components (1,8-cineole and limonene) as a volatile fraction exhibited effective control of *Tyrophagus putrescentiae*, reaching 100% death of mites; this study evaluated pine oils from *P. pinea*, *P. halepensis*, *P. pinaster*, and *Pinus nigra*; three oils showed good acaricidal control (>50% of death mites), but only the oil of *P. pinea* had the ability to kill all mites, while the essential oil of *P. nigra* had poor performance (Macchioni et al. 2002). Also, poultry red mites (*Dermanyssus gallinae*) were eliminated by pine oil (*P. sylvestris* L.) at 0.21 mg/cm²; the $LD_{50}$ was 0.11 mg/cm² (George et al. 2010). Mites cause several problems to stored products, human health, some foods, cultivars, and animals, and pine oil may be an option to control or repel them; nevertheless, more studies are still necessary.

### 14.7.4 Copepod Control

Potential applications in order to control *Lernaea cyprinacea* (copepod, water crustacean) in fish like *Brycon lundii* Reinhardt (freshwater teleostei) and *Leporinus piau* (freshwater fish) have been evaluated utilizing *Pinus elliottii* oil. A steamed oil fraction and a chloroform fraction were effective in killing 100% of the parasites after 18 h; crude resin required 24 h, while pure compounds such as α- and β-pinene were less effective and required 30 h at 0.5 mg/L. On the other hand, a toxic dose of the resin was at least 200 times higher than that for *L. piau*, highlighting the use of pine resin or derived products from pine oil to control dangerous parasites of pisciculture (Tóro et al. 2003).

The insecticidal effects of pine oil and its compounds vary according to intrinsic and extrinsic factors; the main factors include pine species, tree age, chemotypes, geographic conditions (altitude, rainfall, season, temperature, etc.), when the resin or wood was collected, insect species, and age of insect. Despite this, reports presented here suggest that pine oil has several potential applications as an insecticide.

### 14.7.5 Fungal Control

Several plants, fruits, crops, and wood diseases are attributed to fungi. *Aspergillus*, *Penicillium*, *Fusarium*, *Rhizopus*, *Botrytis*, *Colletotrichum*, and *Sclerotinia* are the main fungal pest genera. Published studies have focused on evaluating the inhibitory or fungicidal activity of pine oil in model systems (broth or agar culture medium) against certain molds;

however, no reports were found regarding its application on cultivars or food systems. Pine oil exhibited fungicidal activity against *Fusarium oxysporum*, *Aspergillus niger*, and *Aspergillus flavus* when commercial pine oil from *P. silvestris* L. was utilized; tests were performed in broth (Tullio et al. 2006). On the other hand, inhibitory activity of pine oil (on agar) against the main indoor environmental air molds (*A. niger* at 1.5%, *Cladosporium cladosporioides* at 0.5%, *Penicillium chrysogenum* at 2.5%, *Aureobasidium pullulans* at 0.75%, and *Paecilomyces variotti* at 1.2% [v/v]) was reported by Motiejūnaitė and Pečiulytė (2004) when a volatile fraction of commercial pine oil of *P. sylvestris* L. was evaluated. Commercial pine oils of *Pinus mungo* var. *pulmilio* and *P. sylvestris* also inhibited *Penicillium citrinum*, *Penicillium griseofulvum*, *Penicillium expansum*, *Penicillium crustosum*, and *Penicillium brevicompactum* when tested oils were evaluated by the disc diffusion technique on model systems (Felšöciová et al. 2015). Inhibitory activity was stronger for *P. sylvestris* than *P. mungo* for studied molds; thus, the composition of pine oil determined its antifungal activity. More research is necessary in order to understand the pine oil fungicidal activity, as well as to suggest its application on final products.

## 14.7.6 Bactericide

According to the U.S. Environmental Protection Agency (USEPA), approved pesticide uses for pine oil include bacteria targets such as *Brevibacterium ammoniagenes*, *Escherichia coli*, *Klebsiella pneumoniae*, *Pseudomonas aeruginosa*, *Salmonella choleraesuis*, *Salmonella typhi*, *Salmonella typhosa*, *Serratia marcescens*, *Shigella sonnei*, *Staphylococcus aureus*, *Streptococcus faecalis*, and *Streptococcus pyogenes* (USEPA 2006).

Current reports regarding bactericidal effects of pine oil confirm or propose potential applications of the essential oil. Antibacterial properties of pine oil have been tested under different assays (liquid or solid culture media); for example, pine oil from *P. elliottii* has been suggested as an antimicrobial against a wide range of bacteria: *Staphylococcus epidermidis* ATCC 149990 (25 µg/ml), *S. aureus* ATCC 29213 (200 µg/ml), *Enterococcus faecium* NCTC 717 (100 µg/ml), *Staphylococcus capitis* ATCC 27840 (100 µg/ml), *Enterococcus faecalis* ATCC 14990 (100 µg/ml), and *Staphylococcus haemolyticus* ATCC 29970 (100 µg/ml). The minimal inhibitory concentration varied from 25 to 200 µg/ml when pine oil was tested in tryptic soy broth (TSB) by Leandro et al. (2014); notably, tested bacteria included multidrug-resistant strains, so pine oil could be an important asset in the fight against these microorganisms. Commercial pine oil from *P. brutia* also displayed antimicrobial activity against food-borne bacteria when 50 µl of essential oil was tested by disc agar diffusion against *E. coli* ATCC 25922 (5.25 mm), *Salmonella* Paratyphi A NCTC 13 (6.00 mm), *K. pneumoniae* ATCC 700603 (20.50 mm), *Yersinia enterocolitica* NCTC 11175 (15.50 mm), *P. aeruginosa* ATCC 27853 (15.00 mm), *Aeromonas hydrophila* NCIMB 1135 (24.50 mm), *Campylobacter jejuni* ATCC 33560 (22.25 mm), *E. faecalis* ATCC 29212 (21.25 mm), and *S. aureus* ATCC 29213 (28.25 mm) by Ozogul et al. (2015). In addition, pine oil from *P. pinaster* Aiton was able to inhibit or kill *Actinomadura madurae* at 124.8 or 150.0 µl/ml, respectively, in TSB medium when assayed by the microdilution method (Stojkovic et al. 2008).

## 14.7.7 Repellents

Pine oil has been traditionally used as a mosquito repellent in rural areas. The repellent effect of commercial pine oil from *P. longifolia* was similar to that of citronella oil (the most common commercial repellent); both oils exerted a protective effect of 100% for 11 h against bites of *Anopheles culicifacies* on human skin. Pine oil also repelled 97.4% of bites from *Culex quinquefaciatus*; meanwhile, citronella oil repelled 98.5% on human skin during 9 h. The same pine oil was added to mats; heated mats protected human skin from bites of *An. culicifacies* and *C. quinquefaciatus* at 94.1% (10.3 h) and 92.0% (8.2 h), respectively (Ansari et al. 2005). The strong repellent effect of pine oil against *An. culicifacies* shows its potential application in repellent formulas of commercial products.

Horn flies (*Hematobia irritans* L.) can be repelled by commercial pine oil on pastured cows and heifers; 86.5% of the flies were repelled up to 24 and 12 h, respectively. Pine oil was also applied in a mixture with sunflower oil; after 12 h of repellent activity, $16 \pm 9$ and $4 \pm 3$ flies were observed for each treatment (sunflower oil and sunflower oil + pine oil, respectively). Furthermore, alcohol was probed as a carrier of the tested essential oil; however, a repellent effect was observed only for 8 h and only 69% of flies were repelled (Lachance and Grance 2014). *Cryptorrhynchus laphati* (poplar and willow borer) adult males and females were repelled by pine oil (1%) or α-pinene (1 mol/L) when exposed to their volatile compounds through a Y-tube olfactometer or wind tunnel by Cao et al. (2015). Potential repellent applications can be derived from these reports that highlight the efficacy of pine oils.

### 14.7.8 Security Measures in the Handling and Application of Pine Oil

Security measures for pine oil as a pesticide are similar to those for synthetic products. These include goggles, a respiratory mask, clothing, and boots. If the product splashes the face, eyes, or mouth, immediately rinse under running water (Biocoat Australia 2016).

### 14.7.9 Effects of Pine Oil on Human Health

The natural origin of pine oil is not a reason to use it without care. Essential oils can decrease membrane potential and depolarization of the mitochondrial membranes in eukaryotic cells. The cycling of ionic $Ca^{++}$ and other ionic channels is also affected. Then, changes in pH, proton pump, and the ATP pool can occur. In mammalian cells, essential oils' toxicity causes apoptosis and necrosis (Bakkali et al. 2008). Commercial pine oil may cause slight erythema in 1% of humans, and there is doubtful evidence of mild sensitization when patch tests (12.5% of aqueous solution) are applied (Hercules Inc. 1974). An acute oral toxicity and an acute inhalation toxicity of pine oil have been reported at $LD_{50}$ values in rats of 2.7 g/kg and >3.67 mg/L, respectively, as well as primary eye irritation in rabbits lasting up to 16 days (USEPA 2006). Inhalation exposure to pine oil has a limit of 50 mg/kg/day for rats; pine oil has not been classified as carcinogenic, while neurotoxic effects are not available (USEPA 2006). Based on these and detailed information available at the USEPA website, it can be determined that there is a reasonable certainty that no harm will come to the general population when pine oil is utilized as a pesticide, disinfectant, sanitizer, microbicide or microbistat, virucide, or insecticide, or when it is applied indoors for nonfood, residential, or medical purposes (USEPA 2006).

### 14.7.10 Additional Considerations

Undesirable effects can be derived from pine oil use. The most important are pest resistance derived from recurrent use of pine oil and/or cross-resistance to other compounds utilized as agents of control. On the other hand, lengthy exposures to pine oil or human sensitization can lead to allergies.

There are few reports with regard to pine oil leading to possible resistance in pests; reports are available only for bacteria. Some authors have studied possible relationships between antibacterial cleaners (including pine oil) and bacterial resistance to antibiotics; however, there are no clear conclusions about this. Cole et al. (2003) showed a lack of antibiotic cross-resistance in target bacteria from home antibacterial products, while Price et al. (2002) reported that *S. aureus* resistant to household pine oil cleaner leads to glycopeptide-intermediate *S. aureus* (strains with intermediate levels of vancomycin resistance).

No published reports were found about pine oil as a pesticide and its relation to allergies; however, information from other applications or studies is presented in order to take appropriate care when pine oil is used. Early studies showed an increase in human sensitization to pine oil between 1992 and 1997, from 0.5% to 3.1% when patch tests were applied in 45,005 patients by the Dermatological Centers of Germany and Austria (Treudler et al. 2000); this study concluded that turpentine allergy was found less frequently in men, resulted in fewer symptoms on the hands and more on the legs or face, and impacted subjects older than 60 years more often. Other reports have proven reemerged allergies for turpentine in 24 cases of hand dermatitis in pottery workers in the United Kingdom (Lear et al. 1996). Also, asthma can be induced in some people when recurrent exposure to turpentine occurs, for example, for art painters (Dudek et al. 2009). In Iran, a study showed very few patients (1.5%) were allergic to turpentine (patch test); patients were tested for fragrance contact allergy (Firooz et al. 2010).

## 14.8 Commercial Pesticide Products

Selected commercial products derived from pine oil are available in some countries for weed control. Pine oil–derived products are recommended for organic production. Organic Interceptor™ is the first organic weed control product concentrate. It is recommended at levels of 15%–20% and diluted with water. Its applications are for a wide range of weeds, and the active ingredient in the product is pine oil at 680 g/L (Biocoat Australia 2016). Weed Blitz™ organic herbicide knockdown pine oil weed control (certified as organic, Biological Farmers of Australia [BFA] organic) is nonselective and kills most weeds and their seeds by stripping their outer protective coating. Weed control success depends on the kind of plant, size, and season, among other factors; however, nonwoody weeds respond well to Weed Blitz™ (Green Harvest Organic Gardening Supplies 2016). Its active ingredient is pine oil at 136 g/L. Another certified organic weed control is Organic Weedfree Rapid™, whose active ingredients are pine oil (150 g/L) and fatty acids (Kiwicare 2016). This product is available in two presentations: liquid ready to use and liquid concentrate. Other weed control products with pine oil as their active ingredient are available as commercial products, like Amgrow Weed Blitz Concentrate and Organic Herbicide of Go Natural.

## 14.9 Advantages and Limitations of Pine Oil as a Pesticide

Pine oil, as many other essential oils, is environmentally friendly for pest management because it is natural and its main compounds include volatiles with short half-lives. Pine oil components like α-pinene, β-pinene, camphene, and *trans*-pinane have a half-life in the air between 1.4 and 9.4 h (Regnault-Roger 2013); meanwhile, α-terpineol has a short half-life in air (4 h), 466 h in surface water, and 2 days on soil where there is degradable soil microbiota (Chemwatch 2009). On the other hand, pine oil is a selective pesticide, and thus effects on nontargeted species can be limited.

The main limitations of pine oil are its limited effect on insect pests when compared with other natural products. Approved uses by the USEPA are also limited. In addition, more studies are necessary on its use as a pest control product.

## 14.10  Essential Oil–Based Insecticides (Patents, Commercial Uses, and Approved Uses)

Due to pine oil's bactericidal, repellent, and insecticidal properties, several patents with specific formulations have been filed. Bactericidal claims for a detergent bar with improved stain removal are protected under U.S. Patent 5013486. This patent attributes a bactericidal effect to pine oil; 0.5% of pine oil reduced 90% of bacterial counts, and 1.5%–5% of pine oil achieved reductions of >99% (Joshi 1991). A liquid formula as a surface cleaner was proposed to have antibacterial activity when pine oil was combined with oil-soluble organic acid (benzilic acid, benzoic acid, salicyclic acid, sorbic acid, etc.); the antimicrobial activity was attributed to a synergistic effect of the pine oil and organic acids (Spaulding et al. 1989).

It is claimed that mixtures of pine oil–surfactant–water in ratios of 1:1:128 to 1:1:8 (surfactant can be anionic and nonionic) control the population of certain insects, especially fire ants (Eichhoefer 1990). "Pine oil fire ant insecticide formulations" claim to use pine oil surfactant and bait in solid or fluid form, from 1 part of pine oil and 99 parts of bait to about 99 parts of pine oil and 1 part of bait. Bait can be a solid, liquid, water, paste, syrup, mist, or gas, and can contain a retardant and/or odorant (Eichhoefer 1992).

A pH-modified insect repellent or insecticide soap composed of plant essential oils claims the use of pine oil as part of its formulation. Pine oil is used as one part of six, and five parts from other essential oils are equal parts of eucalyptus and citronella oils. Pine oil is claimed to function as a deodorizer, antibacterial, antiseptic, and disinfectant (Baube 2014).

As an herbicide, U.S. Patent 6759370 claims to use pine oil with an alcohol content of at least 60% by weight, combining with tall oil reacted with alkali to form fatty acid soap; its primary use is as an herbicide on spray application (foam or bubbles) (Innes 2004). Also, a knockdown herbicidal composition that includes pine oil and tea tree oil or eucalyptus oil was patented by Selga and Kiely (1999). The main commercial uses are as disinfectant or weed control product, as previously mentioned in Sections 14.6.3 and 14.8. Approved uses as a pesticide are determined by the USEPA, as pointed out in Section 14.7.6; for its application (outdoor and indoor spaces), refer to Section 14.7.9.

## 14.11  Final Remarks

The essential oil of turpentine is currently used in different products, such as pesticides, food, cleaning products, and paints. However, its composition, properties, and yield depend on many factors that have to be controlled to obtain a uniform product. It is also important to mention that there are few published reports related to the extraction, properties, and uses of turpentine oil.

There are few publications on pine oil properties as a pesticide; however, commercial products are available for organic production in Australia. Despite the little information reported, potential applications of pine oil as a pesticide are of great interest due to its effect on a wide variety of insects, bacteria, and molds. In addition, pine oil as a pesticide can be an alternative to synthetic pesticides in order to reduce environmental issues, although more studies with regard to pine oil effects are still necessary to suggest appropriate applications.

# References

Ansari, M.A., Mittal, P.K., Razdan, R.K., and Sreehari, U. 2005. Larvicidal and mosquito repellent activities of pine (*Pinus longifolia*, family: Pinaceae) oil. *J Vector Borne Dis* 42:95–99.

Ascenzi, J.M. 1996. *Handbook of Disinfectants and Antiseptics*. New York: Marcel Dekker.

Bakkali, F., Averbeck, S., Averbeck, D., and Idaomar, M. 2008. Biological effects of essential oils—A review. *Food Chem Toxicol* 46:446–475.

Barceloux, D.G. 2008. *Medical Toxicology of Natural Substances: Foods, Fungi, Medicinal Herbs, Plants and Venomous Animals*. Hoboken, NJ: John Wiley & Sons.

Baube, H. 2014. pH modified insect repellent/insecticide soap composed of plant essential oils. U.S. Patent US 8647684 B2.

Bidlack, J.E., and Jansky, S.H. 2011. Introduction to seed plants gymnosperms. In *Stern's Introductory Plant Biology*, 413–431. Boston: McGraw Hill Higher Education.

Biocoat Australia Pty Ltd. 2016. Commercial application manual Organic Interceptor™. https://www .yumpu.com/en/document/view/11350726/weed-application-manualqxp-organic-interceptor.

Can Başer, K.H., and Buchbauer, G. 2016. *Handbook of Essential Oils Science, Technology and Applications*. Boca Raton, FL: CRC Press.

Cao, Q.J., Yu, J., Ran, Y.L., and Chi, D.F. 2015. Effects of plant volatiles on electrophysiological and behavioral responses of *Cryptorrhynchus lapathi*. *Entomol Exp Appl* 156:105–116.

Chemwatch. 2009. Material safety data sheet, alpha terpineol sc-291877. http://datasheets.scbt.com /sc-291877.pdf.

Cheung, T.W., and Smialowicz, D. 2000a. Biphenyl based solvents in blooming type hard surface cleaners. U.S. Patent US 6100231 A.

Cheung, T.W., and Smialowicz, D. 2000b. Biphenyl based solvents in blooming type germicidal hard surface cleaners. U.S. Patent US 6075002 A.

Chupin, L., Motillo, C., Charrier-El Bouhtoury, F., Pizzi, A., and Charrier, B. 2013. Characterization of maritime pine (*Pinus pinaster*) bark tannins extracted under different conditions by spectroscopic methods, FTIR and HPLC. *Ind Crop Prod* 49:897–903.

Cole, E.C. et al. 2003. Investigation of antibiotic and antibacterial agent cross-resistance in target bacteria from homes of antibacterial product users and nonusers. *J Appl Microbiol* 95: 664–676.

Díaz, E., Cortiñas, J., Ordóñez, S., Vega, A., and Coca, J. 2004. Selectivity of several liquid phases for the separation of pine terpenes by gas chromatography. *Chromatographia* 60:573–578.

Dorow, P., Weiss, T., Felix, R., and Schmutzler, H. 1987. Effect of a secretolytic and a combination of pinene, limonene and cineole on mucociliary clearance in patients with chronic obstructive pulmonary disease. *Arzneimittelforsch* 37:1378–1381.

Dudek, W., Wittczak, T., Swierczynska-Machura, D., Walusiak-Skorupa, J., and Palczynski, C. 2009. Occupational asthma due to turpentine in art painter—Case report. *Int J Occup Med Environ Health* 22(3):293–295.

Eichhoefer, G.W. 1990. Pine oil ant insecticide. U.S. Patent 4891222.

Eichhoefer, G.W. 1992. Pine oil fire ant insecticide formulations. U.S. Patent 5118506.

Encyclopedia Britannica. 2016 turpentine. *Encyclopedia Britannica.* http://global.britannica.com /topic/turpentine.

FAOSTAT (Food and Agriculture Organization Corporate Statistical Database). 2015. Pesticide use in United States of America. http://faostat3.fao.org/browse/R/RP/E.

Felšöciová, S. et al. 2015. Antifungal activity of essential oils against selected terverticillate penicillia. *Ann Agric Environ Med* 22(1):38–42.

Fermont, H.A. 1981. Extraction of coniferous woods with fluid carbon dioxide and other supercritical fluids. U.S. Patent US 4308200 A.

Firooz, A. et al. 2010. Fragrance contact allergy in Iran. *J Eur Acad Dermatol Venereol* 24:1437–1441.

Fraise, A.P., Maillard, J.Y., and Sattar, S. 2012. *Russell, Hugo and Ayliffe's Principles and Practice of Disinfection, Preservation and Sterilization.* West Sussex, UK: John Wiley & Sons.

Francesconi, R., and Castellari, C. 2001. Densities, viscosities, refractive indices, and excess molar enthalpies of methyl tert-butyl ether + components of pine resins and essential oils at 298.15 K. *J Chem Eng* 46:1520–1525.

George, D.R. et al. 2010. Environmental interactions with the toxicity of plant essential oils to the poultry red mite *Dermanyssus gallinae. Med Vet Entomol* 24:1–8.

Giepen, M., Neto, F.S., and Köpke, U. 2014. Controlling weeds with natural phytotoxic substances (NPS) in direct seeded soybean. In *Proceedings of the 4th ISOFAR Scientific Conference, "Building Organic Bridges" at the Organic World Congress,* Istanbul, Turkey, 469–472.

Green Harvest Organic Gardening Supplies. 2016. Weed organic control. http://greenharvest.com .au/PestControlOrganic/WeedControlProducts.html.

Gryj, A.C., Pachecano, J.A., Connors, T.F., and Suriano, D.F. 1999. Stable liquid cleaners containing pine oil. U.S. Patent US 5962394 A.

Hercules Inc. 1974. Yarmor 302W pine oil for formulation of disinfectants. Bulletin OR-1058. Wilmington, DE: Hercules Inc.

Innes, R.M. 2004. Herbicides. U.S. Patent 6759370.

Jagodzińska, K., Feliczak-Guzik, A., and Nowak, I. 2011. Analytical methods for identification and determination of some cosmetics ingredients *Chemik* 65(2):91–93.

Joshi, D. 1991. Detergent bar with improved stain removing and antibacterial properties. U.S. Patent 5013486.

Kanat, M., and Alma, M.H. 2003. Insecticidal effects of essential oils from various plants against larvae of pine processionary moth (*Thaumetopoea pityocampa* Schiff) (Lepidoptera: Thaumetopoeidae). *Pest Manag Sci* 60(2):173–177.

Kaushik, D., Kumar, A., Kaushik, P., and Rana, A.C. 2012. Analgesic and anti-inflammatory activity of *Pinus roxburghii* Sarg. *Adv Pharm Sci* 2012:1–6.

Kelkar, V.M., Geils, B.W., Becker, D.R. Overby, S.T., and Neary, D.G. 2006. How to recover more value from small pine trees: Essential oils and resins. *Biomass Bioenerg* 30:316–320.

Kiwicare. 2016. Organic Weedfree Rapid. http://www.kiwicare.co.nz/help/product/?sid=organic -weedfree-rapid.

Lachance, S., and Grance, G. 2014. Repellent effectiveness of seven plant essential oils, sunflower oil and natural insecticides against horn flies on pastured dairy cows and heifers. *Med Vet Entomol* 28:193–200.

Leandro, L.F. et al. 2014. Antibacterial activity of *Pinus elliottii* and its major compound, dehydroabietic acid, against multidrug-resistant strains. *J Med Microbiol* 63:1649–1653.

Lear, J.T., Heagerty, A.H.M., Tan, B.B., Smith, A.G., and English, J.S.C. 1996. Transient re-emergence of oil of turpentine allergy in the pottery industry. *Contact Dermatitis* 35:169–172.

Leyva, M. et al. 2009. Utilización de aceites esenciales de pinaceas endémicas como una alternativa en el control del *Aedes aegypti. Rev Cubana Med Trop* 61(3):239–243.

Leyva, M., Marquetti, M.C.F., Tacoronte, G.J.E., Tiomno, T.O., and Montada, D.D. 2010. Efecto inhibidor del aceite de trementina sobre el desarrollo de larvas de *Aedes aegypti* (Diptera: Culicidae). *Rev Cubana Med Trop* 62(3):212–216.

Liazid, A. et al. 2010. Evaluation of various extraction techniques for obtaining bioactive extracts from pine seeds. *Food Bioprod Process* 88:247–252.

Lu, R.Z., and Kloeppel, A.A. 2000a. Blooming type disinfecting cleaning compositions. U.S. Patent US 6110295 A.

Lu, R.Z., and Kloeppel, A.A. 2000b. Blooming type cleaning compositions including a system of amphoteric and nonionic surfactants. U.S. Patent US 6066606 A.

Lu, R.Z., and Smialowicz, D.T. 1999. Blooming pine oil containing compositions. U.S. Patent US 5985819 A.

Lucia, A. et al. 2007. Larvicidal effect of *Eucalyptus grandis* essential oil and turpentine and their major components on *Aedes aegypti* larvae. *J Am Mosq Control Assoc* 23(3):299–303.

Macchioni, F. et al. 2002. Acaricidal activity of pine essential oils and their main components against *Tyrophagus putrescentiae*, a stored food mite. *J Agric Food Chem* 50:4586–4588.

Manske, S.D., and McPherson, M.S. 2002. Pine oil cleaning composition. U.S. Patent US 6465411 B2.

McLaren, D., Fridman, M., and Bonilla, J. 2012. Effects of pine oil, sugar and covers on germination of serrated tussock (*Nassella trichotoma*) and kangaroo grass (*Themeda triandra*) in a pot trial. In *Eighteenth Australasian Weeds Conference*, Victoria, Australia, 333–335.

Merchant, P., Chokshi, K.K., and Kowalik, R.M. 2000. Cleaning composition containing pine oil extenders. U.S. Patent US 6010998 A.

Mercier, B., Prost, J., and Prost, M. 2009. The essential oil of turpentine and its major volatile fraction (α- and β-pinenes): A review. *Int J Occup Med Environ Health* 22(4):331–342.

Missio, A.L. et al. 2016. The effect of oleoresin tapping on physical and chemical properties of *Pinus elliottii* wood. *Sci For* 43(107):721–732.

Motiejūnaitė, O., and Pečiulytė, D. 2004. Fungicidal properties of *Pinus sylvestris* L. for improvement of air quality. *Medicina (Kaunas)* 40(8):787–794.

Nagnan, P., and Clement, J.L. 1990. Terpenes from the maritime pine *Pinus pinaster*: Toxins for subterranean termites of the genus *Reticulitermes* (Isoptera: Rhinotermitidae)? *Biochem Syst Ecol* 18(1):13–16.

National Institute of Environmental Health Sciences. 2002. Toxicological summary for turpentine. Research Triangle Park, NC: National Institute of Environmental Health Sciences. https://ntp .niehs.nih.gov/ntp/htdocs/chem_background/exsumpdf/turpentine_508.pdf.

National Institute for Occupational Safety and Health. 2015. Turpentine. Atlanta, GA: Centers for Disease Control and Prevention. http://www.cdc.gov/niosh/ipcsneng/neng1063.html.

Ntiamoah, Y.A., Borden, J.H., and Pierce, H.D., Jr. 1996. Identity and bioactivity of oviposition deterrents in pine oil for the onion maggot, *Delia antique. Entomol Exp Appl* 79:219–226.

Oerke, E.C. 2006. Crop losses to pests. *J Agric Sci* 144(01):31–43.

Ortuño Sánchez, M.F. 2006. *Manual práctico de aceites esenciales, aromas y perfumes*. Murcia, Spain: Aiyana Ediciones.

Ozogul, Y., Kuley, E., Ucar, Y., and Ozogul, F. 2015. Antimicrobial impacts of essential oils on food borne-pathogens. *Recent Pat Food Nutr Agric* 7(1):53–61.

Pakdel, H., Sarron, S., and Roy, C. 2001. r-Terpineol from hydration of crude sulfate turpentine oil. *J Agric Food Chem* 49:4337–4341.

Panda, H. 2011. *Spirit Varnishes Technology Handbook with Testing and Analysis*. Delhi: Asia Pacific Business Press.

Pandey, A., Höfer, R., Taherzadeh, M., Nampoothiri, M., and Larroche, C. 2015. *Industrial Biorefineries and White Biotechnology*. Amsterdam: Elsevier.

Papajannopoulos, A.D., Song, Z.Q., Liang, Z.Q., and Spanos, J.A. 2001. GC-MS analysis of oleoresin of three Greek pine species. *Eur J Wood Wood Prod* 59:443–446.

Price, C.T.D., Singh, V.K., Jayaswal, R.K., Wilkinson, B.J., and Gustafson, J.E. 2002. Pine oil cleaner-resistant *Staphylococcus aureus*: Reduced susceptibility to vancomycin and oxacillin and involvement of SigB. *Appl Environ Microbiol* 68(11):5417–5421.

Regnault-Roger, C. 2013. Essential oils in insect control. In *Natural Products: Phytochemistry, Botany and Metabolism of Alkaloids, Phenolics and Terpenes*, ed. K.G. Ramawat and J.-M. Mérillon, 4087–4107. Berlin: Springer-Verlag.

Richter, A.F. 1997. Pine oil hard surface cleaning compositions. U.S. Patent US 5591708 A.

Richter, A.F. 1998. Pine oil hard surface cleaning compositions. U.S. Patent US 5728672 A.

Richter, A.F., and Taraschi, F.A. 1997. Germicidal pine oil cleaning compositions. U.S. Patent US 5629280 A.

Rodrigues-Corrêa, K.C.S., Lima, J.C., and Fett-Neto, A.G. 2012. Pine oleoresin: Tapping green chemicals, biofuels, food protection, and carbon sequestration from multipurpose tress. *Food Energy Secur* 1(2):81–93.

Rodrigues-Corrêa, K.C.S., Lima, J.C., and Fett-Neto, A.G. 2013. Oleoresins from pine: Production and industrial uses. In *Natural Products: Phytochemistry, Botany and Metabolism of Alkaloids, Phenolics and Terpenes*, ed. K.G. Ramawat and J.-M. Mérillon, 4037–4060. Berlin: Springer-Verlag.

Sanz Bascuñana, E. 2015. *Aromaterapia*. Barcelona: Editorial Hispano Europea.

Seabra, I.J., Dias, A.M.A., Braga, M.E.M., and Sousa, H.C. 2012. High pressure solvent extraction of maritime pine bark: Study of fractionation, solvent flow rate and solvent composition. *J Supercit Fluid* 62:135–148.

Selga, J., and Kiely, W.A. 1999. Herbicidal composition and method. U.S. Patent 5998335.

Spaulding, L., Rebarber, A., and Wiese, E. 1989. Broad spectrum antimicrobial system for hard surface cleaners. U.S. Patent 4867898.

Stanislowski, A.G., and England, B.J. 1991. Odor controlling animal litter with pine oil. U.S. Patent US 5016568 A.

Stanislowski, A.G., and England, B.J. 1993. Odor controlling animal litter with pine oil. U.S. Patent US 5189987 A.

Stanislowski, A.G., England, B.J., and Ratcliff, S.D. 1992a. Combined odor controlling animal litter. U.S. Patent US 5135743 A.

Stanislowski, A.G., England, B.J., and Ratcliff, S.D. 1992b. Combined odor controlling animal litter. U.S. Patent US 5183655 A.

Stojkovic, D. et al. 2008. Susceptibility of three clinical isolates of *Actinomodura madurae* to α-pinene, the bioactive agent of *Pinus pinaster* turpentine oil. *Arch Biol Sci Belgrade* 60(4):697–701.

Stolerman, I. 2010. *Encyclopedia of Psychopharmacology*. Berlin: Springer-Verlag.

Sukarno, A., Hardiyanto, E.B., Marsoem, S.N., and Na'iem, M. 2015. Oleoresin production, turpentine yield and components of *Pinus merkusii* from various Indonesian provenances. *J Trop Forest Sci* 27(1):136–141.

Tongnuanchan, P., and Benjakul, S. 2014. Essential oils: Extraction, bioactivities, and their uses for food preservation. *J Food Sci* 79(7):1231–1249.

Tóro, R.M. et al. 2003. Activity of the *Pinus elliottii* resin compounds against *Lernaea cyprinacea* in vitro. *Vet Parasitol* 118:143–149.

Treudler, R. et al. 2000. Increase in sensitization to oil of turpentine: Recent data from a multicenter study on 45,005 patients from the German-Austrian Information Network of Departments of Dermatology (IVDK). *Contact Dermatitis* 42:68–73.

Tullio, V. et al. 2006. Antifungal activity of essential oils against filamentous fungi determined by broth microdilution and vapour contact methods. *J Appl Microbiol* 102:1544–1550.

Tümen, I., and Reunanen, M. 2010. A comparative study on turpentine oils of oleoresins of *Pinus sylvestris* L. from three districts of Denizli. *Rec Nat Prod* 4:224–229.

Turek, C., and Stintzing, F.C. 2011. Application of high-performance liquid chromatography diode array detection and mass spectrometry to the analysis of characteristic compounds in various essential oils. *Anal Bioanal Chem* 400:3109–3123.

Ulukanli, Z. et al. 2014. Chemical composition, antimicrobial, insecticidal, phytotoxic and antioxidant activities of Mediterranean *Pinus brutia* and *Pinus pinea* resin essential oils. *Chin J Nat Med* 12(1):901–910.

Urfer, A.D., and Lazarowitz, V.L. 1994. Pine-oil containing hard surface cleaning composition. U.S. Patent US 5308531 A.

USEPA (U.S. Environmental Protection Agency). 2006. Re-registration eligibility decision for pine oil (case 3113). https://www3.epa.gov/pesticides/chem_search/reg_actions/reregistration /red_PC-067002_2-Oct-06.pdf.

Usha, R.P. 2012. Fumigant and contact toxic potential of essential oils from plant extracts against stored product pests. *J Biopest* 5(2):120–128.

WHO Pesticide Evaluation Scheme. 2011. Generic risk assessment model for insecticides used for larviciding rev. 1. WHO Library Cataloguing-in-Publication Data, WA 240. Available at: http://apps.who.int/iris/bitstream/10665/44675/1/9789241502184_eng.pdf?ua=1.

Wiyono, B., Tachibana, S., and Tinambunan, D. 2006a. Chemical compositions of the pine resin, rosin and turpentine oil from West Java. *J Forest Res* 3(1):7–17.

Wiyono, B., Tachibana, S., and Tinambunan, D. 2006b. Chemical composition of Indonesian *Pinus merkussi* turpentine oils, gum oleoresins and rosins from Sumatra and Java. *Pak J Biol Sci* 9(1):7–14.

# 15

# Orange Oil

Rosaria Ciriminna, Francesco Meneguzzo, and Mario Pagliaro

## CONTENTS

## 15.1 Introduction

Orange oil is an essential oil (EO) produced by cells within the rind of the orange fruit, which is extracted as a by-product of orange juice production by centrifugation due to numerous domestic, industrial, and medicinal uses [1]. Being nontoxic, nonirritating, and nonsensitizing, orange oil is granted the generally recognized as safe (GRAS) status in the United States by the Food and Drug Administration (GRAS 182.20). EO is mostly composed of (greater than 90%) d-limonene, a highly valued terpene [2]. Although present in minor amounts, other components are important to enhance, for example, the plant antifungal and antibacterial properties with respect to pure d-limonene, as first reported in 1993 by Singh and colleagues [3].

First registered as an insecticide in the United States in 1958 (and later as an antibacterial in 1971), d-limonene extracted from orange oil by distillation has been among the first natural pesticide ingredients used in environmentally friendly pest control. We note here that *plant pest* means any organism causing a detrimental effect on the plant's health and vigor, including fungi, bacteria, viruses, molds, insects, mites, and nematodes, but excluding mammals, fish, and birds.

Today, the great potential of *Citrus sinensis* essential oil in crop protection against insect pests is well established, although perhaps little known among farmers (the pesticide users) and even chemical research practitioners (the pesticide developers) [4].

Orange essential oil is furthermore highly effective as a contact insecticide against ants, roaches, palmetto bugs, fleas, silverfish, and many home and garden insects, including houseflies and their larvae and pupae [5]. Several insecticide products are available in the marketplace, along with the first crop biopesticides.

Owing to their low toxicity, nonpersistence in the environment, and good repellent, insecticidal, and growth-reducing activities on a variety of insects [6], in a trajectory common to several other functional products, driven by health and environmentally conscious users and consumers, pest and insect repellents, originally developed via synthetic organic chemistry in the early 1920s, have evolved to rediscover low-risk essential oils [7].

Several thorough studies on botanical pesticides have been published [8], including recent accounts that assess the current and future situation of biopesticides, especially focusing on their potential within the European pesticide legislative framework [9]. In 2008, Isman, whose research group in Canada has pioneered the field, argued that the greatest benefits from botanicals might be achieved in developing countries where human pesticide poisonings are prevalent [10], whereas their use in industrialized countries would be more restricted to pest control in and around homes and gardens, in food storage facilities, and on companion animals.

Reviewing recent research and industrial achievements, in this chapter we summarize the main features of orange oil as a biopesticide. We conclude by suggesting arguments for which the employment of this valued natural product as an environmentally friendly biopesticide will be a central tenet of the emerging bioeconomy not only in organic food production, but also in conventional agriculture, where it will replace a number of synthetic pesticides.

## 15.2 Composition, Toxicity, and Biological Activity

In general, the composition varies with the cultivated species of the *Citrus* genus, geographical origin, harvesting (and thus weather) conditions, and extraction method. For example, the limonene content in the same Valencia orange crop may range from 91.4% in vacuum-distilled concentrates [11] to 97.0% in cold-pressed concentrates [12]. From now on, we will refer to sweet orange oil, namely, the oil obtained from *C. sinensis* (sweet orange). Note that *Citrus* is a large genus beyond *C. sinensis* that originated from a backcross hybrid between pummelo and mandarin [13]; it includes several major cultivated species, such as *Citrus reticulata* (tangerine and mandarin), *Citrus limon* (lemon), *Citrus grandis* (pummelo), and *Citrus paradisi* (grapefruit).

Figure 15.1 displays the chemical structures of selected components of orange essential oil. Along with dominant d-limonene (>90%), the principal components of orange oil are monoterpenes ($\alpha$-pinene, sabinene, and $\beta$-myrcene), followed by oxygenated compounds such as alcohols (linalool and $\alpha$-terpineol) and aldehydes (geranial, neral, and citronellal), including long-chain aliphatic aldehydes like decanal and octanal [14].

The orange color is due to the presence of polymethoxyflavones, such as sinestetin, tangeretin, quercetogetin, nobiletin, and heptamethoxyflavone (Table 15.1), which, like other flavonoids, show strong antifungal activity [15], besides playing an important role as antioxidants and enzyme inhibitors.

Remarkably, from the applicative viewpoint, significant differences exist in the oil composition between orange oils obtained using biological cultivation and those obtained through traditional cultivation using organophosphorus and organochlorine pesticides [14]. In particular, the content of aliphatic aldehydes is much lower (0.22%) in orange oil obtained from traditionally grown fruits than that (0.79%) found in oil extracted from organically grown fruits. These aldehydes, especially decanal and octanal, characterize the olfactory notes of sweet orange oils. Furthermore, the amount of terpene aldehydes neral and geranial, the so-called "citral" substances that determine lemon oil olfactory peculiarity and the economic value of the oil in the perfume market, was considerably higher in oil from organically grown crops (2.90%) than in oil from conventionally grown fruits (2.41%).

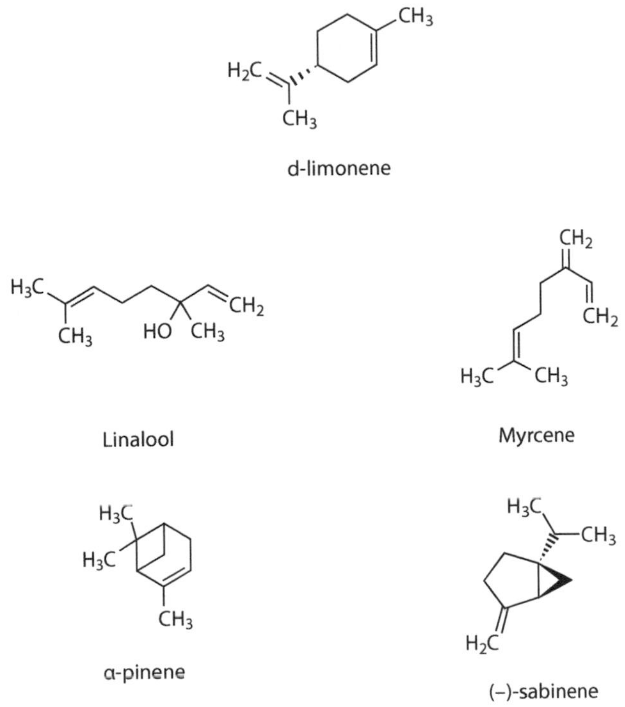

**FIGURE 15.1**
Chemical structures of the main components of orange oil.

**TABLE 15.1**

Polymethoxyflavone Content in Sweet Orange Oil Isolated in Spain

| Flavonoid | Concentration (g/L) |
|---|---|
| Sinensetin | 0.1 |
| Quercetogetin | 0.1 |
| Nobiletin | 1.0 |
| Tangeretin | 0.5 |
| Heptamethoxyflavone | 2.5 |

Source: Reproduced from Del Río, J. A. et al., *J. Agric. Food Chem.*, 46, 4423–4428, 1998. With permission.

Renewable d-limonene extracted from orange oil occurs naturally in food and is therefore classified by the U.S. Food and Drug Administration as a GRAS food additive to baked goods, ice cream products, gelatins, puddings, and chewing gum. With a few exceptions, its mammalian toxicity is low and its environmental persistence is very short. It is practically nontoxic to mammals [16] and birds, and slightly toxic to freshwater species, both fish and invertebrates.

The constituents of orange oil are quickly degraded in the open environment, thanks to rapid oxidation. In general, d-limonene has poor solubility in water (13.8 mg/L), and it is highly resistant to microbial biodegradation in water and soil due to its well-known antimicrobial activity. It rapidly volatilizes from both dry and moist soil to the atmosphere, where it rapidly undergoes gas phase oxidation reactions with photochemically produced hydroxyl radicals, ozone, and at night, nitrate radicals, with calculated half-lives for these processes on the order of <2 h. A similar fate is expected for the other components, indicating little or no persistence in the environment, and thus no bioaccumulation or biomagnification.

The aerial oxidative degradation compounds of d-limonene, and limonene 1,2-oxide in particular, are respiratory irritants [17], although having short lives, and may cause adverse skin reactions. Hence, during plant protection with orange oil, applicators will use proper equipment observing the labeling recommendations aimed at minimizing exposure from dermal contact or inhalation during delivery by spraying or fogging. For example, public authorities in Canada require that when applying a new orange oil–based insecticide formulation for commercial environments, applicators "wear long pants, a long-sleeved shirt, shoes plus socks, goggles or face shield, and chemical-resistant gloves [18]."

As mentioned above, the plant protection properties of EOs, and orange essential oil in particular, have long been known. It is perhaps not surprising that in 2009 researchers in Greece reported that orange oil is an excellent pesticide against neonates of the Mediterranean fruit fly, namely, the *Ceratitis capitata*, causing frequent infestation of citrus fruits plantations [19]. Administering larvae with diets containing orange oil caused rapid mortality, with $LC_{50}$ values ranging from 7 to 11 ml/g.

In 1991, Coats and colleagues were the first to ascribe to neurotoxicity the significant biological effects and symptoms observed—hyperactivity, loss of coordination, tremors, trembling, and paralysis of legs, followed by convulsions and death—when insects came into contact with monoterpenoids. "Symptoms of acute poisoning of insects by orange oil monoterpenoids are similar to those effected by some neurotoxic compounds [20]." The same team in 1988 had reported that d-limonene found in the essential oil of various citrus leaves and fruit peels exhibits significant insect control properties [21]. Today, we know that monoterpenoid compounds exert their activities through neurotoxic effects involving several mechanisms that affect multiple targets, thereby more effectively disrupting cellular activity and biological processes of insects [6].

In 2015, Kourimska and colleagues in Czechia completed a landmark study for future applications of orange oil to crop protection. The research team tested *C. sinensis* essential oil in wheat protection against *Oulema melanopus* L. (*Coleoptera: Chrysomelidae*) [4], a small cereal leaf beetle feeding on various grasses that can cause economically important damage, especially on wheat and barley.

Tested against *O. melanopus* larvae and adults under laboratory conditions, the oil showed remarkably high direct contact toxicity against the larvae, causing mortality of 85% during 48 h (Table 15.2). The larvae cause damage to the growing plant during the last stage of development. A high dissipation rate of the oil from treated plants (concentration lower than 0.01 g/kg 5 min after application of the oil) clearly points to its low persistency (Figure 15.2), and thus to its environmental and food safety.

Adding relevance to the use of orange oil as a biopesticide that goes well beyond use in homes and gardens, in 2013 researchers in Greece discovered that the oil is an excellent pesticide against the vine mealybug, a grape vine pest that is affecting an increasing number of grape vine–growing areas worldwide [22].

In detail, bioassays were conducted in the laboratory by spraying an aqueous solution of the EO obtained via hydrodistillation of orange or lemon peels (using 1% Tergitol as an emulsifier) directly on grape leaves bearing field representative clusters of *Planococcus ficus*.

The results clearly show the high insecticidal activity and the lack of any phytotoxic effect on grape vine by citrus orange and lemon oils. The $LC_{50}$ and $LC_{90}$ mortality data

**TABLE 15.2**

Mortality of *Oulema melanopus* Larvae and Adults after Apical
Application of *Citrus sinensis* Essential Oil

| Exposition Time (h) | Larvae (%) | Adults (%) |
|---|---|---|
| 1 | 12.5 | 0 |
| 24 | 42.5 | 0 |
| 48 | 85.0 | 0 |

*Source:* Reproduced from Zarubova, L. et al., *Acta Agric. Scand. Sect. B,*
65, 89–93, 2015. With permission.

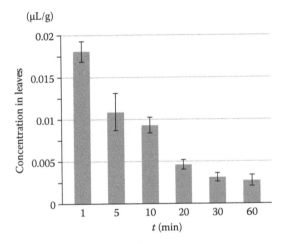

**FIGURE 15.2**
Concentration of *C. sinensis* essential oil after application on wheat leaves. Bars indicate standard error. (Image
courtesy of the Florida Chemical Company; Lenka Zarubova et al. *Acta Agric Scand B Plant Soil Sci,* 2015.)

values (lethal concentration where 50% and 90% of the population is destroyed, respectively) in the two pest life stages studied were significantly lower than the corresponding values of the reference paraffin oil, pointing to a higher toxic effect of the orange oil than that of the reference product.

Referring to the previous work of Hollingsworth, dating back to 2005 [23], the team emphasized the importance of a good emulsion when mixing the oil with water in the presence of the required surfactant in order to prevent the two stages of phase separation of the formulation, and thus variability of the EO insecticidal effect on the target pest and cause of phytotoxicity to the plant. In detail, Hollingsworth identified a 1% limonene, 0.75% APSA-80 (a nonionic surfactant functioning as a spray adjuvant), and 0.1% Silwet L-77 (a superspreading surfactant based on a trisiloxane ethoxylate surfactant) microemulsion as a suitable spray biopesticide providing superior control of mealybugs when sprayed on green scales on potted gardenia plants, averaging 95% mortality, in comparison with 89% and 88% mortality on plants sprayed with insecticidal soap or horticultural oil, respectively.

The relevance of the right choice of the surfactants in formulating orange oil as a biopesticide is demonstrated by the successful introduction of PREV-AM, the first orange oil–based biopesticide commercialized on a global scale.

In detail, PREV-AM is the trade name of the orange oil formulation in water with borax (sodium borate) and biodegradable surfactants, such as, originally, ethoxylated alcohols and alcohol ethoxy sulfate [24]. It is owned by Oro Agri, a South African company with production sites in Africa, Europe, and the United States. Manufactured in Spain and California in large and increasing amounts, PREV-AM Plus is a biopesticide containing 60 g/L orange oil, along with borax and a blend of surfactants, such as ethoxylated alcohols, alkyl glucoside polysaccharides, urea, and parabens, as preservatives [25]. Lately, it has been used as a multipurpose contact insecticide, fungicide, and acaricide in many of the world's countries for the eco-friendly treatment of several pest-mediated diseases, such as powdery mildew fungus (*oidium tuckery*) affecting grape crops [26]. The formulation provides highly effective contact activity against various insects, fungi, and mites, acting as an alternative to conventional pesticides.

In 2015, the product gained approval by Italy's Health Ministry for the treatment of a number of plant pest-induced diseases [27]. For example, it fights against the vector of the *Xylella fastidiosa* bacterium that is severely hitting olive orchards in Italy's Puglia. Formulated in the PREV-AM composition, orange oil has shown particularly good efficacy in killing the *Philaneus spurmarius* insect that acts as a vector of the plant pathogenic bacterium that lives inside the xylem vessels of the host plant, where it forms a biofilm that disrupts the passage of water and nutrients within the plant. Applied at an optimal rate of 8 L/ha, the biopesticide desiccates the cuticles of the insects that are carriers of *Xylella bacteria*, leading to suffocation and death.

Assessed by acute toxicity tests carried out in the open field, the toxicity of PREV-AM to predatory mites on strawberry (tested in 2013 in California) and on grape (tested in 2009 in Czechia) and to honeybees (tested in 2013 in Germany) is low [28]. Indeed, the formulation is approved for organic farming, even though acute toxicity assessment alone cannot fully predict the actual impact of biopesticides on nontarget parasitoids, requiring full consideration of sublethal effects and longer-term population dynamics in order to assess pesticide risks [29].

Pointing to the usefulness of orange oil as a postharvest freshness-saving agent, researchers in Korea recently showed that orange oil is also suitable as a new fumigant for the control of two grain storage insects, adults of the maize weevil (*Sitophilus zeamais*) and the red flour beetle (*Tribolium castaneum*), against which it shows strong fumigant and contact activity [30]. Except for linalool, which shows exceptionally high toxicity, the toxicity of the oil in contact and fumigant toxicity tests (Table 15.3) was higher than that of its

**TABLE 15.3**

Fumigant Toxicity of Basil Oil, Orange Oil, and Their Components against *Sitophilus zeamais* and *Tribolium castaneum* Adults, 24 h

| EO/Terpene | *S. zeamais* $LC_{50}$ (mg/ml) | *T. castaneum* $LC_{50}$ (mg/ml) |
|---|---|---|
| Orange oil | 0.106 | 0.130 |
| Limonene | 0.122 | 0.171 |
| Linalool | 0.016 | 0.023 |
| β-Myrcene | 0.274 | 0.275 |
| α-Pinene | 0.264 | 0.273 |

*Source:* Reproduced from Ministero della Salute, Decreto Autorizzazione in deroga per situazioni di emergenza fitosanitaria, ai sensi dell'art, 53, paragrafo 1, del regolamento (CE) n. 1107/2009, del prodotto fitosanitario denominato PREV-AM PLUS reg. n. 16379, contenente la sostanza attiva olio essenziale di arancio dolce, May 13, 2015. See http://www.salute.gov.it/imgs/C_17_pagineAree_1110_listaFile_itemName _35_file.pdf.

constituents (α-pinene, 0.54%; sabinene, 0.38%; β-myrcene, 1.98%; d-limonene, 96.5%; and linalool, 0.6%). Suggesting a respiratory mode of action, the toxicity was mainly exerted by fumigant action via the vapor phase.

Orange oil is highly effective in killing and repelling entire ant and termite colonies, preventing reinfestation. Indeed, orange oil showed both contact and fumigant toxicity, dissolving and dispersing the pheromone trail left behind by ants so that new ants cannot follow it back to the original nesting spot. For example, dispersing orange oil (~92% d-limonene) at concentrations as low as 5 ppm (v/v) in the vapor phase causes practically complete (96%) mortality to the Formosan subterranean termite (*Isoptera: Rhinotermitidae*) within 5 days [31]. Furthermore, termites did not tunnel any longer through glass tubes fitted with sand treated with 0.2%–0.4% orange oil.

## 15.3 Open Challenges

Several home insecticides based on water-based formulated orange oil are available in the marketplace, especially in North America, for instance, *Orange Guard*, *Ortho Home Defense Indoor Insect Killer*, *Concern Citrus Home Pest Control*, *Safer Fire Ant Killer*, and *Citrex Fire Ant Killer*. Yet, to the best of our knowledge, only one pesticide to protect plants against weeds, pests, and diseases has been commercialized.

Hence, it is natural to ask, why are there so few products commercially available in 2017, when the sustainability issue has emerged as a central global problem?

One might think that the procedure for regulatory approval of plant protection products is too expensive. Yet, the United States reduced and released the regulation process for GRAS biopesticides, establishing a simplified procedure, as early as 1996. In only 5 years, this single policy act led to a large diversity of EO-based insecticide products available to users in the U.S. marketplace [6].

In Europe, in 2009 the European Food Safety Authority published its review of orange oil risk assessment [17], which was used in 2013 by the European Commission to approve the PREV-AM biopesticide containing orange oil for control of sweet potato whitefly, *Bemisia tabaci*, on field pumpkin (*Cucurbita pepo*) and for control of greenhouse whitefly on tomato. The same PREV-AM nonpersistent formulation provides insecticide, fungicide, and acaricide activity on a wide range of crops, and thanks to this company, orange oil is now present in the European approved pesticide substance list. Reduced and flexible regulation for essential oil products that not only do not have a history of adverse effects, but also show significant health beneficial effects, is thus in place.

In our viewpoint, there are three reasons for which orange oil pesticides are not yet massively marketed. First, strong competition exists from readily available synthetic pesticides in a global market historically controlled by large chemical companies [32]. It is revealing, in this respect, to learn that recently the world's largest chemical company applied and got registration in Canada for its orange oil–based insecticides, *MotherEarth Botanical Crawling Insect Killer* and *ProCitra-DL Botanical Crawling Insect Killer*, to control "cockroaches, spiders, crickets, millipedes, centipedes, flour beetles, cluster flies, ticks, fleas, bed bugs and Asian lady beetles on contact" in commercial and domestic environments, respectively [18]. Both products contain 10% d-limonene (technical grade), and when applied to indoor commercial facilities, they are required not to be in operation.

Second, the orange oil supply is limited and the price is increasing. This is especially true for food-grade orange oil, as obviously a biopesticide manufacturing company will not buy orange oil contaminated with synthetic pesticides.

Even though orange oil is among the cheapest essential oils, at around $10 per kilogram, its price rose from around $3 per kilogram in 2010 to around $8 per kilogram in less than a year (due to the Deep Water Horizon oil spill in the United States), and then to $15 per kilogram in 2014 due to a drop in supplies after a drought in Brazil, the main producer of orange oil. Furthermore, driven by a steady rise in the demand for "naturals," the demand for orange oil as an alternative cleaning solvent, as an aromatic flavor in food and beverages, and as a relaxing fragrance in personal care, aromatherapy, cosmetic, and perfume applications [33] is causing almost weekly price increases [34].

A manufacturer of an orange oil–based biopesticide would then ensure safe supply of the oil, including the possibility to self-extract the oil from the waste orange peel, as producers of rosemary EO as a novel food antioxidant have recently done [35]. The latter established new extraction plants in Morocco and other countries where large plantations of rosemary exist, and started to target a market dominated by manufacturers of synthetic phenolic antioxidants since the 1930s, namely, from the same years in which synthetic pesticides started to be massively used in agriculture [36].

The third reason is technical and lies in the highly volatile nature of orange oil whose components quickly evaporate. Orange oil terpenes, indeed, are not stable and are easily oxidized and deteriorated when exposed to high temperature, oxygen, and humidity [37], namely, the conditions typical of open fields. Twenty years ago, Kim and Morr were among the first to encapsulate orange oil in gum arabic microcapsules using a spray-drying technique [38]. Several other conventional encapsulation techniques have been employed, including advanced sol-gel technology for the entrapment of d-limonene in silica microcapsules [39]. However, most efforts were aimed at developing either new food or cosmetic functional ingredients.

Much room for improvement remains for the development of solid pesticides based on stable microencapsulated orange oil subject to optimized controlled release in the field environment. Indeed, whereas the open literature provides few examples of microencapsulated orange oil for pest control, several patents claim excellent pesticide properties of the microencapsulated oil. For example, applied research has shown that the efficacy of these encapsulated volatile oils increases if a nonvolatile agent is used to carry the encapsulated volatile EO [40]. Upon application to the plant or fruit, the EO and the nonvolatile agent remain in contact, the evaporation rate is reduced, and a synergestic effect enhancing the exerted protective action is observed. The Israeli company *Botanocap*, for instance, develops pesticides for crop protection with microencapsulation and slow release of essential oils based on synergistic technology [41].

## 15.4 Outlook and Conclusions

Looking at the future of EO-based botanicals, in 2012, Vincent and colleagues in Canada noted that "clearly, if the public wishes to have access to botanical products for home, garden, organic agriculture, or greenhouse use, then governments must be lobbied to consider reduced registration processes more seriously [6]." "Furthermore," the team insisted, "open field evaluation was needed as most research has focused on lab and greenhouse experiments."

Five years later, many things have changed. Looking at insecticide applications, it is instructive to review the feedback posted online by hundreds of customers commenting on the performance and price of one commercial orange oil–based insecticide on the website of one of the world's largest online stores [42]. Out of 339 reviews, 270 were positive and only 69 critical. The product, one customer complained, was "too expensive," being priced at $29.97 for almost 1 L. Yet, the oil not only killed ants in homes and gardens, but also effectively repelled dogs and cats, leaving a pleasant, delicate smell of oranges. "I feel much much safer this way, and my wife loves how the yard smells when I'm killing fire ants," wrote one customer.

Pointing to the practical and commercial relevance of citrus oil insecticides, in 2014 BASF entered the world's richest market, North America, with two products aimed at domestic and commercial built environments.

Even more significantly from an economic and environmental perspective is the fact that the first multipurpose biopesticide for a variety of crops entered the market, replacing several different products for pest control with a single product (PREV-AM) showing little toxicity, high efficacy, and an excellent environmental fate.

In order to make a difference, however, orange oil–based biopesticides will have to penetrate the pesticide market for crop pest control in China, Brazil, and India, namely, the world's leading agriculture countries, which, incidentally, are among the top five leading countries in the production of orange fruits.

In brief, orange essential oil will be used not only to formulate plant protection products for organic farming or for integrated pest management, but also to replace synthetic pesticide products for protecting a wide number of crops and plants in both economically developed and developing countries.

Relying on successful registration processes in Europe and North America, new and long-established chemical companies will start to manufacture large-volume products for large-scale commercial agriculture, differentiating their new environmentally friendly products via the formulation chemistry, some of which will also use nanochemistry to develop microencapsulated orange oil.

The broad spectrum of action of orange oil will also allow postharvest application of these new biopesticides, to enhance the storage of produce and valued crops, by preventing pest-induced deterioration. This, *inter alia*, will contribute to ending what Clark has rightly called a significant example of waste of our societies [43], as most waste orange peel obtained from orange juice production ends up in landfills, without extracting the valued essential oil, pectin, and hemicellulose comprising the orange peel. Providing a critical review of the scientific, economic, and environmental state of affairs, this study will hopefully accelerate this transition.

## Acknowledgments

This study is dedicated to Gino Fazio, AMG Energia (Palermo), for all he has done for one of us (MP) during his presidency. Thanks to Dr. Sabino Lorusso (Nufarm Italia) for valued information about the PREV-AM Plus formulation and its actual performance.

# References

1. K. Bauer, D. Garbe, H. Surburg, *Common Fragrance and Flavor Materials*, Wiley VCH, Weinheim, 2001.
2. R. Ciriminna, M. Lomelli, P. Demma Carà, J. Lopez-Sanchez, M. Pagliaro, Limonene: A versatile chemical of the bioeconomy, *Chem. Commun.* 2014, *50*, 15288–15296.
3. G. Singh, R. K. Upadhyay, C. S. Narayanan, K. P. Padmkumari, G. P. Rao, Chemical and fungitoxic investigations on the essential oil of *Citrus sinensis* (L.) Pers., *J. Plant Dis. Protect.* 1993, *100*, 69–74.
4. L. Zarubova, L. Kourimska, M. Zouhar, P. Novy, O. Douda, J. Skuhrovec, Botanical pesticides and their human health safety on the example of *Citrus sinensis* essential oil and *Oulema melanopus* under laboratory conditions, *Acta Agric. Scand. Sect. B* 2015, *65*, 89–93.
5. P. Kumar, S. Mishra, A. Malik, S. Satya, Insecticidal evaluation of essential oils of *Citrus sinensis* L. (Myrtales: Myrtaceae) against housefly, *Musca domestica* L. (Diptera: Muscidae), *Parasitol. Res.* 2012, *110*, 1929–1236.
6. C. Regnault-Roger, C. Vincent, J. Thor Arnason, Essential oils in insect control: Low-risk products in a high-stakes world, *Annu. Rev. Entomol.* 2012, *57*, 405–424.
7. C. Peterson, J. Coats, Insect repellents—Past, present and future, *Pestic. Outlook* 2001, *12*, 154–158.
8. Y. Akhtar, M. B. Isman, Plant natural products for pest management: The magic of mixtures, in *Advanced Technologies for Managing Insect Pests*, ed. I. Ishaaya, S. R. Palli, A. R. Horowitz, Springer Science + Business Media, Dordrecht, 2012, chap. 11.
9. J. J. Villaverde, P. Sandín-España, B. Sevilla-Morán, C. López-Goti, J. Luis Alonso-Prados, Biopesticides from natural products: Current development, legislative framework, and future trends, *Bioresources* 2016, *11*, 5618–5640.
10. M. B. Isman, Botanical insecticides: For richer, for poorer, *Pest Manag. Sci.* 2008, *64*, 8–11.
11. J. Pino, M. Sánchez, R. Sánchez, E. Roncal, Chemical composition of orange oil concentrates, *Nahrung* 1992, *36*, 539–542.
12. R. L. Colman, E. D. Lund, M. G. Moshonas, Composition of orange essence oil, *J. Food Sci.* 1969, *34*, 610–611.
13. Q. Xu et al. The draft genome of sweet orange (*Citrus sinensis*), *Nat. Genet.* 2013, *45*, 59–66.
14. A. Verzera, A. Trozzi, G. Dugo, G. Di Bella, A. Cotroneo, Biological lemon and sweet orange essential oil composition, *Flavour Fragr. J.* 2004, *19*, 544–548.
15. J. A. Del Río, M. C. Arcas, O. Benavente-García, A. Ortuño, Citrus polymethoxylated flavones can confer resistance against *Phytophthora citrophthora*, *Penicillium digitatum*, and *Geotrichum species*, *J. Agric. Food Chem.* 1998, *46*, 4423–4428.
16. U.S. Environmental Protection Agency, *Reregistration eligibility decision (RED): Limonene*, EPA 738-R-94-034, Office of Prevention, Pesticides, and Toxic Substances, Washington, DC, 1994.
17. OJEU (*Official Journal of the European Union*), Commission decision of 8 June 2009 recognising in principle the completeness of the dossier submitted for detailed examination in view of the possible inclusion of orange oil in Annex I to Council Directive 91/414/EEC (notified under document number C(2009) 4232), *Off. J. Eur. Union* 2009, 52, L145/47.
18. Health Canada, *Registration decision RD2015-23, d-Limonene*, Health Canada Ottawa, December 4, 2015.
19. D. P. Papachristos, A. C. Kimbaris, N. T. Papadopoulos, M. G. Polissiou, Toxicity of citrus essential oils against *Ceratitis capitata* (Diptera: Tephritidae) larvae, *Ann. Appl. Biol.* 2009, *155*, 381–389.
20. J. R. Coats, L. L. Karr, C. D. Drewes, Toxicity and neurotoxic effects of monoterpenoids: In insects and earthworms, *ACS Symp. Ser.* 1991, *449*, 305–316.
21. L. L. Karr, J. R. Coats, Insecticidal properties of d-limonene, *J. Pestic. Sci.* 1988, *13*, 2287–2290.
22. F. Karamaouna, A. Kimbaris, A. Michaelakis, D. Papachristos, M. Polissiou, P. Papatsakona, E. Tsora, Insecticidal activity of plant essential oils against the vine mealybug, *Planococcus ficus*, *J. Insect. Sci.* 2013, *13*, 142.

23. R. G. Hollingsworth, Limonene, a citrus extract, for control of mealybugs and scale insects, *J. Econ. Entomol.* 2005, *98*, 772–799.

24. E. M. Pullen, Citrus oil compositions and methods of use, WO 2008097553 A2, 2008.

25. E. M. Pullen, D. C. Uys, Methods of reducing phytotoxicity of a pesticide, EP 2420141 A1, 2012.

26. O. Gamberini, S. Lorusso, Olio essenziale di arancio dolce (PREV-AM PLUS): Insetticida/fungicida consentito anche in agricoltura biologica, *Prodotti fitosanitari, Le novità 2016*, Regione Emilia Romagna, Bologna, March 1, 2016.

27. Ministero della Salute, Decreto Autorizzazione in deroga per situazioni di emergenza fitosanitaria, ai sensi dell'art, 53, paragrafo 1, del regolamento (CE) n. 1107/2009, del prodotto fitosanitario denominato PREV-AM PLUS reg. n. 16379, contenente la sostanza attiva olio essenziale di arancio dolce, May 13, 2015. See http://www.salute.gov.it/imgs/C_17_pagineAree_1110_listaFile _itemName_35_file.pdf (accessed August 26, 2016).

28. Oro Agri, Company guide, January 2015.

29. A. Biondi, L. Zappalà, J. D. Stark, N. Desneux, Do biopesticides affect the demographic traits of a parasitoid wasp and its biocontrol services through sublethal effects? *PLoS One* 2013, *8(9)*, e76548.

30. S.-I. Kim, D.-W. Lee, Toxicity of basil and orange essential oils and their components against two coleopteran stored products insect pests, *J. Asia Pac. Entomol.* 2014, *17*, 13–17.

31. A. K. Raina, J. Bland, M. Dollittle, A. Lax, R. Boopathy, M. Lolkins, Effect of orange oil extract on the Formosan subterranean termite (*Isoptera: Rhinotermitidae*), *J. Econ. Entomol.* 2007, *100*, 880–885.

32. V. Pelaez, L. Rodrigues da Silva, E. Borges Araújo, Regulation of pesticides: A comparative analysis, *Sci. Public Policy* 2013, *40*, 644–656.

33. Grand View Research, *Essential oil market analysis by product (orange, corn mint, eucalyptus, citronella, peppermint, lemon, clove leaf, lime, spearmint), by application (medical, food & beverage, spa & relaxation, cleaning & home) and segment forecasts to 2024*, Grand View Research, Boulder, CO, August 2016.

34. N. Murray, Orange oil prices increasing every week, agra-net.com, December 10, 2015.

35. N. Baldwin, Inside rosemary's approval, *World of Food Ingredients*, April/May 2011, pp. 40–41.

36. R. Ciriminna, F. Meneguzzo, R. Delisi, M. Pagliaro, Olive biophenols as new antioxidant additives in food and beverage, *Chemistry Select* 2017, *2(4)*, 1360–1365.

37. S. Ananaram, G. A. Reineccius, Stability of encapsulated orange peel oil, *Food Technol.* 1986, *40*, 88–93.

38. Y. D. Kim, C. V. Morr, Microencapsulation properties of gum arabic and several food proteins: Spray-dried orange oil emulsion particles, *J. Agric. Food Chem.* 1996, *44*, 1314–1320.

39. R. Ciriminna, M. Pagliaro, Sol-gel microencapsulation of fragrances and flavors: Opening the route to sustainable odorants and aromas, *Chem. Soc. Rev.* 2013, *42*, 9243–9250.

40. Formulations containing microencapsulated essential oils, U.S. Patent US 9101143 B2.

41. A. Markus, C. Linder, Advances in the technology of controlled release pesticide formulations, in *Microencapsulation: Methods and Industrial Applications*, ed. S. Benita, Informa Healthcare, Zug, Switzerland, 2005, pp. 55–77.

42. Customer reviews, Nature's Wisdom Orange Oil Concentrate, amazon.com (accessed August 23, 2016).

43. R. Luque, J. H. Clark, Valorisation of food residues: Waste to wealth using green chemical technologies, *Sustain. Chem. Process.* 2013, *1*, 10.

# 16

# Essential Oil of *Cinnamomum cassia* for Pest Control

**Abhishek Niranjan, Alok Lehri, and S.K. Tewari**

## CONTENTS

*Cinnamomum cassia*, called Chinese cassia or Chinese cinnamon, is an evergreen tree originating in southern China, and is widely grown there and other parts of southern and eastern Asia (India, Indonesia, Taiwan, Thailand Laos, Vietnam, and Malaysia). Among several species of *Cinnamomum*, the aromatic bark of *C. cassia* is commercially used as a spice. In the United States, Chinese cassia is the most common type of cinnamon used. The buds are also most commonly used as a spice, especially in India.

## 16.1 Origin and History

The species is one of the oldest spices of the world. The ancient Egyptians used the spice for embalming. The Dutch had a monopoly on the cinnamon trade until 1776. The tree is indigenous to Myanmer, Indo-China, Japan, and China. The principal producer of cassia oil is China, and there it is an age-old cottage industry. *C. cassia* is found mostly in southern China, Vietnam, and Laos and Myanmar (Burma), and is commercially cultivated in Vietnam and China. In China, the main production areas are in the Kwangsi and Kwangtung provinces in south China. The area under cultivation is approximately 35,000 ha. It gives an annual production of approximately 28,000 tons of cassia bark. The United Arab Emirates is the major buyer of cassia bark and cassia leaf oil. *C. cassia* was also described under *Cinnamomum aromaticum* Nees. The correct name for Chinese cassia is *C. cassia* (Linn.) Bercht. & Presl.

## 16.2 Habit and Habitat

Common names of cassia (*Cinnamomum cassia*)

> Chinese: Kuei, rou gui pi, kweior kui
> Arabic: Darseen, kerfee, salikha
> English: Chinese cassia, bastard cinnamon, Chinese cinnamon
> Dutch: Kassie, bastaardkaneel, valsekaneel
> Finnish: Talouskaneli, kassia FrenchCasse, canefice, canelle de Chine
> Estonian: Hiinakaneelipuu
> German: Chinesisches zimt, kassie
> Icelandic: Kassia
> Hungarian: Kasszia, fahejkasszia, kinaifahej
> Italian: Cassia, cannelladella Cina
> Japanese: Kashia keihi, bokei
> Russian: Korichnojederevo
> Laotian: Sa chouang
> Norwegian: Kassia

Spanish: Casia, canela de la China
Thai: Ob choey
Swedish: Kassia
Urdu: Taj

It is a tree. It grows up to 10–15 m tall, with grayish color and hard bark, elongated leaves that are 10–15 cm long, and reddish color at the young stage. *C. cassia* Presl (family: Lauraceae) evidently originates in southern China, but it is now widely cultivated in tropical or subtropical areas, for example, Indonesia, India, Laos, Thailand, Malaysia, and Vietnam. It has so far been unsuccessful outside China, but considering its importance, it is now being tried in different parts of India. The tree is native to China and cultivated in southeastern China for extraction of essential oil. It is also grown in Afghanistan and northern China. All parts of the tree are fragrant, and it yields the cassia oil of commerce. The plant grows well on the slopes of hills. These are propagated from seeds or shoot cuttings. After the trees attain a height of 2–3 m in 5–6 years, the trunk is cut close to the ground. This encourages new shoots to spring up. When the twigs attain 4 cm in diameter, they are cut into pieces. The bark is separated from the wood and the cut pieces, from which the bitter outer layer, that is, epidermis is scraped off, and inner bark is collected and dried in the sun. This inner bark is highly valued and is used in expensive Chinese medicine. In India, this evergreen aromatic tree grows up to 18 m in height and 1.5 m in width, and is reported to grow wild in isolated places in Mizoram and is also cultivated to a limited extent in Tamil Nadu. The bark is gray, smooth, and thick. The leaves are oblong elliptic, dark shiny green, 15 × 7.5 cm with three prominent nerves from the base. There are small pale yellow flowers in lax, spreading, terminal, and auxiliary panicles. The fruits are black pulpy, aromatic, elliptic drupes with a single seed in the cup of the calyx lobe. Indian information on this plant is very scarce due to its highly restricted distribution. However, in China the plant is extensively cultivated and considerable international trade exists in this species (Asolkar et al. 1994; Supriya 2005).

## 16.3 Botanical Description and Morphology

Botanical name: *Cinnamomum cassia*
Order: Laurales
Family: Lauraceae
Genus: *Cinnamomum*
Species: *cassia* (Nees & T. Nees)

It is an evergreen tree of 10–15 m or more in height with smooth bark, small pale yellow flowers, and ellipsoidal sappy fruit about the size of a pea. The tree is long cylindrical and much branched, 30–75 cm long, with the thick end being 0.3–1 cm in diameter. It is physically reddish-brown, along with longitudinal ridges, fine wrinkles, dotted leaf scars, branch scars, bud scars, and dotted lenticels. Is has a hard texture and is easily breakable. Slices are 2–4 mm thick, with a cut surface of bark showing reddish-brown wood of yellowish-white to pale yellowish-brown subsquare pith. Its odor is characteristically aromatic, and its taste is sweet and slightly pungent, relatively strong for bark. Morphologically leaves are three-nerved, oval oblong and narrow at each end, up to 15 cm long. Flowers are borne in silky panicles, pale yellow in color, and small. The tree bark is pale and smooth with corky patches. It is a small tree, the bark of which forms the cassia bark or Chinese cassia or Chinese cinnamon. Morphologically leaves are opposite, glabrous upper, minutely hairy lower, with microscopic hairs, oblong lanceolate, three-ribbed from nearly 5 mm above the base; side veins are ascending to the apex, exstipulate. The length–breadth value is 4.26. The inflorescence of cassia is an axillary panicle (panicled cyme), exceeding the leaves; it is multiple flowered, with a long peduncle, and flowers have a long pedicel and minute hairs. The fruit is an ovoid one-seeded berry, seated in an enlarged perianth cup having truncate perianth lobes. It flowers from October to December.

### 16.3.1 Cell Inclusions in the Leaf Tissue

Leaf tissues of *C. cassia* and other species contain cellular inclusions (oil globules, mucilage, and calcium oxalate crystals). Tannic substances are found in the epidermal and mesophyll cells, and appear as deep brown or black deposits. Yellow or golden-yellow deposits of mucilage occur in the mesophyll cells. Such deposits are abundant in *C. cassia*. Lysigenous cavities containing oil occur within the mesophyll. Oil globules are found extensively distributed in the leaf and petiolar cells. In *C. cassia*, mucilage and tannin cells are present in the ground tissue. The shape of the vascular bundle is quite characteristic. Both ends of the vascular bundle are curved toward the center (incurving), thereby presenting a wide arc-shaped appearance. Sclereids lie scattered around the vascular strand. Xylem cell rows are surrounded by a thick continuous zone of sclerenchyma at the abaxial side.

### 16.3.2 Bark Anatomy

Cinnamon bark is of great economic importance, but very little work has been done regarding its structure and development. Santos (1930) conducted earlier studies on bark anatomy, while Chaudhuri and Kayal (1971), Bamber and Summerville (1979), Shylaja (1984), and Shylaja and Manilal (1992) have carried out more recent studies. These studies have shown differences in the bark structure in cinnamon, especially with regard to the distribution of bast fibers, sclerified tissue, stone cells, and secretory cells. Shylaja (1984) and Shylaja and Manilal (1992) reported on the anatomy of the fresh bark of *C. cassia*. Externally, the bark is gray in color. Bark tissues are characterized by secretion cells, containing mucilage or essential oil droplets, but the frequency and distribution of such cells vary in the four species. Bark is characterized by islands of sclerenchyma in the pericyclic region, which are connected by a continuous band of stone cells in three of the species (*C. cassia*). This band is a demarcating zone between the extrapericyclic region (comprising phellem and phelloderm) and the secondary phloem. Phloem ray cells of both young and old barks are filled with a dark, golden, or brown substance in *C. cassia*. Islands of sclerenchyma in pericyclic regions are characteristic features of *Cinnamomum* barks. The distribution of bast fibers is another characteristic feature of the barks. In *C. cassia*, such fibers are rare and sparsely distributed. The cortical region of young bark and the phellem and regions of the phelloderm of older bark contain a large number of cells with brown tannic deposits. Such cells are frequent in *C. cassia*. Oil globules are abundant in the outer bark tissues of *C. cassia*. Crystalline inclusions, mostly in the form of raphides, are distributed sparsely in *C. cassia*. Raphides are needle or spindle shaped, occurring alone or in groups in the phloem tissues.

### 16.3.3 Bark Regeneration

Cassia bark is obtained from standing trees (in some parts of China and Vietnam); therefore, fast bark regeneration is very important. Major efforts should be made to improve the bark regeneration after bark removal. Hong and Yetong (1997) have used the application of a bark regeneration fluid, followed by covering with plastic film, which resulted in 80% regeneration of bark. That bark is thicker, with higher oil and cinnamaldehyde contents. The composition of the regeneration fluid is not known.

### 16.3.4 Floral Morphology, Biology, and Breeding Behavior

Flowers of cinnamon are produced in lax terminal or axillary panicles and are nearly equal to or, to some extent, longer than the leaves. The flowers are abundant with long green-whitish peduncles. The flowers are softly hairy, actinomorphic, bracteate, bisexual, perigynous, and trimerous. Six free perianth in two whorls of three each, 9 + 3 stamens in four whorls of three each; the outer two whorls are introrse and glandless, the third whorl is extrorse and flanked by two prominent glands, and the fourth whorl is represented by glandless, sagittate, stalked staminodes. Valvular dehiscence is shown by fertile stamens. The filaments of stamens and staminodes are embedded with minute hairs. The ovary is superior, unilocular, with a single pendulous anatropous ovule. Its long style ends in a stigma that belongs to a dry, papillate class (Heslop-Harrison and Shivanna 1977). Flowering starts by November and lasts until the early part of March. Normally, flower development takes 14–15 days from the stage of visible initiation. The male and female phases are temporally separated (protogynous dichogamy), thereby ensuring outcrossing. Due to protogynous dichogamy, maturing of the female phase occurs before that of the male phase. Both phases are separated by almost a day, during which time the flower closes completely. Every flower opens twice in two stages, as reported by Joseph (1981). In stage 1 on the first day, after the flower opens, its stigma is whitish and fresh and appears to be receptive. There is an absence of anther dehiscence, and the stamens of the first and third whorls are fused. After nearly 5 hours, the flower closes. Stage 2 starts the next day, when the flower opens again. The stigma appears shriveled and not receptive. The stamens of the third whorl are now separated from those of the first and now adhere to the pistil. After the opening of flower, the anthers dehisce 30–60 minutes. After approximately 5 hours, the flower again closes and does not open again.

### 16.3.5 Cassia Buds

The cassia bud consists of dried immature fruits that vary from approximately 6 to 14 mm in length and about 4 to 6 mm in the width of the cup. The immature fruit is present in the calyx cup. Dried cassia buds have a sweet, warm, pungent taste, similar to that of cassia bark.

### 16.3.6 Calyx Tube

The calyx tube consists of the following parts:

- An epidermis of minute cells, rectangular or nearly so, with thick outer walls.
- A cortex consisting of parenchyma cells, isodiametric, about 45 μ in diameter; secretion cells, nearly 66 μ in diameter; and occasional stone cells. The cell materials are light to dark brown in color.

- A pericycle distinguished by a ring of fibers and ring that are broken by the parenchymatous tissue of pith rays. The fibers are thick walled and up to about 66 μ in diameter.
- Fibrovascular tissue similar to that of the calyx lobes.
- A pith of isodiametric, parenchymatous cells and occasional stone cells.

### 16.3.7 Pericarp

The pericarp is made up of the following tissues: (1) a sclerenchymatous epicarp, the cells of which have thick outer walls and thin inner walls, and (2) a mesocarp made of stone cells, secretion cells, parenchyma cells, and vascular tissue. Stone cells differ in size and shape and lie beneath the epicarp. The major portion of the mesocarp is occupied by parenchyma cells. Cell contents differ in color from light to dark brown.

## 16.4 Cultivation

Commercially, it is known as *Cassia lignea* in the Chinese and European markets. All the plant parts are aromatic, although only its bark and leaves are used commercially. The main chemical constituent of bark and leaf oil is cinnamaldehyde. During cultivation, the trees are coppiced from time to time to keep the height to 3–4 m. Trees of 15–20 years old contain a central bole of 15–20 m. As branches are produced at an acute angle, usually the tree acquires a conical shape. The bark is extensively used as a spice, as well as an important Chinese medicine. For medicinal use, bark from wildly grown matured trees is preferred. For use as a spice, the internal bark from older trees, as well as bark from young shoots, is used, which is scraped or peeled out and dried. The inner bark possesses 1.5%–4% essential oil, in which 98% cinnamaldehyde is present. The bark obtained from higher altitudes, known as *kwangsi* bark, is of better quality, with higher oil content than that of plants grown at lower altitudes (*kwantung* bark). The immature fruits are often dried and sold in the market as cassia buds. In southern China, the plant is grown on terraced hillsides at altitudes of 90–300 m, under a moderately cool climate with an annual rainfall of 1285 mm. It is usually propagated by cutting. The plant sometimes grows from seeds in which the seedlings are raised in nursery beds and transplanted at the age of 1 or 2 years old.

## 16.5 Crop Improvement

### 16.5.1 Genetic Resources and Varieties

China developed 35 lines of Chinese cassia (*C. cassia*) from the open pollinated progenies of cassia trees introduced during the early 1950s.

### 16.5.2 Improvement of Chinese cassia

The crop improvement program in Chinese cassia mainly focuses on (1) clone selection of high quality (high essential oil content having a high percentage of cinnamaldehyde), (2) maximum bark recovery, and (3) maximum growth and regeneration capacity. With these aims, cassia germplasms were maintained and evaluated for morphological and quality parameters at the Indian Institute of Spices Research (IISR), Calicut, India (Krishnamoorthy et al. 1999, 2001). Bark oil in various lines varied from 1.2% to 4.9%, bark oleoresin from 6.0% to 10.5%, and leaf oil from 0.4% to 1.6%. The cinnamaldehyde content of leaf oil varied from 40% to 86%, and that of bark oil from 61% to 91%. The coefficient of variation was found to be high for bark and leaf oil (26.76% and 26.74%, respectively). Based on the quality and other characteristics, four promising lines were identified. Among them, two lines have high bark oleoresin (10.2% and 10.5%, respectively) and two others have high bark oil (4.7% and 4.9%, respectively) and high cinnamaldehyde in bark oil (91.0% and 90.5%, respectively). Little crop improvement work has gone into the Chinese cassia in China or Vietnam, where the production is located. This is true of the Indonesian cassia as well. In all these countries, cassia cinnamon is treated, managed, and exploited as a forest tree crop.

## 16.6 Climate

It grows in temperate climates and prefers terraced hillsides at altitudes of about 90–300 m. Those growing at high latitudes yield better-quality bark.

## 16.7 Propagation

Cassia is generally propagated by cuttings that are planted between February and April. For propagation, seeds are sown in nursery beds. In Chinese plantations, cassia is grown as a coppiced bush.

## 16.8 Micropropagation of Cinnamon

Micropropagation of *C. cassia* was reported by Huang et al. (1998) and Nirmal Babu et al. (1997). They induced callus from petiole explants cultured in Linsmaier and Skoog (LS) agar medium, which contains numerous combinations of growth regulators (1965). Callus tissues were further grown in LS medium containing indole-3-acetic acid (IAA) and 6-benzyl adenine napthalene acetic acid (NAA) and benzyladenine (BA). Callus was further subcultured with 0.2% gellan gum at 25°C in the dark. High-performance liquid chromatography (HPLC) analysis of the callus tissues developed found them to contain procyanin B2, epicatechin, procyanidin B4, and procyanidin C1, which are precursors of condensed tannins. The vanillin–HCl test showed the presence of a large amount of high-molecular-weight condensed tannins in the callus cultures. Yazaki and Okuda (1993) reported that condensed tannins are produced in tissue cultures of *C. cassia*.

## 16.9 Harvesting Techniques

The leaf and bark are harvested. When the plants attain an age of 5–7 years and height of 1.5–2.7 m, they are cut down close to the ground level and a number of new shoots come up. The bark of harvested branches (nearly 4 cm thick) is removed in lengths of 30–60 cm for leaf oil extraction, and small twigs with leaves are harvested.

## 16.10 Grading and Processing

The outer bark (epidermal layer) is somewhat bitter in taste and is scrapped off, and the inner bark is dried in the sun until it turns brown and curls into a hollow tube or quill. Leaf oil has a composition similar to that of the bark generally; cassia leaf is distilled for oil, and cassia bark is ground and used as a spice.

## 16.11 Various Products of *C. cassia*

Three products from the Chinese cassia are bark; oil from the twigs, leaves, and bark; and cassia buds, which are the dried, unripe fruits. The yield of oil is 0.31%, but as much as 0.77% has also been reported. Chinese cassia bark is sweet and aromatic, resembling Ceylon cinnamon in flavor, but is rather less delicate and sometimes slightly astringent. It is less uniformly thin and darker in color. The bark, in both whole and ground form, is used as a flavoring agent, especially for processed foods. The ground form is used in the flavoring of bakery products, sauces, pickles, puddings, curry powder, some beverages, and confectionery, and is a preferred spice for chocolate manufacture. It is also used as a disinfectant.

## 16.12 Extraction of Oil

The oil is extracted from leaves and twigs, which yield from 0.30% to 0.70%, depending on the season, age, and quality of the material. Crude oil obtained by distillation contains resin, kerosene, and traces of lead. This crude oil is rectified before use. The odor and flavor of rectified oil are powerful and characteristic for cassia bark, which is yellow to slightly brownish in color, with the smell and taste of cinnamon. The chemical constituents present in the oil are cinnamic aldehyde, cinnamylacetae, phenylpropyl acetate, salicylaldehyde, salicylic acid, benzaldehyde, cinnamic acid, methoxy benzaldehyde, and coumarin. The major chemical compound in the oil is cinnamaldehyde, which varies from 80% to 95%. The oil is an important flavoring agent of different types of beverages and food products, like confectionery, baked goods, cakes, beverages, desserts, table sauces, soft drinks, pickles, candies, chewing gums, and meat. The oil is also employed in the preparation of cosmetics and perfumes and the scenting of soaps. The oil is mixed with olive oil to rub into the skin and massage into the scalp. Shylaja (1984) and Ravindran et al. (1992) have analyzed flavonoids, terpenoids, and steroids and found much variation between *C. cassia* and other species of *Cinnamomum*. They are chemically very distinct among themselves and from other species. The authors also carried out a centroid clustering analysis of the flavonoid–triterpenoid data, which shows the independent clustering of species.

In *Cinnamomum*, the complexity in the flavonoid pattern is the result of o-methylation, which is considered an advanced characteristic in flavonoid evolution. It was shown that flavones have replaced the simpler flavonols, which, again, is an advanced characteristic in the evolutionary history of flavonoids (Crawford 1978). Fujita (1967) attempted a subspecific classification of several species based on the composition of volatile oil. Such infraspecific variability might have evolved as a result of interbreeding, and forces like mutations, segregation, and isolation mechanisms acting on the populations.

## 16.13 Chemical Constituents

The bark contains lyoniresional-3-o-B-D-glucopyranoside; syrigaresinol; cinnamic aldehyde; cyclic glycerol 1,3-acetal and its cis-isomer, 5,7,3-trimethyl(–)epicatechin; 5,7-dimethyl-3,4-di-o-methylene(+)epicatechin; 3-o-8-C and 6C-B-D glucopyranosides; cinnamtannins A1, A2, A3, and A4; procyanidins A02B-1, B-2, B-5, B-7 and C-1 and B-D; and glucopyranosides of procyanidins B28 and B26C. Cinnamtannin A1 and procyanidin B5 show antiviral activity (Anonymous 1992). The bark yields essential oil (0.3% on a fresh basis) containing cinnamaldehyde (82.2%) and eugenol (1.5%). Other constituents present in the oil are cinnamic acid, cinnamyl acetate, benzaldehyde, methyl salicylate, hydrocinnamaldehyde, α-pinene, 1,8-cineol, linalool, α-terpineol, and guiacol. According to the Indian Pharmacopoeia, the bark should contain volatile oil, not less than 1.0% ash, not more than 5.0% (acid insoluble, 2.0%), and foreign organic matter, not more than 1.0%. The bark oil resembles cassia oil from leaves and twigs, but it contains less o-methoxycinnammaldehyde than cassia oil. It is sometimes used in soaps, perfumes, spices, essence, and beverages (Anonymous 1992). The leaves and twigs on steam distillation yield a pale yellow or brown essential oil commercially known as cassia oil, which becomes darker and thicker during storage. The leaves alone give a lower yield than a mixture of leaves and twigs. All parts of the plant (stem, bark, and leaves) contain an essential oil, having cinnamaldehyde as the major constituent. The leaves from northeast India yield an essential oil (0.4% on a fresh basis) containing cinnamaldehyde (71.3%) and eugenol (10.1%) as major constituents. However, the leaf oil from Nigeria contains benzyl benzoate (90%) and cinnamaldehyde (4.0%) as major constituents. According to the Indian Pharmacopoeia, the oil should have the following characteristics: specific gravity at 25°C, 1.045–1.063 and refractive index

($\eta_D$) 1.6020–1.6135, and 7.5% total aldehyde. The presence of α- and β-pinene, 1,8-cineol, hexanol, hex-3-en-1-guiacol, fenchone, hydroxycinnamaldehyde, and several other phenolics is also reported in the leaf oil. The oil of Chinese cassia is one of the important flavoring agents and is used widely in all kinds of food products and beverages, such as meat, table sauce, cakes, baked goods, confectionery, desserts, candies, chewing gums, pickles, and soft drinks. The oil's flavor is similar to but harsher than that of celon cinnamon bark oil, for which it is employed as a substitute to a limited extent. Oil of cassia is also used in perfumes, and to a restricted extent in the scenting of soaps. Although most of the plant yields oil containing up to 90% *trans*-isomers of cinnamaldehyde, some varieties yield oil containing the cis-isomer; this oil has a very spicy, pungent, and lasting fragrance and may prove more useful for cola-type drinks. The demand for oil, however, has been affected by synthetic cinnamaldehyde. It is also used as a stimulant in amenorrhea and as an antiseptic and germicide (Anonymous 1992). The dried cassia buds (unripe fruits) possess a faint cinnamon-like odor and warm, sweet, pungent flavor. They are used as a spice, much like cloves, and are added to sweet pickles and other dishes. The flowers and unripe fruits yield 1.5%–1.9% and 0.5%–2.0%, respectively, of an essential oil containing 80%–90% cinnamaldehyde as its major constituent. The oil is used as an aromatic spice. The fruits are eaten by birds, which disperse the seeds.

## 16.14 Physicochemical Properties of *Cinnamomum cassia* Oil

The crude oil is thick and dark in color. The rectified one has a yellow to slightly brownish color (Table 16.1).

## 16.15 Pharmacological Activities

During the Vedic period in India, herbal drugs as traditional medicines were used for the treatment of various diseases. Hence, use of herbal medicines, in spite of the great advances observed in modern medicine in recent decades, makes an important contribution to health care. Cinnamon is an ancient spice used in many countries. It consists of the

**TABLE 16.1**

Physicochemical Properties of Bark Oil and Leaf Essential Oil of *Cinnamomum cassia*

|                        | Bark Oil                                          | Leaf Oil                       |
| ---------------------- | ------------------------------------------------- | ------------------------------ |
| Specific gravity       | 1.010–1.030 at (20°)<br>1.046–1.059 at (25°/25°)  | 1.030–1.050 at (20°)           |
| Optical rotation (25°C) | −0.40 to +0.30                                   | 0°36′–0°40′                    |
| Refractive index (20°) | 1.6045–1.6135<br>1.529–1.537                       | 1.529–1.537 at (20°)           |
| Aldehyde content       | 88%–99%                                            | 4%                             |
| Eugenol content        | 77.3%–90.5%                                        | 4%–10%                         |

dried inner bark of *C. cassia*. The twigs and bark of this species are widely consumed in Asia as a spice and normally used as traditional Chinese medicine for treating gastritis, dyspepsia, blood circulation disturbances, and inflammatory diseases. Chinese Cassia has been used in Chinese traditional medicine as a diaphoretic, antipyretic, and analgesic. It is used as a purgative, carminative, and astringent. It is also useful as a cardiac stimulant, refrigerant, and diuretic; in resolving the viscosity of phlegm of the lungs in asthma and mucus from the throat and trachea; as an intensifier of eyesight; in neuralgia; for the stimulation of liver functions; in uterine pains, ascites, and edema; and as a powerful antidote for poison. The bark is one of the constituents of a Unani preparation (*jawarish jalinoos*) used in the treatment of diseases of the stomach, intestines, and urinary bladder. Its constant use is reported to prevent premature graying of hair. The bark is also one of the constituents of a Chinese drug (*saiko keishi to*) that has been used as an antiepileptic and sedative agent and in other neurological diseases requiring a tranquilizing effect. The bark markedly reduces blood pressure in experimental rats (Anonymous 1992). Aqueous extract of the bark shows significant antiallergic activity in guinea pigs. The extract also shows potent antiulcerogenic activity in rats, and it is comparable in activity to cimetidine. A number of diterpenes (i.e., cinnzeylanin; anhydrocinnzeylanin; cinnzeylanol; anhydrocinnzeylanol; cinncassiols A, B, C, D-1, D-4, and E; and glycosides of cinncassiols ABD-1 and D-4) have been isolated from the antiallergic fraction of the aqueous extract. Three antiulcerogenic compounds, cassioside ($C_{20}H_{32}O_9$), cinnamoside ($C_{24}H_{38}O_{12}$), and 3,4,5-trimethoxypenol B-D apiofuranosyl–(1–6)-B-D glucopyranside, have been isolated from the bark. The aqueous extract prevents an increase of protein level in urine when administered orally to rats with nephritis. A solution of the bark shows local anesthetic action on nerve fibers (Anonymous 1992). Various pharmacological activities, like antioxidant, antimicrobial, antibacterial, antifungal, anticancer (colorectal cancer), antidiabetic, antiobesity, and antileukemia activities, in cinnamon species have been reported (Meena Vangalapati et al. 2012; Varsha 2012). It has also been reported for activities against Parkinson's and Alzheimer's diseases (Syed et al. 2015).

### 16.15.1 Antioxidant Activity

Cheng-Hong et al. (2012) investigated the antioxidant activities of various parts (barks, buds, and leaves) of *C. cassia*. The plant parts were extracted with ethanol and supercritical fluid extraction (SFE). For the antioxidant activity comparison, $IC_{50}$ values of extracts from SFE and ethanol in the DPPH scavenging assay were 0.562–10.090 and 0.072–0.208 mg/ml; the Trolox equivalent antioxidant capacity (TEAC) values varied from 6.789 to 58.335 mmole Trolox/g and 133.039 to 335.779 mmoleTrolox/g, respectively. The amount of total flavonoid contents ranged from 0.031 to 1.916 g/100 g dry weight (DW) of materials and 2.030 to 3.348 g/100 g DW, and the total phenolic contents varied from 0.151 to 2.018 g/100 g DW and 6.313 to 9.534 g/100 g DW in SFE and ethanol extracts, respectively. Based on the results, the ethanol extracts of *Cinnamomum* bark shows potential value as an antioxidant substitute, and this study also provides a better technique to extract the natural antioxidant substances from *C. cassia*.

### 16.15.2 Antityrosinase and Antimelanogenic Activities

Chou et al. (2013) reported that essential oil of *C. cassia* and its major constituent, cinnamaldehyde, could effectively inhibit melanin production in B16 melanoma cells. Their observations indicated that essential oil and cinnamaldehyde inhibited α-melanocyte-stimulated hormone (MSH)-induced melanogenesis via tyrosinase inactivation, and oxidative stress was simultaneously suppressed in B16 melanoma cells. Both essential oil and cinnamaldehyde are generally recognized as safe (GRAS), although cinnamaldehyde is a potent skin sensitizer. Therefore, essential oil and cinnamaldehyde are potentially employed as effective skin-whitening agents and as antioxidants for future development under complementary and alternative medicine–based aromatherapy.

### 16.15.3 Antidiabetic Activity

Hoehn and Stockert (2012) reported that 18 type II diabetics (9 women and 9 men) participated in a 12-week trial that consisted of two parts, a 3-week control phase, followed by a 9-week experimental phase where half of the subjects received 1000 mg of *C. cassia*, while the other half were given 1000 mg of a placebo pill. All the subjects that were in the group of cinnamon showed a decrease in their blood sugar levels. The authors found it statistically significant. In the cinnamon group, the subjects had an average overall decrease in their blood sugar levels of about 30 mg/dl, which is comparable to that of oral medications available for diabetes. All subjects concerned had appropriate diabetic diets and maintained that diet for the entire 12-week study.

## 16.16 Pesticidal and Insecticidal Activity

Essential oils are volatile-natural complex secondary metabolites having a strong odor with a lower density than water. Approximately 3000 essential oils are known, out of which 300 have been commercialized for pesticidal potential and in the cosmetics, pharmaceuticals, and perfume industries (Nur et al. 2014). Several authors have studied the chemical composition of the essential oil of *C. cassia*. The essential oil was used against several stored product insects and house dust mites by applying contact and fumigant toxicity. Moreover, the essential oil of *C. cassia* exhibits larvicidal activity and strong repellency against several mosquitoes. Xin et al. (2014) reported that the essential oil displayed contact toxicity against adult *Lioscelisbos trychophila*, with a median lethal concentration ($LC_{50}$) of 55.68 µg/cm$^2$, as well as fumigant toxicity ($LC_{50}$, 1.33 mg/L air). *Trans*-cinnamaldehyde exhibited strong contact and fumigant toxicity, having $LC_{50}$ values of 43.40 µg/cm$^2$ and 1.29 mg/L air, respectively. The booklice (*L. trychophila* Badonnel; Psocoptera: Liposcelidae) are frequently found in stored product grains and in great numbers in amylaceous products. Infestations of stored product insects can be controlled by insecticidal treatment of commodities and surfaces or fumigation, which has led to problems such as increasing costs of pesticide application, disturbances of the environment, pest resurgence, pest resistance to pesticides, lethal effects on nontargeted organisms, and direct toxicity to the users. These problems increased the need to develop novel selective insect control alternatives. Several researchers have investigated and confirmed that a variety of essential oils not only repel insects, but also contain activity of contact and fumigant toxicity against stored product pests, as well as exhibit feeding inhibition or harmful effects on the reproductive system of insects. Essential oils or their constituents may provide an alternative substitute for currently used fumigants and insecticides to control stored food insects. Nur et al. (2014) identified a total of 35 components, accounting for 97.44% of the essential oil of *C. cassia*. The principal compounds in the essential oil analyzed by gas chromatography–mass spectrometry

(GC-MS) were *trans*-cinnamaldehyde (49.33%), acetophenone (6.94%), *trans*–cinnamic acid (5.45%), and ciscinnamaldehyde (4.44%), followed by o-methoxycinnamaldehyde (3.48%), coumarin (3.42%), and (E)–cinnamyl alcohol (3.21%). Nur et al. (2014) reported repellency and insecticidal activity of cinnamon oil obtained by the hydrodistillation method. It was directly exposed to ants. Different concentrations of fabricated repellency paper were used to test the repellency and insecticidal activity of cinnamon essential oil. The authors concluded that extraction through solvent is an effective method of cinnamon essential oil extraction with positive repellent and insecticidal activity on ants. In both the repellency and insecticidal activity of ants, cinnamon essential oil shows a positive result, repelling and killing ants at particular concentrations. The highest concentration of essential oil gives the highest mortality and repellency percentage. The oil is effective and environmentally safe for ant control. *Repellent* is defined by Fried et al. (2007) as material that causes any insect to turn away. Repellents have been used to prevent insects or, specifically, ants from harming or irritating human routine life, as they always swarm over food, particularly food that has been left on tables or uncovered. Ant repellent is used to prevent the ants from invading houses and spoiling foods. Chemicals are effective but equally harmful, causing different diseases in the population, children, and pets. One example, N,N-diethyl-3-methylbenzamide (DEET), which has been used to either repel or kill ants, is a carcinogenic chemical. Therefore, cinnamon oil was reported to be an alternative natural repellent to replace the carcinogenic chemical repellent. Youssif and Shaal (2011) reported mosquitocidal activity of volatile oils of *C. cassia* against *Aedes caspius* mosquitoes.

### 16.16.1 Antilarval Activity

Gomaa et al. (2014) reported activity against *Dermestes maculates*. It is considered to be among the pests that caused the most damage to Egyptian mummies. Petroleum ether, hexane, acetone, chloroform, and ethanol extracts from *C. cassia* were analyzed for their insecticidal activities against *D. maculatus* larvae obtained from Egyptian mummies. *C. cassia* was extracted by five separate solvents. Among all extracts, a chloroform extract of cassia was the most effective against *D. maculatus* larvae. The series of intensity of activity starts from the chloroform extract and is followed by petroleum ether, hexane, acetone, and ethanol. Chloroform extract at any concentration results in complete mortality after a period that does not exceed 5 days, and 8, 10, 13, and 16 days with petroleum ether, hexane, acetone, and ethanol, respectively.

### 16.16.2 Nematicidal Activity

The nematicidal activity of two cassia (*C. cassia*) oils (especial and true) and their compounds (e.g., *trans*-cinnamaldehyde and *trans*–cinnamic acid) was examined by a direct contact bioassay by Jeong et al. (2007) toward adult *Bursaphelenchus xylophilus*. The results were compared with those of 34 related compounds. After 24 hours, the $LC_{50}$ values of two cassia oils (0.084–0.085 mg/ml) and four cinnamon oils (0.064–0.113 mg/ml) were found to be toxic toward adult *B. xylophilus*. Out of 45 tested chemical compounds, *trans*-cinnamaldehyde (0.061 mg/ml) was found to have maximum active nematicide, followed by ethyl cinnamate, methyl-*trans*-cinnamaldehyde, methyl cinnamate, and allyl cinnamate (0.114–0.195 mg/ml). Potent nematicidal activity was also shown by 4-methoxycinnamonitrile, *trans*-4-methoxycinnamaldehyde, *trans*-2-methoxycinnamaldehyde, ethyl-cyanocinnamate, cinnamonitrile, and cinnamyl bromide (range 0.224–0.502 mg/ml). Structure–activity relationships show a correlation of types of functional groups, saturation, and carbon skeleton that plays an exclusive role in determining the toxicities to adult *B. xylophilus*. Cassia and cinnamon oils and test compounds were found to be potential nematicides or leads for the control of pine wilt disease caused by *B. xylophilus*.

### 16.16.3 Antignawing Activity

Hee-Kwon et al. (1999) reported antignawing activity of cinnamom cortex (the dried bark of *C. cassia*)–derived materials against laboratory-reared mice. Cinnamaldehyde from the bark exhibited potent repellent activity.

### 16.16.4 Antimicrobial Activity

Syed et al. (2015) reported that the methanol extracts of *C. cassia* showed minimum inhibitory concentration (MIC) values of 13.3 mg/ml when treated against *Fusarium moniliforme* and *Phyllosticta caricae*. The acetone extracts of *C. cassia* had MIC values of 8.3 and 10 mg/ml, respectively, when tested against *Botrytis cinerea* and *Glomerella cingulata*. Growth of *Aspergillus niger*, *B. cinerea*, *F. moniliforme*, and *P. caricae* was inhibited significantly by the hot water extracts of *C. cassia*, with MIC values of 10.0, 11.7, 5.0, and 6.7 mg/ml, respectively. Nguyen et al. (2009) reported that methanol extracts of cinnamon and their substrate fractions are useful as a fungicide against *Rhizoctonia solani*. This author stated that cinnamon (*C. cassia*) extract at 250, 500, and 750 mg/ml concentrations showed antibacterial activity against *Enterobacter aeruginosa*, *Escherichia coli*, *Staphylococcus aureus*, and *Staphylococcus epidermidis*, and ether, chloroform, acetone, and ethanol extracts from *Cinnamomum* against *Bacillus subtilis*. Nazia et al. (2006) studied the antibacterial activity of aqueous infusion, decoction, and essential oil of *C. cassia* (cinnamon bark) using a standard disk diffusion method. This author analyzed it against 178 bacterial strains belonging to 12 different genera of bacterial populations isolated from the oral cavity of 250 specimens of healthy individuals between 2 and 85 years of age. Generally, all types of tested bacterial strains were inhibited by oil of *C. cassia*, except *Salmonella* Paratyphi B, which exhibits a 99.4% antibacterial effect compared with an aqueous decoction (70.2%) and infusion (52.2%). The aqueous decoction of *C. cassia* results in high antibacterial activity against *Streptococcus oralis* and *Streptococcus sanguis* (23 mm), *Micrococcus roseus* (21 mm), *Streptococcus intermedius* (20 mm), and *Streptococcus mutans* (17 mm), whereas *Pseudomonas aeruginosa*, *Salmonella* Paratyphi A, *Salmonella* Paratyphi B, *Klebsiella ozaenae*, *Citrobacter* sp., *E. aerogenes*, *S. aureus*, *Plesiomonas shigelloides*, and *Alcaligens* sp. were not inhibited. Biavati et al. (1997) studied the anitmicrobial effects and found a good inhibitory effect against gram-positive and gram-negative microbial strains by different plant extracts and their essential oils. The aqueous infusion of *C. cassia* inhibits microbial strains of *M. roseus*, *Streptococcus anginosus*, *S. sanguis*, *S. oralis*, *S. mutans*, *Streptococcus salivarius*, *Klebsiella pneumoniae*, *S. intermedius*, *Streptococcus morbillorum*, *Flavobacterium* sp., and *Streptococcus uberis*.

## 16.17 Conclusions

*C. cassia* is a good source of essential oil, having pesticidal, insecticidal, larvicidal, and mosquitocidal activity. These activities are attributed to the presence of cinnamaldehyde as a principal active compound. Apart from the above activities, the parts of the tree are also used in various pharmacological activities. These biological activities make this plant economically important. Very little information is available worldwide, opening a new source of research and development in various areas.

# References

Anonymous. 1992. *The Wealth of India*. Vol. III. Council of Scientific and Industrial Research, New Delhi.

Asolkar, L.V., Kakkar, K.K., and Chakre, O.J. 1994. Glossary of Indian Medicinal Plants with Active Principles Part—I (A–K), Publication and Information Directorate (Council of Scientific and Industrial Research), New Delhi.

Bamber, R.K., and Summerville, R. 1979. Taxonomic significance of sclerified tissue in the bark of Lauraceae. *IAWE Bull.* 4: 69–75.

Biavati, B., Franzoni, S., and Ghazvinizadeh, H. 1997. Antimicrobial and antioxidant properties of plant essential oils. Essential oils. Basic and Applied Research. Proceedings of the 27th International Symposium on Essential Oils, Vienna, Austria, September 8–13. 326–331.

Chaudhuri, R.H.N., and Kayal, R.N. 1971. Pharmacognostic studies on the stem barks of four species of *Cinnamomum*. *Bull. Bot. Surv. India* 13: 94–104.

Cheng-Hong, Y., Rong-Xian, L., and Li-Yeh, C. 2012. Antioxidant activity of various parts of *Cinnamomum cassia* extracted with different extraction methods. *Molecules* 17: 7294–7304.

Chou, S.-T., Chang, W.-L, Chang, C.-T., Hsu, S.-L., Lin, Y.-C., and Shih, Y. 2013. *Cinnamomum cassia* essential oil inhibits α-MSH-induced melanin production and oxidative stress in murine B16 melanoma cells. *Int. J. Mol. Sci.* 14: 19186–19201.

Crawford, D.J. 1978. Flavanoid chemistry and angiosperm evolution. *Bot. Rev.* 44: 431–456.

Fried, H.L., Khazan, D., and Morales, M.N. 2007. U.S. Patent No. 7, 201, 926. Washington, DC: U.S. Patent and Trademark Office. Frienkel, R.K., and Traczyk, T.N. The phospholipases A of epidermis. *J. Invest. Dermatol.* 74:169–173.

Fujita, Y. 1967. Classification and phylogeny of the genus *Cinnamomum* in relation to the constituent essential oils. *Bot. Mag. Tokyo* 80: 261–271.

Gomaa, A.M., Abdel-Rahman, E., and Fathy, A. 2014. Insecticidal activity of *Cinnamomum cassia* extractions against the common Egyptian mummies insect pest (*Dermestes maculatus*). *Int. J. Conserv. Sci.* 5: 355–368.

Hee-Kwon, L., Hoi-Seon, L., and Young-Joon, A. 1999. Antignawing factor derived from *Cinnamomum cassia* bark against mice. *J. Chem. Ecol.* 25: 1131–1139.

Heslop-Harrison, Y., and Shivanna, K.R. 1977. The receptive surface of the angiosperm stigma. *Ann. Bot.* 41: 1233–1258.

Hoehn, A.N., and Stockert, A.L. 2012. The effects of *Cinnamomum cassia* on blood glucose values are greater than those of dietary changes alone. *Nutr. Metab. Insights* 5: 77–83.

Hong, L., and Yetong, C. 1997. Regeneration of bark of *Cinnamomum cassia* Presl. after girdling. *J. Plant Resour. Environ.* 6(3): 1–7.

Huang, L.C., Huang, B.L., and Murashige, T. 1998. A micropropagation protocol for *Cinnamomum camphora*. *In Vitro Cell. Dev. Biol. Plant* 34: 141–146.

Jeong-Ok, K., Sang-Myung, L., Yil-Seong, M., Sang-Gil, L., and Young-Joon, A. 2007. Nematicidal activity of cassia and cinnamon oil compounds and related compounds toward *Bursaphelenchus xylophilus* (Nematoda: Parasitaphelenchidae). *J. Nematol.* 39: 31–36.

Joseph, J. 1981. Floral biology and variation in cinnamon. In S. Vishveshwara (ed.), *Proceedings of Placrosym IV*, Kasaragod, India, pp. 431–434.

Krishnamoorthy, B., Zachariah, T.J., Rema, J., and Mathews, P.A. 1999. Evaluation of selected Chinese cassia, *Cinnamomum cassia* accessions for chemical quality. *J. Spices Aromat. Crops* 8: 215–217.

Krishnamoorthy, B., Zachariah, T.J., Rema, J., and Mathews, P.A. 2001. High quality cassia selections from IISR, Calicut. *Spice India* 14(6): 2–4.

Meena Vangalapati, S.S.N., Surya Prakash, D.V., and Sumanjali, A. 2012. A review on pharmacological activities and clinical effects of cinnamon species. *Res. J. Pharm. Biol. Chem. Sci.* 3: 653–663.

Nazia, M., Ahmed, C., and Perween, T. 2006. Anti-microbial activity of *Cinnamomum cassia* against diverse microbial flora with its nutritional and medicinal impacts. *Pak. J. Bot.* 38: 169–174.

Nguyen, V.N., Nguyen, D.-M.-C., Seo, D.-J., Park, R.-D., and Jung, W.-J. 2009. Antimycotic activities of Cinnamon-derived compounds against Rhizoctonia solani in vitro. *BioControl* 54: 697–707.

Nirmal Babu, K., Ravindran, P.N., and Peter, K.V. 1997. Protocols for micropropagation of spices and aromatic crops. Indian Institute of Spices Research, Calicut, Kerala, p. 35.

Nur, N.K., Syarifah, N., Azlina, S.I., Masdar, N.D., Hamid, F.A., and Nawawi, W.I. 2014. Extraction and potential of cinnamon essential oil towards repellency and insecticidal activity. *Int. J. Sci. Res. Publications* 4: 2250–3153.

Ravindran, S., Krishnaswamy, N.R., Manilal, K.S., and Ravindran, P.N. 1992 Chemotaxonomy of *Cinnamomum* Schaeffer occurring in Western Ghats. *J. Indian Bot. Soc.* 71: 37–41.

Santos, J.K. 1930. Leaf and bark structures of some cinnamon trees with special reference to Philippine trees. *Philipp. J. Sci.* 43: 305–365.

Shylaja, M. 1984. Studies on Indian cinnamomum. PhD thesis, University of Calicut.

Shylaja, M., and Manilal, K.S. 1992. Bark anatomy of four species of *Cinnamomum* from Kerala. *J. Spices Aromat. Crops* 1: 84–87.

Supriya, K.B. 2005. *Handbook of Aromatic Plants*. Pointer Publisher, Jaipur, India.

Syed, F.Z., Muhammad, A., Jibran, S.M., and Makoto, K. 2015. Diverse pharmacological properties of *Cinnamomum cassia*: A review. *Pak. J. Pharm. Sci.* 28: 1433–1438.

Varsha, J.B. 2012. A review on pharmacological activities of *Cinnamomum cassia* Blume. *Int. J. Green Pharm.* 102–108.

Xin, C.L., Jun, C., Na Na, Z., and Zhi, L.L. 2014. Insecticidal activity of essential oil of *Cinnamomum cassia* and its main constituent, *trans*-cinnamaldehyde, against the booklice, *Liposcelisbos trychophila*. *Trop. J. Pharm. Res.* 13: 1697–1702.

Yazaki, K., and Okuda, T. 1993. *Cinnamomum cassia* Blume (cinnamon): In vitro culture and the production of condensed tannins. In Y.P.S. Bajaj (ed.), *Biotechnology in Agriculture and Forestry. Medicinal and Aromatic Plants V.* Vol. 24. Springer-Verlag, Berlin, 122–131.

Youssif, R.S., and Shaal, E.A. 2011. Mosquitocidal activity of some volatile oils against *Aedes caspius* mosquitoes. *J. Vector Borne Dis.* 48: 113–115.

# 17

# Essential Oil of *Thymus vulgaris* L. for Pest Control

Abhishek Niranjan, Alok Lehri, and S.K. Tewari

## CONTENTS

*Thymus vulgaris* is a flowering plant of the mint family Lamiaceae. It grows up to 15–30 cm tall by 40 cm wide. Thyme is cultivated in most European countries, including France, Switzerland, Spain, Italy, Bulgaria, Portugal, and Greece. The yield and quality of oil obtained from the thyme plant varies in line with the genetic variations. The essential oil distilled from the thyme plant is mostly used in preservation (meat and butter) and in making chewing gum and ice cream. Thyme oil is also known for its antibacterial, anti-inflammatory, antiviral, antioxidant, antifungal, and insecticidal activities.

## 17.1 Classification

Kingdom: Plantae

Class: Magnoliopsida

Order: Lamiales

Family: Lamiaceae

Genus: *Thymus*

Species: *vulgaris*

Scientific name: *Thymus vulgaris* L.

## 17.2 Vernacular Names

The vernacular names of *T. vulgaris* L. are wild thyme, creeping thyme, mountain thyme, common thyme, farigola, garden thyme, herbatimi, herbathymi, mother of thyme, red thyme, rubbed thyme, thick leaf thyme, thym, thymian, thyme, timi, and tomillo.

## 17.3 Origin and Distribution of *Thymus vulgaris* L.

The general name for the many herb varieties of the *Thymus* species is thyme. Each variety is native to Europe and Asia. Garden thyme or common thyme is considered the principal type, and is used commercially for ornamental and flowering purposes. Thyme is also distributed to the western Mediterranean region, extending to southeastern Italy. In Greek, the name *thyme* was first given to the plant as a derivative of a word that meant "to fumigate," because of its use as incense, for its balsamic odor. Others derive the name from the Greek words *thyo*, meaning "perfume," and *thumus*, meaning "courage." In ancient and medieval days, the plant was held to be a great source of invigoration, with its pleasant qualities inspiring courage. Another source quotes its use by the Sumerians as long ago as 3500 BC, and to the ancient Egyptians, who called it tham.

## 17.4 Production of *Thymus vulgaris* L.

In South Africa, some producers distill thyme for essential oil production. Most of the thyme produced is for the fresh and dried market. Yields of *T. vulgaris* for fresh herb production can be 5–6 t/ha, and for dry herb, production can be 2 t/ha. Thyme will yield about 15 tons of plant material per hectare per year, at an oil recovery rate of 0.5%–1%, or 75 to 150 kg/ha/year under irrigation conditions. Under dry land conditions, considerable variation in yields is found. Internationally, thyme is produced from cultivated and wild harvested plants in most European countries, including Switzerland, France, Spain, Bulgaria, Italy, Portugal, and Greece. The quality and yield of the essential oil vary according to the genetic material of the plant, harvesting at crop maturity, the environment, and the distillation process. About 90% of the production of thyme oil for world trade is obtained from Spain. In southern Europe, farmers and herb growers benefit from the longer growing season, according to climate advantages. Therefore, most thyme is produced in Europe. Cultivation of the herb for trade mainly occurs in Spain, France, Italy, and Bulgaria. In hot summer conditions, essential oil yields approximately 1.0% (10 ml of oil/kg fresh thyme) from wild thyme, while it may decrease to 0.10% in the winter. As per variety, essential oil yields from cultivated material range from 0.05% to 0.50%. However, herbage yields under cultivation far exceed production in the wild, so more oil is produced per hectare in cultivated crops. Selected cultivars in Switzerland yield 3% essential oil from fresh herbage of more than 15 t/ha. Thyme can be grown in most provinces of South Africa. The Western Cape is climatically the closest to the Mediterranean area, where thyme grows naturally in the wild. The plant is very adaptable and therefore performs well in other climatic zones, for example, Kwa Zulu-Natal, Gauteng, and Mpumalanga. It does well in drier parts, such as the Karoo, but irrigation is needed.

## 17.5 Botanical Description of the Plant

### 17.5.1 Stem

The stem is perennial, aromatic, and a subshrub, 20–30 cm in height, with ascending, quadrangular, grayish-brown to purplish-brown lignified and twisted stems bearing oblong-lanceolate to ovate-lanceolate grayish green leaves with pubescence on the lower surface. The flowers have a pubescent calyx and a bilobate, pinkish or whitish corolla, and are borne in verticillasters. The fruit is made of four brown ovoid nutlets. It has both horizontal and upright habits. With age, the stems become woody.

### 17.5.2 Leaves

Leaves of thyme are tiny, usually 2.5–5 mm in length, and differ considerably in shape and hair covering, depending on the cultivar, with each species having a slightly different scent. It is sessile or has a very short petiole. The lamina is entire, tough, and covered on both surfaces by a gray to greenish-gray indumentum; lanceolate to ovate, the edges are markedly rolled up toward the abaxial surface. The adaxial surface of the midrib is depressed and is very prominent. The calyx is green, regular with violet spots, and is tubular. At the end are two lips, of which the upper one is bent back and has three lobes on its end; the lower is longer and has two hairy teeth. After flowering, the calyx tube is closed by a crown of long, stiff hairs. The corolla, about twice as long as the calyx, is usually brownish in the dry state and is slightly bilabiate. *T. vulgaris* leaves are oval to oblong in shape and somewhat fleshy. Leaves are almost stalkless, with margins curved inward and highly aromatic. The fragrance of its leaves is the result of an essential oil, which gives it its flavoring value for culinary purposes, and is the source of its medicinal properties.

### 17.5.3 Flowers

The flowers terminate the branches in whorls. The calyx is tubular, striated, closed at the mouth with small hairs, and divided into two lips, the uppermost cut into three teeth and the lower into two. The corolla consists of a tube about the length of the calyx, spreading at the top into two lips of a pale purple color, the upper lip erect or turned back and notched at the end, and the under lip longer and divided into three segments.

### 17.5.4 Seeds

The seeds are round and very small and retain their germinating power for 3 years.

### 17.5.5 Thyme Cultivars

The genus *Thymus* consists of approximately 215 species with several hybrids. Three principal varieties, the broad-leaved, narrow-leaved, and variegated varieties, are usually grown for use. The narrow-leaved type, with tiny, gray-green leaves, is more aromatic than the broad-leaved thyme (winter or German thyme). Broader leaves than the ordinary garden thyme differentiates it from others; the fragrant lemon thyme has a lemon flavor. It is not curved at the margins, and is marked as a variety of *Thymus serpyllum* (the wild thyme). Dominating all thyme with a strong flavor is silver thyme. The most cultivated thymes used for culinary and essential oil extraction are as follows:

- *Thymus vulgaris*: Common thyme; prostrate form; yellow, silver, and variegated foliage available; used in cooking
- *Thymus zygis*: Like above; mostly distilled for essential oil
- *Thymus* × *citriodorus*: Lemon thyme; upright form; golden and variegated silver foliage available; strong lemon scent

Varico, a robust cultivar, has an erect growth form with grayish-blue foliage and a great herbage yield. It contains thymol concentrations up to 50% and higher, as well as more than a 3% essential oil yield. It is immune to frost and can be propagated with seeds. Other promising new cultivars are currently being developed in various countries. Approximately 66 different species and hybrids have been selected for the color of their leaves and flowers, and are mainly used as ornamental shrubs.

## 17.6 Soil and Climatic Requirements

Thyme grows well in temperate to warm, dry, sunny climates. It does not grow well under shade. It needs full sunlight for best growth. Extreme moisture makes thyme more susceptible to rot diseases. Rainfall in the Mediterranean region is 500–1000 mm/year, mainly in winters, where thyme is cultivated in the majority. Thyme grows well in appropriately drained soils with a pH of 5.0–8.0. Thyme species do excellent in coarse, rough soils. Although thyme grows easily, especially in calcareous light, dry, stony soils, it can be cultivated in heavy wet soils, but it becomes less aromatic.

## 17.7 Cultivation Practices

### 17.7.1 Propagation

Propagation of thyme proceeds from seeds, layering, and stem cuttings. Plantations are enhanced by dividing the plants at their roots. Seeds are sown in the spring below 6 mm or less from the soil level. Seeds germinate in about 2 weeks. Seeds planted in trays take 6–8 weeks to reach transplant readiness. The seedlings are transplanted outdoors after the danger of frost has passed. If established and growing well before winter, the small plants can withstand frost. The source of thyme seed has to be known, as there are possibilities of hybridization. To have homogeneous plants, it is advisable to make cuttings. Thyme grows easily from 5 to 10 cm cuttings taken in the spring. Promoting hormones for roots may be beneficial. Take care not to use this method if any harmful soil organism is present.

## 17.8 Soil Preparation

Thyme can be planted successfully in very shallow soil, where other crops cannot grow. High-quality yield products are obtained from herbal and essential oil crops grown in natural soils. Soil analyzed at the laboratory can be checked for mineral deficiencies and excesses, carbon ratios, and organic status. A soil analysis will lead farmers to correct the nutrient composition of the soil. It provides the crop with optimum growing conditions, like balanced mineral status and correct pH. Soil fertility levels should be within acceptable ranges. Soil parameters like pH and minerals contained should be corrected accordingly to cultivate *T. vulgaris*. Fertilizer use has to be planned according to whether the crop will be grown inorganically or organically. Organic soil preparation practices are encouraged to ensure that the organic matter and soil microorganisms are present.

### 17.8.1 Pest Control

The volatile oils of the plants have pest-repellent properties; therefore, pests on thyme are not very frequent. However, whitefly, scale, and spider mites may infest the plants. For prospective producers of herbal and essential oil crops, some of following pest control guidelines are recommended.

- The preliminary option is to choose natural pest control measures.
- A pest management program should be strictly followed.
- Regular inspection of the crop is needed.
- Major problems can be prevented by early detection and management of pest control.
- Exact identification of pests for natural beneficial predators is necessary.
- Introduce and use biological controls, natural predators, parasites, nematodes, fungi, bacteria, and beneficial microorganisms. Use of chemicals that kill such organisms should be avoided.
- Other organic methods, such as insecticidal soaps, reflective mulches, traps, plant extracts, and handpicking of pests, water sprays, and vacuum may be used.

Effective controls may be used that target specific taxonomic groups, eating habits, or life stages: insecticidal soaps, horticultural oils, pheromones, and growth-regulating natural substances such as neem oil. For pest control, knowledge of certain herbs that repel or attract insects can be used in companion plantings. If organic practices will be used, make sure that products are certified for use. For more information on the identification of insects and for recommended controls, contact agricultural institutes. In wetter environments with improper soil drainage, *Rhizoctonia* root rot can cause problems. Thyme plants can get a few diseases, like rust, *Alternaria* blight, and *Botrytis*.

## 17.9  Harvesting

Thyme is harvested for essential oil once per year, during the late summer, when flowering begins. In certain conditions, two harvests per year are achievable. For dried produce, harvest stems and leaves just as flowering begins, cutting the entire plant back to about 10–15 cm above the ground. For fresh produce, harvest only the tips of the branches, so the plants are strong enough to produce enough young shoots. To produce a uniform product, the dried product should be processed to remove the leaves from the stems, and then sieved to remove dirt. Numerous methods exist for drying, from natural sunrays to human-made sophisticated driers. The use of sun-drying methods results in poor-quality essential oil. Artificial drying methods result in better control of product quality. A forced-airflow drier is a suitable system to dry better-quality leaves. To reduce loss of flavor through volatilization of essential oil, and to maintain a good green color, thyme should be dried at temperatures lower than 40°C. After drying, the leaves should be separated further from the stems, sieved, and graded. Fresh product has to be clean of foreign material and look fresh and crispy, with a good color and flavor.

### 17.9.1  Grading of *Thymus vulgaris*

Grading of dried thyme depends on the quality requirements prescribed ISO 6754:1996. The standard prescribes certain requirements of the finished product. The essential oil content of the dried herb is an important factor contributing to the flavor intensity. To meet the requirements, whole thyme leaves should contain a minimum of 0.5% essential oil, which equals 5 ml/kg dried herb, and ground thyme should contain at least 0.2% essential oil. For essential oil production, there is a different range of chemotypes occurring in thyme. There are at least six different chemotypes of importance: thymol, carvacrol, linalol, geraniol, thuyan-4-ol, and α-terpinyl acetate. The most frequent are thymol and carvacrol, which are generally extracted from plants growing near sea level, and linalool, which is generally extracted from plants occurring at higher altitudes. The geraniol, thuyan-4-ol, and α-terpinyl acetate chemotypes are rare and found mixed with the first three chemotypes. There is a current recovery in the demand for thymol used in the pharmaceutical industry, owing to its powerful properties as a disinfectant.

## 17.10 Storage

Essential oil of thyme should be stored in a cool, dry area until it is used. Oil should be kept in dark, airtight glass bottles and not be exposed to heat or heavy metals. Once opened, the vial of essential oil should be refrigerated, and tightly closing the cap will prolong its shelf life. Deterioration begins if the liquid is much darker or more viscous than normal.

## 17.11 Extraction of Oil

Dried and ground parts of plants are cut into small pieces and subjected to hydrodistillation for 3 hours using a Clevenger-type apparatus; the oils obtained are dried using anhydrous sodium sulfate. Essential oil yielded from the air-dried plant parts of *T. vulgaris* was 1.6%. For extraction of essential oil, Ivan (2013) used different green solvents, namely, ethanol, limonene, and ethyl lactate, to extract thymol from thyme plants. Ethyl lactate and limonene are agrochemical solvents, easily biodegradable, with polarities in the range of acetonitrile and hexane, respectively. Both solvents are generally recognized as safe (GRAS) and approved by the U.S. Food and Drug Administration as pharmaceutical and food additives. Further, the high solubility of thymol in ethyl lactate has been recently determined and reported by the authors. They used pressurized liquid extraction (PLE) in an ASE 350 system with the three green liquid solvents at different extraction temperatures (60°C, 130°C, and 200°C), employing *T. vulgaris* as the model thyme variety. Then, the extraction of thymol from other thyme varieties (*T. zygis* and *Thymus citriodorus*) was studied. The extraction yield and thymol recovery obtained from the different extracts were quantified and compared. The three green solvents have shown good capacity to extract thymol from thyme plants.

## 17.12 Chemical Composition of the Essential Oil

The essential oil from *T. vulgaris* shows a high content of oxygenated monoterpenes (56.53%) and low contents of monoterpene hydrocarbons (28.69%), sesquiterpene hydrocarbons

(5.04%), and oxygenated sesquiterpenes (1.84%). The predominant compound among the essential oil components is thymol (51.34%), while the amount of all other components of the oil is less than 19%. Thyme oil monoterpene hydrocarbon content is made up of p-cymene and g-terpinene. Alcohols such as linalool, a-terpineol, and thujan-4-ol are also present (Thompson et al. 2003; Khan and Abourashed 2010). Gas chromatography–mass spectrometry (GC-MS) analysis identified the two primary constituents of thyme oil as thymol (57.8%) and p-cymene (28.6%), with all other constituents making up <5% of the total. The aerial parts of *T. vulgaris* collected from Western Ghats of India were analyzed for essential oil by gas chromatography. Forty-eight compounds were detected, among which 36 compounds, constituting 98.63% of the oil, were identified, with thymol (61.6%), p-cymene (11.2%), ϒ-terpinene (7.4%), methyl thymol (3.9%), methyl carvacrol (3.3%), and β-caryophyllene (2.3%) being the major chemical constituents (Syamasundar 2008). El-Nekeety et al. (2011) reported that the oil contains carvarcrol (45 mg/g), thymol (24.7 mg/g), β-phellandrene (9.7 mg/g), humuline (3.1 mg/g), α-phellandrene (2.3 mg/g), and myrcene (2.1 mg/g). However, α- and β-pinene, myrcene, α-thyjone, tricyclene, 1,8-cineole, and β-sabinene were found in very low concentrations. The density and refraction index of the essential oil of thyme at 20°C are 0.944 g/ml and 1.507, respectively (Viuda-Martos et al. 2008).

## 17.13 Essential Oil Market

The world major market for thyme essential oils is the United States, Japan, and Europe. Production continues to be concentrated in Europe, with seven of the world's largest essential oil processing firms. In the United States, the major users of essential oils are the soft drink companies. Japan accounts for 10% of the world demand. The Canadian market is dominated by the U.S. perfume and flavoring industry. France dominates the world perfumery market, and Switzerland is one of the leaders in the pharmaceutical field. Britain and India are known to feature strongly in the flavoring sector. Most countries import all their dried thyme mainly from Spain and Morocco, the major world producers. Most bulk dried herbs are produced in countries with low labor costs, so the challenge to producers in South Africa is to produce crops of superior quality at a competitive price. The essential oil component, thymol, has a wide range of uses in the manufacturing of liqueurs, perfumes, pharmaceutical products, and toilet articles. The essential oil of thyme is used to preserve processed meat and butter, and in making chewing gum, ice cream, candy, and Benedictine liqueur.

## 17.14 Various Properties of Thyme Oil

### 17.14.1 Medicinal Properties

Thyme is prescribed with other herbs for asthma and hay fever, and is often used to treat worms in children. Thyme has been thought of as an antiseptic, antimicrobial, astringent, anthelmintic, carminative, and tonic. Thyme is incredibly useful in cases of assorted intestinal infections and infestations, like hookworms, ascarids, gram-positive and gram-negative bacterium, fungi, and yeasts, as well as *Candida albicans*. Its active constituent, thymol, is active against enterobacteria and cocci bacteria. Thyme may also improve liver functioning, and act as an appetite stimulant. It can be used in the treatment of cartilaginous tube, bronchial, and urinary infections. Used as a gargle, thyme is helpful in the treatment of laryngitis and inflammation. It is used for skin issues like oily skin, sciatica, acne, dermatitis, and bug bites. The thymol-rich *T. vulgaris* essential oil, known as "red thyme oil," has strong antiseptic activity; the linalool-rich essential oil, known as "garden thyme," has potent antiparasitic and antifungal properties; and the thujanol-rich essential oil, known as "sweet thyme," has antiviral properties. These essential oils are used in aromatherapy to stimulate the mind, strengthen memory and concentration, and calm the nerves. "White thyme oil" is also used, and it is milder on the skin. Applied to the skin, thyme relieves bites and stings and rheumatic aches and pains (Prasanth et al. 2014). Other authors have reported the medicinal uses of thyme species (Zarzuelo and Crespo 2002; Saleh et al. 2015).

### 17.14.2 Anti-Inflammatory Activity

*T. vulgaris* oil is a combination of monoterpenes. The main compounds of this oil are the natural terpenoid thymol and its phenol chemical compound carvacrol (Namsa et al. 2009), which have antioxidative, antimicrobial, medicinal drug, antitussive, antispasmodic, and antibacterial effects. Terpenoids, flavonoid aglycones, flavonoid glycosides, and synthetic resin acids were additionally found in *Thymus* spp.

### 17.14.3 Antibacterial Activity

The essential oils obtained from *T. vulgaris* L. harvested at four biological process stages were evaluated for their biological activity and chemical components. The thyme volatile oils were studied for their inhibition effects against nine strains of gram-negative bacteria and six strains of gram-positive bacteria. The bioimpedance methodology was used for antibacterial activity of the essential oils, and the parameter chosen for outlining and quantifying the antibacterial activity of the thyme oils was the detection time. The plate-counting technique was used to study the inhibitory effect by direct exposure. All the thyme essential oils examined had a significant bacteriostatic activity against the microorganisms tested. This activity was additionally marked against the gram-positive bacteria. The oil from full thyme flowers was the most effective at stopping the growth of the microorganism species examined. The oils tested were conjointly shown to possess smart antibacterial activity by direct contact against the gram-negative microorganism. The main component of the essential oil of thyme, thymol, is active against *Salmonella* and *Staphylococcus* bacteria. The antiseptic and tonic properties of thyme make it a useful tonic for the immune system in chronic, especially fungal, infections, as well as an effective remedy for chest infections such as bronchitis, whooping cough, and pleurisy. Thyme and thyme oil have been used as fumigants, disinfectants, and mouthwashes. The pleasant-tasting infusion can be taken for minor throat and chest infections, and the fresh leaves may be chewed to relieve sore throats (Viuda-Martos et al. 2008).

### 17.14.4 Antiviral Properties

Nolkemper et al. (2006) conducted an experiment with aqueous extracts from species of the Lamiaceae family. These were examined for their antiviral activity against herpes simplex virus (HSV). Extracts from thyme (*T. vulgaris*) showed inhibitory activity against herpes simplex virus types 1 and 2 (HSV-1 and HSV-2), and an acyclovir-resistant strain of HSV-1 was tested *in vitro* on RC-37 cells in a plaque reduction assay.

## 17.14.5 Antioxidant Properties

Antioxidants are those compounds that inhibit the oxidation of different molecules. Oxidation is a chemical process that transfers electrons or hydrogen from a substance to an oxidizing agent. This reaction produces free radicals. In turn, these radicals begin chain reactions. Once the chain reaction happens in a cell, it causes damage or death to the cell. Antioxidants stop these chain reactions by removing free radical intermediates, and inhibit different oxidation reactions. The leafy parts of thyme and its oil are utilized in foods for flavor, aroma, and preservation, and additionally in folk medicines. El-Nekeety et al. (2011) conducted an experiment to work out the elements of *T. vulgaris* L. oil and evaluate the protecting effects of this oil against aflatoxin-induced oxidative stress in rats. Treatment with aflatoxins alone disturbs the lipid profile in blood serum, decreases the total antioxidant capability, and increases creatinine, uric acid, and nitric oxide in blood serum and lipid peroxidation in the liver. Amiri (2012) reported that thyme has potential antioxidant activity.

## 17.15 Insecticidal Activity

The insecticidal activity of thyme volatile oil, thymol, and carvacrol was evaluated in the laboratory against completely different larval stages of lesser mealworm. The first and later larval stages were reared on diets containing one or two acetone solutions of tested compounds. The insecticidal activity of thyme volatile oil and pure monoterpenes against *Alphitobius diaperinus* larvae relied on the dose and age of the larvae. The growth of younger larvae was considerably affected, whereas that of the older larval stage was less influenced, and only by pure oil components. In young larvae, the application of 1% thyme oil, thymol, and carvacrol caused mortality of 50.0%, 86.67%, and 85%, respectively. Choi et al. (2003) tested the efficacy of oil against greenhouse whitefly adults, nymphs, and eggs. At the highest rate of $9.3 \times 10^{-3}$ concentration, thyme oil caused 100% mortality in adults and 88% mortality in eggs. However, there were many other oils in this study, including clove and peppermint, which provided greater mortality when tested at lower rates. In a study by Shaaya et al. (1991), thyme oil had highly toxic fumigant toxicity against the stored grain pest sawtoothed grain beetle (*Oryzaephilus surinamensis*). Among constituents tested in the same study, carvacrol, linalool, and a-terpineol were also highly toxic. Machial et al. (2010) tested 17 essential oils for their toxic effect on two Lepidopteran species, the oblique-banded leafroller (*Choristoneura rosaceana*) and the cabbage looper (*Trichoplusia ni*), in which thyme oil was the second most toxic on first-instar *C. rosaceana* larvae, with 64% mortality at a concentration of 5.0 µl/ml, superseded only by the 97% mortality with patchouli oil at the same concentration. Despite thyme oil's potency on *C. rosaceana*, it was one of the least toxic essential oils on first-instar *T. ni*. Leila et al. (2014) reported that the greatest toxicity was observed at 250 µg/L of essential oil, with the $LC_{50}$ values of 134.1 µg/L after 24 hours using essential oil of *T. transcaspicus*. It exhibited strong insecticidal activity against *Anopheles stephensi*, which can be attributed to its constituents, especially carvacrol and thymol phenols. Maqtari et al. (2011) and Marino and Bersani (1999) studied the composition of essential oil and its potent effect on antimicrobial activity.

### 17.15.1 Antilarval Activity

The essential oil of *T. vulgaris* L. was studied by Elham et al. (2014) for its toxicity and physiological effects on the lesser mulberry pyralid *Glyphodes pyloalis* Walker in controlled conditions. The leaf disc method was used to study acute toxicity; the effects of $LC_{10}$, $LC_{30}$, and $LC_{50}$ on the feeding efficiency of fourth-instar larva; and biochemical indices. The essential oil doses of $LC_{10}$, $LC_{30}$, and $LC_{50}$ were estimated to be 0.107%, 0.188%, and 0.279% for *T. vulgaris*. The authors found that *T. vulgaris* was more toxic than other essential oils. The essential oil sublethal dose $LC_{30}$ affected the nutritional indices of fourth-instar larvae of *G. pyloalis*. The essential oils reduced total protein, carbohydrate, and lipid. Some concentrations of essential oils changed the activity level of α-amylase, protease, lipase, general esterases, and glutathione S-transferase (GST), but others showed no effect on these enzymes. It was concluded that the essential oil concentrations used were toxic to *G. pyloalis* and showed irreversible effects on key metabolic processes; therefore, the used essential oil concentrations may be considered alternatives to the classic pest control agents. Fouad et al. (2016) evaluated the potential larvicide of essential oils from aromatic *T. vulgaris* cultivated in northeast Morocco on the malaria vector *Anopheles labranchiae*. The sample showed larvicidal activity against larvae in stages 3 and 4. The *T. vulgaris* essential oil demonstrated an $LC_{50}$ of the order of 351.63 μg/ml and an $LC_{90}$ of 621.34 μg/ml. According to the authors, the results open interesting perspectives for the application of essential oil of *T. vulgaris* in the production of biocides. Szczepanik et al. (2012) has reported the essential oil of thyme to be a powerful antilarval agent.

### 17.15.2 Acaricidal Activity

Mansour et al. (2015) have investigated acaricidal effects of thyme essential oil on *Dermanyssus gallinae*. The authors used a filter paper contact test. Thirty live adult female mites were exposed to different concentrations of each examined compound for 2, 4, 6, and 24 hours. The mortality rate of mites at each concentration and time was recorded, and each treatment was performed in triplicate. Thyme essential oil showed acaricidal activities, and the effects of exposure time and concentration on mortality rate were significant ($p < 0.05$). The highest mortality was achieved at 24 hours. The authors suggested that essential oil of thyme is a potent green pesticide.

### 17.15.3 Insect-Repellent Activity

*T. vulgaris* is a good companion crop that repels cabbage fly, whitefly, and aphids. Only a few thyme species are used as landscape ornamentals (Chintalchere et al. 2013). Thyme is excellent for rock gardens. Creeping thyme tolerates occasional foot traffic and can be used between stepping stones along garden paths. Thyme can also be utilized as an edging or border plant in herb gardens.

### 17.15.4 Antifungal Activity

Shazia and Muzafar (2011) reported that volatiles of thyme oil were extremely effective in reducing gray mold and soft rot incidence in strawberry fruits caused by *Botrytis cinerea* and *Rhizopus stolonifer*, respectively.

## 17.16 Conclusion

*T. vulgaris* is a good source of essential oil, having pesticidal, insecticidal, larvicidal, and mosquitocidal activity. These activities are attributed to the presence of thymol as a principal active compound. Apart from the above activities, the plant possesses various pharmacological activities, such as antioxidant, anti-inflammatory, and antimicrobial activities. All types of activities are attributed to the presence of biological compounds of therapeutic use. These biological activities make this plant economically important. Scanty information is available worldwide, opening a new vista of research and development in various areas.

## References

Amiri, H. 2012. Essential oils composition and antioxidant properties of three *Thymus* species. *Evid. Based Complement. Alternat. Med.* 2012: 728065.

Chintalchere, J.M., Lakare, S., and Pandit, R.S. 2013. Bioefficacy of essential oils of *Thymus vulgaris* and *Eugenia caryophyllus* against housefly, *Musca domestica* L. *Bioscan* 8: 1029–1034.

Choi, W.-I., Lee, E.H., Choi, B.R., Park, H.M., and Ahn, Y.J. 2003. Toxicity of plant essential oils to *Trialeurodes vaporariorum* (Homoptera: Aleyrodidae). *Hort. Entomol.* 96: 1479–1484.

Elham, Y., Jalal, J.S., and Jalil, H. 2014. Effect of *Thymus vulgaris* L. and *Origanum vulgare* L. essential oils on toxicity, food consumption, and biochemical properties of lesser mulberry pyralid *Glyphodes pyloalis* Walker (Lepidoptera: Pyralidae). *J. Plant Prot. Res.* 54(1).

El-Nekeety, A.A., Mohamed, S.R., Hathout, A.S., Hassan, N.S., Aly, S.E., and Abdel-Wahhab, M.A. 2011. Antioxidant properties of *Thymus vulgaris* oil against aflatoxin-induce oxidative stress in male rats. *Toxicon* 57: 984–991.

Fouad, E.T., Raja, G.S., and Abdelhakim, E.O.L. 2016. Larvicidal activity of essential oils of *Thymus vulgaris* and *Origanum majorana* (Lamiaceae) against the malaria vector *Anopheles labranchiae* (Diptera: Culicidae). *Int. J. Pharm. Pharm. Sci.* 8: 372–376.

Ivan, A., David, V.B., Roumiana, P.S., Guillermo, R., Elena, I., and Tiziana, F. 2013. Extraction of thymol from different varieties of thyme plants using green solvents. Presented at III Iberoamerican Conference on Supercritical Fluids, Cartagena de Indias, Colombia.

Khan, I.A., and Abourashed, E.A. 2010. *John Leung's Encyclopedia of Common Natural Ingredient: Used in Food, Drugs, and Cosmetics.* 3rd ed. Wiley & Sons, Hoboken, NJ.

Leila, D., Kamal, R.A., Mohammad, R., Mohammad, R.A., and Javad, B. 2014. Insecticidal activity of the essential oil of *Thymus transcaspicus* against *Anopheles stephensi*. *Asian Pac. J. Trop. Biomed.* 4: S589–S591.

Machial, C.M., Shikano, I., Smirle, M., Bradbury, R., and Isman, MB. 2010. Evaluation of the toxicity of 17 essential oils against *Choristoneura rosaceana* (Lepidoptera: Tortricidae) and *Trichoplusia ni* (Lepidoptera: Noctuidae). *Pest Manag. Sci.* 66: 1116–1121.

Mansour, E., Ali, M., Gholam, A.K., and Afkhami-Goli, A. 2015. *In vitro* acaricidal effects of thyme essential oil, tobacco extract and carbaryl against *Dermanyssus gallinae* (Acari: Dermanyssidae). *Sci. Parasitol.* 16: 89–94.

Maqtari, M.A.A., Akghalibi, S.M., and Alhamzy, E.H. 2011. Chemical composition and antimicrobial activity of essential oil of *Thymus vulgaris* from Yemen. *Turkish J. Biochem.* 36: 342–349.

Marino, M., and Bersani, C. 1999. Antimicrobial activity of the essential oils of *Thymus vulgaris* L. measured using a bioimpedometric method. *J. Food Prot.* 62: 1017–1023.

Namsa, N.D., Tag, H., Mandal, M., Kalita, P., and Das, A.K. 2009. An ethnobotanical study of traditional anti-inflammatory plants used by the Lohit community of Arunachal Pradesh, India. *J. Ethnopharmacol.* 125: 234–245.

Nolkemper, S., Reichling, J., Stintzing, F.C., Carle, R., and Schnitzler, P. 2006. Antiviral effect of aqueous extracts from species of the Lamiaceae family against herpes simplex virus type 1 and type 2 in vitro. *Planta Med.* 72: 1378–1382.

Prasanth, R.V., Ravi, V.K., Varsha, P.V., and Satyam, S. 2014. Review on *Thymus vulgaris* traditional uses and pharmacological properties. *Med. Aromat. Plants* 3: 1–3.

Saleh, H., Azizollah, J., Ahmadreza, H., and Raham, A. 2015. The application of medicinal plants in traditional and modern medicine: A review of *Thymus vulgaris*. *Int. J. Clin. Med.* 6: 635–642.

Shaaya, E., Ravid, U., Paster, N., Juven, B., Zisman, U., and Pissarve, V. 1991. Fumigant toxicity of essential oils against four major stored-product insects. *J. Chem. Ecol.* 17: 499–504.

Shazia, S., and Muzafar, G.W. 2011. Essential oil composition of *Thymus vulgaris* L. and their uses. *J. Res. Dev.* 11: 83–94.

Syamasundar, K.V., Srinivasulu, B., Stephen, A., Ramesh, S., and Rao, R.R. 2008. Chemical composition of volatile oil of *Thymus vulgaris* L. from Western Ghats of India. *J. Spices Aromat. Crops* 17: 255–258.

Szczepanik, M., Zawitowska, B., and Szumny, A. 2012. Insecticidal activities of *Thymus vulgaris* essential oil and its components (thymol and carvacrol) against larvae of lesser mealworm, *Alphitobius diaperinus* Panzer (Coleoptera: Tenebrionidae). *Allelopathy J.* 30: 129–142.

Thompson, J., Chalcha, J., Michet, A., Linhart, Y., and Ehlers, B. 2003. Qualitative and quantitative variation in monoterpene co-occurrence and composition in the essential oil of *Thymus vulgaris* Chemotypes. *J. Chem. Ecol.* 29: 859–880.

Viuda-Martos, M., Ruiz-Navajas, Y., Fernández-López, J., and Pérez-Álvarez, J.A. 2008. Antibacterial activity of different essential oils obtained from spices widely used in Mediterranean diet. *Int. J. Food Sci. Technol.* 43: 526–531.

Zarzuelo, A., and Crespo, E. 2002. The medicinal and nonmedicinal uses of thyme. In *Thyme: The Genus Thymus*. Stahl-Biskup, E., and Saez, F., Eds. London: Taylor & Francis, 263–292.

# 18

# Castor Oil

**R.T. Gahukar and Sheetal Mital**

## CONTENTS

## 18.1 Introduction

Castor as an oil crop has gained importance in the global market, with its demand increasing annually by 3%–5% (Anjani 2012). This point is pertinent to recent development in the biodiesel feedstock supply, and of significant importance in the industrial production of pharmaceuticals. Therefore, the seed price is attractive for farmers who are inclined to cultivate castor crop in place of nonremunerative crops. Consequently, a lot of field research has been undertaken on oil extraction methods and the physicochemical properties for improving oil quality. As a pesticide, castor oil can be an ideal eco-friendly and cheap control measure against field pests and disease pathogens. Its use as medicinal product, however, needs intensive studies for human safety. Only a few studies have reported the isolation of essential oils from castor oil and their use compared with medicinal and aromatic plants. Thus, literature on this subject is scanty, and not easily available in compiled form. To close up this gap, the current review gathers information on the botany of the plant, compares oil extraction methods and physicochemical properties, and elaborates on oil utilization in crop and seed protection and also in pharmaceutical and industrial production. In Table 18.1, the local names for the castor bean plant used in different countries are given.

## 18.2 Botany of the Plant

The castor plant (castor bean or castor oil plant), *Ricinus communis* L., belongs to the family Euphorbiaceae (order Euphorbiales). The genus *Ricinus* is named after a Latin word for "tick," probably because its seed has markings and a lump at the end that resembles a certain tick. The origin of the plant is the Ethiopian region, but it has spread widely to East Africa, China, Thailand, South America, the Mediterranean basin, and India. Now, it is found throughout the world in tropical countries and a few temperate countries, where it is cultivated on a small scale as a border crop or as the sole crop on poor soils in semiarid zones, and it has become an abundant weed in the United States (Nangbes et al. 2013). It can grow well in all types of soil. However, well-drained soil with moisture retentive capacity, such as sandy loam soil, is ideal. It adapts to varied climates and grows fast even with a low availability of water and plant nutrients. However, insufficient nitrogen supply to plants results in reduced seed yield (Weiss 2000).

**TABLE 18.1**

Local Names for the Castor Bean Plant Used in Different Countries

| Local Name | Country |
|---|---|
| Kerua, kerroa, charua | Arabia |
| Cherva, higuera del diablo infernal, tartago | Argentina |
| Bafureira, baga, carrapateira, mamona, mamono, ricino | Brazil |
| Pi ma, yuen kin tse, ta ma tse | China |
| Kai-dudu-deu | Cochin China |
| Higuerila | Costa Rica |
| Higuereta | Cuba |
| Ricin, bois de carapat, palma christi, paume dieu | France |
| Wunderbaum, ricinusol | Germany |
| Aporano | Ancient Greece |
| Kiki, kroton, mbacibo' | Modern Greece |
| Bupurura | Guinea Bisau (Manjaco tribe) |
| Buorai | Guinea Bisau (Biafada tribe) |
| Djague-djague | Guinea Bisau (Crioulo tribe) |
| Djacula | Guinea Bisau (Futa-Fula tribe) |
| Torra, entogai | Guinea Bisau (Balanta tribe) |
| Castor oilseed, palma christi, castor bean | Great Britain and the United States |
| Caffe da olio, erba da latte, erba lattaria, erba venaria, fagiolo d'india, fagiolo romano, fico d'inferno, girasole, girasole maggiore, girasole piccolo, mano aperta, meo, mirasole, palma christi, riccino, ricino, ricino comune, ricino minore, ricino volgare, scatapuzia, zecca | Italy |
| African coffee tree | Africa |
| Armanata | Falkland Islands |
| Higuerila, heguerilla, higuerillo, tartago, tlapatl | Mexico |
| Wonderolie | Holland |
| Tartago, castor | Paraguay |
| Higuereta | Dominican Republic |
| Ricinusa | Ancient Rome |
| Catoputia major | Russia |
| Eranda, erando, erumba, arand, erand, andi | India (Sanskrit) |
| Palma christi | Panama |
| Tartago | Venezuela |

*Source:*  U.S. Department of Agriculture, Natural Resources Conservation Service, *Ricinus communis* L. castorbean, PLANTS Database, National Plant Data Center, Baton Rouge, 2006, http://plants.usda.gov; Rana, M. et al., *Int. J. Pharmtech. Res.*, 4(4), 1706–1711, 2012.

Castor is basically a long-day plant, but it is adaptable to various photoperiods (12–18 h) and pH values (4.5–8.3) (Salihu et al. 2014). Also, plants can tolerate a wide range of annual temperatures (7°C–27.8°C) and annual precipitation (20–429 cm) (Salihu et al. 2014). Because of its genetic characteristics, the plant greatly varies in its appearance and growth characteristics. For example, it may be a nonhardy fast-growing suckering perennial shrub, often developing into a small tree (12 m height), or a short-lived dwarf annual shrub. It is a common annual crop on marginal land and coastal sandy belts, where it can reach a height of 2–3 m and withstand even sandy and saline environments. The improved hybrids and varieties are generally dwarf plants that have been developed especially for high oil content.

### 18.2.1 Root

The tall plant has a well-developed taproot of a few meters, with substantial laterals and secondary roots. Roots of a dwarf plant reflecting a particular variety or cultural system show a less apparent root system. In low-rainfall areas, a poor root system is associated with slow aerial growth affecting overall plant development (Weiss 2000). A well-developed root system allows the plant to take maximum soil moisture and build plant resistance to drought. Also, it allows the plant to tap necessary nutrients for accumulating biomass, which is mostly correlated with yield performance (Weiss 2000). Therefore, planting castor in soft and loose soil (such as sandy loam) is advantageous for getting a better crop.

### 18.2.2 Stem

The stem is round in shape and red or purple in color, and sometimes covered with a waxy bloom that gives a red or green stem a blush appearance. In aged plants, the stem color turns to gray at the base. The stem has many branches, but only primary branches give rise to secondary branches, and this sequence is continued over the whole plant life. The stem of the dwarf plant remains solid, whereas in tall plant it becomes hollow after considerable height. The presence of plastids in the stem at the juvenile stage enhances the photosynthetic activity. Nodes are well developed, from each of which a leaf arises. The node at which the first racemes appear is a characteristic of quick maturity of the plant. In dwarf hybrids, it usually occurs after 6–12 nodes, but it can vary from 6 to 45 nodes in segregating populations (Weiss 2000).

### 18.2.3 Leaf

The leaf is large (about 10–60 cm long), often dark glossy green, light green, or reddish, with a long petiole. In some varieties, leaves start off as dark-reddish-purple or bronze when young, but gradually change to a dark green, sometimes with a reddish tinge as they mature. In other varieties, they are green from the seedling stage, whereas in still others, a pigment masks the green color of the chlorophyll-bearing parts (Weiss 2000). Leaf development and expansion is not affected by sunlight, but by soil moisture. Leaves are of a palmate type, with 5–11 lobes with toothed edges and prominent veins on the underside. They are alternate, except for two opposite leaves at the node immediately above the cotyledons (Weiss 2000).

### 18.2.4 Flower

Plants can produce flowers over a long period under favorable climatic conditions. Flowers are borne on inflorescences forming a pyramidal raceme or spikes on main and lateral branches. Flowers may be monoecious (male and female), pistillate (only female), or interspersed on the inflorescence. Male flowers are yellowish-green with prominent creamy stamens and are carried in void pikes up to 15 cm long. They are borne at the tips of the spikes to occupy the underportion of the spike with no corolla, but have a green calyx deeply cut into three to five segments enclosing numerous branched yellow stamens. Female flowers occupy the upper portion of the spike and likewise have no corolla (Weiss 2000). The three narrow segments of the calyx are reddish, and the ovary in the center is crowned by deeply divided red threadlike styles. There is wide variation between the flowers, the ratio of male to female flowers, and the number of fertile female flowers. Female flowers have prominent stigmas and open before the male flowers in most of the varieties, while a reverse process occurs in others. Depending on the variety, the period of opening of female flowers, as well as that of male flowers, is 3–7 days (Weiss 2000). Generally, male flowers shed most of the viable pollens between 1 and 2 days after opening, and pollens shed from 2–3 h before sunrise to late afternoon, with a peak at midmorning. Pollen shedding is common in a temperature range of 26°C–29°C and relative humidity of 60%. Stigmas can remain receptive for a period of 7–10 days after opening (Weiss 2000).

### 18.2.5 Fruit

Fruit is a schizocarp globular spiny capsule with three cells, each of which splits open at maturity into separate parts and then breaks away explosively, shattering the seeds. Some varieties produce capsules with rudimentary spines, while others have soft, flexible, and nonirritant spiny capsules. However, some varieties produce spiny irritant capsules. After fertilization, a capsule is formed in 3–7 days. Racemes (indeterminate inflorescence) are conical, cylindrical, or oval with varied capsule arrangements, which can be compact, semicompact, or loose. The color of the capsule is mostly light green. The period of capsule maturity varies from 140 to 160 days depending on the variety (Weiss 2000). The lowest flowering racemes usually mature first; the others follow in sequence up to the stem. Ripening of fruits along the racemes is sometimes uneven in some wild varieties, and the period between the first and last mature fruits may be several weeks.

When ripe, capsules become hard and brittle and shatter mostly at maturity. In some varieties, the whole capsule falls from a desiccated raceme, with the seed remaining enclosed, or the capsule may split to release seeds (Weiss 2000). Strong capsules tend to preclude mechanical hulling, while very soft capsules become difficult to hull without damaging seeds (Salihu et al. 2014).

### 18.2.6 Seed

The castor seed has warty appendages called "caruncle," which is a type of elaiosome (fleshy structures attached to seeds). The caruncle promotes the dispersal of the seed. The seed is elongated, oval, or square, covered with thin, brittle, and mottled testa enclosing a white kernel of varied length (up to 250 mm) and breadth (5–15 mm). Seed color varies considerably; for example, it can be white, dark brownish–red, brown, dark chocolate, red, or black. The seed weight varies from 9 to 100 g for 100 seeds, depending on the number of seeds produced. The period of dormancy extends to several months in some varieties, while some seeds can be sown with normal germination after harvest. Dormancy can be broken by soaking seeds for 2 h in water or by removing the caruncle and piercing testa at the side. Germination is epigeal, with cotyledons coming out above the soil, expanding as green leaves (Weiss 2000). Bigger seeds germinate earlier than the smaller seeds.

## 18.3 Seed Oil

Castor oil is composed of 0.7% moisture, 48.8% fat, 7.2% protein, 11.6% carbohydrates, and 10.6% ash (Mbah et al. 2014). Several phytochemicals, such as cineole, 2-octanol, terpene-4-ol, limonene, subinene, pinene, terpinene, and the methyl groups tannin, phenol, alkaloid, oxalate, phytate, saponin, cyanogenic glycoside, and flavonoid, have been isolated or extracted, and only tannin is abundant (0.35%) (Momoh et al. 2012). Also, five types of triacylglycerols have been identified by Salimon et al. (2010) in castor oil from Malaysia: triricinolein (84.1%), diricinoleoylstearoylglycerol (8.2%), diricinoleoyloleoyl-glycerol (5.6%), diricinoleoyllinoleoylglycerol (1.2%), and diricinoleoylpalmitoylglycerol (0.9%). Castor oil is rich in triglycerides, mainly ricinolein and ricin (water-soluble toxin). Castor oil has the following unique properties.

1. It is colorless to very pale yellow, and tasteless or with mild taste.
2. Ricinoleic acid in castor oil is highly unusual because of its unique hydroxyl fatty acid ($C_{18}H_{34}O_3$), structurally known as cis-12-hydroxyoctadeca-9-enoic acid, with 18-carbon hydroxylated fatty acid having one double bond that is composed of triglycerides (esters) containing a 3-carbon alcohol (glycerol) and three 18-carbon (or 16-carbon) fatty acids (Ogunniyi 2006). Cvengros et al. (2006) showed that the hydroxyl group of ricinoleic acid affects the density and viscosity of the oil. No other commercial seed oil has such a high predominance of a single fatty acid with a high energy value and biofuel potential.
3. Oil uniformity and consistency, and comparatively, its high specific gravity, are important properties for its industrial use.
4. It is nontoxic and biodegradable, with a high oxidative stability of 44 h and a low cloud point of –14°C (making it useful even in cold weather) and can be stored up to 1 year if refrigerated (Abdelaziz et al. 2014).
5. Its easy solubility in alcohols at room temperature and limited solubility in aliphatic petroleum solvents facilitate several chemical reactions.
6. It is the most promising renewable raw material for the chemical and polymer industries.
7. It has a low energy requirement for oil production, for example, 56.8 GJ/ha. Castor seed consumes only 19% of the total energy, whereas 39% of energy is consumed in oil extraction and refining and 42% in biodiesel production (Severino et al. 2012).
8. Castor oil differs from other oils with its acetyl or hydroxyl value and comparatively high viscosity, making it an excellent emollient and lubricant (lubricity as low as 2g/kg) over a wide range of temperatures, and it has the capacity to wet and disperse rapidly.

## 18.4 Methods of Oil Extraction

Three methods of oil extraction have been used by various researchers and villagers. Among these, chemical extraction is common, particularly in industrial production.

### 18.4.1 Wet Extraction

In Nigeria, wet extraction is popularly employed as tradition by women, with only a 19% oil yield (Oluwole et al. 2012). In this method, castor beans are crushed and oil is extracted by using hot water or steam. This method, although economical, is not effective, and therefore extensive research is needed for its improvement and validation so that resource-poor villagers will be able to use it efficiently.

## 18.4.2 Mechanical Extraction

In this process, mechanical compressors, including the hydraulic press and continuous screw press, are employed to extract oil at room temperature. Some modifications by increasing the temperature or cold press have also been tested. A general method used by Perdomo et al. (2013) is described below.

A sample of 200 g of castor seeds is placed in a mechanical extruder. Seeds are compressed with the hydraulic press for 2 min at a pressure of 490 kPa, and then to 748 kPa for 2 min. Later, the cake is reused and the process is repeated. Extracted oil is centrifuged twice for 16 min at 1300 revolutions per minute (rpm) in order to clean the oil and remove suspended solids.

## 18.4.3 Chemical (Soxhlet) Extraction

### 18.4.3.1 Processing

Before extraction, castor seeds are processed to get the maximum yield of quality oil. The preliminary operations include cleaning, drying, dehulling, winnowing, and grinding. By handpicking or using a sieve, castor seeds are cleaned and separated from foreign materials such as dirt. Cleaned seeds are packed in net bags and dried in a greenhouse (Perdomo et al. 2013) or sun-dried for 4–5 days in the open until the shells split and the seeds shed. In order to further reduce the moisture content, seeds are dried in an oven at 60°C for 7 h (Akpan et al. 2006), 90°C for 6 h (Mgudu et al. 2012), 95°C for 7 h (Perdomo et al. 2013), or 105°C for 1 h (Salimon et al. 2010). These seeds are put in a desiccator for about 30 min, removed, and reweighed every 2 h until a constant weight is obtained.

Moisture content in seeds can be affected by plant morphology, specific biomass composition, seed immaturity, and conditions of seed storage (Perdomo et al. 2013). In any case, low moisture content in extracted oil is an indicator of good shelf life (Abitogun et al. 2008). The moisture content in seeds is calculated per the following formula: $(W1 - W0)/W0 \times 100$, where $W1$ and $W0$ are the initial weight before drying and final weight after drying, respectively. The winnowing is done in the tray to blow away seed cover. The cleaned seeds are then ground by using hand machines or simply crushed with a mortar and pestle into a paste or cake to make extraction easy and rapid.

### 18.4.3.2 Extraction

Researchers used different methods to extract oil with chemical solvents in a Soxhlet apparatus. A common method and modifications are described below.

A solvent is poured into a round-bottom flask in which paste is placed and then inserted in the center of the extractor inside the heating mantle. A condenser is connected to the extractor. The Soxhlet is heated at 60°C, and when solvent boils, the vapor rises through the vertical tube into the condenser at the top. The condensed liquid drips on a filter paper thimble. The extract seeps through the pores of the thimble and fills the siphon tube, where it flows back down into the round-bottom flask. This operation takes around 30–45 min. The extract is removed from the tube, dried in the oven, and cooled in the desiccator.

This operation is repeated several times to get the desired quantity of oil. At the end of the extraction, the resulting mixture containing the oil is heated at 70°C to recover or remove solvent from the oil, and the percentage of oil extracted is determined. The solvent can also be evaporated using a rotary evaporator (Salimon et al. 2010).

Oil can be extracted by various chemical solvents (hexane, cyclohexane, ethyl acetate, methanol, isopropanol, ethanol, pentane, and petroleum ether). Among them, ethyl acetate and methanol have been extensively used with oil contents of 56.0% and 55.2%, respectively (Dasari and Goud 2014). Meneghetti et al. (2006) reported that biodiesel can be obtained by transesterification of castor oil with either ethanol or methanol, with a similar yield of fatty acid esters; however, methanol as a transesterification agent is rapid. Methanol as a solvent and sodium hydroxide as a catalyst can replace ethanol and potassium hydroxide, respectively (Dasari and Goud 2014). Also, despite the high moisture content, great reduction in viscosity is possible with the transesterification reaction (Thomas et al. 2013). Recently, Compton et al. (2015) suggested solvent-less esterification with ferulic acid. The feruloytated oil is a ultraviolet (UV) (280–361 nm) absorbent and antioxidant, and therefore can be a potential candidate for incorporation into lipid bilayers to protect liposomes and their contents from reactive oxygen species. Similarly, a liquid industrial waste from distillery (known as feint) can be an effective bioresource substitute for commercial solvents for the extraction of oil used in resin production (Akartha and Anusiem 1996). Conclusively, mechanical extraction (hot pressing using a hydraulic press), followed by chemical extraction (solvent), is common. Oil extracted (%) is calculated as (oil/sample size) × 100 (Akpan et al. 2006; Mgudu et al. 2012). The difficulty is that even with the same method used by several researchers, there is still a difference in oil yield, which is attributed to use of immature seeds and different conditions of seed storage (Meneghetti et al. 2006).

The data of two extraction methods are shown in Table 18.2. With mechanical extraction, a higher content of 49.8% of free fatty acids (FFAs) was obtained compared with wet extraction (vs. 33.4%), but the content of ricinoleic acid was less (64.81%) in mechanical extraction than in chemical extraction (76.06%). Viscosity is much higher with mechanical extraction at room temperature (Perdomo et al. 2013). Density becomes important for estimating the production cost and process design for large-scale industrial production of castor oil. Generally, the density of extracted oil is a little greater with mechanical extraction than with chemical extraction (Table 18.2). Otherwise, the oil yield obtained by Soxhlet extraction is 1.5 times greater than that obtained by mechanical extraction. Since relative

**TABLE 18.2**

Effect of Two Extraction Methods on Oil Yield, Content of Fatty Acids, and Oil Density

| Content/Method | Soxhlet Extraction | Mechanical Extraction (Room Temperature) | Mechanical Extraction (60°C) |
|---|---|---|---|
| Oil yield (%) | 40.175–56.218 | – | 26.314–36.597 |
| Free fatty acids (%) | 0.0291–0.0557 | 0.0272–0.1504 | 0.0420–0.0950 |
| Palmitic acid (%) | 0.15 | 0.30 | 0.22 |
| Stearic acid (%) | 1.03 | 6.42 | 10.16 |
| Oleic acid (%) | 1.39 | 4.61 | 3.80 |
| Linoleic acid (%) | 9.17 | 23.28 | 1.75 |
| Linolenic acid (%) | 0.36 | 0.58 | 0.16 |
| Ricinoleic acid (%) | 76.06 | 64.81 | 83.99 |
| Density (g/cm$^3$) | 0.931–0.949 | – | 0.956–0.958 |

*Source:* Perdomo, F.A. et al., *Curr. Sci.*, 110(10), 1890–1892, 2013.

solubility and solvent chemical affinity are important factors in the extraction process, chemical extraction results in cleaner oil, as solvent hexane is washed out during processing. If the cold-press method is used in mechanical compression, oil has a low acid value and iodine value, with lighter color, and a high saponification value compared with Soxhlet-extracted oil (Okullo et al. 2012).

## 18.5 Refining of Extracted Oil

Refining is needed to improve certain oil properties. For example, there is a reduction at high temperature in the viscosity, peroxide value, acid value, saponification value, and iodine value, and an increase in the pour point (Table 18.3). Generally, viscosity is reduced after a transesterification reaction, irrespective of the initial high viscosity and high moisture content (Akpan et al. 2006; Thomas et al. 2013). After removing stones and other impurities, clay is ground and mixed with water. To activate, 2 M HCl is added to the clay and the mixture is boiled at 100°C for 2 h. The mixture is washed with water and then dried and ground (Akpan et al. 2006). The extracted oil is degummed by adding boiling water, and the mixture is stirred for 2 min and allowed to stand in the separating funnel. The aqueous layer is removed and the procedure is repeated until all gum is removed (Akpan et al. 2006). The next step is neutralization. For this process, 60 g of degummed oil is poured into a beaker and heated to 80°C, after which 40 ml of 0.1 M NaOH is added and stirred to a uniform solution. Sodium chloride is added at 10% (w/w) so that formed soap settles down. This material is transferred into a separating funnel and allowed to stand for 1 h. Soap is separated from the oil, and hot water is added again and again to the oil solution until soap is fully removed. Neutralized oil is then drawn off into a beaker (Akpan et al. 2006).

**TABLE 18.3**

Certain Properties of Crude Oil and Refined Oil

| Oil Properties | Nigeria | | Sudan | |
|---|---|---|---|---|
| | Crude | Refined | Crude | Refined |
| Viscocity at 28°C | 9.424 | 6.484 | – | – |
| Viscocity at 40°C | – | – | 234.0 | 209.6 |
| Viscocity at 100°C | – | – | 18.79 | 18.30 |
| Ester value | – | – | 178.09 | 177.65 |
| Peroxide value | – | – | 6.93 | 5.90 |
| Pour point (°C) | – | – | 5.0 | 7.0 |
| pH | 6.11 | 6.34 | – | – |
| Acid value (mg of NaOH/g of oil) | 1.148 | 0.869 | 1.231 | 0.916 |
| Saponification value (mg of KOH/g of oil) | 185.83 | 181.55 | 179.33 | 178.56 |
| Iodine value (g of $I_2$/100 g of oil) | 87.72 | 84.8 | 86.98 | 84.23 |
| Specific gravity | – | – | 0.963 | 0.960 |

*Source:* The Nigeria data are from Akpan, U.G. et al., *Leonardo J. Sci.*, 8, 43–52, 2006. The Sudan data are from Abdelaziz, A.I.M. et al., *J. Chem. Eng.*, 2(1), 1–4, 2014.

For bleaching, 50 g of neutralized oil is poured into a beaker and heated to 90°C. Activated clay is added at 15% (w/w), and the mixture is stirred continuously for 30 min until the temperature rises to 110°C for another 30 min. The whole content is filtered hot in an oven at 70°C (Akpan et al. 2006; Abdelaziz et al. 2014). Further, to obtain pure oil, 20 g of oil is warmed at 35°C and 15 ml of concentrated sulfuric acid (98% pure) is added. The mixture is allowed to react with constant stirring, and then is washed with hot distilled water and left to stand for 2 h, after which water is removed and the sulfuric acid ester formed is finally neutralized with 10 ml of 0.1 M sodium hydroxide (Akpan et al. 2006). After refining and bleaching, oil becomes colorless or slightly yellowish, from the original pale straw color.

Equally important for industrial use are the various reactions of castor oil and characteristics of oil grades (Tables 18.4 and 18.5).

**TABLE 18.4**

Various Reactions of Castor Oil

|  | Nature of Reaction | Added Reactants | Type of Products |
|---|---|---|---|
| Ester linkage | Hydrolysis | Acid, enzyme, or Twitchell reagent catalyst | Fatty acids, glycerol |
|  | Esterification | Monohydric alcohols | Esters |
|  | Alcoholysis | Glycerol glycols, pentaerythritol, and other compounds | Mono- and diglycerides, monoglycols, etc. |
|  | Saponification | Alkalies, alkalies plus metallic salts | Soluble soaps, insoluble soaps |
|  | Reduction | Na reduction | Alcohols |
|  | Amidation | Alkyl amines, alkanolamines, and other compounds | Amine salts, amides |
| Double bond | Oxidation, polymerization | Heat, oxygen, cross-linking agents | Polymerized oils |
|  | Hydrogeneration | Hydrogen (moderate pressure) | Hydroxystearates |
|  | Epoxidation | Hydrogen peroxide | Epoxidized oils |
|  | Halogeneration | $Cl_2$, $Br_2$, $I_2$ | Halogenated oils |
|  | Addition reactions | S, maleic acid | Polymerized oils, factice |
|  | Sulfonation | $H_2SO_4$ | Sulfonated oils |
| Hydroxyl group | Dehydration, hydrolysis, distillation | Catalyst (plus heat) | Dehydrated castor oil, octadecadienoic acid |
|  | Caustic fusion | NaOH | Sebacic acid, capryl alcohol |
|  | Pyrolysis | High heat | Undeclenic acid, heptaldehyde |
|  | Halogenation | $PCl_5$, $POCL_3$ | Halogenated castor oils |
|  | Alkoxylation | Ethylene and/or propylene oxide | Alkoxylated castor oils |
|  | Esterification | Acetic, phosphoric, maleic, and phthalic anhydrides | Alkyl and alkylaryl esters, phosphate esters |
|  | Urethane reaction | Isocyanates | Urethane polymers |
|  | Sulfation | $H_2SO_4$ | Sulfated castor oil (Turkey red oil) |

**TABLE 18.5**

Characteristics of Castor Oil Grades

| Properties | Cold-Pressed Oil | Solvent-Extracted Oil | Dehydrated Oil | Methanol-Extracted Oil |
|---|---|---|---|---|
| Specific gravity | 0.961–0.963 | 0.957–0.963 | 0.926–0.937 | 0.961 |
| Acid value | 3 | 10 | 62 | 0.91 |
| Iodine value (Wij) | 82–88 | 80–88 | 125–145 | 89 |
| Saponification value | 179–185 | 177–182 | 185–188 | 185 |

## 18.6 Methods of Oil Analysis for Physicochemical Properties

This is a very important aspect of castor oil for its utilization. Therefore, intensive research has been undertaken on the extraction and characterization of castor seed oil in tropical countries, particularly Brazil (Conceicao et al. 2007), India (Sridhar et al. 2010), Malaysia (Salimon et al. 2010), Mexico (Perdomo et al. 2013), Nigeria (Akpan et al. 2006; Abitogun et al. 2008; Nangbes et al. 2013), Pakistan (Chakrabarti and Ahmad 2008), Sudan (Abdelaziz et al. 2014), and South Africa (Mgudu et al. 2012).

Different methods have been used, the most common being those recommended by the Association of Official Analytical Chemists, British Pharmacopoeia, and American Society for Testing and Materials (ASTM). Stubiger et al. (2003) found that the techniques of high-performance liquid chromatography (HPLC) and mass spectrometry (MS) are satisfactory for detecting and quantifying fatty acids. Later, Cvengros et al. (2006) suggested vapor pressure osmometry (VPO) to determine molecular weight accurately. General methodologies for studying the physicochemical properties of castor oil are discussed below.

### 18.6.1 Acid Value

This is the quantity of potassium hydroxide expressed in milligrams that is needed to neutralize the acidic constituents or free fatty acids in a gram of castor oil. Diethyl ether and ethanol, 25 ml each, are mixed in a 250 ml beaker. This mixture is added to 10 g of oil in a 250 ml conical flask, and a few drops of phenolphthalein are added to the mixture, which is titrated with 0.1 M NaOH to the end point with continuous shaking until the mixture becomes dark pink. An increase in acid value is due to hydrolysis of triacylglycerol (an important aspect for long shelf life of castor oil). The acid value is calculated as 2× the free fatty acid value (Akpan et al. 2006; Mgudu et al. 2012).

### 18.6.2 Color

Generally, the color of castor oil is determined by using a lovibund tintometer and half-inch cell. The pale-yellow color is generally attributed to the low acid value (Mgudu et al. 2012).

### 18.6.3 Fatty Acid Profile

The fatty acid profile is important because of the dependence of biodiesel properties on the structure and type of fatty acid alkyl esters. This profile can be used to check the level of oxidative deterioration of the oil by enzymatic and chemical oxidation. For example, heating (Soxhlet extraction or mechanical extraction at 60°C) increases the FFA percent, and thereby the acid value, compared with extraction done at room temperature (Table 18.6). Perdomo et al. (2013) quantified free fatty acid content by taking 100 µl of acid and mixing with 1 ml of NaOH methanolic solution. Oil samples are heated to 100°C for 25 min, and 6 ml of HCl methanolic solution is placed into the solution and heated again up to 80°C for 10 min, followed by the addition of 3.75 ml of an equimolar hexane–methyl tert-butyl ether solution. The upper phase is removed and mixed with 9 ml of NaOH solution, and its volume is measured. Fatty acids are analyzed by using a HP 5890 gas chromatograph equipped with software. The identification of peaks is performed by comparing the

**TABLE 18.6**

Physicochemical Properties of Castor Oil from Different Provenances

| Parameters | Malaysia | Pakistan | Sudan | Nigeria (Plateau State) | Nigeria (Osun State) | ASTM |
|---|---|---|---|---|---|---|
| Acid value (mg NaOH/g of oil) | 4.9 | – | 0.916 | 14.8 | 14.8 | 0.4–4.0 |
| Color (unit) | – | – | – | 14.00 | 14.00 | – |
| Congealing temperature (°C) | – | – | – | – | –18 | –21.7 |
| Copper corrosion (1–3 scale, 3 being corrosive) | – | – | – | – | 1 | – |
| Density (g/cm³, 20°C) | – | 0.9584 | – | 0.948 | – | – |
| Ester value | – | – | 177.6 | – | – | – |
| Fire point (°C) | – | – | – | 256 | 256 | – |
| Flash point (°C) | – | 310 | 305 | 225 | 225 | 320 |
| Free fatty acids (%) | 3.4 | – | – | 7.4 | 7.4 | 0.3–2.0 |
| Iodine value (mg/g) | 84.5 | – | 84.2 | 58.64 | 58.39 | 82–88 |
| Lipid content (%) | 43.3 | – | – | – | – | – |
| Moisture content (%) | 0.2 | – | – | 0.30 | 0.30 | 3.16–3.72 |
| Molecular weight | 937.7 | – | – | – | – | – |
| pH | – | – | – | 5.8 | 5.8 | – |
| Peroxide value (mEq/kg) | – | – | 5.9 | 158.6 | 178.0 | – |
| Refractive index (25°C–30°C) (20°C–25°C) | 1.47 | – | – | 1.792 | 1.792 | 1.476–1.179 |
| Saponification value (mg KOH/g oil) | 182.9 | – | 178.6 | 180.77 | 178.0 | 175–187 |
| Smoke point (°C) | – | – | – | 215 | 215 | – |
| Specific gravity (29°C–25°C) | – | – | 0.96 | – | 0.948 | 0.957–0.968 |
| Tert-butyl nitrite | – | – | 0.34 | – | – | – |
| Turbidity (Jackson turbidity unit) | – | – | – | 5.0 | 5.0 | – |
| Viscocity (mm²/s, 25°C) | – | – | – | – | – | 6.3–6.8 |
| Viscocity (mm²/s, 40°C) | 332 | 239 | 209 | 0.425 | 0.425 | 240.12 |

*Source:* The Malaysia data are from Salimon, J. et al., *Sains Malaysiana*, 39(5), 761–764, 2010. The Pakistan data are from Chakrabarti, M.H., and Ahmad, R., *Pak. J. Bot.*, 40(3), 1153–1157, 2008. The Sudan data are from Abdelaziz, A.I.M. et al., *J. Chem. Eng.*, 2(1), 1–4, 2014. The Nigeria (Plateau State) data are from Nangbes, J.O. et al., *Int. J. Eng. Sci.*, 2(9), 105–109, 2013. The Nigeria (Osun State) data are from Abitogun, A. et al., *Internet J. Nutr. Wellness*, 8(2), 2008.

retention times with standard methyl ricinoleate and other fatty acid methyl esters. The lower content of FFA is generally a result of impurities in crude oil.

Mgudu et al. (2012) used another method. A mixture of 12.5 ml of diethyl ether + 12.5 ml of ethanol is taken in a beaker and 5 g of oil is in a conical flask. A few drops of phenolphthalein are added. The mixture is titrated with 0.1 M NaOH with constant shaking until a dark pink color appears. A volume of 0.1 M NaOH is noted. Here, 100 ml of 0.1 M NaOH is equal to 2.83 g of oleic acid. They proposed a formula for calculating the percentage of fatty acid content (%) as (Vo/Wo) × 2.83 × 100, where Vo is the volume of 0.1 M NaOH (100 ml of 0.1 M NaOH = 2.83 g of oleic acid) and Wo is the sample weight.

### 18.6.4 Iodine Value

This value is the measure of the degree of unsaturation. The method used by Akpan et al. (2006) is described for estimating iodine value. An oil sample of 0.4 g is weighed in a conical flask, and 20 ml of carbon tetrachloride is added to dissolve the oil. Then, 25 ml of Dam's reagent is added to the flask using a safety pipette in a fume chamber. The content is vigorously stirred after putting a stopper in the flask and placing it in the dark for 2 h, 30 min. At the end of the period, 20 ml of 10% aqueous potassium iodide and 125 ml of water are added using a measuring cylinder. The content is titrated with 0.1 M sodium thiosulfate solution until the yellow color disappears. A few drops of 1% starch indicator are added, and the titration is continued by adding thiosulfate drops until the blue coloration disappears after vigorous shaking. The same procedure is used for a blank sample and the iodine value is calculated as $12.69C(V1 - V2)/M$, where C is the concentration of sodium thiosulfate, V1 is the volume of sodium thiosulfate used for the blank, V2 is the volume of sodium thiosulfate used for determination, and M is the mass of the sample.

### 18.6.5 Value of pH

Mgudu et al. (2012) proposed a method in which 2 g of oil is poured into a clean dry beaker and 13 ml of hot distilled water is added. The mixture is stirred slowly and then cooled in a cold water bath to 25°C. The pH electrode is first standardized with a buffer solution and then submerged into the oil–water mixture. The formula for estimating the pH was not given by these researchers.

### 18.6.6 Refractive Index

This value is an indication of the level of saturation of the oil. A few drops of the sample are transferred into the glass slide of the refractometer coupled with a thermometer, calibrated specimen, and light source. Water at 30°C is circulated around the glass slide to keep a uniform temperature. Through the eyepiece of the refractometer, the dark portion viewed is adjusted to be in line with the intersection of the cross. At no parallax error, the pointer on the scale points to the refractive index. This procedure is repeated and the mean value is recorded as refractive index (Akpan et al. 2006).

### 18.6.7 Saponification Value

Akpan et al. (2006) described the following method of saponification. In a conical flask, 2 g of cake + 25 ml of 0.1 N ethanolic potassium hydroxide are taken. This mixture is boiled gently for 60 min with continuous stirring. A reflex condenser is placed on the flask containing this mixture, and a few drops of phenolphthalein indicator are added to the warm solution and then titrated with 0.5 M HCl to the end point until the pink color of the indicator disappears. The same procedure is followed for a blank test. The formula for calculating the saponification value is 56.1 N(V0 – V1)/M, where V0 is the volume of the solution used for the blank test, V1 is the volume of the solution used for determination, N is the actual normality of HCl used, and M is the mass of the sample.

### 18.6.8 Specific Gravity

A clean and dry-density bottle or conical flask of 25 ml capacity is weighed and then filled in with 10 ml of oil. After inserting a stopper, the bottle is reweighed (W1). Oil is substituted with water after washing and drying the bottle, and weighed (W2). The formula proposed by Akpan et al. (2006) and Mgudu et al. (2012) is Sp. gr. = (W1 – W0)/W2 – W0 = mass of the substance/mass of an equal volume of water.

### 18.6.9 Triacylglycerols

Salimon et al. (2010) used high-performance liquid chromatography. The mobile phase was acetone–acetonitrile (63.5:36.5), and the flow rate, column temperature, detector temperature, and analysis time were 1 ml/min, 30°C, 40°C, and 30 min, respectively. Oil samples (2 ml containing 0.1 ml of oil dissolved in mobile phase solvent) were injected, and each peak was identified by comparing it with the standard sample based on the equivalent carbon number.

### 18.6.10 Viscosity

For estimating oil viscosity, Akpan et al. (2006) have proposed a rapid system. A clean, dry viscometer with a flow time above 200 s for the fluid is used. The sample is filtered through a fine mesh screen to eliminate dust and other solid material. The viscometer is charged with the sample by inverting the tube in the thinner arm into the liquid sample, and suction force is drawn up to the upper timing mark of the viscometer, after which the instrument is tuned to its normal vertical position. The meter is placed into a holder and inserted into a constant temperature bath at 29°C; about 10 min is allowed for the sample to reach the bath temperature of 29°C. The suction force is then applied to the thinner arm to draw the sample slightly above the upper timing mark. The afflux time is noted by timing the flow of the sample from the upper mark to the lower mark.

Perdomo et al. (2013) used a VM 3000 Stabinger viscometer at 40°C and 5 ml of oil sample and reported fourfold greater viscosity with mechanical extraction at room temperature than with mechanical extraction at 60°C or chemical extraction. The greater value is generally due to suspended particles present in the crude oil.

## 18.7 Comparison of Physicochemical Properties from Different Provenances

Data for some important physicochemical properties of refined oil from four countries and those recommended by ASTM are given in Table 18.6. The properties differ considerably from one country to another country, and from one region to another region within a country. Besides, the factors affecting seed quality, methods used for oil extraction, and methods for estimating physicochemical properties are major contributing factors for this variation. Therefore, standardization of these methods is needed for making oil marketable at a cheaper rate than petroleum-based fuels. Since the data on properties are often fragmentary, comparison is difficult and recommendations cannot be made. However, it gives an idea of content from different geographic origins. It is also useful to improve these properties in order to bring them closer to the standard of ASTM.

## 18.8 Factors Affecting Oil Content and Physicochemical Properties

### 18.8.1 Cultural Practices of Castor Cultivation

In India, delayed sowing was found to be promising with respect to oil quality characteristics and fatty acid composition. On the other hand, the content of ricinoleic acid was not influenced by the date of sowing. Higher oil content under dry cultivation than in irrigated fields has been reported (Ramanjaneyulu et al. 2013). In Egypt, castor cultivation with foliar spraying of K-amino fertilizers, irrigations (75% of water requirement), and/or 100% of the water evapotranspiration regime was found appropriate to produce maximum oil (Mohammed and Mursey 2015). These examples show that suitable cultural practices can be one of the factors for high oil content in castor plants, although these criteria may differ in each crop-growing area.

### 18.8.2 Castor Genotypes

In Brazil, Ramos et al. (1984) surveyed 36 castor varieties with a large variability in oil content and fatty acid composition. Similarly, 11 genotypes in Mexico (Armendaziz et al. 2015) and 31 genotypes in India (Kallamadi et al. 2015) were tested for oil content with wide variation. Perdomo et al. (2013) analyzed oil extracted from seven cultivars from Queretaro state alone in Mexico and reported minimum and maximum oil contents of 31.5% and 56.21%, respectively. The content of fatty acids also varied considerably, for example, palmitic acid (0.00%–0.41%), stearic acid (0.04%–1.58%), oleic acid (0.32%–2.08%), linoleic acid (1.9%–21.69%), linolenic acid (0.05%–0.86%), and ricinoleic acid (74.68%–95.49%). A natural mutant with a high content of oleic acid (cv. A-74) and another mutant with a low content of ricinoleic acid (cv. OLE-1) showed great potential for castor oil content (Velasco et al. 2005). In both cultivars, there was a 10% increase in oil at plant maturity, with no difference in the content of palmitic, linoleic, and linolenic acids (Velasco et al. 2005). Ricinoleic acid content raised to >80% at 21 days after pollination in one mutant and to 75% at 78 days in another mutant. However, the content of gamma-tocopherol was similar in both cultivars and alpha-, beta-, and delta-tocopherol raised at maturity to 605–785 mg/kg of dry seeds (Velasco et al. 2005). The acid value (0.66%–3.88%) and saponification values (166.5%–195.2%) differed significantly among local cultivars from Nigeria, whereas the iodine value, specific gravity, viscosity, refractive index, pH, and peroxide value did not differ among cultivars (Oluwole et al. 2015).

Several methods, including genetic markers, *in vitro* propagation, and genetic transformation, have been successfully exploited to screen cultivars for obtaining high oil yield (Singh et al. 2015). Genetic diversity can also be assessed with a dendogram prepared with the help of simple sequence repeat (SSR) and amplified fragment length polymorphism (AFLP) (Pecina-Quintero et al. 2013). These techniques are useful to breed and select cultivars with desired traits of high oil content from local, regional, or national-level collections. Currently, there are 50 gene banks of castor genotypes across the world (Anjani 2012), which facilitate exchange and distribution of seed for different countries.

### 18.8.3 Geographical Origin

Each region has its own characteristic features, such as climate, soil, crop cultivation practices, and varietal preference. Therefore, oil content differs considerably from one country to another, for example, 33.4%–49.8% in Mexico (Perdomo et al. 2013), 32% in Sudan (Abdelaziz et al. 2014), 43.69%–52.78% in Tanzania (Omari et al. 2015), 35%–46% in Iran (Alizeralu et al. 2011), and 48% in Nigeria (Abitogun et al. 2008). There is wide variation in physiochemical properties among oils from different countries, for example, acid value (0.44–1.97 mg NaOH/g), free fatty acids (0.22%–0.99%), peroxide value (10.87–13.73 mEq/g), saponification value (165.5–187.5 mg/KOH/g), iodine value (78.15–83.42 g of $I_2$/100 g), specific gravity (0.945–0.985), refractive index (1.468–1.473), pH (5.7–6.3), and viscosity (9.0–9.3) for oil from Nigeria (Abitogun et al. 2008), and moisture content (0.3%–1.2%), refractive index (1.404–0.430), saponification value (164–179 mg KOH/g), acid value (0.2–0.9 mg NaOH/g), iodine value (75–86 g of $I_2$/100 g), and peroxide value (0–0.5 mEq/kg) for oil from Iran (Alizeralu et al. 2011).

In the case of fatty acids, Table 18.7 shows that a higher content (>90%) of the main component (ricinoleic acid) is from Brazil, Pakistan, and Tanzania. The exchange of seeds of castor genotypes with great oil potential should therefore be encouraged.

### 18.8.4 Temperature of Extraction

According to Canvin (1965), the temperature of chemical extraction does not have any effect on the oil content and fatty acid composition. On the contrary, temperature did affect these contents in mechanical extraction. For example, Perdomo et al. (2013) reported an oil

**TABLE 18.7**

Content (%, w/w) of Fatty Acids in Castor Oil from Five Countries

| Fatty Acid | Malaysia | Brazil | Pakistan | Nigeria (Plateau State) | Nigeria (Osun State) | Tanzania (6 Regions) |
|---|---|---|---|---|---|---|
| Linoleic | 7.3 | 4.4 | 4.4 | 0.61 | 0.61 | 2.9–4.8 |
| Linolenic | 0.5 | 0.2 | 0.2 | 0.33 | 0.30 | – |
| Olei | 5.5 | 2.8 | 2.8 | 2.28 | 2.28 | 1.4–5.1 |
| Palmitic | 1.3 | 0.7 | 0.7 | 0.46 | 0.46 | 2.3–4.8 |
| Ricinoleic (of total acids) | 84.2 | 90.2 | 90.2 | 83.97 | 81.94 | 83.5–92.3 |
| Stearic | 1.2 | 0.9 | 0.9 | 0.52 | 0.50 | 1.1–4.2 |
| Saturated acids | 2.5 | 1.6 | – | – | – | 3.8–11.7 |
| Unsaturated acids | 97.5 | 97.6 | – | – | – | 88.3–96.2 |

*Source:* The Malaysia data are from Salimon, J. et al., *Sains Malaysiana*, 39(5), 761–764, 2010. The Brazil data are from Conceicao, M.M. et al., *Renew. Sustain. Energy Rev.*, 11, 964–975, 2007. The Pakistan data are from Chakrabarti, M.H., and Ahmad, R., *Pak. J. Bot.*, 40(3), 1153–1157, 2008. The Nigeria (Plateau State) data are from Nangbes, J.G. et al., *Int. J. Eng. Sci.*, 2(9), 105–109, 2013. The Nigeria (Osun State) data are from Abitogun, A. et al., *Internet J. Nutr. Wellness*, 8(2), 2008. The Tanzania data are from Omari, A. et al., *Green Sustain. Chem.*, 5(4): 154–163, 2015.

content of 40.17%–56.22% in chemical extraction and a lower quantity (26.31%–36.595%) in mechanical extraction, even at 60°C. A greater content at 60°C of stearic acid (10.16% vs. 6.42% at room temperature) and ricinoleic acid (83.99% vs. 64.81% at room temperature), and a smaller amount of palmitic acid (0.30% vs. 0.22% at room temperature), oleic acid (4.61% vs. 3.8% at room temperature), linoleic acid (1.75% vs. 23.28% at room temperature), and linolenic acid (0.16% vs. 0.58% at room temperature), has been reported by Perdomo et al. (2013). Similarly, a maximum oil yield of 41.67% (oil recovery of 75.76%) from crushed seed was obtained at 90°C and 135 kPa for a pressing time of 12 min (Olaniyan 2010). Better oil recovery is the effect of temperature inducing a rupture in triacylglycerols in the oil and generating a rise in free fatty acids and diacylglycerols (Perdomo et al. 2013). Oil viscosity decreases when heated, but the yield is reduced (Perdomo et al. 2013). In the cold-press method, oil has a low acid value and iodine value, with lighter color, and a high saponification value compared with Soxhlet (heat-treated)–extracted oil (Okullo et al. 2012).

### 18.8.5 Pretreatment of Castor Seeds

In Nigeria, as a traditional method, dehulled or undehulled castor seeds are boiled, roasted, or simply heated as raw seeds before oil extraction (Oluwole et al. 2015). When seeds were pretreated with methanol, there is a significant increase in oil yield (50% from 40%) and pour point (–21°C from –15°C), and a reduction in acid value (0.925 from 3.92 mg KOH/g) (Dasari and Goud 2014). Mgudu et al. (2012) used microwave heating before extraction with six treatments (119 W for 30, 50, and 120 s and 280 W for 30, 50, and 120 s, frequency of 2450 MHz) and control (0 W, 0 s), and reported a significant increase in oil yield (44.34% vs. 39.5% in control), and no significant change in the content of FFA (0.339% vs. 0.336% in control) and pH (7.7–8.1 vs. 7.58 in control). The records on refractive index, specific gravity, and color did not show any particular trend, and differences between treatments and control samples were not mentioned (Mgudu et al. 2012). Microwave treatment decreases the moisture content of the seeds, thereby making them more fragile and their tissues rupture easily, their cell membranes disintegrate, and porosity is increased. Also, the process is quick, requires less energy, and preserves most thermolabile compounds from oxidative deterioration (Mgudu et al. 2012). Considering these criteria, microwave preheating can be suggested as a substitute for conventional oven heating. Perdomo et al. (2013) also recommended heating the seeds before compressing in both methods to make industrial oil production economic and efficient.

Other factors, including solute-to-solvent ratio and extraction time, can be important for oil extraction (Dasari and Goud 2014). Ideally, a leaching time of 2 h at 50°C for 0.05 g of solvent per milliliter of solute has been suggested by Mbah et al. (2014). These parameters need intensive studies, as data are not available on their impact.

## 18.9 Uses of Castor Oil

### 18.9.1 Medicinal Uses

The uses described in this section are from various informal sources of literature (product brochures, company bulletins, etc.) and websites. Therefore, they should not be taken as recommendations, particularly for public health. Medical practitioners should also verify some of these uses for future prospects because various grades of castor oil are commercially available with different acid values, moisture levels, colors, and purities. Extensive review by Atkinson (2015) enumerated recipes of castor oil for beauty and health. Therefore, only salient features are discussed below.

1. Castor oil stimulates the lymphatic system and strengthens the immune system by increasing white blood cells (WBCs) that help fight against infections, because oil increases the count of T-11 cells and production of lymphocytes in the blood and initiates more antibodies, as well as kills bacteria, fungi, viruses, and even cancer cells.

2. Traditionally, castor oil is used in the treatment of constipation and dysentery. When taken orally, oil acts as a laxative to relieve one from constipation, as it acts on the digestive track in about 4–6 h. The mode of action is that ricinoleic acid is released in the intestine and the digestive system is relaxed, facilitating bowel movement. It is also given to children orally for deworming.

3. Oil is antibacterial, antiviral, and antifungal. In an *in vitro* assay, Momoh et al. (2012) assessed essential oils of castor against 14 species of bacteria and 6 species of fungi and reported that bacteria are more susceptible than fungi. The minimum inhibitory concentration was 6.25–12.50 mg/ml for bacteria and 12.5–25 mg/ml for fungi. When compared with commercial products (erythromycin, ampiclox, rifampin, etc.), castor essential oils were found to be less effective. Castor oil is routinely used in microbial infection in the urinary bladder and vagina.

4. Being anti-inflammatory and analgesic, it is used to treat joint pains, sore muscles, and nerve inflammation. Thus, oil is the cheapest, safest, and most effective treatment option for all inflammatory, degenerative, and malignant disorders (Isah et al. 2006). Better joint pain management is done by placing a hot water bath after applying oil and by repeating this process several times for relief from arthritis. In Africa, seed kernels or hulls are boiled in milk and water. This decoction is used as a traditional remedy to relieve arthritis, low back pain, and sciatica. In the case of acne, oil helps to break up clogged skin glands and pores. It also helps to sanitize skin because it contains undecylenic acid, which kills bacteria and viruses causing acne (Salihu et al. 2014).

5. Before delivery, pregnant women are asked to drink a few tablespoons of castor oil because it induces labor by pushing uterus contraction (Kelly et al. 2013). In fact, ricinoleic acid activates prostaglandin EP3 receptors in the uterus, and thus delivery becomes easier (Tunaru et al. 2012).

6. Oil is added as an ingredient in skin care products and cosmetics (lipstick, shampoo, and soap). Application on sensitive skin is nontoxic, nonirritating, and safe because it is a cell membrane stabilizer, a valuable detergent for degreasing the matrix, and a mitochondrial cleanser (Isah et al. 2006). Application of a mixture of castor oil + baking soda dissolves corns, cysts, and moles due to the fatty acid content of the oil. It is also applied on skin against dermatitis, wound healing, acne, sunburn, keratosis, wrinkles, ringworm, and other skin infections (Salihu et al. 2014). Protection of burns and wounds from infections helps to prevent infections like dry skin warts, boils, athlete's foot, and chronic itching. Oil acts as good skin moisturizer.

7. Oil mixed with either coconut oil or almond oil initiates hair growth and thickens eyebrows and eyelashes, as it boosts blood circulation to the follicles, leading to

faster growth (Salihu et al. 2014). Castor oil is an effective cure for bald patches, and probably makes the hairs dark, because it contains omega-6 essential fatty acids, which are responsible for healthy hair (Salihu et al. 2014). In fact, this occurs in the Odisha State in India, where ladies apply castor oil to their hair almost daily to make it healthy and prevent premature shedding.

8. Castor oil is a remedy for health ailments, such as multiple sclerosis, cerebral palsy, migraine, and menstrual disorders; oral care; proper lactation; and birth control, and a medication for HIV-positive patients and those with sciatica and back pain. The American Cancer Society has suggested that the oil-based commercial product Cremophor® be used in chemotherapy against cancerous tumors, but it may cause allergy.

There are a few constraints in using castor oil. For example, skin reactions and gastrointestinal upset may occur. Oil is broken down in the small intestine into ricinoleic acid, which acts as an irritant to the intestinal lining and causes digestive discomfort due to diarrhea, cramps, irritable bowel, ulcers, diverticulitis, hemorrhoids, colitis, and prolapses. In the case of maternal labor, discomfort due to nausea is possible. Oil being thick, its application to the skin leaves a sticky feeling.

## 18.9.2 Industrial Uses

### 18.9.2.1 Biodiesel Production

Castor oil is a valuable renewable potential source of raw material for the production of biodiesel in several countries (Ogunniyi 2006; Mutlu and Meier 2010), and an array of reviews and research papers have been published on this subject. Biodiesel (also known as methyl esters of fatty acids) is nontoxic, biodegradable, and an excellent substitute for petroleum-based diesel fuel. The hydroxyl group renders ricinoleic acid as a better valuable feedstock than other oils. Since cold-flow properties are better due to the higher unsaturated fatty acid concentration, castor oil can be blended with petroleum diesel for better performance.

High solubility in methanol makes it an ideal oil for biodiesel production using esterification–neutralization–transesterification (ENT) requiring a minimum amount of catalyst and heating, and thereby reducing the cost of production (Bello and Makanju 2011). The ENT process can yield a high quantity of methyl esters with good biodiesel properties (Silitonga et al. 2016) by reducing viscosity and improving fuel quality for application in the compression ignition (Chakarbarti and Ahmad 2008). Further, transesterification can be optimized by adjustments on the type and ratio of catalysts and reagents, reaction time, catalytic system, temperature, and purification system. By using the ENT method, Armendaziz et al. (2015) recorded an oil yield of 27–431.7 kg/ha in Mexico, with ricinoleic acid methyl ester content up to 84.7%–89.2%. Generally, the best results of biodiesel production have been obtained by using (1) ethanol instead of methanol, (2) an acid catalyst instead of traditional catalysts (vegetable oils), (3) microwave heating, and (4) cosolvents.

Since castor oil has a high energy value and positive fuel properties due to fatty acid composition, its blend can be mixed with diesel at 10% (v/v) (Berman et al. 2011). Fatty acids increase oil lubricity and can therefore replace soybean, sunflower, and canola oilseeds as feedstock for biodiesel (Drown et al. 2001). It is also possible to blend it with regular diesel or biodiesel made with other lipid feedstocks (cottonseed or soybean biodiesel) up to 200 g/kg and with normal diesel at 400 g/kg (Saribiyik et al. 2010). It is a common motor lubricant for internal combustion engines due to its high boiling and low melting points, and resistance to heat compared with petroleum-based oils (Ogunniyi 2006; Mutlu and Meier 2010).

Castor oil is being increasingly used in the pharmaceutical and chemical industries to manufacture high-value products, but it becomes uneconomical to use it for biodiesel with the high cost of feedstock. Castor oil biodiesel may pose difficulties in internal combustion engines because of its high density, viscosity, and hydroscopic properties. These constraints, however, are minor, considering its easy availability and potential for biodiesel production.

### 18.9.2.2 Other Industrial Uses

Recent consumer awareness toward green consumerism has made food industries search for preservatives for safe, healthy, and nutritious food. Essential oils offer an appropriate solution to this (Prakash and Kiran 2016). As such, castor oil is used in food industries as an additive and flavoring agent because glycerol, popularly known as glycerin, is a sugar alcohol that has a sweet taste and low toxicity. Dietary intake per the International Castor Oil Association is as high as 10% for 90 days without any ill effects. The Joint FAO/ WHO Committee on Food Additives has established an acceptable daily oil intake of up to 0.7 mg/kg of body weight, for example, 1 tablespoon for adults and 1 teaspoonful for children. Oil in a processed form, known as polyglycerol polyricinoleate, is used in chocolate bar manufacturing as a less expensive substitute for cocoa butter and as a mold inhibitor to prevent rotting in rice, wheat, and pulses (Wilson et al. 1998). In the future, there is a great scope for essential oils in the marketing chain by improving current food products by using nanoencapsulation, edible coatings, and controlled-release systems.

Thomas et al. (2015) discussed the industrial perspectives of castor oil. Oil serves as a raw material in paint and nylon industries (Thomas et al. 2015). The Turkey red oil (sulfonated castor oil) is used to manufacture industrial lubricants, hydraulic and brake fluids, plastics, and detergents, and in the treatment of leather and other industrial products, such as nylon-6, nylon-10 from sebacic acid, and nylon-11 from 11-amino-undecanoic acid (Isah 2006). The essential oils of nine plants were assessed against two fungi, *Aspergillus niger* Tiegh. and *Geotrichum candidum* Link, to protect wooden structures. Among them, castor oil showed the lowest effectiveness at all doses (5, 10, 20, 30, and 50 ppm). However, a paint containing oil can be an eco-friendly approach in buildings and indoor environments (Verma et al. 2011).

Unsaturated polyester resin (UPR) can be fabricated from oil through blending pentaerythritol glyceride maleates with petroleum-based UPR (Liu et al. 2014). This resin is suitable for a liquid molding process with less shrinking and better mechanical and thermal properties (Liu et al. 2014). Glycerol in castor oil is converted to acrolein and epichlorohydrin and used as a raw material for epoxy resins and for the manufacturing of polyols for flexible foams and, to a lesser extent, rigid polyurethane foams (Thomas et al. 2015). Glycerol is also used to produce nitroglycerin (an essential heart medication) and ingredients of dynamite (smokeless gun powder) and other explosives, fuel components (hydrogen gas production and conversion to ethanol), and chemical products (citric acid, ethylene, and propylene glycol) (Thomas et al. 2015). Detergent manufactured from castor oil is biodegradable and better than synthetic detergents (Isah 2006).

Apart from the above-mentioned major uses, castor oil is an ingredient in the manufacture of aviation fuels, transparent typewriter and printing inks, soaps, textile dyes, coatings, polishes, varnishes, lacquers, grease, hydraulic fluids, linoleum, and coatings of fabrics due to its semidrying property (Ogunniyi 2006; Perdomo et al. 2013; Thomas et al. 2015). Oil serves as bio-based renewable monomers and polymers that are used in the manufacturing of polyurethanes, polyesters, and polyamides (Mutlu et al. 2010).

## 18.10 Castor Oil as Pesticide

Castor oil proved effective against insect pests and diseases of agricultural crops and stored commodities, including food grains. Table 18.8 shows various modes of action of castor oil. Insects and disease pathogens affecting human health are also discussed in this section, with an objective of facilitating an effective control program using castor oil as one of the components.

### 18.10.1 Insect Pests and Diseases of Field Crops

In greenhouse trials, a castor oil–based detergent at 1.5 ml/L in aqueous solution or 10% mixture in water was added to pesticides for spraying a strawberry crop in Brazil. The application resulted in a better control of ants (*Atta* spp.), leaf-feeding beetles (*Epilachna* spp.), red spider mite (*Tetranychus urticae* Koch), anthracnose (*Colletorichum gloeosporioides* [Penz.] Penz. & [Sacc.]), gray mold rot (*Botrytis cinera* Pers.), and bacterial blight or leaf spot (*Xanthomonas fragariae* Kennedy & King) (Galhiane et al. 2012). The residue of pesticides (deltamethrin, folpet, tebuconazle, abamectin, and mancozeb) was at a lower level than when pesticides were used without oil. In a 3-year survey in India, it was found that spraying of a mixture of 3% castor oil + butter milk + extract of fenugreek, betel vine, and onion is practiced by tribal communities to control insect pests on peanut and pulses (Mohapatra et al. 2009). These examples show that castor oil can be a potential synergist for organic and synthetic pesticides. The diamondback moth, *Plutella xylostella* L., was successfully controlled under both laboratory and semifield conditions (Kodjo et al. 2011). In the laboratory, complete mortality of third-instar larvae was achieved with 10% oil emulsion (topical application and ingestion methods) compared with 2%–5% mortality in control and 82%–88% mortality with 5% chlorpyriphos (Dursban®) dusting. Lowest adult emergence (42.3%) and the highest abnormal adults (48%–79%) were also recorded. In field caged cabbage plants, larval mortality (57.7%), adults without deformities (23.7%), and longevity (8–9 days) were reported with 5% castor oil. The corresponding data on untreated plants were 10% mortality, 98% adults without deformity, and longevity of 12–13 days (Kodjo et al. 2011). In yam (*Dioscorea* sp.), infection on tubers caused by two fungi, *Aspergillus flavus* Tiegh and *Fusarium verticillioides* (Sacc.) is common. In an *in vitro* test, crude oil lowered the growth and development of fungal mycelium and showed potential in disease control (Makun et al. 2011).

In a green gram (*Vigna radiata* [L.]) pot culture experiment, the damage of the root-knot nematode, *Meloidogyne incognita* (Kofoid & White) Chitwood, was studied by Wani and Bhat (2012). They found that the soil application of urea coated with castor oil at 0.06 g/pot significantly reduced the root-knot index (1.3 vs. 4.6 in control, rated on a 0–5 scale, 5 ≥ 100 galls/root) and improved the root nodule index (3.3 vs. 1.1 in control, rated on a 1–5 scale, 5 = 100 nodules/root). However, urea coated with neem-based Nimin® gave better results. Similarly, in laboratory and greenhouse experiments on cucumber plants, castor seed oil at 1000 ppm inhibited egg hatching of *M. incognita*, 72 h after treatment, by 71.3% and reduced egg hatching by 16%, thus resulting in low gall density (Katooli et al. 2010).

**TABLE 18.8**

Examples of Field and Storage Pests and Disease Pathogens Controlled with Castor Oil

| Crop/Commodity | Pest/Pathogen Species (Family) | Mode of Action | References |
|---|---|---|---|
| Strawberry (F) | *Epilachna* sp. (Coccinellidae) | SY | Galhiane et al. 2012 |
| | *Tetranychus urticae* (Tetranychidae) | SY | Galhiane et al. 2012 |
| | *Colletotrichum gloesporioides* (Glomerellaceae) | SY | Galhiane et al. 2012 |
| | *Botrytis cinerea* (Sclerotiniaceae) | SY | Galhiane et al. 2012 |
| | *Xanthomonas fragariae* (Xanthomonadaceae) | SY | Galhiane et al. 2012 |
| Cabbage (GH) | *Plutella xylostella* (Plutellidae) | IGR, IN | Kodjo et al. 2011 |
| Bean (S) | *Zabrotes subfasciatus* (Chrysomelidae) | IN | Mushobozy et al. 2009 |
| Cowpea (S) | *Callosobruchus chinensis* (Bruchidae) | OD, OV, IN | Bhargava and Meena 2002; Neog and Singh 2012 |
| Cowpea (S) | *Callosobruchus maculatus* (Bruchidae) | IN, IGR, IN, OD, OV, RE | Haghtalab et al. 2009; Shinde et al. 2011; Fouad 2013a; Yahaya et al. 2013 |
| Cowpea (S) | *Acanthoscelides obtectus* (Bruchidae) | IGR, IN, OD, OV | Nana et al. 2014 |
| Chickpea (S) | *Callosobruchus maculatus* (Bruchidae) | IN | Paceco et al. 1995 |
| Maize (S) | *Sitiphilus zeamais* (Curculionidae) | IN | Wale and Assegie 2015 |
| Coffee (S) | *Hypothenemus hampei* (Curculionidae) | IN | Mohapatra et al. 2009; Celestino et al. 2016 |
| Green gram (GH) | *Meloidogyne incognita* (Heteroderidae) | NE | Wani and Bhat 2012 |
| Green gram (S) | *Callosobruchus hinensis* (Bruchidae) | IN | Singh and Yadav 2003 |
| Green gram (S) | *Callosobruchus maculatus* (Bruchidae) | IGR, OD, RE | Ratnasekara and Rajapakse 2009 |
| Pigeon pea (S) | *Callosobruchus chinensis* (Bruchidae) | IN | Kadve et al. 2016 |
| Wheat (S) | *Trogoderma granarium* (Dermestidae) | IGR, IN | Jakhar and Jat 2010 |
| | *Sitophilus oryzae* (Curculionidae) | IN | Meghwal et al. 2012 |
| | *Rhyzopertha dominica* (Bostrichidae) | IN | Patel and Vekaria 2013 |

*Note:* F, field; GH, greenhouse; IGR, insect growth regulator; IN, insecticidal/toxic; NE, nematicidal/toxic; OD, oviposition deterrent; OV, ovicidal; RE, repellent; S, storage; SY, synergist.

### 18.10.2  Insect Pests of Stored Food Grains

In the laboratory, a pulse beetle or adzuki bean beetle, *Callosobruchus chinensis* (L.), was controlled with seed treatment of castor oil at 1 ml/100 g of seeds of cowpea with 80.7% mortality in adults 3 days after treatment (DAT). It inhibited oviposition (26.5 eggs/female vs. 79.4 eggs in control), reduced egg viability up to 61.7%, and reduced F1 progeny by 85% (Bhargava and Meena 2002). Recently, Kadve et al. (2016) reported the effectiveness of pigeon pea seed treatment with castor oil at 5 ml/kg, in terms of fecundity (66 eggs/100 seeds vs. 122 eggs in untreated seeds), adult emergence (26.6% vs. 88.3% in control), and resulting seed damage (26.3% vs. 85% in control) and seed weight loss (9.3% vs. 20.7% in control). The best treatment, however, was seed treatment with deltamethrin 2.8EC at 0.04 ml/kg, with significantly reduced pest fecundity, adult emergence, seed damage, and weight loss, and better seed germination (90% compared with 81% with castor oil and 50% in untreated seeds) (Kadve et al. 2016). In a trial with the powder (5%, w/w) of leaves or the seed coat of 10 plants, 8 vegetable oils (1%, w/w), and malathion dust (1%, w/w) tested for 2 years against pulse beetle (Neog and Singh 2012), castor oil performed well with 3.33%–4.33% infestation against 56.7%–61.3% in control. In another test, castor oil at seven doses (4, 5, 6, 7, 8, 9, and 10 ml/kg of seeds) with an exposure of 24–120 h showed that insect mortality was dependent on dose and exposure period, with a maximum mortality of 86.7% at 9 ml/kg of seeds with an $LC_{95}$ of 10.9 ml/kg for 72 h in the cowpea seed beetle, *Callosobruchus maculatus* (F.), in stored cowpea (Haghtalab et al. 2009), and a mortality of 99.1% at 9 ml/kg with an $LC_{95}$ value of 2.95 ml/kg for 120 h in the common bean weevil, *Acanthoscelides obtectus* (Say), in bean (*Phaseolus vulgaris* L.) (Nana et al. 2014).

A mixture of castor oil + neem oil (both 2%) in a 1:1 proportion used for treating cowpea seeds at 2 ml/50 g of seeds significantly reduced fecundity of *C. maculatus* at 14 days after treatment (4.7 eggs/50 g of seeds vs. 41.7 eggs in control). When castor oil was used alone, the fecundity was 7.63 eggs/50 g of seeds versus 44.8 eggs in control (Yahaya et al. 2013). Castor oil applied at 5 ml/kg seed inhibited egg laying (38 eggs/200 g of seeds vs. 161 eggs in control); reduced egg hatching (60% vs. 100% in control), adult longevity (1.1 days vs. 9.7 days in control), the infestation rate after 3 months of storage (9.65% vs. 51.5% in control), and adult emergence (4.5 adults/100 g of seeds vs. 131 adults in control); and prolonged the development period (33 days vs. 25 days in control) (Shinde et al. 2011). At a higher dose at 10 ml/kg of seeds, castor oil gave protection to stored green gram up to 280 days, with significant mortality in *C. chinensis* grubs (Singh and Yadav 2003).

Castor oil applied at 1 ml/100 g of seed gave 100% protection for 150 days in chickpea from *C. maculatus*, and for 90 days from the bean beetle, *Callosobruchus phaseoli* (Gyllenhal), in bean (Paceco et al. 1995). A mixture of essential oils + acetone on filter paper was tested in the laboratory against *C. maculatus* at 0.01%, 0.1%, and 1%. In a free choice at 1%, castor oil showed only 1.5% repellency compared with cinnamon oil (47.5% repellency) after 4 h of treatment (Fouad 2013a). In another experiment, essential oils of camphor, castor, cinnamon, mustard, and clove oil at 0.5%, 1%, 2%, and 4% against the faba bean beetle, *Bruchidius incarnatus* (Boheman), were tested (Fouad 2013b). From toxicity and repellent activities, the superiority of cinnamon (*Cinnamomum zeylanicum*) oil was confirmed, whereas castor oil not effective as either a repellent or an insecticidal, even at the highest concentration of 4% (e.g., repellency of 12.1% with castor oil vs. 54.1% with cinnamon oil recorded 1 h after exposure). In a test on oil vapors at 200 µl/20 seeds of green gram, Ratnasekara and Rajapakse (2009) observed significant reduction in the rate of oviposition and adult emergence in *C. maculatus* due to effective repellent action of castor oil, but it was inferior to that of mustard oil, *Solanum indicum* L. or *Calophyllum inophyllum* L. In cowpea stored in a silo in Africa, damage by two bruchids (*C. maculatus* and *Bruchidius atrolineatus* [Pic.]) was significantly reduced by treating seeds with castor oil at 6 ml/kg (Yakubu et al. 2012).

When stored seeds of wheat were treated at 1 ml/100 g of seeds, the duration of larval and pupal development of the khapra beetle, *Trogoderma granarium* Everts, was significantly prolonged (32.6 days vs. 25.1 days in untreated seeds) (Jakhar and Jat 2010). This treatment also reduced adult emergence to 31.3% from 64% in the control and recorded threefold less damage in treated grains (13% vs. 39% in control) (Jakhar and Jat 2010). The rice weevil, *Sitophilus oryzae* (L.), attacking stored wheat was best controlled by mustard oil (75% mortality and 4.53% grain damage), followed by castor oil (61.7% mortality and 7.6% grain damage) (Meghwal et al. 2012). Similarly, only 88% of adults emerged in castor oil–treated grains. In Ethiopia, mortality of the maize weevil, *Sitophilus zeamais* (Mots.), with 2 ml of oil/kg of stored maize was 53% 1 h after treatment, but it increased to >85% when the dose was increased; the corresponding $LD_{50}$ was 2.04 ml (Wale and Assegie 2015). In Brazil, in a control trial against the coffee berry borer, *Hypothenemus hampei* (Ferrai), castor oil applied at 3% (v/v) resulted in 53.7% mortality with an $LC_{50}$ of 3.49% against 40.8% mortality and an $LC_{50}$ of 6.71% in water-sprayed berries (Celestino et al. 2016). Mohapatra et al. (2009) reported from a 3-year survey in India an effective control for the coffee borer with a seed treatment of a mixture of 3% castor oil + 3% peanut oil at 10 ml/kg of seeds.

Castor oil at a concentration of 10 mg/ml proved effective as a strong repellent (58.66%) against the red flour beetle, *Tribolium castaneum* (Herbst), with an $LD_{50}$ value of 5.52 mg/cm for 24 h, with a dose of 2.04 mg/cm$^2$ resulting in significant reduction in egg hatching, survival rate of larvae, and adult emergence (Islam et al. 2014). On the contrary, Jenan (2014) reported 100% mortality in first- through third-instar larvae of the confused flour beetle, *Tribolium confusum* (Val), and khapra beetle due to fumigant action of 10% castor oil in acetone, against only 40% mortality in acetone (control). The seed protectant capacity of essential oil of castor due to contact toxicity ($LD_{50}$ of 615.28 µg/cm$^2$) and fumigant toxicity ($LD_{50}$ of 5.52 and 4.05 mg/cm 24 and 48 h after treatment) was demonstrated against *T. castaneum* by Islam et al. (2014). Khalequzzaman and Choudhury (2003) evaluated four oils, each mixed with pirimophos–methyl in a 1:10 proportion, and reported the maximum toxic effect against four strains of *T. castaneum*, with an $LD_{50}$ of 0.264–0.355 µg/cm$^2$ for pesticide and 9.48–42.97 µg/cm$^2$ for castor oil. The maximum synergism evaluated by a cotoxicity coefficient value was for neem oil (4908.53), followed by sesame oil (434.11), castor oil (295.24), and soybean oil (232.93). This test demonstrated the use of castor oil as a potential synergist.

Wheat seeds were treated with crude castor oil against the lesser grain borer, *Rhyzopertha dominica* (Fb.). There was significantly lower grain damage (9.9% vs. 13% in control) and less seed weight loss (5.6% vs. 11.4% in control), but deltamethrin 2.8EC (0.01 ml/kg) proved to be the most effective treatment (Patel and Vekaria 2013). Infestation of the Mexican bean beetle, *Zabrotes subfasciatus* (Boh.) was controlled by treating bean seeds with castor oil at 5 ml/kg. This treatment was as effective as the oil of sesamum, oil palm, cotton, or maize, but it was inferior to malathion dust (2%) applied to seeds at 0.5 g/kg (Mushobozy et al. 2009). From the examples cited, it can be concluded that crude castor oil applied at various doses either for spraying field crops or treating stored food grains could effectively control insects, nematodes, and plant pathogens.

### 18.10.3 Insects Affecting Human Health

Castor oil is an effective repellent against mosquitoes. It is applied on the body or embodied in lotion, cream, paste, and other preparations, either to facilitate application or for a more lasting effect (Patel et al. 2012). The commercial product Bite Blocker Xtreme®, containing 8% castor oil + 3% soybean oil + 6% geranium oil, has given protection up to 163 min against the flood water mosquito, *Psorophora columbiae* (Dyar & Knob.), in the United States (Qualls et al. 2011). By using a misting system, Cilek et al. (2011) compared essential oils of castor with commercial repellents against two caged mosquito species: dengue/Asian tiger mosquito, *Aedes albopunctus* (Skuse), and home mosquito, *Culex quinquefasciatus* (Say). Castor oil (3%) applied at a dose of 4 ml/L resulted in lower mortality, but repellency was equal compared with two commercial products, Riptide® containing 5% pyrethrin (9 ml/L) and Ecoexempt® containing 18% rosemary oil (9 ml/L). Castor oil is an effective repellent to household insects, including cockroaches, which spread human diseases (Patel et al. 2012). In a laboratory trial, ricinoleic acid esters from castor oil showed cytoplasmic changes that inhibited the development of oocytes and proved to be an effective acaricide against the brown dog tick, *Rhipicephalus sanguineus* (Latreille), which is a recognized vector of many pathogens (Arnosti et al. 2010).

## 18.11 Advantages and Limitations of Oil as a Pesticide

Castor cultivation is easy and profitable, and therefore seed availability in large quantities around the year is possible by planting on large areas. Phytotoxicity due to vegetable and organic oils on agricultural crops has not been reported (Galhiane et al. 2012; Nagar et al. 2012). In crop protection, plant-derived products are used due to their comparatively low cost of preparation, low toxicity to natural enemies of crop pests, safety to humans, biodegradability, and having no record of development of resistance in pests (Gahukar 2014). Moreover, oil affects physiological, nutritional, and behavioral processes, resulting in insect mortality, and shows diverse effects, such as antifeedant, oviposition-deterrent, ovicide, growth regulation and fecundity reduction, and fungal mycelium effects, as well as inhibition of the hatching of insect eggs and root penetration of nematodes. The potential of oil against insect vectors of human diseases plays a significant role in public health. In some instances, the addition of oil to chemicals or other products increased pest mortality and not only showed promising results in crop protection through synergism, but also reduced treatment cost (Gahukar 2014).

There are some limitations that may hinder use of oil. In crop and seed protection, the quality of the crude oil reflects its performance in pest control. For example, while listing pests of stored castor seeds in Africa, Salihu et al. (2014) mentioned the infestation of three insects (*T. castaneum*, the cigarette beetle; *Lasioderma serricorne* [Fb.], and the tropical warehouse moth, *Ephestia cautella* [Walker]) that resulted in deterioration of the quality of seeds, and thereby the quality of oil, making it unsuitable as a pesticide. Similarly, reports from India (Lalithakumari et al. 1971) and Africa (Negedu et al. 2014) showed that oil quality was considerably deteriorated by fungi, *Aspergillus tamarii* (Kita) and *Aspergillus* spp. This fact was evidenced from a significant increase in the content of the moisture, crude protein, ash, crude fiber, and peroxide value, and a significant reduction in the content of the total fat and soluble sugar (Negedu et al. (2014). Thus, stored seeds attacked by insects and disease pathogens become unsuitable for oil extraction.

Seed germination was lowered by 20%–30% when castor oil was used to treat the seed of pulses in India (Raghuwani and Kapadia 2003; Sanappa and Acharya 2014) and maize seeds in Ethiopia (Wale and Assegie 2015). Castor oil treatment at 5 ml/kg to maize seeds against disease pathogens resulted in seed germination up to 64% against 80% with captan at 3 g/kg, but seed vigor was much better in seedlings from oil-treated seeds (Wani et al. 2014).

In the tropics, residual persistence of castor oil on crop plants is affected due to the high temperature in the summer and runoff from plants due to heavy rains during the rainy season. Further, slow toxic action and less effectiveness compared with chemicals make farmers reluctant to use castor oil in pest control. In countries where poor farmers cannot buy costly synthetic pesticides, castor oil–based "ready-to-use" products are not available in the local market. Also, infrastructure for storing seeds, technical know-how for extraction, and quality testing are lacking at the village level, as reported by Morse et al. (2002) for Nigeria. Also, nonstandardization of extraction methods and determination of the physical, chemical, and engineering properties of the oil, fatty acid fractions, and derived chemical feedstocks make biodiesel incompetent in commercial chains.

## 18.12 Conclusions

For extracting castor oil of reasonably high quality, heathy seeds are needed. Therefore, storage conditions should be kept ideal for long shelf life. Considering yield and quality parameters for ultimate utilization of castor oil in industries, chemical extraction proved efficient compared with wet or mechanical extraction. Since castor oil is readily available through large-scale cultivation of castor crop that farmers can cultivate without much technical help and money, its utilization as an insecticide needs to be promoted as seed treatment to reduce pest infestation in stored commodities. In the future, intensive studies may show promising results against pests affecting human health. Of course, there are some limitations that need the attention of researchers and drug manufacturers.

Biodiesel production shows a significant perspective, particularly to reduce greenhouse gas (GHG) emission (Severino et al. 2012). Biodiesel engines have a lower impact on environment than those operating on petroleum fuel (Chakrabarti and Ahmad 2008). These aspects are important to avoid GHG emissions in the environment. Castor oil replacing the feedstock of seeds of cotton, soybean, sunflower, and canola oil should be encouraged, as their consumption as human food and animal feed has increased in recent years.

In food industries, new preservatives based on essential oils should be tested for their toxicology, environmental effect, and product chemistry to ensure their quality and safety to human health. From this perspective, collaboration between the scientific community, food industries, and regulatory authorities is needed at local and national levels.

## References

Abdelaziz, A.I.M., Elamin, I.H.M., Gasmelseed, G.A., Abdalla, B.K. 2014. Extraction, refining and characterization of Sudanese castor seed oil. *J. Chem. Eng.* 2(1): 1–4. http://researchpub.org/journal /jce/number/vol2-no1/vol2-no1-1/pdf.

Abitogun, A., Alademeyin, O., Oloye, D. 2008. Extraction and characterization of castor seed oil. *Internet J. Nutr. Wellness* 8(2).

Akartha, O., Anusiem, A.C.I. 1996. A bioresource solvent for extraction of castor oil. *Ind. Crops Prod.* 5: 273–77.

Akpan, U.G., Jimoh, A., Mohammed, A.D. 2006. Extraction, characterization and modification of castor seed soil. *Leonardo J. Sci.* 8: 43–52.

Alizeralu, A.F.N., Shirzad, H., Harzati, S. 2011. The effect of climatic factors on the production and quality of castor oil. *Nature Sci.* 9(4): 15–19.

Anjani, K. 2012. Castor genetic resources: A primary genepool for exploitation. *Ind. Crops Prod.* 35: 1–14.

Armendaziz, J., Lapuerta, M., Zavala, F., Garcia-Zambrano, E., Ojeda, M.C. 2015. Evaluation of eleven genotypes of castor oil plant (*Ricinus communis* L.) for the production of biodiesel. *Ind. Crops Prod.* 77: 484–90.

Arnosti, A., Brienza, P.D., Furquim, K., Mathias, M.I.C. 2010. Effects of ricinoleic acid esters from castor oil (*Ricinus communis*) on the vitellogenesis of *Rhipcephalus sanguineus* (Latreille, 1806) (Acari: Ixodidae) tick. *Exp. Parasitol.* 127(2): 575–80.

Atkinson, A. 2015. *Essential Oils for Beauty, Wellness and the Home: 100 Natural Non-Toxic Recipes for the Beginner and Beyond.* New York: Skyhorse Publishing.

Bello, E.I., Makanju, A. 2011. Production, characterization and evaluation of castor oil biodiesel as alternative fuel for diesel engines. *J. Emerg. Trends Eng. Appl. Sci.* 2(3): 525–30.

Berman, P., Nizri, S., Wiesman, Z. 2011. Castor oil biodiesel and its blends as alternative fuel. *Biomass Bioenergy* 35(7): 2861–66.

Bhargava, M.C., Meena, B.L. 2002. Efficacy of some vegetable oils against pulse beetle, *Callosobruchus chinensis* (Linn.) on cowpea, *Vigna unguiculata* (L.). *Indian J. Plant Prot.* 30(1): 46–50.

Canvin, D.T. 1965. The effect of temperature on the oil content and fatty acid composition of the oils from several oil seed crops. *Can. J. Bot.* 43(1): 63–69.

Celestino, F.N., Pratissoli, D., Machado, L.C., Santos, H.J.G., Jr., Queiroz, V.T., Mordgan, L. 2016. Control of coffee berry borer, *Hypothenemus hampei* (Ferrai) (Coleoptera: Curculionidae: Scolytinae) with botanical insecticides and mineral oils. *Acta Sci. Agron.* 38(1): 1–8.

Chakrabarti, M.H., Ahmad, R. 2008. Trans-esterification studies on castor oil as a first step towards its use in biodiesel production. *Pak. J. Bot.* 40(3): 1153–57.

Cilek, J.E., Hallmon, C.F., Johnson, R. 2011. Efficacy of several commercially formulated essential oils against caged female *Aedes albopunctus* and *Culex quinquefasciatus* when operationally applied via an automatic-timed insecticide application system. *J. Am. Mosquito Contr. Assoc.* 27(3): 252–55.

Compton, D.L., Laszio, J.A., Evans, K.O. 2015. Phenylpropanoid esters of lesquerella and castor oil. *Ind. Crops Prod.* 63: 9–16.

Conceicao, M.M., Candeia, R.A., Silva, F.C., Bezerra, A.F., Fernandes, V.J., Jr., Souza, A.G. 2007. Thermo-analytical characterization of castor oil biodiesel. *Renew. Sustain. Energy Rev.* 11: 964–75.

Cvengros, J., Paligova, J., Cvengrosova, Z. 2006. Properties of esters base on castor oil. *Eur. J. Lipid Sci. Technol.* 108(8): 629–35.

Dasari, S.R., Goud, V.V. 2014. Effect of pretreatment on solvents extraction and physicochemical properties of castor seed oil. *J. Renew. Sustain. Energy* 6: 063108.

Drown, D.C., Harper, K., Frame, E. 2001. Screening vegetable oil alcohol esters as fuel lubricity enhancers. *J. Am. Oil Chem. Soc.* 78(6): 579–84.

Fouad, H.A. 2013a. Effect of five essential oils as repellents against the cowpea weevil, *Callosobruchus maculatus* (F.). *Bull. Environ. Pharm. Life Sci.* 2(5): 23–27.

Fouad, H.A. 2013b. Bioactivity of essential oils of medicinal plants against *Bruchidius incarnatus* (Boheman 1833). *Res. J. Agric. Environ. Manage.* 2(6): 136–41.

Gahukar, R.T. 2014. Potential and utilization of plant products in pest control. In *Insect Pest Management: Current Concepts and Ecological Perspectives*, ed. D.P. Abrol, 125–139. New York: Elsevier.

Galhiane, M.S., Rissato, S.R., Santos, L.S. et al. 2012. Evaluation of the performance of a castor oil-based formulation in limiting pesticide residues in strawberry crops. *Quimica Nova (Sao Paulo)* 35(2).

Haghtalab, N., Shayesteh, N., Aramideh, S. 2009. Insecticidal efficacy of castor and hazelnut oils in stored cowpea against *Callosobruchus maculatus* (F.) (Coleoptera: Bruchidae). *J. Biol. Sci.* 9(2): 178–79.

Isah, A.G. 2006. Production of detergent from castor oil. *Leonardo Electronic J. Pract. Technol.* 9: 153–60.

Islam, M.R., Song, C.H., Jeong, Y.T. et al. 2014. Bioprotectant capacities of essential oil from *Ricinus communis* (L.) against stored product red flour beetle, *Triboium castanium* (Herbst). *J. Parasitol. Photon* 104: 150–56.

Jakhar, B., Jat, S.L. 2010. Efficacy of plant oils as grain protectants against beetle, *Trogoderma granarium* Everts in wheat. *Indian J. Entomol.* 72(3): 205–8.

Jenan, U. 2014. Fumigant toxicity of *Ricinus communis* L. oil on adults and larvae of some stored product insects. *J. Nat. Sci. Res.* 4(4): 26–29.

Kadve, P.S., Gawande, R.W., Deotale, R., Rathod, K.B., Lavhe, N.V. 2016. Evaluation of different organic products against pigeon pea pulse beetle, *Callosobruchus chinensis* L. *Pestology* 40(5): 36–41.

Kallamadi, P.R., Ranga Rao Nadigotla, V.P.R., Mulpuri, S. 2015. Molecular diversity in castor (*Ricinus communis* L.). *Ind. Crops Prod.* 66: 271–81.

Katooli, N., Moghadam, E.M., Tahen, A., Nasrollahnejad, S. 2010. Management of root-knot nematode (*Meloidogyne incognita*) on cucumber with the extract and oil of nematicidal plants. *Int. J. Agric. Res.* 5: 582–86.

Kelly, A.J., Kavanagh, J., Thomas, J. 2013. Castor oil, bath and/or enema for cervical priming and induction of labour. *Cochrane Database Syst. Rev.* 7: CD003099.

Khalequzzaman, M., Choudhury, F.D. 2003. Evaluation of mixtures of plant oils as synergists for pirimophos-methyl in mixed formulation against *Tribolium castaneum* (Herbst). *J. Biol. Sci.* 3(3): 347–59.

Kodjo, T.A., Gbenonchi, M., Sadate, A., Homi, A., Dieudonne, G.Y.M., Komla, S. 2011. Bio-insecticidal effects of plant extracts and oil emulsions of *Ricinus communis* L. (Malpighiales: Euphorbiaceae) on the diamondback, *Plutella xylostella* L. (Lepidoptera: Pluttellidae) under laboratory and semi-field conditions. *J. Appl. Biosci.* 43: 2899–914.

Lalithakumari, D., Vidhyasekharan, P., Govindaswamy, C.V., Doraiswami, S. 1971. Reduction in oil content of castor seed due to storage fungi. *Curr. Sci.* 40: 273.

Liu, C., Li, J., Lei, W., Zhou, Y. 2014. Development of biobased unsaturated polyester resin containing highly fractionalized castor oil. *Ind. Crops Prod.* 52: 329–37.

Makun, H.A., Anjorin, S.T., Adeniran, L.A. et al. 2011. Antifungal activities of *Jatropha curcas* and *Ricinus communis* seeds on *Fusarium verticillioides* and *Aspergillus flavus* in yam. *ARPN J. Agric. Biol. Sci.* 6(6): 22–27.

Mbah, G.O., Amulu, N.F., Onyiah, M.I. 2014. Effects of process parameters on the yield of oil from castor seed. *Am. J. Eng. Res.* 3(5): 179–86.

Meghwal, H.P., Bajpai, N.K., Chaudhary, H.R. 2012. Efficacy of vegetable oils as grain protectant against *Sitophilus oryzae* (L.) in stored maize. *Indian J. Entomol.* 74(1): 9–12.

Meneghetti, S.M.P., Meneghetti, M.R., Wolf, C.R. et al. 2006. Biodiesel from castor oil: A comparison of ethynolysis versus methynolysis. *Energy Fuels* 20(5): 2262–65.

Mgudu, L., Muzenda, E., Kabula, J., Belaid, M. 2012. Microwave-assisted extraction of castor oil. Presented at International Conference on Nanotechnology and Chemical Engineering, Bangkok, December, 21–22.

Mohammed, M.H., Mursey, H.M. 2015. Improving quantity and quality of castor bean oil for biodiesel growing under severe conditions in Egypt. *Energy Procedia* 68: 117–21.

Mohapatra, P., Ponnurasan, N., Narayanaswamy, P. 2009. Tribal pest control practices of Tamil Nadu for sustainable agriculture. *Indian J. Trad. Know.* 8(2): 218–24.

Momoh, A.O., Oladunmoye, M.K., Adebolu, T.T. 2012. Evaluation of the antimicrobial and phytochemical properties of oil from castor seeds (*Ricinus communis* Linn.). *Bull. Environ. Pharm. Life Sci.* 1(10): 21–27.

Morse, S., Ward, A., McNamara, N., Denholm, I. 2002. Exploring the factors that influence the uptake of botanical insecticides by farmers: A case study of tobacco based products in Nigeria. *Exp. Agric.* 38: 469–79.

Mushobozy, D., Nganilevanu, G., Ruheza, S., Swella G. 2009. Plant oils as common bean (*Phaseolus vulgaris*) seed protectants against infestation by the Mexican bean weevil, *Zabrotes subfasciatus* (Boh.). *J. Plant Prot. Res.* 49(1): 35–40.

Mutlu, H., Meier, M.A.R. 2010. Castor oil as a renewable resource for the chemical industry. *Eur. J. Lipid Sci. Technol.* 112(1): 10–30.

Nagar, A., Singh, S.P., Singh, Y.P., Singh, R., Meena, H., Nagar, R. 2012. Bioefficacy of vegetable and organic oils, cakes and plant extracts against mustard aphid, *Lipaphis erysimi* (Kalt.). *Indian J. Entomol.* 74: 114–19.

Nana, P., Nchu, F., Bikomo, R.M., Kutima, H.L. 2014. Efficacy of vegetable oils against dry bean beetles. *Afr. Crop Sci. J.* 22(3): 175–80.

Nangbes, J.G., Nvak, J.B., Buba, W.M., Zukdimma, A.N. 2013. Extraction and characterization of castor (*Ricinus communis*) seed oil. *Int. J. Eng. Sci.* 2(9): 105–9.

Negedu, A., Joseph, A.B., Umoh, U.J., Atawodi, S.E., Rai, M. 2014. Biodeterioration of stored castor (*Ricinus communis*) seeds by *Aspergillus tamarii*. *Nusantara Biosci.* 6(2): 126–31.

Neog, P., Singh, H.K. 2012. Efficacy of plant powders and vegetable oils against *Callosobruchus chinensis* (L.) on stored green gram. *Indian J. Entomol.* 74(3): 267–73.

Ogunniyi, D.S. 2006. Castor oil: A vital industrial raw material. *Bioresour. Technol.* 97(9): 1086–91.

Okullo, A., Temu, A.K., Ogwok, P., Ntalikwa, J.W. 2012. Physico-chemical properties of biodiesel from jatropha and castor oils. *Int. J. Renew. Energy Res.* 2: 47–56.

Olaniyan, A.M. 2010. Effect of extraction conditions on the yield and quality of oil from castor bean. *J. Cereals Oilseeds* 1(2): 24–33.

Oluwole, F.A., Abdulrahim, D.T., Aviara, N.A., Nana, S. 2012. Traditional method of extracting castor oil. *Continental J. Eng. Sci.* 7(2): 6–10.

Oluwole, F.A., Aviara, N.A., Umar, B., Mohammed, A.B. 2015. Influence of variety and pre-treatment on oil properties by mechanically expressed castor oil. *Global Adv. Res. J. Eng. Technol. Innov.* 4(1): 1–9.

Omari, A., Mgani, Q.A., Mubofu, E.B. 2015. Fatty acid profile and physico-chemical parameters of castor oils in Tanzania. *Green Sustain. Chem.* 5(4): 154–63.

Paceco, I.A., de Castro, M.F.P.P.M., de Paula, D.C., Lourencao, A.L., Bolonhezi, S., Barbieri, M.K. 1995. Efficacy of soybean and castor oils in the control of *Callosobruchus maculatus* (F.) and *Callosobruchus phaseoli* (Gyllenhal) in stored chick peas. *J. Biol. Sci.* 9(2): 175–79.

Patel, A.V., Vekaria, M.V. 2013. Management of lesser grain borer, *Rhyzopertha dominica* (F.) on wheat. *Indian J. Entomol.* 75(4): 347–48.

Patel, E.K., Gupta, A., Oswal, R.J. 2012. A review on mosquito repellent methods. *Int. J. Pharm. Chem. Biol. Sci.* 2(3): 310–17.

Pecina-Quintero, V., Anaya-Lopez, J.L., Nunez-Colin, C.A. et al. 2013. Assessing the genetic diversity of castor bean from Chiapas, Mexico using SSR and AFLP markers. *Ind. Crops Prod.* 41: 134–43.

Perdomo, F.A., Acosta-Osorio, A.A., Herrera, G. et al. 2013. Physicochemical characterization of seven Mexican *Ricinus communis* L. seeds & oil contents. *Biomass Bioenergy* 48: 17–24.

Prakash, B., Kiran, S. 2016. Essential oils: A traditionally realized natural resource for food preservation. *Curr. Sci.* 110(10): 1890–92.

Qualls, W.A., Xu, R.D., Hott, J.A., Smith, M.L., Moeller, J.J. 2011. Field evaluations of commercial repellents against the flood water mosquito, *Psorophora columbiae* (Diptera: Culicidae) in St. Johns County, Florida. *J. Med. Entomol.* 48(6): 1247–49.

Raghuwani, B.R., Kapadia, M.N. 2003. Efficacy of different vegetable oils as seed protectants of pigeon pea against *Callosobruchus maculatus* (Fab.). *Indian J. Plant Prot.* 31(1): 115–18.

Ramanjaneyulu, A.V., Reddy, A.V., Madhavi, A. 2013. The impact of sowing date and irrigation regime on castor (*Ricinus communis* L.) seed yield, oil quality characteristics and fatty acid composition during post rainy season in South India. *Ind. Crops Prod.* 44: 25–31.

Ramos, L.C.D., Tango, J.S., Savi, A., Lcal, N.R. 1984. Variability for oil and fatty acid composition in castor bean varieties. *J. Am. Oil Chem. Soc.* 61: 1841–1843.

Rana, M., Dhamija, H., Prashar, B., Sharma, S. 2012. *Ricinus communis* L.—A review. *Int. J. Pharmtech. Res.* 4(4): 1706–11.

Ratnasekara, D., Rajapakse, R.H.S. 2009. Repellent properties of plant oil vapours in pulse beetle, *Callosobruchus maculatus* (Fab.) (Coleoptera: Bruchidae) in stored green gram (*Vigna radiata* Walp.). *Trop. Agric. Res. Ext.* 12(1): 13–16.

Salihu, B.Z., Gana, A.K., Apuyor, B.O. 2014. Castor oil plant (*Ricinus communis* L.): Botany, ecology and uses. *Int. J. Sci. Res.* 3(5): 1334–41.

Salimon, J., Noor, D.A.M., Nazrizawati, A.T., Firdaus, M.Y.M., Noraishah, A. 2010. Fatty acid composition and physicochemical properties of Malaysian castor bean *Ricinus communis* L. seed oil. *Sains Malaysiana* 39(5): 761–64.

Sanappa, M.B., Acharya, M.F. 2014. Efficacy of botanical oils in the control of cowpea bruchid, *Callosobruchus maculatus* Fab. in stored cowpea. *Pestology* 38(2): 44–47.

Saribiyik, O.Y., Ozcanli, M., Serin, H., Serin, S., Aydir, K. 2010. Biodiesel production from *Ricinus communis* oil and its blends with soybean biodiesel. *J. Mech. Eng.* 56: 811–16.

Severino, L.S., Auld, D.L., Baldanzi, M. et al. 2012. A review on the challenges for increased production of castor. *Agron. J.* 104(4): 853–80.

Shinde, P.S., Godase, S.K., Mule, R.S., Jalgaonkar, V.N. 2011. Use of some surface protectants against pulse beetle, *Callosobruchus maculatus* Fab. (Coleoptera: Bruchidae) infesting stored cowpea. *Pestology* 35(1): 38–43.

Silitonga, A.S., Masjuki, H.H., Ong, H.C., Kusumo, T.Y.F., Mahila, T.M.I. 2016. Synthesis and optimization of *Hevea brasiliensis* and *Ricinus communis* as feedstock for biodiesel production: A comparative study. *Ind. Crops Prod.* 85: 274–86.

Singh, A.S., Kumari, S., Modi, A.R., Gajera, B.B., Narayanan, S., Kumar, N. 2015. Role of conventional and biotechnological approaches in genetic improvement of castor (*Ricinus communis* L.). *Ind. Crops Prod.* 74: 55–62.

Singh, V., Yadav, D.S. 2003. Efficacy of different oils against pulse beetle, *Callosobruchus chinensis* in green gram, *Vigna radiata* and their effect on germination. *Indian J. Entomol.* 65(2): 281–86.

Sridhar, B.S., Binna, K.V., Anita, M.V., Paramjeet, K.B. 2010. Optimization and characterization of castor seed oil. *Leonardo J. Sci.* 17(9): 59–70.

Stubiger, G., Pittenauer, E., Allmaier, G. 2003. Characterization of castor oil by on-line and off-line non-aqueous reverse-phase high performance liquid chromatography-mass spectrometry (APCI and UV/MALDI). *Phytochem. Analy.* 14(6): 337–46.

Thomas, A., Matthaus, B., Fiebig, H.J. 2015. Fats and fatty oils. In *Ullman's Encyclopedia of Industrial Chemistry*, ed. B. Elvers, 1–84. Weinheim: Wiley-VCH Verlag.

Thomas, T.P., Birney, D.M., Auld, D.L. 2013. Optimizing esterification of safflower, cottonseed, castor and used cottonseed oils. *Ind. Crops Prod.* 41: 102–6.

Tunaru, S., Althoff, T.F., Nusing, R.M., Diener, M., Offermanns, S. 2012. Castor oil induces laxation and uterus contraction via ricinoleic activating prostaglandin EP3 receptors. *Proc. Nat. Acad. Sci. USA* 109(2): 9179–84.

U.S. Department of Agriculture, Natural Resources Conservation Service. 2006. *Ricinus communis* L. castorbean. PLANTS Database. Baton Rouge: National Plant Data Center. http://plants.usda .gov.

Velasco, L., Rajas-Barros, P., Fernandez-Martinez, J.M. 2005. Fatty acid and tocopherol accumulation in the seeds of a high oleic acid castor mutant. *Ind. Crops Prod.* 22: 201–6.

Verma, R.K., Chaurasia, L., Kumar, M. 2011. Antifungal activity of essential oils against selected building fungi. *Indian J. Nat. Prod. Resour.* 2(4): 448–51.

Wale, M., Assegie, H. 2015. Efficacy of castor bean oil (*Ricinus calamus* L.) against maize weevils (*Sitophilus zeamais* Mots.) in northwestern Ethiopia. *J. Stored Prod. Res.* 63: 38–41.

Wani, A.H., Bhat, M.Y. 2012. Control of root-knot nematode, *Meloidogyne incognita* by urea coated with nimin or other natural oils on mung, *Vigna radiata* (L.) R. Wilczek. *J. Biopesticides* 5(Suppl.): 255–58.

Wani, A.H., Joshi, J., Titov, A., Tomar, D.S. 2014. Effect of seed treatment and packing materials on seed quality parameters of maize (*Zea mays* L.) during storage. *J. Appl. Res.* 4(4): 40–44.

Weiss, E.A. 2000. *Oilseed Crops.* 2nd ed. Oxford: Blackwell Scientific Ltd.

Wilson, R., van Schie, B.J., Howes, D. 1998. Overview of the preparation, use and biological studies on polyglycerol polyrecinoleate (PGPR). *Food. Chem. Toxicol.* 36: 711–18.

Yahaya, M.M., Bandiya, H.M., Yahaya, M.A. 2013. Efficacy of selected seed oils against the fecundity of *Callosobruchus maculatus* (F.) (Coleoptera: Bruchidae). *Experiment* 8(4): 513–21.

Yakubu, B.L., Mbonu, O.A., Nda, A.J. 2012. Cowpea (*Vigna unguiculata*) pest control methods in storage and recommended practices for efficiency: A review. *J. Biol. Agric. Healthcare* 2(2): 27–33.

# 19

# Chamomile Oil

Ompal Singh, Zakia Khanam, and Leo M.L. Nollet

## CONTENTS

## 19.1 Introduction

Chamomile (*Matricaria chamomilla* L.) is one of the important medicinal herbs that is native to southern and eastern Europe. It is also grown in Germany, Hungary, France, Russia, Yugoslavia, and Brazil. It was introduced to India during the Mughal period. Now it is grown in Punjab, Uttar Pradesh, Maharashtra, Jammu, and Kashmir. The plant is found in North Africa, Asia, North and South America, Australia, and New Zealand [1]. Hungary is the main producer of the plant biomass. In Hungary, it grows abundantly in poor soils and is a source of income to poor inhabitants of these areas. Flowers are exported to Germany in bulk for distillation of the oil [2]. In India, the plant had been cultivated in Lucknow for about 200 years, and it was introduced in Punjab about 300 years ago. It was introduced in Jammu in 1957 by Handa et al. [3]. The plant was first introduced in the alkaline soils of Lucknow in 1964–1965 by Chandra et al. [4,5]. There is no demand for blue oil as such at present in India. However, flowers of chamomile are in great demand. Presently, two firms cultivating chamomile exist: M/S Ranbaxy Labs Limited, New Delhi, and M/S German Remedies. The latter is the main grower of chamomile flowers. Chamomile has been used in herbal remedies for thousands of years. It was known in ancient Egypt, Greece, and Rome [6]. This herb has been believed by Anglo-Saxons to be one of the nine sacred herbs given to humans by the Lord [7]. The chamomile drug is included in the pharmacopoeia of 26 countries [8]. It is an ingredient of several traditional, unani, and homeopathy medicinal preparations [9–12]. As a drug, it finds use in flatulence, colic, hysteria, and intermittent fever [13]. The flowers of *M. chamomilla* contain the blue essential oil at 0.2%–1.9% [14,15], which finds a variety of uses. Chamomile is used mainly as an anti-inflammatory and antiseptic medicine, but also as an antispasmodic and mild sudorific [16]. It is used internally mainly as a tisane (infuse 1 tablespoonful of the drug in 1 L of cold water and do not heat) for disturbance of the stomach associated with pain, for sluggish digestion, and for diarrhea and nausea; it is used more rarely and very effectively for inflammation of the urinary tract and for painful menstruation. Externally, the drug in powder form may be applied to wounds slow to heal; for skin eruptions and infections, such as shingles and boils; for haemorrhoids; and for inflammation of the mouth, throat, and eyes [17]. Tabulated products from chamomile flower extracts are marketed in Europe and used for various ailments. Chamomile tea eye washing can induce allergic conjunctivitis. Pollen of *M. chamomilla* contained in these infusions is the allergen responsible for these reactions [18]. Antonelli quoted from writings of several doctors of the sixteenth and seventeenth centuries that chamomile was used in those times in intermittent fevers [19]. Gould et al. have evaluated the hemodynamic effects of chamomile tea in patients with cardiac disease [20]. It was found in general that the patients fell into a deep sleep after taking the beverage. Pasechnik reported that infusion prepared from *M. chamomilla* exercised a marked stimulatory action on the secretary function of the liver [21]. Gayar and Shazli reported toxicity of acetone extract of *M. chamomilla* against larvae of *Culex pipens* L. [22]. The other pharmacological properties include anti-inflammatory, antiseptic, carminative, healing, sedative, and spasmolytic activities [23]. However, *M. chamomilla* has exhibited both positive and negative bactericidal activity with *Mycobacterium tuberculosis*, *Salmonella typhimurium*, and *Staphylococcus aureus*. The international demand for chamomile oil has been steadily growing. As a result, the plant is widely cultivated in Europe and has been introduced in some Asian countries for the production of its essential oil. *M. chamomilla* L., *Anthemis nobilis* L., and *Ormenis multicaulis* Braun Blanquet and Maire, belonging to the family Asteraceae, are a natural and major source of "blue oil" and flavonoids. The oil is used as a mild sedative and for digestion [20,24–29], besides being antibacterial and fungicidal in action [20]. In addition to pharmaceutical uses, the oil is extensively used in perfumery, cosmetics, and aromatherapy, as well as in the food industry [27,30–33]. Gowda et al. studied that the essential oil present in the flower heads contains azulene and is used in perfumery, cosmetic creams, hair preparations, skin lotions, toothpastes, and fine liquors [34]. The dry flowers of chamomile are also in great demand for use in herbal tea and baby massage oil, for promoting the gastric flow of secretion, and for the treatment of cough and cold [35]. The use of herbal tea preparations eliminated colic in 57% of infants [36]. Because of its extensive pharmacological and pharmaceutical properties, the plant possesses great economic value and is in great demand in European countries.

## 19.2 Botany of the Plant

True chamomile is an annual plant with thin spindle-shaped roots only, penetrating flatly into the soil. The branched stem is erect and heavily ramified, and grows to a height of 10–80 cm. The long and narrow leaves are bi- to tripinnate. The flower heads are placed separately; they have a diameter of 10–30 mm and are pedunculate and heterogamous. The golden-yellow tubular florets with five teeth are 1.5–2.5 mm long, always ending in a glandulous tube. The 11–27 white plant flowers are 6–11 mm long, 3.5 mm wide, and arranged concentrically. The receptacle is 6–8 mm wide, flat in the beginning and conical, cone shaped later, hollow—the latter being a very important distinctive characteristic of

*Matricaria,* and without paleae. The fruit is a yellowish-brown achene. The true chamomile is very often confused with plants of the genera *Anthemis.* Special attention has to be paid to avoid confusion with *Anthemis cotula* L., a poisonous plant with a revolting smell. In contrast to true chamomile, *A. cotula,* similar to *Anthemis arvensis* L. and *Anthemis austriaca* Jacq., has setiform, prickly pointed paleae, and a filled receptacle. The latter species are nearly odorless [37]. Although the systematic status is quite clear nowadays, there are a number of inaccuracies concerning the names. Apart from misdeterminations and confusion, the synonymous use of the names *Anthemis, Chamomilla,* and *Matricaria* leads to uncertainty with regard to the botanical identification. Moreover, the nomenclature is complicated by the fact that Linnaeus made mistakes in the first edition of his *Species Plantarum* that he corrected later on. The best-known botanical name for true chamomile is *Matricaria recutita* (syn. *M. chamomilla* and *Chamomilla recutita* (L.) Rauschert), belonging to the genus *Chamomilla* and family Asteraceae [37]. *M. chamomilla* is a diploid species ($2n = 18$), allogamous in nature, exhibiting wide segregation as a commercial crop. Chamomile, a well-known old-time drug, is known by an array of names, such as baboonig, babuna, babuna camornile, babunj, German chamomile, Hungarian chamomile, Roman chamomile, English chamomile, camomilla, flos chamomile, single chamomile, sweet false chamomile, pinheads, and scented mayweed, suggesting its widespread use [38,39]. The three plants, namely, *A. nobilis* Linn, *Corchorus depressus* Linn, and *M. chamomilla* Linn, are reported under one unani name, *babuna,* at different places in the literature. This created a lot of confusion and misuse of the drug as an adulterant and so forth. Ghauri et al. conducted a detailed taxonomic and anatomical study and concluded that babuna belongs to the family Compositae (Asteraceae) and the correct scientific name of babuna is *M. chamomilla* L. [40].

## 19.3 Methods of Extraction of Chamomile Oil

Extraction of essential oils was performed using hydrodistillation of dried samples of flower heads using a Clevenger-type apparatus over 3 h. The oils were dried over sodium sulfate. Early efforts at extraction used alcohol and a fermentation process. New methods of essential oil extraction are entering the mainstream of aromatherapy, offering new choices in oils never before available. With the new labels of supercritical $CO_2$, along with the traditional steam and hydrodistillations, "absolutes," and cold pressing, a little education for the aromatherapy enthusiast can go a long way in essential oil selection. Is one process better than another? Does one produce a nicer-smelling oil, or one with greater aromatherapeutic value? It turns out that essential oil production, like winemaking, is an art form, as well as a science. The way in which oils are extracted from plants is important because some processes use solvents that can destroy the therapeutic properties. Some plants, particularly flowers, do not lend themselves to steam distilling. They are too delicate, or their fragrance and therapeutic essences cannot be completely released by water alone. These oils will be produced as absolutes—and while not technically considered essential oils, they can still be of therapeutic value. Jasmine oil and rose oil, in particular, are delicate flowers whose oils are often found in absolute form. The value of the newer processing methods depends greatly on the experience of the distiller, as well as the intended application of the final product. Each method is important, and has its place in the making of aromatherapy-grade essential oils. Some of the few methods for extractions of essential oils are given below:

- *Maceration*: Maceration actually creates more of an "infused oil" than an essential oil. The plant matter is soaked in vegetable oil, heated, and strained, at which point it can be used for massage.

- *Cold pressing*: Cold pressing is used to extract the essential oils from citrus rinds such as orange, lemon, grapefruit, and bergamot. This method involves the simple pressing of the rind at about 120°F to extract the oil. The rinds are separated from the fruit, are ground or chopped, and are then pressed. The result is a watery mixture of essential oil and liquid that will separate given time. Little, if any, alteration from the oil's original state occurs—these citrus oils retain their bright, fresh, uplifting aromas, like that of smelling a wonderfully ripe fruit. It is important to note that oils extracted using this method have a relatively short shelf life, so make or purchase only what you will be using within the next 6 months.

- *Solvent extraction*: A hydrocarbon solvent is added to the plant material to help dissolve the essential oil. When the solution is filtered and concentrated by distillation, a substance containing resin (resinoid) or a combination of wax and essential oil (known as concrete) remains. From the concentrate, pure alcohol is used to extract the oil. When the alcohol evaporates, the oil is left behind. This is not considered the best method for extraction, as the solvents can leave a small amount of residue behind, which could cause allergies and effect the immune system.

- *Enfleurage*: This is an intensive and traditional way of extracting oil from flowers. The process involves layering fat over the flower petals. After the fat has absorbed the essential oils, alcohol is used to separate and extract the oils from the fat. The alcohol is then evaporated and the essential oil collected.

- *Hydrodistillation*: Some processes become obsolete to carry out extraction, like hydrodistillation, which is often used in primitive countries. The risk is that the still can run dry, or be overheated, burning the aromatics and resulting in an essential oil with a burnt smell. Hydrodistillation seems to work best for powders (i.e., spice powders and ground wood) and very tough materials, like roots, wood, or nuts.

- *Supercritical fluid extraction*: This modern technique involves the use of carbon dioxide as the "solvent" that carries the essential oil away from the raw plant material. Carbon escapes in its gaseous form, leaving the essential oil behind.

- *Turbo distillation extraction*: Turbo distillation is suitable for hard-to-extract or coarse plant material, such as bark, roots, and seeds. In this process, the plants soak in water and steam is circulated through this plant and water mixture. Throughout the entire process, the same water is continually recycled through the plant material. This method allows faster extraction of essential oils from hard-to-extract plant materials.

- *Steam distillation*: Most commonly, the essence is extracted from the plant using a technique called distillation. One type of distillation places the plants or flowers on a screen. Steam is passed through the area and becomes "charged" with the essence. The steam then passes through an area where it cools and condenses. This mixture of water and essential oil is separated and bottled. Since plants contain such a small amount of this precious oil, several hundred pounds may be needed to produce a single ounce.

## 19.4 Methods of Analysis of Chamomile Oil

Gas chromatography (GC) is one of the best techniques to identify the constituents of an essential oil. When properly used, it can easily detect and identify major components of essential oils, and give some indications of the quality and authenticity of the oil. The technique does have limitations, however. Many minor components of essential oils (<0.01%) do not register on GC detector systems. The separation of essential oil components is usually carried out by GC with fused-silica capillary columns. The properties and conditions of columns used are variable, depending on the polarity of the components to be separated. It is advantageous to use a more selective phase for a given separation, as the overlapping of peaks in the final chromatogram is often a significant drawback of chromatographic techniques in natural samples. The discovery of chiral phases (mostly based on cyclodextrin derivatives) allows the resolution of enantiomers of volatile components. These phases can give different elution sequences for a polarity range and provide a distinct advantage in identification because of large changes in solute relative retention times. The information obtained from high-resolution GC analysis of the volatile fraction of essential oils must be sufficient to determine whether the product is genuine. If the product is adulterated, the kind and level of adulteration must be detected. Therefore, a selective and accurate separation is absolutely necessary in the case of industrial analysis. On the other hand, GC sometimes permits the separation and further identification of some components of the nonvolatile residue as well.

## 19.5 Composition of Chamomile Oil

Different cultivars have different amounts of active components. However, their chemical composition is affected by the local ecological conditions and the cultivation method [41]. A study of the main sesquiterpenes of chamomile essential oil revealed the major components to be chamazulene (19.9%), α-bisabolol (20.9%), A and B α-bisabolol oxides (21.6% and 1.2%, respectively), and β-farnesene (3.1%). Among the minor components was spathulenol [42]. An Iranian experiment studied four cultivars of German chamomile, Bodegold (tetraploid), Germania (diploid), Bona (diploid), and Goral (tetraploid). The results showed that the plant heights of Goral and Bodegold were significantly higher than those of Germania and Bona. Goral produced the highest anthodia yield. The lowest dry anthodia yield was produced by Bona. The highest essential oil content (0.627% w/w) was extracted from Bona in the first harvest, but Germania produced the lowest essential oil content (0.627% w/w) at the third harvest. The chamazulene content of the cultivars ranged between 9.6% and 14% [43]. The essential oils of *M. recutita* L. cultivated in Estonia were isolated, and 37 components were identified. The main components were α-bisabolol oxide A (20%–33%) and B (12%–85%), bisabolone oxide A (14%–75%), (E)-farnesene (13%–45%), α-bisabolol (14%–85%), chamazulene (7%–55%), and en-yn-dicycloether (22%–175%) [44]. A recent investigation in Estonia indicated that the main constituents of the essential oils were as follows: α-bisabolol oxide A (39.4%), bisabolone oxide A (13.9%), (Z)-en-yne-dicycloether (11.5%), α-bisabolol oxide B (9.9%), α-bisabolol (5.6%), and chamazulene (4.7%) [45]. A study regarding the responses of young plants of diploid and tetraploid *M. chamomilla* cultivars to abiotic stress (within an interval from 6 h before to 54 h after spraying the leaf rosettes with aqueous CuCl$_2$ solution) revealed that the content of herniarin in the treated plants rose approximately three times. The highest amounts of umbelliferone in stressed plants exceeded 9 times and 20 times those observed in control plants of the tetraploid and diploid cultivars, respectively. Due to stress, the concentration of en-yn-dicycloether in leaves decreased by more than 40% [46]. An Iranian study in Isfahan indicated essential oil components of German chamomile isolated by hydrodistillation of the aerial parts of the plant. Sixty-three components were characterized, representing 86.21% of the total oil components detected; α-bisabolol oxides A (25.01%) and B (9.43%), spathulenol (8.49%), *cis*-en-yn-dicycloether (7.42%), and α-24-bisabolene oxide A (7.17%) were the major constituents of the oil. Chamazulene represented 3.28%, and α-bisabolol 6.01% [47].

## 19.6 Physical and Chemical Properties of Oil

The chemical and physical properties of *Chamomilla recutita* (*matricaria*) flower oil are included in Table 19.1. Information on the other 10 ingredients was not found, nor was unpublished information provided.

## 19.7 General Use of Chamomile Oil

Traditionally, chamomile has been used for centuries as an anti-inflammatory, antioxidant, mild astringent, and healing medicine [50]. As a traditional medicine, it is used to treat wounds, ulcers, eczema, gout, skin irritations, bruises, burns, canker sores, neuralgia, sciatica, rheumatic pain, hemorrhoids, mastitis, and other ailments [51,52]. Externally, chamomile has been used to treat diaper rash; cracked nipples; chicken pox; ear and eye infections; disorders of the eyes, including blocked tear ducts; conjunctivitis; nasal inflammation; and poison ivy [53,54]. Chamomile is widely used to treat inflammations of the skin and mucous membranes, and for various bacterial infections of the skin, oral cavity

**TABLE 19.1**

Chemical and Physical Properties of Chamomile

| Properties | *Chamomilla recutita* (*matricaria*) Oil |
| --- | --- |
| Form | Deep blue or blue-green liquid with strong, characteristic odor |
| Specific gravity | Between 0.910 and 0.950 |
| Refractive index solubility | Soluble in most fixed oils and in propylene glycol; insoluble in glycerine and in mineral oil |
| Acid value | Between 5 and 50 mg of KOH/g of oil |
| Ester value | Between 65 and 155 mg of KOH/g of oil |
| Saponification number | ≈43 |
| UV absorption maximum | 285 nm |

*Source:* Pauli, A., *Int. J. Essent. Oil*, 2(2), 60–68, 2008; U.S. Pharmacopeial Convention, *Food Chemicals Codex*, 6th ed., U.S. Pharmacopeial Convention, Rockville, MD, 2009.

and gums, and respiratory tract. Chamomile in the form of an aqueous extract has been frequently used as a mild sedative to calm nerves and reduce anxiety, and to treat hysteria, nightmares, insomnia, and other sleep problems [55]. Chamomile has been valued as a digestive relaxant and has been used to treat various gastrointestinal disturbances, including flatulence, indigestion, diarrhea, anorexia, motion sickness, nausea, and vomiting [56,57]. Chamomile has also been used to treat colic, croup, and fevers in children [58]. It has been used as an emmenagogue and a uterine tonic in women. It is also effective in arthritis, back pain, bedsores, and stomach cramps.

## 19.8 Use as a Pesticide

Chamomile oil is volatile, which means it has the ability to vaporize at room temperature. One of the methods to determine the suitability of the chamomile drug is to determine the amount of essential oil or volatile oil content in it. To determine the total volatile component in the chamomile flowers, the flowers are steam distilled in a Clevenger-type apparatus. The volume of the oil distilled per 100 g of flowers is expressed in percentage. This percentage indicates the total volatile oil content. The lousicidal and repellent effects of five essential oils, camphor (*Cinnamomum camphora*), onion (*Allium cepa*), peppermint (*Mentha piperita*), chamomile (*M. chamomilla*), and rosemary (*Rosmarinus officinalis*) oils, were investigated for the first time against the buffalo louse, *Haematopinus tuberculatus*, and flies infesting water buffaloes in Qalyubia Governorate, Egypt [59]. For *in vitro* studies, filter paper contact bioassays were used to test the oils and their lethal activities were compared with that of *d*-phenothrin. Four minutes posttreatment, the median lethal concentration (LC50) values were 2.74%, 7.28%, 12.35%, 18.67%, and 22.79% for camphor (*C. camphora*), onion (*A. cepa*), peppermint (*M. piperita*), chamomile (*M. chamomilla*), and rosemary (*R. officinalis*) oils, respectively, whereas for *d*-phenothrin, it was 1.17%. The lethal time (LT50) values were 0.89, 2.75, 15.39, 21.32, 11.60, and 1.94 min after treatment with 7.5% camphor, onion, peppermint, chamomile, rosemary, and *d*-phenothrin, respectively. All the materials used except rosemary, which was not applied, were ovicidal to the eggs of *H. tuberculatus*. Despite the results of the *in vitro* assays, the *in vivo* treatments revealed that the pediculicidal activity was more pronounced with oils. All treated lice were killed after 0.5–2 min, whereas with *d*-phenothrin, 100% mortality was reached only after 120 min. The number of lice infesting buffaloes was significantly reduced 3, 6, 4, 6, and 9 days after treatment with camphor, peppermint, chamomile, onion, and *d*-phenothrin, respectively. Moreover, the oils and *d*-phenothrin significantly repelled flies, *Musca domestica*, *Stomoxys calcitrans*, *Haematobia irritans*, and *Hippobosca equina*, for 6 and 3 days posttreatment, respectively. No adverse effects were noted on either animals or pour-on operators after exposure to the applied materials.

Seven essential oils of *C. camphora*, *Cymbopogon winterianus*, *M. chamomilla*, *Mentha viridis*, *Prunus amygdalus* var. *amara*, *R. officinalis*, and *Simmondsia chinensis* were evaluated in the laboratory for their toxicities and repellent effectiveness against adults of the sawtoothed grain beetle, *Oryzaephilus surinamensis* (L.), and rust-red flour beetle, *Tribolium castaneum* (Herbst) [60]. Five concentrations of every essential oil (0.125%, 0.25%, 0.5%, 0.75%, and 1%) were tested. Adult beetles were exposed to the treated wheat for 2 weeks. Percent mortality was recorded after 3 days, 1 week, and 2 weeks from exposure. The repellent action of the previous essential oils was also studied using the same concentrations used in toxicity tests. Results showed that complete mortality of *O. surinamensis* was achieved by *M. viridis*,

*M. chamomilla,* and *C. camphora* at concentrations more than 0.5%, although 1% *P. amygdalus* or *C. winterianus* resulted in complete mortality of *T. castaneum* after 2 weeks of exposure. Conversely, *R. officinalis* was the least toxic to both insect species. The rest of the essential oils gave adequate toxicity to both insect species. A pronounced increase of mortality was observed for most of the essential oils with increasing time of exposure. *T. castaneum* was less susceptible to tested oils than *O. surinamensis.* Moreover, *M. chamomilla* exhibited a high repellency of 81.94% and 84.73% at 1% concentration against *O. surinamensis* and *T. castaneum,* respectively.

The lesser grain borer, *Rhyzopertha dominica,* is a major insect pest of stored grain in the tropics. Vegetable oils (chamomile, sweet almond, and coconut) at 2.5, 3.5, 5.0, 7.0, and 10.0 ml/kg were tested against *R. dominica* (F.) in wheat grain. All bioassays were conducted at 30°C and 65% ± 2% relative humidity (RH). Treatments with vegetable oils at a high dose (10.0 ml/kg) achieved more than 95% control within 24 h of exposure to freshly treated grain. There was little difference between the three oils in their effect. The persistence of the oils in grains was tested at short-term storage intervals (48, 72, and 96 h) and intermediate-term intervals (10, 20, and 30 days) after treatments. The activity of all products decreased with storage period. Seed viability was reduced by a high dose rate (10.0 ml/kg) of oil treatments. The potential of acaricidal activity of chamomile flower extract was studied against engorged *Rhipicephalus annulatus* tick under laboratory condition [61]. For this purpose, the engorged females of *R. annulatus* were exposed to twofold serial dilutions of chamomile flower extract (0.5%, 1.0%, 2.0%, 4.0%, and 8.0%) using the "dipping method" *in vitro.* The engorged ticks were immersed in different plant dilutions (five ticks for each dilution) for 1 min, and they were immediately incubated in separate Petri dishes for each replicate at 26°C and 80% relative humidity. The mortality rate for each treatment was recorded 5 days after incubation.

The mortality rate caused by different dilutions of chamomile flower extract ranged from 6.67% to 26.7%, whereas no mortality was recorded for the nontreated control group. The mass of produced eggs varied from 0.23 g (in 8.0% solution) to 0.58 g (in control), with no statistical differences between the treatments and control ($p > 0.05$). Also, the chamomile flower extract in the highest concentration used (8.0%) caused 46.67% failure in egg laying in engorged females, while no failure was observed for the nontreated control group. Macroscopic observations indicated that in effective concentrations of plant (4.0% and 8.0%), patchy hemorrhagic swelling appeared on the skin of treated ticks. Pesticide residues are found in chamomile drug due to agricultural practices during cultivation. The pesticide residues usually have compounds such as chlorinated hydrocarbons or sulfur-containing dithiocarbamate or organophosphorus. These can be detected by column chromatography (CC) and GC. A purified chamomile extract is specifically prepared for the detection of the pesticide residues in it by GC. Abdel-Gawad et al. [62] suggested that the insecticide [14]C-ethion could be satisfactorily eliminated from the essential oil of chamomile if adsorbents such as calcium oxide and sawdust were added during the distillation process. Such methods that can efficiently eliminate pesticide residues need to be developed.

## 19.9 Antibacterial Activity

The extract and essential oil of Roman chamomile flower head showed antibacterial activity against *Porphyromonas gingivalis.* The antimicrobial effects were evaluated by the disk diffusion method. The results indicated that the means of the inhibition zone for chamomile extract and essential oil were 13.33 ± 3.4 and 20.5 ± 0.5, respectively [63]. Two hydroperoxide compounds isolated from *A. nobilis* showed a medium antibacterial activity [64]. The antimicrobial activity of an essential oil of the flower of *A. nobilis* from Provence (France) was tested against various strains of Gram-positive bacteria (*S. aureus* and *Enterococcus faecalis*) and Gram-negative bacteria (*Escherichia coli*, *Pseudomonas aeruginosa*, *Proteus vulgaris*, *Klebsiella pneumoniae*, and *Salmonella* sp.), as well as against the yeast *Candida albicans* using a modified agar dilution and agar diffusion method. In addition, some pure main and minor compounds (chemical composition obtained by means of GC and GC–mass spectrometry [MS] measurements), such as isobutyl angelate (32.1%), 2-methylbutyl angelate (16.2%), isobutyl isobutyrate (5.3%), methyl 2-methylbutyrate (1.9%), prenyl acetate (1.4%), 2-methylbutyl 2-methylbutyrate (1.2%), and 2-methylbutyl acetate (1.2%), were also studied for their antimicrobial effects. The Roman chamomile sample showed high antimicrobial activity against all strains of tested microbes. A similar result was found for 2-methylbutyl 2-methylbutyrate, 2-methylbutyl acetate, and prenyl acetate [65]. The volatile oil of *A. nobilis* showed activity against Gram-positive bacteria, especially *Bacillus subtilis*, *Bacillus anthracis*, *Micrococcus glutamicus*, *Bacillus saccharolyticus*, *Bacillus thuringiensis*, *Sarcina lutea*, *Bacillus*, *Lactobacillus plantarum*, *S. aureus*, *Staphylococcus* sp., and *Lactobacillus casei*, whereas the oil showed no activity against Gram-negative bacteria species, including *Salmonella* group B, *Citrobacter* sp., *Enterobacter* sp., *E. coli*, *Pseudomonas* sp., *Salmonella saintpaul* and *Salmonella weltevreden*. The volatile oil also inhibited the growth of dermatophytons, *Alternaria* sp., *Aspergillus fumigatus*, and *Aspergillus parasiticus*. Volatile oil was inactive against *C. albicans*, *Cryptococcus neoformans*, *Histoplasma capsulatum*, and *Aspergillus niger*. Hydroperoxides (Z-2-methyl-2-butyric acid-(2-hydroperoxy-2-methyl-3-butenyl) ester and Z-2-methyl-2-butyric acid-(3-hydroperoxy-2-methylidenebutyl) ester), isolated from the ethanolic extract of the *A. nobilis* flowers, showed antibacterial activity against *E. coli*, *P. aeruginosa*, and *E. faecalis*. The minimum inhibitory concentration (MIC) values of the first compound were 256 µg/ml against *E. coli* and 512 µg/ml against *P. aeruginosa*. The MIC values of the second compound were 512 and 128 µg/ml against the same microorganisms, respectively [66]. Volatile oil of *A. nobilis* showed high activity against the whitefly (*Trialeurodes vaporariorum*) nymphs at 0.0047 and 0.0093 µg/ml using an impregnated filter paper test, whereas it was ineffective against the adult or egg forms.

# References

1. Ivens GM. Stinking mayweed. *NZJ Agric* 1979;138:21–3.
2. Svab J. New aspects of cultivating chamomile. *Herba Polonica* 1979;25:35–9.
3. Handa KL, Chopra IC, Abrol BK. Introduction of some of the important exotic aromatic plants in Jammu and Kashmir. *Indian Perfumer* 1957;1:42–9.
4. Chandra V. Cultivation of plants for perfumery industry at Lucknow. *Indian Perfumer* 1973;16:40–4.
5. Chandra V, Singh A, Kapoor LD. Experimental cultivation of some essential oil bearing plants in saline soils, *Matricaria chamomilla* L. *Perfum Essent Oil Rec* 1968;59:871.
6. Issac O. *Recent Progress in Chamomile Research—Medicines of Plant Origin in Modern Therapy*. 1st ed. Prague: Prague Press, 1989.

7. Crevin JK, Philpott J. *Herbal Medicine Past and Present*. 1st ed. Durham, NC: Duke University Press, 1990.
8. Pamukov D, Achtardziev CH. *Natural Pharmacy* [in Slova]. 1st ed. Bratislava, Slovakia: Priroda, 1986.
9. Das M, Mallavarapu GR, Kumar S. Chamomile (*Chamomilla recutita*): Economic botany, biology, chemistry, domestication and cultivation. *J Med Aromat Plant Sci* 1998;20:1074–109.
10. Kumar S, Das M, Singh A, Ram G, Mallavarapu GR, Ramesh S. *J Med Aromat Plant Sci* 2001;23:617–23.
11. Lawrence BM. Progress in essential oils. *Perfume Flavorist* 1987;12:35–52.
12. Mann C, Staba EJ. The chemistry, pharmacology and commercial formulations of chamomile. In Craker LE, Simon JE, eds., *Herbs, Spices and Medicinal Plants—Recent Advances in Botany, Horticulture and Pharmacology*. Phoenix: Oryx Press, 1986, pp. 235–80.
13. Tyihak E, Sarkany-Kiss J, Verzar-Petri G. Phytochemical investigation of apigenin glycosides of *Matricaria chamomilla*. *Pharmazie* 1962;17:301–4.
14. Bradley P, ed. *The British Herbal Compendium*. 1st ed. London: British Herbal Medicine Association, 1992.
15. Mann C, Staba EJ. The chemistry, pharmacology and commercial formulations of chamomile. In Craker LE, Simon JE, eds., *Herbs, Spices and Medicinal Plants—Recent Advances in Botany, Horticulture and Pharmacology*. Binghamton, NY: Haworth Press, 2002, pp. 235–80.
16. Mericli AH. The lipophilic compounds of a Turkish *Matricaria chamomilla* variety with no chamazuline in the volatile oil. *Int J Crude Drug Res* 1990;28:145–7.
17. Fluck H. *Medicinal Plants and Authentic Guide to Natural Remedies*. 1st ed. London: W. Foulsham and Co., 1988.
18. Subiza J, Subiza JL, Marisa A, Miguel H, Rosario G, Miguel J et al. Allergic conjunctivitis to chamomile tea. *Ann Allergy* 1990;65:127–32.
19. Antonelli G. *Matricaria chamomilla* L. and *Anthemis nobilis* L. in intermittent fevers in medicines. *Biol Abstr* 1928;6:21145.
20. Gould L, Reddy CV, Compreht FF. Cardiac effect of chamomile tea. *J Clin Pharmacol* 1973;13:475–9.
21. Pasechnik LK. Cholagogic action of extracts prepared from wild chamomile (*Matricaria chamomilla*). *Farmakol Toksikol* 1996;29:468–9.
22. Gayar F, Shazli A. Toxicity of certain plants to *Culex pipens* L. larvae (Diptera: Culicidae). *Bull Soc Entomol (Egypt)* 1968;52:467–75.
23. Salamon I. Chamomile: A medicinal plant. *J Herbs Spices Med Plants* 1992;10:1–4.
24. Das M, Mallavarapu GR, Kumar S. Isolation of a genotype bearing fascinated capitula in chamomile (*Chamomilla recutita*). *J Med Aromat Plant Sci* 1999;21:17–22.
25. Gasic O, Lukic V, Adamovic R, Durkovic R. Variability of content and composition of essential oil in various chamomile cultivators (*Matricaria chamomilla* L.). *Herba Hungarica* 1989;28:21–8.
26. Lal RK, Sharma JR, Misra HO. Vallary: An improved variety of German chamomile. *Pafai J* 1996;18:17–20.
27. Lal RK, Sharma JR, Misra HO, Singh SP. Induced floral mutants and their productivity in German chamomile (*Matricaria recutita*). *Indian J Agric Sci* 1993;63:27–33.
28. Shahi NC. Traditional cultivation of Babunah *Chamomilla recutita* (L.) Rauschert syn. *Matricaria chamomilla* L. Lucknow. *Bull Medico-ethnobotanical Res* 1980;1:471–7.
29. Sharma A, Kumar A, Virmani OP. Cultivations of German chamomile—A review. *Curr Res Med Aromat Plants* 1983;5:269–78.
30. Anonymous. A superior variety of German chamomile identified. *CIMAP Newsl* 1993;20:13.
31. Earle L. *Vital Oils*. 1st ed. London: Ebury Press, 1991.
32. Masada Y. *Analysis of Essential Oils by Gas Chromatography and Mass Spectrometry*. 1st ed. New York: John Wiley & Sons, 1976.
33. Misra N, Luthra R, Singh KL, Kumar S, Kiran L. Recent advances in biosynthesis of alkaloids. In Nanishi K, Methcohn O, eds., *Comprehensive Natural Product Chemistry (CONAP)*. Oxford: Elsevier, 1999, pp. 25–69.

34. Gowda TNV, Farooqi AA, Subbaiah T, Raju B. Influence of plant density, nitrogen and phosphorus on growth, yield and essential oil content of chamomile (*Matricaria chamomilla* Linn.). *Indian Perfumers* 1991;35:168–72.

35. Anonymous. Azulene in pharmacy and cosmetics. *Dragoco Rep* 1969;16:23–5.

36. Weizman ZVI, Alkrinawi S, Goldfarb DAN, Bitran G. Efficacy of herbal tea preparation in infantile colic. *J Pediatr* 1993;122:650–2.

37. Franz Ch, Bauer R, Carle R, Tedesco D, Tubaro A, Zitterl-Eglseer K. Study on the assessments of plants/herbs, plant herb extracts and their naturally or synthetically produced components as additives for use in animal production. CFT/EFSA/FEEDAP/2005/01, 2005, pp. 155–69. http://www.Agronavitor.cz/UserFiles/File/Agronavitor/kvasnickva_feedap_reportplantsherbs.pdf.

38. Franke R. Plant sources. In Franke R, Schilcher H, eds., *Chamomile: Industrial Profiles.* 1st ed. Boca Raton, FL: CRC Press, 2005, pp. 39–42.

39. Leung A, Foster S. *Encyclopedia of Common Natural Ingredients Used in Food, Drugs, and Cosmetics.* 2nd ed. New York: John Wiley & Sons, 1996.

40. Ghauri IG, Malih S, Ahmed I. Correct scientific name of "babuna" used widely as a drug in unani system of medicine. *Pak J Sci Ind Res* 1984;27:20–3.

41. Franke R, Schilcher H. *Chamomile Industrial Profile.* Boca Raton, FL: CRC Press, 2005, pp. 49–82.

42. Costescu CI, Hadaruga NG, Rivis A, Hadaruga DI, Lupea AX, Parvu D. Antioxidant activity evolution of some *Matricaria chamomilla* L. extracts. *J Agroaliment Processes Technol* 2008; 14:417–32.

43. Azizi M. Study of four improved cultivars of *Matricaria chammomilla* L. in climatic condition of Iran. *Iran J Med Aromat Plants* 2006;22:386–96.

44. Orav V, Kailas T, Ivask K. Volatile constituents of *Matricaria recutita* L., from Estonia. *Proc Estonian Acad Sci Chem* 2001;50:39–45.

45. Raal A, Kaur H, Orav A, Arak E, Kailas T, Muuricepp M. Content and composition of essential oils in some Asteraceae species. *Proc Estonian Acad Sci Chem* 2011;60:55–63.

46. Eliasoba A, Repcak M, Pastirova A. Quantitative changes of secondary metabolites of *Matricaria chamomilla* by a biotic stress. *Z Naturforsch* 2004;59:543–8.

47. Shams-Ardakani M, Ghannadi A, Rahimzadeh A. Volatile constituents of *Matricara chamomilla* L. from Isfahan, Iran. *Iran J Pharm Sci* 2006;2:57–60.

48. Pauli A. Relationship between lipophilicity and toxicity of essential oils. *Int J Essent Oil* 2008;2(2):60–8.

49. U.S. Pharmacopeial Convention. *Food Chemicals Codex.* 6th ed. Rockville, MD: U.S. Pharmacopeial Convention, 2009.

50. Weiss RF. *Herbal Medicine.* Beaconsfield, UK: Beaconsfield Publishers, 1988, pp. 22–8.

51. Rombi M. *Cento Piante Medicinali.* Bergamo, Italy: Nuovo Insttuto d'Arti Grafiche, 1993, pp. 63–5.

52. Awang-Dennis VC. *The Herbs of Choice: The Therapeutic Use of Phytomedicinals.* New York: CRC Press/Taylor & Francis Group, 2006, p. 292.

53. Martens D. Chamomile: The herb and the remedy. *J Chiropr Acad Homeopathy* 1995;6:15–8.

54. Newall CA, Anderson LA, Phillipson JD. *Herbal Medicine: A Guide for Health Care Professionals.* Vol. 296. London: Pharmaceutical Press, p. 996.

55. Forster HB, Niklas H, Lutz S. Antispasmodic effects of some medicinal plants. *Planta Med* 1980;40:309–19.

56. Crotteau CA, Wright ST, Eglash A. Clinical inquiries; what is the best treatment for infants with colic? *J Fam Pract* 2006;55:634–6.

57. Sakai H, Misawa M. Effect of sodium azulene sulfonate on capsaicin-induced pharyngitis in rats. *Basic Clin Pharmacol Toxicol* 2005;96:54–5.

58. Peña D, Montes de Oca N, Rojas S. Anti-inflammatory and anti-diarrheic activity of *Isocarpha cubana* Blake. *Pharmacologyonline* 2006;3:744–9.

59. Khater HF, Ramadan MY, El-Madawy RS. Lousicidal, ovicidal and repellent efficacy of some essential oils against lice and flies infesting water buffaloes in Egypt. *Vet Parasitol* 2009;164(2–4):257–66.

60. Al-Jabr AM. Toxicity and repellency of seven plant essential oils to *Oryzaephilus surinamensis* (Coleoptera: Silvanidae) and *Tribolium castaneum* (Coleoptera: Tenebrioidae). *Sci J King Faisal Univ (Basic Appl Sci)* 2006;7(1):1427H.

61. Pirali-Kheirabadi K, Razzaghi-Abyaneh MJ. Biological activities of chamomile (*Matricaria chamomile*) flowers' extract against the survival and egg lying of the cattle fever tick (*Acari ixodidae*). *Zhejiang Univ Sci B* 2007;8(9):693–6.

62. Abdel-Gawad H, Abdel Hameed RM, Elmmesalamy AM, Hegazi B. 2011. Distribution and elimination of $^{14}$C-ethion insecticide in chamomile flowers and oil. *Phosphorus Sulfur Silicon Relat. Elem.* 186(10):2122–34.

63. Khare CP. *Indian Medicinal Plants—An Illustrated Dictionary.* Berlin: Springer Science and Business Media, 2007, p. 55.

64. Bail S, Buchbauer G, Jirovetz L, Denkova Z, Slavchev A, Stoyanova A, Schmidt E, Geissler M. Antimicrobial activities of Roman chamomile oil from France and its main compounds. *J Essent Oil Res* 2009;21(3):283–6.

65. Hänsel R, Keller K, Rimpler H, Schneider G, eds. *Hagers Handbuch der Pharmazeutischen Praxis. Drogen A–D.* Berlin: Springer-Verlag, 1993, pp. 808–17.

66. Choi WI, Lee SG, Park HM, Ahn YJ. Toxicity of plant essential oils to *Tetranychus urticae* (Acari: Tetranychidae) and *Phytoseiulus persimilis* (Acari: Phytoseiidae). *J Econ Entomol* 2004;97:553–8.

# 20

## Neem Oil

Zakia Khanam, Hanan M. Al-Yousef, Ompal Singh, and Irshad Ul Haq Bhat

## CONTENTS

## 20.1 Botany

### 20.1.1 Origin and Distribution

Neem, botanically known as *Azadirachta indica* A. Juss, belongs to the Meliaceae (mahogany) family (Anon, 2011). It is indigenous to India and found in tropical and subtropical regions like Pakistan, Bangladesh, Sri Lanka, and Mayanmar. The Siwalik hills, dry forests of Andhra Pradesh, Karnataka, and Tamil Nadu (India) are the main habitat of the wild population (Hashmat et al., 2012). It thrives well in the dry regions of the northwest, and approximately 50% of the tree population of India is reported in Uttar Pradesh. The Science and Technology Panel of the International Development National Research

Council (1992) has documented that around 60% of the total neem population of the world inhabits India (Tinghui et al., 2001). It is also grown and naturalized in Southeast Asian (Thailand, Indonesia, Peninsular Malaysia, the Philippines, and Singapore) and West African countries, as well as Australia and Saudi Arabia. More recently, it has been familiarized to the Caribbean and various zones of America (Parotta, 2001).

## 20.1.2 Taxonomical Classification

The neem plant is taxonomically classified as (Girish and Shankara, 2008; Anon, 2011)

Kingdom: Plantae
Division: Tracheophyta
Class: Magnoliopsida
Order: Sapindles
Family: Meliaceae
Subfamily: Melioideae
Tribe: Melieae
Genus: *Azadirachta*
Species: *indica*

*A. indica* is synonymous with *Melia azadirachta* L. and *Antelaea azadirachta* (L.) Adelb (Anon, 2011).

## 20.1.3 Vernacular Names

*A. indica* has many local names, depending on the languages used in a country. The appellations given in the different languages of India and other countries are shown in Table 20.1.

## 20.1.4 Plant Description

*A. indica* belongs to Meliaceae, a family of dicots mostly represented by trees and shrubs. The family includes about 51 genera and 550 species, with many of them prized for their wood, edible fruits, and medicinal and ornamental qualities (Wiart, 2006). It is a small to medium-sized evergreen tree with a height of 15 m (30 m maximum), having a large rounded crown (10–20 m) with spreading branches and a branchless bole (7.5 m, diameter 90 cm). The bark of the tree is thick, fissured, dark gray to red (inside) in color, and it possesses a gummy colorless sap. The leaves are long (20–40 cm), alternate, pinnate, exstipulate, and glabrous with a light green hue. The leaves have two pairs of basal glands with a subglabrous petiole (2–7 cm) and above, channeled rachis. Each leaf comprises 8–19 serrated, proximally alternate, ovate to lanceolate leaflets. Inflorescence is axillary clustered multiflowered thyrsus (150–250 flowers) with a length of 15–30 cm and minute caducous bracts. Flowers of the tree are small (1 cm in diameter), white or pale yellow, and sweet smelling. They are actinomorphic, pentamerous, and bisexual or unisexual male on the same plant. The calyx of the flowers is imbricate, ovate, thin, and puberulous from inside, while petals are free, spreading, imbricate, spathulate and ciliolate from inside. Fruits are single (maximum of two) and small (1–2 cm) in size. They are greenish to yellow in color and an ellipsoidal seeded drupe. The tree has a thin exocarp, pulpy mesocarp,

**TABLE 20.1**

Vernacular Names of *A. indica*

| Country | Language: Vernacular Names |
|---|---|
| India | Bengali: Nim, nimgachh<br>Guajarati: Danujhada, limbado, limbra, limdo<br>Hindi: Balnimb, neem, nim, nimb, nind, vempu, veppam<br>Kannada: Bemu, bevinamara, bivu, kaybevu<br>Punjabi: Bakam, drekh, nim<br>Sanskrit: Arista, nimba, nimbah, picumarda<br>Tamil: Vepa, veppu, veppam, vembu |
| Indonesia | Indonesian: Mind, intaran, membha, imba, mempheuh, mimba<br>Javanese: Mimba, imba |
| Malaysia | Malay: Sadu, baypay, mambu, veppam |
| Myanmar | Burmese: Bowtamaka, thinboro, tamarkha, tamar, tamaka, tamabin |
| Nepal | Nepali: Neem |
| Thailand | Thai: Sadao, kadao, sadao India, khwinin, saliam, cha-tang |
| Vietnam | Vietnamese: Saafu daau, sàu-dàu |
| England, Canada, America | English: Persian lilac, neem tree, bastard tree, Indian lilac, bead tree, margosa tree, cornucopia, Indian cedar |
| France | French: Margousier, margosier, neem, nim, azadirac de l'Inde |
| Saudi Arabia | Arabic: Nim, neem |

*Source:* Parotta, J. A., *Healing Plants of Peninsular India*, CABI Publishing, New York, 2001, pp. 495–496.

and cartilaginous endocarp. Seeds are an unwinged, oval, or spherical structure with thin testa. The tree has a profound taproot system with widespread lateral roots. It may form suckers if roots encounter some damage (Hearne 1975; Csurhes, 2008; Hashmat et al., 2012).

## 20.2 Methods of Extraction of Oil

Neem essential oil is usually prepared from the seed kernels and is well known for its high insecticidal and medicinal value (Lokanadhan et al., 2012). The fruit, flower, and leaves are minor sources of neem essential oil (Narsing Rao et al., 2014). According to a survey, only 20% of the seeds are being harvested due to scattered growth of neem trees in India. Out of it, India produces approximately 8300 tonnes of neem oil annually. In general, the neem oil yield reported from seeds varies from 25% to 45% (Anya et al., 2012; Ismadji et al., 2012). Mechanical pressing, solvent extraction, and more recently, supercritical fluid extraction (SFE) are among the numerous methods to extract neem seed oil (NSO) (Liauw et al., 2008). In the mechanical method, cold pressing or temperature-controlled pressing can be employed to procure oil from the neem seed kernel via physical crushing. Approximately 82% of the neem oil can be recovered by the mechanical extraction method. Although mechanical pressing is a frequently used technique, neem oil acquired with this process has poor quality due to it low azadirachtin content (1427 ± 51 ppm, 25.3%), suggesting nonselectivity of the extraction process. Moreover, oil produced by mechanical pressing is turbid, containing a considerable amount of water and metal content, and hence it has cheap market value (Adeeko and Ajibola, 1990; Lalea and Abdulrahman, 1999). According to a study conducted by Nitièma-Yefanova et al. (2012), the oil yield by cold pressing has a positive effect on increased kernel compression, reduced particle size, and decreased cage loading. It was concluded that the best oil yield (40.3% ± 0.0) from ground kernels can be obtained by cold pressing at 25°C, 33.7 MPa ± 2.9 pressure, and quarter-cage loading. However, while the cold-pressing method is easy, economical, and solvent-free, to obtain high-quality oil, intensive purification steps are needed, which often reduce active components of oil and may cause technical and monetary constraints for commercial-scale production (Ismadji et al., 2012).

The solvent extraction method is generally the preferred choice for obtaining neem oil. It furnishes a high yield and clear oil compared with the mechanical extraction method. In the Soxhlet extraction method, neem oil percentage recovery corresponded to 92.3%–99.1%, with the azadirachtin content (4658.4 ± 92.5 ppm) three times more than that obtained from the cold-pressed extraction procedure. Furthermore, the conventional solvent extraction method has a relatively low operational cost and is economical compared with the other modern methods of extraction, like supercritical fluid extraction (Liauw et al., 2008; Adewoye and Ogunleye, 2012; Ismadji et al., 2012). Hexane is the most commonly used solvent in the extraction of seed oil, including neem seeds, due to its suitable functional properties, that is, nonpolar nature, which facilitate the high solubility of hydrocarbons, lipids, and glycerides at moderate temperatures. Also, hexane is inexpensive and unreactive with oil (Ayoola et al., 2014). However, it is enlisted among 189 hazardous air pollutants of the Clean Air Act, and is being watched as both a "criteria pollutant" and a "hazardous air pollutant." Thus, interest has been generated among researchers to discover an alternative, nonflammable, efficient, less hazardous, and environmentally friendly solvent. Liauw et al. (2008) compared $n$-hexane and ethanol solvents as a medium of neem oil extraction. The maximum oil yields obtained were 41.11% and 44.29% with ethanol and $n$-hexane at a low extraction temperature (50°C), respectively. At the same temperature, petroleum ether (42.60%) demonstrated higher efficiency than ethanol (39.17%) for neem oil extraction (Satyanandam et al., 2011). Later, it was anticipated that ethanol may be a good substitute for hexane, as there is a possibility to enhance oil extractability using ethanol at high temperatures (Liauw et al., 2008; Ayoola et al., 2014). This can be easily achieved because of ethanol's tendency to withstand high temperatures (80°C), which is not possible in the case of hexane due to its highly flammable behavior. Also, there were several reports that showed that ethanol–hexane mixtures have the potential for high neem seed oil extraction compared with hexane alone. In one study, ethanol–hexane mixtures utilized for oil extraction by the Soxhlet extraction method in ratios of 60/40, 50/50 and 40/60 (v/v) furnished oil yields of 44%, 43%, and 41.2%, respectively. The ethanol–hexane mixtures gave a better yield than 100% hexane (40.25%) at 55°C over a 6 h duration (Ayoola et al., 2014). Similarly, Edres (2014) extracted neem oil through the Soxhlet extraction method using only $n$-hexane and obtained a very low yield of 17.60%.

The high oil yield by the solvent extraction method depends not only on the type of solvent, but also on various other physical factors, such as solvent composition, volume of solvent, temperature, sample–solvent ratio, and material size. Liauw et al. (2008) has optimized oil extraction from ground neem seeds, categorized into three types of particle size ranges (i.e., 0.85–1.40 mm, 0.71–0.85 mm, and 0.425–0.71 mm). The results revealed that the maximum oil yield was obtained at 0.425–0.71 mm seed sizes. Adewoye and Ogunleye (2012) improved neem oil extraction with regard to the three aspects of extraction, that is, solvent composition ($n$-hexane and ethanol), temperature, and time, by a central composite design (CCD). The maximum predicted percentage yield was obtained as 43.48%

with 80.77% *n*-hexane at 34.93°C over a 6 h duration. Okonkwo et al. (2013) reported the influence of agitation, type of impeller, and time of contact on NSO yield using a pilot solvent extraction plant. In their research, the maximum percentage yield was found to be 36.86% within 40 min with a flat-blade turbine impeller type A1 operating at 84 rpm. At 50°C, food-grade ethanol was used as a medium for extraction with 0.425–0.710 mm seed particle size.

Among solvent extraction procedures, the Soxhlet method is conventionally performed for oil extraction (Ayoola et al., 2014). Because of its low yield and to meet demand, recently, microwave-assisted extraction (MAE) was manipulated for optimization of neem oil extraction by Doehlert design. Parameters such as time, temperature, and solvent-to-biomass ratio were studied to obtain an optimized yield of neem oil (Nde et al., 2015). It was observed that within 24 min, 80% of the oil was extracted at 80°C and a solvent–biomass ratio of 3:1. This method was found to be more efficient and faster than the conventional method (10 h), and without significantly affecting oil quality. This attributed to the penetrative behavior of microwaves, as shown by scanning electron microscopy (SEM) analysis, which revealed structural deterioration of the neem seed kernels.

Furthermore, SFE has been of great interest for being an efficient and effective method for the recovery of essential oils and active metabolites from solid matrices. It produces high-quality oil with no solvent residues, but its operating and investment costs are high. Beyond a critical temperature (31°C) and pressure (74 bars), carbon dioxide ($CO_2$) exists as a liquid and finds use in SFE to extract the active ingredients of neem seeds (Martinelli et al., 1991). The density and solvation of the supercritical $CO_2$ (SC-$CO_2$) at 200 bars is analogous to that of hexane; hence, it behaves as a nonpolar solvent and can dissolve triglycerides at concentrations up to 1% mass. Unlike other solvents, $CO_2$, being a green, nonhazardous, inert, and affordable solvent, is perfectly adapted to extract essential oils without interfering with its active components (Sapkale et al., 2010). Contrary to the solvent extraction method, SFE is environmentally friendly and produces no waste, and removal of solvent (SC-$CO_2$) from the oil or analyte does not require rigorous heating, which can be achieved by releasing pressure, leaving almost no trace of $CO_2$ in the oil yield. By using SC-$CO_2$, extraction of neem essential oil and its active triterpenoids has been reported elsewhere (Mongkholkhajornsilp et al., 2005; Zahedi et al., 2010).

The solvation properties of SC-$CO_2$ can be tailored by adding cosolvents such as ethanol and methanol. Johnson and Morgan (1997) reported SC-$CO_2$ extraction of neem seed oil and its triterpenoids (azadirachtin, nimbin, and salannin) at 328.15 K with methanol to enhance the selectivity and extraction performance. There have been studies that have suggested that the solvation properties of SC-$CO_2$ can be modulated by altering temperature and pressure as well, to obtain a high oil yield. The adjustable solvent power offers high extraction selectivity. Ismadji et al. (2012) demonstrated the extraction of neem oil and active triterpenoid compounds from seed kernel without cosolvent through changeable pressures (10–35 MPa) and temperatures (313.15, 323.15, and 333.15 K). It was observed that the neem oil yield and active molecules extracted were temperature and pressure dependent. The maximum percentage recovery of neem oil was reported as $6.67 \pm 0.12$ ($\times 10^3$ kg) at the highest studied temperature (333.15 K) and pressure (35 MPa). This was explained by the high pressure greatly increasing the solvent density (solvating power) of $CO_2$, thus enhancing the solubility of the solute. Moreover, the solvent-to-solute ratio ($CO_2$ to neem seed kernels) has a significant impact on the oil yield at supercritical conditions, as revealed by Ambrosino et al. (1999).

## 20.3 Methods of Analysis of Oil

Essential oils are widely used for medicinal, cosmetic, culinary, perfumery, and agricultural products. The geographical location, climatic conditions, soil properties, chemotype, developmental stages, and so forth, affect the distribution of components and manifest quantitative changes in the oil composition. However, chemical fingerprinting or key active components help to determine the purity as well as potency of the oil. Due to the high market price of premium essential oils, they are adulterated, diluted, or substituted with poor-quality oils, cheap terpenes, and low-density petroleum fractions. Therefore, analyses of oil to check adulterant and purity is of utmost importance to consumers and manufacturers for safety and quality control.

Essential oils are complex mixtures of secondary metabolites, responsible for their unique properties and aroma. The characterization of oil for their chemical composition is vital for industrial and economic purposes, where organic chemistry plays a fundamental role. The growth of chromatographic techniques has made significant progress in the study of the chemical composition of essential oils, for example, high-performance liquid chromatography (HPLC), gas chromatography (GC), gas chromatography–mass spectrometry (GC-MS), and capillary electrophoresis (Forim et al., 2010; Djenontin et al., 2012). GC is considered one of the best methods for analyzing oils, for both qualitative and quantitative determination, due to its simplicity, rapidity, and efficiency. Generally, conventional GC analyses require 30–60 m columns and 30–60 min to furnish a chromatogram. The high efficiency of chromatography and quantitative determination of the important groups of a compound were achieved by baseline separation. It is imperative that all reliable analytical methods have prior validation. The criteria for authentication usually include linearity, specificity, accuracy, precision, robustness, recovery, limits of quantification (LOQs) and detection (LODs), and repeatability in HPLC methods. Along with these chromatographic techniques, development of detectors acted as a powerful tool for accurate analyses of essential oils (Forim et al., 2012).

The interesting feature of neem essential oil is that it is potently aromatic and viscous, and predominantly constitutes fatty acids, along with hydrocarbons (Djenontin et al., 2012). In order to obtain fatty acid profile of neem oil, fatty acid methyl ester (FAME) derivatives were injected into a GC equipped with a flame ionization detector (FID) with a glycol succinate column (Hossain, 2005). For identification of individual fatty acids, both internal and external standards were used. Recently, Djenontin et al. (2012) reported the neem oil profile by GC-FID, involving both identification and quantitation of its components. Among fatty acids, prominent concentrations of oleic (43.5%), linoleic (18.7%), palmitic (17.8%), and stearic (17.4%) acids were observed. During analysis, a HP-INNO Wax capillary column and helium as the carrier gas were used. Furthermore, the composition of neem essential oil was explored by both chromatographic and spectral techniques, such as quantitative thin-layer chromatography (TLC), GC, GC-MS, and $^{13}C$ nuclear magnetic resonance (NMR). They revealed that fatty acids were the main component of NSO. Other compounds include triacylglycerols, sterols, $n$-alkanes, aromatics, esters, sulfur, and nitrogenous compounds and terpenoids (Kurose and Yatagai, 2005; Momchilova et al., 2007). The fatty acid of neem oil was also studied by high-speed countercurrent chromatography (HSCCC) (Gossé et al., 2005). They successfully resolved different fatty acids and unsaponified organic compounds via the HSCCC technique.

In neem essential oil, azadirachtin is a good-quality control biomarker compound that possibly simplifies the demand of various equipment, time, and the cost required for extensive analyses, to check the purity of the oil and its industrial formulations. The amount of azadirachtin is usually detected in various neem oil products by a simple, sensitive, and selective HPLC technique (Ambrosino et al., 1999; Sidhu et al., 2003). Forim et al. (2010) developed a method for simultaneous quantification of two important limonoids of NSO, that is, azadirachtin and 3-tigloylazadirachtol, by using HPLC equipped with an ultraviolet (UV)–visible detector. The chromatographic analyses were performed isocratically with a ratio of 35:65 (v/v) of acetonitrile and water as the mobile phase, and the UV wavelength was selected at 217 nm (maximum) to measure low-concentration marker compounds. The study was aimed at controlling the quality and promotion of Brazilian neem seed and NSO. In another study, analysis was performed by a reverse phase Spherisorb C-18 ODS 5 µm column with an acetonitrile–water gradient system to determine the azadirachtin content in neem formulations and oil (Sundaram and Curry, 1993). Similarly, a fast preconcentration method was established for azadirachtin A, azadirachtin B, nimbin, and salannin determination in neem oil. It comprises solid phase extraction with graphitized carbon, followed by quantification using HPLC-UV with an upper limit of quantification, 100 µg/ml (Ramesh and Balasubramanian, 1999).

Moreover, HPLC combined with a mass spectrometry (MS) detector has been utilized by researchers to study the azadirachtin content of neem oil extracted from insecticidal formulations (Ambrosino et al., 1999; Barrek et al., 2004). MS is a more precise tool for the identification of compounds than a conventional UV attachment. In the study, atmospheric pressure chemical ionization (APCI) and electrospray ionization (ESI) were used as ionization sources, and these enabled rapid identification of neem oil limonoids. The ESI source assisted in the identification of the largest number of structures (azadirachtin A, azadirachtin D, azadirachtin I, deacetylnimbin, de-acetylsalannin, nimbin, and salannin) (Barrek et al., 2004). In another study on seed oil, HPLC analysis coupled with a UV diode array detector (DAD) indicated the presence of sterols, where the major sterol was β-sitosterol (77.7%) (Djenontin et al., 2012). Similar studies on sterols and tocopherol of NSO were carried out by GC and HPLC, respectively (Djibril et al., 2015). Although the above-discussed chromatographic techniques combined with detectors may be utilized for more precise screening, identification, and quantification of essential oils, they are expensive for regular analyses. Lalla et al. (2003) developed a cheap, simple, accurate, and specific method to analyze NSO using azadirachtin as a biomarker compound. The method employs high-performance thin-layer chromatography (HPTLC) for quantitative study of the oil and can be exploited for routine analyses. More recently, HPTLC was carried out to investigate the composition of NSOs obtained from three production sites in Italy. The study exhibited a marked difference in chemical composition and variation among NSOs, with respect to the limonoids (Benelli et al., 2015).

## 20.4 Composition of Oil

Neem essential oil from flower and leaves is a minor source of volatile oil (0.08%), composed mainly of caryophyllene (85%) (Narsing Rao et al., 2014). NSO is a main source of volatiles, being composed of essential oil and fatty acids (Djenontin et al., 2012). NSO is subjected to extensive phytochemical studies due to its strong biological, agricultural, and medicinal properties (Lokanadhan et al., 2012). The chemical composition of NSO is very complex and rich in terpenoids, limonoids, and volatile sulfur compounds (Ricci et al., 2009). Until now, more than 300 compounds have been isolated from various parts of *A. indica* (Gossé et al., 2005). However, NSO alone recounted more than 100 determined biologically active compounds (Benelli et al., 2015). The key chemical constituents reported from neem essential oil can be divided into individual classes, as described in the next sections.

### 20.4.1 Hydrocarbons

El-Hawary et al. (2013) has studied essential oils obtained from neem leaves and flowers using the hydrodistillation method. The constituents of leaf and flower essential oil were analyzed by GC-MS. The main hydrocarbons (85.36%) detected in leaf oil were β-elemene (33.39%), γ-elemene (9.89%), germacrene D (9.72%), caryophyllene (6.8%), and bicyclogermacrene (5.23%). The oxygenated compounds were mainly sesquiterpene oxide (5.04%). However, flower oil hydrocarbons were composed primarily of pentacosane (18.58%), tetracosane (10.65%), β-germacrene (9.73%), β-caryophyllene (5.84%), and dodecene (4.54%). The principal oxygenated compounds (28.3%) were octadecanol (16.7%), verdiflorol (5.32%), farnesol (1.63%), and α-terpineol (1.51%). In addition, Narsing Rao et al. (2014) showed that volatile oil obtained from neem flower powder by hydrosteam distillation contains the sesquiterpenes caryophyllene (56.03%), caryophyllene oxide (17.41%), and α-caryophyllene (humulene) (12.10%) as major components. The other minor components were identified as copaene, bicyclo[5,2,0]non-1-ene, cyclohexene, 2H-indene 2-one, cyclohexane, and 1,2-benzenedicarboxylic acid. The neem seed essential oil has been studied by Kurose and Yatagai (2005), along with other essential oils of *Azadirachta* species. GC and GC-MS have detected hexadecanoic acid (34.0%), oleic acid (15.7%), 5,6-dihydro-2,4,6-triethyl-(4H)-1,3,5-dithiazine (11.7%), methyl oleate (3.8%), and eudesm 7(11)-en-4-ol (2.7%) as major oil constituents. The minor components spotted were *n*-alkanes, aromatics, esters, sulfur and nitrogen compounds, and terpenoids.

### 20.4.2 Fatty Acids

NSO is a major source of fatty acid and is mainly composed of oleic acid (50%–60%), palmitic acid (13%–15%), stearic acid (14%–19%), linoleic acid (8%–16%), and arachidic acid (1%–3%). Oleic acid, linoleic acid, and α-linoleic acid are the principal ω-9, ω-6, and ω-3 fatty acids, respectively, in NSO (Mongkholkhajornsilp et al., 2005; Ismadji et al., 2012). Besides, palmitic (31.76%), linoleic (18.57%), linolenic (12.64%), oleic (9.74%), arachidonic (7.38%), and docosatrienoic (5.7%) acids were also reported in flower oil (Narsing Rao et al., 2014). Nevertheless, Momchilova et al. (2007) identified and quantified 13 fatty acids and 25 triacylglycerols by amalgamation of chromatographic and spectral techniques.

### 20.4.3 Limonoids

The important bioactive compounds of NSO belong to the limonoid class of triterpenoids, such as azadirachtin (azadirachtin A), salannin, salannol, nimbin, nimbinin, nimbidin, nimbidiol, nimolicinol, gedunin, 3-tigloylazadirachtol (azadirachtin B), epoxyazadiradione, 17β-hydroxyazadiradione, 1-tigloyl-3-acetyl-11-hydroxymeliacarpin (azadirachtin D),

1α,2α-epoxy-17β-hydroxyazadiradione, 1α,2α-epoxynimolicinol, and 7-deacetylnimolicinol (Hallur et al., 2002; Ismadji et al., 2012). In 1942, Siddiqui reported bitter principles, nimbin, nimbinin, and nimbidin, where nimbidin was the major bitter principle of NSO. All these plant metabolites are well known for their effective biological properties against insects and pests; among them, azadirachtins (0.3%–0.6%) are the most active component of neem essential oil (Brahmachari, 2004). The concentration of triterpenoid secondary metabolite in neem seeds is dependent on the geographical location of the plant grown (Sidhu et al., 2003).

### 20.4.3.1 Azadirachtins

Azadirachtins are the most celebrated and studied active principles of neem oil due to their deterrent, antiovipositional, antifeedant, growth-disrupting, growth-regulating, fecundity, and fitness-reducing properties against insects and various kinds of arthropods (Ambrosino et al., 1999; Morgan, 2009). They are a group of closely related isomers that belong to the steroid-like tetranortriterpenoid class, called azadirachtin A to H (Rembold et al., 1984, 1987). Among all azadirachtins identified so far, azadirachtin A (azadirachtin) is a highly appreciated and interesting compound, as it is considered the most potent and principal agent for controlling insects (Sinha et al., 1999). Hence, it acts as a biomarker for standardization of neem oil and commercial insecticidal formulations (Sundaram and Curry, 1993; Sidhu et al., 2003). In 1968, Butterworth and Morgan isolated azadirachtin ($C_{35}H_{44}O_{16}$, MW 720, m.p., 160°C), and its synthesis was published 22 years after the discovery, due to its complicated structure (Gossé et al., 2005). The first total synthesis was given by Stevenley in 2007 (Veitch et al., 2007). The azadirachtin content in crude neem oil varies (100–4000 ppm), depending on the extraction technique, seed quality, environment, and genetic factors (Ambrosino et al., 1999; Ismadji et al., 2012). Neem seed kernel from Bali, Indonesia, is reported to contain up to 6200 ppm of azadirachtin.

### 20.4.4 Sterols

The major sterols in NSO reported were β-sitosterol, stigmasterol, campesterol, and fucosterol (Momchilova et al., 2007). The total tocopherols (298 ppm) indicate α-tocopherol and γ-tocopherol as major components (30.8% and 62.3%, respectively) (Djenontin et al., 2012).

### 20.5 Physical and Chemical Properties of Oil

Neem oil is normally a golden-yellow, yellowish-brown, reddish-brown, dark brown, greenish-brown, or bright red liquid. It has an unpleasant strong and offensive odor. The smell of the oil is a partial combination of peanut and garlic. The obnoxious odor of neem oil is ascribed to the presence of sulfur-containing volatile compounds (Dasa Rao and Seshadri, 1941). It has an acrid taste, which is attributed to several triterpenoids present in it. It is a nondrying oil and, due to its hydrophobic nature, needs appropriate surfactants for proper emulsification during industrial application (Mongkholkhajornsilp et al., 2005; Usman et al., 2013; Edres, 2014). The quality of the oil depends on its composition, which in turn affects the properties. Since neem oil mainly contains fatty acids as one of the active components, it is commonly analyzed for its quality by determining the saponification (SV), acid (AV), and iodine (IV) values, and so forth. Table 20.2 represents standard physicochemical properties of NSO (Okonkwo et al., 2013; Djibril et al., 2015).

The temperature has a considerable effect on the quality of NSO, as it decreases with an increase in temperature (Satyanandam et al., 2011). The poor quality of NSO can be confirmed by increased acid, saponification, and peroxide values (PVs) and decreased iodine value with high temperature (Liauw et al., 2008; Satyanandam et al., 2011). A high acid value indicates the presence of a high amount of free fatty acids in oil due to the degradation caused by hydrolysis at high temperature. The lipase enzymes in oil are responsible for the hydrolysis of triglycerides into free fatty acid and glycerol. Since the optimum temperatures for enzymes is 30°C–40°C, the extraction and storage temperatures are the chief concern in maintaining the quality of NSO (Khraisha, 2000; Choe and Min, 2006). The polarity of the extraction solvent also influences the AV, as low polarity caused the solvent to efficiently extract free fatty acids, thus escalating its acid value. According to Erakhrumen (2011), bio-oil with a high acid value can be exploited as a preservative for lignocellulose to enhance wood durability and properties.

The saponification value signifies the average molecular weight of the oil triglycerides. As high temperature causes lipid degradation, it reduces the average molecular weight of the NSO. The reduction in average molecular weight leads to a reduction in the viscosity of the oil. Also, the specific gravity (SG) or density of NSO decreases with an increasing temperature of exposure, and results in reduced viscosity and increased flow of oil. The peroxide value of oil implies rancidity and increases with temperature (Adeeko and Ajibola, 1990). Rancidity can be caused by hydrolysis, oxidation, or microbes. It gives an unpleasant smell to the oil due to the degradation of glycerides or formation of aldehydes and ketones. Oil with a high AV has poor resistance to peroxidation, particularly during storage. Increased extraction and storage temperature or long improper storage may lead to rancidity of NSO, thereby reducing its oxidative stability (Mongkholkhajornsilp et al., 2005). The iodine value expresses the degree of unsaturation in oil. The high temperature degrades the bonding, thereby leading to a decreased iodine value. Moreover, IV represents the drying nature of oil, where iodine values greater than 140 g/100 g and IVs less than 125 g/100 g are characteristics of drying and nondrying oil, respectively (Wicks et al., 1992). Unlike temperature, the polarity of the extraction solvent did not have any influence on SV and IV (Liauw et al., 2008; Erakhrumen, 2011). The results of some recent research on the physical and chemical properties of NSO are presented in Table 20.3.

**TABLE 20.2**

Standard Physicochemical Properties of Neem Seed Oil

| Properties | Value |
|---|---|
| Odor | Garlic |
| Specific gravity at 30°C | 0.908–0.934 |
| Viscosity at 37.8°C (mm²/s) | 49.79 |
| Refractive index at 30°C | 1.4615–1.4705 |
| pH | 5.7–6.5 |
| Iodine value | 65–80 g/g |
| Acid value | 40 mg of KOH/g |
| Saponification value | 175–205 mg of KOH/g |

**TABLE 20.3**

Properties of NSO Obtained by Various Methods

| Extraction Method | AV (mg/g) | SV (mg of KOH/g) | IV (g/100 g) | PV (mg/g) | Reference |
|---|---|---|---|---|---|
| Cold pressing | 18.24 | 172.88 | 93.11 | 1.42 | Erakhrumen, 2011 |
| Solvent (hexane–ethanol, 60:40) | 12.90 | 199.99 | | | Ayoola et al., 2014 |
| Solvent (Soxhlet extractor, cyclohexane) | 10.2 | 200 | 72.82 | | Djibril et al., 2015 |
| Solvent (Soxhlet extractor, hexane) | 17.40 | 186.4 | 58.20 | 78.40 mEq/g | Zaku et al., 2012 |
| Solvent (hexane) | 1.411 g/g | 176.64 mg/g | 89.35 g/g | | Sodeinde and Samuel, 2014 |

## 20.6 General Uses of Oil

### 20.6.1 Medicine

The therapeutic application of neem had been identified by Indians 4000 years ago, since the Vedic period. In Sanskrit, the neem tree is recognized as "Arishtha," meaning "reliever of sickness." It is regarded as the "village dispensary" (Brahmachari, 2004). Traditionally, all parts of this divine tree have been utilized against various human ailments. According to Ayurvedic medicine, NSO has been used for the treatment and control of leprosy, syphilis, eczema, chronic ulcer, and intestinal helminthiasis (Hashmat et al., 2012). In the Siddha system of medicine, a preparation containing NSO, called *onan cutar tailam*, is used for epilepsy. However, apart from being practiced in traditional (Ayurvedic, Siddha, and Unani) medicine, there are several modern scientific investigations that have been carried out on NSO that have justified its therapeutic usage. NSO and its active components have displayed several pharmacological activities, including anti-inflammatory, antiarthritic, antipyretic, hypoglycemic, diuretic, spermicidal, antifungal, antibacterial, antigastric ulcer, antiviral, and antipsoriasis activities (Brahmachari, 2004).

### 20.6.2 Agriculture and Public Health

Besides curative abilities, neem oil has been reputed as a source of naturally occurring pesticide and insecticide (Ahmad et al., 2015; Jhalegar et al., 2015; Rodrigues et al., 2015; Sridharan et al., 2015). Utilization of NSO products as agrochemicals is remarkable. Pusa neem golden urea (PNGU), a urea–neem oil adduct, is one example of a neem-based agrochemical applied as a nitrification inhibiting agent. Moreover, NSO is also used for public health, mainly as an antilouse and antimalarial agent (Al-Quraishy et al., 2015). Neem oil mixed with coconut oil acts as an effective mosquito repellent (Brahmachari, 2004).

### 20.6.3 Personal Care Products

NSO and its components are utilized in toiletries and cosmetics such as skin care (soap, eczema cream, antiseptic cream, nail care, and balm), hair care (shampoo and hair oils), oral hygiene (toothpaste), and other household products (insect repellent spray and lotion, candles, wax, and lubricants) (Hashmat et al., 2012; Shetty et al., 2016). Purified NSO is also exploited in nail polish and other cosmetics preparation (Shetty et al., 2016).

## 20.7  Pesticidal Uses of Oil

Neem oil is recognized as a powerful biopesticide and may offer a solution to global agricultural, environmental, and public health problems. The NSO allelochemicals are reported to have feeding and oviposition deterrence, repellency, growth disruption, reduced fitness, and sterility activities, and hence have been widely used in agricultural pest control (Brahmachari, 2004). In NSO, high concentrations of bioinsecticide limonoids are reported, mainly azadirachtin A, azadirachtin B, nimbin, and salannin (Stark and Walter, 1995). The most potent limonoid in NSO, azadirachtin, primarily acts as an insect repellent and insect growth regulator (IGR). Its structure is similar to that of insect hormones, "ecdysones," responsible for metamorphosis in insects. It is active at minute concentrations (1–10 ppm) and responsible for hindering the action of ecdysones, thus preventing the larvae from shedding their exoskeletons. Thus, azadirachtin alters their life cycle and inhibits the development of immature insects (Lokanadhan et al., 2012; Radwan and El-Shiekh, 2012). Also, NSO exhibits antifeedent and oviposition deterrent activity (Benelli et al., 2015). Antifeedent activity, credited to azadirachtin, nimbin, salannin, epoxyazadiradione, and melandriol, causes antiperistalitic movement in the alimentary canal and initiates a vomiting sensation in the insect (Esparza-Díaz et al., 2011). The nauseated feeling and inability to swallow do not allow insects to feed on NSO-treated surfaces. It checks feeding in approximately 200 types of insects at concentrations of 10–100 ppm. Similarly, NSO sprayed during storage does not allow female insects to lay eggs (Lokanadhan et al., 2012).

The broad-spectrum activity coupled with non-toxicity to mammals brands NSO as a perfect candidate for biopesticide treatment. As it is effective without being unacceptably hazardous to users and the environment, positive steps have been taken by the government to promote the use of neem biopesticide in agriculture and public health programs (Kaushik, 2004). The use of biopesticides is assumed to be a significant component of integrated pest management (IPM) for the realization of sustainable agriculture, due to their economic viability and eco-friendly nature. Hence, a tremendous amount of research has been conducted in the last few decades to exploit neem's pesticidal potential in agriculture and the public health sector (Table 20.4).

## 20.8  Advantages as a Pesticide

Neem essential oil plays an important role in pest management, and interest in neem pesticides has grown during the last few decades, as numerous pesticides have been restricted due to environment and food safety issues (Stark and Walter, 1995). The widespread understanding of the undesirable effects of synthesized pesticides on plants, soil, and nontargeted creatures has shifted interest toward readily available sources of biopesticides, that is, botanical pesticides. They are safe, degradable, and cheap (Brahmachari, 2004; Lokanadhan et al., 2012). The growing acceptence of neem essential oil–based pesticides has generated a great

**TABLE 20.4**

Recent Research on Neem Essential Oil for Agricultural and Public Health Pest Control

| Effected Organism | Pathogen/Disease | Treatment | Effects | Reference |
|---|---|---|---|---|
| *Mangifera indica* L. (Taimour mango) | Powdery mildew and malformation | Neem oil (1%) | Powdery mildew disease severity index (DSI) was 18.17%, more efficient against mango malformation | Ismail, 2016 |
| Humans | *Aedes aegypti* (dengue mosquito) | *Metarhizium anisopliae* (entomopathogenic fungi) + neem oil (0.001%) | 12% survival | Gomes et al., 2015 |
| Cowpea (Brazil) | *Spodoptera eridania* (southern armyworm) | Neem oil (0.35% and 0.7%) | Reduced leaf consumption | Rodrigues et al., 2015 |
| Brinjal | *Leucinodes orbonalis guenee* (shoot and fruit borer) | Neemarin (neem oil) (3 L/ha) | Reduced shoot and fruit damage | Singh and Sachan, 2015 |
| Humans | *Pediculus humanus capitis* (head louse) | Antilouse shampoo Licener® (shampoo + neem oil) | Oxygen uptake is prohibited in 3–10 min | Al-Quraishy et al., 2015 |
| Cowpea | *Maruca vitrata* | Multinucleopolyhedrovirus (MaviMNPV) + neem oil | Induced MaviMNPV infection in *M. vitrata* populations | Sokame et al., 2015 |
| Humans | *Anopheles arabiensis* (Ethiopian malaria mosquito) | Neem oil (20%) | More than 70% repellency (protection) in 3 h | Abiy et al., 2015 |
| Kinnow mandarin (*Citrus nobilis* × *Citrus deliciosa*) | *Penicillium digitatum* and *Penicillium italicum* | Neem essential oil | Influenced overall acceptability of postharvest crop | Jhalegar et al., 2015 |
| Cultivated crops | *Helicoverpa armigera* (moth) | Neem oil (1% emulsifiable concentrate [EC] azadirachtin) | Reduction in fecundity, reproductive rates, immature development | Ahmad et al., 2015 |
| Humans | *Aedes albopictus* (Asian tiger mosquito), filariasis | Neem seed oil | Larvicidal toxicity and field oviposition deterrence | Benelli et al., 2015 |
| Cotton | Cotton pest | *Beauveria bassiana* (entomopathogenic fungi) + neem oil | Damaged reproductive organs of pests | Togbé et al., 2015 |
| Kale plant | *Brevicoryne brassicae* (cabbage aphid) | Neem oil (1%) | Less cabbage aphid population | Pissinati and Ventura, 2015 |
| Okra (*Abelmoschus esculentus* L. Moench) | *Bemisia tabaci* (whitefly) | Mineral oil + neem oil (2%) | 95% mortality (after 48 h) | Sridharan et al., 2015 |

*(Continued)*

**TABLE 20.4 (CONTINUED)**

Recent Research on Neem Essential Oil for Agricultural and Public Health Pest Control

| Effected Organism | Pathogen/Disease | Treatment | Effects | Reference |
|---|---|---|---|---|
| Western white pine | *Zootermopsis augusticollis* (dampwood termite) | Neem oil | Rapid mortality | Fatima and Morrell, 2015 |
| Cashew trees | *Toxoptera odinae* (cashew aphid) | Neem oil | Killed 72.7%–78.9% of aphid population | Ambethgar, 2015b |
| Stone fruits | *Monilinia fructicola* | Neem oil (3.53 g/L) | 50% inhibition of mycelial growth | Lalancette and McFarland, 2015 |
| Watermelon | *Aphis gossypii* (watermelon aphid) | Neem oil | Dose-dependent decrease in population growth rate | Souza et al., 2015 |
| Humans | *Cimex lectularius* L. (bed bug) | Commercial neem oil | Killed 100% bed bugs | Feldlaufer and Ulrich, 2015 |
| Humans | *Aedes aegypti* (dengue mosquito) | Leaf essential oil | Effective against first-instar larvae and pupal stage (48 h) | Nasir et al., 2015a |
| Humans (contaminated food) | *Penicillium verrucosum* and *Penicillium nordicum* (produce ochratocin) | Neem essential oil (15 µl/ml) | 100% and 77.52%–92.49% inhibition, respectively | Koteswara Rao et al., 2015 |
| Coconut | *Aceria guerreronis* (coconut mite) | Neem oil (3%) | 31.31% reduction in mite population | Balaji and Hariprasad, 2015 |
| Cultivated crops | *Helicoverpa armigera* (moth/cotton bollworm) | PONNEEM (neem + pongam oils, 1:1 ratio) (20 ppm) | Feeding deterrence and genotoxicity | Packiam et al., 2015 |
| Humans | *Aedes aegypti* and *Aedes albopictus* | Neem essential oil (10% solution with canola oil) | Insect repellency of 246 ± 15.78 and 256 ± 14.87 min | Nasir et al., 2015b |
| *Jasminum auriculatum* | *Aceria jasmini* (eriophyid mite) | Neem oil (30 mL/L) | Reduced mite population | Devi et al., 2015 |
| Tomato | Whitefly and leaf miner | Neem oil (2.5 L/ha) | Effective after 20 days of transplanting | Chavan et al., 2015 |
| Cashew | *Ferrisia virgata* (Cockerell) | Neem oil | Effective to a limited extent after two rounds of spraying at 7-day intervals | Ambethgar, 2015a |
| Okra | *Bemisia tabaci* Genn. (whitefly), yellow vein mosaic | Neem oil | 8.89% of disease intensity | Kumar et al., 2015 |
| Tomato | *Tuta absoluta* Meyrick (tomato moth) | Neem seed oil | Significant effect recorded | Salem and Abdel-Moniem, 2015 |
| *Phaseolus vulgaris* L. (dry bean plants) | *Bemisia tabaci* Genn. biotype B | Neem oil | High nymphal mortality (>81%) was achieved | de Almeida Marques et al., 2015 |

prospect for producers to exploit the same for commercial gain, and plentiful researches have been directed toward the safety and efficacy evaluation of neem pesticides (Boeke et al., 2004; Anis Joseph et al., 2010; Vethanayagam and Rajendran, 2010). NSO pesticide has exhibited very low toxicity toward most vertebrates, and no significant adverse effect on the ecosystem has been observed (Sinha et al., 1999; Ismadji et al., 2012). A clinical study conducted on adults (156) and children (110) did not show any side effects on them after 1 year of exposure to neem oil (1%) (Brahmachari, 2004). Among all the neem-based products, NSO is considered an extremely safe insecticide to protect stored seeds for human consumption (Boeke et al., 2004). Thus, NSO is generally recognized as safe (GRAS) by the U.S. Food and Drug Administration (FDA) and Environmental Protection Agency (EPA) for use in food products. Also, they exempted NSO from the requirement of a maximum allowable pesticide limit on agricultural products. Neem pesticides are eco-friendly and environmentally friendly because their components rapidly biodegrade in the presence of sunlight and do not leave any residue on terrestrial and aquatic environment (Radwan and El-Shiekh, 2012). The half-life of azadirachtin (the main active component of NSO) on plants, soil, and water has been reported as 1–2.5 days, 3–44 days, and 48 min to 4 days, respectively. The remaining components of NSO are disintegrated by microbes present in the soil and water bodies. Additionally, the treated pests do not develop any resistance for NSO after prolonged use, as it modifies their life cycle instead of killing them. A remarkable feature of neem oil pesticides is their selective approach toward pests, as they do not harm beneficial insects, predators, parasites, and species that assist in pollination. Instead, they target only the chewing and sucking-type insects feeding on plants and animals, respectively (Radwan and El-Shiekh, 2012). Moreover, NSO can nourish and condition the soil and should be used, along with other pesticides and oils, for more efficacy (Lokanadhan et al., 2012).

## 20.9 Limitations as a Pesticide

Although NSO is a GRAS biopesticide, side effects in humans, animals, fish, and some nontargeted insect species have been reported in a number of isolated cases (Table 20.5) (El-Badawi et al., 2015). NSO is slightly irritating to the eyes, skin, and stomach due to the presence of azadirachtin (Shetty et al., 2016). In some instances, NSO in humans showed acute toxicity after oral administration (Boeke et al., 2004). Lai et al. (1990) reported that an even small amount of NSO can cause toxic encephalopathy, where vomiting, drowsiness, tachypnea, recurrent generalized seizures, leukocytosis, and metabolic acidosis are the main symptoms. In one extreme report, a child expired after oral administration of NSO for cough treatment. Autopsy findings revealed alterations in the liver and kidney of the child, consistent with Reye's syndrome (Sinniah et al., 1982). These incidences imply that the dosage taken as a drug was very high for human consumption, and the safe dose should be below ±0.20 ml/kg of body weight (bw), as calculated from animal studies. Recently, Shetty et al. (2016) reported a case of an adult female who committed suicide by consuming NSO. In rats and rabbits, NSO causes damage to the central nervous system (CNS) and lungs at a lethal dose ($LD_{50}$) of 14.1 and 24.0 ml/kg of bw, respectively (Gangopadhyay, 1994). Many studies indicated an antifertility effect of NSO in humans, attributed to the salanin compound (Brahmachari, 2004). *In vitro* studies on neem essential oil exhibited spermicidal activity in rats (0.25 mg/ml), rhesus monkeys, and humans (25 mg/ml) (Riar et al., 1990). In females, NSO administration in rats

**TABLE 20.5**

Adverse Effect of NSO on Humans and Animals

| Effected Organism | Exposure | Dose | Observed Effects | Reference |
|---|---|---|---|---|
| Humans (old female) | Ingestion | | Multiple organ failure, toxic encephalopathy | Shetty et al., 2016 |
| Ceraeochrysa claveri (predator insect) | Intake of neem oil-contaminated prey, Diatraea saccharalis eggs | 0.5%, 1%, and 2% neem oil | Sublethal effect: Cytotoxic effects in the adult midgut | Scudeler et al., 2016 |
| Oreochromis niloticus (Nile tilapia fish) | Treated with NSO | 112.5 ppm | Interfered with the antioxidant defense system; decrease in GST, CAT, and SOD | El-Badawi et al., 2015 |
| Rats | Ingestion | 2.0, 3.3, 4.6 ml/ kg of bw | Subacute toxicity: Antifertility in females | Dhaliwal et al., 1998 |
| Mice | Ingestion | 1.0–28.2 g/kg of bw | Acute toxicity | Tandan et al., 1995 |
| Humans (child) | Oral droplets | 5 ml | Toxic encephalopathy | Lai et al., 1990 |
| Rabbits, rats | Ingestion | 10–80 ml/kg of bw | Acute toxicity | Gandhi et al., 1988 |
| Humans (child) | Ingestion | 12 ml | Changes in the liver and kidneys | Sinniah et al., 1982 |

*Note:* CAT, catalase; GST, glutathione-S-transferase; SOD, superoxide dismutase.

suggested high abortive effects during early pregnancy (Lal et al., 1987). Since NSO is used to treat stored seeds against insects, much controversy exists on the claims of the negative effect on treated seeds. According to Naik and Dumbre (1985), NSO, being bitter in nature, affects the taste and influences the germination of treated seeds. On the contrary, no such adverse change in the taste of treated seeds was reported elsewhere (Boeke et al., 2004). Also, NSO can easily turn rancid and be contaminated by aflatoxins, which might pose additional health risks to consumers (Sinniah et al., 1982). Although neem oil poisoning is rare, it may provoke deleterious changes in vital organs of the organism, if exposed directly or indirectly. Hence, precautions must be taken while it is administered as a pest management tool.

## 20.10 Essential Oil–Based Insecticides

Neem has undergone extensive research as a bioinsecticide from the past three decades, specifically in the United States and other European countries. Ample scientific evidence has encouraged the formulation of several commercial products based on either NSO or its most active component, azadirachtin. The first marketable neem insecticide, Margosan O, was produced in the United States (W.R. Grace and Co., Columbia, Maryland). It is composed of 0.25% azadiractin and 3%–5% neem oil and has received exemption from the U.S. EPA (Radwan and El-Shiekh, 2012). Neemguard (W.R. Grace and Co.) is another commercial insecticide, consisting of formulated neem oil from neem seed kernel (Stark and Walter, 1995). Later, several neem formulations were prepared from neem oil and have found wide usage as a bioinsecticide for organic cultivation. Neem oil pesticide resists a wide variety of pests including mealy bugs, beet armyworms, aphids, cabbage worms, thrips, white-flies, mites, fungus gnats, beetles, moth larvae, mushroom flies, leaf miners, caterpillars, locusts, nematodes and the Japanese beetles (Table 20.4) (Ahmad et al., 2015; Jhalegar et al., 2015; Rodrigues et al., 2015; Sridharan et al., 2015). In India, neem oil–based insecticide is commercially manufactured, and applied in cotton, vegetables, fruit trees, coffee, tea, rice farming, and so forth (Ambrosino et al., 1999). The other well-known commercial neem-based formulations with azadirachtin as an active ingredient are Neemix (W.R. Grace and Co.), Nimbecidine, Neemgold, Econeem Plus, Econeem, Soluneem, Limonool, FortuneAza, and NeemAzal-F.

## 20.11 Conclusions

Neem essential oil–based pesticide is environmentally benign. It is selectively toxic, does not bioaccumulate, and has short persistence in the ecosystem, hence making it an ideal candidate for an integrated pest management program. Various scientific studies have shown that severe side effects were only encountered when NSO was consumed directly in large amounts. However, direct and indirect contact with children and lactating and pregnant women should be avoided. Although there is a meager possibility of neem essential oil, applied as pesticide, entering the food chain of humans, most NSO degrades rapidly—before it reaches the consumer. Since a few studies have indicated the possibility of adverse effects, care should be taken in the administration of NSO as a pest-controlling agent. Furthermore, NSO is easily contaminated with aflatoxin; thus, the aflatoxin concentration in NSO-based pesticides should be properly controlled to avoid an unnecessary risk factor. As far as neem essential oil extraction is concerned, the solvent extraction method is the preferred choice due to its low operational and economic cost. Among modern methods of extraction, SFE is of great interest for being a green, efficient, and effective method for the recovery of essential oils and active metabolites.

## References

Abiy, E., Gebre-Michael, T., Balkew, M., Medhin, G. (2015). Repellent efficacy of DEET, MyggA, neem (*Azedirachta indica*) oil and chinaberry (*Melia azedarach*) oil against *Anopheles arabiensis*, the principal malaria vector in Ethiopia. *Malaria Journal*, 14(187), 1–6.

Adeeko, K. A., Ajibola, O. O. (1990). Processing factors affecting yield and quality of mechanically expressed groundnut oil. *Journal of Agricultural Engineering*, 45, 31–43.

Adewoye, T. L., Ogunleye, O. O. (2012). Optimization of neem seed oil extraction process using response surface methodology. *Journal of Natural Sciences Research*, 2(6), 66–76.

Ahmad, S., Ansari, M. S., Muslim, M. (2015). Toxic effects of neem based insecticides on the fitness of *Helicoverpa armigera* (Hübner). *Crop Protection*, 68, 72–78.

Al-Quraishy, S., Abdel-Ghaffar, F., Mehlhorn, H. (2015). Head louse control by suffocation due to blocking their oxygen uptake. *Parasitology Research*, 114(8), 3105–3110.

Ambethgar, V. (2015a). Field evaluation of some insecticides against white-tailed mealy bug, *Ferrisia virgata* (Cockerell) infesting cashew. *Acta Horticulturae*, 1080, 469–472.

Ambethgar, V. (2015b). Management of cashew aphid, *Toxoptera odinae* van der Goot (Homoptera: Aphididae) with some insecticides and neem products. *Acta Horticulturae*, 1080, 473–476.

Ambrosino, P., Fresa, R., Fogliano, V., Monti, S. M., Ritieni, A. (1999). Extraction of azadirachtin A from neem seed kernels by supercritical fluid and its evaluation by HPLC and LC/MS. *Journal of Agricultural and Food Chemistry*, 47(12), 5252–5256.

Anis Joseph, R., Premila, K. S., Nisha, V. G., Rajendran, S., Sarika Mohan, S. (2010). Safety of neem products to tetragnathid spiders in rice ecosystem. *Journal of Biopesticides*, 3(1), 88–89.

Anon. (2011). Meliaceae of North America Update. Database, version 2011. Updated for ITIS by the Flora of North America Expertise Network, in connection with an update for USDA PLANTS (2007–2010).

Anya, U. A., Chioma, N. N., Obinna, O. (2012). Optimized reduction of free fatty acid content on neem seed oil, for biodiesel production. *Journal of Basic and Applied Chemistry*, 2(4), 21–28.

Ayoola, A. A., Efeovbokhan, V. C., Bafuwa, O. T., David, O. T. (2014). A search for alternative solvent to hexane during neem oil extraction. *International Journal of Science*, 4(4), 66–70.

Balaji, K., Hariprasad, Y. (2015). Efficacy of botanicals on the management of coconut mite *Aceria guerreronis* (Keifer) (Acaridae: Eriophyidae). *Journal of Biopesticides*, 8(1), 13–18.

Barrek, S., Paisse, O., Grenier-Loustalot, M.-F. (2004). Analysis of neem oils by LC-MS and degradation kinetics of azadirachtin-A in a controlled environment: Characterization of degradation products by HPLC-MS-MS. *Analytical and Bioanalytical Chemistry*, 378(3), 753–763.

Benelli, G., Bedini, S., Cosci, F., Toniolo, C., Conti, B., Nicoletti, M. (2015). Larvicidal and ovideterrent properties of neem oil and fractions against the filariasis vector *Aedes albopictus* (Diptera: Culicidae): A bioactivity survey across production sites. *Parasitology Research*, 114(1), 227–236.

Boeke, S. J., Boersma, M. G., Alink, G. M., van Loon, J. J. A., van Huis, A., Dicke, M., Rietjens, I. M. C. M. (2004). Safety evaluation of neem (*Azadirachta indica*) derived pesticides. *Journal of Ethnopharmacology*, 94, 25–41.

Brahmachari, G. (2004). Neem—An omnipotent plant: A retrospection. *ChemBioChem*, 5, 408–421.

Chavan, R. D., Yeotikar, S. G., Gaikwad, B. B., Dongarjal, R. P. (2015). Management of major pests of tomato with biopesticides. *Journal of Entomological Research*, 39(3), 213–217.

Choe, E., Min, D. B. (2006). *Mechanisms and Factors for Edible Oil Oxidation: Comprehensive Reviews in Food Science and Food Safety*. Chicago: Institute of Food Technologists, Vol 5.

Csurhes, S. (2008). Pest plant risk assessment: Neem tree *Azadirachta indica*. Queensland: Department of Primary Industries and Fisheries.

Dasa Rao, C. J., and Seshadri, T. R. (1941). Fatty acids of neem oil. Andhra Pradesh: Department of Chemistry and Chemical Technology, pp. 161–167.

de Almeida Marques, M., Quintela, E. D., Mascarin, G. M., Fernandes, P. M., Arthurs, S. P. (2015). Management of *Bemisia tabaci* biotype B with botanical and mineral oils. *Crop Protection*, 66, 127–132.

Devi, M., Umapathy, G., Asokan, G. (2015). Efficacy of newer insecticides against *Aceria jasmini* in *Jasminum auriculatum*. *Journal of Entomological Research*, 39(3), 237–241.

Dhaliwal, P. K., Roop, J. K., Guraya, S. S., Dhawan, A. K. (1998). Antifertility activity of neem seed oil in cyclic female rats [abstract]. In *Proceedings of the International Conference on Ecological Agriculture: Towards Sustainable Development*, Chandigarh, India, pp. 340–346.

Djenontin, T. S., Wotto, V. D., Avlessi, F., Lozano, P., Sohounhloué, D. K. C., Pioch, D. (2012). Composition of *Azadirachta indica* and *Carapa procera* (Meliaceae) seed oils and cakes obtained after oil extraction. *Industrial Crops and Products*, 38(1), 39–45.

Djibril, D., Mamadou, F., Gérard, V., Geuye, M. D. C., Oumar, S., Luc, R. (2015). Physical characteristics, chemical composition and distribution of constituents of the neem seeds (*Azadirachta indica* A. Juss) collected in Senegal. *Research Journal of Chemical Sciences*, 5(7), 52–58.

Edres, A. E. E. A. B. E. (2014). Extraction of neem oil from neem seeds (2013– 2014). Thesis, Department of Chemistry, College of Science, Sudan University of Science and Technology, Sudan.

El-Badawi, A. A., Gaafar, A. Y., Abbas, H. H., Authman, M. M. N. (2015). Toxic effects of neem seeds oil on Nile Tilapia (*Oreochromis niloticus*) and application of different trials of control. *Research Journal of Pharmaceutical, Biological and Chemical Sciences*, 6(1), 645–658.

El-Hawary, S. S., El-Tantawy, M. E., Rabeh, M. A., Badr, W. K. (2013). Chemical composition and biological activities of essential oils of *Azadirachta indica* A. Juss. *International Journal of Applied Research in Natural Products*, 6(4), 33–42.

Erakhrumen, A. A. (2011). Selected physical and chemical properties of mechanically extracted neem seed oil sourced as a preservative for ligno-cellulose in south western Nigeria. *Foreign Studies in China*, 13(4), 263–269.

Esparza-Díaz, G., Villanueva-Jiménez, J. A., López-Collado, J., Osorio-Acosta, F. (2011). Multi-insecticide extractive technology of neem seeds for small growers. *Tropical and Subtropical Agroecosystems*, 13, 409–415.

Fatima, R., Morrell, J. J. (2015). Ability of plant-derived oils to inhibit dampwood termite (*Zootermopsis augusticollis*) activity. *Maderas: Cienciay Tecnologia*, 17(3), 685–690.

Feldlaufer, M. F., Ulrich, K. R. (2015). Essential oils as fumigants for bed bugs (Hemiptera: Cimicidae). *Journal of Entomological Science*, 50(2), 129–137.

Forim, M. R., das Graças, M. F., da Silva, F., Fernandes, J. B. (2012). Secondary metabolism as a measurement of efficacy of botanical extracts: The use of *Azadirachta indica* (neem) as a model. In Perveen, F. (ed.), *Insecticides—Advances in Integrated Pest Management*. Rijeka, Croatia: InTech Publishing, pp. 367–390.

Forim, M. R., Da Silva, M. F. D. G. F., Cass, Q. B., Fernandes, J. B., Vieira, P. C. (2010). Simultaneous quantification of azadirachtin and 3-tigloylazadirachtol in Brazilian seeds and oil of *Azadirachta indica*: Application to quality control and marketing. *Analytical Methods*, 2(7), 860–869.

Gandhi, M., Lal, R., Sankaranarayanan, A., Banerjee, C. K., Sharma, P. L. (1988). Acute toxicity study of the oil from *Azadirachta indica* seed (neem oil). *Journal of Ethnopharmacology*, 23, 39–51.

Gangopadhyay, S. (1994). International Conference on Current Progress in Medicinal and Aromatic Plant Research, India, December 30, p. 99.

Girish, K., Shankara, B. S. (2008). Neem—A green treasure. *Electronic Journal of Biology*, 4(3), 102–111.

Gomes, S. A., Paula, A. R., Ribeiro, A., Moraes, C. O. P., Santos, J. W. A. B., Silva, C. P., Samuels, R. I. (2015). Neem oil increases the efficiency of the entomopathogenic fungus *Metarhizium anisopliae* for the control of *Aedes aegypti* (Diptera: Culicidae) larvae. *Parasites and Vectors*, 8(669), 1–8.

Gossé, B., Amissa, A. A., Adjé, F. A., Niamké, F. B., Ollivier, D., Ito, Y. (2005). Analysis of components of neem (*Azadirachta indica*) oil by diverse chromatographic techniques. *Journal of Liquid Chromatography and Related Technologies*, 28(14), 2225–2233.

Hallur, G., Sivramakrishnan, A., Bhat, S. V. (2002). Three new tetranortriterpenoids from neem seed oil. *Journal of Natural Products*, 65(8), 1177–1179.

Hashmat, I., Azad, H. Ahmed, A. (2012) Neem (*Azadirachta indica* A. Juss)—A nature's drugstore: An overview. *International Research Journal of Biological Sciences*, 1(6), 76–79.

Hearne, D. A. (1975). Trees for Darwin and northern Australia. Canberra: Department of Agriculture, Forestry and Timber Bureau.

Hossain, M. A. (2005). Examples of the development of pharmaceutical products from medicinal plants, In *Neem Seed Oil*. Bangladesh: Bangladesh Council of Scientific and Industrial Research (BCSIR), vol. 10, pp. 59–63.

Ismadji, S., Kurniawan, A., Ju, Y. H., Soetaredjo, F. E., Ayucitra, A., Ong, L. K. (2012). Solubility of azadirachtin and several triterpenoid compounds extracted from neem seed kernel in supercritical $CO_2$. *Fluid Phase Equilibria*, 336, 9–15.

Ismail, O. M. (2016). Effect of spraying 'Taimour' mango trees with neem and lemon grass oils on fruit set. *Research Journal of Pharmaceutical, Biological and Chemical Sciences*, 7(1), 259–264.

Jhalegar, M. D. J., Sharma, R. R., Singh, D. (2015). *In vitro* and *in vivo* activity of essential oils against major postharvest pathogens of Kinnow (*Citrus nobilis* × *C. deliciosa*) mandarin. *Journal of Food Science and Technology*, 52(4), 2229–2237.

Johnson, S., Morgan, E. D. (1997). Supercritical fluid extraction of oil and triterpenoids from Neem seeds. *Phytochemical Analysis*, 8(5), 228–232.

Kaushik, N. (2004). *Biopesticides for Sustainable Agriculture: Prospects and Constraints*. New Delhi: Energy and Resources Institute, p. 74.

Khraisha, Y. H. (2000). Retorting of oil shale followed by solvent extraction of spent shale: Experiment and kinetic analysis. *Journal of Energy Sources*, 22, 347–355.

Koteswara Rao, V., Girisham, S., Reddy, S. M. (2015). Inhibitory effect of essential oils on growth and ochratoxin A production by *Penicillium* species. *Research Journal of Microbiology*, 10(5), 222–229.

Kumar, H., Singh, R., Gupta, V., Zutshi, S. K. (2015). Performance of different germplasm, plant extracts and insecticides against yellow vein mosaic of okra (OYVMV) under field conditions. *Vegetos*, 28(1), 31–37.

Kurose, K., Yatagai, M. (2005). Components of the essential oils of *Azadirachta indica* A. Juss, *Azadirachta siamensis* Velton, and *Azadirachta excelsa* (Jack) Jacobs and their comparison. *Journal of Wood Science*, 51(2), 185–188.

Lai, S. M., Lim, K. W., Cheng, H. K. (1990). Margosa oil poisoning as a cause of toxic encephalopathy. *Singapore Medical Journal*, 31, 463–465.

Lal, R., Gandhi, M., Sankaranarayanan, A., Mathur, V. S., Sharma, P. L. (1987). Antifertility effect of *Azadirachta indica* oil administered per os to female albino rats on selected days of pregnancy. *Fitoterapia*, 58, 239–241.

Lalancette, N., McFarland, K. A. (2015). Effect of biorational fungicides on *in vitro* growth of *Monilinia fructicola*. *Acta Horticulturae*, 1084, 563–567.

Lalea, N. E. S, Abdulrahman, H. T. (1999). Evaluation of neem (*Azadirachta indica* A. Juss) seed oil obtained by different methods and neem powder for the management of *Callosobruchus maculatus* (f.) (Coleoptera: Bruchidae) in stored cowpea. *Journal of Stored Products Research*, 35(2), 135–143.

Lalla, J. K., Hamrapurkar, P. D., Patil, P. S. (2003). Azadirachtin as biomarker compound in HPTLC assay of seed and seed oil of *Azadirachta indica* A. Juss. *Journal of Planar Chromatography—Modern TLC*, 16(4), 311–314.

Liauw, M. Y., Natan, F. A., Widiyanti, P., Ikasari, D., Indraswati, N., Soetaredjo, F. E. (2008). Extraction of neem oil (*Azadirachta indica* A. Juss) using *n*-hexane and ethanol: Studies of oil quality, kinetic and thermodynamic. *Journal of Engineering and Applied Sciences*, 3(3), 49–54.

Lokanadhan, S., Muthukrishnan, P., Jeyaraman, S. (2012). Neem products and their agricultural applications. *Journal of Biopesticides*, 5, 72–76.

Martinelli, E., Schulz, K., Mansoori, G. A. (1991). Supercritical fluid extraction/retrograde condensation (SFE/RC) with applications in biotechnology. *Supercritical Fluid Technology*, 451–478.

Momchilova, S., Antonova, D., Marekov, I., Kuleva, L., Nikolova-Damyanova, B., Jham, G. (2007). Fatty acids, triacylglycerols, and sterols in neem oil (*Azadirachta Indica* A. Juss) as determined by a combination of chromatographic and spectral techniques. *Journal of Liquid Chromatography and Related Technologies*, 30(1), 11–25.

Mongkholkhajornsilp, D., Douglas, S., Douglas, P. L., Elkamel, A., Teppaitoon, W., Pongamphai, S. (2005). Supercritical CO2 extraction of nimbin from neem seeds—A modelling study. *Journal of Food Engineering*, 71, 331–340.

Morgan, E. D. (2009). Azadirachtin, a scientific gold mine. *Bioorganic & Medicinal Chemistry*, 17(12), 4096–4105.

Naik, R. L., Dumbre, R. B. (1985). Effect of some vegetable oils used as surface protectants against *Callosobruchus maculatus* on storability and qualities of cowpea. *Bulletin of Grain Technology*, 23, 33–39.

Narsing Rao, G., Prabhakara Rao, P. G., Satyanarayana, A. (2014). Chemical, fatty acid, volatile oil composition and antioxidant activity of shade dried neem (*Azadirachta indica* L.) flower powder. *International Food Research Journal*, 21(2), 807–813.

Nasir, S., Batool, M., Hussain, S. M., Nasir, I. B., Hafeez, F., Debboun, M. (2015a). Bioactivity of oils from medicinal plants against immature stages of dengue mosquito *Aedes aegypti* (Diptera: Culicidae). *International Journal of Agriculture and Biology*, 17(4), 843–847.

Nasir, S., Batool, M., Qureshi, N. A., Debboun, M., Qamer, S., Nasir, I., Bashir, R. (2015b). Repellency of medicinal plant extracts against dengue vector mosquitoes, *Aedes albopictus* and *A. aegypti* (Diptera: Culicidae). *Pakistan Journal of Zoology*, 47(6), 1649–1653.

Nde, D. B., Boldor, D., Astete, C. (2015). Optimization of microwave assisted extraction parameters of neem (*Azadirachta indica* A. Juss) oil using the Doehlert's experimental design. *Industrial Crops and Products*, 65, 233–240.

Nitièma-Yefanova, S., Son, G., Yé, S., Nébié, R. H. C., Bonzi-Coulibaly, Y. (2012). Optimization of the parameters of cold extraction of the oil of *Azadirachta indica* A. Juss and effects on some chemical characteristics of the extracted oil. *Biotechnology, Agronomy and Society and Environment*, 16(4), 423–428.

Okonkwo, P. C., Mukhtar, B., Usman, J. G. (2013). Approach to evaluation of solvent extraction of oil from neem seed. *International Journal of Applied Science and Technology*, 3(6), 76–85.

Packiam, S. M., Emmanuel, C., Baskar, K., Ignacimuthu, S. (2015). Feeding deterrent and genotoxicity analysis of a novel phytopesticide by using comet assay against *Helicoverpa armigera* (Hübner) (Lepidoptera: Noctuidae). *Brazilian Archives of Biology and Technology*, 58(4), 487–493.

Parotta, J. A. (2001). *Healing Plants of Peninsular India*. New York: CABI Publishing, pp. 495–496.

Pissinati, A., Ventura, M. U. (2015). Control of cabbage aphid, *Brevicoryne brassicae* (L.) using kaolin and neem oil. *Journal of Entomology*, 12(1), 48–54.

Radwan, O. A., El-Shiekh, Y. W. A. (2012). Degradation of neem oil 90% EC (azadirachtin) under storage conditions and its insecticidal activity against cotton leaf worm *S. littoralis*. *Researcher*, 4(3), 77–83.

Ramesh, A., Balasubramanian, M. (1999). Rapid preconcentration method for the determination of azadirachtin A and B, nimbin and salannin in neem oil samples by using graphitised carbon solid phase extraction. *Analyst*, 124(1), 19–21.

Rembold, H., Forster, H., Czoppelt, Ch. (1987). Structure and biological activity of azadirachtin A and B. In Schmutterer, H., Ascher, K. R. S. (eds.), *Natural Pesticides from Neem Tree (Azadirachta indica* A. Juss) *and Other Tropical Plants*. GTZ: Eschborn, Germany, pp. 149–160.

Rembold, H., Forster, H., Czoppelt, Ch., Rao, J. P., Sieber, K. P. (1984). The azadirachtins, a group of insect growth regulators from the neem tree. In Schmutterer, H., Ascher, K. R. S. (eds.), *Natural Pesticides from Neem Tree (Azadirachta indica* A. Juss) *and Other Tropical Plants*. GTZ: Eschborn, Germany, pp. 153–162.

Riar, S., Devakumar, C., Ilavazhagan, G., Bardhan, J., Kain, A. K., Thomas, P., Singh, R., Singh, B. (1990). Volatile fraction of neem oil as a spermicide. *Contraception*, 42(4), 479–487.

Ricci, F., Berardi, V., Risuleo, G. (2009). Differential cytotoxicity of MEX: A component of neem oil whose action is exerted at the cell membrane level. *Molecules*, 14(1), 122–132.

Rodrigues, N. E. L., da Silva, A. G., de Souza, B. H. S., Costa, E. N., Ribeiro, Z. A., Boiça Júnior, A. L. (2015). Effects of cowpea cultivars and neem oil on attractiveness, feeding, and development of *Spodoptera eridania* (Cramer) (Lepidoptera: Noctuidae). *Idesia*, 33(4), 65–74.

Salem, S. A., Abdel-Moniem, A. S. H. (2015). Evaluation of non-traditional approaches for controlling tomato moth, *Tuta absoluta* Meyrick (Lepidoptera, Gelechiidae), a new invasive pest in Egypt. *Archives of Phytopathology and Plant Protection*, 48(4), 319–326.

Sapkale, G. N., Patil, S. M., Surwase, U. S., Bhatbhage, P. K. (2010). Supercritical fluid extraction. *International Journal of Chemical Sciences*, 8(2), 729–743.

Satyanandam, T., Rosaiah, G., Babu, K., Rao, Y. S., Rao, K. R. S. S. (2011). Variations in quantity and quality of neem oil (*Azadirachta indica* A. Juss) extracted by petroleum ether and ethanol in select areas of Guntur, Andhra Pradesh, India. *Current Trends in Biotechnology and Pharmacy*, 5(1), 1029–1037.

Scudeler, E. L., Garcia, A. S. G., Padovani, C. R., Pinheiro, P. F. F., Santos, D. C. D. (2016). Cytotoxic effects of neem oil in the midgut of the predator *Ceraeochrysa claveri*. *Micron*, 80, 96–111.

Shetty, P., Kumar, A., Nayak, V. C., Patil, N. B., Avinash, A., Shashidhara, S., Karthik Rao, N., Rao, R. (2016). A rare case of neem oil ingestion as a suicidal modality. *Research Journal of Pharmaceutical, Biological and Chemical Sciences*, 7(1), 1253–1255.

Siddiqui, S. (1942). A note on the isolation of three new bitter principles from the nim oil. *Current Science*, 11, 278–279.

Sidhu, O. P., Kumar, V., Behl, H. M. (2003). Variability in neem (*Azadirachta indica*) with respect to azadirachtin content. *Journal of Agricultural and Food Chemistry*, 51(4), 910–915.

Singh, M., Sachan, S. K. (2015). Comparative efficacy of some biopesticides against shoot and fruit borer, *Leucinodes orbonalis* Guenee in Brinjal. *Plant Archives*, 15(2), 805–808.

Sinha, S., Murthy, P. N. S., Rao, C. V. N., Ramaprasad, G., Sitaramaiah, S., Kumar, D. G., Savant, S. K. (1999). Simple method for enrichment of azadirachtin from neem seeds. *Journal of Scientific and Industrial Research*, 58, 990–994.

Sinniah, D., Baskaran, G., Looi, L. M., Leong, K. L. (1982). Reye-like syndrome due to margosa oil poisoning: Report of a case with postmortem findings [abstract]. *American Journal of Gastroenterology*, 77, 158–161.

Sodeinde, O. A., Samuel, O. (2014). Optimization of neem oil extraction from its seed using response surface methodology. In *Separations Division 2014—Core Programming Area at the 2014 AIChE Annual Meeting*, Atlanta, United States, vol. 1, pp. 417–424.

Sokame, B. M., Tounou, A. K., Datinon, B., Dannon, E. A., Agboton, C., Srinivasan, R., Pittendrigh, B. R., Tamò, M. (2015). Combined activity of *Maruca vitrata* multi-nucleopolyhedrovirus, MaviMNPV, and oil from neem, *Azadirachta indica* A. Juss and *Jatropha curcas* L., for the control of cowpea pests. *Crop Protection*, 72, 150–157.

Souza, C. R., Sarmento, R. A., Venzon, M., Dos Santos, G. R., De Silveira, M. C. A. C., Tschoeke, P. H. (2015). Lethal and sublethal effects of neem on *Aphis gossypii* and *Cycloneda sanguinea* in watermelon. *Acta Scientiarum-Agronomy*, 37(2), 233–239.

Sridharan, S., Shekhar, K. C., Ramakrishnan, N. (2015). Bioefficacy, phytotoxicity, and biosafety of mineral oil on management of whitefly in okra. *International Journal of Vegetable Science*, 21(1), 28–35.

Stark, J. D., Walter, J. F. (1995). Neem oil and neem oil components affect the efficacy of commercial neem insecticides. *Journal of Agriculture and Food Chemistry*, 43(2), 507–512.

Sundaram, K. M. S., Curry, J. (1993). High performance liquid chromatographic method for the analysis of azadirachtin in two commercial formulations and neem oil. *Journal of Environmental Science and Health, Part B*, 28(2), 221–241.

Tandan, S. K., Gupta, S., Chandra, S., Lal, J., Singh, R. (1995). Safety evaluation of *Azadirachta indica* seed oil, a herbal wound dressing agent. *Fitoterapia*, 66, 69–72.

Tinghui, X., Wegener, M., O'Shea, M., Deling, M. (2001). World distribution and trade in neem products with reference to their potential in China. Presented at AARES 2001 Conference of Australian Agricultural and Resource Economics Society, Adelaide, January 22–25, 2001.

Togbé, C. E., Haagsma, R., Zannou, E., Gbèhounou, G., Déguénon, J. M., Vodouhê, S., Kossou, D., van Huis, A. (2015). Field evaluation of the efficacy of neem oil (*Azadirachta indica* A. Juss) and *Beauveria bassiana* (Bals.) Vuill. in cotton production. *Journal of Applied Entomology*, 139(3), 217–228.

Usman, J. G., Okonkwo, P. C., Mukhtar, B. (2013). Design and construction of pilot scale process solvent extraction plant for neem seed oil. *Nigerian Journal of Technology (NIJOTECH)*, 32(3), 528–537.

Veitch, G. E., Beckmann, E., Burke, B., Boyer, A., Masten, S. L., Ley, S. V. (2007). Synthesis of azadirachtin: A long but successful journey. *Angewandte Chemie (International Edition in England)*, 46(40),7629–7632.

Vethanayagam, S. M., Rajendran, S. M. (2010). Bioefficacy of neem insecticidal soap (NIS) on the disease incidence of bhendi, *Abelmoschus esculentus* (L.) Moench under field conditions. *Journal of Biopesticides*, 3(1), 246–249.

Wiart, C., ed. (2006). Medicinal plants of Asia and the pacific. In *Medicinal Plants Classified in the Family Meliaceae*. Boca Raton, FL: Taylor & Francis Group, CRC Press, chap. 30

Wicks, Z. W., Jr., Jones, F. N., Pappas, S. P. (1992). *Organic Coatings: Science and Technology*. New York: John Wiley & Sons, pp. 133–143.

Zahedi, G., Elkamel, A., Lohi, A. J. (2010). Genetic algorithm optimization of supercritical fluid extraction of nimbin from neem seeds. *Food Engineering*, 97(2), 127–134.

Zaku, S. G., Emmanual, S. A., Isa, A. H., Kabir, A. (2012). Comparative studies on the functional properties of neem, jatropha, castor, and moringa seeds oil as potential feed stocks for biodiesel production in Nigeria. *Global Journal of Science Frontier Research Chemistry*, 12(7), 23–26.

# 21

## Pyrethrum Oils

**Basil K. Munjanja**

### CONTENTS

### 21.1 Introduction

Synthetic pesticides are an invaluable component of agriculture, which enhance agricultural productivity because in their absence, heavy losses caused by pests would be incurred. For instance, in the United Kingdom wheat yields rose from 2.5 t/ha in 1948 to 7.5 t/ha in 1997 (Cooper and Dobson 2007). As a result, an astronomic increase has been recorded in their use, and to date, more than 1000 substances have been registered as pesticides (Tomlin 2003). However, not all the pesticide reach the target (Pimentel 1992), because they find their way into environmental compartments where their residues can be detected at parts per million, parts per billion, or parts per trillion levels, depending on the persistence of the pesticides (Ccanccapa et al. 2016). Because of that, many regulations, such as the Food Quality Protection Act, have been put in place to control their registration (Dayan et al. 2009). In addition, there has been a paradigm shift toward exploring the use of biopesticides as a viable option to alleviating pesticide pollution. Biopesticides are mainly classified as plant extracts, microorganisms, pheromones, and genes.

Plant-based pesticides, which are also known as botanical pesticides, have gained increasing popularity because they are "green pesticides," which reduce the pest population, while at the same time being environmentally compatible (Koul et al. 2008). They were developed into pesticides by noticing their traditional use in crop protection, checking their efficacy, and consequently identifying the active ingredients (Ntalli and Menkissoglu-Spiroudi 2011). However, their major problem is the great difference in the composition and quality of the plant extracts, which can be natural or due to the extraction technique employed (Miresmailli and Isman 2014). For this reason, they have failed to outcompete synthetic pesticides in the field of plant protection. Moreover, it has been suggested that their greatest benefit can be realized in developing countries, where the plants are locally available at a low cost, compared with in industrialized nations, where legislation is very stringent and they cannot be registered easily (Isman 2008).

Examples of these include neem (*Azadirachta indica* A. Juss), pyrethrum (*Chrysanthemum cinerariaefolium*), and nicotine (*Nicotiana tabacum*) (Singh et al. 2010). They can be classified according to their modes of action as antifeedants, attractants, antimicrobials, fumigants, contact toxicants, or repellents (Akhtar and Isman 2012; Miresmailli and Isman 2014). Pyrethrum is an example of an antifeedant that disturbs the feeding process of insects (Akhtar and Isman 2012).

The insecticidal value of pyrethrum flower was discovered by the Chinese more than 2000 years ago (Singh et al. 2010). However, its use was only fully realized after an American trader, Jumticoff, discovered its use in the control of lice by the Caucuses tribes (Schleier and Peterson 2011). Since then, production increased in the Dalmatia Coast, followed by Japan (Anon 2010). However, after World War II, increased production was observed in Kenya. By the 1960s, Kenya supplied more than 90% of the world's pyrethrum. However, the production dropped around 2004, and since that time, the island state of Tasmania in Australia has dominated with 65% of the world production (Monda 2014). In addition, production has steadily increased in countries like Rwanda, China, and Tanzania, as shown by Figure 21.1.

Currently, pyrethrum plants are found in countries such as Kenya (Wandahwa et al. 1996), India (Bhakuni et al. 2007), Tanzania, Ecuador, Brazil, Russia, Japan (Srivastava et al. 2010), and Australia (Morris et al. 2006). The plantations in Tasmania, Australia, have become the second largest producers after Kenya, producing plants almost similar to those found in East Africa (Isman 2006). However, the plants obtained in India produce

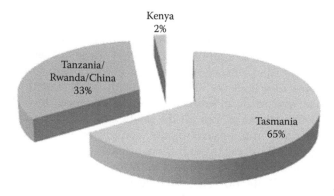

**FIGURE 21.1**
World production of pyrethrum flowers 2011–2012 (pale refined extract 50%). (From BRA, MGK, Ulverstone, Tasmania.)

an oil that has a different composition from those found in Kenya (Bhakuni et al. 2007). Nevertheless, it is imperative to note that the Tasmanian plantations have dominated the world market because of a thriving breeding program, compared with the East African ones (Li et al. 2011). However, despite the production in different regions of the world, the yield obtained after processing is very small compared with the amounts of dried flowers produced. For instance, 20,000 metric tonnes of dried flowers was observed to give a potential yield of 500 tonnes of 50% pale extract (Glynne-Jones 2001). For this reason, there have also been recent efforts to increase the production by *in vitro* production, although some of these techniques have proved not to be viable because they cannot be used on a large scale (Hitmi et al. 2000).

Owing to the intensive production of this plant in different parts of the world, it is imperative that we critically discuss the pyrethrum plant and the oils extracted from it. Special attention is first given to the botany of the plant. After that, the composition of the oil and its physicochemical properties are evaluated. The general and pesticidal uses are discussed as well, with their merits and demerits highlighted. Finally, an evaluation of the strengths and weaknesses of the analytical and extraction techniques is made.

## 21.2 Botany of the Plant

Pyrethrum oil is extracted from the dried flowers of the plant *C. cinerariaefolium* (*Tanacetum cinerariaefolium*) (Majoni and Munjanja 2015). The plant, which belongs to the Asteraceae family (Gallo et al. 2017), is small and perennial, with a daisy-like appearance and white petals, and is reported to have originated from Yugoslavia (Wandahwa et al. 1996). The flowering process in the plant is slowed down by low-photon flux density, regardless of temperature. Moreover, high day temperatures of up to 25°C, combined with low-photon flux intensity, have been observed to prevent pyrethrum from flowering in otherwise inductive conditions (Brown and Menary 1994). For this reason, it is grown in highlands of both tropical and subtropical regions of the world, or in lowland regions with temperate climates that induce blooming (Li et al. 2011).

The active ingredients from the pyrethrum flower, which are known as pyrethrins, do not usually exceed 2% of the dry mass (Kiriamiti et al. 2003b). Of this amount, more of it is found in the seed part of the flower than in the flower part. Moreover, when dried, the flower extracts are observed to have a high pyrethrin content (Roncevic et al. 2014). Thus, the drying process carried out before extraction of pyrethrins does not change the pyrethrin content (Morris et al. 2006). However, when a pyrethrum flower is stored for prolonged periods at high temperatures, after harvesting, the pyrethrins may degrade (Atkinson et al. 2004).

## 21.3 Composition of Oil

The oil, which has insecticidal action, is made up of pyrethrum esters of chrysanthemic acid, and pyrethric acid to three alcohols, namely, pyrethrolone, cinerolone, and jasmololone. This leads to the formation of two fractions commonly known as pyrethrin I (P I), which comprises

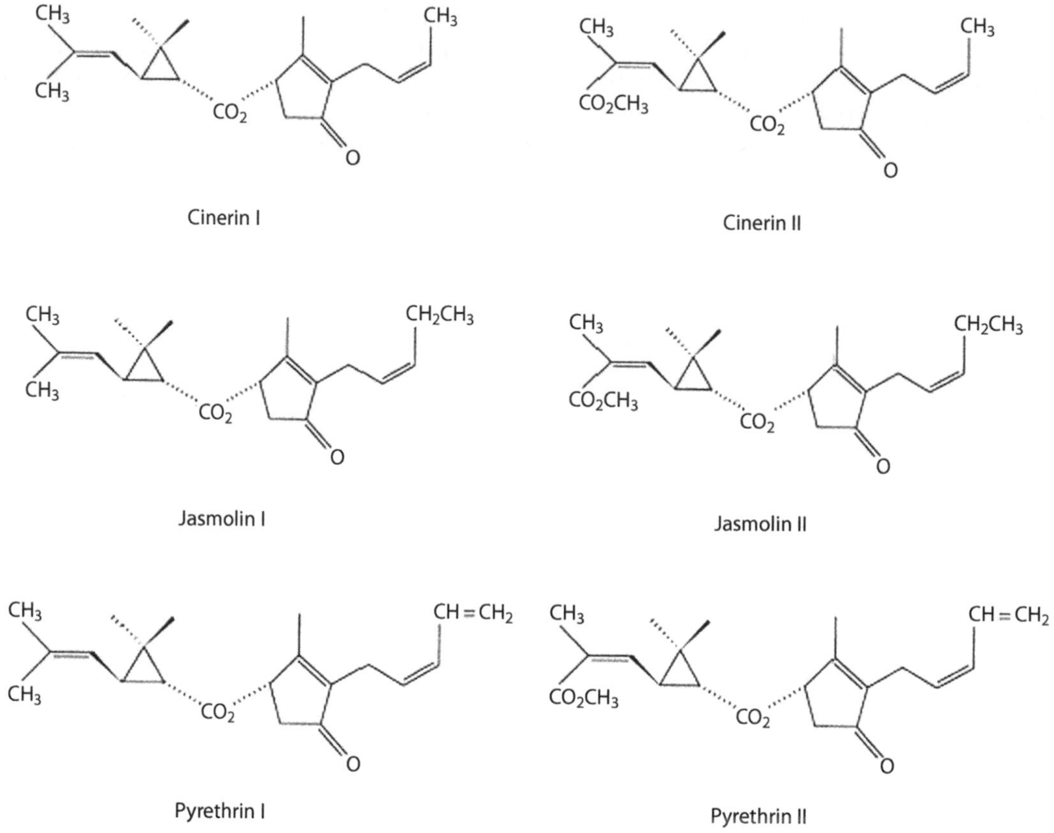

**FIGURE 21.2**
Structure of pyrethrins. (Reproduced from Henry, C.W. et al., *J. Chromatogr. A*, 863, 89–103, 1999. With permission.)

three esters of chrysanthemic acid: pyrethrin I, cinerin I, and jasmolin I. Pyrethrin II (P II) comprises three esters of pyrethric acid: pyrethrin II, cinerin II, and jasmoline II (Wang et al. 1997). P I compounds have a single ester group, linked to both the tricyclic and pentyl ring systems, differing from each other in the alkene group attached to the five-member ring (Henry et al. 1999). On the other hand, P II compounds have an additional ester moiety attached to the three-member ring, making them more polar than the former. The acid moiety in P I is thought to be formed from D-glucose via 2-C-methyl-D-erytritol 4-phosphate, whereas the alcohol moiety is possibly synthesized from linolenic acid (Matsuda et al. 2005). The relative amounts of the pyrethrins, whose structures are shown in Figure 21.2, depend on factors such as plant genotype, geographical origin and time of harvest, soil conditions, and climate (Li et al. 2014).

## 21.4 Physicochemical Properties of Oil

The physicochemical properties of pyrethrin oils shown in Table 21.1 may help to explain their behavior in the environment. Generally, pyrethrins are nonpolar and have low

**TABLE 21.1**

Physicochemical Properties of Pyrethrins

| Chemical Compound | Pyrethrin I | Cinerin I | Jasmolin I | Pyrethrin II | Cinerin II | Jasmolin II | Reference |
|---|---|---|---|---|---|---|---|
| Chemical formula | $C_{21}H_{28}O_3$ | $C_{20}H_{28}O_3$ | $C_{21}H_{30}O_3$ | $C_{22}H_{28}O_5$ | $C_{21}H_{28}O_5$ | $C_{22}H_{30}O_5$ | Head 1973 |
| Molecular weight | 328.4 | 316.4 | 330.4 | 372.4 | 360.4 | 374.4 | Head 1973 |
| Boiling point (°C) | 170 | 136–138 | – | 200 | 182–184 | | Tomlin 2003 |
| Vapor pressure (mmHg) | $2.02 \times 10^{-5}$ | $1.1 \times 10^{-6}$ | $4.8 \times 10^{-7}$ | $3.9 \times 10^{-7}$ | $4.6 \times 10^{-7}$ | $1.9 \times 10^{-7}$ | Tomlin 2003 |
| Water solubility (mg/L) | 0.35 | 3.62 | 0.60 | 125.6 | 1038 | 214.8 | Crosby 1995 |
| $K_{ow}$ (log) | 5.62 | 4.77 | 5.43 | 3.56 | 2.71 | 3.37 | Crosby 1995 |
| BCF | 11,000 | 2,500 | 4,700 | 300 | 70 | 210 | Crosby 1995 |
| Volatilization ($\mu g/cm^3/h$) | 0.89 | 1.98 | 1.18 | 0.65 | 1.38 | 1.80 | Crosby 1995 |
| Henry's law constant | $4.3 \times 10^{-3}$ | $9.4 \times 10^{-4}$ | $3.3 \times 10^{-3}$ | $8.9 \times 10^{-6}$ | $2.2 \times 10^{-6}$ | $1.4 \times 10^{-5}$ | Crosby 1995 |

solubility in water (Crosby 1995). P I has lower solubility than P II, as shown by Table 21.1. It also exhibits low volatility, as shown by the low vapor pressures (Tomlin 2003). Finally, the high bioconcentration factors (BCFs) of P I compared with P II can be attributed to their low solubility in water and high octanol–water coefficients. The BCF is a ratio of the concentration of a chemical in an animal to the concentration of that chemical in the environment (Gunasekara 2004).

## 21.5 General and Pesticidal Uses

Pyrethrin formulations can be in a liquid, dust, or aerosol form, depending on the application. They can be dissolved in water or alcohol, but to increase their toxicity, petroleum is used (Anadon et al. 2009). They find great use in households, where they are used to control human lice, and in mosquito-repellent coils. In addition, they can be used on crops to control insects with great lethality. Applications of their use as veterinary medicines in dogs and cats have also been reported, and they are mainly used as shampoos (Anadon et al. 2009). Finally, they have been reported to be effective against grain weevils (Biebel et al. 2003).

In most of these applications, their efficacy is enhanced by the use of synergists such as piperonyl butoxide (Tomar et al. 1979) and sesamol ethers (Devakumar et al. 1985). Piperonyl butoxide can exist as a synthetic or natural, with some countries preferring the use of the latter to the former in formulations because it is "green" (Cavoski et al. 2011). For instance, piperonyl butoxide synergized pyrethrins were observed to have enhanced activity as grain protectants against four liposcelidid psocids (*Liposcelis bostrychopila* Badonnel, *Liposcelis entomophila* [Enderlein], *Liposcelis decolor* [Pearman], and *Liposcelis paeta*), which were known to be difficult to control in Australia (Nayak 2010).

## 21.6 Toxicity

Pyrethrin insecticides block the volt-gated sodium channels in nerve axon insects, resulting in a knockdown effect, hyperactivity, and convulsions (Isman 2006). However, the knockdown effect of P I has a lower time than P II. Nonetheless, P II is easily metabolized by insects (Kiriamiti et al. 2003b). For this reason, P I and P II are used synergistically to achieve effective pest control (Winney 1979). A study was carried out to determine the relative toxicities of the pyrethrins to female houseflies (*Musca domestica* L.) (Sawicki et al. 1962). It was found that at 20°C, pyrethrum extract was 1.0; pyrethrin II, 1.3–1.5; pyrethrin I, 0.9–1.0; cinerin II, 0.5–0.6; and cinerin I, 0.4–0.5. In a more recent study, to compare the antifeedent activities of polygodial and pyrethrins against those of white flies (*Bemisia tabaci*) and aphids (*Myzus persicae*), the former was 2–20 times less deterrent than the latter, depending on the insect species (Prota and Bouwmeester 2014). The *Plasmodium falciparum* is the causative agent of malaria, and pyrethrins were also found to have antiplasmodial activity against it, with P II being the most selective antiplasmodial compound (Hata et al. 2011). However, the disadvantage observed with pyrethrins is that some insects are not affected because they render them harmless (Atkinson et al. 2004).

In rats, the toxicity of pyrethrin oils was largely observed to occur by decreasing the ATPase activity by up to 40%, and this was attributed to the presence of piperonyl butoxide (Kakko et al. 2000). Pyrethrin toxicity in humans is very low, although taking large amounts may lead to convulsions (Proudfoot 2005).

## 21.7 Metabolic Fate

Pyrethrin pesticides have low mammalian toxicity, because they are rapidly metabolized (Majoni and Munjanja 2015). For instance, moderate toxicity was observed for rats (rat oral acute $LD_{50}$ values ranging from 350 to 500 mg/kg). The metabolic pathway in mammals is thought to take place by oxidation of the alcohol and acid moieties of P I. Alternatively, metabolism may take place by hydrolysis of the methyl ester groups (Yamamoto et al. 1969). In rats, cinerin I and jasmoline I were metabolized by hydroxylation of the methyl and methylene groups (Class et al. 1990). In humans, absorption takes place more quickly through the gut than through the skin. Nonetheless, the active components are metabolized rapidly by the liver (Proudfoot 2005).

## 21.8 Environmental Fate

Despite their advantages, pyrethrins have been replaced by synthetic pyrethroids because of their poor stability in sunlight, reduced efficacy, and high production costs (Katsuda 1999). Pyrethrins degrade in sunlight; for instance, in sunlight the (Z)-pent-2-enyl side chain of the rethrolone moiety changed to the (E)-isomer (Kawano and Yanagihara 1980). However, a study carried out a few years ago showed that the photodegradation of pyrethrins may be slowed down by adding cyclodextrins (Biebel et al. 2003). In another study, the use of sunscreen agents was reported as a possible alternative to decrease the pho-todegradation of pyrethrins in formulations (Minello et al. 2005). Pyrethrins were also observed to be degraded in the presence of ultraviolet (UV) light regardless of humidity or the presence of oxygen (Blackith 1952). However, according to another study carried out, the stabilizing effect of piperonyl butoxide to pyrethrins exposed to ultraviolet light was also discovered (Donaldson and Stevenson 1960).

According to another study, temperature is another factor that accelerates the degrada-tion of pyrethrins (Atkinson et al. 2004). This was exemplified by the fact that the pyre-thrins decreased by 26%, 65%, and 68%, respectively, as the temperature increased from 20°C to 60°C and finally 100°C. The authors suggested that the concentration did not reach zero because of the plant structure.

The fate of pyrethrins in plants has been studied intensively by many authors, who have shown that they degrade quickly. For instance, a recent study showed that the half-lives of pyrethrins on field-grown tomatoes and bell pepper fruits did not exceed 2 hours (Antonius 2004). In another field experiment, the fate of pyrethrins in peaches was investi-gated, and the half-life of P I was found to be 2.3 days, and that of P II 6.6 days (Angioni et al. 2005). It is important to note that the half-lives increased in formulations where pipero-nyl butoxide was used. Moreover, the findings from both studies corroborate the fact that pyrethrins degrade quickly under field conditions, especially when light is present. In contrast, the absence of sunlight was shown to decrease the degradation rate of pyrethrins. This is exemplified by a study carried out to determine the degradation of pyrethrin resi-dues on stored durum wheat after postharvest treatment. It was found that in the absence of light, pyrethrins were stable for 22 days and took 8 months to dissipate completely (Caboni et al. 2007).

Residues of pyrethrins on potato leaves and in soil under field conditions were also determined using high-performance liquid chromatography (HPLC) coupled to UV detec-tion. Residues of pyrethrin I in compost treatments (0.056 µg/g) were higher than in no-mulch treatments (0.026 µg/g) (Antonius et al. 2001). In addition, P I bound more strongly to soils because it has a very large $K_{oc}$.

Concentrations of pyrethrins on an average of 36.1 ng/L have been reported in runoff water after they had been applied 11 days in advance (Antonius et al. 1997). In a separate study carried out, pyrethrins were detected in surface waters collected from five tributar-ies of the San Francisco Bay, California, at concentrations of less than 8.96 ng/L (Woudneh and Oros 2006).

## 21.9 Methods of Extraction

Pyrethrins are extracted from pyrethrum flowers to obtain oleoresin, which contains pyre-thrum I and II as the major components; other chemical components found in the extracts include carotenoids, sesquiterpenes, sesquiterpenoid lactones, flavonoids, n-alkanes, and various fatty acids (Casida 1973; Head 1973). These components can give rise to possible interference during analysis (Head 1968). For this reason, extraction techniques should produce an extract with high recovery of pyrethrin esters. The recovery rate may depend on the solvent used; for instance, nonpolar solvents like hexane give a very high content of total pyrethrins in the final extract (Nagar et al. 2015). Examples of techniques that have been used to extract pyrethrins include ultrasonic extraction (USE), Soxhlet extraction, and supercritical fluid extraction (SFE) (Otterbach and Wenclawiak 1999). A similarity between Soxhlet extraction and USE is that both of them use energy, which facilitates the continuous extraction of analytes by mass transfer, and thus the analyte leaches out in two successive elutions (Nagar et al. 2015). Other methods that have been used are maceration in a solvent such as n-hexane and cyclically pressurized extraction, also known as rapid solid–liquid dynamic extraction (RSDLE) (Gallo et al. 2017).

### 21.9.1 Soxhlet Extraction

Soxhlet extraction involves placing the sample in an extractor, and subsequently distilling, while introducing fresh portions of solvent at intervals (Turiel and Martin-Esteban 2008). The advantage of this technique is that good extraction results are obtained because the extraction solvent is recycled continuously at high temperatures (Nagar et al. 2015).

In an early study that was carried out to explore the possibility of using Soxhlet extraction for extraction of pyrethrin oils from pyrethrum flowers, it was observed that use of a warm solvent gave erroneous results for P II. On the other hand, cold extraction gave low values of P II, and it eliminated the variations that were obtained with different grades of ligroin (Mitchell and Tresadern 1949). The amount of pyrethrins extracted using this technique also depends on the solvent used. For instance, acetonitrile gave the best results in a recent study carried out (Nagar et al. 2015).

### 21.9.2 Supercritical Fluid Extraction

Supercritical fluid extraction involves the use of a supercritical fluid such as carbon dioxide to extract solid samples. The advantage of using a supercritical fluid is that extraction of analytes is possible even in pores that are not easily accessible (Turiel and Martin-Esteban 2008). In addition, it can be coupled to gas chromatography (GC) (Wenclawiak and Otterbach 2000) or HPLC. The coupling of SFE to these chromatographic techniques offers extraction of unlimited sample volumes (Pol and Wenclawiak 2003). The variables that affect the extraction efficiency include pressure, temperature, and particle size (Kiriamiti et al. 2003b). Other advantages include being environmentally friendly because of the absence of solvents. Notwithstanding the above advantages, SFE is very expensive, and it is not universally applicable (Gallo et al. 2017).

A method was developed to extract P I and P II from pyrethrum flower, and the extraction efficiencies under various conditions were examined (Pan et al. 1995). It was found that the most effective extractions of P I and P II were at 40°C and 1200 psi. In addition, the extraction efficiencies of the technique were better than those obtained with n-hexane. However, this is in contrast to a more recent study that was carried out to explore the supercritical fluid extraction of pyrethrins from pyrethrum flowers, where the extract was similar to that obtained using hexane in terms of waxes and oils (Kiriamiti et al. 2003b). Notwithstanding the above fact, the same research group improved the quality of the extract by carrying out the fractionation in two steps (Kiriamiti et al. 2006). However, the ideal was not reached because the separators were too small. In another study, the same research group showed that supercritical carbon dioxide can be used to purify oleoresin that had been previously extracted with hexane (Kiriamiti et al. 2003a). The extract obtained was of high quality and contained a high concentration of pyrethrins.

Studies have been carried out to compare SFE with other extraction techniques, such as USE and Soxhlet extraction (Otterbach and Wenclawiak 1999). Both studies have shown that the quality of the extract was better than that of USE. Recently, the technique has also been compared with maceration and cyclic pressurized extraction, and the extraction efficiency was comparable to that of the other techniques (Gallo et al. 2017).

### 21.9.3 Ultrasonic Extraction

USE makes use of mechanical waves to alter the physical and chemical properties of a matrix by cavitations, thus releasing the extractant from the matrix (Luque-Garcia and Luque de Castro 2003). Sonication is usually carried out in a sonication bath or sonication probes (Fenoll et al. 2011).

A study was carried out to compare the extraction efficiency of four extraction techniques: percolation, agitation with heat, Soxhlet extraction, and USE. It was discovered that energy-assisted extraction techniques like agitation and USE increased the extractive yield by 20%–50% (Nagar et al. 2015). Moreover, the best results for sonication were found using acetonitrile as the extraction solvent.

## 21.10 Methods of Analysis

### 21.10.1 Supercritical Fluid Chromatography

Supercritical fluid chromatography (SFC) uses a supercritical fluid such as carbon dioxide as the mobile phase. This gives it a competitive edge over other chromatographic techniques because it is fast, environmentally friendly, and highly efficient (El-Saeid and Khan 2010). Factors affecting chromatographic separation efficiency in SFC include pressure gradients, density, and temperature (Wenclawiak et al. 1998). Another important thing to consider before analysis can be done is calibration. However, this may not always be problematic. Hence, allethrin can be used as a reference standard because it has a structure that is similar to that of the compounds (Wenclawiak and Otterbach 2000).

SFC with positive pressure and negative temperature gradients was successfully used to separate pyrethrins (Wenclawiak et al. 1998).

### 21.10.2 Gas Chromatography

Gas chromatography is used to analyze analytes that can be volatilized by coupling to a selective detector such as a flame ionization detector (FID) (Class 1991) or electron capture detector (ECD) (Berger-Preib et al. 1997). Gas chromatographic methods have been used for a long time to analyze pyrethrum extracts with great precision (Kawano et al. 1974). A study revealed that use of shorter columns and a thinner stationary phase is not sufficient to eliminate degradation of pyrethrins (Wieboldt et al. 1989). It is of paramount importance to note that when carrying out the analyses of pyrethrins, elution temperatures should not exceed 200°C, because hot injection ports cause tautomerization of P I and P II and poor peak shapes (Wieboldt et al. 1989). GC analysis using a temperature less than 210°C was successfully used to separate pyrethrins without any degradation (Class 1991).

A trend observed is the use of gas chromatography–mass spectrometry (GC-MS) to separate, quantify, and identify the pyrethrins (Wenclawiak et al. 1997; Cai et al. 2013). GC-MS coupled to SFE at high temperature and a catalyst was used to quantify and identify the pyrethrin esters, with high quantitative conversion (Wenclawiak et al. 1997). In a recent study, GC-MS was successfully used to determine the amounts of the chemical compounds in flower extracts, by use of peak area normalization (Cai et al. 2013).

Despite the usefulness of GC-based methods in the separation and quantitation of pyrethrins, their use has decreased over the years because pyrethrins degrade at high temperatures. For this reason, HPLC-based methods can be used.

### 21.10.3 High-Performance Liquid Chromatography

Normal phase high-performance liquid chromatography (NP-HPLC) involves the use of a stationary phase that is more polar than the mobile phase. In an earlier study, NP-HPLC was used to separate the pyrethrin esters at 229 nm, with some interference from UV-absorbing material in the extract (McEldowney and Menary 1988). However, in a more recent study NP-HPLC was successfully used to quantify the components of pyrethrum extract, without any interference from the sample matrix (Essig and Zhao 2001a). The same research group separated and characterized pyrethrum extract standard by semipreparative NP-HPLC, obtaining an assay content that was close to that reported by the AOAC International titration method (Essig and Zhao 2001b).

Reverse phase high-performance liquid chromatography (RP-HPLC), which relies on nonpolar stationary phases such as C18 and C8 (Botitsi et al. 2011), can be used to separate and purify pyrethrum extracts (Wei et al. 2006). A study was carried out to determine pyrethrins in pyrethrum extracts by RP-HPLC with diode array detection at 240 nm (Wang et al. 1997). Well-resolved peaks were obtained at 240 nm, as shown by Figure 21.3. Changing the wavelength to 230 nm gave poor resolution between cinerin I and pyrethrin I. In another study, different stationary phases were compared for the separation efficiency using RP-HPLC (Nagar et al. 2015). It was found that the C18 column (3.9 × 150 mm, 5 µm) resolved the pyrethrin mixtures well with great sensitivity, as shown by Figure 21.4.

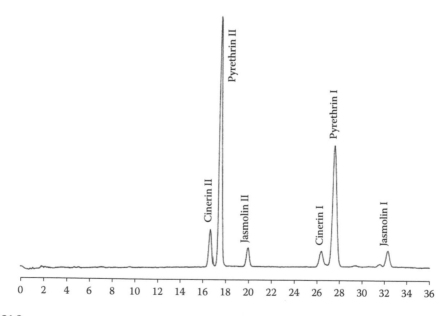

**FIGURE 21.3**
Reverse phase HPLC chromatogram of pyrethrum extract. (Reproduced from Wang, I. et al., *J. Chromatogr. A*, 766, 277–281, 1997. With permission.)

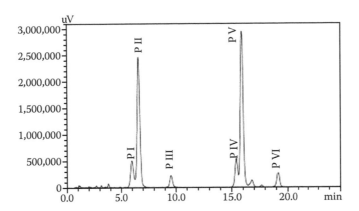

**FIGURE 21.4**
HPLC chromatogram of standard pyrethrins. P I, cinerin II; P II, pyrethrin II; P III, jasmolin II; P IV, cinerin I; P V, pyrethrin; P VI, jasmolin I. (Reproduced from Nagar, A. et al., *Ind. Crops Prod.*, 76, 955–960, 2015. With permission.)

A recent advancement in HPLC is the use of ultra-high-performance liquid chromatography (UHPLC), which has reduced particle size and higher flow rates for increased speed, with improved resolution and signal-to-noise ratio (Leandro et al. 2006). UHPLC with photodiode array detection and aerosol detection was used to detect pyrethrins with great sensitivity and precision (Thomas et al. 2015).

Another recent advance is the use of nanoliquid chromatography, which utilizes flow rates of nanoliters per minute, and therefore provides high sensitivity because of lower chromatographic dilution and higher efficiency (Asimakopoulos et al. 2015). This technique has previously been used in the analysis of synthetic pesticides using various stationary phases (Cappielo et al. 2003; Buonasera et al. 2009). Recently, reverse phase nanoliquid chromatography coupled to direct electron ionization–mass spectrometry (nanoLC–direct EI MS) was successfully used to detect and quantify pyrethrins in pyrethrum extracts with high sensitivity and precision (Cappielo et al. 2012).

### 21.10.4 High-Performance Capillary Electrophoresis

The pyrethrin esters were separated using micellar electrokinetic chromatography (MEKC) by comparing two pseudostationary phases, namely, sodium dodecyl sulfate (SDS) and polymeric sodium N-undecyl sulfate (poly-SUS) (Henry et al. 1999). The advantage of this technique over HPLC is the shorter analysis times obtained for both pseudostationary phases. However, the use of a pseudostationary phase can be a problem. Hence, other techniques that use a stationary phase, such as capillary electrochromatography (CEC), can be explored.

### 21.10.5 Capillary Electrochromatography

Capillary electrochromatography is a technique that separates analytes between the mobile and stationary phases by electro-osmotic flow (Dittman and Rozing 1996). CEC was successfully used to separate six pyrethrin esters with reduced runtimes of up to 16 minutes (Henry et al. 2001). However, according to the study, the technique shows low sensitivity for pyrethrins that are in low concentration. This may require an additional concentration step.

## 21.11 Conclusions

There is now ample evidence to prove that the continuous use of synthetic pesticides leads to environmental degradation. Hence, the use of plant-based pesticides, which are environmentally friendly, has been explored. The pyrethrins derived from the pyrethrum plant, apart from being environmentally friendly are very effective as insecticides, producing a rapid knockdown effect. Another distinct advantage is that they are readily metabolized in mammals such as rats and humans, only causing toxicity if taken in large amounts.

However, they also have a major disadvantage of rapidly degrading in the presence of sunlight. Synergists and sunscreen agents have been successfully used to improve their half-lives under field conditions. For this reason, they remain one of the most effective plant-based insecticides.

It is worth mentioning that numerous advances have been made in both extraction and analysis techniques. Extraction techniques have evolved from laborious techniques such as Soxhlet extraction, which require large amounts of organic solvents, to the use of green techniques such as SFE, which have high recoveries. In addition, chromatographic techniques have been successfully used to separate and quantify the pyrethrins in extract, with good sensitivity and precision. A notable improvement is the use of mass spectrometry to identify the individual pyrethrins without any interference.

In conclusion, much research has been published on pyrethrin-related topics such as extraction and analysis techniques, and metabolic and environmental fate. Moreover, the use of stabilizers and synergists to increase their photostability is remarkable, and this may lead to their continued use. However, in terms of efficacy, they may not exactly match the synthetic pesticides.

## References

Akhtar, Y, and MB Isman. 2012. Plant Natural Products for Pest Management: The Magic of Mixtures. In *Advanced Technologies for Managing Insect Pests*, ed. I Ishaaya, 231–247. Dordrecht: Springer Science Business Media.

Anadon, A, MR Martinez-Larranaga, and MA Martinez. 2009. Use and Abuse of Pyrethrins and Synthetic Pyrethroids in Veterinary Medicine. *Veterinary Journal* 182: 7–20.

Angioni, A, F Dedola, EV Minello, A Barra, P Cabras, and P Caboni. 2005. Residues and Half Lives of Pyrethrins on Peaches after Field Treatments. *Journal of Agricultural and Food Chemistry* 53: 4059–4063.

Anon. 2010. The Power Daisy. www.pyrethrum.com/About_Pyrethrum/Agronomy.html.

Antonius, GF. 2004. Residues and Half Lives of Pyrethrins on Field-Grown Pepper and Tomato. *Journal of Environmental Science and Health Part B* 39: 491–503.

Antonius, GF, ME Byers, and WC Kerst. 1997. Residue Levels of Pyrethrins and Piperonyl Butoxide in Soil and Runoff Water. *Journal of Environmental Science and Health Part B* 32: 621–644.

Antonius, GF, JC Snyder, and GA Patel. 2001. Pyrethrins and Piperonyl Butoxide Residues on Potato Leaves and in Soil under Field Conditions. *Journal of Environmental Science and Health Part B* 36: 261–271.

Asimakopoulos, AG, A Bletsou, K Kannan, and NS Thomaidis. 2015. Recent Developments in Liquid Chromatography-Mass Spectrometry: Advances in Liquid Chromatographic Separations and Ionization Techniques/Interfaces. In *Mass Spectrometry for the Analysis of Pesticide Residues and Their Metabolites*, ed. D Tsipi, H Botitsi, and A Economou, 113–130. 1st ed. Hoboken, NJ: John Wiley & Sons.

Atkinson, BL, AJ Blackman, and H Faber. 2004. The Degradation of Natural Pyrethrins in Crop Storage. *Journal of Agricultural and Food Chemistry* 52: 280–287.

Berger-Preib, E, K Levsen, and A Preib. 1997. Analysis of Individual Natural Pyrethrins in Indoor Matrices by HRGC/ECD. *Journal of Separation Science* 20: 284–288.

Bhakuni, RS, AP Kahol, SP Singh, and A Kumar. 2007. Composition of North Indian Pyrethrum (*Chrysanthemum cinerariaefolium*) Flower Oil. *Journal of Essential Oil Bearing Plants* 10: 31–35.

Biebel, R, E Rametzhofer, H Klapal, D Polheim, and H Viernstein. 2003. Action of Pyrethrum-Based Formulations against Grain Weevils. *International Journal of Pharmaceutics* 256: 175–181.

Blackith, RE. 1952. Stability of Contact Insecticides. I. Ultraviolet Photolysis of the Pyrethrins. *Journal of the Science of Food and Agriculture* 3: 219–224.

Botitsi, H, SD Garbis, A Economou, and D Tsipi. 2011. Current Mass Spectrometry Strategies for the Analysis of Pesticides and Their Metabolites in Food and Water Matrices. *Mass Spectrometry Reviews* 30: 907–939.

Brown, PH, and RC Menary. 1994. Flowering in Pyrethrum (*Tanacetum cinerariaefolium* L.). Environmental Requirements. *Journal of Horticultural Science* 69: 877–884.

Buonasera, K, G D'Orazio, S Fanali, P Dugo, and L Mondello. 2009. Separation of Organophosphorus Pesticides by Using Nano-Liquid Chromatography. *Journal of Chromatography A* 1216: 3970–3976.

Caboni, P, EV Minello, M Cabras, A Angioni, G Sarias, F Dedola, and P Cabras. 2007. Degradation of Pyrethrin Residues on Stored Durum Wheat after Postharvest Treatment. *Journal of Agricultural and Food Chemistry* 55: 832–835.

Cai, TT, M Ye, ZY Li, LM Fan, YG Zha, and J Wang. 2013. Investigation of the Main Chemical Compounds in Pyretrum Extract Obtained by Supercritical Fluid Extraction. *Advanced Materials Research* 781–784: 737–740.

Cappielo, A, G Famiglini, F Mangani, P Palma, and A Siviero. 2003. Nano-High-Performance Liquid Chromatography-Electeon Ionization Mass Spectrometry Approach for Environmental Analysis. *Analytica Chimica Acta* 493: 125–136.

Cappielo, A, B Tirillini, G Famiglini, H Trufelli, V Termopoli, and C Flender. 2012. Determination of Natural Pyrethrins by Liquid Chromatography-Electron Ionisation-Mass-Spectrometry. *Phytochemical Analysis* 23: 191–196.

Casida, JE, ed. 1973. *Pyrethrum—The Natural Insecticide*. New York: Academic Press.

Cavoski, I, P Caboni, and Y Miano. 2011. Natural Pesticides and Future Perspectives. In *Pesticides in the Modern World—Pesticide Use and Management*, ed. M Stoytcheva, 163–190. Rijeka, Croatia: Intech.

Ccanccapa, A, A Masia, V Andreu, and Y Pico. 2016. Spatio-Temporal Patterns of Pesticide Residues in the Turia and Jucar Rivers (Spain). *Science of the Total Environment* 540: 200–210.

Class, T. 1991. Optimized Gas Chromatographic Analysis of Natural Pyrethrins and Pyrethroids. *Journal of Separation Science* 14: 48–51.

Class, T, T Ando, and J Casida. 1990. Pyrethroid Metabolism—Microsomal Oxidase Metabolites of (S-) Bioallethrin and the Six Natural Pyrethrins. *Journal of Agricultural and Food Chemistry* 38: 529–537.

Cooper, J, and H Dobson. 2007. The Benefits of Pesticides to Mankind and the Environment. *Crop Protection* 26(9): 1337–1348.

Crosby, DG. 1995. Environmental Fate of Pyrethrins. In *Pyrethrum Flowers; Production, Chemistry, Toxicology, and Uses*, ed. JE Casida and GB Quistad, 194–213. New York: Oxford University Press.

Dayan, FE, CL Cantrell, and SO Duke. 2009. Natural Products in Crop Protection. *Bioorganic and Medicinal Chemistry* 17: 4022–4034.

Devakumar, C, VS Saxena, and SK Mukerjee. 1985. New Sesamol Ethers as Pyrethrum Synergists. *Agricultural and Biological Chemistry* 49: 725–730.

Dittman, MM, and GP Rozing. 1996. Capillary Electrochromatography—A High-Efficiency Micro-Separation Technique. *Journal of Chromatography A* 744: 63–74.

Donaldson, JM, and JH Stevenson. 1960. The Stabilising Effect of Piperonyl Butoxide on Pyrethrins Exposed to Ultraviolet Light. *Journal of the Science of Food and Agriculture* 11: 370–373.

El-Saeid, MH, and HA Khan. 2010. Analysis of Pesticides in Food Samples by Supercritical Fluid Chromatography. In *Handbook of Pesticides: Methods of Pesticide Residue Analysis,* ed. LML Nollet and HS Rathore, 93–113. Boca Raton, FL: CRC Press.

Essig, K, and Z Zhao. 2001a. Method Development and Validation of a High-Performance Liquid Chromatographic Method for Pyrethrum Extract. *Journal of Chromatographic Science* 39: 473–480.

Essig, K, and Z Zhao. 2001b. Preparation and Characterization of a Pyrethrum Extract Standard. *LCGC* 19: 722–730.

Fenoll, J, P Hellín, CM Martínez, P Flores, and S Navarro. 2011. Determination of 48 Pesticides and Their Main Metabolites in Water Samples by Employing Sonication and Liquid Chromatography-Tandem Mass Spectrometry. *Talanta* 85 (2): 975–982.

Gallo, M, A Formato, D Ianniello, A Andolfi, E Conte, M Ciaravolo, V Varchetta, and D Naviglio. 2017. Supercritical Fluid Extraction of Pyrethrins from Pyrethrum Flowers (*Chrysanhthemum cinerariifolium*) Compared to Traditional Maceration and Cyclic Pressurization Extraction. *Journal of Supercritical Fluids* 119: 104–112.

Glynne-Jones, A. 2001. Pyrethrum. *Biopesticides* 12: 195–198.

Gunasekara, AS. 2004. Environmental Fate of Pyrethrins. Sacramento: California Environmental Protection Agency.

Hata, Y, S Zimmermann, M Quitschau, M Kaiser, M Hamburger, and M Adams. 2011. Antiplasmodial and Antitrypanosomal Activity of Pyrethrins and Pyrethroids. *Journal of Agricultural and Food Chemistry* 59: 9172–9176.

Head, SW. 1968. Fatty Acid Composition of Extract from Pyrethrum Flowers (*Chrysanthemum cinerariaefolium*). *Journal of Agricultural and Food Chemistry* 16: 762–765.

Head, SW. 1973. Carotenoid Constituents of Pyrethrum Flowers. *Journal of Agricultural and Food Chemistry* 21: 999–1001.

Henry, CW, III, ME Mccarroll, and IM Warner. 2001. Separation of the Insecticidal Pyrethrin Esters by Capillary Electrochromatography. *Journal of Chromatography A* 905: 319–327.

Henry, CW, III, SA Shamsi, and IM Warner. 1999. Separation of Natural Pyrethrum Extracts Using Micellar Electrokinetic Chromatography. *Journal of Chromatography A* 863: 89–103.

Hitmi, A, A Coudret, and C Barthomeuf. 2000. The Production of Pyrethrins by Plant Cell and Tissue Cultures of *Chrysanthemum cinerariaefolium* and *Tagetes* Species. *Critical Reviews in Plant Sciences* 19: 69–89.

Isman, MB. 2006. Botanical Insecticides, Deterrents, and Repellents in Modern Agriculture and an Increasingly Regulated World. *Annual Reviews of Entomology* 51: 45–66.

Isman, MB. 2008. Botanical Insecticides—For Rich, for Poorer. *Pesticide Management Science* 64: 8–11.

Kakko, I, T Toimela, and H Tahti. 2000. Piperonyl Butoxide Potentiates the Synaptosome ATPase Inhibiting Effect of Pyrethrin. *Chemosphere* 40: 301–305.

Katsuda, Y. 1999. Development and Future Prospects for Pyrethroid Chemistry. *Pesticide Science* 55: 775–782.

Kawano, Y, and K Yanagihara. 1980. Examination of the Conversion Products of Pyrethrins and Allethrin Formulations Exposed to Sunlight by Gas Chromatography and Mass Spectrometry. *Journal of Chromatography A* 198: 317–328.

Kawano, Y, K Yanagihara, and A Bevenue. 1974. Analytical Studies of Pyrethrin Formulations by Gas Chromatography III. Analytical Results on Insecticidally Active Components of Pyrethrins from Various World Sources. *Journal of Chromatography A* 90: 119–128.

Kiriamiti, HK, S Camy, C Gourdon, and J Condoret. 2003a. Supercritical Carbon Dioxide Processing of Pyrethrum Oleoresin and Pale. *Journal of Agricultural and Food Chemistry* 51: 880–884.

Kiriamiti, HK, S Camy, J Gourdon, and S Condoret. 2003b. Pyrethrins Extraction from Pyrethrum Flowers Using Carbon Dioxide. *Journal of Supercritical Fluids* 26: 193–200.

Kiriamiti, HK, S Sarmat, and C Nzila. 2006. Fractionation of Crude Pyrethrum Extract Using Supercritical Carbon Dioxide. In *Proceedings from the International Conference on Advances in Engineering and Technology*, 339–346. Entebbe, Uganda.

Koul, O, S Walia, and GS Dhaliwal. 2008. Essential Oils as Pesticides: Potential and Constraints. *Biopesticides International* 4: 63–84.

Leandro, CC, P Hancock, RJ Fussell, and BJ Keely. 2006. Comparison of Ultra-Performance Liquid Chromatography and High Performance Liquid Chromatography for the Determination of Priority Pesticides in Baby Foods by Tandem Quadrupole Mass Spectrometry. *Journal of Chromatography A* 1103: 94–101.

Li, J, MA Jongsma, and C Wang. 2014. Comparative Analysis of Pyrethrin Content Improvement by Mass Selection, Family Selection and Polycross in Pyrethrum [*Tanacetum cinerariifolium* (Trevir.) Sch Bip.] Populations. *Industrial Crops and Products* 53: 268–273.

Li, J, LY Yin, MA Jongsma, and CY Wang. 2011. Effect of Light, Hydropriming and Abiotic Stress on Seed Germination, and Shoot and Root Growth of Pyrethrum (*Tanacetum cinerariifolium*). *Industrial Crops and Products* 34: 1543–1549.

Luque-Garcia, JL, and MD Luque de Castro. 2003. Ultrasound: A Powerful Tool for Leaching. *TrAC Trends in Analytical Chemistry* 22: 41–47.

Majoni, S, and BK Munjanja. 2015. Metabolism of Biopesticides. In *Biopesticides Handbook*, ed. LMM Nollet and HS Rathore, 25–50. Boca Raton, FL: CRC Press.

Matsuda, K, Y Kikuta, A Haba, K Nakayama, Y Katsuda, A Hatanaka, and K Komai. 2005. Biosynthesis of Pyrethrin I in Seedlings of *Chrysanthemum cinerariaefolium*. *Phytochemistry* 66: 1529.

McEldowney, AM, and RC Menary. 1988. Analysis of Pyrethrins in Pyrethrum Extracts by High-Performance Liquid Chromatography. *Journal of Chromatography A* 447: 239–243.

Minello, EV, F Lai, MT Zonchello, M Melis, M Russo, and P Cabras. 2005. Effect of Sunscreen and Antioxidant on the Stability of Pyrethrin Formulations. *Journal of Agricultural and Food Chemistry* 53: 8302–8305.

Miresmailli, S, and MB Isman. 2014. Botanical Insecticides Inspired by Plant-Herbivore Chemical Interactions. *Trends in Plant Science* 19: 29–35.

Mitchell, WM, and FH Tresadern. 1949. The Analysis of Pyrethrum Flowers. *Journal of Chemical Technology and Biotechnology* 68: 221–225.

Monda, JM. 2014. Pyrethrum Sector in Kenya—Current Status. http://projects.nri.org/options/images /Current_status_of_pyrethrum_sector_in_Kenya.pdf.

Morris, SE, NW Davies, PH Brown, and T Groom. 2006. Effects of Drying Conditions on Pyrethrins Content. *Industrial Crops and Products* 23: 9–14.

Nagar, A, A Chatterjee, LU Rehman, A Ahmad, and S Tandon. 2015. Comparative Extraction and Enrichment Techniques for Pyrethrins from Flowers of *Chrysanthemum cinerariaefolium*. *Industrial Crops and Products* 76: 955–960.

Nayak, MK. 2010. Potential of Piperonyl Butoxide-Synergised Pyrethrins against Psocids (Psocoptera: Liposcelididae) for Stored-Grain Protection. *Pest Management Science* 66: 295–300.

Ntalli, NG, and U Menkissoglu-Spiroudi. 2011. Pesticides of Botanical Origin: A Promising Tool in Plant Protection. In *Pesticides—Formulations, Effects, Fate*, ed. M Stoytcheva, 1–24. Rijeka, Croatia: Intech.

Otterbach, A, and BW Wenclawiak. 1999. Ultrasonic/Soxhlet/Supercritical Fluid Extraction Kinetics of Pyrethrins from Flowers and Allethrin from Paper Strips. *Fresenius Journal of Analytical Chemistry* 365: 472–474.

Pan, WHT, C Chang, T Su, F Lee, and MS Fuh. 1995. Preparative Supercritical Fluid Extraction of Pyrethrin I and II from Pyrethrum Flower. *Talanta* 42: 1745–1749.

Pimentel, D. 1992. Environmental and Economic Costs of Pesticide Use. *Bioscience* 42: 750–760.

Pol, J, and BW Wenclawiak. 2003. Direct On-Line Continuous Supercritical Fluid Extraction and HPLC of Aqueous Pyrethrins Solution. *Analytical Chemistry* 75: 1430–1435.

Prota, N, and HJ Bouwmeester. 2014. Comparative Antifeedent Activities of Polygodial and Pyrethrins against Whiteflies (*Bemisia tabaci*) and Aphids (*Myzus persicae*). *Pest Management Science* 70: 682–688.

Proudfoot, AT. 2005. Poisoning due to Pyrethrins. *Toxicological Reviews* 24: 107–113.

Roncevic, S, LP Svedruzic, and I Nemet. 2014. Elemental Composition and Chemometric Characterization of Pyrethrum Plant Materials and Insecticidal Flower Extracts. *Analytical Letters* 47: 627–640.

Sawicki, RM, M Elliot, JC Gower, M Snarey, and EM Thain. 1962. Insecticidal Activity of Pyrethrum Extract and Its Four Insecticidal Constituents against houseflies. I. Preparation and Relative Toxicity of the Pure Constituents; Statistical Analysis of the Action of Mixtures of These Components. *Journal of the Science of Food and Agriculture* 13: 172–185.

Schleier, JJ, III, and RKD Peterson. 2011. Pyrethrins and Pyrethroid Insecticides. In *Green Trends in Insect Control*, ed. O Lopez and JG Fernandez-Bolanos, 94–131. Vol. 11. London: Royal Society of Chemistry.

Singh, KK, A Tomar, and HS Rathore. 2010. Scope and Limitations of Neem Products and Other Botanicals in Plant Protection: A Perspective. In *Handbook of Pesticides: Methods of Pesticide Residue Analysis*, ed. LML Nollet and HS Rathore, 67–89. Boca Raton, FL: CRC Press.

Srivastava, S, P Goyal, and MM Srivastava. 2010. Pesticides: Past, Present, and Future. In *Handbook of Pesticides: Methods of Pesticide Residue Analysis*, ed. LML Nollet and HS Rathore, 47–65. Boca Raton, FL: CRC Press.

Thomas, D, A Wong, JA Glinski, and I Acworth. 2015. Determination of Pyrethrins in Pyrethrum Oil Extracts by UHPLC-UV-CAD. *Planta Medica* 82-PA2.

Tomar, SS, ML Maheshwari, and SK Mukerjee. 1979. Syntheses and Synergistic Activity of Some Pyrethrum Synergists from Dillapiole. *Agricultural and Biological Chemistry* 143: 1479–1483.

Tomlin, C. 2003. *The Pesticide Manual—A World Compendium*. 13th ed. Alton, UK: British Crop Protection Council.

Turiel, E, and A Martin-Esteban. 2008. Sample Handling of Pesticides in Food and Environmental Samples. In *Analysis of Pesticides in Food and Environmental Samples*, ed. JL Tadeo, 36–58. Boca Raton, FL: CRC Press.

Wandahwa, D, E Van Ranst, and P Van Damme. 1996. Pyrethrum (*Chrysanthemum cinerariaefolium* Vis) Cultivation in West Kenya: Origin, Ecological Conditions and Management. *Industrial Crops and Products* 5: 307–322.

Wang, I, V Subramanian, R Moorman, J Burleson, and J Ko. 1997. Direct Determination of Pyrethrins in Pyrethrum Extracts by Reversed-Phase High Performance Liquid Chromatography with Diode Array Detection. *Journal of Chromatography A* 766: 277–281.

Wei, D, Z Li, G Wang, Y Yang, Y Li, and Y Xia. 2006. Separation and Purification of Natural Pyrethrins by Reversed Phase High Performance Liquid Chromatography. *Chinese Journal of Analytical Chemistry* 34: 1776–1779.

Wenclawiak, BW, M Krappe, and A Otterbach. 1997. In Situ Transesterification of the Natural Pyrethrins to Methyl Esters by Heterogenous Catalysis Using a Supercritical Fluid Extraction System and Detection by Gas-Chromatography-Mass Spectrometry. *Journal of Chromatography A* 785: 263–267.

Wenclawiak, BW, and A Otterbach. 2000. Carbon-Based Quantitation of Pyrethrins by Supercritical-Fluid Chromatography. *Journal of Biochemical and Biophysical Methods* 43: 197–207.

Wenclawiak, BW, A Otterbach, and M Krappe. 1998. Capillary Supercritical Fluid Chromatography of Pyrethrins and Pyrethroids with Positive Pressure and Negative Temperature Gradients. *Journal of Chromatography A* 799: 265–273.

Wieboldt, RC, KD Kempfert, DW Later, and ER Campbell. 1989. Analysis of Pyrethrins Using Capillary Supercritical Fluid Chromatography and Capillary Gas Chromatography with Fourier Transform Infrared Detection. *Journal of Separation Science* 12: 106–111.

Winney, R. 1979. Performance of Pyrethroids as Domestic Insecticides. *Pyrethrum Post* 13: 132–136.

Woudneh, MB, and DR Oros. 2006. Quantitative Determination of Pyrethroids, Pyrethrins, and Piperonyl Butoxide in Surface Water by High-Resolution Gas Chromatography/High Resolution Mass Spectrometry. *Journal of Agricultural and Food Chemistry* 54: 6957–6962.

Yamamoto, I, EC Kimmel, and J Casida. 1969. Oxidative Metabolism of Prethroids in Houseflies. *Journal of Agricultural and Food Chemistry* 17 (6): 1227–1236.

# 22

# Pennyroyal Oil as a Green Pesticide

**N.C. Basantia and Hamir Singh Rathore**

## CONTENTS

## 22.1 Introduction

Pennyroyal (*Mentha pulegium* L.) is known by several common names, such as pennyroyal, mint, and poleo. Pennyroyal is native to Ireland, across southern and central Europe, and in the Ukraine (Tutin et al. 1976). A European folk name for the plant is "grows-in-the-ditch" (Polonin 1969). Repeatedly introduced in North America since European settlement, pennyroyal is now found naturalized in wildlands throughout the world (Grieve, 1959). These plants also thrive in Asia, Iran, the United Arab Emirates, Ethiopia, Brazil, Tunisia, and Portugal. This plant also grows spontaneously in humid parts of Iran. The worth of *Mentha* is evident, as its essential oils (EOs) and dried and fresh plant material are in daily use in confectionary, beverages, bakeries, cosmetics, pharmaceuticals, and pesticides (Shaikh et al., 2014). Essential oils of this herb have been used in traditional medicine in many countries. The oils are also reported to have potential for use to prevent oxidation of unsaturated fatty acids and increase a product's shelf life. Pennyroyal is a good alternative for synthetic antioxidants and is used as flavors in many foods. Besides food products, pennyroyal essential oils can be used in detergents and soaps, dental products, and insect and tick repellent agents.

## 22.2 Botany of the Plant

### 22.2.1 Taxonomic Classification (Shah et al., 2011)

Kingdom: Plantae
Subkingdom: Tracheobionta
Superdivision: Spermatophyta
Division: Magnoleophyta
Class: Magnoliliopsida
Order: Lamiales
Family: Lamiaceae/Labiatae
Genus: *Mentha* L.
Species: *pulegium* (pennyroyal)
Botanical name: *Mentha pulegium* (pennyroyal)

### 22.2.2 Habitat and Distribution

*M. pulegium* L. is an herbaceous perennial plant group wildly found in humid areas of the plains and mountains. Pennyroyal is native to Ireland, across the southern and central European countries, in North Africa, Ethiopia, Brazil, Tunisia, Portugal, and the Arabic countries. Spain and Morocco are main pennyroyal oil producer countries. This plant also grows spontaneously in humid parts of Iran.

### 22.2.3 Botanical Description of Plant

*M. pulegium* L. is an aromatic perennial herbaceous plant reaching up to 40 cm. These plants are characterized by a uniformly prostrate to upright habit while flowering.

### 22.2.4 Leaves

The leaves of *M. pulegium* are petiolate, ovate to suborbicular, obtuse to cuneate, and entire to serrate.

### 22.2.5 Flower

The flower is of an inflorescence shape and verticillate, with four to many flowers.

---

## 22.3 Methods of Extraction of Oil

The common methods to extract essential oil from *M. pulegion* are (1) hydrodistillation (HD), steam distillation, and steam and water distillation; (2) Soxhlet extraction; (3) supercritical fluid extraction (SFE); (4) microwave-assisted hydrodistillation; and (5) ultrasound-assisted extraction (UAE).

### 22.3.1 Distillation

Distillation is the extracting oil process that converts volatile liquid (essential oils) into a vapor state and then condenses the vapor into a liquid state. In this process, the botanic material is completely immersed in water and the still is brought to boil. It is used to protect the oils to a certain degree since the surrounding water acts as a barrier to prevent overheating. When the condensed material cools down, the water and essential oil are separated, and the oil is decanted for use as an essential oil. As water distillation tends to be small, the operation takes a long time to accumulate much oil (Ranjitha and Vijiyalakshmi, 2014). In the steam distillation method, the botanical material is placed in a still and steam is forced over the material. The hot steam is used to release the aromatic molecules from the plant material. The steam forces the pockets open, and then the molecules of these volatile oils escape from the plant material and evaporate into the steam. The steam containing the essential oil is passed through the cooling system to condense the steam, which forms a liquid form of essential oil. Then, water is separated. The major advantage of steam distillation is that the temperature never goes above 100°C, so temperature-sensitive compounds can be distilled. The disadvantage is that not many compounds can be steam distilled—usually only aromatic ones (Masango, 2005; Ranjitha and Vijiyalakshmi, 2014). The hydrodiffusion method is similar to the steam distillation process. The main difference between these two methods is that the steam is introduced into the still. In the case of hydrodiffusion, steam is fed into the top, onto the botanical material, instead of from bottom, as in normal steam distillation. The steam containing the essential oil is passed through a cooling system to condense the steam, which forms a liquid of essential oil, and then water is then separated. The main advantages of this method are that less steam is used, the processing time is shorter, and there is a higher yield (Ranjitha and Vijiyalakshmi, 2014).

### 22.3.2 Soxhlet Extraction

Soxhlet extraction is a general and well-established technique that surpasses other conventional extraction techniques in performance, except for limited fields of application, for example, the extraction of thermolabile compounds. Most of the solvent extraction units worldwide are based on Soxhlet principles with recycling of solvents. Basically, the equipment consists of a drug holder extractor, a solvent storage vessel, a reboiler kettle, a condenser, a breather system, and supporting structures like a boiler, a refrigerated chilling unit, and a vacuum unit (William, 2007). This technique is based on the choice of solvent coupled with heat or agitation. In this process, the circulation of solvents causes the displacement of transfer equilibrium by repeatedly bringing fresh solvent into contact with the solid matrix. This method maintains a relatively high extraction temperature and no filtration of extract is required (Shams et al., 2015).

However, the limitation of this technique is that there is a possibility of thermal decomposition of thermolabile targeted compounds because the extraction usually occurs at the boiling point of the solvent for a long time.

### 22.3.3 Microwave Extraction Method

Solvent-free microwave extraction is used to separate the essential oil from plant material. The method involves placing the sample in a microwave reactor without any addition of organic solvent or water. The internal heating of the water within the sample distends its cells and leads to rupture of the glands and oleiferous receptacles. This process frees essential oil, which is evaporated by the *in situ* water of plant material.

A cooling system outside the microwave oven continuously condenses the vapors, which are collected in specific glassware. The excess of water is refluxed back to the extraction vessel in order to restore the *in situ* water to the sample.

The microwave isolation offers a net advantage in terms of yield and better oil composition. Furthermore, it is environmentally friendly. In this method, low-boiling-point hydrocarbon compounds undergo decomposition (Marie et al., 2004; Ranjitha and Vijiyalakshmi, 2014).

### 22.3.4 Supercritical Fluid Extraction

Supercritical fluid extraction is used for the extraction of flavors and fragrances. SFE is a separation technology that uses supercritical fluid as solvent. Every fluid is characterized by a critical point, which is defined in terms of the critical temperature and critical pressure. Fluids cannot be liquefied above the critical temperature regardless of the pressure applied, but may reach the density close to the liquid state. A substance is considered to be a supercritical fluid when it is above its critical temperature and critical pressure. Several compounds have been examined as SFE solvents (e.g., hexane, pentane, butane, nitrous oxide, sulfur hexafluoride, and fluorinated hydrocarbon).

However, the main supercritical solvent used is carbon dioxide. Carbon dioxide (critical condition 30.9°C and 73.8 bar) is cheap, environmentally friendly, and generally recognized as safe. Supercritical carbon dioxide is also attractive because of its high diffusivity and easily tunable solvent strength. Another advantage is that carbon dioxide is gaseous at room temperature and ordinary pressure, which makes analyte recovery very simple and solvent-free (Taylor, 1996; Ranjitha and Vijiyalakshmi, 2014).

### 22.3.5 Ultrasound-Assisted Extraction

The mechanical effect of ultrasound accelerates the release of organic compounds within the plant body due to cell wall disruption, mass transfer intensification, and easier access of the solvent to the cell content. Ultrasound-assisted extraction (UAE) is reported to be one of the important techniques for extracting valuable compounds from the vegetable material (Vilkhu et al. 2008). General ultrasonic devices are the ultrasonic cleaning bath and ultrasonic probe system.

The efficiency of UAE depends on various factors, such as the nature of the tissue being extracted, the location of the component to be extracted, the treatment of the tissue prior to extraction, the effect of ultrasonics, the surface mass transfer, and intraparticle diffusion.

UAE can extract analytes under a concentrated form and free from any contaminants or artifacts. It also demonstrates advantages in terms of yield, selectivity, operating time, energy input, and preservation of thermolabile compounds (Shams et al., 2015).

## 22.4 Composition of Oil

The essential oil contains a complex mixture consisting mainly of oxygenated monoterpenes such as menthone, pulegone, neomenthol, and 8-hydroxy-$\partial$-4(5)-p-menthen-3-one. In some studies of *M. pulegium* essential oil, however, pulegone does not occur as a major constituent. In the essential oil of *M. pulegium* from Austria, piperitone was found to be the main component, followed by limonene, menthone, and neomenthone, while pulegone was not even detected. This type of oil is sometimes described as the piperitone–piperitenone chemotype. Chemotypes are defined as individual plants producing distinctive dominant secondary compounds (Santesson, 1968; Keefover-Ring et al., 2009). These are usually qualitative designations and arbitrary. According to the essential composition, the existence of nine chemotypes has been reported for *Mentha* plants. The main components of the nine chemotypes of the genus *Mentha* are (1) pulegone/isopulegone, (2) piperitone/piperitone oxide, (3) geraniol and/or geranyl acetate, (4) linalool and/linalool acetate, (5) carvone/dihdrocarvone, (6) menthofuran, (7) menthone/isomenthone/menthol isomers, (8) piperitone oxide/piperitone oxide, and (9) pulegone/menthone/isomenthone.

There is great variability in the chemical composition of *M. pulegium* essential oil among the studies performed so far (Pino et al., 1996). The variation of yields and chemical composition of essential oils depend on several factors, such as the method of extraction, the used plant parts, the products and reagents used for extraction, the environment, the plant genotype, the geographical origin, the harvest period of the plant, the degree of drying, the drying conditions, the temperature and drying time, and the presence of parasites, viruses, and weeds (Karousou et al., 2005; Kelen et al., 2008). The major components of pennyroyal oil (*M. pulegon*) reported in different studies are given in Table 22.1.

## 22.5 Properties of Pennyroyal Oil

### 22.5.1 Physicochemical Properties

The volatile oil obtained by steam distillation of *M. pulegium* is a light yellow liquid having a mint-like odor. As per the Committee on Food Chemicals Codex (2004), the physicochemical properties of pennyroyal oil are as given in Table 22.2.

**TABLE 22.1**

Major Components of Pennyroyal Oil (*M. pulegon*)

| Serial No. | Compound | Origin | Parts of Plant | Sample Period | Extraction Method | Content % | Reference |
|---|---|---|---|---|---|---|---|
| 1 | Pulegon | Iran | Aerial parts | Summer | Hydrodistillation | 37.8 | Aghel et al., 2004 |
| | | Jammu and Kashmir, India | Aerial parts | July 2001–2003 | Hydrodistillation | 42.9–45.4 | Agnihotri et al., 2005 |
| | | Austria | Plant | nd | nd | 25.1 | Zwaving and Smith, 1971 |
| | | Algeria | Aerial parts | nd | Hydrodistillation | 4.4–87.3 | Beghidja et al., 2007 |
| | | Turkey | nd | Summer, 1993 | Steam distillation | 20.5 | Mueller-Riebau et al., 1995 |
| | | Portugal | Flowers and leaves | July | Hydrodistillation | 78.3–80.9 | Reis-Vasco et al., 1999 |
| | | Greece | Aerial parts | nd | nd | 0.1–90.7 | Kokkini et al., 2004 |
| | | Morocco | Leaves | nd | nd | 6.5 | Elhoussine et al., 2010 |
| | | Egypt | nd | nd | nd | 88.1 | Aziz and Craker, 2009 |
| | | Brazil | nd | nd | nd | 28.4–61.4 | Oliveira et al., 2011 |
| 2 | Piperitone | Iran | Aerial parts | Summer | Hydrodistillation | 0.0–97.2 | Aghel et al., 2004 |
| | | Jammu and Kashmir, India | Aerial parts | July 2001–2003 | Hydrodistillation | 0.1–26.7 | Agnihotri et al., 2005 |
| | | Austria | Plant | nd | nd | 70.0 | Zwaving and Smith, 1971 |
| | | Algeria | Aerial parts | nd | Hydrodistillation | 0.1–19.2 | Beghidja et al., 2007 |
| | | Turkey | nd | Summer, 1993 | Steam distillation | nd | Mueller-Riebau et al., 1995 |
| | | Portugal | Flowers and leaves | July | Hydrodistillation | nd | Reis-Vasco et al., 1999 |
| | | Greece | Aerial parts | nd | nd | 0.1–39.8 | Kokkini et al., 2004 |
| | | Morocco | Leaves | nd | nd | 35.6 | Elhoussine et al., 2010 |
| | | Egypt | nd | nd | nd | nd | Aziz and Craker, 2009 |
| | | Brazil | nd | nd | nd | nd | Oliveira et al., 2011 |

*(Continued)*

**TABLE 22.1 (CONTINUED)**

Major Components of Pennyroyal Oil (*M. pulegon*)

| Serial No. | Compound | Origin | Parts of Plant | Sample Period | Extraction Method | Content % | Reference |
|---|---|---|---|---|---|---|---|
| 3 | Piperitone oxide | Iran | Aerial parts | Summer | Hydrodistillation | nd | Aghel et al., 2004 |
| | | Jammu and Kashmir, India | Aerial parts | July 2001–2003 | Hydrodistillation | 0.1–26.7 | Agnihotri et al., 2005 |
| | | Austria | Plant | nd | nd | nd | Zwaving and Smith, 1971 |
| | | Algeria | Aerial parts | nd | Hydrodistillation | 0.1–19.2 | Beghidja et al., 2007 |
| | | Turkey | nd | Summer, 1993 | Steam distillation | nd | Mueller-Riebau et al., 1995 |
| | | Portugal | Flowers and leaves | July | Hydrodistillation | nd | Reis-Vasco et al., 1999 |
| | | Greece | Aerial parts | nd | nd | 0.1–39.8 | Kokkini et al., 2004 |
| | | Morocco | Leaves | nd | nd | 21.2 | Elhoussine et al., 2010 |
| | | Egypt | nd | nd | nd | nd | Aziz and Craker, 2009 |
| | | Brazil | nd | nd | nd | nd | Oliveira et al., 2011 |
| 4 | Isomenthone | Iran | Aerial parts | Summer | Hydrodistillation | nd | Aghel et al., 2004 |
| | | Jammu and Kashmir, India | Aerial parts | July 2001–2003 | Hydrodistillation | 3.8–4.0 | Agnihotri et al., 2005 |
| | | Austria | Plant | nd | nd | nd | Zwaving and Smith, 1971 |
| | | Algeria | Aerial parts | nd | Hydrodistillation | nd | Beghidja et al., 2007 |
| | | Turkey | nd | Summer, 1993 | Steam distillation | nd | Mueller-Riebau et al., 1995 |
| | | Portugal | Flowers and leaves | July | Hydrodistillation | nd | Reis-Vasco et al., 1999 |
| | | Greece | Aerial parts | nd | nd | 4.3–28.6 | Kokkini et al., 2004 |
| | | Morocco | Leaves | nd | nd | nd | Elhoussine et al., 2010 |
| | | Egypt | nd | nd | nd | 5.8 | Aziz and Craker, 2009 |
| | | Brazil | nd | nd | nd | nd | Oliveira et al., 2011 |

*Note:* nd, not determined.

**TABLE 22.2**

Physicochemical Properties of Pennyroyal Oil

| Serial No. | Parameter | Value | Reference |
|---|---|---|---|
| 1 | Angular rotation | +18 to + 25° | FCC, 1996 |
| 2 | Specific gravity at 20°C | 0.928–0.940 | FCC, 1996 |
| 3 | Refractive index at 20°C | 1.483–1.488 | FCC, 1996 |
| 4 | Solubility | 1 ml is soluble in 2 ml of 70% alcohol | FCC, 2004 |

## 22.5.2 Biological Properties

Pennyroyal essential oil possesses various useful properties, which can be classified as follows:

1. Antimicrobial properties
2. Aromatic characteristics
3. Antioxidant properties
4. Insecticidal properties
5. Medicinal characteristics

### 22.5.2.1 Antimicrobial Properties

Antimicrobial properties of pennyroyal essential oil can be due to the existence of some groups like pulegone, menthone, and neo-menthone. The EO can disrupt the structure of different layers in polysaccharides, fatty acids, and phospholipids of the bacterial membrane by changing the permeability of the cellular membrane and damaging the bacterial cell wall (Nobakht et al., 2011; Teixeira et al., 2012). Different researchers have demonstrated that pennyroyal essential oil has a strong antimicrobial property against microorganisms such as *Salmonella typhimurium*, *Listeria monocytogenes*, *Escherichia coli*, *Bacillus cereus*, *Clostridium perfrigens*, *Staphylococcus aureus*, *Helicobacter pylori*, *Pseudomonas aeruginosa*, *Klebsiela pneumoniae*, and *Brochothrix thermosphacta*. Among these microorganisms, the most and least antimicrobial properties have been observed for *S. typhimurium* and *B. thermosphacta*, respectively (Teixeira et al., 2012).

### 22.5.2.2 Aromatic Characteristics

Pennyroyal essential oils are cyclohexenes and aromatic. Pulegone is the original compound of these oils that has a specific mint aroma (from intense to balsamic and pungent). However, the quality and quantity of the aroma vary, depending on the variety, growth conditions, weather conditions, amount of sun, and lack of water. High temperature, lack of water, and intense sun in summers make the herb produce more monoterpenes (like menthone and isomenthone). Pulegone, besides its isomers isopulegone and piperiton, can have an important role in producing the mint aroma. Some other compounds, such as 3-octanol and noneone, have been observed in pennyroyal, so that both of them, and especially 3-octanol, have a huge role in the aromatic characteristics (Díaz-Maroto et al., 2007). The method of volatiles extraction has a considerable impact on the quality and quantity of the aroma. Pennyroyal essential oil extracted by supercritical fluid extraction is better than that extracted by the other methods (Reis-Vasco et al., 1999).

### 22.5.2.3 Antioxidant Properties

Pennyroyal essential oils have an antioxidant property due to the existence of phenolic compounds such as flavonoids, phenolic acids, tannins, and phenolic diterpenes. The main components are pulegone and menthone. The essential oil of *M. pulegium* exerted a strong antioxidant activity that is almost equal to that of butylated hydroxytoluene (Ouakouak et al., 2015).

### 22.5.2.4 Insecticidal and Insect Repellency Properties

Essential oils show good potential against insect and mite pests. They have shown effectiveness by fumigation and topical application, besides having antifeedant and repellent properties (Regnault-Roger, 1997). The essential oil of pennyroyal wards off fleas, ants, lice, mosquitoes, ticks, and moths (Koul et al., 2008). Pennyroyal oil has effective insecticidal activity due to its chemical composition in general due to monoterpenes but also due to the synergestic effect of some minor consituents (Zekri et al., 2013). The components, mainly pulegone, piperitenone, thymol, limonene, and pionene, show important insecticidal activity against several pests (Frazios et al., 1997; Boughdad et al., 2011). The fumigant effect of this oil against adults of *Sitophilus oryzae* is due to pulegone (Zekri et al., 2013). The toxic effect of *M. pulegium* essential oil against *S. oryzae* is through inhibition of neurotransmitters (acetylcholinesterase and octopamine) (Price and Berry, 2006; López and Pascual-Villalobos, 2010; Zekri et al., 2013; Enan et al., 1998; Hollingworth et al., 1964).

Zekri et al. (2013) evaluated the insecticidal activity of *M. pulegium* essential oil against stored cereal pests. Pennyroyal oil has shown an important fumigant effect against *S. oryzae* (L.) adults. At 2 µl of essential oil per liter of air, insects have been totally decimated on the fourth day. The calculated lethal concentration ($LC_{50}$) and $LC_{99}$ vary according to the time. They decrease gradually as the fumigation time increases. The extreme lethal concentrations vary from 2.65 to 0.044 µl/L of air and from 143.9 to 0.518 µl/L of air, respectively. The effectiveness of the pennyroyal oil may be attributed to its chemical composition generally and monoterpene particularly. The fumigant effect of this oil against adults of *S. oryzae* could be explained by the high content of pulegone and piperitenone without ignoring the synergistic role of the minor compounds. The essential oils extracted from the mint species *M. pulegium* and *Mentha spicata*, together with their main constituents, pulegone, menthone, and carvone, were tested for insecticidal and genotoxic activities on *Drosophila melanogaster* (Franzios et al., 1997). The EOs of both aromatic plants showed strong insecticidal activity, while only the oil of *M. spicata* exhibited a mutagenic one. Among the constituents studied, the most effective insecticide was found to be pulegone, while the most effective for genotoxic activity was menthone.

### 22.5.2.5 Medicinal Characteristics

Pennyroyal essential oil has been used as an abortifacient and may induce menstruation (Conway and Slocumb, 1979). It has been useful for treatments in some conditions and diseases. Pennyroyal can be consumed as an antiflatulent, antitussive, diuretic, or anti-inflammatory. It is used for the treatment of infant's chronic diarrhea, asthma, and candidiasis. It has also been applied as an anticonvulsive, antiemetic, antinausea, antistress, heart stimulant, or sedative medication in traditional medicine (Vallnce and Edin, 1955; Dietz and Bolton, 2010; Nobakht et al., 2011; Da Rocha et al., 2012).

The consumption of pennyroyal essential oil in high to very high doses in humans can cause disorders, such as gastritis, convulsions, disseminated intravascular coagulation, hepatotoxicity, pulmonary toxicity, renal failure, or central nervous system (CNS) toxicity; coma; and even death (Anderson et al., 1996; Sztajnkrycer et al., 2003; Da Rocha et al., 2012).

Only a few investigations have been conducted to survey pennyroyal essential oil's stability over time. The effects of various factors on essential oil in general are discussed here to understand the storage stability of pennyroyal oil.

- *Light*: Ultraviolet light and visible light are considered to accelerate the autoxidation process. Monoterpenes have been seen to degrade rapidly under the influence of light (Misharina et al., 2003).

- *Temperature*: Ambient temperature crucially influences essential oil stability in several respects. Generally, chemical reactions accelerate with temperature and the rate of reaction expressed by the Arrhenius equation. Hence, both autoxidation and decomposition of hydroperoxide advance with increasing temperature. Contrarily, lower temperature favors the solubility of oxygen in liquids, which in turn may negatively affect essential oil stability. Concerning hydroperoxide stability, it was revealed that at low or moderate temperature, hydroperoxide formation predominated the decomposition rate, while the opposite is true at 50°C (Aidos et al., 2002).

- *Oxygen availability*: Oxygen consumption during storage of different monoterpenes has been recorded, and changes in composition as well as physicochemical properties of essential oil were generally more pronounced in half-filled containers than when only little or no headspace was present (El-Nikeety et al., 1998). Oil oxidation accelerates with the concentration of dissolved oxygen, which in turn depends on oxygen partial pressure in the headspace, as well as ambient temperature.

- *Metal contaminants*: Upon distillation in primitive stills or during storage in metallic containers, impurities of metals can be released into essential oils. Equal to light and heat, heavy metals, especially copper and ferrous ions, are considered to promote autoxidation in particular if hydroperoxides are already present.

Altogether, reliable and comprehensive studies on pennyroyal essential oil storage are rarely found, and concrete specifications on appropriate storage conditions, as well as shelf life, have not been clearly defined to date.

## 22.6 General Uses

The aerial parts of this plant contain a wide diversity of secondary metabolites, such as tannins, resins, pectins, and essential oils. Fresh or dried leaves and flowering tops are commonly used for their healing and culinary properties.

## 22.7 Advantage of Pennyroyal Oil as a Pesticide

The constituents of pennyroyal oil are selective and have little or no harmful effect on the environment and the nontarget organisms. Essential oils have received much attention as potentially useful bioactive compounds against insects, showing a broad spectrum of activity. They have low mammalian toxicity, degrade rapidly in the environment, and are locally available. The aromatic characteristic of essential oils present a variety of functions for the plants, together with repelling or attracting insects utilizing chemical components in the oil as defense resources. Essential oil could be a useful alternative to synthetic insecticides in the production of organic food.

## 22.8 Constraints of Pennyroyal as a Pesticide

In terms of specific constraints, the efficacy of these materials falls short when compared with synthetic pesticides. Essential oils also require somewhat greater application rates (as high as 1% active ingredients) and may require frequent reapplication when used outdoors. The commercial application of plant essential oil–based pesticide has challenges, like sufficient quantities of plant material, standardization and refinement of the pesticide product, protection of technology, and regulatory approval. In addition, as the chemical profile of plant species can vary naturally depending on geographic, genetic, climatic annual, or seasonal factors, pesticide manufacturers have to take additional steps to ensure that their product will perform consistently. All this requires substantial costs, and smaller companies are not willing to invest the required funds unless there is a high probability of recovering the costs through some form of market exclusivity. Finally, once all these issues are addressed, regulatory approval is required (Mohan et al., 2011).

## 22.9 Conclusion

Pennyroyal oil consists of a diverse array of bioactive compounds and exhibits a wide range of activities, such as antimicrobial, antioxidative, and insecticidal activities. Therefore, this oil has great potential not only in food, pharmaceuticals, and cosmetics, but also in insect repellents, insecticides, and a variety of other ways to control a large number of pests. Pennyroyal oil has huge potential as an alternative to synthetic pesticides in stored cereal and crop protection, although the efficacy of these oil constituents is comparatively less than that of the synthetic pesticides. However, it is gaining momentum as far as environmental pollution and human health are concerned. It is expected that the innovative formulations of pesticides based on this essential oil will find their greatest commercial applications in urban pest control, vector control vis-à-vis human health, and pest control in agriculture, and will help in organic food production systems.

## References

Aghel, N., Yamini, Y., Hadjiakhoondi, A., Pourmortazavi, S.M. Supercritical carbon dioxide extraction of *Mentha pulegium* L. essential oil. *Talanta* 62(2), 407–411, 2004.

Agnihotri, K., Agarwal, S.G., Dhar, L., Thappa, R.K., Baleshwar, M., Kapahi, B.K., Saxena, R.K., Gazi, G. Essential oil composition of *Mentha pulegium* L. growing wild in the north-western Himalayas, India. *Flavour Fragr. J.* 20(6), 607–610, 2005.

Aidos, I., Lourenclo, S., Van der Padt, A., Luten, J.B., and Boom, R.M. Stability of crude herring oil produced from fresh byproducts influence of temperature during storage. *J. Food Sci.* 67, 3314–3320, 2002.

Anderson, I.B., Mullen, W.H., Meeker, J.E. Pennyroyal toxicity: Measurement of toxic metabolite levels in two cases and review of the literature. *Ann. Intern. Med.* 124, 726–734, 1996.

Aziz, E.E., Craker, L.E. Essential oil constituents of peppermint, pennyroyal, and apple mint grown in a desert agrosystem. *J. Herbs Spices Med. Plants* 15(4), 361–367, 2009.

Beghidja, N., Bouslimani, N., Benayache, F., Benayache, S., Chalchat, J.C. Composition of the oils from *Mentha pulegium* grown in different areas of the east of Algeria. *Chem. Nat. Compd.* 43(4), 481–483, 2007.

Boughdad, A., Elkasimi, R., and Kharchafi, A. Activité insecticide des huiles essentielles de *Mentha Sur Callosobrochus maculatus* (F) (Coleoptera, Bruchidea). AFPP—Neuvième Conférence Internationale sur les ravageurs en Agriculture. Montpellier, October 26–27, 2011, 9 p.

Committee on Food Chemicals Codex. Food Chemicals Codex, 5th edition. Washington, DC: National Academy Press, 2004.

Conway, G.A., Slocumb, J.C. Plants used as abortifacients and emmenagogues by Spanish New Mexicans. *J. Ethnopharmacol.* 1(3): 241–261, 1979.

Da Rocha, M.S., Dodmane, P.R., Arnold, L.L., Pennington, K.L., Anwar, M.M., Adams, B.R., Taylor, S.V., Wermes, C., Adams, T.B. Mode of action of pulegone on the urinary bladder of F344 rats. *Toxicol. Sci.* 128, 1–8, 2012.

Díaz-Maroto, C., Castillo, N., Castro-Vázquez, L., González-Viñas, M., Pérez-Coello, S. Volatile composition and olfactory profile of pennyroyal (*Mentha pulegium* L.) plants. *Flavour Fragr. J.* 22, 114–118, 2007.

Dietz, B.M., Bolton, J.L. Biological reactive intermediates (BRIs) formed from botanical dietary supplements. *Chem. Biol. Interact.* 192, 72–80, 2010.

Elhoussine, D., Zineb, B., Abdellatif, B. GC/MS analysis and antibacterial activity of the essential oil of *Mentha pulegium* grown in Morocco. *Res. J. Agric. Biol. Sci.* 6(3), 191–198, 2010.

El-Nikeety, M.M.A., El-Akel, A.T.M., Abd El-Hady, M.M.I., Badei A.Z.M. Changes in physical properties and chemical constituents of parsley herb volatile oil during storage. *Egypt. J. Food Sci.* 26(28), 35–49, 1998.

Enan, E., Beigler, M., Kende, A. Insecticidal actions of terpenes and phenols to cockroaches: Effect on octopamine receptors. In Proceedings of the International Symposium on Plant Protection, Ghent, Belgium, 1998.

Franzios, G., Mirotsou, M., Hatziapostolou, E., Kral, J., Scouras, Z.G., Mavragani-Tsipidou, P. Insecticidal and genotoxic activities of mint essential oils. *J. Agric. Food Chem.* 45, 2690–2694, 1997.

Grieve, M. *A Modern Herbal*. Hafner, New York, 1959.

Hollingworth, R.M., Johnstone, E.M., Wright, N. Aspect of the biochemistry and toxicology of octopamine in arthropods. In Magee, P.S., Kohn, G.K., Menn, J.J. (eds.), *Pesticide Synthesis through Rational Approaches*. ACS Symposium Series 255. Washington, DC: American Chemical Society, 1984, pp. 103–125.

Karousou, R., Koureas, D., Kokkini, S. Essential oil composition is related to the natural habitats: *Coridothymus capitatus* and *Satureja thymbra* in NATURA 2000 sites of Crete. *Phytochem.* 66: 2668–2673, 2005.

Keefover-Ring, K., Thompson, J.D., Linhart, Y.B. Beyond six scents: Defining a seventh *Thymus vulgaris* chemotype new to southern France by ethanol extraction. *Flavour Fragr. J.* 24, 117–122, 2009.

Kelen, M., Tepe, B. Chemical composition, antioxidant and antimicrobial properties of the essential oils of three Salvia species from Turkish flora. *Bioresource Technol.* 99, 4096–4104, 2008.

Kokkini, S., Hanlidou, E., Karousou, R., Lanaras, T. Clinical variation of *Mentha pulegium* essential oils along the climatic gradient of Greece. *J. Essent. Oil Res.* 16(6), 588–593, 2004.

Koul, O., Walia, S., Dhaliwal, G.S. Essential oil as green pesticides: Potential and constraints. *Biopestic. Int.* 4, 63–84, 2008.

López, M.D., Pascual-Villalobos, M.J. Mode of inhibition of acetylcholinesterase by monoterpenoids and implications for pest control. *Ind. Crops Prod.* 31, 284–288, 2010.

Masango, P. Cleaner production of essential oils by steam distillation. *J. Cleaner Prod.* 29(1), 171–176, 2005.

Misharina, T.A., Polshkov, A.N., Ruchkina, E.L., Medvedeva, I.B. Changes in the composition of essential oil of marjoram during storage. *Appl. Biochem. Microbiol.* 39, 311, 2003.

Mohan, M., Haider, S.Z., Andola, H.C., Purohit, V.K. Essential oils as green pesticides; for sustainable agriculture. *Res. J. Pharm. Biol. Chem. Sci.* 2(4), 100–106, 2011.

Mueller-Riebau, F., Berger, B., Yegen, O. Chemical composition and fungitoxic properties to phytopathogenic fungi of essential oils of selected aromatic plants growing wild in Turkey. *J. Agric. Food Chem.* 43(80), 2262–2266, 1995.

Nobakht, A., Norani, J., Safamehr, A. The effects of different amounts of *Mentha pulegium* L. (pennyroyal) on performance, carcass traits, hematological and blood biochemical parameters of broilers. *J. Med. Plant. Res.* 5, 3763–3768, 2011.

Oliveira, R.A., Sa, I.C.G., Duarte, L.P., Oliveira, F.F. Constituintes voláteis de *Mentha pulegium* L. e *Plectranthus amboinicus* (Lour.) Spreng. *Rev. Bras. Plant. Med.* 13(2), 165–169, 2011.

Pino, J.A., Rosado, A., Fuentes, V. Chemical composition of the essential oil of *Mentha pulegium* L. from Cuba. *J. Essent. Oil Res.* 8(3), 295–296, 1996.

Polunin, O. *Flowers of Europe*. Oxford University Press, London, 1969.

Price, D.N., and Berry, M.S. Comparison of effects of octopamine and insecticidal essential oils on activity in the nerve cord, foregut, and dorsal unpaired median neurons of cockroaches. *J. Insect Physiol.* 52, 309–319, 2006.

Regnault-Roger, C. The potential of botanical essential oils for insect pest control. *Integrated Pest Manag. Rev.* 2, 25–34, 1997.

Quakouak, H., Chohra, M., Denane, M. Chemical composition, antioxidant activities of essential oil of *Mentha pulegium* L. *Int. Lett. Nat. Sci.* 39, 49–55, 2015.

Ranjitha, J., and Vijiyalakshmi, S. Facile methods for the extraction of essential oil from the plant species—A review. *Int. J. Pharm. Sci. Res.* 5(4), 1107–1115, 2014.

Reis-Vasco, E.M.C., Coelho, J.A.P., Palavra, A.M.F. Comparison of pennyroyal oils obtained by supercritical $CO_2$ extraction and hydrodistillation. *Flavour Fragr. J.* 14(3), 156–160, 1999.

Santesson, R., *Svensk Some aspects of lichen taxonomy, Naturvetenskap* 21, 176–184, 1968.

Shah, G., Shri, R., Panchal, V., Sharma, N., Singh, B., Mann, A.S. Scientific basis for the therapeutic use of *Cymbopogon citrates* Stapf (lemongrass). *J. Adv. Pharm. Technol. Res.* 2, 3–8, 2011.

Shaikh, S., Yaacob, H.B., Rahim, Z.H.A. Prospective role in treatment of major illnesses and potential benefits as a safe insecticide and natural food preservative of mint (*Mentha* spp.): A review. *Asian J. Biomed. Pharm. Sci.* 4(35), 1–12, 2014.

Shams, K.A., Nahal, S.A., Ibrahim, A.S., Mohamed-Elamir, F.H., Mostafa, M.E., Faiza, M.H. Green technology: Economically and environmentally innovative methods fo extraction of medicinal and aromatic plants (MAP). *J. Chem. Pharm. Res.* 7(5), 1050–1074, 2015.

Sztajnkrycer, M.D., Otten, E.J., Bond, G.R., Lindsell, C.J., Goetz, R.J. Mitigation of pennyroyal oil hepatotoxicity in the mouse. *Acad. Emerg. Med.* 10, 1024–1028, 2003.

Taylor, L.T. *Supercritical Fluid Extraction*. John Wiley & Sons, New York, 1996.

Teixeira, B., Marques, A., Ramos, C., Batista, I., Serrano, C., Matos, O., Neng, N.R., Nogueira, J.M.F., Saraiva, J., Nunes, M.L. European pennyroyal (*Mentha pulegium*) from Portugal: Chemical composition of essential oil and antioxidant and antimicrobial properties of extracts and essential oil. *Ind. Crop Prod.* 36, 81–87, 2012.

Vallnce, W.B., Edin, M.B. Pennyroyal poisoning, a fatal case. *Lancet* 269, 850–851, 1955.

Vilkhu, K., Mawson, R., Simons, L., Bates, D. Applications and opportunites for ultrasound assisted extraction in the food industry—A review *Innov. Food Sci. Emerg. Technol.* 9, 161–169, 2008.

William, B.J. The origin of the Soxhlet Extractor. *J. Chem. Educ.* 84(12): 1913, 2007.

Zekri, N., Amlich, S., Boughdad, A., El Belghiti, A.M., Zair, T. Phytochemical study and insecticidal activity of *Mentha pulegium* L. oils from Morocco against *Sitophilus oryzae*. *Mediterr. J. Chem.* 2(4), 607–619, 2013.

Zwaving, J.H., Smith, D. Composition of the essential oil of Austrian *Mentha pulegium*. *Phytochemistry* 10(8), 1951–1953, 1971.

# 23

# Pesticidal Activity of Different Essential Oils

Leo M.L. Nollet

## CONTENTS

In the different chapters of Section II, "Essential and Vegetable Oils," a number of essential oils (EOs) are discussed in depth. Of course, the number of essential oils or plant species is much greater. In this chapter, first the pesticidal activities of some genera, *Lantana*, *Allium*, *Mentha*, and *Juniperus*, are detailed. Next, recent studies on essential oils of a wide range of plants are discussed. Finally, special attention is given to Chinese herbs.

## 23.1 *Lantana*

Dua et al. [1] investigated the insecticidal activity of essential oil isolated from the leaves of *Lantana camara* against mosquito vectors. The essential oil was isolated from the leaves of *L. camara* using the hydrodistillation method. A bioassay test was carried out by the World Health Organization's (WHO) method for determination of adulticidal activity against mosquitoes. $LD_{50}$ values of the oil were 0.06, 0.05, 0.05, 0.05, and 0.06 mg/cm$^2$, while $LD_{90}$ values were 0.10, 0.10, 0.09, 0.09, and 0.10 mg/cm$^2$ against *Aedes aegypti*, *Culex quinquefasciatus*, *Anopheles culicifacies*, *Anopheles fluvialitis*, and *Anopheles stephensi*, respectively. $KDT_{50}$ values of the oil were 20, 18, 15, 12, and 14 min, and $KDT_{90}$ values were 35, 28, 25, 18, and 23 min against *Ae. aegypti*, *C. quinquefasciatus*, *An. culicifacies*, *An. fluviatilis*, and *An. stephensi*, respectively, on 0.208 mg/cm$^2$ impregnated paper. Studies on the persistence of essential oil of *L. camara* on impregnated paper revealed that it has more adulticidal activity for longer periods at low storage temperature. Gas chromatography–mass spectrometry (GC-MS) analysis of the essential oil showed 45 peaks. Caryophyllene (16.37%), eucalyptol (10.75%), α-humelene (8.22%), and germacrene (7.41%) were present in major amounts and contributed 42.75% of the total constituents.

The biological properties of verbascoside are numerous and include antimicrobial activities. Leaf extracts of *Lippia javanica* Spreng. and *L. camara* Linné (Verbenaceae) were partially purified using column chromatography and high-speed centrifugal countercurrent chromatography, the latter yielding fractions with higher purity (71%) than those from a single-column chromatographic separation (38%–44% pure) [2]. Verbascoside remained stable upon heating, but was completely decomposed after 4 h exposure of the extract to sunlight. Compared with the other storage conditions, the compound was best preserved in a dry form in the dark. Analysis by high-performance liquid chromatography revealed that the verbascoside content of plant parts of *L. camara* from natural populations was highly variable, both within and between populations. However, several specimens produced high levels of the compound (Hazyview, Plant 3 [83.0 mg/g dry weight], Magoebaskloof 2, Plant 5 [64.8 mg/g], and White River, Plant 2 [64.0 mg/g]), suggesting that *L. camara* is an excellent source of verbascoside. Extracts of the plant displayed effective *in vivo* inhibition of *Penicillium digitatum* on oranges.

Diaz Napal et al. [3] screened the activity of plant extracts derived from 89 species native to Argentina, against the leaf-cutting ant *Acromyrmex lundi* (Guérin) and its mutualistic fungus, *Leucoagaricus gongylophorus*, through a pickup assay and bioautography, respectively. The pickup assay revealed moderate to strong antiforaging activity for just over 13.5% of the assayed species, including complete ant foraging inhibition for *Aristolochia argentina*, *Flourensia oolepis*, *Gaillardia megapotamica*, *Lantana grisebachii*, and *Lithrea molleoides*. Most plant extracts were well tolerated by fungi, with only 12.3% of the species tested showing some degree of fungus growth inhibition. Among these, *A. argentina*, *F. oolepis*, and *Pterocaulon alopecuroides* were the strongest inhibitors, whereas *Baccharis flabellata*, *Dalea elegans*, and *Zanthoxylum coco* revealed a more moderate activity. Only *A. argentina* and *F. oolepis* extracts showed strong antiforaging effects and affected fungus growth at the same time. Values of $IC_{50}$ and MIC indicated that extracts inhibiting ant foraging at lower concentrations did not necessarily also inhibit fungus growth at lower doses. The active principle of *A. argentina*, on both ant foraging and fungal growth, was identified as argentilactone.

Aromatic plants (i.e., *Citrus* spp., *Eucalyptus* spp., *L. camara*, *Ocimum* spp., and *L. javanica*) are commonly used in all the African studies of repellents against mosquito vectors [4]. Native people know three major methods of using repellent plants: (1) production of repellent smoke from burning plants, (2) hanging plants inside the house or sprinkling leaves on the floor, and (3) use of plant oils, juices from crushed fresh parts of the plants, or various prepared extracts applied on uncovered body parts.

The bioactivity of the essential oil extracted by hydrodistillation from *L. camara* leaves was assessed under laboratory conditions [5]. The composition of *L. camara* essential oil included large amounts of sesquiterpene, mainly β-caryophyllene (35.70%) and caryophyllene oxide (10.04%). The tested essential oil showed good fumigant activity within 1 week of exposure for all tested doses. Moreover, a remanence study confirmed that the oil was efficient during 2 weeks.

## 23.2 *Allium*

Junnila et al. [6] tested the efficacy of attractive toxic sugar bait (ATSB) with garlic oil microencapsulated in β-cyclodextrin as an active ingredient against *Aedes albopictus* in suburban Haifa, Israel. Two 3-acre gardens with high numbers of *Ae. albopictus* were selected for perimeter spray treatment with ATSB and ASB (bait containing no active ingredient). Baits were colored with food dye to verify feeding of the mosquitoes. The mosquito population was monitored by human landing catches and sweep net catches in the surrounding vegetation.

Experiments lasted for 44 days. Treatment occurred on day 13. The mosquito population collapsed about 4 days after treatment and continued to drop steadily for 27 days, until the end of the study. At the experimental site, the average pretreatment landing rate was 17.2 per 5 min. Two days posttreatment, the landing rate dropped to 11.4, and continued to drop to an average of 2.6 during the following 26 days. During the same period, the control population was stable. Few sugar-fed females (8%–10%) approached a human bait, and anthrone tests showed relatively small amounts of sugar within their crop or gut. Around 60%–70% of males caught near our human bait were sugar positive, which may indicate that the males were feeding on sugar for mating-related behavior. From the vegetation treated with the toxic bait, significantly fewer (about 10%–14%) males and females stained by ATSB than the ASB-treated control were recovered. This may indicate that the toxic baits alter the resting behavior of the poisoned mosquitoes within the vegetation. Almost no *Ae. albopictus* females (5.2 ± 1.4) approached human bait after treatment with ATSB.

Fresh garlic cloves were steam distilled to obtain the essential oil [7]. The garlic oil was tested for toxicity against the eggs, larvae, and adults of *Tribolium castaneum* and adults of *Sitophilus zeamais*. *T. castaneum* egg mortality increased with garlic oil concentration, with complete kill of eggs being achieved at 4.4 mg/cm$^2$, using the filter paper impregnation bioassay. The eggs were the most susceptible stage, followed by adults, 10-day-old larvae, and older larvae. *T. castaneum* adults were more susceptible to garlic oil than *S. zeamais* adults, with KD$_{50}$ values of 1.32 and 7.65 mg/cm$^2$, respectively. When rice and wheat were treated with garlic oil, eggs that were laid in the media failed to produce F1 progeny at concentrations of 2000 ppm in rice for *T. castaneum* and 5000 ppm in wheat for *S. zeamais*. The weights of F1 adults of *T. castaneum* and *S. zeamais* in treated media were not significantly different from those of the controls.

In order to develop biological control of aphids by a "push–pull" approach, intercropping using repellent emitting plants was developed in different crop and associated plant models [8]. Garlic is one of the potential plants that could be inserted in crops to decrease the pest occurrence in neighboring crop plots. Field works were conducted in wheat fields in Langfang Experimental Station, Hebei Province, China, from October 2009 to July 2010 during the wheat developmental season. The effect of wheat intercropping with garlic, but also the volatiles emission on the incidence of the English grain aphid, *Sitobion avenae*, was assessed. Natural beneficial occurrence and global yields in two winter wheat varieties that were susceptible or resistant to cereal aphid were also determined, comparing with control plots without the use of garlic plant intercrop or semiochemical releaser in the fields. *S. avenae* was found to be lower in garlic oil blend (GOB) treatment, diallyl disulfide (DD) treatment, and wheat–garlic intercropping (WGI) treatment that compared the control plots for both varieties. Both intercropping and application of volatile chemicals emitted by garlic could improve the population densities of natural enemies of cereal aphid, including ladybeetles and mummified aphids. The ladybeetle population densities in WGI and GOB and the mummified aphid densities in WGI and DD were significantly higher than those in control fields for both varieties. There were significant interactions between cultivars and treatments to the population densities of *S. avenae*. The 1000-grain weight and yield of wheat were also increased compared with the control.

Water-distilled essential oil from the dried bulbs of *Allium chinense* (Liliaceae) was analyzed by GC-MS [9]. Eighteen compounds, accounting for 98.4% of the total oil, were identified, and the main components of the essential oil of *A. chinense* were methyl allyl trisulfide (30.7%), dimethyl trisulfide (24.1%), methyl propyl disulfide (12.8%), and dimethyl disulfide (9.6%), followed by methyl allyl disulfide (3.4%) and methyl propyl trisulfide (3.6%). The essential oil exhibited contact toxicity against the booklice (*Liposcelis bostrychophila*) with a lethal concentration for 50% (LC$_{50}$) value of 441.8 µg/cm$^2$, while the two major constituents, dimethyl trisulfide and methyl propyl disulfide, had LC$_{50}$ values of 153.0 and 738.0 µg/cm$^2$ against the booklice, respectively. The essential oil of *A. chinense* possessed strong fumigant toxicity against the booklice with an LC$_{50}$ value of 186.5 µg/L, while methyl allyl trisulfide (LC$_{50}$ = 90.4 µg/L) and dimethyl trisulfide (LC$_{50}$ = 114.2 µg/L) exhibited stronger fumigant toxicity than methyl propyl disulfide (LC$_{50}$ = 243.4 µg/L) and dimethyl disulfide (LC$_{50}$ = 340.8 µg/L) against the booklice.

Dried bulbs of *A. sativum* were extracted with different solvents and evaluated for insecticidal, antimicrobial, and antioxidant activities.

Aqueous and methanol extracts showed the highest insecticidal activity (mortality rate of 81% and 64%, respectively) against the larvae of *Spodoptera litura* at a concentration of 1000 ppm [10]. With regard to antimicrobial activity, aqueous extract exhibited antibacterial activity against gram-positive (*Bacillus subtilis* and *Staphylococcus aureus*) and gram-negative (*Escherichia coli* and *Klebsiella pneumonia*) strains and antifungal activity against *Candida albicans*. While methanol extract showed antimicrobial activity against all the tested microorganisms except two (*S. aureus* and *C. albicans*), the extracts of hexane, chloroform, and ethyl acetate did not show any antimicrobial activity. The minimum inhibitory concentration of aqueous and methanol extracts against tested bacterial and fungal strains was 100–150 µg/ml. Antioxidant activity of the bulb extracts was evaluated in terms of inhibition of free radicals by 2,2'-diphenly-1-picrylhydrazyl. Aqueous and methanol extracts exhibited strong antioxidant activity (80%–90% of the standard).

Essential oil of *Allium macrostemon* was obtained by hydrodistillation and analyzed by gas chromatography (GC) and GC-MS [11]. The activities of the essential oil and its two major constituents were evaluated, using WHO procedures, against the fourth-instar larvae of *Ae. albopictus* for 24 h, and larval mortalities were recorded at various essential oil or compound concentrations ranging from 9.0 to 150 µg/ml.

The essential oil of *A. macrostemon* exhibited larvicidal activity against the early fourth-instar larvae of *Ae. albopictus* with an LC$_{50}$ value of 72.86 µg/ml. The two constituent compounds, dimethyl trisulfide and methyl propyl disulfide, possessed strong larvicidal activity against the early fourth-instar larvae of *Ae. albopictus*, with LC$_{50}$ values of 36.36 and 86.16 µg/ml, respectively.

### 23.3 *Mentha*

EO of Algerian *Mentha pulegium* L. leaves is obtained by steam distillation with a yield of 1.45% ± 0.01% and analyzed by GC-MS, where 39 compounds were identified [12]. The analysis has revealed that EO contains pulegone (70.66%) and neo-menthol (11.21%) as the major compounds. Antimicrobial study showed good activity tested against 11 bacteria (3 g+ and 8 g–) and 2 yeasts, and also showed good inhibitory (minimum inhibitory concentration [MIC]) and bactericidal (minimum bactericidal concentration [MBC]) properties.

Insecticide assessment was carried out against the food pest *Sitophilus granarius* (L.). Three toxicity tests of EO were performed using three different methods: contact, inhalation, and ingestion. Also, assessment of the toxicity of *M. pulegium* EO was carried out using leaves at different statuses (fresh, dried, and semidried), ground and nonground. Inhalation and ingestion methods were implemented on wheat seeds that showed high efficiency (100% mortality), and a lethal dose (LD$_{50}$ = 9.11 ± 2.53 µl/ml) was obtained using the contact method. Leaves in various statuses did not show any efficiency except crushed ones.

The essential oils of *M. pulegium* L. (MPE) and *Mentha rotundifolia* (L.) Huds (MRE), which are growing in Algeria, were prepared by hydrodistillation and their chemical compositions investigated by GC-MS [13]. The oils were tested for their antimicrobial activity using disc diffusion and spot assays, and antioxidant activity using the 2,2-Azinobis(3-ethylbenzothiazoline-6-sulfonic) acid (ABTS) test and Kit Radicaux Libres® (KRL) biological assay. Also, contact toxicity, fumigant toxicity, and repellency tests of these essential oils were evaluated against adults of *Rhyzopertha dominica* (F.) (the principal pest of wheat). The major components found in MPE are pulegone (70.4%), neo-menthol (13.4%), neo-menthol acetate (3.5%), and menthone (2.7%). On the other hand, MRE provided *trans*-piperitone epoxide (30.2%), piperitone oxide (8.7%), thymol (4.5%), germacrene D (3.5%), and terpinen-4-ol (2.7%) as major ingredients. MRE exhibited a stronger antimicrobial effect and antioxidant activity in the KRL test than MPE. In the contact assay, $DL_{50}$ values of MRE and MPE were 3.3 and 6.9 µl/ml, respectively. The fumigant toxicity assay of MPE and MRE showed mortality ratios of 39.2 and 44.3%, respectively, at the dose of 2 µl/ml. Moreover, at this dose and after 30 min exposure time, the repellent effect showed death rates of 46.03% and 47.54% for MPE and MRE, respectively.

See also Chapter 4 of this book.

## 23.4 *Juniperus*

The essential oil of *Juniperus procera* was evaluated against the larvae of *Anopheles arabiensis* under the laboratory and semifield conditions by adopting the World Health Organization standard protocols [14]. The larval mortality was observed for 24 h postexposure.

The essential oil of *J. procera* demonstrated varying degrees of larvicidal activity against *An. arabiensis*. The $LC_{50}$ and $LC_{90}$ values of *J. procera* were 14.42 and 24.65 mg/L, respectively, under laboratory conditions. However, under semifield conditions the $LC_{50}$ and $LC_{90}$ values of *J. procera* were 24.51 and 34.21 mg/L. The observations clearly showed that larval mortality rate is completely time and dose dependent compared with the control.

Sindhu et al. [15] report a new bioassay "syringe test" (modified larval immersion test [LIT]) for *in vitro* evaluation of acaricidal activity of crude plant extracts. Prepared syringes, containing eggs of tick, were incubated until 14 days after hatching of eggs, when the bioassay was performed on the larvae. $LC_{50}$, $LC_{90}$, and $LC_{99}$ values were calculated for each tested product. Ninety-five percent confidence intervals for $LC_{50}$ were very narrow, indicating a high degree of repeatability for the new bioassay on larvae of *Riphicephalus* (*Boophilus*) *microplus*. Bioassays were applied to six crude aqueous-methanol extracts from five plants (*Acacia nilotica*, *Buxus papillosa*, *Fumaria parviflora*, *Juniperus excelsa*, and *Operculina turpethum*), of which three showed discernible effects. Twenty-four hours postexposure, $LC_{99}$ values were 11.9% (w/v) for *F. parviflora* and 20.8% (w/v) and 29.2% (w/v) for *B. papillosa* and *A. nilotica*, respectively. After six days of exposure, these values were 9.1% (w/v), 9.2% (w/v), and 15.5% (w/v) for *F. parviflora*, *A. nilotica*, and *B. papillosa*, respectively.

A laboratory-based study has been conducted to evaluate the repellency of the Ethiopian ethnomedicinal plant Tedh (vernacular name [local native language, Amharic], *J. procera* [Cupressaceae]) against the Afro-tropical malarial vector *A. arabiensis* Patton at four different concentrations: 1.0, 1.5, 2.5, and 5.0 mg/cm² [16]. Experimentation on the percentage of protection in relation to the dosage has been performed. The tested concentrations of the essential oil of *J. procera* exhibited various degrees of repellency in terms of percentage of repellency and complete protection time against female *An. arabiensis*, that is, 1.0, 1.5, 2.5, and 5.0 mg/cm² (64.10% [92 min], 68.10% [125 min], 72.20% [190 min], and 80.60% [311 min], respectively). Student's *t*-test results showed a statistically significant difference between the treated and control groups.

The acaricidal effect of seven essential oils was examined *in vitro* against the cattle tick (*R. microplus*) [17]. Engorged female ticks were manually collected in farms of southern Brazil and placed into Petri dishes (*n* = 10) in order to test the following oils: juniper (*Juniperus communis*), palmarosa (*Cymbopogon martinii*), cedar (*Cedrus atlantica*), lemongrass (*Cymbopogon citratus*), ginger (*Zingiber officinale*), geranium (*Pelargonium graveolens*), and bergamot (*Citrus aurantium* var. *bergamia*) at concentrations of 1%, 5%, and 10% each. A control group was used to validate the tests containing Triton X-100 only. Treatment effectiveness was measured considering inhibition of tick oviposition (partial or total), egg's weight, and hatchability. *C. martinii*, *C. citratus*, and *C. atlantica* essential oils showed an efficacy higher than 99% at all concentrations tested. In addition, *J. communis*, *Z. officinale*, *P. graveolens*, and *C. aurantium* var *bergamia* oils showed efficiencies ranging from 73% to 95%, depending on the concentration tested, where higher concentrations showed greater efficacy.

## 23.5 Different EOs

The essential oils from accessions of *Lippia sidoides* (Verbenaceae) were characterized by GC and GC-MS and investigated for their acaricidal activity against the two-spotted spider mite (*Tetranychus urticae*) [18]. Twenty-nine compounds were identified with potential acaricidal activity. Glass receptacles were used as test chambers. For each dose and exposure time combination, three replicates were used. Each replicate consisted of 30 adult females of *T. urticae*, 10 mites in each leaf disc of *Canavalia ensiformis* placed in a Petri dish. Increasing amounts of oil or terpene were applied on a blotting paper strip, fixed on the inner surface of the glass recipient cover, corresponding to 2, 4, 6, 8, and 10 µl/L of air, respectively. Exposure periods were 24, 48, and 72 h. Data obtained in these experiments were submitted to probit analysis. The essential oil of *L. sidoides*, thymol, and carvacrol exhibited potent acaricidal activity against *T. urticae*.

Laboratory bioassays on insecticidal activity of essential oils extracted from six Mediterranean plants (*Achillea millefolium*, *Lavandula angustifolia*, *Helichrysum italicum*, *Foeniculum vulgare*, *Myrtus communis*, and *Rosmarinus officinalis*) were carried out against the larvae of the Culicidae mosquito *Ae. albopictus* [19]. The chemical composition of the six EOs was also investigated. All tested oils had insecticidal activity, with differences in mortality rates as a function of both oil and dosage. At the highest dosage (300 ppm), EOs from *H. italicum*, *A. millefolium*, and *F. vulgare* caused higher mortality than the other three oils, with mortality rates ranging from 98.3% to 100%. *M. communis* EO induced only 36.7% larval mortality at the highest dosage (300 ppm), a value similar to those recorded at the same dosage by using *R. officinalis* and *L. angustifolia* (51.7% and 55%, respectively). The analyzed EOs had a higher content of monoterpenoids (80%–99%) than sesquiterpenes (1%–15%), and they can be categorized into three groups on the basis of their composition. Few EOs showed the hydrocarbon sesquiterpenes, and these volatile compounds were generally predominant in comparison with the oxygenated forms, which were detected in lower quantities only in *H. italicum* (1.80%) and in *M. communis* (1%).

The efficacy of the essential oil and various organic extracts from flowers of *Cestrum nocturnum* L. was evaluated for controlling the growth of some important phytopathogenic fungi [20]. The oil (1000 ppm) and the organic extracts (1500 μg/disc) revealed antifungal effects against *Botrytis cinerea*, *Colletotrichum capsici*, *Fusarium oxysporum*, *Fusarium solani*, *Phytophthora capsici*, *Rhizoctonia solani*, and *Sclerotinia sclerotiorum* in the growth inhibition range of 59.2%–80.6% and 46.6%–78.9%, respectively, and their MIC values ranged from 62.5 to 500 μg/ml and 125 to 1000 μg/ml. The essential oil had a remarkable effect on spore germination of all the plant pathogens with concentration- and time-dependent kinetic inhibition of *P. capsici*. Further, the oil displayed remarkable *in vivo* antifungal effect up to 82.4%–100% disease suppression efficacy on greenhouse-grown pepper plants.

The acaricidal activity of a volatile essential oil hydrodistillate of *Satureja thymbra* L. (Lamiaceae) and its major constituents, carvacrol and γ-terpinene, was evaluated against field-collected unfed adult *Hyalomma marginatum* [21]. The distillate was tested against this tick species at 5, 10, 20, and 40 μl/L, while the two major components were each tested at 10 μl/L. Generally, tick mortality to the *S. thymbra* distillate increased with concentration and exposure time. Ticks exposed to vapors from cotton wicks containing at least 40 μl/L resulted in complete (100%) mortality at 3 h. The lower concentrations provided ≥90% mortality at 3 h posttreatment, with complete mortality at 24 h. Knockdown was observed only in the carvacrol and γ-terpinene treatments. Ticks exposed to carvacrol-treated wicks produced >93% knockdown at 3 h, but at 24 h, approximately 57% were dead. The γ-terpinene treatment produced ≥90% knockdown at 105 min through 3 h, but at 24 h, only about 87% of the ticks were dead.

Prakash et al. [22] report the essential oil of *Ocimum gratissimum* as a plant-based preservative and recommend its application as a nontoxic antimicrobial and antiaflatoxigenic agent against fungal and aflatoxin contamination of spices, as well as a shelf-life enhancer in view of its antioxidant activity. The EO exhibited antifungal activity against fungal isolates from some spices and showed better efficacy as a fungitoxicant than the prevalent fungicide Wettasul-80. The EO showed antioxidant activity through 2,2-diphenyl-1-picrylhydrazyl (DPPH) free radical scavenging and β-carotene–linoleic acid bleaching assay. Methyl cinnamate (48.29%) and γ-terpinene (26.08%) were recorded as the major components of the oil through GC-MS analysis. The EO was found to be nonmammalian toxic, showing a high $LD_{50}$ (11622.67 μl/kg) during oral toxicity on mice.

The bioactivity of the essential oil extracted by hydrodistillation from the seed of *Coriandrum sativum* was assessed under laboratory conditions for its biological activity against *S. granarius* in chickpea grains [23]. The components of the essential oil were identified through GC and GC-MS. The identity of 17 constituent compounds of the essential oil was confirmed, and their relative proportion was determined. Linalool has the highest percentage composition in the *C. sativum* seed essential oil (73.11%). All essential oil dosages showed a significant level of toxicity to the insect after 5 days.

*Hesperozygis ringens* (Lamiaceae) is a strongly aromatic plant employed popularly for its antiparasitic properties [24]. The leaves afforded 4% essential oil, constituted mainly by pulegone (86%). Laboratory tests were carried out to determine the toxicity of the essential oil species on engorged females and larvae of the cattle tick *R. microplus* using the adult immersion test (AIT) and the LIT. It was observed that the essential oil at the concentration of 50 and 25 μl/ml inhibited the egg laying significatively in relation to the controls, and the eggs from these treated females were affected by the oil; the hatching was inhibited in 95% and 30%, respectively. In the LIT, it was verified that the $LC_{99.9}$, $LC_{50}$, and $LC_1$ were 0.541, 0.260, and 0.015 μl/ml, respectively. Pulegone, isolated from the oil, showed a similar effect on the females and the larvae, indicating that it is responsible for the oil activity.

In order to identify natural products for plant disease control, the essential oil of star anise (*Illicium verum* Hook. f.) fruit was investigated for its antifungal activity on plant pathogenic fungi [25]. The fruit essential oil obtained by hydrodistillation was analyzed for its chemical composition by GC and GC-MS. *trans*-Anethole (89.5%), 2-(1-cyclopentenyl)-furan (0.9%), and *cis*-anethole (0.7%) were found to be the main components among 22 identified compounds, which accounted for 94.6% of the total oil. Both the essential oil and *trans*-anethole exhibited a strong inhibitory effect against all test fungi, indicating that most of the observed antifungal properties were due to the presence of *trans*-anethole in the oil, which could be developed as a natural fungicide for plant disease control in fruit and vegetable preservation.

Essential oil of *Plectranthus amboinicus* was studied for its chemical composition and larvicidal potential against the malarial vector mosquito *An. stephensi* [26]. In total, 26 compounds were identified by GC and GC-MS. The major chemical compound was carvacrol (28.65%), followed by thymol (21.66%), α-humulene (9.67%), undecanal (8.29%), γ-terpinene (7.76%), ρ-cymene (6.46%), caryophyllene oxide (5.85%), α-terpineol (3.28%), and β-selinene (2.01%). The larvicidal assay was conducted to record the $LC_{50}$ and $LC_{90}$ values, and the larval mortality was observed after 12 and 24 h of exposure. The $LC_{50}$ values of the oil were 33.54 ppm (after 12 h) and 28.37 ppm (after 24 h). The $LC_{90}$ values of the oil were 70.27 ppm (after 12 h) and 59.38 ppm (after 24 h). The results of the present study showed that the essential oil of *P. amboinicus* is one of the inexpensive and eco-friendly sources of natural mosquito larvicidal agent to control or reduce the population of malarial vector mosquitoes.

The aim of the research of Liu et al. [27] was to determine larvicidal activity of the essential oil derived from roots of *Saussurea lappa* (Compositae) and the isolated constituents against the larvae of the Culicidae mosquito *Ae. albopictus*. The essential oil of *S. lappa* roots was obtained by hydrodistillation and analyzed by GC and GC-MS. A total of 39 components of the essential oil of *S. lappa* roots were identified. The essential oil has a higher content (79.80%) of sesquiterpenoids than monoterpenoids (13.25%). The principal compounds in *S. lappa* essential oil were dehydrocostus lactone (46.75%), costunolide (9.26%), 8-cedren-13-ol (5.06%), and α-curcumene (4.33%). Dehydrocostus lactone and costunolide exhibited strong larvicidal activity against *Ae. albopictus* with $LC_{50}$ values of 2.34 and 3.26 µg/ml, respectively, while the essential oil had an $LC_{50}$ value of 12.41 µg/ml.

The objectives of the study of Tian et al. [28] were to determine the antifungal activity *in vitro* of the essential oil extracted from the seeds of dill (*Anethum graveolens* L.) and to evaluate its antifungal activity *in vivo* as a potential food preservative. The antifungal activity of this oil was tested by a poisoned food technique against *Aspergillus flavus*, *Aspergillus oryzae*, *Aspergillus niger*, and *Alternaria alternata*. The wet and dry mycelium weights of the tested fungi were also determined in a liquid culture to evaluate the antifungal activity. The minimum inhibitory concentration of oil for the four tested fungi was found to be 2.0 µl/ml, and the mycelial growth inhibition was determined at day 9. The effect of the essential oil on inhibition of decay development on cherry tomatoes was tested *in vivo* by exposing inoculated and control fruit to essential oil vapor at 120 and 100 µl/ml concentrations, respectively.

The essential oil composition and *in vitro* antioxidant and antimicrobial activity of the essential oil and methanol extract of *Salvia eremophila* were evaluated in the research of Ebrahimabadi et al. [29]. GC and GC-MS analysis of the plant essential oil resulted in the identification of 28 compounds representing 99.24% of the oil. Borneol (21.83%), α-pinene (18.80%), bornyl acetate (18.68%), and camphene (6.54%) were detected as the major components, consisting of 65.85% of the oil. The plant essential oil and methanol extract were also subjected to screenings for the evaluation of their antioxidant activities using DPPH and β-carotene–linoleic acid tests. While the plant essential oil showed only weak antioxidant activities, its methanol extract was considerably active in both DPPH ($IC_{50}$ = 35.19 µg/ml) and β-carotene–linoleic acid (inhibition percentage 72.42%) tests. The plant was also screened for its antimicrobial activity, and good to moderate inhibitions were recorded for its essential oil and methanol extract against most of the tested microorganisms.

The aspects of the antifungal activity of essential oil of laurel (*Laurus nobilis*) obtained by means of a supercritical carbon dioxide (SFE-$CO_2$) technique against postharvest spoilage fungi have been studied by tests performed under *in vitro* and *in vivo* conditions [30]. The determination of the main active substances was carried out by gas chromatography analysis: laurel oil was characterized by a high content (≥10%) of 1,8-cineole, linalool, terpineol acetate, and methyl eugenol and a low content (<10%) of linalyl acetate, eugenol, sabinene, β-pinene, and α-terpineol. The inhibition of the mycelial growth of *B. cinerea*, *Monilinia laxa*, and *P. digitatum* was evaluated *in vitro* at concentrations of 200, 400, 600, 800, and 1000 µg/ml. *M. laxa* was totally inhibited by application of the oil at the lowest concentration, *B. cinerea* was completely inhibited at the highest concentration, and a fungistatic action was observed in both cases. *P. digitatum* was only partially inhibited at all the concentrations. The activity of the oil, placed in the form of spray on the fruit skin at concentrations of 1, 2, and 3 mg/ml, was studied by biological tests. Both curative and protective activities of the oil have been evaluated on peaches, kiwifruits, oranges, and lemons artificially inoculated with *M. laxa*, *B. cinerea*, and *P. digitatum*, respectively. A very good antifungal activity was found on kiwifruits and peaches when the oil was placed before the inoculation at a concentration of 3 mg/ml (68% and 91% of decay inhibition, respectively). The same activity was found on peaches when the oil was placed after the infection (76% of decay inhibition). The application of the oil did not cause any phytotoxic effect and kept any fruit flavor, fragrance, or taste.

Mishra et al. [31] report on fungal deterioration of five herbal raw materials and the antifungal, antiaflatoxigenic, and antioxidant efficacy of Jamrosa essential oil and its two major components. Herbal raw materials were found to be associated with 14 fungal species, including strains of aflatoxin-producing *A. flavus*. Jamrosa EO and its major components Z-citral and linalyl acetate were assessed against the highest aflatoxin $B_1$ ($AFB_1$)–producing strain, *A. flavus* LHPA$_9$. The Jamrosa EO MIC for *A. flavus* LHPA$_9$, as well as the concentration that suppressed aflatoxin $B_1$ production, was 0.4 µl/ml. This EO was found to be more efficacious than its major components individually, as well as in combination. Z-citral inhibited $AFB_1$ completely at 1.0 µl/ml, while linalyl acetate did so at 0.7 µl/ml. The combination of both compounds completely inhibited $AFB_1$ production at 0.8 µl/ml. Free radical scavenging activities ($IC_{50}$) of EO, Z-citral, linalyl acetate, and a combination of both compounds were 86, 94, 217, and 158 µl/ml, respectively.

The essential oil of *Deverra scoparia* Coss. & Durieu was investigated for its acaricidal activity against a worldwide pest, the two-spotted spider mite, *T. urticae* Koch (Acari: Tetranychidae) [32]. The essential oil was analyzed by fast GC and GC-MS. The activities of its individual and blended constituents were determined. This study showed that female mortality increased with increasing *D. scoparia* oil concentrations, with $LD_{50}$ and $LD_{90}$ values at 1.79 and 3.2 mg/L, respectively. A reduction in fecundity had already been observed for concentrations of 0.064, 0.08, and 0.26 mg/L *D. scoparia* essential oil. Ten major components, comprising 98.52% of the total weight, were identified; α-pinene was the most abundant constituent (31.95%), followed by sabinene (17.24%) and Δ3-carene (16.85%). The 10 major constituents of *D. scoparia* oil were individually tested against *T. urticae* females.

The most potent toxicity was found with α-pinene, Δ3-carene, and terpinen-4-ol. The presence of all constituents together in the artificial mixture caused a significant decrease in the number of eggs laid by females, at 0.26 mg/L (11 eggs), compared with the control (50 eggs). The toxicity of blends of selected constituents indicated that the presence of all constituents was necessary to reproduce the toxicity level of the natural oil.

Essential oil of *Satureja hortensis* isolated via hydrodistillation was investigated against 1- to 7-day-old adults of the red flour beetle, *T. castaneum* (Herbst); 12- to 14-day-old larvae of the Mediterranean flour moth, *Ephestia kuehniella* (Zell.); and Indian meal moth, *Plodia interpunctella* (Hübner) [33]. Repellency of this oil on all three pest species' adults was also studied. After 48 h of exposure, the $LC_{50}$ value for *T. castaneum* was 192.35 µl/L. $LC_{50}$ values were calculated as 80.9 µl/L and 139.8 µl/L after 9 h for *E. kuhniella* and *P. interpunctella*, respectively. *S. hortensis* oil showed more contact toxicity against *P. interpunctella* ($LC_{50}$ = 0.19 µl/cm$^2$) than *E. kuehniella* ($LC_{50}$ = 0.27 µl/cm$^2$). Repellency of this oil on all the insect species was high. The relationship between exposure time and oil concentration on the mortality of all species indicated that mortality was increased by increasing the oil concentration and exposure time.

The essential oil from *Hyptis suaveolens* L. (Lamiaceae) was analyzed by GC and gas chromatography–electron impact mass spectroscopy (GC-EIMS), and 66 constituents were identified [34]. *H. suaveolens* EO contains sabinene (34%), β-caryophyllene (11.2%), terpinolene (10.7%), β-pinene (8.2%), limonene (5.8%), and 4-terpineol (2.5%) as major constituents. Moreover, *H. suaveolens* EO and its major constituents were evaluated for their repellent activity against adults of the granary weevil *S. granarius* (L.) (Coleoptera: Dryophthoridae) in Petri dish tests and in pitfall bioassays. Data showed that *H. suaveolens* EO possesses a noticeable repellent activity against *S. granarius* in both testing methods. Furthermore, in all trials good repellence rates of terpinolene, β-pinene, and sabinene were found, in particular at lower dosages.

Zapata and Smagghe [35] report on the repellent activity, as well as contact and fumigant toxicity, of four essential oils extracted from the leaves and bark of *Laurelia sempervirens* and *Drimys winteri* against an important stored product insect pest: the red flour beetle, *T. castaneum*. The four oils tested had a very strong repellent activity toward *T. castaneum* when tested in a filter paper arena test. After 4 h exposure, >90% repellency was achieved with *L. sempervirens* oils at low concentrations of 0.032 µl/cm$^2$, while for *D. winteri* oils, concentrations of 3–10 times higher were needed to achieve this activity. Oils of both *L. sempervirens* and *D. winteri* were found to be toxic toward *T. castaneum* when applied topically or by fumigation. $LD_{50}$ values by topical application of *L. sempervirens* oils were from 39 to 44 µg/mg of insect; for *D. winteri* oils, these were from 75 to 85 µg/mg of insect. By fumigation, $LC_{50}$ values for *L. sempervirens* oils were 1.6–1.7 µl/L air, while these were 9.0–10.5 µl/L air for *D. winteri* oils. In addition, with *L. sempervirens* oils, 50% of the tested beetles were killed at 100 µl/L air within 3.0–4.4 h, while with *D. winteri* oils, the $LT_{50}$ values were 6.1–7.4 h.

Oregano essential oil (*Origanum onites*) was applied at two doses, 0.55 and 0.75 µl/cm$^3$, and two exposure times, 3 and 6 h, as a disinfectant for hatching eggs [36]. The formaldehyde-treated eggs were used as a positive control, and untreated eggs were used as a negative control. After chemical analysis, the main constituents of oregano essential oil were carvacrol, linalool, para-cymene, and γ-terpinene. The lowest microbial counts on eggs were obtained from oregano essential oil. Microbial inhibition increased with the increasing essential oil concentrations. Essential oil exposure times had no significant effects on microbial counts. Essential oil fumigation lowered middle embryonic mortality and the discarded chick rate, but increased early and late embryonic mortalities compared with formaldehyde treatment. Essential oil doses significantly affected late embryonic mortality, the discarded chick rate, the contamination rate, hatchability of the fertile egg, the body weight at 21 and 42 days, body weight gain, and total feed consumption. But, early and middle embryonic mortality were not significantly affected by treatments.

The cowpea weevil, *Callosobruchus maculatus* (F.), is a major pest of cowpea *Vigna unguic-ulata* (L.) Walp. in storage units, making the grains unsuitable for consumption [37]. The aim of this study was to obtain and chemically identify the components of essential oils extracted from fruit peels of *Citrus latifolia* Tanaka, *Citrus reticulata* Blanco, *Citrus sinensis* L. Osbeck, and *Citrus paradisi* Macf., as well as to determine the contact and fumigant toxicity of these oils and their repellent effect on *C. maculatus* adults. GC-MS analysis identified 45 compounds in the essential oils; the major components were described as follows: *C. latifolia* (limonene 57.7%, γ-terpinene 17.2%, β-pinene 12.3%, and α-pinene 2.0%), *C. sinensis* (limonene 93.8% and myrcene 2.1%), *C. reticulata* (limonene 94.2% and myrcene 1.6%), and *C. paradisi* (limonene 94.2% and myrcene 1.8%). In the contact toxicity tests using treated cowpeas, the $LC_{50}$ values ranged from 943.9 to 1037.7 ppm, with the lowest value for *C. latifolia* and the highest for *C. sinensis*. The number of eggs and newly emerged adults was inversely proportional to the essential oil concentration increase. In the fumigant toxicity test, $LC_{50}$ values ranged from 10.2 to 12.98 μl/L air, with *C. latifolia* showing the best results. In the repellency test, the essential oils were classified as neutral at all concentrations. The percentages of oviposition reduction ranged from 29.74% to 71.66%, while reduction in emergence varied from 15.43 to 85.31.

The purpose of the review of Pavela [38] was to evaluate the current research on using EOs as potential larvicides based on their chemical composition and biological efficacy. The selected plants (their EOs), as the case may be, were therefore required to meet two essential conditions: (1) $LC_{50} \leq 100$ ppm and (2) their chemical composition had to be known.

In total, 122 plant species from 26 families were selected from the available literature. However, more than two-thirds of the plants (68.8%) were from only five families: Lamiaceae, Cupressaceae, Rutaceae, Apiaceae, and Myrtaceae.

Considering the above-estimated $LC_{50}$ value as the main criterion of efficacy, 77 showed $LC_{50} < 50$ ppm. Some of these efficient EOs were obtained from aromatic plants also grown commercially on relatively large areas, with good cultivation technology (e.g., *Pimpinella anisum*, *C. sativum*, *F. vulgare*, *Mentha longifolia*, *Ocimum basilicum*, *Thymus* spp., *Eucalyptus* spp., and *Piper* spp.). Such plants could become a suitable source of active substances for potential botanical larvicides. Only seven plants (*Blumea densiflora*, *Auxemma glazioviana*, *Callitris glaucophylla*, *Cinnamomum microphyllum*, *Cinnamomum mollissimum*, *Cinnamomum rhyncophyllum*, and *Zanthoxylum oxyphyllum*) can be considered significantly most efficient, given that $LC_{50} < 10$ ppm has been estimated for their EOs. These EOs contained less common substances, predominantly from the group of sesquiterpenes, aromatic acids, and ketones.

## 23.6 Chinese Herbs

In the screening program for new agrochemicals from Chinese medicinal herbs, *Murraya exotica* was found to possess insecticidal activity against the maize weevil, *S. zeamais*, and red flour beetle, *T. castaneum* [39]. The essential oil of aerial parts of *M. exotica* was obtained by hydrodistillation and investigated by GC and GC-MS. The main components of *M. exotica* essential oil were spathulenol (17.7%), α-pinene (13.3%), caryophyllene oxide

(8.6%), and α-caryophyllene (7.3%). Essential oil of *M. exotica* possessed fumigant toxicity against *S. zeamais* and *T. castaneum* adults with $LC_{50}$ values of 8.29 and 6.84 mg/L, respectively. The essential oils also show contact toxicity against *S. zeamais* and *T. castaneum* adults with $LD_{50}$ values of 11.41 and 20.94 mg/adult, respectively.

In the same screening program for new agrochemicals from Chinese medicinal herbs and wild plants, essential oil of *Kadsura heteroclite* stems was found to possess strong toxicities against the root-knot nematode, *Meloidogyne incognita*, and the maize weevil, *S. zeamais* [40]. The essential oil of *K. heteroclita* was extracted via hydrodistillation and analyzed by gas chromatography–flame ionization detector (GC-FID) and GC-MS. A total of 46 components of the essential oil were identified. The main components of the essential oil were α-eudesmol (17.56%), 4-terpineol (9.74%), δ-cadinene (9.27%), and δ-cadinol (6.32%), followed by δ-4-carene (4.78%) and calarene (4.01%). The essential oil exhibited strong nematicidal activity against *M. incognita*, with an $LC_{50}$ value of 122.94 μg/ml. The essential oil possessed contact toxicity against *S. zeamais* adults, with an $LD_{50}$ value of 25.57 μg/adult, and also showed pronounced fumigant toxicity against *S. zeamais* ($LC_{50} = 14.04$ mg/L air).

This screening for bioactive principles from several Chinese medicinal herbs showed that the essential oil of *Cymbopogon distans* aerial parts possessed strong repellency against the booklouse, *L. bostrychophila*, and the red flour beetle, *T. castaneum* [41]. A total of 36 components of the essential oil were identified by GC and GC-MS. *trans*-Geraniol (16.54%), (*R*)-citronellal (15.44%), (+)-citronellol (11.51%), and α-elemol (9.06%) were the main components of the essential oil, followed by β-eudesmol (5.71%) and (+)-limonene (5.05%). Geraniol and citronellol were strongly repellent against *L. bostrychophila*, whereas citronellal and limonene exhibited weak repellency against the booklouse. Geraniol and citronellol exhibited comparable repellency against the booklouse relative to the positive control, DEET. Moreover, geraniol and citronellol exhibited stronger repellency against the red flour beetle than DEET, whereas the two other compounds showed the same level of repellency against the red flour beetle compared with DEET.

In the screening program of Chu et al. [42] for new agrochemicals from local wild plants, essential oil of *Artemisia vestita* (Asteraceae) was found to possess strong insecticidal activity against the maize weevil, *S. zeamais*. Essential oil of aerial parts of *A. vestita* was obtained from hydrodistillation and was investigated by GC and GC-MS. The main components of essential oil were grandisol (40.29%), 1,8-cineol (14.88%), and camphor (11.37%). The essential oil of *A. vestita* possessed strong fumigant toxicity against *S. zeamais* adults, with an $LC_{50}$ value of 13.42 mg/L air. The essential oil of *A. vestita* also showed contact toxicity against *S. zeamais* adults with an $LD_{50}$ value of 50.62 mg/adult.

The essential oil of *Rhododendron anthopogonoides* flowering aerial parts possesses significant toxicity against maize weevils, *S. zeamais* [43]. A total of 37 components were identified in the essential oil, and the main constituents of the essential oil were 4-phenyl-2-butanone (27.22%), nerolidol (8.08%), 1,4-cineole (7.85%), caryophyllene (7.63%), and γ-elemene (6.10%), followed by α-farnesene (4.40%) and spathulenol (4.19%). Repeated bioactivity-directed chromatographic separation on silica gel columns resulted in the isolation of three compounds: 4-phenyl-2-butanone, 1,4-cineole, and nerolidol. 4-Phenyl-2-butanone shows pronounced contact toxicity against *S. zeamais* ($LD_{50} = 6.98$ mg/adult) and was more toxic than either 1,4-cineole or nerolidol ($LD_{50} = 50.86$ and 29.30 mg/adult, respectively) against the maize weevils, while the crude essential oil had an $LD_{50}$ value of 11.67 mg/adult. 4-Phenyl-2-butanone and 1,4-cineole also possessed strong fumigant toxicity against the adults of *S. zeamais* ($LC_{50} = 3.80$ and 21.43 mg/L), while the crude essential oil had an $LC_{50}$ value of 9.66 mg/L.

Liu et al. [44] found that *Artemisia capillaris* and *A. mongolica* possess insecticidal activity against the maize weevil, *S. zeamais*. The essential oils of aerial parts of the two plants were obtained by hydrodistillation and investigated by GC and GC-MS. The main components of *A. capillaris* essential oil were 1,8-cineole (13.75%), germacrene D (10.41%), and camphor (8.57%). The main constituents of *A. mongolica* essential oil were α-pinene (12.68%), germacrene D (8.36%), and γ-terpinene (8.17%). Essential oils of *A. capillaris* and *A. mongolica* possess fumigant toxicity against *S. zeamais* adults with $LC_{50}$ values of 5.31 and 7.35 mg/L, respectively. The essential oils also show contact toxicity against *S. zeamais* adults, with $LD_{50}$ values of 105.95 and 87.92 mg/adult, respectively.

The essential oil of *Atractylodes chinensis* (DC.) Koidz was found to possess strong insecticidal activity against the common vinegar fly, *Drosophila melanogaster* L. [45]. The essential oil was extracted via hydrodistillation, and its constituents were determined by GC-MS analysis.

The main components of *A. chinensis* essential oil were β-eudesmol (21.05%), β-selinene (11.75%), γ-elemene (7.16%), and isopetasam (5.36%). Bioactivity-directed chromatographic separation on repeated silica gel columns led to the isolation of five compounds: atractylon, α-elemol, β-eudesmol, hinesol, and β-selinene. β-Selinene, α-elemol, and hinesol showed pronounced contact toxicity against *D. melanogaster* adults, with $LD_{50}$ values of 0.55, 0.65, and 0.71 μg/adult, respectively. Atractylon and β-eudesmol were also toxic to the fruit flies ($LD_{50}$ = 1.63 and 2.65 μg/adult, respectively), while the crude oil had an $LD_{50}$ value of 2.44 μg/adult.

# References

1. V.K. Dua, A.C. Pandey, A.P. Dash. Adulticidal activity of essential oil of *Lantana camara* leaves against mosquitoes. *Indian Journal of Medical Research*, 2010, 131(3), 434–439.
2. J.-N. Oyourou, S. Combrinck, T. Regnier, A. Marston. Purification, stability and antifungal activity of verbascoside from *Lippia javanica* and *Lantana camara* leaf extracts. *Industrial Crops and Products*, 2013, 43, 820–826.
3. G.N. Diaz Napal, L.M. Buffa, L.C. Nolli, M.T. Defagó, G.R. Valladares, M.C. Carpinella, G. Ruiz, S.M. Palacios. Screening of native plants from central Argentina against the leaf-cutting ant *Acromyrmex lundi* (Guérin) and its symbiotic fungus. *Industrial Crops and Products*, 2015, 76, 275–280.
4. R. Pavela, G. Benelli. Ethnobotanical knowledge on botanical repellents employed in the African region against mosquito vectors—A review. *Experimental Parasitology*, 2016, 167, 103–108.
5. S. Zoubiri, A. Baaliouamer. GC and GC/MS analyses of the Algerian *Lantana camara* leaf essential oil: Effect against *Sitophilus granarius* adults. *Journal of Saudi Chemical Society*, 2012, 16(3), 291–297.
6. A. Junnila, E.E. Revay, G.C. Müller, V. Kravchenko, et al. Efficacy of attractive toxic sugar baits (ATSB) against *Aedes albopictus* with garlic oil encapsulated in beta-cyclodextrin as the active ingredient. *Acta Tropica*, 2015, 152, 195–200.
7. S.H. Ho, L. Koh, Y. Ma, Y. Huang, K.Y. Sim. The oil of garlic, *Allium sativum* L. (Amaryllidaceae), as a potential grain protectant against *Tribolium castaneum* (Herbst) and *Sitophilus zeamais* Motsch. *Postharvest Biology and Technology*, 1996, 9(1), 41–48.
8. H.-B. Zhou, J.-L. Chen, Y. Liu, F. Francis, E. Haubruge, C. Bragard, J.-R. Sun, D.-F. Cheng. Influence of garlic intercropping or active emitted volatiles in releasers on aphid and related beneficial in wheat fields in China. *Journal of Integrative Agriculture*, 2013, 12(3), 467–473.

9. X.C. Liu, X.N. Lu, Q.Z. Liu, Z.L. Liu. Evaluation of insecticidal activity of the essential oil of *Allium chinense* G. Don and its major constituents against *Liposcelis bostrychophila* Badonnel. *Journal of Asia-Pacific Entomology*, 2014, 17(4), 853–856.

10. B. Meriga, R. Mopuri, T. Murali Krishna. Insecticidal, antimicrobial and antioxidant activities of bulb extracts of *Allium sativum*. *Asian Pacific Journal of Tropical Medicine*, 2012, 5(5), 391–395.

11. X.C. Liu, Q. Liu, L. Zhou, Z.L. Liu. Evaluation of larvicidal activity of the essential oil of Allium macrostemon Bunge and its selected major constituent compounds against Aedes albopictus (Diptera: Culicidae). *Parasites & Vectors*, 2014, 7, 184.

12. M. Abdelli, H. Moghrani, A. Aboun, R. Maachi. Algerian *Mentha pulegium* L. leaves essential oil: Chemical composition, antimicrobial, insecticidal and antioxidant activities. *Industrial Crops and Products*, 2016, 94(30), 197–205.

13. F. Brahmi, A. Abdenour, M. Bruno, P. Silvia, P. Alessandra, F. Danilo, Y.-G. Drifa, E. Mahmoud Fahmi, M. Khodir, C. Mohamed. Chemical composition and in vitro antimicrobial, insecticidal and antioxidant activities of the essential oils of *Mentha pulegium* L. and *Mentha rotundifolia* (L.) Huds growing in Algeria. *Industrial Crops and Products*, 2016, 88, 96–105.

14. K. Karunamoorthi, A. Girmay, S. Fekadu. Larvicidal efficacy of Ethiopian ethnomedicinal plant *Juniperus procera* essential oil against Afrotropical malaria vector *Anopheles arabiensis* (Diptera: Culicidae). *Asian Pacific Journal of Tropical Biomedicine*, 2014, 4(1), S99–S106.

15. Z. Sindhu, N.N. Jonsson, Z. Iqbal. Syringe test (modified larval immersion test): A new bioassay for testing acaricidal activity of plant extracts against *Rhipicephalus microplus*. *Veterinary Parasitology*, 2012, 188(3–4), 362–367.

16. K. Karunamoorthi, A. Girmay, S. Fekadu Hayleeyesus. Mosquito repellent activity of essential oil of Ethiopian ethnomedicinal plant against Afro-tropical malarial vector *Anopheles arabiensis*. *Journal of King Saud University—Science*, 2014, 26(4), 305–310.

17. R. Pazinato, A. Volpato, M.D. Baldissera, R.C.V. Santos, D. Baretta, R.A. Vaucher, J.L. Giongo, A.A. Boligon, L. Moura Stefani, A. Schafer Da Silva. In vitro effect of seven essential oils on the reproduction of the cattle tick *Rhipicephalus microplus*. *Journal of Advanced Research*, 2016, 7(6), 1029–1034.

18. S.C.H. Cavalcanti, E. dos S. Niculau, A.F. Blank, C.A.G. Câmara, I.N. Araújo, P.B. Alves. Composition and acaricidal activity of *Lippia sidoides* essential oil against two-spotted spider mite (*Tetranychus urticae* Koch). *Bioresource Technology*, 2010, 101(2), 829–832.

19. B. Conti, A. Canale, A. Bertoli, F. Gozzini, L. Pistelli. Essential oil composition and larvicidal activity of six Mediterranean aromatic plants against the mosquito *Aedes albopictus* (Diptera: Culicidae). *Parasitology Research*, 2010, 107(6), 1455–1461.

20. S.M. Al-Reza, A. Rahman, Y. Ahmed, S.C. Kang. Inhibition of plant pathogens *in vitro* and *in vivo* with essential oil and organic extracts of *Cestrum nocturnum* L. *Pesticide Biochemistry and Physiology*, 2010, 96(2), 86–92.

21. H. Cetin, J.E. Cilek, E. Oz, L. Aydin, O. Deveci, A. Yanikoglu. Acaricidal activity of *Satureja thymbra* L. essential oil and its major components, carvacrol and γ-terpinene against adult *Hyalomma marginatum* (Acari: Ixodidae). *Veterinary Parasitology*, 2010, 170(3–4), 287–290.

22. B. Prakash, R. Shukla, P. Singh, P.K. Mishra, N.K. Dubey, R.N. Kharwar. Efficacy of chemically characterized *Ocimum gratissimum* L. essential oil as an antioxidant and a safe plant based antimicrobial against fungal and aflatoxin B$_1$ contamination of spices. *Food Research International*, 2011, 44(1), 385–390.

23. S. Zoubiri, A. Baaliouamer. Essential oil composition of *Coriandrum sativum* seed cultivated in Algeria as food grains protectant. *Food Chemistry*, 2010, 122(4), 1226–1228.

24. V.L. Sarda Ribeiro, J. Campiol dos Santos, S.A.L. Bordignon, M.A. Apel, A.T. Henriques, G.L. van Poser. Acaricidal properties of the essential oil from *Hesperozygis ringens* (Lamiaceae) on the cattle tick *Riphicephalus* (*Boophilus*) *microplus*. *Bioresource Technology*, 2010, 101(7), 2506–2509.

25. Y. Huang, J. Zhao, L. Zhou, J. Wang, Y. Gong, X. Chen, Z. Guo, Q. Wang, W. Jiang. Antifungal activity of the essential oil of *Illicium verum* fruit and its main component *trans*-anethole. *Molecules* 2010, 15(11), 7558–7569.

26. A. Senthilkumar, V. Venkatesalu. Chemical composition and larvicidal activity of the essential oil of *Plectranthus amboinicus* (Lour.) Spreng against *Anopheles stephensi*: A malarial vector mosquito. *Parasitology Research*, 2010, 107(5), 1275–1278.

27. Z.L. Liu, Q. He, S.S. Chu, C.F. Wang, S.S. Du, Z.W. Deng. Essential oil composition and larvicidal activity of *Saussurea lappa* roots against the mosquito *Aedes albopictus* (Diptera: Culicidae). *Parasitology Research*, 2012, 110(6), 2125–2130.

28. J. Tian, X. Ban, H. Zeng, B. Huang, J. He, Y. Wang. *In vitro* and *in vivo* activity of essential oil from dill (*Anethum graveolens* L.) against fungal spoilage of cherry tomatoes. *Food Control*, 2011, 22(12), 1992–1999.

29. A.H. Ebrahimabadi, A. Mazoochi, F.J. Kashi, Z. Djafari-Bidgoli, H. Batooli. Essential oil composition and antioxidant and antimicrobial properties of the aerial parts of *Salvia eremophila* Boiss. from Iran. *Food and Chemical Toxicology*, 2010, 48(5), 1371–1376.

30. U. De Corato, O. Maccioni, M. Trupo, G. Di Sanzo. Use of essential oil of *Laurus nobilis* obtained by means of a supercritical carbon dioxide technique against post harvest spoilage fungi. *Crop Protection*, 2010, 29(2), 142–147.

31. P.K. Mishra, R. Shukla, P. Singh, B. Prakash, A. Kedia, N.K. Dubey. Antifungal, anti-aflatoxigenic, and antioxidant efficacy of Jamrosa essential oil for preservation of herbal raw materials. *International Biodeterioration & Biodegradation*, 2012, 74, 11–16.

32. S. Attia, K.L. Grissa, G. Lognay, S. Heuskin, A.C. Mailleux, T. Hance. Chemical composition and acaricidal properties of *Deverra scoparia* essential oil (Araliales: Apiaceae) and blends of its major constituents against *Tetranychus urticae* (Acari: Tetranychidae). *Journal of Economic Entomology*, 2011, 1220–1228.

33. M. Maede, I. Hamzeh, D. Hossein, A. Majid, R.K. Reza. Bioactivity of essential oil from *Satureja hortensis* (Laminaceae) against three stored-product insect species. *African Journal of Biotechnology*, 2011, 10(34).

34. G. Benelli, G. Flamini, A. Canale, I. Molfetta, P.L. Cioni, B. Conti. Repellence of *Hyptis suaveolens* whole essential oil and major constituents against adults of the granary weevil *Sitophilus granarius*. *Bulletin of Insectology*, 2012, 65(2), 177–183.

35. N. Zapata, G. Smagghe. Repellency and toxicity of essential oils from the leaves and bark of *Laurelia sempervirens* and *Drimys winteri* against *Tribolium castaneum*. *Industrial Crops and Products*, 2010, 32(3), 405–410.

36. G. Copur, M. Arslan, M. Duru, M. Baylan, S. Canogullari, E. Aksan. Use of oregano (*Origanum onites* L.) essential oil as hatching egg disinfectant. *African Journal of Biotechnology*, 2010, 9(17), 2531–2538.

37. K. de Andrade Dutra, J. Vargas de Oliveira, D.M. do Amaral Ferraz Navarro, D.R. e Silva Barbosa, J.P. Oliveira Santos. Control of *Callosobruchus maculatus* (FABR.) (Coleoptera: Chrysomelidae: Bruchinae) in *Vigna unguiculata* (L.) WALP. with essential oils from four *Citrus* spp. plants. *Journal of Stored Products Research*, 2016, 68, 25–32.

38. R. Pavela. Essential oils for the development of eco-friendly mosquito larvicides: A review. *Industrial Crops and Products*, 2015, 76, 174–187.

39. W.Q. Li, C.H. Jiang, S.S. Chu, M.X. Zuo, Z.L. Liu. Chemical composition and toxicity against *Sitophilus zeamais* and *Tribolium castaneum* of the essential oil of *Murraya exotica* aerial parts. *Molecules*, 2010, 15(8), 5831–5839.

40. H.Q. Li, C.Q. Bai, S.S. Chu, L. Zhou, S.S. Du, Z.L. Liu, Q.Z. Liu. Chemical composition and toxicities of the essential oil derived from *Kadsura heteroclita* stems against *Sitophilus zeamais* and *Meloidogyne incognita*. *Journal of Medical Plant Research*, 5(19), 4943–4948, 2011.

41. J.S. Zhang, N.N. Zhao, Q.Z. Liu, Z.L. Liu, S.S. Du, L. Zhou, Z.W. Deng. Repellent constituents of essential oil of *Cymbopogon distans* aerial parts against two stored-product insects. *Journal of Agricultural and Food Chemistry*, 2011, 59(18), 9910–9915.

42. S.S. Chu, Q.R. Liu, Z.L. Liu. Insecticidal activity and chemical composition of the essential oil of *Artemisia vestita* from China against *Sitophilus zeamais*. *Biochemical Systematics and Ecology*, 38(4), 2010, 489–492.

43. K. Yang, Y.X. Zhou, C.F. Wang, S.S. Du, Z.W. Deng, Q.Z. Liu, Z.L. Liu. Toxicity of *Rhododendron anthopogonoides* essential oil and its constituent compounds towards *Sitophilus zeamais*. *Molecules*, 2011, 16(9), 7320–7330.
44. Z.L. Liu, S.S. Chu, Q.R. Liu. Chemical composition and insecticidal activity against *Sitophilus zeamais* of the essential oils of *Artemisia capillaris* and *Artemisia mongolica*. *Molecules*, 2010, 15(4), 2600–2608
45. S.S. Chu, G.H. Jiang, Z.L. Liu. Insecticidal compounds from the essential oil of Chinese medicinal herb *Atractylodes chinensis*. *Pest Management Science*, 2011, 67(10), 1253–1257.

# Section III

# Different Aspects of Essential Oils

# 24

## Essential Oils and Synthetic Pesticides

Vasakorn Bullangpoti

## CONTENTS

## 24.1 Introduction

The environmental problems caused by the overuse of synthetic pesticides have been of concern in recent years. Thus, many scientists now need to research new, highly selective and biodegradable pesticides to solve problems and develop techniques that can be used to reduce pesticide use while maintaining crop yields (Koul et al., 2008). Essential oil compounds are an alternative to synthetic pesticides as a means to reduce the negative impacts to human health and the environment. They are safe and ecofriendly. Moreover, they are more compatible with environmental components than synthetic pesticides (Isman and Machial, 2006).

Essential oil compounds are defined as any volatile oil compounds that have strong aromatic components and that give a distinctive odor, flavor, or scent to a plant (Pavela, 2015). These are the by-products of plant metabolism and are commonly referred to as volatile plant secondary metabolites. Essential oil compounds are produced by more than 17,500 aromatic plant species commonly belonging to many angiospermic families, for example, Lamiaceae, Rutaceae, Myrtaceae, Zingiberaceae, and Asteraceae (Regnault-Roger et al., 2012). Most essential oils comprise monoterpenes—compounds that contain 10 carbon atoms often arranged in a ring or in acyclic form—as well as sesquiterpenes, which are hydrocarbons comprising 15 carbon atoms. The compounds of essential oil compounds are synthesized and stored in complex secretary structures, that is, glandular trichomes, secretory cavities, and resin ducts, and are present as droplets of fluid in the leaves, stem, flowers, fruits, bark, or roots of plants (Fahn, 2000).

The aromatic characteristics of essential oils provide various functions for the plants, including (1) attracting or repelling insects, (2) protection from heat or cold, and (3) utilizing chemical constituents in the oil as defense materials (Koul et al., 2008; Pavela, 2015).

In the next section are some developments where essential oils have been projected as pesticides in some recent research publications, along with their potential and constraints emphasized.

## 24.2 Essential Oil Compounds and Their Efficacy as a Pesticide

Essential oil compounds are complex natural mixtures. They are characterized by two or three major components at fairly high concentrations (20%–70%) compared with other components present in trace amounts, and generally, these major components determine the biological properties of the essential oils (Pavela, 2015). The components include two groups of distinct biosynthetical origin. The main group is composed of terpenes, terpenoids, while the other is composed of aromatic and/or aliphatic constituents, all characterized by low molecular weights. Essential oil compounds are synthesized in the cytoplasm and plastids of plant cells via malonic acid, mevalonic acid, and methyl-d-erythritol-4-phosphate (MEP) pathways.

Terpenes are hydrocarbons made up of several units of isoprene (C5), while terpenoids are terpenes that have been biochemically modified via enzymes that add oxygen molecules and move or remove a methyl group (Burt, 2004). Monoterpenes are formed from the coupling of two isoprene units (C10). They are the most represented molecules, constituting about 90% of the essential oil compounds, and allow for a great variety of structures (acyclic, monocyclic, and bicyclic) and several functions (carbures, e.g., myrcene, p-cimene, and camphene; alcohols, e.g., linalool, menthol, and borneol; aldehydes, e.g., geranial and citronellal; ketones, e.g., carvone, pulegone, and camphor; esters, e.g., linalyl acetate, menthyl, and citronellyl acetate; ethers, e.g., 1,8-cineole and menthofurane; peroxides, e.g., ascaridole; phenols, e.g., thymol and carcacrol; etc.). Examples of plants containing these compounds include bay leaves, cannabis, thyme, parsley, hops, laurel, tea tree, mugwort, sweet basil, wormwood, and rosemary.

Sesquiterpenes are formed from the assembly of three isoprene units (C15). The extension of the chain increases the number of cyclizations and thus leads to a great variety of structures. The functions of sesquiterpenoids are similar to those of monoterpenoids (cabures, e.g., azulene, β-caryophyllene, and elemenes; alcohols, e.g., bisabol, cedrol, and patchoulol; ketones, e.g., nootkatone, germacrone, and turmerones; etc.). The principal plant sources for these compounds include angelica, bergamot, celery, mint, orange, rosemary, and sage (Banthorpe, 1991).

Derived from phenylpropane, aromatic compounds occur less frequently than terpenes. The biosynthetic pathways concentrating the terpenes and phenylpropanic derivatives are generally separated in plants but may coexist in some, with one major pathway taking over. The functions of aromatic compounds are similar to those of monoterpenoids and/or sesquiterpenoids and comprise, for example, aldehydes (cynnamal-dehyde), alcohols (cinnamic alcohol), phenols (chavicol and eugenol), methoxy derivatives (anethole, estragole, and methyleugenol), and methylene dioxy compounds (apiole, myristine, and safrole). The principal plant sources for these compounds include anise, cinnamon, clove, and fennel (Grayson, 2000).

Essential oils are usually obtained via steam distillation of aromatic plants. There are many plant species from various families, which show pesticidal efficacy. There are many mixtures that have been considered for use as insecticides (Dev and Koul, 1997). However, the greatest pesticidal efficiency is limited to six families: Lamiaceae (14.72%), Rutaceae (12.88%), Cupressaceae (10.43%), Apiaceae (9.20%), Myrtaceae (8.59%), and Asteraceae (6.13%) (Table 24.1).

**TABLE 24.1**

Most Common Substances Contained in Essential Oil Compounds of Plants

| Family | Compound |
|---|---|
| Apiaceae | Limonene, α- and β-pinene, sabinene, *trans*-anethole, carvone |
| Asteraceae | 1,8-Cineole, linalool, limonene |
| Cupressaceae | α-Pinene, 3-carene, sabinene, limonene |
| Lamiaceae | 1,8-Cineole, thymol, p-cymene, cavacol, euglenol, linalool |
| Myrtaceae | 1,8-Cineole, p-cymene |
| Rutaceae | Limonene, α- and β-pinene |

Those substances represented in the essential oil compounds by more than 10% are considered majority substances. Although each of the essential oil compounds contain at least 10 substances and every essential oil compound is unique due to its complex composition, only 1–5 substances can be considered as majority (Pavela, 2015).

In addition, it can be seen that some substances commonly occur in multiple plant species of the same family, and can thus be considered typical for individual families, although sometimes the same compounds can be found in other families. In this chapter, not only the six families mentioned above are considered; various families looking at the most common substances contained in essential oil compounds of plants and their efficacy as pesticides are reviewed (Table 24.2).

Considering Table 24.2, toxicity is indicated as having the greatest efficacy as a fumigant and as contact insecticidal activity for a wide range of pests. Some of these efficient essential oil compounds were obtained from aromatic plants that are grown commercially on relatively large areas, with good cultivation technology (e.g., *Pimpinella anisum*, *Coriandrum sativum*, *Foeniculum vulgare*, *Mentha longifolia*, *Ocimum basilicum*, *Thymus* spp., *Eucalyptus* spp., and *Piper* spp.). Such plants could become a suitable source of active substances for potential botanical pesticides. These essential oil compounds contain less common substances, predominantly from the group of sesquiterpenes (guaiol, β-bisabolol, δ-cadinol, germacrene D, and β-caryophyllene), aromatic acids, and ketones (Table 24.2).

In fact, the concentration needed to achieve mortality depends on many factors, such as the life stage of the pest, ambient temperature, the capacity of the substances to penetrate the cuticle, and the mechanism of action (Pavela et al., 2009; Rattan, 2010). Although many papers have focused on the pesticidal efficacy of essential oil compounds, little information is available on their mechanism of action against insects. The relation between pesticidal effect and chemical composition of the essential oil is difficult to determine because the interactions among compounds can influence the activity of the mixture. In principle, every aromatic compound contained in an essential oil compound is unique in terms of its structure and biological activity. Various mechanisms of action of individual substances, many of which still remain unknown, may in their combinations not only provide a significant increase in effectiveness, but also prevent the development of pest resistance (Rattan, 2010).

The mode of action of essential oil compounds against some pests has been recorded in some publications, including as a neurotoxic by interfering with the neuromodulator octopamine (OA) (Kostyukovsky et al., 2002) and by GABA-gated chloride channels (Priestley et al., 2003). The neurotoxicity of several monoterpenoids (d-limonene, myrcene, terpineol, linalool, and pulegone) was evaluated against the house fly as well as on the German cockroach (Coats et al., 1991). Toxicity from essential oils or their constituents in insects and other arthropods points to a neurotoxic mode of action; the most prominent symptoms are

**TABLE 24.2**

Most Common Substances Contained in Essential Oil Compounds of Plants as Pesticides

| Family | Plant Genus and Species | Major Constituents | $LC_{50}$ or $LD_{50}$ | Target Organism | Method | Reference |
|---|---|---|---|---|---|---|
| Amaranthaceae | *Chenopodium ambrosioides* | α-Terpinene, p-cymene | 66.81 ppm | *P. xylostella* (third instar) | Antifeed | Wei et al. (2015) |
| | | | 6.142 mg/L air | *P. xylostella* (third instar) | Fumigant | Wei et al. (2015) |
| | | | 2.916 µl/larva | *P. xylostella* (third instar) | Contact | Wei et al. (2015) |
| Amaryllidaceae | *Allium cepa* | Disulfide dipropyl, p-cymene | 35 ppm | *Ae. aegypti* | Contact | Leyva et al. (2009) |
| | | | 1.11 µg/g | *Schistocerca gregaria* (third instar) | Topical | Mansour et al. (2015) |
| | | | 20.153 µg/g | *S. gregaria* (third instar) | Contact | Mansour and Abdel-Hamid (2015) |
| | *Allium macrostemon* | Methyl propyl disulfide, dimethyl trisulfide | 73 ppm | *Aedes albopictus* | Contact | Liu et al. (2014) |
| | *Allium monanthum* | Dimethyl trisulfide, dimethyl tetrasulfide | 23 ppm | *Ae. aegypti* | Contact | Moon et al. (2011) |
| | *Allium victorialis* L. var. *platyphyllum* | Allyl methyl disulfide, dimethyl trisulfide | 24 ppm | *Ae. aegypti* | Contact | Chung et al. (2011a) |
| Anacardiaceae | *Pistacia terebinthus* L. subsp. *palaestina* (Boiss.) Engler | α-Pinene, cyclopentane | 59 ppm | *Culex pipiens* | Contact | Cetin et al. (2011) |
| | *Spondias purpurea* | Caryophylleneoxide, α-cadinol | 39 ppm | *Ae. aegypti* | Contact | Lima et al. (2011) |
| Annonaceae | *Cananga odorata* (Lam.) Hook. F & Thomson | Benzyl acetate, linalool, methyl benzoate | 52 ppm | *Ae. aegypti* | Contact | Vera et al. (2014) |
| Annonaceae | *Cananga odorata* (Lam.) Hook. F. & Thomson | Benzyl acetate, linalool, methyl benzoate | 52 ppm | *Ae. aegypti* | Contact | Vera et al. (2014) |
| | *Guatteria blepharophylla* Mart. | Caryophylleneoxide | 58 ppm | *Ae. aegypti* | Contact | Aciole et al. (2011) |
| | *Guatteria friesiana* Erkens & Maas | Eudesmols | 52 ppm | *Ae. aegypti* | Contact | Aciole et al. (2011) |
| Apiaceae | *Angelica purpuraefolia* | β-Phellandrene, nerolidol | 31 ppm | *Ae. aegypti* | Contact | Park et al. (2010) |

*(Continued)*

**TABLE 24.2 (CONTINUED)**

Most Common Substances Contained in Essential Oil Compounds of Plants as Pesticides

| Family<br>Plant Genus<br>and Species | Major Constituents | LC$_{50}$ or LD$_{50}$ | Target Organism | Method | Reference |
|---|---|---|---|---|---|
| *Apium graveolens* | d-Limonene, 4-chloro-4,4-dimethyl-3-(1-imidazolyl)-valerophenone, 1-dodecanol | 42–59 ppm | *Ae. aegypti* | Contact | Pitasawat et al. (2007), Nagella et al. (2012a) |
| | | 59 ppm | *Anopheles dirus* | Contact | Pitasawat et al. (2007) |
| *Bupleurum fruticosum* | Pinene, carvone, limonene | 64 ppm<br>54 ppm<br>72 ppm | *C. pipiens*<br>*Ae. aegypti*<br>*An. dirus* | Contact | Evergetis et al. (2009)<br>Pitasawat et al. (2007) |
| *Carum carvi* | Carvone, limonene | 54 ppm | *Ae. aegypti* | Contact | Pitasawat et al. (2007) |
| | | 72 ppm | *An. dirus* | Contact | Pitasawat et al. (2007) |
| *Conopodium capillifolium* | Pinene, sabinene | 68 ppm | *C. pipiens* | Contact | Evergetis et al. (2009) |
| *Coriandrum sativum* | Linalool | 20 ppm | *Ae. aegypti* | Contact | Nagella et al. (2012b) |
| *Cuminum cyminum* | Cuminic aldehyde | 1.54 µg/g | *S. gregaria* (third instar) | Topical | Mansour et al. (2015) |
| | | 18.284 µg/g | *S. gregaria* (third instar) | Contact | Mansour and Abdel-Hamid (2015) |
| *Elaeoselinum asclepium* | α-Pinene, sabinene | 96 ppm | *C. pipiens* | Contact | Evergetis et al. (2009) |
| *Ferulago carduchorum* | (Z)-Ocimene, α-pinene | 12 ppm | *Anopheles stephensi* | Contact | Golfakhrabadi et al. (2015) |
| *Foeniculum vulgare* | *trans*-Anethole | 49 ppm | *A. aegypti* | Contact | Pitasawat et al. (2007) |
| | | 35 ppm | *An. dirus* | Contact | Pitasawat et al. (2007) |

*(Continued)*

**TABLE 24.2 (CONTINUED)**

Most Common Substances Contained in Essential Oil Compounds of Plants as Pesticides

| Family | Plant Genus and Species | Major Constituents | LC$_{50}$ or LD$_{50}$ | Target Organism | Method | Reference |
|---|---|---|---|---|---|---|
| | *Heracleum sphondylium* ssp. *pyrenaicum* (Lam.) Bonnier & Layens | Octyl acetate, limonene | 77 ppm | *C. pipiens* | Contact | Evergetis et al. (2009) |
| | *Oenanthe pimpinelloide* | γ-Terpinene, o-cymene | 40 ppm | *C. pipiens* | Contact | Evergetis et al. (2009) |
| | *Pimpinella anisum* | trans-Anethole | 9.3 µl/L 1.9 µl/L 29 µl/L | *Culex quinquefasciatus* *C. quinquefasciatus* *Daphnia magna* | Spray Fumigant Contact | Pavela et al. (2014) |
| | *Petroselinum sativum* | Apiol | 1.34 µg/g | *S. gregaria* (third instar) | Topical | Mansour et al. (2015) |
| | | | 63.406 µg/g | *S. gregaria* (third instar) | Contact | Mansour and Abdel-Hamid (2015) |
| | *Seseli montanum* ssp. *tomasinii* (Reichenb. fil.) Archangeli | α-Pinene, sabinene, β-phellandrene | 86 ppm | *C. pipiens* | Contact | Evergetis et al. (2009) |
| Apocynaceae | *Cionura erecta* (L.) Griseb. | Cedren-9-one, α-cadinol, eugenol, α-muurolene | 77 ppm | *An. stephensi* | Contact | Mozaffari et al. (2014) |
| Araliaceae | *Dendropanax morbifera* Leveille | γ-Elemene, tetramethyltricyclohydrocarbon, β-selinene, α-zingibirene | 62 ppm | *Ae. aegypti* | Contact | Chung et al. (2009) |
| Asteraceae | *Artemisia gilvescens* Miquel. | 1,8-Cineole, camphor, germacrene D | 49 ppm | *Anopheles anthropophagus* | Contact | Zhu and Tian (2013) |
| | *Blumea densiflora* D.C. | Borneol, germacrene D, β-caryophyllene | 10 ppm | *An. anthropophagus* | Contact | Zhu and Tian (2013) |
| | *Blumea mollis* (D. Don) Merr. | Linalool, γ-elemene, copaene, estragole, alloocimene | 71 ppm | *C. quinquefasciatus* | Contact | Senthilkumar et al. (2008) |
| | *Tagetes erecta* | Piperitone | 79 ppm | *Ae. aegypti* | Contact | Marques et al. (2011) |
| | *Tagetes filifolia* | trans-Anethole | 47 ppm | *Ae. aegypti* | Contact | Ruiz et al. (2011) |

*(Continued)*

**TABLE 24.2 (CONTINUED)**

Most Common Substances Contained in Essential Oil Compounds of Plants as Pesticides

| Family | Plant Genus and Species | Major Constituents | LC$_{50}$ or LD$_{50}$ | Target Organism | Method | Reference |
|---|---|---|---|---|---|---|
| | *Tagetes lucida* | Methyl chavicol | 66 ppm | *Ae. aegypti* | Contact | Vera et al. (2014) |
| | *Tagetes minuta* | *trans*-Ocimenone | 52 ppm | *Ae. aegypti* | Contact | Ruiz et al. (2011) |
| | *Tagetes patula* | Limonene, terpinolene | 13 ppm | *Ae. aegypti* | Contact | Dharmagadda et al. (2005) |
| | | | 22 ppm | *C. quinquefasciatus* | Contact | Dharmagadda et al. (2005) |
| | | | 12 ppm | *An. stephensi* | Contact | Dharmagadda et al. (2005) |
| | *Matricaria chamomilla* | Franesene | 1.59 µg/g | *S. gregaria* (third instar) | Topical | Mansour et al. (2015) |
| | | | 62.389 µg/g | *S. gregaria* (third instar) | Contact | Mansour and Abdel-Hamid (2015) |
| | *Tagetes erecta* | Piperitone | 79 ppm | *Ae. aegypti* | Contact | Marques et al. (2011) |
| Boraginaceae | *Auxemma glazioviana* | -Bisabolol, α-cadinol | 3 ppm | *Ae. aegypti* | Contact | José et al. (2004) |
| | *Cordia curassavica* | α-Pinene | 97 ppm | *Ae. aegypti* | Contact | Santos et al. (2006) |
| | *Cordia leucomalloides* | γ-Cadinene, (E)-caryo phyllene | 63 ppm | *Ae. aegypti* | Contact | Santos et al. (2006) |
| Cupressaceae | *Callitris glaucophylla* | Guaiol, citronellic acid | 0.7 ppm | *Ae. aegypti* | Contact | Shaalan et al. (2006) |
| | | | 0.2 ppm | *Culex annulirostris* | Contact | Shaalan et al. (2006) |
| | *Cryptomeria japonica* | 16-Kaurene, elemol, eudesmol, sabinene | 28–56 ppm, 51–56 ppm | *Ae. aegypti*, *Ae. albopictus* | Contact, Contact | Cheng et al. (2009), Cheng et al. (2009) |
| | *Cunninghamia konishii* | Cedrol | 85 ppm, 189 ppm | *Ae. aegypti*, *Ae. albopictus* | Contact, Contact | Cheng et al. (2013), Cheng et al. (2013) |
| | *Cupressus arizonica* | *trans*-Muurola-3,5-diene, *cis*-14-nor-muurol-5-en-4-one | 64 ppm | *Ae. albopictus* | Contact | Cheng et al. (2013) |
| | *Cupressus benthamii* | Limonene, umbellulone | 37 ppm | *Ae. albopictus* | Contact | Giatropoulos et al. (2013) |

*(Continued)*

**TABLE 24.2 (CONTINUED)**

Most Common Substances Contained in Essential Oil Compounds of Plants as Pesticides

| Family | Plant Genus and Species | Major Constituents | LC$_{50}$ or LD$_{50}$ | Target Organism | Method | Reference |
|---|---|---|---|---|---|---|
| | Cupressus macrocarpa | Sabinene, α-pinene, terpinen-4-ol | 54 ppm | Ae. albopictus | Contact | Giatropoulos et al. (2013) |
| | Cupressus sempervirens | α-Pinene, δ-3-carene | 54 ppm | Ae. albopictus | Contact | Giatropoulos et al. (2013) |
| | Cupressus torulosa | α-Pinene, δ-3-carene | 57 ppm | Ae. albopictus | Contact | Giatropoulos et al. (2013) |
| | Chamaecyparis formosensis | Myrtenol, myrtenal | 38 ppm / 35 ppm | Ae. aegypti / Ae. albopictus | Contact / Contact | Kuo et al. (2007) / Kuo et al. (2007) |
| | Chamaecyparis lawsoniana | Limonene, oplopanonyl acetate, beyerene | 47 ppm | Ae. albopictus | Contact | Giatropoulos et al. (2013) |
| | Juniperus communis | α-Pinene, sabinene, γ-3-Carene | 65 ppm | C. pipiens | Contact | Vourlioti-Arapi et al. (2012) |
| | Juniperus drupacea | α-Pinene, limonene | 26 ppm | C. pipiens | Contact | Vourlioti-Arapi et al. (2012) |
| | Juniperus foetidissima | Sabinene, 4-methyl-1-(1-methyl ethyl)-3-cyclohexen-1-ol | 53 ppm | C. pipiens | Contact | Vourlioti-Arapi et al. (2012) |
| | Juniperus oxycedrus L. ssp. oxycedrus | Myrcene, germacrene-D, α-pinene | 55 ppm | C. pipiens | Contact | Vourlioti-Arapi et al. (2012) |
| | Juniperus oxycedrus L. subsp. macrocarpa (Sm.) Ball. | α-Pinene | 65 ppm | C. pipiens | Contact | Vourlioti-Arapi et al. (2012) |
| | Juniperus phoenicea | α-Pinene, β-3-carene, γ-phellandrene, α-terpinyl acetate | 55 ppm | Ae. albopictus | Contact | Giatropoulos et al. (2013) |
| | Tetraclinis articulata | α-Pinene, bornyl acetate | 70 ppm | Ae. albopictus | Contact | Giatropoulos et al. (2013) |
| Euphorbiaceae | Croton argyrophylloides | β-Trans-guaiene | 94 ppm | Ae. aegypti | Contact | Lima et al. (2013) |
| | Croton nepetaefolius | Methyleugenol | 66 ppm | Ae. aegypti | Contact | Lima et al. (2013) |
| | Croton sonderianus | Spathulenol | 55 ppm | Ae. aegypti | Contact | Lima et al. (2013) |

*(Continued)*

**TABLE 24.2 (CONTINUED)**

Most Common Substances Contained in Essential Oil Compounds of Plants as Pesticides

| Family | Plant Genus and Species | Major Constituents | LC$_{50}$ or LD$_{50}$ | Target Organism | Method | Reference |
|---|---|---|---|---|---|---|
| | Croton zehntneri | (E)-Anethole | 25–56 ppm | Ae. aegypti | Contact | Lima et al. (2013), Santos et al. (2007), Morais et al. (2006) |
| Fabaceae | Copaifera multijuga | β-Caryo phyllene, α–humulene | 18 ppm / 128 ppm | Ae. aegypti / An. darlingi | Contact / Contact | Trindade et al. (2013) |
| | Hymenaea courbaril | α-Copaene, spathulenol, β-silinene | 15 ppm | Ae. aegypti | Contact | Aguiar et al. (2010) |
| | Psoralea corylifolia | Caryophyllene oxide, phenol, 4-(3,7-dimethyl-3-ethenylocta-1,6-dienyl), caryophyllene | 63 ppm | C. quinquefasciatus | Contact | Dua et al. (2013) |
| Geraniaceae | Pelargonium radula | Cetronellol geranial | 1.54 µg/g | S. gregaria (third instar) | Topical | Mansour et al. (2015) |
| | | | 22.719 µg/g | S. gregaria (third instar) | Contact | Mansour and Abdel-Hamid (2015) |
| Hypericaceae | Hypericum scabrum | α-Pinene | 82 ppm | C. pipiens | Contact | Cetin et al. (2011) |
| Lamiaceae | Coleus aromaticus | Thymol | 72 ppm | Culex tritaeniorhynchus | Contact | Govindarajan et al. (2013a) |
| | | | 76 ppm | Ae. albopictus | Contact | Govindarajan et al. (2013a) |
| | | | 60 ppm | Anopheles subpictus | Contact | Govindarajan et al. (2013a) |
| | Dracocephalum kotschyi | Limonene | 4.4 µl/L | Myzus persicae | Contact | Jalaei et al. (2015) |
| | Hyptis martiusii | γ-3-Carene, 1,8-cineole | 18 ppm / 27 ppm | Ae. aegypti / C. quinquefasciatus | Contact / Contact | Costa et al. (2004) / Costa et al. (2005) |
| | Lavandula gibsoni | α-Terpinolen, thymol | 48 ppm / 62 ppm / 54 ppm | Ae. aegypti / An. stephensi / C. quinquefasciatus | Contact / Contact / Contact | Kulkarni et al. (2013) / Kulkarni et al. (2013) / Kulkarni et al. (2013) |
| | Lippia sidoides | Thymol, α-felandreno | 20–25 ppm / 17 ppm | Ae. aegypti / C. quinquefasciatus | Contact / Contact | Lima et al. (2013) / Costa et al. (2005) |

(Continued)

**TABLE 24.2 (CONTINUED)**

Most Common Substances Contained in Essential Oil Compounds of Plants as Pesticides

| Family | Plant Genus and Species | Major Constituents | LC$_{50}$ or LD$_{50}$ | Target Organism | Method | Reference |
|---|---|---|---|---|---|---|
| | *Mentha longifolia* | Piperitenone oxid | 17 ppm | *C. quinquefasciatus* | Contact | Pavela et al. (2014) |
| | *Mentha spicata* | Carvone, *cis*-carveol, limonene | 62 ppm | *C. quinquefasciatus* | Contact | Govindarajan et al. (2011) |
| | | | 52 ppm | *Ae. aegypti* | Contact | Govindarajan et al. (2011) |
| | | | 49 ppm | *An. stephensi* | Contact | Govindarajan et al. (2011) |
| | *Ocimum americanum* | (E)-Methyl-cinnamate | 67 ppm | *Ae. aegypti* | Contact | Cavalcanti et al. (2004) |
| | *Ocimum basilicum* | Linalool, eugenol, methyl eugenol | 1.54 µg/g | *S. gregaria* (third instar) | Topical | Mansour et al. (2015) |
| | | | 19.541 µg/g | *S. gregaria* (third instar) | Contact | Mansour and Abdel-Hamid (2015) |
| | | | 14 ppm | *C. tritaeniorhynchus* | Contact | Govindarajan et al. (2013b) |
| | | | 11 ppm | *Ae. albopictus* | Contact | Govindarajan et al. (2013b) |
| | | | 9 ppm | *An. subpictus* | Contact | Govindarajan et al. (2013b) |
| | *Ocimum gratissimum* | Eugenol, 1,8-cineole | 60 ppm | *Ae. aegypti* | Contact | Cavalcanti et al. (2004) |
| | *Ocimum sanctum* | Methyleugenol | 85 ppm | *Ae. aegypti* | Contact | Gbolade and Lockwood (2008) |
| | *Origanum vulgare* | Terpinene 1-ol-4 | 1.56 µg/g | *S. gregaria* (third instar) | Topical | Mansour et al. (2015) |
| | | | 53.333 µg/g | *S. gregaria* (third instar) | Contact | Mansour and Abdel-Hamid (2015) |

*(Continued)*

**TABLE 24.2 (CONTINUED)**

Most Common Substances Contained in Essential Oil Compounds of Plants as Pesticides

| Family | Plant Genus and Species | Major Constituents | LC$_{50}$ or LD$_{50}$ | Target Organism | Method | Reference |
|---|---|---|---|---|---|---|
| | *Plectranthus amboinicus* | Carvacrol | 52 ppm | *Ae. aegypti* | Contact | Lima et al. (2011) |
| | | | 55 ppm | *Anopheles gambiae* | Contact | Kweka et al. (2012) |
| | *Plectranthus molli* | Piperitone oxide, fenchone | 25 ppm | *Ae. aegypti* | Contact | Kulkarni et al. (2013) |
| | | | 33 ppm | *An. stephensi* | Contact | Kulkarni et al. (2013) |
| | | | 29 ppm | *C. quinquefasciatus* | Contact | Kulkarni et al. (2013) |
| | *Pulegium vulgare* | Pulegone, carvone | 64 ppm | *C. quinquefasciatus* | Contact | Pavela et al. (2014) |
| | *Rosmarinus officinalis* | 2-Methoxy-3-(2-propenyl)-phenol, 1,8-cineole, camphor | 0.057 µl/L air: 72 h | *S. oryzae* | Fumigation | Kiran and Prakash (2015) |
| | | | 0.039 µl/L air: 36 h | *Oryzaephilus surinamensis* | Fumigation | Kiran and Prakash (2015) |
| | *Satureja hortensis* | γ-Terpinene, carvacrol | 38 ppm | *C. quinquefasciatus* | Contact | Yu et al. (2013) |
| | *Thymus bovei* | Thymol, p-cymene | 36 ppm | *C. quinquefasciatus* | Contact | Pavela (2009) |
| | | | 1300 µg/g | *S. gregaria* (third instar) | Topical | Mansour et al. (2015) |
| | *Thymus capitatus* | Thymol carvacrol | 1038 µg/g | *S. gregaria* (third instar) | Topical | Mansour et al. (2015) |
| | *Thymus leucospermus* | p-Cymene | 34 ppm | *C. pipiens* | Contact | Pitarokili et al. (2011) |
| | *Thymus satureoides* | Thymol, borneol | 43 ppm | *C. quinquefasciatus* | Contact | Pavela (2009) |
| | *Thymus teucrioides* subsp. Candilicus | p-Cymene, γ-terpinene, thymol | 23 ppm | *C. pipiens* | Contact | Pitarokili et al. (2011) |
| | *Thymus vulgaris* | Thymol, p-cymene | 1128 µg/g | *S. gregaria* (third instar) | Topical | Mansour et al. (2015) |
| | | | 33 ppm | *C. quinquefasciatus* | Contact | Pavela (2009) |
| | *Vitex agnus castus* | *trans*-β-Caryophyllene, 1,8-cineole | 83 ppm | *C. pipiens* | Contact | Cetin et al. (2011) |

*(Continued)*

**TABLE 24.2 (CONTINUED)**

Most Common Substances Contained in Essential Oil Compounds of Plants as Pesticides

| Family | Plant Genus and Species | Major Constituents | $LC_{50}$ or $LD_{50}$ | Target Organism | Method | Reference |
|---|---|---|---|---|---|---|
| Lauraceae | *Cinnamomum impressicostatum* | Benzyl benzoate, α-phellandrene | 11 ppm | *Ae. aegypti* | Contact | Lima et al. (2011) |
| | *Cinnamomum microphyllum* | Benzyl benzoate | 7 ppm | *Ae. aegypti* | Contact | Jantan et al. (2005) |
| | *Cinnamomum mollissimum* | Benzyl benzoate | 10 ppm | *Ae. aegypti* | Contact | |
| | *Cinnamomum osmophloeum* | Cinnamaldehyde, cinnamyl acetate | 36 ppm | *Ae. aegypti* | Contact | Cheng et al. (2004) |
| | *Cinnamomum pubescens* | Benzyl benzoate | 13 ppm | *Ae. aegypti* | Contact | Jantan et al. (2005) |
| | *Cinnamomum rhyncophyllum* | Benzyl benzoate | 6 ppm | *Ae. aegypti* | Contact | Jantan et al. (2005) |
| | *Cinnamomum scortechinii* | β-Phellandrene, linalool | 22 ppm | *Ae. aegypti* | Contact | Jantan et al. (2005) |
| | *Lindera obtusiloba* | α-Copaene, β-caryophyllene | 24 ppm | *Ae. aegypti* | Contact | Chung and Moon (2011) |
| Meliaceae | *Guarea humaitensis* | Caryophyllene epoxide, humulene epoxide II | 48 ppm | *Ae. aegypti* | Contact | Magalhães et al. (2010) |
| | \**Guarea scabra* | *cis*-Caryophyllene, α-*trans*-bergamotene | 98 ppm | *Ae. aegypti* | Contact | Magalhães et al. (2010) |
| Myrtaceae | *Eucalyptus astringens* | Pinene, camphene | 13.91–17.58 μl/L air | *Rhyzopertha dominica* | Fumigation | Hamdi et al. (2015) |
| | | | 121.46–180.432 μl/L air | *Callosobruchus maculatus* | Fumigation | Hamdi et al. (2015) |
| | | | 274.16–293.15 μl/L air | *Tribolium castaneum* | Fumigation | Hamdi et al. (2015) |
| | *Eucalyptus camaldulensis* | 1,8-Cineole, α-terpinyl acetate | 31 ppm 55 ppm | *Ae. aegypti* *Ae. albopictus* | Contact Contact | Cheng et al. (2009) Cheng et al. (2009) |
| | *Eucalyptus citriodora* | Citronellal, citronellol, α-humulene isopulegol | 71 ppm | *Ae. aegypti* | Contact | Vera et al. (2014) |
| | *Eucalyptus dunnii* | 1,8-Cineole, γ-terpinene | 25 ppm | *Ae. aegypti* | Contact | Lucia et al. (2008) |
| | *Eucalyptus grandis* | α-Pinene, β-pinene, 1,8-cineole | 32 ppm | *Ae. aegypti* | Contact | Lucia et al. (2007) |

*(Continued)*

**TABLE 24.2 (CONTINUED)**

Most Common Substances Contained in Essential Oil Compounds of Plants as Pesticides

| Family | Plant Genus and Species | Major Constituents | LC$_{50}$ or LD$_{50}$ | Target Organism | Method | Reference |
|---|---|---|---|---|---|---|
| | *Eucalyptus gunnii* | 1,8-Cineole, p-cymene, β-phellandrene | 21 ppm | *Ae. aegypti* | Contact | Lucia et al. (2008) |
| | *Eucalyptus lehmani* | Pinene, camphene | 11.51–16.06 μl/L air | *R. dominica* | Fumigation | Hamdi et al. (2015) |
| | | | 80.43–144.82 μl/L air | *C. maculatus* | Fumigation | Hamdi et al. (2015) |
| | | | 225.35–282.06 μl/L air | *T. castaneum* | Fumigation | Hamdi et al. (2015) |
| | *Eucalyptus saligna* | 1,8-Cineole, p-cymene | 22 ppm | *Ae. aegypti* | Contact | Lucia et al. (2008) |
| | *Eucalyptus tereticornis* | β-Phellandrene, 1,8-cineole | 22 ppm | *Ae. aegypti* | Contact | Lucia et al. (2008) |
| | *Eugenia melanadenia* | 1,8-Cineole | 85 ppm | *Ae. aegypti* | Contact | Aguilera et al. (2003) |
| | *Pimenta racemosa* | Terpinem-4-ol, 1,8-cineole | 27 ppm | *Ae. aegypti* | Contact | Leyva et al. (2009) |
| | *Psidium guajava* | 1,8-Cineole, β-caryophyllene | 25 ppm | *Ae. aegypti* | Contact | Lima et al. (2011) |
| | *Psidium rotundatum* | 1,8-Cineole, α-pinene | 63 ppm | *Ae. aegypti* | Contact | Aguilera et al. (2003) |
| | *Syzigium aromaticum* | Eugenol | 21 ppm | *Ae. aegypti* | Contact | Costa et al. (2005) |
| | | | 25 ppm | *C. quinquefasciatus* | Contact | Cardoso and Lemos (2005) |
| Pinaceae | *Pinus brutia* | α-Pinene, β-pinene | 67 ppm | *Ae. albopictus* | Contact | Koutsaviti et al. (2014) |
| | *Pinus halepensis* | β-Caryophyllene | 70 ppm | *Ae. albopictus* | Contact | Koutsaviti et al. (2014) |
| | *Pinus stankewiczii* | Germacrene D, α-pinene, β-pinene | 82 ppm | *Ae. albopictus* | Contact | Koutsaviti et al. (2014) |
| Piperaceae | *Piper aduncum* | β-Pinene | 57 ppm | *Ae. aegypti* | Contact | Leyva et al. (2009) |
| | *Piper auritum* | Safrole | 17 ppm | *Ae. aegypti* | Contact | Leyva et al. (2009) |
| | *Piper capense* | δ-Cadinene | 34 ppm | *An. gambiae* | Contact | Matasyoh et al. (2011) |

*(Continued)*

**TABLE 24.2 (CONTINUED)**

Most Common Substances Contained in Essential Oil Compounds of Plants as Pesticides

| Family | Plant Genus and Species | Major Constituents | $LC_{50}$ or $LD_{50}$ | Target Organism | Method | Reference |
|---|---|---|---|---|---|---|
| | *Piper hostmanianum* | Asaricin, myristicin | 54 ppm | *Ae. aegypti* | Contact | Morais et al. (2007) |
| | *Piper klotzschianum* | 1-Butyl-3,4-methylenedioxybenzene, limonene, α-phellandrene | 13 ppm | *Ae. aegypti* | Contact | Nascimento et al. (2013) |
| | *Piper marginatum* | (Z)-Asarone, patchouli alcohol | 23 ppm | *Ae. aegypti* | Contact | Autran et al. (2009) |
| | *Piper permucronatum* | Dillapiole, myristicin | 36 ppm | *Ae. aegypti* | Contact | Morais et al. (2007) |
| Pittosporaceae | *Pittosporum tobira* | Undecane, L-limonene, 4-methyl-1,3-pentadiene | 58 ppm | *Ae. aegypti* | Contact | Chung et al. (2010) |
| Poaceae | *Cymbopogon citartus* | n.d. | 33.1 µl/L air | *O. surinamensis* | Fumigation | Lambrano et al. (2015) |
| | | | >604 µl/L air | *Sitophilus zeamais* | Fumigation | Lambrano et al. (2015) |
| | *Cymbopogon flexuosus* | Geranial, neral | 17 ppm | *Ae. aegypti* | Contact | Vera et al. (2014) |
| | *Cymbopogon nardus* | n.d. | 46.9 µl/L air | *O. surinamensis* | Fumigation | Lambrano et al. (2015) |
| | | | >604 µl/L air | *S. zeamais* | Fumigation | Lambrano et al. (2015) |
| | *Cymbopogon martinii* | n.d. | 37.2 µl/L air | *O. surinamensis* | Fumigation | Lambrano et al. (2015) |
| | | | 159 µl/L air | *S. zeamais* | Fumigation | Lambrano et al. (2015) |
| Rutaceae | *Citrus aurantium* | Limonene | 2.94 µl/L air | *Bemisia tabaci* | Fumigation | Zarrad et al. (2015) |
| | | | 39 ppm | *C. pipiens* | Contact | Michaelakis et al. (2009) |
| | *Citrus hystrix* | β-Pinene, d-limonene, terpinene-4-ol | 30 ppm | *Ae. aegypti* | Contact | Sutthanont et al. (2010) |

*(Continued)*

**TABLE 24.2 (CONTINUED)**

Most Common Substances Contained in Essential Oil Compounds of Plants as Pesticides

| Family | Plant Genus and Species | Major Constituents | $LC_{50}$ or $LD_{50}$ | Target Organism | Method | Reference |
|---|---|---|---|---|---|---|
| | *Citrus limon* | Limonene | 30 ppm | *C. pipiens* | Contact | Michaelakis et al. (2009) |
| | *Citrus reticulata* | d-Limonene, γ-terpinene | 15 ppm | *Ae. aegypti* | Contact | Sutthanont et al. (2010) |
| | *Citrus sinensis* | Limonene | 20 ppm | *Ae. aegypti* | Contact | Vera et al. (2014) |
| | | | 51 ppm | *C. pipiens* | Contact | Michaelakis et al. (2009) |
| | *Clausena excavata* | Safrole, terpinolene | 37 ppm | *Ae. aegypti* | Contact | Cheng et al. (2009) |
| | | | 41 ppm | *Ae. albopictus* | Contact | Cheng et al. (2009) |
| | *Chloroxylon swietenia* | Geijerene, limonene, germacrene D | 16 ppm | *Ae. aegypti* | Contact | Kiran et al. (2006) |
| | | | 14 ppm | *An. stephensi* | Contact | Kiran et al. (2006) |
| | *Feronia limonia* | Estragole, β-pinene | 15 ppm | *An. stephensi* | Contact | Senthilkumar et al. (2013) |
| | | | 11 ppm | *Ae. aegypti* | Contact | Senthilkumar et al. (2013) |
| | | | 22 ppm | *C. quinquefasciatus* | Contact | Senthilkumar et al. (2013) |
| | *Murraya tetramera* | α-Cedrene caryophyllene, γ-elemene, α-selinene, α-eudesmol | 0.13–0.63 nL/cm² | *T. castaneum* | Fumigation | You et al. (2015) |
| | *Murraya euchrestifolia* | α-Cedrene caryophyllene, clemene, α-selinene, α-eudesmol | 0.63 nl/cm² | *T. castaneum* | Fumigation | You et al. (2015) |
| | *Murraya koenigii* | α-Cedrene caryophyllene, elemene, α-selinene, α-eudesmol | 0.63 nl/cm² | *T. castaneum* | Fumigation | You et al. (2015) |
| | *Murraya kwangsiensis* | α-Cedrene caryophyllene, elemene, α-selinene, α-eudesmol | 3.15–15.73 nl/cm² | *T. castaneum* | Fumigation | You et al. (2015) |
| | *Murraya exotica* | α-Cedrene caryophyllene, elemene, α-selinene, α-eudesmol | 3.15–15.73 nl/cm² | *T. castaneum* | Fumigation | You et al. (2015) |
| | *Murraya alata* | α-Cedrene caryophyllene, elemene, α-selinene, α-eudesmol | <0.13 nl/cm² | *T. castaneum* | Fumigation | You et al. (2015) |

*(Continued)*

**TABLE 24.2 (CONTINUED)**

Most Common Substances Contained in Essential Oil Compounds of Plants as Pesticides

| Family | Plant Genus and Species | Major Constituents | LC$_{50}$ or LD$_{50}$ | Target Organism | Method | Reference |
|---|---|---|---|---|---|---|
| | *Ruta chalepensis* | 2-Undeca none, 2-nonanone | 22 ppm | *Ae. aegypti* | Contact | Ali et al. (2013) |
| | | | 15 ppm | *Anopheles quadrimaculatus* | Contact | Conti et al. (2013) |
| | | | 35 ppm | *Ae. albopictus* | Contact | Conti et al. (2013) |
| | *Swinglea glutinosa* | β-Pinene, piperitenone, α-pinene | 65 ppm | *Ae. aegypti* | Contact | Vera et al. (2014) |
| | *Toddalia asiatica* | Geraniol, d-limonene, isopimpinellin, 4-vinylguaiacol | 69 ppm | *Ae. albopictus* | Contact | Liu et al. (2012) |
| | *Zanthoxylum armatum* | Linalool, limonene | 54 ppm | *Ae. aegypti* | Contact | Tiwary et al. (2007) |
| | | | 58 ppm | *An. stephensi* | Contact | Tiwary et al. (2007) |
| | | | 49 ppm | *C. quinquefasciatus* | Contact | Tiwary et al. (2007) |
| | *Zanthoxylum limonella* | d-Limonene, terpinen-4-ol | 24 ppm | *Ae. aegypti* | Contact | Pitasawat et al. (2007) |
| | | | 57 ppm | *An. dirus* | Contact | Pitasawat et al. (2007) |
| | *Zanthoxylum oxyphyllum* | Methyl heptyl ketone, methyl nonyl ketone | 7 ppm | *Ae. aegypti* | Contact | Borah et al. (2012) |
| | *Zanthoxylum rhoifolium* | β-Blemene | ~1% | *B. tabaci* | Contact | Christofoli et al. (2015) |
| Scrophulariaceae | *Capraria biflora* | α-Humulene, *trans*-cariophillene | 73 ppm | *Ae. aegypti* | Contact | Souza et al. (2012) |
| | *Stemodia maritima* | β-Caryophyllene, caryophyllene oxide | 23 ppm | *Ae. aegypti* | Contact | Arriaga et al. (2007) |
| Valerianaceae | *Valeriana fauriei* | Bornyl acetate | 34 ppm | *Ae. aegypti* | Contact | Chung et al. (2011b) |
| Verbenaceae | *Lantana camara* | (Z)-Caryo phyllene, bicyclogermacrene, (E)-caryophyllene | 42 ppm | *Ae. aegypti* | Contact | Costa et al. (2010) |

*(Continued)*

**TABLE 24.2 (CONTINUED)**

Most Common Substances Contained in Essential Oil Compounds of Plants as Pesticides

| Family | Plant Genus and Species | Major Constituents | $LC_{50}$ or $LD_{50}$ | Target Organism | Method | Reference |
|---|---|---|---|---|---|---|
| | *Lippia alba* | Carvone, limonene | 15.2–16.7 µl/ml | *S. zeamais* | Repellency | Peixoto et al. (2015) |
| | | | 19.7–28.7 µl/ml | *T. castaneum* | Repellency | Peixoto et al. (2015) |
| | | | 44 ppm | *Ae. aegypti* | Contact | Vera et al. (2014) |
| | *Lippia origanoides* | Carvacrol, p-cymene | 53 ppm | *Ae. aegypti* | Contact | Pavela (2015) |
| | *Lippia sidoides* | Thymol | 63 ppm | *Ae. aegypti* | Contact | Cavalcanti et al. (2004) |
| Zingiberaceae | *Alpinia purpurata* | n.d. | 41.4 µl/L | *S. zeamais* | Fumigant | Soledade de Lira et al. (2015) |
| | *Curcuma aromatica* | 1H-3a,7-methanoazulene, curcumene | 36 ppm | *Ae. aegypti* | Contact | Choochote et al. (2005) |
| | *Curcuma zedoaria* | 1,8-Cineole, p-cymene, α-phellandrene | 31 ppm | *Ae. aegypti* | Contact | Pitasawat et al. (2007) |
| | | | 29 ppm | *An. dirus* | Contact | Pitasawat et al. (2007) |
| | *Kaempferia galanga* | 2-Propeonic acid, pentadecane, ethyl-p-methoxycinnamate | 53 ppm | *Ae. aegypti* | Contact | Sutthanont et al. (2010) |
| | *Zingiber officinale* | Zingiberene, citronellol β-sesguiphel landrene | 46 ppm | *Ae. aegypti* | Contact | Moon et al. (2011) |
| | *Zingiber zerumbet* | α-Humulene, zerumbone | 48 ppm | *Ae. aegypti* | Contact | Sutthanont et al. (2010) |

*Note:* n.d., no data.

hyperactivity, followed by hyperexcitation, leading to rapid knockdown and immobilization (Enan, 2001). The inhibition of acetylcholinesterase (AChE) also plays a key role in modulating pesticidal activity. Several essential oil compounds have been shown to be inhibitors of AChE against different insect species. For example, essential oil compounds of *Zingiber officinale* were found to alter behavior and memory in the cholinergic system (Felipe et al., 2008), while linalool was identified as an inhibitor of acetylcholinesterase (Ryan and Byrne, 1988).

Another possible target suggested for essential oil compounds is interference with GABA-gated chloride channels in insects (Rattan, 2010). Thujone has been classified as a neurotoxic insecticide that acts on $GABA_A$ receptors. Thujone is a competitive inhibitor of [3H]EBOB binding (i.e., of the noncompetitive blocker site of the GABA-gated chloride channel) and is a reversible modulator of the $GABA_A$ receptor (Hold et al., 2000). It has been suggested that thymol potentiates $GABA_A$ receptors through an unidentified binding site (Priestley et al., 2003). The silphinenes antagonize the action of GABA on insect neurons (Bloomquist et al., 2008; Rattan, 2010).

Octopamine is also a target for essential oil compound activity in insects. OA is a naturally occurring, multifunctional, biogenic amine, which plays key roles as a neurotransmitter, neuromodulator, and neurohormone in the invertebrate system, with a physiological function comparable to that of norepinephrine in vertebrates (Enan, 2001). The acute and sublethal behavioral effects of essential oil compounds on insects are consistent with an octopamenergic target site in insects, which acts by blocking octopamine receptors (Rattan, 2010).

In addition, essential oils and their constituents could affect biochemical processes, which specifically disrupt the endocrinological balance of insects. They may act as insect growth regulators, disrupting the normal process of morphogenesis (Reynolds, 1987; Rattan, 2010). However, it is thus apparent that the substances contained in essential oil compounds exhibit not only various mechanisms of action, but also various levels of capacity to penetrate an insect's cuticle and enter its body, which is directly related to the capacity to provide an insecticidal effect.

Although sesquiterpenes seem to provide better efficacy than some monoterpenes, recent research indicates that the ability of mutual synergistic or antagonistic action may also play a very important role in terms of efficacy, which is probably related to the ability of mutual complementarism of the mechanism of action. Complex mixtures containing substances with different mechanisms of action can significantly hinder the ability of insects to intoxicate substances in the body.

For example, the author team has reported some substances that do not cause mortality alone or show low mortality, but when mixed with other compounds, such substances exert a significant effect on the resulting efficacy as a synergist. This is the case for a pulegone together with thymol and 1,8-cineole (Kumrungsee et al., 2014), or as shown in *Spodoptera littoralis* larvae, borneol by itself causes only very low acute toxicity; however, when mixed with other monoterpenes, it significantly increases their efficacy (Pavela, 2014).

## 24.3 Essential Oil Compounds as a Synthetic Pesticide

The environmental problems caused by the overuse of pesticides have been of concern to both scientists and the public in recent years. It has been estimated that about 2.5 million tons of pesticides are used on crops each year, and the worldwide damage caused by pesticides reaches $100 billion annually. The reasons for this are twofold: (1) the high toxicity and nonbiodegradable properties of pesticides and (2) the residues in soil, water resources, and crops that affect public health. Thus, on the one hand, one needs to search for new, highly selective, and biodegradable pesticides to solve the problem of long-term toxicity to mammals, while on the other hand, one must study the environmentally friendly pesticides and develop techniques that can be used to reduce pesticide use while maintaining crop yields. Natural products are an excellent alternative to synthetic pesticides as a means of reducing the negative impacts on human health and the environment. The move toward green chemistry processes and the continuing need for developing new crop protection tools with novel modes of action make discovery and commercialization of natural products as green pesticides an attractive and profitable pursuit that is commanding attention. The concept of "green pesticides" refers to all types of nature-oriented and beneficial pest control materials that can contribute to reducing the pest population and increasing food production. They are safe and eco-friendly. They are more compatible with the environmental components than synthetic pesticides (Isman and Machial, 2006).

As noted in the discussion on efficacy above, until now, plant essential oil compounds have been produced commercially from several botanical sources by many companies. The oils are generally sometimes composed of complex mixtures of monoterpenes, biogenetical related phenols, and sesquiterpenes, and examples include 1,8-cineole, the major constituent of oils from rosemary and eucalyptus; eugenol from clove oil; thymol from garden thyme; menthol from various species of mint; asarones from calamus; and carvacrol and linalool from many plant species.

As lipophilicity plays a key role in modulating pesticidal activity, the association between lipophilic compounds and protein deactivation or enzyme inhibition may be a reasonable explanation for this fact (Ryan and Byrne, 1988). This was confirmed in a chemometric study applied to active compounds such as terpenes and phenylpropanoids, and the activity was strongly correlated with independent variables having a hydrophobic profile (Scotti et al., 2013).

Double bonds are important in the pesticidal activity of natural molecules because hydrogenation of these bonds decreases the lipophilic character of these compounds, restricting their passage through the larvae cuticle (Lomonaco et al., 2009). For example, Lucia et al. (2007) and Perumalsamy et al. (2009) found that α-pinene, which possesses an exocyclic double bond, is more toxic to *Aedes aegypti* larvae than β-pinene, which has an endocyclic double bond.

As there is little information available about the mechanism of essential oil action against insects, the relation between pesticidal effect and the chemical composition of the essential oil has been the focus of some scientific work. Thus, some chemists try syntheses by changing the structural composition of essential oil compounds to study the possible mechanism. Other objectives are to increase the activity, as well as economic gain, as then there is no need to use and/or be reliant on the natural source to extract the essential oil compounds. All these objectives have become active fields for research in drug design. For example, Li et al. (2015) described finding potential pesticides against *Plutella xylostella* based on essential oils by synthesizing the oriented chiral esters in their structure. The preliminary results revealed that synthesized compounds showed significantly improved insecticidal activities compared with the essential oil molecules used at the start.

## 24.4 Advantages of Essential Oil Compounds as a Pesticide

The advantage of using essential oil compounds as pesticides is that they are moderately toxic to mammals (Table 24.3), but with few exceptions, the oils themselves or products based on oils are mostly less toxic to mammals, birds, and fish (Stroh et al., 1998). The toxicity of essential oil compounds based on ecological information is shown in Table 24.4, therefore justifying their placement under acceptable green pesticides. Owing to their volatility, essential oils have limited persistence under field conditions; therefore, although natural enemies are susceptible via direct contact, predators and parasitoids reinvading a treated crop 1 or more days after treatment are unlikely to be poisoned by residue contact, as often occurs with conventional insecticides (Koul et al., 2008). In addition, there are some specific modes of action of neurotoxin between invertebrate and vertebrate that cause differing toxicities between mammals and invertebrate pests.

In fact, pesticides derived from plant essential oils do have several important benefits. Pesticides based on plant essential oils or their constituents have demonstrated efficacy against a range of stored product pests, domestic pests, blood-feeding pests, and certain soft-bodied, agricultural pests, as well as against some plant pathogenic fungi responsible for pre- and postharvest diseases. They may be applied as fumigants, granular formulations, or direct sprays, with a range of effects, from lethal toxicity to repellence and/or oviposition deterrence in insects. These features indicate that pesticides based on plant essential oils could be used in a variety of ways to control a large number of pests.

Oil-in-water microemulsions are being developed as a nanopesticide delivery system to replace the traditional emulsifiable concentrates (oils), to reduce the use of organic solvent, and to increase the disparity and penetration properties of the droplets. The advantages of using pesticide oil-in-water microemulsions for improving the biological efficacy and reducing the dosage of pesticides would be a useful strategy in green pesticide technology.

Due to their volatile nature, there is a much lower level of risk to the environment with essential oil compounds than with current synthetic pesticides. Predator, parasitoid, and pollinator insect populations will be less impacted because of the minimal residual activity, making essential oil–based pesticides compatible with integrated pest management programs such as in the research of Yotavong et al. (2015), where the toxicity on the parasitoid was less than that on the host. It is also clear that resistance will develop more slowly to essential oil–based pesticides owing to the complex mixtures of constituents that characterize many of these oils (Koul et al., 2008). Ultimately, it is in developing countries where the source plants are endemic that these pesticides may have their greatest impact in an integrated pest management strategy. It is expected that these pesticides will find their greatest commercial application in urban pest control, public health, veterinary health, vector control vis-à-vis human health, and protection of stored commodities. In agriculture, these pesticides will be most useful for protected crops (e.g., greenhouse crops), for high-value row crops, and within organic food production systems where few alternative pesticides are available (Koul et al., 2008). Koul et al. (2008) noted the advantages of using essential oil compounds: (1) there are changing consumer preferences toward the use of natural as opposed to synthetic products; (2) there are expanding niche markets, where quality is more important than price; (3) there is strong growth in demand for essential oils and plant extracts; (4) there is potential to extend the range of available products, including new product development through biotechnology; and (5) there is production of essential oils and plant extracts from low-cost, developing countries.

**TABLE 24.3**

Mammal Toxicity of Some Essential Oil Compounds

| Compound | Animal Tested | Route | $LD_{50}$ (mg/kg) | Reference |
|---|---|---|---|---|
| 2-Acetonaphthone | Mice | Oral | 599 | Sigma-Aldrich (2012a) |
| | Dog | Intravenous | 500 | Koul (2005) |
| Apiol anisaldehyde | Rat | Oral | 1,510 | Koul (2005) |
| | Rabit | Dermal | 5,000 | Thermo Scientific (2015) |
| *trans*-Anethole | Rat | Oral | 2,090 | FAO (1999) |
| Carene | Rat | Oral | 4,800 | FAO (1999) |
| (+)Carvone | Rat | Oral | 1,640 | FAO (1999) |
| 1,8-Cineole | Rat | Oral | 2,480 | FAO (1999); Sigma-Aldrich (2013d) |
| Cinnamaldehyde | Guinea pig | Oral | 1,160 | FAO (1999) |
| | Rat | Oral | 2,220 | Koul (2005) |
| Citral | Rat | Oral | 4,960 | FAO (1999) |
| | Rabbit | Dermal | 2,250 | Sigma-Aldrich (2013a) |
| Citronellol | Mice | IMS | 4,000 | FAO (1999) |
| | Rat | Oral | 3,450 | Sigma-Aldrich (2015a) |
| | Rabbit | Dermal | 2,650 | Sigma-Aldrich (2015a) |
| Clove oil | Rat | Oral | 3,720 | FAO (1999) |
| Cumin oil | Rat | Oral | 2,500 | FAO (1999) |
| Dillapiol | Rat | Oral | 1,000–1,500 | Koul (2005) |
| iso-Eugenol | Rat | Oral | 1,560 | FAO (1999) |
| Geraniol | Rat | Oral | 3,600 | FAO (1999) |
| | Rabbit | Dermal | >5,000 | Sigma-Aldrich (2014a) |
| 3-Isothujone | Mice | Subcutaneous | 442.2 | Koul (2005) |
| Limonene | Rat | Oral | 4,600 | FAO (1999) |
| | Rabbit | Dermal | >5,000 | Sigma-Aldrich (2015b) |
| Linalool | Rat | Oral | 2,790 | FAO (1999) |
| | Rabit | Dermal | 5,610 | Koul (2005) |
| Maltol | Rat | Oral | 2,330 | Koul (2005) |
| | Mice | Oral | 550 | Sigma-Aldrich (2015c) |
| | Rabbit | Oral | 1,620 | Sigma-Aldrich (2015c) |
| | Mice | Subcutaneous | 820 | Sigma-Aldrich (2015c) |
| Menthol | Rat | Oral | 3,180 | FAO (1999) |
| 4-Methoxyphenol | Rat | Dermal | >2,000 | Sigma-Aldrich (2015d) |
| Methyl chavicol | Rat | Oral | 1,820 | FAO (1999) |
| Methyl eugenol | Rat | Oral | 810 | Sigma-Aldrich (2012b) |
| Myrcene | Rat | Oral | >11,390 | Sigma-Aldrich (2014b) |
| | Rabbit | Dermal | >5,000 | Sigma-Aldrich (2014b) |
| Origanum oil | Rat | Oral | 1,850 | FAO (1999) |
| | Rabbit | Dermal | 320 | Sigma-Aldrich (2013b) |
| α-Pinene | Rat | Oral | 3,700 | Sigma-Aldrich (2014c) |
| | Rabbit | Dermal | >5,000 | Sigma-Aldrich (2014c) |
| (+)-Pulegone | Mice | Intraperitoneal | 150 | FAO (1999) |
| γ-Terpinene | Rat | Oral | 3,650 | Sigma-Aldrich (2012c) |
| Terpinen-4-ol | Rat | Oral | 4,300 | FAO (1999) |
| Thujone | Mice | Subcutaneous | 87.5 | Koul (2005) |
| | Rat | Intraperitoneal | 120 | FAO (1999) |
| Thymol | Mice | Oral | 1,050–1,200 | EFSA (2012) |
| | Rat | Oral | 980 | FAO (1999) |
| | Rat | Dermal | >2,000 | EFSA (2012) |
| | Guinea pig | Oral | 880 | EFSA (2012) |

*Note:*  IMS, intramuscular stimulation.

**TABLE 24.4**

Ecological Information on Some Essential Oil Compounds

| Compound | Animal Tested | Method | Result | Reference |
|---|---|---|---|---|
| 1,8-Cineole | *Pimephales promelas* | Static test | 96 h LC50 = 102 mg/L | Sigma-Aldrich (2015f) |
| Citronellol | *Leuciscus idus* | Static test | 96 h LC50 = 10–22 mg/L | Sigma-Aldrich (2015a) |
| | *Daphnia magna* | OECD test Guideline 202 | 48 h LC50 = 17 mg/L | Sigma-Aldrich (2015a) |
| | *Algae* | Static test | 72 h LC50 = 2.4 mg/L | Sigma-Aldrich (2015a) |
| Euglenol | *Danio rerio* | Static test | 96 h LC50 = 13 mg/L | Sigma-Aldrich (2013c) |
| | *Daphnia magna* | OECD test Guideline 202 | 48 h LC50 = 1.13 mg/L | Sigma-Aldrich (2013c) |
| Geraniol | *Danio rerio* | Static test | 96 h LC50 = 22 mg/L | Sigma-Aldrich (2014a) |
| | *Daphnia magna* | OECD test Guideline 202 | 48 h LC50 = 10.8 mg/L | Sigma-Aldrich (2014a) |
| | *Desmodesmus subspicatus* | Growth inhibitor | 72 h LC50 = 13.1 mg/L | Sigma-Aldrich (2014a) |
| Linalool | *Oncorphynchus mykiss* | Static test | 96 h LC50 = 27.8 mg/L | Sigma-Aldrich (2015e) |
| | *Daphnia magna* | Immobilization | 48 h EC50 = 59 mg/L | Sigma-Aldrich (2015e) |
| | *Desmodesmus subspicatus* | Static test | 96 h LC50 = 156.7 mg/L | Sigma-Aldrich (2015e) |
| Limonene | *Pimephales promelas* | OECD test Guideline 203 | 96 h LC50 = 0.72 mg/L | Sigma-Aldrich (2015b) |
| | *Daphnia magna* | OECD test Guideline 202 | 48 h LC50 = 0.36 mg/L | Sigma-Aldrich (2015b) |
| 4-Methoxyphenol | *Oncorhynchus mykiss* | OECD test Guideline 202 | 96 h LC50 = 28.5 mg/L | Sigma-Aldrich (2015d) |
| | *Daphnia magna* | OECD test Guideline 202 | 48 h LC50 = 3 mg/L | Sigma-Aldrich (2015d) |
| | *Pseudokirchneriella subcapitata* | OECD test Guideline 201 | 72 h EC50 = 54.7 mg/L | Sigma-Aldrich (2015d) |
| Methyl eugenol | *Oncorhynchus mykiss* | N/A | 96 h LC50 = 6 mg/L | Sigma-Aldrich (2012b) |
| α-Pinene | *Daphnia magna* | OECD test Guideline 202 | 48 h EC50 = 48 mg/L | Sigma-Aldrich (2014c) |
| Pulegone | *Daphnia pulex* | N/A | 48 h EC50 = 24.4 mg/L | Sigma-Aldrich (2015f) |

*Note:* N/A, no testing method described.

## 24.5 Limitations of Essential Oil Compounds as a Pesticide

In spite of the considerable research effort in many laboratories throughout the world and an ever-increasing volume of scientific literature on the pesticidal properties of essential oils and their constituents, surprisingly few pest control products based on plant essential oils have appeared in the marketplace. This may be a consequence of regulatory barriers to commercialization (i.e., cost of toxicological and environmental evaluations), or the fact that the efficacy of essential oils toward pests and diseases is not as apparent or obvious as that seen with currently available products.

Many countries still face some problems in the liberal use of essential oil compounds as biopesticides. There are several reasons why farmers are reluctant to adopt biopesticides to replace synthetic chemicals. Essential oil compounds generally act more slowly than synthetic pesticides, killing arthropod pests over a longer time, rather than having an immediately apparent knockdown effect, so the pests can still damage the crops after application. Moreover, essential oils also require somewhat greater application rates (as high as 1% active ingredient) and may require frequent reapplication when used outdoors. Because of this, farmers still prefer to use and rely on the quick-acting synthetic pesticides, causing high pest mortality shortly after application.

In addition, as the chemical profile of plant species can vary naturally, depending on geographic, genetic, climatic, annual, or seasonal factors, pesticide manufacturers must take additional steps to ensure that their products will perform consistently. All this requires substantial cost, and smaller companies may not be willing to invest the required funds unless there is a high probability of recovering the costs through some form of market exclusivity (e.g., patent protection).

Lastly, the poor quality of some of the noncommercial biopesticide products is also a cause for concern (Warburton et al., 2002). A relevant issue is the illegal importation of unregistered products that may contain extremely low levels of active agents and sometimes a cocktail of several agents, so that their use requires high volumes. In addition, some commercial products, which do not pass the registration process, are still being sold in the market, which creates doubts about ineffective products for control, and thus people are reluctant to use them.

Therefore, cost-competitiveness and product quality and performance of essential oils are the two major barriers standing in the way of their adoption. Overcoming these barriers will require more effort in technology transfer, so that research results are more readily translated into field use. Even more importantly, it will require additional funding for the optimization of local production strategies, farmer education, continuing research, and even subsidization of biopesticide retail costs, so that their availability to farmers is ensured.

## Acknowledgments

I would like to thank the KURDI service for their kind help in checking the English grammar in an earlier version and for their useful suggestions. In addition, I thank my family and especially my husband, Philippe Milon, for their encouragement. Thanks also to Dr. Eric Wajnberg and Dr. Opender Koul for their encouragement, which enabled me to start writing this chapter. Finally, thanks to Laika and Tako for being with me whenever I was writing this chapter.

## References

Aciole, S.D.G., Piccoli, C.F., Duquel, J.E., Costa, E.V., Navarro-Silva, M.A., Marques, F.A., Maia, B.L.N.S., Pinheiro, M.L.B., Rebelo, M.T. 2011. Insecticidal activity of three species of *Guatteria* (Annonaceae) against *Aedes aegypti* (Diptera: Culicidae). *Rev. Colomb. Entomol.* 37: 262–268.

Aguiar, J.C.D., Santiago, G.M.P., Lavor, P.L., Veras, H.N., Ferreira, Y.S., Lima, M.A., Arriaga, A.M. et al. 2010. Chemical constituents and larvicidal activity of *Hymenaea courbaril* fruit peel. *Nat. Prod. Commun.* 5: 1977–1980.

Aguilera, L., Navarro, A., Tacoronte, J.E., Leyva, M., Marquetti, M.C. 2003. Efecto letal de myrtaceas cubanas sobre *Aedes aegypti* (Diptera: Culicidae). *Rev. Cubana Med. Trop.* 55: 100–104.

Ali, A., Demirci, B., Kiyan, H.T., Bernier, U.R., Tsikolia, M., Wedge, D.E., Khan, I.A., Başer, K.H., Tabanca, N. 2013. Biting deterrence, repellency, and larvicidal activity of *Ruta chalepensis* (Sapindales: Rutaceae) essential oil and its major individual constituents against mosquitoes. *J. Med. Entomol.* 50: 1267–1274.

Arriaga, A.M.C., Rodrigues, F.E.A., Lemos, T.L.G., Oliveira, M.C.F., Lima, J.Q., Santiago, G.M.P., Braz-Filho, R., Mafezoli, J. 2007. Composition and larvicidal activity of essential oil from *Stemodia maritima* L. *Nat. Prod. Commun.* 2: 1237–1239.

Autran, E.S., Neves, I.A., Silva, C.S.B., Santos, G.K.N., Câmara, C.A.G., Navarro, D.M.A.F. 2009. Chemical composition, oviposition deterrent and larvicidal activities against *Aedes aegypti* of essential oils from *Piper marginatum* Jacq. (Piperaceae). *Bioresour. Technol.* 100: 2284–2288.

Banthorpe, D.V. 1991. Classification of terpenoids and general procedures for their characterization. In B.V. Charlwood, D.V. Banthorpe (eds.), *Methods in Plant Biochemistry*. Vol. 7. Academic Press, London, pp. 1–41.

Bloomquist, J.R., Boina, D.R., Chow, E., Carlier, P.R., Reina, M., Gonzalez-Coloma, A. 2008. Mode of action of the plant-derived silphinenes on insect and mammalian $GABA_A$ receptor/chloride channel complex. *Pestic. Biochem. Phys.* 91: 17–23.

Borah, R., Saikia, K., Talukdar, A.K., Kalita, M.C. 2012. Chemical composition and biological activity of the leaf essential oil of *Zanthoxylum oxyphyllum*. *Planta Med.* 78: 100.

Burt, S. 2004. Essential oils: Their antibacterial properties and potential applications in foods—A review. *Int. J. Food Microbiol.* 94: 223–253.

Cardoso, A.L.H., Lemos, T.L.G. 2005. Chemical-biological study of the essential oils of *Hyptis martiusii*, *Lippia sidoides* and *Syzigium aromaticum* against larvae of *Aedes aegypti* and *Culex quinquefasciatus*. *Braz. J. Pharmacogn.* 15: 304–309.

Cavalcanti, E.S.B., Morais, S.M., Lima, M.A.A., Santana, E.W.P. 2004. Larvicidal activity of essential oils from Brazilian plants against *Aedes aegypti* L. *Mem. Inst. Oswaldo Cruz* 99: 541–544.

Cetin, H., Yanikoglu, A., Cilek, J.E. 2011. Larvicidal activity of selected plant hydrodistillate extracts against the house mosquito, *Culex pipiens*, a West Nile virus vector. *Parasitol. Res.* 108: 943–948.

Cheng, S.S., Lin, J.Y., Tsai, K.H., Chen, W.J., Chang, S.T., 2004. Chemical composition and mosquito larvicidal activity of essential oils from leaves of different *Cinnamomum osmophloeum* provenances. *J. Agric. Food Chem.* 52: 4395–4400.

Cheng, S.S., Chua, M.T., Chang, E.H., Huang, C.G., Chen, W.J., Chang, S.T. 2009. Variations in insecticidal activity and chemical compositions of leaf essential oils from *Cryptomeria japonica* at different ages. *Bioresour. Technol.* 100: 465–470.

Cheng, S.S., Lin, C.Y., Chung, M.J., Liu, Y.H., Huang, C.G., Chang, S.T. 2013. Larvicidal activities of wood and leaf essential oils and ethanolic extracts from *Cunninghamia konishii* Hayata against the dengue mosquitoes. *Ind. Crop Prod.* 47: 310–315.

Choochote, W., Chaiyasit, D., Kanjanapothi, D., Rattanachanpichai, E., Jitpakdi, A., Tuetun, B., Pitasawat, B. 2005. Chemical composition and anti-mosquito potential of rhizome extract and volatile oil derived from *Curcuma aromatica* against *Aedes aegypti* (Diptera: Culicidae). *J. Vector Ecol.* 30: 302–309.

Christofoli, M., Costa, E.C.C., Bicalho, K.U., de C. Domingues, V., Peixoto, M.F., Alves, C.C.F., Araújo, W.L., de M. Caza. C. 2015. Insecticidal effect of nanoencapsulated essential oils from *Zanthoxylum rhoifolium* (Rutaceae) in *Bemisia tabaci* populations. *Ind. Crop Prod.* 70: 301–308.

Chung, I.M., Moon, H.I. 2011. Composition and immunotoxicity activity of essential oils from *Lindera obtusiloba* Blume against *Aedes aegypti* L. *Immunopharmacol. Immunotoxicol.* 33(1): 146–149.

Chung, I.M., Kim, E.H., Moon, H.I. 2011b. Immunotoxicity activity of the major essential oils of *Valeriana fauriei* Briq against *Aedes aegypti* L. *Immunopharmacol. Immunotoxicol.* 33: 107–110.

Chung, I.M., Seo, S.H., Kang, E.Y., Park, S.D., Park, W.H., Moon, H.I. 2009. Chemical composition and larvicidal effects of essential oil of *Dendropanax morbifera* against *Aedes aegypti* L. *Biochem. Syst. Ecol.* 37: 470–473.

Chung, I.M., Seo, S.H., Kang, E.Y., Park, W.H., Moon, H.I. 2010. Larvicidal effects of the major essential oil of *Pittosporum tobira* against *Aedes aegypti* (L.). *J. Enzyme Inhib. Med. Chem.* 25: 391–393.

Chung, I.M., Song, H.K., Yeo, M.A., Moon, H.I. 2011a. Composition and immunotoxicity activity of major essential oils from stems of *Allium victorialis* L. var. *platyphyllum Makino* against *Aedes aegypti* L. *Immunopharmacol. Immunotoxicol.* 33: 480–483.

Coats, R., Karr, L.L., Drewes, C.D. 1991. Toxicity and neurotoxic effects of monoterpenoids in insects and earthworms. In P. Hedin (ed.), *Natural Occurring Pest Bioregulators.* American Chemical Society Symposium Series 449. pp. 305–316. Retrieved from: http://lib.dr.iastate.edu/ent_pubs/377/.

Conti, B., Leonardi, M., Pistelli, L., Profeti, R., Ouerghemmi, I., Benelli, G. 2013. Larvicidal and repellent activity of essential oils from wild and cultivated *Ruta chalepensis* L. (Rutaceae) against *Aedes albopictus* Skuse (Diptera: Culicidae), an arbovirus vector. *Parasitol. Res.* 112: 991–999.

Costa, J.G.M., Pessoa, O.D.L., Menezes, E.A., Santiago, G.M.P., Lemos, T.L.G. 2004. Composition and larvicidal activity of essential oils from heartwood of *Auxemma glazioviana* Taub. (Boraginaceae). *Flavour Fragr. J.* 19: 529–531.

Costa, J.G.M., Rodrigues, F.F.G., Angélico, E.C., Silva, I., Mota, M.L., Santos, N.K.A., Cardoso, A.L.H., Lemos, T.L.G. 2005. Chemical-biological study of the essential oils of *Hyptis martiusii, Lippia sidoides* and *Syzigium aromaticum* against larvae of *Aedes aegypti* and *Culex quinquefasciatus.* *Braz. J. Pharmacogn.* 15: 304–309.

Costa, J.G.M., Rodrigues, F.F.G., Sousa, E.O., Junior, D.M.S., Campos, A.R., Coutinho, H.D.M., Lima, S.G. 2010. Composition and larvicidal activity of the essential oils of *Lantana camara* and *Lantana montevidensis.* *Chem. Nat. Compd.* 46: 313–315.

Dev, S., Koul, O. 1997. *Insecticides of Natural Origin.* Harwood Academic Publishers, Amsterdam.

Dharmagadda, V.S.S., Naik, S.N., Mittal, P.K., Vasudevan, P. 2005. Larvicidal activity of *Tagetes patula* essential oil against three mosquito species. *Bioresour. Technol.* 96: 1235–1240.

Dua, V.K., Kumar, A., Pandey, A.C., Kumar, S. 2013. Insecticidal and genotoxic activity of *Psoralea corylifolia* Linn. (Fabaceae) against *Culex quinquefasciatus* Say, 1823. *Parasites Vector* 6: 30.

EFSA (European Food Safety Authority). 2012. Conclusion on the peer review of the pesticide risk assessment of the active substance thymol. *EFSA J.* 10(11): 2916.

Enan, E. 2001. Insecticidal activity of essential oils: Octopaminergic sites of action. *Comp. Biochem. Physiol. C Toxicol. Pharmacol.* 130: 325–337.

Evergetis, E., Michaelakis, A., Kioulos, E., Koliopoulos, G., Haroutounian, S.A. 2009. Chemical composition and larvicidal activity of essential oils from six Apiaceae family taxa against the West Nile virus vector *Culex pipiens. Parasitol. Res.* 105: 117–124.

Fahn, A. 2000. Structure and function of secretory cells. *Adv. Bot. Res.* 31: 37–75.

FAO (Food and Agriculture Organization). 1999. The use of spices and medicinals as bioactive protectants for grains. *Agric. Serv. Bull.* 137: 201–213.

Felipe, C.F.B., Fonsêca, K.S., Barbosa, A.L.R., Bezerra, J.N.S., Neto, M.A., França Fonteles, M.M., Barros Viana, G.S. 2008. Alterations in behavior and memory induced by the essential oil of *Zingiber officinale* Roscoe (ginger) in mice are cholinergic-dependent. *J. Med. Plant Res.* 2: 163–170.

Gbolade, A.A., Lockwood, G.B. 2008. Toxicity of *Ocimum sanctum* L. essential oil to *Aedes aegypti* larvae and its chemical composition. *J. Essent. Oil Bear. Plants* 11: 148–153.

Giatropoulos, A., Pitarokili, D., Papaioannou, F., Papachristos, D.P., Koliopoulos, G., Emmanouel, N., Tzakou, O., Michaelakis, A. 2013. Essential oil composition, adult repellency and larvicidal activity of eight Cupressaceae species from Greece against *Aedes albopictus* (Diptera: Culicidae). *Parasitol. Res.* 112: 1113–1123.

Golfakhrabadi, F., Khanavi, M., Ostad, S.N., Saeidnia, S., Vatandoost, H., Abai, M.R., Hafizi, M., Yousefbeyk, F., Rad, Y.R., Baghenegadian, A., Ardekani, M.R.S. 2015. Biological activities and composition of *Ferulago carduchorum* essential oil. *J. Arthropod-Borne Dis.* 9: 104–115.

Govindarajan, M., Sivakumar, R., Rajeswari, M., Yogalakshmi, K. 2011. Chemical composition and larvicidal activity of essential oil from *Mentha spicata* (Linn.) against three mosquito species. *Parasitol. Res.* 110: 2023–2032.

Govindarajan, M., Sivakumar, R., Rajeswary, M., Veerakumar, K. 2013a. Mosquito larvicidal activity of thymol from essential oil of *Coleus aromaticus* Benth. against *Culex tritaeniorhynchus, Aedes albopictus,* and *Anopheles subpictus* (Diptera: Culicidae). *Parasitol. Res.* 112: 3713–3721.

Govindarajan, M., Sivakumar, R., Rajeswary, M., Yogalakshmi, K. 2013b. Chemical composition and larvicidal activity of essential oil from *Ocimum basilicum* (L.) against *Culex tritaeniorhynchus, Aedes albopictus* and *Anopheles subpictus* (Diptera: Culicidae). *Exp. Parasitol.* 134: 7–11.

Grayson, D.H. 2000. Monoterpenoids. *Nat. Prod. Rep.* 17: 385–419.

Hamdi, S.H., Hedjal-Chebheb, M., Kellouche, A., Khouja, M.L., Boudabous, A., Jemâa, J.M.B. 2015. Management of three pests' population strains from Tunisia and Algeria using *Eucalyptus* essential oils. *Ind. Crop Prod.* 74: 551–556.

Hold, K.M., Sirisoma, N.S., Ikeda, T., Narahashi, T., Casida, J.E. 2000. alpha-Thujone (the active component of absinthe): g-Aminobutyric acid type A receptor modulation and metabolic detoxification. *Proc. Natl. Acad. Sci. USA* 97: 3826–3831.

Isman, M.B., Machial, C.M. 2006. Pesticides based on plant essential oils: From traditional practice to commercialization. In M. Rai, M.C. Carpinella (eds.), *Naturally Occurring Bioactive Compounds.* Elsevier, Amsterdam, pp. 29–44.

Jalaei, Z., Fattahi, M., Aramideh, S. 2015. Allelopathic and insecticidal activities of essential oil of *Dracocephalum kotschyi* Boiss. from Iran: A new chemotype with highest limonene-10-al and limonene. *Ind. Crop Prod.* 73: 109–117.

Jantan, I., Yalvema, M.F., Ahmad, N.W., Jamal, J.A. 2005. Insecticidal activities of the leaf oils of eight *Cinnamomum* species against *Aedes aegypti* and *Aedes albopictus. Pharm. Biol.* 43: 526–532.

Kiran, S.R., Bhavani, K., Devi, P.S., Rao, B.R.R., Reddy, K.J. 2006. Composition and larvicidal activity of leaves and stem essential oils of *Chloroxylon swietenia* DC against *Aedes aegypti* and *Anopheles stephensi. Bioresour. Technol.* 97: 2481–2484.

Kiran, S., Prakash, B. 2015. Toxicity and biochemical efficacy of chemically characterized *Rosmarinus officinalis* essential oil against *Sitophilus oryzae* and *Oryzaephilus surinamensis. Ind. Crop Prod.* 74: 817–823.

Kostyukovsky, M., Rafaeli, A., Gileadi, C., Demchenko, N., Shaaya, E. 2002. Activation of octopaminergic receptors by essential oil constituents isolated from aromatic plants: Possible mode of action against insect pests. *Pest Manag. Sci.* 58: 1101–1106.

Koul, O. 2005. *Insect Antifeedants.* CRC Press, Boca Raton, FL.

Koul, O., Walia, S., Dhaliwal, G.S. 2008. Essential oils as green pesticides: Potential and constraints. *Biopestic. Int.* 4(1): 63–84.

Koutsaviti, K., Giatropoulos, A., Pitarokili, D., Papachristos, D., Michaelakis, A., Tzakou, O. 2014. Greek *Pinus* essential oils: Larvicidal activity and repellency against *Aedes albopictus* (Diptera: Culicidae). *Parasitol. Res.* 114: 583–592.

Kulkarni, R.R., Pawar, P.V., Joseph, M.P., Akulwad, A.K., Sen, A., Joshi, S.P. 2013. *Lavandula gibsoni* and *Plectranthus mollis* essential oils: Chemical analysis and insect control activities against *Aedes aegypti, Anopheles stephensi* and *Culex quinquefasciatus. J. Pest Sci.* 86: 713–718.

Kumrungsee, N., Pluempanupat, W., Koul, O., Bullangpoti, V. 2014. Toxicity of essential oil compounds against diamondback moth, *Plutella xylostella,* and their impact on detoxification enzyme activities. *J. Pest. Sci.* 87: 721–729.

Kuo, P.M., Chu, F.H., Chang, S.T., Hsiao, W.F., Wang, S.Y. 2007. Insecticidal activity of essential oil from *Chamaecyparis formosensis* Matsum. *Holzforschung* 61: 595–599.

Kweka, E.J., Senthilkumar, A., Venkatesalu, V. 2012. Toxicity of essential oil from Indian borage on the larvae of the African malaria vector mosquito, *Anopheles gambiae. Parasites Vector* 5: 277.

Lambrano, R.H., Castro, N.P., Gallardo, K.C., Stashenko, E., Verbel, J.O. 2015. Essential oils from plants of the genus *Cymbopogon* as natural insecticides to control stored product pests. *J. Stored Prod. Res.* 62: 81–83.

Leyva, M., Marquetti, M.C., Tacoronte, J.E., Scull, R., Tiomno, O., Mesa, A., Montada, D. 2009. Actividad larvicida de aceites esenciales de plantas contra *Aedes aegypti* (L.) (Diptera: Culicidae). *Rev. Biomed.* 20: 5–13.

Li, H., Chen, C., Cao, X. 2015. Essential oils-oriented chiral esters as potential pesticides: Asymmetric syntheses, characterization and bio-evaluation. *Ind. Crop Prod.* 76: 432–436.

Lima, G.P.G., Souza, T.M., Freire, G.P., Farias, D.F., Cunha, A.P., Ricardo, N.M.P.S., Morais, S.M., Carvalho, A.F.U. 2013. Further insecticidal activities of essential oils from *Lippia sidoides* and *Croton* species against *Aedes aegypti* L. *Parasitol. Res.* 112: 1953–1958.

Lima, M.A.A., de Oliveira, F.F.M., Gomes, G.A., Lavor, P.L., Santiago, G.M.P., Nagao-Dias, A.T., Arriaga, A.M.C., Lemos, T.L.G., de Carvalho, M.G. 2011. Evaluation of larvicidal activity of the essential oils of plants species from Brazil against *Aedes aegypti* (Diptera: Culicidae). *Afr. J. Biotechnol.* 10: 11716–11720.

Liu, X., Liu, Q., Zhou, L., Liu, Z. 2014. Evaluation of larvicidal activity of the essential oil of *Allium macrostemon Bunge* and its selected major constituent compounds against *Aedes albopictus* (Diptera: Culicidae). *Parasites Vector* 7: 184.

Liu, X.C., Dong, H.W., Zhou, L., Du, S.S., Liu, Z.L. 2012. Essential oil composition and larvicidal activity of *Toddalia asiatica* roots against the mosquito *Aedes albopictus* (Diptera: Culicidae). *Parasitol. Res.* 112: 1197–1203.

Lomonaco, D., Santiago, G.M.P., Ferreira, Y.S., Arriaga, A.M.C., Mazzetto, S.E., Melec, G., Vasapollo, G. 2009. Study of technical CNSL and its main components as new green larvicides. *Green Chem.* 11: 31–33.

Lucia, A., Audino, P.G., Seccacini, E., Licastro, S., Zerba, E., Masuh, H. 2007. Larvicidal effect of *Eucalyptus grandis* essential oil and turpentine and their major components on *Aedes aegypti* larvae. *J. Am. Mosq. Control Assoc.* 23: 299–303.

Lucia, A., Licastro, S., Zerba, E., Masuh, H. 2008. Yield, chemical composition, and bioactivity of essential oils from 12 species of *Eucalyptus* on *Aedes aegypti* larvae. *Entomol. Exp. Appl.* 129: 107–114.

Magalhães, L.A.M., Lima, M.P., Marques, M.O.M., Facanali, R., Pinto, A.C.S., Tadei, W.P. 2010. Chemical composition and larvicidal activity against *Aedes aegypti* larvae of essential oils from four *Guarea* species. *Molecules* 15: 5734–5741.

Mansour, S., Abdel-Hamid, N. 2015. Residual toxicity of bait formulations containing plant essential oils and commercial insecticides against the desert locust, *Schestocerca gregaria* (Forskäl). *Ind. Crop Prod.* 76: 900–909.

Mansour, S.A., El-Sharkawy, A.Z., Abdel-Hamid, N.A. 2015. Toxicity of essential plant oils, in comparison with conventional insecticides, against the desert locust, *Schistocerca gregaria* (Forskål). *Ind. Crop Prod.* 63: 92–99.

Marques, M.M.M., Morais, S.M., Vieira, Í.G.P., Vieira, M.G.S., Silva, A.R.A., Almeida, R.R., Guedes, M.I.F. 2011. Larvicidal activity of *Tagetes erecta* against *Aedes aegypti*. *J. Am. Mosq. Control Assoc.* 27: 156–158.

Matasyoh, J.C., Wathuta, E.M., Kariuki, S.T., Chepkorir, R. 2011. Chemical composition and larvicidal activity of *Piper capense* essential oil against the malaria vector, *Anopheles gambiae*. *J. Asia-Pacific Entomol.* 14: 26–28.

Michaelakis, A., Papachristos, D., Kimbaris, A., Koliopoulos, G., Giatropoulos, A., Polissiou, M.G. 2009. *Citrus* essential oils and four enantiomeric pinenes against *Culex pipiens* (Diptera: Culicidae). *Parasitol. Res.* 105: 769–773.

Moon, H.I., Cho, S.B., Kim, S.K. 2011. Composition and immunotoxicity activity of essential oils from leaves of *Zingiber officinale* Roscoe against *Aedes aegypti* L. *Immunopharmacol. Immunotoxicol.* 33: 201–204.

Morais, S.M., Cavalcanti, E.S.B., Bertini, L.M., Oliveira, C.L.L., Rodrigues, J.R.B., Cardoso, J.H.L. 2006. Larvicidal activity of essential oils from Brazilian *Croton* species against *Aedes aegypti* L. *J. Am. Mosq. Control Assoc.* 22: 161–164.

Morais, S.M., Facundo, V.A., Bertini, L.M., Cavalcantia, E.S.B., Júniorb, J.F.A., Ferreirab, S.A., Britod, E.S., Netod, M.A.S. 2007. Chemical composition and larvicidal activity of essential oils from *Piper* species. *Biochem. Syst. Ecol.* 35: 670–675.

Mozaffari, E., Abai, M.R., Khanavi, M., Vatandoost, H., Sedaghat, M.M., Moridnia, A., Saber-Navaei, M., Sanei-Dehkordi, A., Rafi, F. 2014. Chemical composition, larvicidal and repellency properties of *Cionura erecta* (L.) Griseb. against malaria vector, *Anopheles stephensi* Liston (Diptera: Culicidae). *J. Arthropod-Borne Dis.* 8: 147–155.

Nagella, P., Ahmad, A., Kim, S.J., Chung, I.M. 2012a. Chemical composition, antioxidant activity and larvicidal effects of essential oil from leaves of *Apium graveolens*. *Immunotoxicology* 34: 205–209.

Nagella, P., Kim, M.Y., Ahmad, A., Thiruvengadam, M., Chung, I.M. 2012b. Chemical constituents, larvicidal effects and antioxidant activity of petroleum ether extract from seeds of *Coriandrum sativum* L. *J. Med. Plants Res.* 6: 2948–2954.

Nascimento, J.C., David, J.M., Barbosa, L.C.A., Paula, V.F., Demuner, A.J., David, J.P., Conserva, L.M., Ferreira, J.C., Jr., Guimarães, E.F. 2013. Larvicidal activities and chemical composition of essential oils from *Piper klotzschianum* (Kunth) C DC. (Piperaceae). *Pest Manag. Sci.* 69: 1267–1271.

Park, Y.J., Chung, I.M., Moon, H.I. 2010. Effects of immunotoxic activity of the major essential oil of *Angelica purpuraefolia* Chung against *Aedes aegypti* L. *Immunopharmacol. Immunotoxicol.* 32: 611–613.

Pavela, R. 2009. Larvicidal effects of some Euro-Asiatic plants against *Culex quinquefasciatus* Say larvae (Diptera: Culicidae). *Parasitol. Res.* 105: 887–892.

Pavela, R. 2014. Acute, synergistic and antagonistic effects of some aromatic compounds on the *Spodoptera littoralis* Boisd. (Lep.: Noctuidae) larvae. *Ind. Crops Prod.* 60: 247–258.

Pavela, R. 2015. Essential oils for the development of eco-friendly mosquito larvicides. *Ind. Crop Prod.* 76: 174–187.

Pavela, R., Kaffkova, K., Kumsta, M. 2014. Chemical composition and larvicidal activity of essential oils from different *Mentha* L. and *Pulegium* species against *Culex quinquefasciatus* Say (Diptera: Culicidae). *Plant Prot. Sci.* 50: 36–42.

Peixoto, M.G., Bacci, L., Blank, A.F., Araújo, A.P.A., Alves, P.B., Silva, J.H.S., Santos, A.A., Oliveira, A.P., da Costa, A.S., de F. Arrigoni-Blank, M. 2015. Toxicity and repellency of essential oils of *Lippia alba* chemotypes and their major monoterpenes against stored grain insects. *Ind. Crop Prod.* 71: 31–36.

Perumalsamy, H., Kim, N.J., Ahn, A.J. 2009. Larvicidal activity of compounds isolated from *Asarum heterotropoides* against *Culex pipiens* pallens, *Aedes aegypti*, and *Ochlerotatus togoi* (Diptera: Culicidae). *J. Med. Entomol.* 46: 1420–1423.

Pitarokili, D., Michaelakis, A., Koliopoulos, G., Giatropoulos, A., Tzakou, O. 2011. Chemical composition, larvicidal evaluation, and adult repellency of endemic Greek *Thymus* essential oils against the mosquito vector of West Nile virus. *Parasitol. Res.* 109: 425–430.

Pitasawat, B., Champakaew, D., Choochote, W. 2007. Aromatic plant derived essential oil: An alternative larvicide for mosquito control. *Fitoterapia* 78: 205–210.

Priestley, C.M., Williamson, E.M., Wafford, K.A., Satelle, D.B. 2003. Thymol, a constituent of thyme essential oil, is a positive allosteric modulator of human $GABA_A$ receptors and a homo-oligomeric GABA receptor from *Drosophila melanogaster. Br. J. Pharmacol.* 140: 1363–1372.

Rattan, R.S. 2010. Mechanism of action of insecticidal secondary metabolites of plant origin. *Crop Prot.* 29: 913–920.

Regnault-Roger, C., Vincent, C., Arnason, J.T. 2012. Essential oils in insect control: Low-risk products in a high-stakes world. *Ann. Rev. Entomol.* 57: 405–424.

Reynolds, S.E. 1987. The cuticle, growth and moulting in insects: The essential background to the action of acylurea insecticides. *Pest Manag. Sci.* 20: 131–146.

Ruiz, C., Cachay, M., Domínguez, M., Velásquez, C., Espinoza, G., Ventosilla, P., Rojas, R. 2011. Chemical composition, antioxidant and mosquito larvicidal activities of essential oils from *Tagetes filifolia, Tagetes minuta* and *Tagetes elliptica* from Perú. *Planta Med.* 77-PE30.

Ryan, M.F., Byrne, O. 1988. Plant-insect coevolution and inhibition of acethylcholinesterase. *J. Chem. Ecol.* 14: 1965–1975.

Santos, R.P., Nunes, E.P., Nascimento, R.F., Santiago, G.M.P., Menezes, G.H.A., Silveira, E.R., Pessoa, O.D.L. 2006. Chemical composition and larvicidal activity of the essential oils of *Cordia leucomalloides* and *Cordia curassavica* from the northeast of Brazil. *J. Braz. Chem.* Soc. 17: 1027–1030.

Scotti, L., Scotti, M.T., Silva, V.B., Santos, S.R.L., Cavalcanti, S.C.H., Mendonça, F.J.B., Jr. 2013. Chemometric studies on potential larvicidal compounds against *Aedes aegypti. Med. Chem.* 9: 1–10.

Senthilkumar, A., Jayaraman, M., Venkatesalu, V. 2013. Chemical constituents and larvicidal potential of *Feronia limonia* leaf essential oil against *Anopheles stephensi, Aedes aegypti* and *Culex quinquefasciatus. Parasitol. Res.* 112: 1337–1342.

Senthilkumar, A., Kannathasan, K., Venkatesalu, V. 2008. Chemical constituents and larvicidal property of the essential oil of *Blumea mollis* (D. Don) Merr. against *Culex quinquefasciatus*. *Parasitol. Res.* 103: 959–962.

Shaalan, E.A.S., Canyon, D.V., Bowden, B., Younes, M.W.F., Abdel-Wahab, H., Mansour, A.H. 2006. Efficacy of botanical extracts from *Callitris glaucophylla*, against *Aedes aegypti* and *Culex annulirostris* mosquitoes. *Trop. Biomed.* 23: 180–185.

Sigma-Aldrich. 2012a. 2-Acetonaphthone. Safety Data Sheet. Sigma-Aldrich Co., St. Louis, MO.

Sigma-Aldrich. 2012b. Methyl eugenol. Safety data sheet. Sigma-Aldrich Co., St. Louis, MO.

Sigma-Aldrich. 2012c. γ-Terpinene. Safety data sheet. Sigma-Aldrich Co., St. Louis, MO.

Sigma-Aldrich. 2013a. Citral. Safety data sheet. Sigma-Aldrich Co., St. Louis, MO.

Sigma-Aldrich. 2013b. Oraganum oil. Safety data sheet. Sigma-Aldrich Co., St. Louis, MO.

Sigma-Aldrich. 2013c. Oraganum oil. Safety data sheet. Sigma-Aldrich Co., St. Louis, MO.

Sigma-Aldrich. 2013d. 1,8-cineole. Safety data sheet. Sigma-Aldrich Co., St. Louis, MO.

Sigma-Aldrich. 2014a. Geraniol. Safety data sheet. Sigma-Aldrich Co., St. Louis, MO.

Sigma-Aldrich. 2014b. Mycene. Safety data sheet. Sigma-Aldrich Co., St. Louis, MO.

Sigma-Aldrich. 2014c. α-Pinene. Safety data sheet. Sigma-Aldrich Co., St. Louis, MO.

Sigma-Aldrich. 2015a. Citronellol. Safety data sheet. Sigma-Aldrich Co., St. Louis, MO.

Sigma-Aldrich. 2015b. (R)-(+)-Limonene. Safety data sheet. Sigma-Aldrich Co., St. Louis, MO.

Sigma-Aldrich. 2015c. Maltol. Safety data sheet. Sigma-Aldrich Co., St. Louis, MO.

Sigma-Aldrich. 2015d. 4-Methoxyphenol. Safety data sheet. Sigma-Aldrich Co., St. Louis, MO.

Sigma-Aldrich. 2015e. Linalool. Safety data sheet. Sigma-Aldrich Co., St. Louis, MO.

Sigma-Aldrich. 2015f. (+)-Pulegone. Safety data sheet. Sigma-Aldrich Co., St. Louis, MO.

Soledade de Lira, C., Pontual, E.V., de Albuquerque, L.P., Paiva, L.M., Paiva, P.M.G., de Oliveira, J.V., Napoleaõ, T.H., do A.F. Navarro, D.M. 2015. Evaluation of the toxicity of essential oil from *Alpinia purpurata* inflorescences to *Sitophilus zeamais* (maize weevil). *Crop Prot.* 71: 95–100.

Souza, L.G.S., Almeida, M.C.S., Monte, F.J.Q., Santiago, G.M.P., Braz-Filho, R., Lemos, T.L.G., Gomes, C.L., Nascimento, R.F. 2012. Chemical constituents of *Capraria biflora* (Scrophulariaceae) and larvicidal activity of essential oil. *Quim. Nova* 1012: 2258–2262.

Stroh, J., Wan, M.T., Isman, M.B., Moul, D.J. 1998. Evaluation of the acute toxicity to juvenile Pacific coho salmon and rainbow trout of some plant essential oils, a formulated product, and the carrier. *Bull. Environ. Contam. Toxicol.* 60: 923–930.

Sutthanont, N., Choochote, W., Tuetun, B., Junkum, A., Jitpakdi, A., Chaithong, U., Riyong, D., Pitasawa, B. 2010. Chemical composition and larvicidal activity of edible plant-derived essential oils against the pyrethroid-susceptible and -resistant strains of *Aedes aegypti* (Diptera: Culicidae). *J. Vector Ecol.* 35: 106–115.

Thermo Scientific. 2015. p-Anisaldehyde. Safety data sheet. Thermo Fisher Scientific, Waltham, MA.

Tiwary, M., Naik, S.N., Tewary, D.K., Mittal, P.K., Yadav, S. 2007. Chemical composition and larvicidal activities of the essential oil of *Zanthoxylum armatum* DC (Rutaceae) against three mosquito vectors. *J. Vector Borne Dis.* 44: 198–204.

Trindade, F.T.T., Stabeli, R.G., Pereira, A.A., Facundo, V.A., Silva, A.A. 2013. *Copaifera multijuga* ethanolic extracts, oil-resin, and its derivatives display larvicidal activity against *Anopheles darling* and *Aedes aegypti* (Diptera: Culicidae). *Braz. J. Pharmacogn.* 23: 464–470.

Vera, S.S., Zambrano, D.F., Méndez-Sanchez, S.C., Rodríguez-Sanabria, F., Stashenko, E.E., Luna, J.E.D. 2014. Essential oils with insecticidal activity against larvae of *Aedes aegypti* (Diptera: Culicidae). *Parasitol. Res.* 113: 2647–2654.

Vourlioti-Arapi, F., Michaelakis, A., Evergetis, E., Koliopoulos, G., Haroutounian, S.A. 2012. Essential oils of indigenous in Greece six *Juniperus* taxa. *Parasitol. Res.* 110: 1829–1839.

Warburton, H., Ketunuti, U., Grzywacz, D. 2002. Survey of the supply, production and use of microbial pesticides in Thailand. NRI Report 2723. National Resource Institute, University of Greenwich, Chatham, UK.

Wei, H., Liu, J., Li, B., Zhan, Z., Chen, Y., Tian, H., Lin, S., Gu, X. 2015. The toxicity and physiological effect of essential oil from *Chenopodium ambrosioides* against the diamondback moth, *Plutella xylostella* (Lepidoptera: Plutellidae). *Crop Prot.* 76: 68–74.

Yotavong, P., Boonsoong, B., Pluempanupat, W., Koul, O., Bullangpoti, V. 2015. Effects of the botanical insecticide thymol on biology of a braconid, *Cotesia plutellae* (Kurdjumov), parasitizing the diamondback moth, *Plutella xylostella* L. *Int. J. Pest Manag.* 61(2): 171–178.

You, C., Zhang, W., Guo, S., Wang, C., Yang, K., Liang, J., Wang, W., Geng, Z., Du, S., Deng, Z. 2015. Chemical composition of essential oils extracted from six *Murraya* species and their repellent activity against *Ribolium castaneum. Ind. Crop Prod.* 76: 681–687.

Yu, J., Liu, X.Y., Yang, B., Wang, J., Zhang, F.Q., Feng, Z.L., Wang, C.Z., Fan, Q.S. 2013. Larvicidal activity of essential extract of *Rosmarinus officinalis* against *Culex quinquefasciatus. J. Am. Mosq. Control Assoc.* 29(1): 44–48.

Zarrad, K., Hamouda, A.B., Chaieb, I., Laarif, A., Jemâa, J.M. 2015. Chemical composition, fumigant and anti-acetylcholinesterase activity of the Tunisian *Citrus aurantium* L. essential oils. *Ind. Crop Prod.* 76: 121–127.

Zhu, L., Tian, Y. 2013. Chemical composition and larvicidal activity of essential oil of *Artemisia gilvescens* against *Anopheles anthropophagus. Parasitol. Res.* 112: 1137–1142.

# 25

# Essential Oils and Their Constituents as Novel Biorational Molluscicides for Terrestrial Gastropod Pests

Rory Mc Donnell

## CONTENTS

## 25.1 Background

Invasive species are found in most taxonomic groups, and the Mollusca are no exception. Bivalves and both terrestrial and freshwater snails have been linked with many invasive events throughout the world, and these invasions are typically initiated by either deliberate or inadvertent introductions by humans (Cowie and Robinson 2003). In fact, terrestrial mollusks have been associated with the transport of goods by humans for thousands of years (Welter-Schultes 2008). In the 1850s, *Cornu aspersum* was introduced to California as a potential source of food (escargot), and it is now a very serious pest of citrus and ornamentals (Dekle and Fasulo 2011). In more recent times, the deliberate introduction of the carnivorous rosy wolf snail, *Euglandina rosea*, from subtropical North America to control the invasive giant African land snail, *Lissachatina fulica*, on Pacific and Indian Ocean islands, is widely regarded as being an ecological disaster (Civeyrel and Simberloff 1996). This invasion is thought to have resulted in the extinction of greater than 60% of the endemic Hawaiian land snails (Solem 1990) and the extinction of partulid tree snails (family Partulidae) on the French Polynesian island of Moorea (Clarke et al. 1984).

The management of snail and slug pests throughout the world relies heavily on the use of molluscicides, which are underpinned by only four active ingredients (metaldehyde, iron phosphate, sodium ferric EDTA, and methiocarb). However, the activity of these products is very variable and heavily influenced by environmental conditions. Baits containing the active ingredient metaldehyde are most commonly used, but metaldehyde is very toxic to many nontarget organisms, such as dogs and cats (Studdert 1985; Bates et al. 2012). There is hence an urgent need to develop new molluscicidal products that are more reliable and have less impact on nontarget organisms. One such option is the development of biorational products that have plant extracts as their active ingredients.

## 25.2 Studies Investigating Essential Oils and Their Constituents as Novel Molluscicides

Essential oils and/or their constituents are known to have a broad spectrum of activity against insect and mite pests, fungi, and nematodes. For example, certain oils and their constituent terpenes are known to be lethal to the American cockroach (*Periplaneta americana*) (Ngoh et al. 1998), the German cockroach (*Blattella germanica*), and the common housefly (*Musca domestica*) (Coats et al. 1991; Rice and Coats 1994). However, their efficacy against terrestrial mollusks has been largely overlooked, except for a small number of papers that have focused on two general areas: the potential use of essential oils or their constituents as antifeedants and repellents, and their use as novel molluscicides.

### 25.2.1 Essential Oils as Antifeedants and Repellents

Many past studies concerned with the antifeedant and repellent properties of plant extracts did not specifically identify the active compounds involved. For example, Barone and Frank (1999) examined the antifeedant properties of methanol extracts of nine plants on the invasive slug *Arion lusitanicus*. The extracts of only two species (*Saponaria officinalis* and *Valerianella locusta*) significantly deterred feeding in the laboratory, but no attempt was made to isolate and identify the actual repellent compounds involved. Similarily, Ali et al. (2003a) determined that the raw plant materials and extracts of the bark of *Detarium microcarpum*, the bark and leaves of *Ximenia americana*, and the shoots of *Polygonum limbatum* were repellent to the slug *Deroceras reticulatum* in both labotatory and field bioassays, but the active compounds were not identified.

Other studies involving terrestrial mollusks have examined the efficacy of essential oils from various plants, but again, the actual repellent compounds were not determined. For example, Ali et al. (2003b) showed that saw dust coated with the essential oil of *Commiphora guidottii* helped reduce wheat seedling damage by *D. reticulatum* in laboratory bioassays. Likewise, Lindqvist et al. (2010) determined that birch tar oil had repellent properties against the slug *A. lusitanicus* and the snail *Arianta arbustorum*. Although the repellent compounds were not identified, when mixed with Vaseline®, birch tar oil prevented the land snails from crossing over treated Perspex® fences (height 40 cm) for up to several months in the field.

Vokou et al. (1998) investigated the interaction of commonly occurring aromatic plants in the Mediterranean region of Europe and the snail herbivores, *C. aspersum*, *Helix lucorum*, and *Eobania vermiculata*. The authors found that the presence of the crude essential oil blend extracted from *Origanum vulgare* subsp. *hirtum* placed in food at naturally occurring concentrations had a repellent effect on all three snail species, but the essential oil blend from subsp. *vulgare* did not. Although the authors did not identify the active compounds, they did state that the main essential oil constituents, carvacrol and thymol, accounted for ~60% of the constituents in subsp. *hirtum* extracts, while in subsp. *vulgare*, these phenols were only present in trace amounts, which suggests that carvacrol and thymol were the active snail repellent compounds. This theory is supported by the work of Linhart and

Thompson (1995), who demonstrated that foods containing carvacrol and thymol were fed on the least by *C. aspersum*.

Rice et al. (1978) compared the differential palatability of monoterpenoid compositional types of *Satureja douglasii* to the slug *Ariolimax dolichophallus*. The authors found that types containing a high proportion of bicyclic monoterpenoids (e.g., camphene and camphor) were more palatable to the slug than those containing high proportions of *p*-menthane monoterpenoids (e.g., isomenthone, pulegone, and carvone). In the case of individual monoterpenoids characteristic of these types, pulegone was the least palatable, and carvone and isomenthone were intermediate.

Dodds (1996) assessed the effects of plant extracts on the feeding behavior of *D. reticulatum*, and extracts of hemlock (*Conium maculatum*), rock sapphire (*Crithmum maritimum*), and curled chervil (*Anthriscus cerefolium*) caused the greatest reduction in feeding. Two active antifeedants, (+) fenchone and 4-allylanisole, were identified in *A. cerefolium*, and a new antifeedant compound (later identified as RESCO-1 by Clark et al. 1997) was responsible for most of the activity in *C. maculatum*. These compounds were not tested by Dodd (1996) under field conditions.

Scott et al. (1977) examined the antifeedant potential of extracts of 60 plant species against *D. reticulatum*, and two showed high activity. These were the root extract from horseradish (*Armoracia rusticana*) and the leaf extract of scented geranium (*Pelargonium graveolens*). The activity of the former was accounted for by 2-phenylethyl isothiocyanate, but field trials with this compound did not protect wheat seedlings because it was phytotoxic. The active compound in *P. graveolens* was the monoterpenoid alcohol, geraniol, but field use was limited by its instability to oxidation. To follow on from this work, Airey et al. (1989) tested the activity of 30 related compounds, primarily monoterpenoids, but only the bicyclic monoterpenoid ketone, (+) fenchone, was active. Furthermore, (+) fenchone was effective in protecting wheat seeds in boxes, but the high volatility and low persistence of the compound were seen as limiting factors to widespread use. Interestingly, the (−) isomer of fenchone was inactive, suggesting that different isomers of the same compound can have variable toxicity to terrestrial mollusks.

Clark et al. (1997), in designing a standard method for screening materials that influence feeding in *D. reticulatum*, tested a range of plant extracts. The authors found that a methanol extract of tarragon (*Artemisia dracunculus*) reduced consumption by 82% (±7.5), and the monoterpenoid alcohol geraniol at 1.5% concentration reduced feeding by 83% (±4.3). The fenchone enantiomers were also both significantly antifeedant at 1.5%, but only (+) fenchone reduced feeding at 0.5%, suggesting that this isomer is more repellent than the (−) isomer. This supports the earlier findings of Airey et al. (1989) that (−) fenchone is less active than (+) fenchone, at least in relation to *D. reticulatum*.

## 25.2.2 Essential Oils as Molluscicides

Research on essential oils as novel molluscicides has also focused on the activity of crude oil blends or on the activity of actual terpenoid compounds. For example, Mc Donnell et al. (2016) investigated the toxicity of 11 essential oils and d-limonene against the eggs and juveniles of the pest snail *C. aspersum*. Clove bud oil was identified as the most toxic oil ($LC_{50}$ 0.027%), and in glasshouse trials with potted *Hosta* "Royal Standard," emulsions of this oil were not phytotoxic when applied as a drench and foliar spray at a concentration of 0.116%. The authors suggest that eugenol was the likely active compound.

Amirmohammadi et al. (2012) investigated the toxicity of the essential oil of *Artemisia annua* on the slug *Deroceras agrestis* under controlled laboratory conditions by presenting starved test animals with radish leaves (*Rhaphanus sativus*) soaked in various concentrations of the oil and then air dried. The calculated $LC_{50}$ was 5.81%. The authors also showed that the essential oil has physiological impacts on *D. agrestis*. For example, there was an increase in cytochrome P450 monooxygenase, likely because of the detoxifying role of this enzyme. Increases in phosphatases in the test slugs are also likely linked to the breakdown of toxins present in the essential oil.

Ferreira et al. (2009) showed that exposure to thymol (at concentrations of 2 and 5 g/L) in laboratory studies killed 97.50% of the eggs of the snail *Subulina octona*. The authors also investigated the effect of thymol at these concentrations on 10-day-old juvenile snails, but there was no significant difference in mortality with an untreated control. Using 30-day-old snails, mortality (55% after 120 days of exposure) was significantly higher using thymol at 5 g/L compared to the control.

Abdelgaleil (2010) investigated the fumigant and contact toxicity of 11 monoterpenes against adults of the pest snail, *Theba pisana*. (–) Fenchone had the highest toxicity ($LC_{50}$ 2.51 mg/ml), but myrcene (3.88 mg/ml) and 1,8-cinerole (4.17 mg/ml) also exhibited strong fumigant toxicity to the test species. Cuminaldehyde, geraniol, and (–) menthol were not active. However, in the contact bioassays, cuminaldehyde ($LD_{50}$ 28.37 µg/snail) was the most toxic, followed by geraniol ($LD_{50}$ 42.29 µg/snail) and (–) limonene ($LD_{50}$ 60.27 µg/snail). In fact, eight of the tested monoterpenoids were more toxic to adult *T. pisana* than the synthetic molluscicide methiocarb, highlighting their potential importance as novel biorational molluscicides.

Eshra et al. (2016) also evaluated the fumigant toxicity of four plant extracts against *T. pisana*, and as did Abdelgaleil (2010), the authors found that fenchone (isomer not specified) ($LC_{50}$ 3.3 µl/L of air) was most toxic, followed by the essential oil of *Lavandula dentata* ($LC_{50}$ 16.3 µl/L), limonene ($LC_{50}$ 19.8 µl/L), and carvone ($LC_{50}$ 33.2 µl/L). In fact, fenchone was significantly more toxic to the test snail than the other plant extracts. Esra et al. (2016) also determined that fenchone was a significantly better inhibitor of acetylcholinesterase (AChE) than carvone and limonene, but the authors suggested that the inhibition of AChE by monoterpenes is not the primary mode of action for these compounds because of the high concentrations required for inhibition. Fenchone was also twice as toxic to *T. pisana* than the commercial fumigant, methyl bromide, highlighting its potential commercial importance in snail management.

Radwan and El-Zemity (2007) tested the molluscicidal activity of 10 naturally occurring compounds against *T. pisana* and found that thymol ($LD_{50}$ 120.61 µg/snail) was most effective, followed by eugenol ($LD_{50}$ 125.82 µg/snail) and pulegone ($LD_{50}$ 361.79 µg/snail). Interestingly, this is the only study to the best of the author's knowledge that investigated the potential synergism of molluscicidal plant extracts with a synergistic compound. The authors found that the toxicity of thymol and eugenol alone was less than that of methiocarb ($LD_{50}$ 107.34 µg/snail), but mixing these monoterpenes with the synergist piperonyl butoxide (1:2 ratio) more than doubled their toxicity ($LD_{50}$ thymol, 34.96 µg/snail; $LD_{50}$ eugenol, 50.13 µg/snail) and resulted in both compounds being more toxic to *T. pisana* than methiocarb.

Iglesias et al. (2002) tested the ovicidal efficacy of a number of molluscicides, herbicides, insecticides, and some plant extracts against the eggs of *D. reticulatum* in the laboratory. The active ingredient of neem (*Azadirachta indica*) oil, azadirachtin, killed all the slug eggs at a dose of 0.020 mg/cm² after 24 h of exposure, while a dose of carvone at 0.063 mg/cm² also caused 100% egg mortality. In the only other study that examined the effect of azadirachtin on terrestrial gastropods, Ploomi et al. (2009) demonstrated that azadirachtin at concentrations of 0.3% and 0.03% was repellent to *A. arbustorum* in field trials in Estonia.

## 25.3 Synthesis of Previous Studies

Although many of the studies above provide encouraging results, it is important to remember that research involving the use of essential oils and their constituent compounds as novel products in managing terrestrial gastropods is still in its infancy. Nevertheless, by synthesizing the above studies, it is possible to provide some valuable conclusions and directions for future research.

Research on the monoterpene fenchone has shown that different isomers can have varying efficacies against different gastropod species. For example, (+) fenchone is known to be more active in repelling *D. reticulatum* than (–) fenchone (Airey et al. 1989; Clark et al. 1997), but (–) fenchone has a strong fumigant toxicity to *T. pisana* (Abdelgaleil 2010). In addition, the mode of action of individual active compounds can be very different. For example, cuminaldehyde is known to be a contact toxin to *T. pisana* but was not active as a fumigant toxin against the same snail species. These studies highlight the importance of investigating both the fumigant and contact toxicity of the same compound, but also the molluscicial potential of different isomers. It will also be beneficial to remember that some compounds can be repellent at certain concentrations and act as phagostimulants at other concentrations. For example, at concentrations of more than 5%, sucrose increasingly deters feeding, but between 2.5% and 5%, it stimulates feeding in *D. reticulatum* (Henderson et al. 1992). It would be wise to be mindful of this concentration effect when working with other repellent compounds, be they plant extracts or not.

One of the main disadvantages with using essential oils and their constituents in pest management is that they can be phytotoxic at those concentrations required to kill the target pest (Isman 2000). It is therefore surprising that the phytoxicity of candidate oils and compounds has been largely overlooked in previous research. Exceptions include Scott et al. (1977), who demonstrated that the gastropod repellent 2-phenylethyl isothiocyanate was phytotoxic to wheat. Mc Donnell et al. (2016) demonstrated that the concentration of clove bud essential oil necessary to kill the eggs and juveniles of *C. aspersum* was not phytotoxic to *Hosta* "Royal Standard," which highlights the potential value of this oil as a novel biorational molluscicide. However, future trials with clove bud oil should consider phytotoxic effects on a range of plant species.

Another potential disadvantage with using essential oils and their constituents is their cost in comparison with other management approaches. For example, Mc Donnell et al. (2016) showed that drenches of clove bud oil were >100 times more expensive than using a synthetic molluscicide product (Slug Fest® All Weather Formula). However, the high cost of alternative approaches to managing slugs and snails has not been a significant impediment to their use by stakeholders. For example, the biological control product Nemaslug® is a high-cost management strategy for gastropod pests in Europe, but it is widely used by growers to protect a diverse range of crops (Rae et al. 2007). Furthermore, it may be possible to reduce the overall cost of using essential oils or their constituents by using synergists. Synergists have been used commercially for about 50 years and have contributed significantly to improving the efficacy of insecticides (Bernard and Philogène 1993), but to the best of the author's knowledge, only one study has investigated their potential with biorational molluscicidal compounds. Radwan and El-Zemity (2007) demonstrated that inclusion of piperonyl butoxide (1:2 ratio) more than doubled the toxicity of eugenol and more than tripled the toxicity of thymol and cinnamyl aldehyde against *T. pisana*. The addition of the synergist also resulted in both thymol and eugenol being more toxic to the test snail than the commercially available molluscicide methiocarb. Therefore, it may be possible to use synergists to increase the toxicity of essential oils or their constituents, which in turn will decrease the quantity of active ingredient required for effective pest control, thereby reducing the per treatment cost.

Most of the studies on terrestrial gastropods to date have focused on laboratory bioassays, with few considering the efficacy of identified active compounds against pest species under field conditions, which ultimately is required for identifying truly effective products. Those studies that did progress to field testing therefore provide useful data for helping to select priority oils and compounds for more thorough future investigations. For example, clove bud oil proved to be an effective potted plant drench for *C. aspersum* (Mc Donnell et al. 2016), and birch tar oil, when mixed with Vaseline, repelled *A. lusitanicus* and *A. arbustorum* for up to several months in the field (Lindqvist et al. 2010), but the authors of these studies did not identify the active compounds in the oils, and this should be a priority for future research. In addition, (+) fenchone was effective in protecting wheat seeds from *D. reticulatum* (Airey et al. 1989), but the high volatility and low persistence of the compound were seen as limiting factors. However, future work with this monoterpene should consider the use of slow-release systems to prolong field life. Such slow-release systems have been identified for other pesticides (Sinclair 1973; Rudzinski et al. 2002).

Finally, research with insects has shown that there is little overlap among species with respect to the most toxic essential oils and constituents, indicating that these substances are generally active against a range of pests, but interspecific toxicity of oils and their constituents may be highly idiosyncratic (Sarac and Tunc 1995; Isman 2000). Unpublished results by the author suggest that a similar pattern exists for terrestrial gastropods. Therefore, it will be critical to test oils and their active components on a variety of pest snail and slug species before widespread adoption as molluscicides. Also, it may ultimately be prudent to test combinations of two or more active compounds or oils with the hope of targeting multiple pest species with the same biorational product.

## Acknowledgment

I am grateful to Veronica Puig Sanvicens for her comments and suggestions on this chapter.

## References

Abdelgaleil, S.A.M. 2010. Molluscicidal and insecticidal potential of monoterpenes on the white garden snail, *Theba pisana* (Müller) and the cotton leafworm, *Spodoptera litroalis* (Boisduval). *Applied Entomology and Zoology* 45:425–433.

Airey, W.J., Henderson, I.F., Pickett, J.A., Scott, G.C., Stephenson, J.W., Woodcock, C.M. 1989. Novel chemical approaches to mollusc control. In *Proceedings, British Crop Protection Conference—Slugs and Snails in World Agriculture*, 301–307. Guilford, United Kingdom.

Ali, A.Y., Müller, C.T., Randerson, P., Bowen, I.D. 2003a. Molluscicidal and repellent properties of African plants. In *Proceedings, British Crop Protection Conference—Slugs and Snails: Agricultural, Veterinary and Environmental Perspectives*, 135–141.

Ali, A.Y., Müller, C.T., Randerson, P., Bowen, I.D. 2003b. Screening African plants for mollusk repellency. In *Proceedings, British Crop Protection Conference—Slugs and Snails: Agricultural, Veterinary and Environmental Perspectives*, 319–324.

Amirmohammadi, F., Sendi, J.J., Zibaee, A. 2012. Toxicity and physiological effect of essential oil of *Artemisia annua* (Labiatae) on *Agriolimax agrestis* L. (Stylommatophora: Limacidae). *Journal of Plant Protection* 52:185–189.

Barone, M., Frank, T. 1999. Effects of plant extracts on the feeding behavior of *Arion lusitanicus*. *Annals of Applied Biology* 134:341–345.

Bates, N.S., Sutton, N.M., Campbell, A. 2012. Suspected metaldehyde slug bait poisoning in dogs: A retrospective analysis of cases reported to the Veterinary Poisons Information Service. *The Veterinary Record* 171:324.

Bernard, C.B., Philogène B.J. 1993. Insecticide synergists: Role, importance, and perspectives. *Journal of Toxicology and Environmental Health* 38:199–223.

Civeyrel, L., Simberloff, D. 1996. A tale of two snails: Is the cure worse than the disease? *Biodiversity and Conservation* 5:1231–1252.

Clarke, B., Murray, J., Johnson, M.S. 1984. The extinction of endemic species by a program of biological control. *Pacific Science* 38:97–104.

Clark, S.J., Dodds, C.J., Henderson, I.F., Martin, A.P. 1997. A bioassay for screening materials influencing feeding in the field slug, *Deroceras reticulatum* (Müller) (Mollusca: Pulmonata). *Annals of Applied Biology* 130:379–385.

Coats, J.R., Karr, L.L., Drewes, C.D. 1991. Toxicity and neurotoxic effects of monoterpenoids in insects and earthworms. *American Chemistry Society Symposium Series* 449:306–316.

Cowie, R.H., Robinson, G.D. 2003. Pathways of introduction of non-indigenous land and freshwater snails and slugs. In *Invasive Species. Vectors and Management Strategies*, ed. G.M. Ruiz and J.T. Carlton, 93–122. Washington, DC: Island Press.

Dekle, G.W., Fasulo, T.R. 2001. *Brown Garden Snail, Cornu aspersum (Müller, 1774) (Gastropoda: Helicidae)*. Gainesville, FL: Institute of Food and Agricultural Sciences Extension, University of Florida.

Dodd, C.J. 1996. The control of slug damage using plant-derived repellents and antifeedants. In *Proceedings, British Crop Protection Conference—Slug and Snail Pests in Agriculture*, 335–340. Canterbury, United Kingdom.

Eshra, E.H., Abobakr, Y., Abddelgalil, G.M., Ebrahim, E., Hussein, H.I., Al-Sarar, A.S. 2016. Fumigant toxicity and antiacetylcholinesterase activity of essential oils against the land snail, *Theba pisana* (Müller). *Egyptian Scientific Journal of Pesticides* 2:91–95.

Ferreira, P., Gonçalves Soares, G.L., D'ávila, S., de Almeida Bessa, E.C. 2009. The influence of caffeine and thymol on the survival, growth and reproduction of *Subulina octona* (Brugüière, 1789) (Mollusca, Subulinidae). *Brazilian Archives of Biology and Technology* 52:945–952.

Henderson, I.F., Martin, A.P., Perry, N.J. 1992. Improving slug baits: The effects of some phagostimulants and molluscicides on ingestion by the slug, *Deroceras reticulatum* (Müller) (Pulmonata: Limacidae). *Annals of Applied Biology* 121:423–430.

Iglesias, J., Castillejo, J., Ester, A. 2002. Laboratory evaluation of potential molluscicides for the control of eggs of the pest slug *Deroceras reticulatum* (Müller) (Pulmonata: Limacidae), *International Journal of Pest Management* 48:19–23.

Isman, M.B. 2000. Plant essential oils for pest and disease management. *Crop Protection* 19:603–608.

Lindqvist, I., Lindqvist, B., Tiilikkala, K. et al. 2010. Birch tar oil is an effective mollusk repellent: Field and laboratory experiments using *Arianta arbustorum* (Gastropoda: Helicidae) and *Arion lusitanicus* (Gastropoda: Arionidae). *Agricultural and Food Science* 19:1–12.

Linhart, Y.B., Thompson, J.D. 1995. Terpene-based selective herbivory by *Helix aspersa* (Mollusca) on *Thymus vulgaris* (Labiatae). *Oecologia* 102:126–132.

Mc Donnell, R.J., Yoo, J., Patel, K. et al. 2016. Can essential oils be used as novel drench treatments for the eggs and juveniles of the pest snail *Cornu aspersum* in potted plants? *Journal of Pest Science* 89:549–555.

Ngoh, S.P., Hoo, L., Pang, F.Y., Huang, Y., Kini, M.R., Ho, S.H. 1998. Insecticidal and repellent properties of nine volatile constituents of essential oils against the American cockroach, *Periplaneta americana* (L.). *Pesticide Science* 54:261–268.

Ploomi, A., Jõgar, K., Metspalu, L. et al. 2009. The toxicity of neem to the snail *Arianta arbustorum*. *Sodininkystė ir daržininkystė Mokslo darbų* 28:153–158.

Radwan, M.A., El-Zemity, S.R. 2007. Naturally occurring compounds for control of harmful snails. *Pakistan Journal of Zoology* 39:339–344.

Rae, R., Verdun, C., Grewal, P.S., Robertson, J.F., Wilson, M.J. 2007. Biological control of terrestrial molluscs using *Phasmarhabditis hermaphrodita*—Progress and prospects. *Pest Management Science* 63:1153–1164.

Rice, P.J., Coats, J.R. 1994. Insecticidal properties of monoterpenoid derivatives to the house fly (Diptera: Muscidae) and red flour beetle (Coleoptera: Tenebrionidae). *Pesticide Science* 41:195–202.

Rice, R.L., Lincoln, D.E., Langenheim, J.H. 1978. Palatability of a monoterpenoid compositional type of *Satureja douglasii* to a generalist molluscan herbivore, *Ariolimax dolichophallus*. *Biochemical Systematics and Ecology* 6:45–53.

Rudzinski, W.E., Dave, A.M., Vaishnav, U.H., Kumbar, S.G., Kulkarni, A.R., Aminabhavi, T.M. 2002. Hydrogels as controlled release devices in agriculture. *Designed Monomers and Polymers* 5:39–65.

Sarac, A., Tunc, I. 1995. Toxicity of essential oil vapours to stored product insects. *Zeitschrift für Pflanzenkrankheiten und Pflanzenschutz* 102:69–74.

Scott, G.C., Griffiths, D.C., Stephenson, J.W. 1977. A laboratory method for testing seed treatments for the control of slugs in cereals. In *Proceedings, British Crop Protection Conference—Pests and Disease*, 129–134. Brighton, United Kingdom.

Sinclair, R.G. 1973. Slow-release pesticide system. Polymers of lactic and glycolic acids as ecologically beneficial, cost-effective encapsulating materials. *Environmental Science and Technology* 7:955–956.

Solem, A. 1990. How many Hawaiian land snail species are left? And what we can do for them. *Bishop Museum Occasional Papers* 30:27–40.

Studdert, V.P. 1985. Epidemiological features of snail and slug bait poisoning in dogs and cats. *Australian Veterinary Journal* 62:269–272.

Vokou, D., Tziolas, M., Bailey, S.E.R. 1998. Essential-oil mediated interactions between oregano plants and Helicidae grazers. *Journal of Chemical Ecology* 24:1187–1202.

Welter-Schultes, F.W. 2008. Bronze Age shipwreck snails from Turkey: First evidence for oversea carriage of land snails in antiquity. *Journal of Molluscan Studies* 74:79–87.

# 26

# Perspectives on Essential Oil–Loaded Nanodelivery Packaging Technology for Controlling Stored Cereal and Grain Pests

Farah Hossain, Peter Follett, Stephane Salmieri, Khanh Dang Vu,
Majid Jamshidian, and Monique Lacroix

## CONTENTS

## 26.1 Introduction

Protection of stored food crops against damage from insect pests and pathogens is a major concern for the food industry, farmers, public health organizations, and environmental agencies. Insect feeding causes damage to stored grains and processed products by reducing their dry weight and nutritional value (Follett et al. 2013). In addition, insect infestation–induced changes in the storage environment can create warm, moist "hot spots" that provide suitable conditions for storage fungi that cause further losses (Sung et al. 2013; Abou-Elnaga 2015). Molds may cause a decrease in the quantity of fats, carbohydrates, vitamins, and proteins (Lamboni and Hell 2009). Reducing postharvest food losses due to stored product pests is critical to ensure food security for our rapidly expanding global population. Stored product packaging is an important component of the food processing chain and a critical step in reducing postharvest losses, maintaining quality, adding value, and extending the shelf life of food commodities (Opara and Mditshwa 2013).

Active and intelligent packaging are new and innovative concepts that involve the incorporation of active chemical molecules into packaging material (Khan et al. 2014a). When the active chemical exhibits antimicrobial properties, the packaging is known as antimicrobial active packaging (AP). Different kinds of active substances (chemical and natural) have been tested and incorporated into packaging materials, both to improve their functioning and to provide new functions (Sung et al. 2013; Khan et al. 2014b; Salmieri et al. 2014a). However, an increase in consumer desire for natural, local, and organic products is creating new challenges to the use of microbial-suppressive packaging to enhance food preservation (Prakash et al. 2013). Among the various biopesticides that have been developed and commercialized, certain plant-derived essential oils (EOs) show great promise (Viuda-Martos et al. 2007; Tripathi et al. 2009) for incorporation in food packaging to reduce food contamination and spoilage (Balasubramanian et al. 2009). Linalool, thymol, carvacrol, cinnamaldehyde, clove, and basil oil are widely used in food packaging (Bilia et al. 2014). Most of the research to date on active food packaging has used active biocides with broad-spectrum activity and controlled-release characteristics designed to counter pathogenic bacteria such as *Escherichia coli*, *Salmonella*, and *Listeria* (Sung et al. 2013; Takala et al. 2013; Salmieri et al. 2014b; Severino et al. 2014). Adopting similar packaging approaches to inhibit fungal growth and insect infestation may prove useful for stored products.

This review focuses on EO-loaded systems in active packaging intended for insecticidal and fungicidal applications, with a special emphasis on nano-based strategies to develop nanobioactive packaging materials for eventual large-scale application for protecting stored grain products.

## 26.2 Plant-Derived EOs: Potential Alternatives to Synthetic Pesticides

In recent years, plant-derived EOs and their bioactive chemical constituents have gained increased attention due to their beneficial insecticidal (Picard et al. 2012; Hossain et al. 2014b) and antifungal (Prakash et al. 2013; Hossain 2014a, 2016) activities. EOs are mixtures of complex secondary metabolites containing a wide spectrum of strong aromatic components, such as monoterpenes and sesquiterpenes and their oxygenated derivatives (alcohols, aldehydes, esters, ethers, ketones, phenols, and oxides) in various ratios (Ebadollahi and Mahboubi 2011; Salmieri et al. 2014b; Huq et al. 2015). Since EOs are rich in volatile terpenoids and phenolic components, they exhibit a wide spectrum of activity against insects and various microorganisms (Lacroix and Follett 2015; Ghabraie et al. 2016). Monoterpenes such as 1,8-cineole, camphor, carvone, linalool, and geraniol can penetrate insect bodies through breathing and interfere with their physiological functions (Lee et al. 2003). These monoterpenes can also act as neurotoxic compounds, affecting acetylcholinesterase activity or octopamine receptors (Tripathi et al. 2009) (Figure 26.1). Other modes of action, such as membrane disruption and blockage of the tracheal system, may also be involved in the insecticidal activity of EO components. These compounds also can act at the cellular level, disrupting organelles such as mitochondria. Mansour et al. (2012) conducted an elaborate study to elucidate the action of allylisothiocyanate (AITC) oil, which occurs naturally in black mustard (*Brassica nigra*) and Indian mustard (*Brassica juncea*), and on the mitochondria of *Sitophilus oryzae*, *Tribolium confusum*, and *Plodia interpunctella*. They found that AITC induced significant alteration in insect mitochondrial structure, reducing the quantity of cristae and causing vacuolization and rarefaction of the mitochondrial matrix.

Phenolic and terpene components have been reported for possible antifungal modes of action as well (Viuda-Martos et al. 2007). The major target sites of EOs on fungal cells are illustrated in Figure 26.2. Abd-Aiia and Haggag (2013) reported that the cell wall of fungal pathogens is the main target of phenolic compounds, which primarily disrupt the

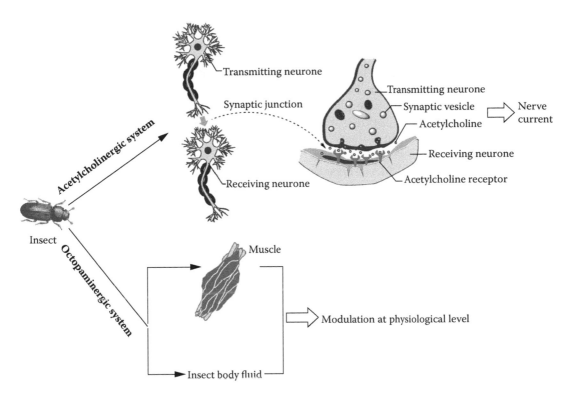

**FIGURE 26.1**
Target sites of EO components in insects. EOs act at the level of the acetylcholinergic nervous system, the action of acetylcholinesterase, which plays an important role in hydrolysing ACh and terminating neurotransmission, thus maintaining neurones in an excited state. Studies have also reported the action of EOs on the octopaminergic nervous system in insects where they disrupt octopamine receptors.

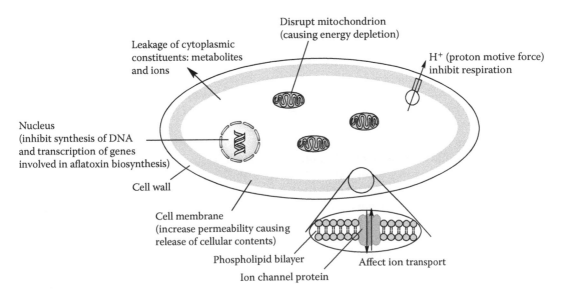

**FIGURE 26.2**
Major sites of action for EO components on fungal cells.

permeability barrier of cell membranes and inhibit respiration. The hydrophobic nature of EOs and their components allows them to penetrate the lipid layer of fungal cell membranes and mitochondria, thereby disrupting their structure and causing energy depletion (Abd-Aiia and Haggag 2013). Tian et al. (2012) reported that EO components like eugenol inhibit respiration and ion transport, and increase membrane permeability, causing release of cellular contents. The physical nature of EOs, which includes low-molecular-weight and pronounced lipophilic tendencies, allows them to penetrate cell membranes more quickly than other substances (Pawar and Thaker 2006). EO components have been shown to inhibit synthesis of DNA and transcription of genes involved in aflatoxin synthesis (Yahyaraeyat et al. 2013). These properties make them attractive alternatives to traditional fumigants, such as phosphine and methyl bromide, which have been reported to exhibit toxic effects against human health and the environment. In addition, insect and fungal species have been shown to develop resistance against traditional fumigants over prolonged exposure, but this is less likely to occur with natural bioactive EOs (Hossain et al. 2014b).

## 26.3 Active EO-Impregnated Packaging against Insects

Insects are typically the most serious pests of dried, stored, durable agricultural commodities (Follett et al. 2013, 2014). Global postharvest grain losses incurred by insect pests are estimated to be up to $17.7 billion per year (C. Oliveira et al. 2014). The most destructive insect pests of stored products are taxonomically found in the orders Coleoptera (beetles) and Lepidoptera (moths), which are worldwide distributed in various climatic conditions (Du et al. 2009). The search for tools to prevent infestations is crucial, and packaging represents a critical step in food quality preservation and the ultimate defense against insect pests. Therefore, considerable efforts should be made for the development of novel insect-proof packages, which are able to resist insect penetration and their potential infestation of food commodities. Active packaging, containing plant-derived EOs, has been designed using many different compounds and found to be effective in reducing food losses resulting from insect infestations (Licciardello et al. 2013). Allahvaisi (2010) investigated the repellent efficacy of EO extracted from *Prunus amygdalus* L. and *Mentha viridis* L. against *Tribolium castaneum*, *Sitophilus granaries*, *Stegobium paniceum*, and *Rhizopertha dominica* pests by spraying the EOs onto the interior surface of polyethylene (PE) packages. The EOs from *P. amygdalus* and *M. viridis* EOs were repellent against *T. castaneum* and *S. granarius*, reducing the contamination of the packaged food by 79% and 64%, respectively, compared with the control. Such a beneficial effect has also been shown on simple paper-based packaging used in grocery stores. Wong et al. (2005) assessed the efficiency of such types of packaging with potential plant extracts of citronella, garlic oil, neem, pine oil, and pyrethrum against *T. castaneum* and reported that out of the five plant extracts, citronella was the most effective in deterring infestation of carton packaging. An application of citronella at 0.2 g m$^{-2}$ of the carton packaging reduced the beetle infestation by 50%, with the repellency activity sustained for at least 16 weeks. The promising use of citronella oil was also evidenced by Licciardello et al. (2013), who showed that packaging coated with EO could inhibit growth of the red flour beetle by 53%–87%. This technology has potential for impregnation of carton packaging, bags, sacks, and containers used for packaging food commodities such as breakfast cereals, confectionery, pet food, grains, and milk

**TABLE 26.1**

Studies Investigating Antifungal and Insecticidal Active Packaging

| Packaging Material/ Polymeric Mixture | Essential Oil | Method of EO Incorporation | Application | Reference |
|---|---|---|---|---|
| Polypropylene | *Cinnamomum zeylanicum* (cinnamon) | Coating | Antifungal | Gutiérrez et al. 2011 |
| Polypropylene, polyethylene/ ethylene | *Cinnamomum zeylanicum* (cinnamon) *Origanum vulgare* oregano *Syzygium aromaticum* (clove) | Film | Antifungal | López et al. 2007a |
| Solid wax paraffin, active paper | Cinnamon | Coating | Antifungal | Rodriguez et al. 2008 |
| PE polymer | *Prunus amygdalus* L. and *Mentha viridis* | Spraying | Insect repellent | Allahvaisi 2010 |
| Carton board | Citronella, garlic oil, neem extract, pine oil, and pyrethrum | Coating | Insect repellent | Wong et al. 2005 |
| Plastic | Citronella, oregano, and rosemary | Coating | Insect repellent | Licciardello et al. 2013 |
| Polyvinyl alcohol | Cinnamon (*Cinnamomum zeylanicum*) | Strip | Insect repellent | Jo et al. 2013 |
| Jute bags | Neem | Coating | Insect repellent | Anwar et al. 2005 |

powder for food preservation. Besides EO-coated PE and paper-based packaging, other approaches, such as use of sachets containing active EOs inside packaging, have also been developed. Jo et al. (2013) used sachets made of polyvinyl alcohol (PVA) loaded with cinnamon EO to test the fumigant and insect repellent activity against *P. interpunctella*. They showed that the cinnamon oil physically diffused throughout the PVA polymer matrix to efficiently repel the insect larvae.

These results clearly show that it is possible to utilize EO-loaded active packaging to control stored food product insect pests. Successful implementation of such packaging would help to minimize loss of stored products. EO-loaded active packaging against insects opens an era of possibilities in contemporary green revolution approaches, as many plant-derived natural products are comparable to or even better than synthetic pesticides. Table 26.1 provides some studies that involved application of active packaging containing EOs against insects.

## 26.4 Active Packaging against Fungi

Another important cause of grain deterioration is fungal infection, which can be caused by high moisture content in crops if they are not adequately dried after harvest. Insect pests are promoters or facilitators of fungal infections (Fathi et al. 2012). Pathogenic species of *Fusarium*, *Aspergillus*, and *Penicillium* are the frequent causal agents of food spoilage and food-borne diseases (Betts et al. 1999; Viuda-Martos et al. 2007). These fungi can contaminate food before and after harvest. Fungal diseases of food crops can be hazardous, as certain fungi produce mycotoxins that can cause severe health problems, such as myco-toxicosis in humans (P. Oliveira et al. 2014) and in farm animals (Zain 2011). Mycotoxins are highly stable compounds that are not destroyed during normal food processing. They can be carcinogenic, mutagenic, genotoxic, teratogenic, neurotoxic, and oestrogenic. The most commonly described effect of acute mycotoxin poisoning is the deterioration of liver or kidney function, which in extreme cases may lead to death (Plooya et al. 2009; Perveen 2012; Hossain et al. 2016). Active food packaging can play an essential role in food process-ing to reduce fungal contamination and proliferation, and ensure food safety. Numerous studies have unravelled the potential of such type of packaging in controlling fungal infestations in food commodities. The use of antimicrobial packaging is more beneficial than directly adding antimicrobial agents onto food. Antimicrobial agents added directly on food surfaces by sprays or drips are not effective enough to inhibit microorganisms in the long term (Turhan 2013).

López et al. (2007a) designed active films made up of polypropylene (PP) and polyethylene–ethylene vinyl alcohol (PE/EVOH) copolymer and used them to incorporate EOs of cin-namon (*Cinnamomum zeylanicum*), oregano (*Origanum vulgare*), clove (*Syzygium aromaticum*), and cinnamon fortified with cinnamaldehyde. They tested these formulations against a wide range of fungi, including *Penicillium islandicum*, *Penicillium roqueforti*, *Penicillium nalgiovense*, *Eurotium repens*, and *Aspergillus flavus*. Their results showed that the PP or PE/EVOH films with incorporated oregano or fortified cinnamon EOs at a concentration of 4% (w/w) were potentially effective as antifungal packaging materials. The packaging sus-tained their antifungal properties over a prolonged period of time and extended the shelf life of the food. Gutiérrez et al. (2011) showed that cinnamon-loaded active packaging con-siderably increased the shelf life of packaged bread and maintained quality up to several months. Their study revealed that the active packaging not only inhibited fungal growth but also maintained the high-quality sensorial properties of gluten-free breads. Simple packaging material such as paper can also be converted to active packing by incorporating bioactive EOs. Rodriguez et al. (2008) evaluated the use of such paper packaging as a smart alternative for protecting bread from fungal infestation. They developed an active paper package based on the incorporation of cinnamon to solid paraffin wax as an active coat-ing. Impregnated cinnamon (4%, w/w) significantly inhibited the mold species *Rhizopus stolonifer* under *in vitro* conditions. When used with an actual bread slice, almost complete inhibition was obtained with 6% cinnamon EO.

Protein-based active films have also been shown to be an option for controlling fungal growth. An active protein-based film developed by Bahram et al. (2014) consisting of whey protein concentrate (WPC) and incorporated cinnamon (1.5%) exhibited a good inhibi-tory effect against *Candida albicans* species. Carbohydrate-based films are equally effective. Cassava starch composite films incorporating cinnamon and clove EO were reported to be effective against *Penicillium commune* and *Eurotium amstelodam* (Souza et al. 2013). In the latter study, the release profiles of cinnamon from the starch films were monitored for 2 h, by ultraviolet–visible spectroscopy at 289 nm. Released amounts of cinnamon were found to vary from 0.88 to 1.19 mg of cinnamon per gram of film.

Other authors have explored the use of sachets containing bioactive agents as smart delivery systems in food packaging against fungal species. Medeiros et al. (2011) used sachets incorporated with EO to preserve mangoes inside paper bags, and assessed their antifungal properties. Sachets containing oregano (*Origanum vulgaris*) and lemongrass (*Cymbopogon citratus*) were tested against *Colletotrichum gloeosporides*, *Lasiodiplodia theobromae*, *Xanthomonas campestris* pv. *Mangiferae indica*, and *Alternaria alternata*. The mangoes were individually wrapped in paper bags containing antimicrobial sachets and maintained at

25°C ± 2°C and relative humidity (RH) of 80% ± 5%, for 9 days. The oregano- and lemongrass-containing sachets were both effective in reducing the growth of the fungi by approximately 2 logs. Another study conducted by Espitia et al. (2011) using active sachets with 20% cinnamon incorporated in polymeric resin were used to study the antifungal effect of the developed matrix in preserving Hawaiian papayas in nonwoven fabric sacks. The treatment with the EO-containing sachets presented a lower growth of filamentous fungi than the control treatment. The cinnamaldehyde release from the polymeric matrix was found to be lineal and gradual during the storage.

Paraffin coating of paper and board has also been evaluated in active packaging systems. Rodriguez-Lafuente et al. (2010) studied the antimicrobial protection and decay retardation of cherry tomatoes in this type of packaging configuration. Bark cinnamon was evaluated against *Alternaria alternata* both *in vitro* and *in vivo* using inoculated cherry tomatoes. Almost total inhibition of the fungus was obtained when 6% of bark cinnamon was applied to the packaging material. Physicochemical parameters such as pH, weight loss, water activity, and color were monitored, and no significant differences were found between the treated and control samples. The maximum transfer of *trans*-cinnamaldehyde and carvacrol to the food was detected after 1 or 2 days of storage. Sensorial analysis performed by expert panelists showed that there were no apparent changes in the cinnamon-based packaged tomatoes.

Active antifungal packaging has been shown to exert its effect mainly through vapor activity. A study conducted by Avila-Sosa et al. (2012) involved the use of Mexican oregano (*Lippia berlandieri* Schauer), cinnamon (*Cinnamomum verum*), or lemongrass (*C. citratus*) incorporated in amaranth, chitosan, and starch edible films. The potential of vapor-induced inhibition of these films was tested against *Aspergillus niger* and *Penicillium digitatum*. It was found that the EOs in the film caused fumigant toxicity against the tested fungal species. For both types of film, a significant increase in the lag phase was observed, as well as a decrease in the maximum specific growth rates of the fungal species. The effectiveness of such active packaging through vapor contact has also been reported by other studies (López et al. 2005, 2007b).

The results of previous reports encourage the study and application of EO-loaded active packaging materials for the safeguarding of packaged food. Although the potential for EO-loaded active packaging is gaining more and more recognition, comprehensive approaches and multidisciplinary research are still necessary in order to implement and commercialize such eco-friendly active packaging technologies (Chulze 2010; Sung et al. 2013). The widespread use of such packaging is currently limited by the lack of scientific studies. Therefore, these results should encourage further investigation considering active substances, doses, modes of application, and target species.

## 26.5 Limits and Challenges for the Application of Plant-Derived EOs in Active Packaging

Although plant-derived EOs have been shown to exhibit highly desirable traits in terms of their insecticidal and antifungal properties, the future potential for their use in active packaging resides in the ability to develop innovative packaging materials that enable their controlled release in food products. EOs bear several features that limit their application, such as their volatility, instability, and insolubility in water (Bilia et al. 2014), and their rapid diffusion into food matrices (Boumail et al. 2013). While the addition of active EOs in packaging may result in immediate inhibition of nondesired microorganisms, survivors may continue to grow, especially when concentrations of the added EOs are depleted as a result of complex interactions with the food matrix or by natural degradation over time, thereby reducing their shelf life. This can effectively lead to the proliferation of antimicrobial-resistant strains (Coma et al. 2003; Chi-Zhang et al. 2004). In order to overcome these potential limitations, nanodelivery systems have been proposed to enhance the efficacy of EO-based formulations in active packaging. Encapsulation of EOs in a nano-based delivery system as a packaging film or diffusion film could provide an alternative media, in such a way that only desired levels of the active component diffuse progressively and come into contact with the food (Boumail et al. 2013; Severino et al. 2015).

## 26.6 EO-Loaded Nanodelivery Systems

Recent advances in nanotechnology have enhanced the efficacy of controlled-release biocide applications through the use of nanoparticles (NPs) (Perlatti et al. 2013). Nanotechnology deals with the application, production, and processing of materials with sizes ranging from a single atom or molecule to particles with 100 molecular diameters (or about 100 nm) (Bilia et al. 2014). The chemical properties of nanoparticles are controlled to promote an efficient assembly of a structure. Such self-assembly or self-organization of nanoparticles ensures a better interaction and mode of action at a target site due to their tunable controlled release and larger surface area (Duncan 2011). As smart delivery systems, they confer more selectivity without hindering the movement of bioactive compounds toward the target pathogen (Perlatti et al. 2013). These features enable the use of smaller amounts of an active compound per area, as long as the formulation provides an optimal concentration of delivery of the target pesticide for a given period of time (Peteu et al. 2010). EO-loaded nanocarriers can be classified as polymer-based and lipid-based nanoparticles. Molecular complexes, such as inclusion complexes with cyclodextrins (CDs), have also been reported (Perlatti et al. 2013). Different strategies employed for EO entrapment in a nano-based system are shown in Figure 26.3.

1. Polymer-based nanocarriers are classified as (a) nanocapsules with a core that is surrounded by a shell of the matrix material and (b) nanospheres with a core that is entrapped within a continuous network of the matrix material (Augustin and Hemar 2009).

2. Lipid-based nanocarriers include phospholipids and emulsions. Phospholipids are amphipathic lipids that can be categorized into liposomes and phytosomes. Liposomes are spherical bilayer vesicles with polar heads facing outward and nonpolar tails pointing to the inner region. The bilayer structure of a liposome allows it to serve as a delivery vehicle for both hydrophilic and lipophilic compounds locating at the center core (Augustin and Hemar 2009). Phytosomes, commonly referred to as phospholipid complexes, are structurally related to liposomes, but phytosomes allow higher compound loading capacity and better chemical stability. Phytosomes are composed of chemically associated molecules of phosphatidylcholine, a bifunctional compound with lipophilic phosphatidyl moiety, and a

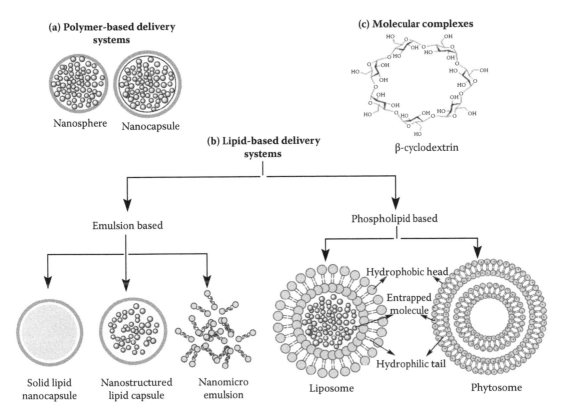

**FIGURE 26.3**
Schematic diagram showing nano-based strategies for EO entrapment. (a) Polymer-based delivery systems. (b) Lipid-based delivery systems. (c) Molecular complexes. (Adapted from Bilia, A. R. et al., *Evid. Based Complement. Altern. Med.* 2014, 651593, 2014; Ting, Y. et al., *J. Funct. Foods*, 7, 112–128, 2014. With modifications.)

hydrophilic choline moiety coupled to a natural active ingredient (Bhattacharya 2009). Flavonoid and terpenoid constituents of plant extracts are able to bind phosphatidylcholine, producing a lipid-compatible molecular complex (Bhattacharya 2009; Ting et al. 2014).

Emulsions are mixtures of two or more originally immiscible liquids, with one dispersed in the other and stabilized by amphipathic molecules. To improve physical functional stability, emulsion-based delivery systems have been designed in various structures, namely, multilayer emulsions, microemulsions, nanoemulsions, solid lipid nanoparticles (SLNs), and self-emulsifying drug delivery systems (SEDDSs). Multilayer emulsions are formed through layer-by-layer adsorption of oppositely charged polyelectrolytes onto a primary emulsion droplet. This enhances the toughness of the interface, and allows more control of core release than conventional unilayer emulsions (McClements et al. 2007). Microemulsions are homogenous, thermodynamically stable isotropic dispersions of nanodroplets with a mean radius of ≥100 nm. In contrast, nanoemulsions are nonequilibrium systems, possessing a relatively high kinetic stability, even for several years, due to their very small size, with a mean radius of ≤100 nm (Bilia et al. 2014). Solid lipid nanoparticles are a class of submicron emulsions prepared using a solid or semisolid lipid core structure. SLNs as a delivery system have been reported to immobilize active elements by the solid particle structure, allowing increased chemical protection, less leakage, and sustained release (Weiss et al. 2008). Self-emulsified drug delivery systems are isotropic mixtures of oils and surfactants and sometimes include cosolvents. SEDDS formulations are an advanced variation of an emulsion-based delivery system that provide a physically stable environment to contain the mixture of bioactives during storage, allowing rapid emulsification upon contact with the aqueous phase (Ingle et al. 2013).

In addition, molecular complexes are developed by the physical association between EOs and cyclodextrins, which are natural macrocyclic oligosaccharides composed of α-(1,4)-linked glucopyranose subunits. They have toroid-shaped structures with rigid lipophilic cavities and a hydrophilic outer surface, capable of enclosing highly hydrophobic molecules inside their cavity (Loftsson and Duchene 2007). A comprehensive list of studies involving the use of nanostructured EO systems applied against insects and fungi is presented in Tables 26.2 and 26.3.

## 26.7 EO Nanosystems in Insecticidal and Fungal Packaging Materials

Nanocomposite polymers, nanoemulsions, and nanoencapsulates have been used in food packaging to enable the slow release of bioactive agents over time, and extend the shelf life of food (Severino et al. 2015). Several studies have explored the application of nanomaterials in food packaging to acquire enhanced protection against insect and fungal pests. Cindi et al. (2015) used chitosan boehmite–alumina nanocomposite films as packaging material in conjunction with thyme oil to control brown rot in peaches (*Prunus persica* L.) during postharvest storage. The nano-based packaging significantly reduced the incidence and severity of brown rot caused by *Monilinia laxa* in artificially inoculated peach fruits at 25°C for 5 days. The active components of thyme oil, namely, thymol (56.43%), β-linalool (37.6%), and caryophyllen (9.47%), were preserved within the headspace of the punnet. This type of packaging preserved the appearance, taste, and natural flavor better than ordinary packaging.

Similar observations were found against fungal species. A recent study conducted by Otoni et al. (2014) involved the formulation of coarse emulsions (diameters of 1.3–1.9 μm) and nanoemulsions (diameters of 180–250 nm) of clove bud (*S. aromaticum*) and oregano (*O. vulgare*) EOs with methyl cellulose (MC) films. The EO emulsions were found to reduce the rigidity and increase the extensibility of the methyl cellulose films. The effects were more pronounced for the nanoemulsions. Although the EOs decreased yeast and mold counts in sliced breads, the nanoencapsulated version of the EOs resulted in a better efficiency by increasing the bioavailability of the EOs. Another study, conducted by Heshmati et al. (2013), showed that nanoparticles in active packaging can also be used to extend the shelf life of luxury and delicate food, such as caviar, which can easily perish at mild or high temperature conditions. They used packaging comprising *Zataria multiflora* EO (0.03%, 0.06%, w/w), nisin (9.18 mg kg$^{-1}$), potassium sorbate (500 and 1000 mg kg$^{-1}$), and low density polyethylene (LDPE) packaging material containing nano-ZnO. They found that the nano-based packaging was able to significantly reduce the fungal growth and preserve the original color of the caviar samples. The results showed considerable Zn migration from the package to the caviar, especially after 40 days of storage. In a recent study, we found similar observations by comparing coarse emulsions (particle size

**TABLE 26.2**

Studies Involving the Use of Nanostructured EO Systems against Insects

| Polymer | Essential Oil | Target Species | Nanodelivery System | Test Condition | Reference |
|---|---|---|---|---|---|
| | *Ageratum conyzoides, Achillea fragrantissima,* and *Tagetes minuta* | *Cowpea beetle, Callosobruchus maculatus* | Nanoemulsion | *In vitro* | Nenaah et al. 2015 |
| Polycaprolactone | *Zanthoxylum rhoifolium* | *Bemisia tabaci* | Nanospheres | *In vitro* | Christofoli et al. 2015 |
| Myristic acid–chitosan | *Carum copticum* | *Sitophilus granarius* (L.) and *Tribolium confusum* (L.) | Nanogel | *In vitro* | Ziaee et al. 2014b |
| Myristic acid–chitosan | Cumin, cyminum cuminum | *Sitophilus granarius* (L.) and *Tribolium confusum* | Nanogels | *In vitro* | Ziaee et al. 2014a |
| Polyethylene glycol (PEG) | Geranium or bergamot | *Tribolium castaneum* and *Rhizopertha dominica* | Nanoparticles | *In vitro* | Werdin González et al. 2014 |
| Polyethylene glycol | Garlic | *Tribolium castaneum* | Nanoparticles | *In vitro* | Yang et al. 2009 |

**TABLE 26.3**

Studies Involving the Use of Nanostructured EO Systems against Fungi

| Polymer | Essential Oil | Target Species | Nanodelivery Systems | Test Condition | Reference |
|---|---|---|---|---|---|
| Chitosan–caffeic acid | *Cuminum cyminum* | *Aspergillus flavus* | Nanogel | *In vitro* | Zhaveh et al. 2015 |
| ZnO-loaded chitosan | *Zataria multiflora* | *Botrytis cinerea* Pers. | Nanoparticle | *In situ* Strawberry | Mohammadi et al. 2015 |
| Chitosan | Thyme | *Aspergillus flavus* | Nanogel | *In situ* Tomato | Khalili et al. 2015 |
| Gold nanoparticle (AuNPs) | *Mentha piperita* | *Aspergillus niger, Aspergillus flavus,* | Nanoparticles | *In vitro* | Thanighaiarassu et al. 2014 |
| Silver | Cinnamon, clove, carom, eucalyptus, coriander, mustard | *Aspergillus, Fusarium, Curvularia, Cladosporium, Phoma* | Nanoparticles | *In vitro* | Jogee et al. 2012 |
| Chitosan–cinnamic acid | *Mentha piperita* | *Aspergillus flavus* | Nanogel | *In situ* Tomato | Beyki et al. 2014 |
| Chitosan | Turmeric | *Candida albicans, Trychophytol mentagrophyte, Fusarium oxysporum, and Penicilium italicum* | Nanoparticles | *In vitro* | Cuc et al. 2014 |

of 270 nm) and nanoemulsions (particle size of 77 nm) of oregano and thyme mixtures that were used to prepare active polymeric biofilms made up of chitosan, methyl cellulose, and polylactic acid (PLA) (Hossain et al. 2016). These film matrices were tested against four fungal species: *A. niger, A. flavus, Aspergillus parasiticus*, and *Penicillium chrysogenum*. Antifungal assays showed that the nanoemulsion-based formulated films exhibited antifungal activities up to three times higher than those of the films designed with coarse emulsions. In addition, the nanoemulsion-based film matrix sustained antifungal activity over 40–72 h, while the films synthesized with the coarse emulsions maintained activity over only a 24 h period. These studies show that nanoactive biodegradable films embedded with active EOs can be customized to produce next-generation food packaging, which can significantly help toward improving food safety.

Nanoparticles can improve the crystallinity and thermal stability of polymeric packaging materials, and thus control the release of EOs in antimicrobial films. Efrati et al. (2014) prepared antifungal films by melt mixing polyethylene copolymers in the presence of oregano-modified montmorillonite nanoclay (NC) and thymol EO. It was found that addition of NC affected the structure and homogeneity of the polymeric crystals. The combination of high-polymer crystallinity during film preparation and chemical affinity between the EO and NC increased the thermal stability of the EO during film processing. The whole packaging system had a multilayer structure with varied densities and polarities. An increase in polarity of the outer packaging layer reduced the desorption of EO due to chemical interactions between the EO and polymer. Such a configuration improved the partitioning of the EO inside the packaging material and led to an enhanced antifungal activity of the nano-based package.

As such, studies have confirmed the beneficial properties that nanoparticles convey to active packaging materials, and the remarkable insecticidal and antifungal activities that they may produce. Nano-based packaging imparts such desirable traits by helping to overcome the limitations of EOs, such as their poor solubility in water, low stability, and high volatility. Nanoencapsulated EOs can also greatly help to preserve the EO-bearing highly volatile property during synthesis of the active packaging material itself, especially at stages involving high-temperature melt processing, while preserving the insecticidal and antifungal activity of the EOs (Cindi et al. 2015; Mohammadi et al. 2015). In addition, as biodegradable packaging systems often lack mechanical stability and have poorer barrier properties than conventional materials, nanomaterials offer the possibility of making up for these limitations.

## 26.8 Release Kinetics Model for EOs from Polymeric Systems

In active packaging systems, the migration of volatile active agents is mainly attributed to the mass transfer phenomenon (Otgonzul 2010). Based on the mass transfer mechanisms involved, release systems can be classified as diffusion controlled, erosion controlled, swelling controlled, or a combination of these. A polymer-based controlled-release device for EOs is a solid-state asymmetrical diffusion system. The main method that can be used to determine the diffusivity of small molecules through solid or semisolid polymer matrices is the determination of sorption or desorption kinetics. The effective diffusivity is calculated from the nonsteady part of the sorption–desorption curve, assuming a Fick's model mechanism and appropriate boundary conditions (Tramón 2014).

Studies have shed light on the physical process of release of EOs embedded in packaging materials. The mathematical modeling of the diffusion process can enable prediction of the bioactive agent release profiles, kinetics, and time during which the agent will remain above the critical inhibiting concentration. Mastromatteo et al. (2009) investigated the release kinetics of thymol from monolayered and multilayered zein-based composite films. The formulations loaded with spelt bran and thymol (35% w/w) (50.24 cm²) were placed in a container with 500 ml distilled water at room temperature, and the concentrations of the released thymol were measured by means of high-performance liquid chromatography (HPLC). The results showed that the release rate of thymol was dependent on the layers, thickness, and fiber amount of the polymer matrix.

Another study, by Kurek et al. (2014), investigated the release of carvacrol from chitosan-based films and coatings at an RH of 2%–96% and temperature in the range of 4°C–37°C for more than 60 days by gas chromatography analysis. They found that the release of carvacrol was significantly enhanced by saturating the system with water vapor and increasing the temperature from 4°C to 37°C. After 60 days, the release was lowest at 0% RH and highest at >96% RH.

EOs can be released either onto the surface or within the headspace between the food and packaging. Compounds released into the gaseous environment within the package can change the atmosphere to prevent ethylene production, lipid oxidation, or growth of microorganisms (Balasubramanian et al. 2009). Various polymer compositions can affect the release rate of the antimicrobials. In addition, incorporation of active substances into the polymer can change polymer chemistry, which in turn can affect its characteristics, such as oxygen permeability, tensile strength, release rate, brittleness, flavor, color, and taste. Therefore, it is important to establish an effective methodology for determining the migration potential of EO-loaded nanodelivery systems from food contact materials (FCMs) to packaged food.

To date, however, studies of nanoparticle migration from composite packaging materials to food remain scarce. According to Hamdani et al. (1997) and Sanches et al. (2008), the migration process is highly dependent on the migrant molecular structure, its ability to form hydrogen bonds with the packaging material and any additive present therein, its affinity with the food, the interaction between the food and composite material, and its molecular size. However, the process of nanoparticle migration can be understood as the mass transfer of NPs through polymeric materials obeying Fick's laws of diffusion. Simon et al. (2008) successfully modeled the migration of NPs from composite materials based on these basic laws of diffusion kinetics to study nonsteady diffusion for different interfaces.

The physicochemical perspective on the potential migration of engineered nanoparticles from packaging to food is based on the average distance traveled by nanoparticles in the polymer matrix. The progress of the migration process can be described using the differential equation below (Equation 26.1), which relates to the migration of an additive from polymeric packaging film (de Abreu et al. 2010).

$$\frac{\partial C_p}{\partial_t} = D_P \frac{\partial^2 C_p}{\partial x^2} \tag{26.1}$$

where $C_p$ (mg kg⁻¹) represents the concentration of the migrant through the composite polymer, $P$, at time $t$ (s) and position $x$ in $P$, with $D_p$ being the diffusion coefficient in $P$ (cm² s⁻¹).

Based on the concept of the potential chemical properties of both nanoparticles and packaging polymers, it is possible to derive the migration potential ($m$) of NPs through composite materials by taking into consideration the density of the polymer, size of the

NPs, density of the NPs in the polymer, and average distance of the nanoparticles from the surface of the polymer (Equation 26.2), and to estimate the number of migrating particles (Equation 26.3) (Wei et al. 2011).

$$m = \sqrt{\frac{K_B T t}{24\pi^2 \eta a}} \tag{26.2}$$

$$n = m S c_0 \tag{26.3}$$

where, in Equation 26.2, $m$ is the migration potential of NPs in the polymer, $K_B$ is the Boltzmann constant ($K_B = 1.3807 \times 10^{-23}$ JK$^{-1}$), $a$ is the radius of NPs, and $T$, $t$, and $\eta$ stand for the absolute temperature, migration time, and dynamic viscosity of the polymer, respectively. In Equation 26.3, $n$, $S$, and $C_0$ represent the amount of migrating nanoparticles, surface area, and initial concentration in the polymer film, respectively. Based on Equations 26.2 and 26.3, $m$ and $n$ for NPs with different radii and polymers with various densities can be easily computed. Theoretical calculations stemming from the above mathematical expressions indicate that the migration of NPs strongly depends on the size of the nanocomposites and density of the matrix. Smaller NPs and less dense polymers will result in higher migration potential (Simon et al. 2008). However, this is based on the assumption that the interphase boundary between the polymer and food does not represent any obstacle for the movement of the NPs, while in fact some NPs may be reflected back to the polymer matrix at the interphase boundary. The mathematical equation accurately models the migration of very small NPs with a radius in the order of magnitude of 1 nm from polymer matrices that have a low viscosity and do not interact with the NPs. These conditions are readily found in the case of nanocomposites of silver with polyolefines (polyethylene and polypropylene). On the other hand, the migration of bigger NPs with relatively high dynamic viscosities, such as nanosilver composites with polyethylene terephthalate, polystyrene, and surface-modified montmorillonite embedded in various polymer matrices, may not be detectable (Wei et al. 2011).

A study conducted by von Goetz et al. (2013) introduced the parameter of the surface area, and according to these authors, in modeling the migration of nanoparticles, constant parameters can be lumped into $a$ and $b$ to take on the form of a power function equation to describe the number of particles migrating from a polymer per surface area (Equation 26.4):

$$f(t) = at^b \tag{26.4}$$

where $f(t)$ relates to the number of nanoparticles migrating from the polymer per surface area ($a$) and $b$ is a calibrated model parameter. By estimating the parameters $a$ and $b$, they attempted to fit a power function to the migrating values for the release of silver from plastic material over time based on Fickian diffusion processes.

The studies conducted so far provide a sound basis for the migration of NPs from composite materials. However, further studies are required for a more in-depth understanding of the mass transport of NPs from nanocomposite film packaging. Similarly important will be to establish an effective methodology for determining the migration potential of NPs from FCM packaged food. In migration tests, the ability of NPs to migrate in a matrix, their dispersion, and their alterations in size and morphologies before and after the migration tests should all be considered to assess the safety of packaging polymers embedded with nanomaterials. The formulation of new models and consolidation of existing ones can help to predict such migration behavior.

## 26.9 Conclusions

Many plant essential oils have been studied that offer novel and effective insecticidal and fungicidal compounds for control of stored product pests. Nano-based controlled delivery systems in packaging using EO or their components are a promising approach for the food industry. Recent investigations into the design and improvement of EO-loaded nano-based active packaging systems are tantalizing, and hold great promise to control stored product pests.

## Acknowledgments

We are thankful to the U.S. Department of Agriculture (USDA). The authors are also thankful to the Ministère de l'Économie, de l'Innovation et des Exportations (MEIE), Quebec, for their support and funding.

## References

Abd-Aiia, M. A., and W. M. Haggag. 2013. Use of some plant essential oils as post-harvest botanical fungicides in the management of anthracnose disease of mango fruits (*Mangi Feraindica* L.) caused by colletotrichum gloeosporioides (Penz). *International Journal of Agriculture and Forestry* 3 (1):1–6.

Abou-Elnaga, Z. S. 2015. Efficacy of extracts of some Egyptian plants against economically important stored grain pest *Sitophilus oryzae* L. *Journal of Entomology and Zoology Studies* 3 (1):87–91.

Allahvaisi, S. 2010. Reducing insects contaminations through stored foodstuffs by use of packaging and repellency essential oils. *Notulae Botanicae Horti Agrobotanici Cluj-Napoca* 38 (3):21–24.

Anwar, M., M. Ashfaq, M.-U. Hasan, and F. M. Anjum. 2005. Efficacy of *Azadirachta Indica* L. oil on bagging material against some insect pests of wheat stored in warehouses at Faisalabad. *Pakistan Entomologist* 27 (1):89–94.

Augustin, M. A., and Y. Hemar. 2009. Nano- and micro-structured assemblies for encapsulation of food ingredients. *Chemical Society Reviews* 38 (4):902–912.

Avila-Sosa, R., E. Palou, M. T. Jiménez Munguía, G. V. Nevárez-Moorillón, A. R. Navarro Cruz, and A. López-Malo. 2012. Antifungal activity by vapor contact of essential oils added to amaranth, chitosan, or starch edible films. *International Journal of Food Microbiology* 153 (1–2):66–72.

Bahram, S., M. Rezaei, M. Soltani, A. Kamali, S. M. Ojagh, and M. Abdollahi. 2014. Whey protein concentrate edible film activated with cinnamon essential oil. *Journal of Food Processing and Preservation* 38 (3):1251–1258.

Balasubramanian, A., L. Rosenberg, K. Yam, and M. Chiknidas. 2009. Antimicrobial packaging: Potential vs. reality—A review. *Journal of Applied Packaging Research* 3 (4):193–221.

Betts, G. D., P. Linton, and R. J. Betteridge. 1999. Food spoilage yeasts: Effects of pH, NaCl and temperature on growth. *Food Control* 10 (1):27–33.

Beyki, M., S. Zhaveh, S. T. Khalili, T. Rahmani-Cherati, A. Abollahi, M. Bayat, M. Tabatabaei, and A. Mohsenifar. 2014. Encapsulation of *Mentha piperita* essential oils in chitosan-cinnamic acid nanogel with enhanced antimicrobial activity against *Aspergillus flavus*. *Industrial Crops and Products* 54:310–319.

Bhattacharya, S. 2009. Phytosomes: The new technology for enhancement of bioavailability of botanicals and nutraceuticals. *International Journal of Health Research* 2 (3):225–232.

Bilia, A. R., C. Guccione, B. Isacchi, C. Righeschi, F. Firenzuoli, and M. C. Bergonzi. 2014. Essential oils loaded in nanosystems: A developing strategy for a successful therapeutic approach. *Evidence-Based Complementary and Alternative Medicine* 2014:651593.

Boumail, A., S. Salmieri, E. Klimas, P. O. Tawema, J. Bouchard, and M. Lacroix. 2013. Characterization of trilayer antimicrobial diffusion films (ADFs) based on methylcellulose-polycaprolactone composites. *Journal of Agricultural and Food Chemistry* 61 (4):811–821.

Chi-Zhang, Y. D., K. L. Yam, and M. L. Chikindas. 2004. Effective control of *Listeria monocytogenes* by combination of nisin formulated and slowly released into a broth system. *International Journal of Food Microbiology* 90 (1):15–22.

Christofoli, M., E. C. C. Costa, K. U. Bicalho, V. de Cássia Domingues, M. F. Peixoto, C. C. F. Alves, W. L. Araújo, and C. de Melo Cazal. 2015. Insecticidal effect of nanoencapsulated essential oils from *Zanthoxylum rhoifolium* (Rutaceae) in *Bemisia tabaci* populations. *Industrial Crops and Products* 70:301–308.

Chulze, S. 2010. Strategies to reduce mycotoxin levels in maize during storage: A review. *Food Additives and Contaminants* 27 (5):651–657.

Cindi, M. D., T. Shittu, D. Sivakumar, and S. Bautista-Baños. 2015. Chitosan boehmite-alumina nanocomposite films and thyme oil vapour control brown rot in peaches (*Prunus persica* L.) during postharvest storage. *Crop Protection* 72:127–131.

Coma, V., A. D. Eschamps, and A. Marchal-Gros. 2003. Bioactive packaging materials from edible chitosan polymers—Antimicrobial activity assessment on dairy-related contaminants. *Journal of Food Science* 68:2788–2792.

Cuc, N. T. K., T. T. K. Dzung, and P. V. Cuong. 2014. Assessment of antifungal activity of turmeric essential oil-loaded chitosan nanoparticles. *Journal of Chemical, Biological, and Physical Sciences* 4 (3):2347–2356.

de Abreu, D. A. P., J. M. Cruz, I. Angulo, and P. P. Losada. 2010. Mass transport studies of different additives in polyamide and exfoliated nanocomposite polyamide films for food industry. *Packaging Technology and Science* 23 (2):59–68.

Du, W. X., C. Olsen, R. Avena-Bustillos, T. McHugh, C. Levin, and M. Friedman. 2009. Effects of allspice, cinnamon, and clove bud essential oils in edible apple films on physical properties and antimicrobial activities. *Journal of Food Science* 74 (7):M372–M378.

Duncan, T. V. 2011. Applications of nanotechnology in food packaging and food safety: Barrier materials, antimicrobials and sensors. *Journal of Colloid and Interface Science* 363:1–24.

Ebadollahi, A., and M. Mahboubi. 2011. Insecticidal activity of the essential oil isolated from *Azilia eryngioides* (Pau) Hedge Et Lamond against two beetle pests. *Chilean Journal of Agricultural Research* 71 (3):406–411.

Efrati, R., M. Natan, A. Pelah, A. Haberer, E. Banin, A. Dotan, and A. Ophir. 2014. The effect of polyethylene crystallinity and polarity on thermal stability and controlled release of essential oils in antimicrobial films. *Journal of Applied Polymer Science* 131 (11).

Espitia, P. J. P., N. F. F. De Soares, L. C. M. Botti, W. A. Da Silva, N. R. De Melo, and O. L. Pereira. 2011. Active sachet: Development and evaluation for the conservation of Hawaiian papaya quality. *Italian Journal of Food Science* 23 (Suppl.):107–110.

Fathi, M., M. R. Mozafari, and M. Mohebbi. 2012. Nanoencapsulation of food ingredients using lipid based delivery systems. *Trends in Food Science & Technology* 23 (1):13–27.

Follett, P. A., K. Snook, A. Janson, B. Antonio, A. Haruki, M. Okamura, and J. Bisel. 2013. Irradiation quarantine treatment for control of *Sitophilus oryzae* (Coleoptera: Curculionidae) in rice. *Journal of Stored Products Research* 52:63–67.

Follett, P. A., A. Swedman, and D. K. Price. 2014. Postharvest irradiation treatment for quarantine control of *Drosophila suzukii* (Diptera: Drosophilidae) in fresh commodities. *Journal of Economic Entomology* 107 (3):964–969.

Ghabraie, M., K. D. Vu, L. Tata, S. Salmieri, and M. Lacroix. 2016. Antimicrobial effect of essential oils in combinations against five bacteria and their effect on sensorial quality of ground meat. *LWT—Food Science and Technology* 66:332–339.

Gutiérrez, L., R. Batlle, S. Andújar, C. Sánchez, and C. Nerín. 2011. Evaluation of antimicrobial active packaging to increase shelf life of gluten-free sliced bread. *Packaging Technology and Science* 24 (8):485–494.

Hamdani, M., A. Feigenbaum, and J. Vergnaud. 1997. Prediction of worst case migration from packaging to food using mathematical models. *Food Additives and Contaminants* 14:499–506.

Heshmati, M. K., N. Hamdami, M. Shahedi, M. A. Hejazi, A. A. Motalebi, and A. Nasirpour. 2013. Impact of *Zataria multiflora* essential oil, nisin, potassium sorbate and LDPE packaging containing nano-ZnO on shelf life of caviar. *Food Science and Technology Research* 19 (5):749–758.

Hossain, F., P. Follett, K. D. Vu, M. Harich, S. Salmieri, and M. Lacroix. 2016. Evidence for synergistic activity of plant-derived essential oils against fungal pathogens of food. *Food Microbiology* 53:24–30.

Hossain, F., P. Follett, K. D. Vu, S. Salmieri, C. Senoussi, and M. Lacroix. 2014. Radiosensitization of *Aspergillus niger* and *Penicillium chrysogenum* using basil essential oil and ionizing radiation for food decontamination. *Food Control* 45:156–162.

Hossain, F., M. Lacroix, S. Salmieri, K. Vu, and P. A. Follett. 2014. Basil oil fumigation increases radiation sensitivity in adult *Sitophilus oryzae* (Coleoptera: Curculionidae). *Journal of Stored Products Research* 59:108–112.

Huq, T., K. D. Vu, B. Riedl, J. Bouchard, and M. Lacroix. 2015. Synergistic effect of gamma (γ)-irradiation and microencapsulated antimicrobials against *Listeria monocytogenes* on ready-to-eat (RTE) meat. *Food Microbiology* 46:507–514.

Ingle, L. M., V. P. Wankhade, T. A. Udasi, and K. K. Tapar. 2013. New approaches for development and characterization of SMEDDS. *International Journal of Pharmacy and Pharmaceutical Science Research* 3 (1):7–14.

Jo, H.-J., K.-M. Park, S. C. Min, J. H. Na, K. H. Park, and J. Han. 2013. Development of an anti-insect sachet using a polyvinyl alcohol–Cinnamon oil polymer strip against *Plodia interpunctella*. *Journal of Food Science* 78 (1):E1713–E1720.

Jogee, S. P., A. P. Ingle, I. R. Gupta, S. R. Bonde, and M. K. Rai. 2012. Detection and management of mycotoxigenic fungi in nuts and dry fruits. *Acta Horticulturae* 963: 69.

Khalili, S. T., A. Mohsenifar, M. Beyki, S. Zhaveh, T. Rahmani-Cherati, A. Abdollahi, M. Bayat, and M. Tabatabaei. 2015. Encapsulation of thyme essential oils in chitosan-benzoic acid nanogel with enhanced antimicrobial activity against *Aspergillus flavus*. *LWT—Food Science and Technology* 60 (1):502–508.

Khan, A., T. Huq, R. A. Khan, B. Riedl, and M. Lacroix. 2014a. Nanocellulose-based composites and bioactive agents for food packaging. *Critical Reviews in Food Science and Nutrition* 54 (2):163–174.

Khan, A., S. Salmieri, C. Fraschini, J. Bouchard, B. Riedl, and M. Lacroix. 2014b. Genipin cross-linked nanocomposite films for the immobilization of antimicrobial agent. *ACS Applied Materials & Interfaces* 6 (17):15232–15242.

Kurek, M., A. Guinault, A. Voilley, K. Galić, and F. Debeaufort. 2014. Effect of relative humidity on carvacrol release and permeation properties of chitosan based films and coatings. *Food Chemistry* 144:9–17.

Lacroix, M., and P. Follett. 2015. Combination irradiation treatments for food safety and phytosanitary uses. *Stewart Postharvest Review* 11 (3):1–10.

Lamboni, Y., and K. Hell. 2009. Propagation of mycotoxigenic fungi in maize stores by post-harvest insects. *International Journal of Tropical Insect Science* 29 (01):31–39.

Lee, S. E., C. J. Peterson, and J. R. Coats. 2003. Fumigation toxicity of monoterpenoids to several stored-product insects. *Journal of Stored Products Research* 39:77–85.

Licciardello, F., G. Muratore, P. Suma, A. Russo, and C. Nerín. 2013. Effectiveness of a novel insect-repellent food packaging incorporating essential oils against the red flour beetle (*Tribolium castaneum*). *Innovative Food Science and Emerging Technologies* 19:173–180.

Loftsson, T., and D. Duchene. 2007. Cyclodextrins and their pharmaceutical applications. *International Journal of Pharmaceutics* 329 (1):1–11.

López, P., C. Sanchez, R. Batlle, and C. Nerin. 2005. Solid- and vapor-phase antimicrobial activities of six essential oils: Susceptibility of selected foodborne bacterial and fungal strains. *Journal of Agricultural and Food Chemistry* 53:6939–6946.

López, P., C. Sánchez, R. Batlle, and C. Nerín. 2007a. Development of flexible antimicrobial films using essential oils as active agents. *Journal of Agricultural and Food Chemistry* 55 (21):8814–8824.

López, P., C. Sánchez, R. Batlle, and C. Nerín. 2007b. Vapor-phase activities of cinnamon, thyme, and oregano essential oils and key constituents against foodborne microorganisms. *Journal of Agricultural and Food Chemistry* 55 (11):4348–4356.

Mansour, E. E., F. Mi, G. Zhang, X. Jiugao, Y. Wang, and A. Kargbo. 2012. Effect of allylisothio-cyanate on *Sitophilus oryzae*, *Tribolium confusum* and *Plodia interpunctella*: Toxicity and effect on insect mitochondria. *Crop Protection* 33:40–51.

Mastromatteo, M., G. Barbuzzi, A. Conte, and M. Del Nobile. 2009. Controlled release of thymol from zein based film. *Innovative Food Science and Emerging Technologies* 10 (2):222–227.

McClements, D. J., E. A. Decker, and J. Weiss. 2007. Emulsion-based delivery systems for lipophilic bioactive components. *Journal of Food Science* 72 (8):R109–R124.

Medeiros, E. A. A., N. F. F. Soares, T. O. S. Polito, M. M. de Sousa, and D. F. P. Silva. 2011. Antimicrobial sachets post-harvest mango fruits. *Revista Brasileira de Fruticultura* 33 (Spec. Issue 1):363–370.

Mohammadi, A., M. Hashemi, and S. M. Hosseini. 2015. Nanoencapsulation of *Zataria multiflora* essential oil preparation and characterization with enhanced antifungal activity for controlling *Botrytis cinerea*, the causal agent of gray mould disease. *Innovative Food Science and Emerging Technologies* 28:73–80.

Nenaah, G. E., S. I. A. Ibrahim, and B. A. Al-Assiuty. 2015. Chemical composition, insecticidal activity and persistence of three Asteraceae essential oils and their nanoemulsions against *Callosobruchus maculatus* (F.). *Journal of Stored Products Research* 61:9–16.

Oliveira, C., A. Auad, S. Mendes, and M. Frizzas. 2014. Crop losses and the economic impact of insect pests on Brazilian agriculture. *Crop Protection* 56:50–54.

Oliveira, P. M., E. Zannini, and E. K. Arendt. 2014. Cereal fungal infection, mycotoxins, and lactic acid bacteria mediated bioprotection: From crop farming to cereal products. *Food Microbiology* 37:78–95.

Opara, U. L., and A. Mditshwa. 2013. A review on the role of packaging in securing food system: Adding value to food products and reducing losses and waste. *African Journal of Agricultural Research* 8 (22):2621–2630.

Otgonzul, O. 2010. Bioactive polymeric systems for food and medical packaging applications. Dizertační práce, Univerzita Tomáše Bati ve Zlíně, Fakulta technologická, Zlín.

Otoni, C. G., S. F. O. Pontes, E. A. A. Medeiros, and N. D. F. F. Soares. 2014. Edible films from methylcellulose and nanoemulsions of clove bud (*Syzygium aromaticum*) and oregano (*Origanum vulgare*) essential oils as shelf life extenders for sliced bread. *Journal of Agricultural and Food Chemistry* 62 (22):5214–5219.

Pawar, V., and V. Thaker. 2006. In vitro efficacy of 75 essential oils against *Aspergillus niger*. *Mycoses* 49 (4):316–323.

Perlatti, B., P. Bergo, M. Silva, J. Fernandes, and M. Forim. 2013. Polymeric nanoparticle-based insecticides: A controlled release purpose for agrochemicals. In *Insecticide-Development of Safer and More Effective Technologies*, ed. S. Trdan. Rijeka, Croatia: Intech Open Access, 523–550.

Perveen, F., ed. 2012. *Insecticides—Advances in integrated pest management*. Rijeka, Croatia: InTech.

Peteu, S., F. Oancea, O. Sicuia, F. Constantinescu, and S. Dinu. 2010. Responsive polymers for crop protection. *Polymers and Polymer Composites* 2:229–251.

Picard, I., R. G. Hollingsworth, S. Salmieri, and M. Lacroix. 2012. Repellency of essential oils to *Frankliniella occidentalis* (Thysanoptera: Thripidae) as affected by type of oil and polymer release. *Journal of Economic Entomology* 105 (4):1238–1247.

Plooya, W. D., T. Regnierb, and S. Combrinckb. 2009. Essential oil amended coatings as alternatives to synthetic fungicides in citrus postharvest management. *Postharvest Biology and Technology* 53:117–122.

Prakash, B., P. Singh, S. Yadav, S. Singh, and N. Dubey. 2013. Safety profile assessment and efficacy of chemically characterized *Cinnamomum glaucescens* essential oil against storage fungi, insect, aflatoxin secretion and as antioxidant. *Food and Chemical Toxicology* 53:160–167.

Rodriguez, A., C. Nerin, and R. Battle. 2008. New cinnamon-based active paper packaging against *Rhizopus stolonifer* food spoilage. *Journal of Agricultural and Food Chemistry* 56: 6364–6369.

Rodriguez-Lafuente, A., C. Nerin, and R. Batlle. 2010. Active paraffin-based paper packaging for extending the shelf life of cherry tomatoes. *Journal of Agricultural and Food Chemistry* 58 (11):6780–6786.

Salmieri, S., F. Islam, R. A. Khan, F. M. Hossain, H. M. Ibrahim, C. Miao, W. Y. Hamad, and M. Lacroix. 2014a. Antimicrobial nanocomposite films made of poly (lactic acid)-cellulose nanocrystals (PLA-CNC) in food applications: Part A—Effect of nisin release on the inactivation of *Listeria monocytogenes* in ham. *Cellulose* 21 (3):1837–1850.

Salmieri, S., F. Islam, R. A. Khan, F. M. Hossain, H. M. Ibrahim, C. Miao, W. Y. Hamad, and M. Lacroix. 2014b. Antimicrobial nanocomposite films made of poly (lactic acid)–cellulose nano-crystals (PLA–CNC) in food applications—Part B: Effect of oregano essential oil release on the inactivation of *Listeria monocytogenes* in mixed vegetables. *Cellulose* 21 (6):4271–4285.

Sanches, A., J. Cruz, R. Franz, and P. Paseiro. 2008. Mass transport studies of model migrants within dry foodstuffs. *Journal of Cereal Science* 48:662–669.

Severino, R., G. Ferrari, K. D. Vu, F. Donsì, S. Salmieri, and M. Lacroix. 2015. Antimicrobial effects of modified chitosan based coating containing nanoemulsion of essential oils, modified atmosphere packaging and gamma irradiation against *Escherichia coli* O157: H7 and *Salmonella* Typhimurium on green beans. *Food Control* 50:215–222.

Severino, R., K. D. Vu, F. Donsì, S. Salmieri, G. Ferrari, and M. Lacroix. 2014. Antibacterial and physical effects of modified chitosan based-coating containing nanoemulsion of mandarin essential oil and three non-thermal treatments against *Listeria innocua* in green beans. *International Journal of Food Microbiology* 191:82–88.

Simon, P., Q. Chaudhry, and D. Bakos. 2008. Migration of engineered nanoparticles from polymer packaging to food—A physicochemical view. *Journal of Food and Nutrition Research* 47 (3):105–113.

Souza, A. C., G. E. O. Goto, J. A. Mainardi, A. C. V. Coelho, and C. C. Tadini. 2013. Cassava starch composite films incorporated with cinnamon essential oil: Antimicrobial activity, microstructure, mechanical and barrier properties. *LWT—Food Science and Technology* 54 (2):346–352.

Sung, S.-Y., L. T. Sina, T.-T. Tee, S.-T. Bee, A. R. Rahmat, W. A. W. A. Rahman, A.-C. Tan, and M. Vikhraman. 2013. Antimicrobial agents for food packaging applications. *Trends in Food Science & Technology* 33:110–123.

Takala, P. N., S. Salmieri, A. Boumail, R. A. Khan, K. D. Vu, G. Chauve, J. Bouchard, and M. Lacroix. 2013. Antimicrobial effect and physicochemical properties of bioactive trilayer polycaprolactone/methylcellulose-based films on the growth of foodborne pathogens and total microbiota in fresh broccoli. *Journal of Food Engineering* 116 (3):648–655.

Thanighaiarassu, R. R., B. Nambikkairaj, R. Devika, D. Raghunathan, and P. Sivamani. 2014. Green synthesis of gold nanoparticles mediated by plant essential oil and its antifungal activity against human pathogenic fungi. *World Journal of Pharmacy and Pharmaceutical Sciences* 3 (12):1752–1768.

Tian, J., X. Ban, H. Zeng, J. He, Y. Chen, and Y. Wang. 2012. The mechanism of antifungal action of essential oil from dill (*Anethum graveolens* L.) on *Aspergillus flavus*. *PloS One* 7 (1).

Ting, Y., Y. Jiang, C.-T. Ho, and Q. Huang. 2014. Common delivery systems for enhancing *in vivo* bioavailability and biological efficacy of nutraceuticals. *Journal of Functional Foods* 7:112–128.

Tramón, C. 2014. Modeling the controlled release of essential oils from a polymer matrix—A special case. *Industrial Crops and Products* 61:23–30.

Tripathi, A. K., S. Upadhyay, M. Bhuiyan, and P. R. Bhattacharya. 2009. A review on prospects of essential oils as biopesticides in insect pest management. *Journal of Pharmacognosy and Phytotherapy* 1:52–63.

Turhan, K. N. 2013. Cellulose based packaging films containing natural antimicrobial agents. *Journal of Hygienic Engineering and Design* 5:13–17.

Viuda-Martos, M., Y. Ruiz-Navajas, J. Fernández-López, and J. Pérez-Álvarez. 2007. Antifungal activities of thyme, clove and oregano essential oils. *Journal of Food Safety* 27 (1):91–101.

von Goetz, N., L. Fabricius, R. Glaus, V. Weitbrecht, D. Günther, and K. Hungerbühler. 2013. Migration of silver from commercial plastic food containers and implications for consumer exposure assessment. *Food Additives and Contaminants: Part A* 30 (3):612–620.

Wei, H., Y. Yanjun, L. N. Tao, and L. W. 2011. Application and safety assessment for nano-composite materials in food packaging. *Materials Science* 56 (12):1216–1225.

Weiss, J., E. A. Decker, D. J. McClements, K. Kristbergsson, T. Helgason, and T. Awad. 2008. Solid lipid nanoparticles as delivery systems for bioactive food components. *Food Biophysics* 3 (2):146–154.

Werdin González, J. O., M. M. Gutiérrez, A. A. Ferrero, and B. Fernández Band. 2014. Essential oils nanoformulations for stored-product pest control—Characterization and biological properties. *Chemosphere* 100:130–138.

Wong, K. K. Y., F. A. Signal, S. H. Campion, and R. L. Motion. 2005. Citronella as an insect repellent in food packaging. *Journal of Agricultural and Food Chemistry* 53:4633–4636.

Yahyaraeyat, R., A. Khosravi, D. Shahbazzadeh, and V. Khalaj. 2013. The potential effects of *Zataria multiflora* Boiss essential oil on growth, aflatoxin production and transcription of aflatoxin biosynthesis pathway genes of toxigenic *Aspergillus parasiticus*. *Brazilian Journal of Microbiology* 44 (2):649–655.

Yang, F. L., X. G. Li, F. Zhu, and C. L. Lei. 2009. Structural characterization of nanoparticles loaded with garlic essential oil and their insecticidal activity against *Tribolium castaneum* (Herbst) (Coleoptera: Tenebrionidae). *Journal of Agricultural and Food Chemistry* 57 (21):10156–10162.

Zain, M. E. 2011. Impact of mycotoxins on humans and animals. *Journal of Saudi Chemical Society* 15 (2):129–144.

Zhaveh, S., A. Mohsenifar, M. Beiki, S. T. Khalili, A. Abdollahi, T. Rahmani-Cherati, and M. Tabatabaei. 2015. Encapsulation of *Cuminum cyminum* essential oils in chitosan-caffeic acid nanogel with enhanced antimicrobial activity against *Aspergillus flavus*. *Industrial Crops and Products* 69:251–256.

Ziaee, M., S. Moharramipour, and A. Mohsenifar. 2014a. MA-chitosan nanogel loaded with *Cuminum cyminum* essential oil for efficient management of two stored product beetle pests. *Journal of Pest Science* 87 (4):691–699.

Ziaee, M., S. Moharramipour, and A. Mohsenifar. 2014b. Toxicity of *Carum copticum* essential oil-loaded nanogel against *Sitophilus granarius* and *Tribolium confusum*. *Journal of Applied Entomology* 138 (10):763–771.

# 27

# Essential Oil Mixtures for Pest Control

Leo M.L. Nollet and Hamir Singh Rathore

## CONTENTS

## 27.1 Introduction

Essential oils are rapidly growing in popularity because they are being used in industries, namely, the food, pharmaceutical, and cosmetic industries; veterinary products; industrial deodorants; and tobacco without any side effects. Pertinent examples are

- *Calendula*: Bright orange candela flowers are also known as marigolds. Calendula essential oil is particularly good for sensitive skin and can be used to reduce the appearance of acne scars. It can also be put in bathwater to soothe psoriasis.

- *Frankincense*: Frankincense is a must-have essential oil in our home. It is used for relaxation, such as in baths, and to help minor cuts and bug bites heal more quickly. It can be used for depression, inflammation, and immunity, and to increase spiritual awareness.

- *Oregano*: It is a well-known flu fighter. This strong-tasting oil has natural antibacterial qualities, so it can help to fight colds and other illnesses. It is taken topically, often by putting a few drops on the tongue—the taste is not pleasant, but many people swear by it during flu season.

- *Chamomile*: Chamomile is often used as a tea. It is particularly well known for its relaxing effects, which is why the tea is popular to drink before bed. Add to the effect with a few drops of the oil on your pillowcase.

- *Grapefruit*: This essential oil has properties similar to those of lemon oil. It is a great choice for people experiencing fatigue, and it is useful in jet lag. It is also a natural antiseptic, so one can add it to homemade household cleansers to keep one's home safe and clean.

Recently, their uses in pest control have been explored. Some essential oils, such as thyme, cloves, salvia, mint, oregano, and pine, possess antibactericidal properties. Others are insecticides, for example

- Against ants: *Mentha spicata* (spearmint) and *Tanacetum*
- Against aphids: Garlic, other *Allium*, coriander, aniseed, and basil
- Against fleas: Lavender, mints, and lemongrass
- Against flies: Rue, citronella, and mint
- Against lice: *M. spicata*, basil, and rue
- Against moths: Mints, hisopo, rosemary, and dill
- Against coleopteran: *Tanacetum*, cumin, wormwood, and thyme
- Against cockroaches: Mint, wormwood, eucalyptus, and laurel
- Against nematodes: Tagetes, salvia, calendula, and asparagus

Essential oils have been used for thousands of years in various cultures for the above-mentioned purposes. They are obtained [1–10] from different parts of plants, including flowers, leaves, barks, roots, resins, and peels. In ancient times, Jews and Egyptians used a simple procedure to obtain essential oils, for example, by soaking the plants or plant parts in oil for a required period and then filtering the oil through a line bag.

Essential oils are volatile (steam volatile) and liquid at room temperature. Their distillates are initially colorless or slightly yellowish. The specific density of most essential oils is less than that of water. They are high-refractory index compounds and have nearly always rotational properties. They are soluble in alcohol, as well as in high-grade alcohol, and in other common organic solvents, such as ether and chloroform. They are slightly soluble in water and are liposoluble. As they are steam volatile, they can be dragged out using steam for their extraction from plants.

Instead of a single essential oil, generally their mixtures are used in many formats for different purposes (daily needs). The admixtures are useful for the following reasons:

1. By mixing two or more essential oils together, their effect increases considerably. For example, lavender and chamomile oils are both good for anti-inflammatory purposes, and by combining them, their effect will be even greater together, as an anti-inflammatory, than either of them used alone.

2. The blend created by mixing two or more oils that have the same therapeutical properties works synergistically and with more density than just a single oil alone.

3. The high concentration of essential oils can produce the opposite effect; that is, concentrated lavender oil can cause restlessness, agitation, and insomnia rather than relaxation. Therefore, a liquid solvent is used to lower to a proper concentration.

4. The additives and diluents are also cost-effective; that is, they minimize the cost of the product.

5. Blended into correct concentrations, they are safe to use, especially for those who are unfamiliar with them.

6. The blended product is ready for immediate use—no fuss and no mess.

However, few studies have reported on the mixing effects of essential oils.

A small number of publications exist in this area [3–10]. These and a few other references are summarized in the following pages. The authors are aware that some references are not proven by scientific tests, but they give an idea of the culture of using essential oils in the garden, at home, and so forth.

## 27.2 Nine Clever Ways to Use Essential Oils in the Garden [10]

### 27.2.1 To Repel Insect Pests

Rosemary oil, peppermint oil, thyme oil, and clove oil are known as potent repellents for many types of pests. A mixture containing equal parts of rosemary, peppermint, thyme, and clove oils (about 10 drops of each) in a spray bottle filled with water has been found to be effective in getting rid of skittering, creeping, crawling, or flying pests.

### 27.2.2 To Suppress Fungus

Oils such as tea tree, neem, rosemary, citronella, oregano, thyme, peppermint, clove, cinnamon, garlic, and onion (1 tablespoon in a cupful of water) and their mixtures (8–10 drops in a cupful of water) are effective in controlling fungus.

### 27.2.3 To Stop Slugs and Snails

A diluted solution (1 teaspoon in a spray bottle filled with water [1 cup]) of cedarwood, hyssop, and pine oils can be sprayed in a ring around plants where slugs and snails are found in order to keep gastropods off the plants. These oils have been claimed to be the best for this purpose. The spray can be recycled if required.

### 27.2.4 To Discourage Vermin

Cotton balls impregnated with one or two drops of fresh peppermint oil have been reported to repel mice and other rodents. The impregnated balls are tucked into the entrances of mouse holes, squirrel nests, and other rodent burrows to persuade rodent residents to relocate. The plugs can be replaced if required.

### 27.2.5 To Dissuade Pets

Cloth strips or small pieces impregnated with diluted (1 teaspoon in 1 cup of water) rosemary oil can be used to keep the neighborhood tomcat from leaving his delightful presents amidst your herbs and veggies. The strips or pieces can be hung between garden rows around plants or around the garden perimeter where the cat likes to dig. From time to time, replacement of the strips or pieces is required. Similarly, impregnated cloth strips or small pieces with black pepper oil may be used to deter larger mammals from the garden. The overuse of black pepper oil may hurt humans as well. A spray of diluted rosemary oil can also be used for the same purposes.

### 27.2.6 Treat Bites and Stings

A cotton ball or pad impregnated with the mixture of two drops of lavender oil, two drops of chamomile oil, one drop of basil oil, and 1 teaspoon of orange apple cider vinegar (ACV) is used to treat bites and stings of bees, wasps, ants, and other insects. A mixture of these oils in jojoba oil in place of ACV can also be applied for the same purpose. The treatment can be made more effective by prior cleaning and dab drying the effective portion, and then applying the impregnated cotton ball.

### 27.2.7 Attract Pollinators

The scent of Neroli (orange blossom) is known to be an irresistible attractant for bees. The essential oils obtained from many small-blossomed flowers, such as lavender, hyssop, marjoram, *Helichrysum*, basil, sage, and rosemary, also possess these characteristics. A spray of lavender, yarrow, catmint, fennel, *Helichrysum*, or sage essential oils gives good results in attracting butterflies to the garden.

### 27.2.8 Enhance the Mood

Mountain rose herbs are known for their calming effects to relax after a long and stressful day.

### 27.2.9 Mosquito Repellent

Citronella oil is one of the widely known mosquito repellents. An admixture of 1 ounce of organic witch hazel and 10 drops of jojoba oil in a small glass spray bottle containing 2 ounces of water is used successfully to repel mosquitoes.

## 27.3 Green Cleaning: 10 Essential Oils That Naturally Repel Insects [3,4,6,7,11]

### 27.3.1 Lavender Oil (*Lavandula angustifolia* [Mill])

Lavender pillows and sachets are used in linen cupboards and chests of drawers to keep away moths and other insects, as well as to have a fresh scent. Lavender oil is well known to provide relaxation and restful sleep. In many places, it is used for its lovely aroma and soothing qualities. Lavender oil may be sprayed from an atomizer or placed in a saucer to keep away ants and insects and to disinfect the air.

### 27.3.2 Basil Oil (*Ocimum basilicum* [L.])

Basil oil sprays are available on the market for repelling mosquitoes, larvae, and dust mites in wet climates. Basil oil is also used for a zesty addition to tomato sauce and to clean green.

### 27.3.3 Thyme (*Thymus vulgaris* [L.])

Thyme has been found to be a highly effective insecticide against houseflies. It is an excellent mosquito repellent as well.

### 27.3.4 Pine Oil (*Pinus sylvestris* [L.])

Pine oil sprays are used to repel mosquitoes, as well as to provide a fresh, forest-like smell.

### 27.3.5 Vetiver Oil (*Vetiveria zizanioides* [Nash])

Vetiver essential oil spray has been found to be very useful in repelling mosquitoes and creating a spicy and Balinese ambience for summer in houses of workers of the vetiver craft industry in Java, Indonesia. In Indonesia, there is a sustainable vertiver essential oil production. It has developed into a viable vetiver industry, producing items such as aromatic mats, baskets, candles, and soaps from the spent root.

### 27.3.6 Bergamot Oil (*Citrus aurantium* [L.] var. *bergamia*)

It has been claimed that bergamot oil is one the favorite oils to use for green cleaning. It is a good insect repellent, and also adds a mood-boosting and fruity lift. Bergamot oil is phototoxic, so it may not be used for pest control. Bergamot oil is wonderful for topical use on insect bites or stings (without sun exposure).

### 27.3.7 Peppermint Oil (*Mentha piperita* [L.])

Peppermint oil is a perfect choice as a natural insecticide and repellent to mosquitoes. Its fresh and minty clean aroma is used in bug spray. Its diluted solution (3 ml/L of water) completely kills larvae of *Culex quinquefasciatus* in 24 hours. Its mixture (peppermint–tea tree Australia: 1:1) is very effective on insect bites and stings. Peppermint oil is a great well-known recipe.

### 27.3.8 Tea Tree Australia Oil (*Melaleuca alternifolia* [Cheel])

Sprays of Australian tea tree oil are a green cleaning powerhouse, as well as antiparasitic. It is capable of destroying or suppressing the growth of parasites such as fleas, leeches, lice, and ticks. Australian tea tree oil is helpful in controlling irritation from bites or stings.

### 27.3.9 Eucalyptus Oil (*Eucalyptus globulus* [Labill])

Eucalyptus oil is a standard part of each and every natural green cleaning kit. It can also be used as an insecticide. Eucalyptus oils obtained from different species of eucalyptus tree have been found to be more effective against *Lutzomyia longipalpis* than other natural products.

### 27.3.10 Lemon Eucalyptus Oil (*Eucalyptus citriodora* [Hook])

It is also used as a natural insecticide, especially by those who love a citrusy smell for summer. Its hydrosol is very effective against mosquitoes. It is safe to use around children and pets as a broad-range insecticide. Lemon eucalyptus oil has been thoroughly studied in Ethiopia as an insect repellent. *E. citriodora* is an excellent resource, as the tree grows quickly and has a fairly high yield of essential oil. This oil is also effective against bites and stings of the summer.

## 27.4 All-Natural Homemade Bug Spray Recipes That Work [12]

Some natural, homemade, and inexpensive bug sprays are summarized in the following sections.

### 27.4.1 Essential Oil Bug Spray

Add witch hazel to fill almost to the top of a spray bottle (8 ounce) that is already half-filled with distilled or boiled water. Add ½ teaspoon vegetable glycerin, and then add 30–50 drops of essential oil or oils (citronella, clove, lemongrass, rosemary, tea tree, cajeput, eucalyptus, cedar, catnip, lavender, cinnamon, and mint). Mix well and store in a cool place. It works well and smells pleasant.

### 27.4.2 Fresh or Dried Herb Bug Spray

Take 3–4 tablespoons of dried herbs total in any combination, from peppermint, spearmint, citronella, lemongrass, catnip, lavender, and so forth. Mix well, cover, boil, and cool, and then strain herbs out. Mix the aliquot with 1 cup of witch hazel or rubbing alcohol and store in a spray bottle in a cool place. Use when desired. It provides a pleasant smell and is very refreshing to the skin.

### 27.4.3 Super Strong Insect Repellent Recipe

The following simple, inexpensive, and wonderful insect repellent spray may be made by mixing ingredients that are commonly available in one's kitchen.

Place 132 ounces of apple cider vinegar and dried herbs (tablespoon each of sage, rosemary, lavender, thyme, and mint) into a large glass jar; seal the jar tightly and shake daily for 2–3 weeks.

Then strain the herbs out and store in a spray bottle or tincture bottles, preferably in the refrigerator. Its diluted solution (50/50 v/v) may be applied on skin to cure insect bites and stings, and the spray can be used for serious bug control. This mixture has been found to be very strong and possess antiviral and antibacterial properties. It can also be used for any illness. The recommended dose is 1 tablespoon in water for adults several times a day and 1 tablespoon in water for children older than 2 several times a day.

## 27.5 Aromatherapy and Pest Control [13]

The ornamental plants can be protected by the recipes described in the book *1001 All-Natural Secrets to a Pest-Free Property* by Dr. Myles H. Bader. These home remedies are very simple and inexpensive to prepare, and their composition can be altered to the individual need. They can be sprayed by using a glass sprayer and can be stored in a cool place. It is also easy to prepare fresh and add mixture when needed. Some of the recipes are described in the next sections.

### 27.5.1  Recipe for Keeping the Aphids Off

A mixture of oil, soap, and essential oils can be made in a ½ gallon of water by adding a squirt of dishwashing soap (Ivory liquid is the mildest), 2 tablespoons of cooking oil, and 10 drops of eucalyptus essential oil. Then, spray the solution on the plants, making sure to spray underneath the leaves, as well as on top. The oil will smother and drown the aphids, and the eucalyptus smell will deter other aphids from coming back. Lemon essential oil can be used, as well as tea tree. The solution can be painted on with a brush if one desires to be more area specific and/or make a smaller batch.

### 27.5.2  Recipe for Controlling Fungus

Mix baking soda (around 2 tablespoons) in a quart of water. Put six drops each of lemon and ginger essential oils in ½ tablespoon of dishwashing liquid and then add this to the baking soda solution and spray directly on the fungus. For mold, use ½ cup of white vinegar in a quart of water, with tea tree, lavender, or any of the citrus essential oils, and spray onto affected areas.

### 27.5.3  Recipe for Spiders, Mites, and Caterpillars

Mix quite a few drops of Tabasco and a small squirt of dishwashing liquid, and add a few drops of spearmint essential oil. Stir thoroughly and spray on the affected areas. In fact, any of the mint oils will be effective in this mix.

### 27.5.4  Recipe for Ants

Take honey or jam in a small container and mix in boric acid. This is very effective to control ants. A mixture of window cleaner with peppermint, cinnamon, or citrus essential oil is very effective in stopping ants moving on their route, while turmeric may be sprinkled around the hole and the ant would not be able to exit over the spice. A few drops of any of the essential oils dripped directly into the hole of the tunnel will ensure that the tunnel is dead.

### 27.5.5  Recipe for Roaches

A solution of Dr. Bronner's peppermint soap with a few drops of citronella essential oil can be sprayed directly on the roach to kill it. To deter roaches and ants from frequenting a place, a solution of borax mixed with crushed black pepper and bay leaves can be sprayed. The spray should be kept away from pets and children.

### 27.5.6  Recipe for Flies

The movement of flies can be restricted by the burning of eucalyptus, clove, or basil oil in the kitchen. A spray of essential oil in water can freshen the room, as well as dissuade pests from lingering. Lemongrass oil is also useful in pest control.

## 27.6 Ten Homemade Organic Pesticides [14]

Long before the invention of harmful chemical pesticides, farmers and householders used multiple remedies for removing insect infestations from their garden plants and house vectors. The following sections offer some favorite, all-natural, inexpensive, organic recipes for pest control.

### 27.6.1 Neem

Add ½ ounce of high-quality organic neem oil and ½ teaspoon of mild organic liquid soap or Dr. Bronner's peppermint soap to 2 quarts of warm water. Stir slowly, and then add to a spray bottle and use immediately.

### 27.6.2 Citrus Oil and/or Cayenne Pepper Mix

Mix 10 drops of citrus essential oil with 1 teaspoon cayenne paper and 1 cup of warm water. Shake well and spray in the affected areas. It is a great organic pesticide that works well on ants.

### 27.6.3 Soap, Orange Citrus, and Water

Mix 3 tablespoons of liquid organic castile soap with 1 ounce of orange oil to 1 gallon of water, shake well, and transfer into a spray bottle. This is a specifically effective treatment against slugs and can be sprayed directly on ants and roaches.

### 27.6.4 Eucalyptus Oil

Eucalyptus can be sprinkled to control flies, bees, and wasps. It is a great natural pesticide to repel insects.

### 27.6.5 Onion and Garlic Spray

Mince one organic clove of garlic and one medium-sized organic onion, add to a quart of water, wait for 1 hour, and then add 1 tablespoon of cayenne pepper and 1 tablespoon of liquid soap to the mix. Transfer to a spray bottle and store in the refrigerator. The mix is stable for 1 week and holds its potency. It works well against common pests.

### 27.6.6 Chrysanthemum Flower Tea

These flowers contain a powerful plant chemical known as pyrethrum that invades the nervous system of insects, rendering them immobile. Boil 100 g of dried flowers into 1 L of water for 20 minutes, strain, cool, and fill a spray bottle. It is stable for up to 2 months. Its effectiveness can be improved by adding some organic neem oil.

## 27.7  Healthy Living: DIY Peppermint Spider Repellent [15]

Chemical pesticides have been found to be good for pest control all over the world. Unfortunately, for healthy living they have been found to be totally unsuitable because they lead to skin allergies, asthma, and serious illness. Avoiding chemicals and using natural products is a key to healthy living. If one has kids and pets, this is one more reason why one should opt for healthier options and avoid commercial pest control products. The key to healthy living can be developed by making our own pest repellent. It is well known that spiders do not like strong-smelling herbs like mint, lavender, and orange. They hate peppermint oil. So, useful and effective pest repellent recipes can be prepared. Examples are given in the following sections.

### 27.7.1  Peppermint Pest-Repelling Recipe

Add 10–15 drops of peppermint essential oil into a spray bottle containing 8–12 ounces of water. Spray around door frames, windows, small cracks, corners of the ceiling, and bathrooms. Use peppermint essential oil without water for a more potent version. Do this once a week, but in the summer, do it twice a week.

### 27.7.2  Mountain Rose Herb Recipe

Mountain rose herbs are a reputable source for essential oils. The repellent recipe can be made as above for fleas, ticks, ants, centipedes, mosquitoes, moths, mice, and so forth.

## 27.8  Organic Pesticide and Fungicide Spray [16]

Some organic pesticide and fungicide sprays can be made as follows:

- Azadirachtin is toxic to many insects and acts as a feeding inhibitor and growth disrupter. It is a biodegradable and natural insecticide.
- *Aloe vera* acts as a source of salicylic acid, which is necessary for activating a plant's immune system to respond to threats.
- Potassium silicate is a natural fungicide, insecticide, and miticide. The essential oils used in the foliar spray have a variety of antifungal and pesticide properties.

### 27.8.1  Neem Recipe

Add ¼ tablespoon neem oil, ½ tablespoon 7.8% potassium silicate solution, and ¼ ounce (1.5 tablespoons or 150 drops) assorted essential oils (ginger, rosemary, clove, peppermint, and eucalyptus) to ¼ gallon (4 cups) of warm water. Shake well to entirely emulsify the mix. Finally, it should be nice and creamy looking, with no oil floating on top. This can be used to control many types of pest and fungus.

### 27.8.2 Aloe Recipe

Add 1 tablespoon *Aloe vera* juice. The juice can be made by shaking powdered *Aloe vera* in water. *Aloe vera* activates the immune system of the plant to give response to any threats.

## 27.9 Pest Control Using Essential Oils for Ants, Mice, Cockroaches, and More [17]

Citronella oil is found in many natural insect repellents. It is distilled from citronella grass, grown mostly in southern Asia. The fragrance is lemony and bearable to most people, but pungent to many insects, especially mosquitoes. Other essential oils possessing insect-repelling properties are cedarwood, lemongrass, eucalyptus, peppermint, pennyroyal, lavender, and bergamot. Common pests and ways to use essential oils to repel them are given in the next sections.

### 27.9.1 Ant Repellent

A common natural way to get rid of ants is by using vinegar spray. Black pepper or cayenne pepper can also be used sprinkled on places where ants show up. Peppermint essential oil spray is useful in the kitchen and bedroom.

### 27.9.2 Aphid Repellent

Mix 10 drops of rosemary and lavender oil both in 1 L (4 cups) of salt water. Spray the affected plants. Other repellents are neem oil spray and calcium powder. Calcium powder is sprinkled around the plant's base.

### 27.9.3 Cockroach Repellent

Impregnated cotton balls with citronella oil can be placed in trouble areas, such as cupboards or under the sink. Peppermint oil and lemongrass oil may also be added to citronella oil to prepare impregnated cotton balls. A spray solution can also be made by mixing 5 drops of cypress essential oil and 10 drops of peppermint oil. Spray the solution whenever and wherever the cockroaches appear.

### 27.9.4 Fly Repellent

Place a handful of dried cloves in a bowl, and sprinkle a few drops of clove oil, lavender oil, and citronella oil or peppermint oil on the dried cloves. Place it in areas where flies may come or linger, such as a front or back door, kitchen, or near the garage.

### 27.9.5 Mosquito Repellent

Put a few drops of citronella oil or lemongrass oil in an oil burner. The oil is to be added again and again in the burner. A citronella candle also works well. A ribbon impregnated with a mixture containing five drops each of citronella oil, lemongrass oil, peppermint oil, and lavender oil can be hung on doorways, in the patio, or in a window for repelling mosquitoes.

### 27.9.6 Mice Repellent

Place cotton balls impregnated with peppermint oil, eucalyptus oil, or spearmint oil in locations where mice may be entering the house, areas of an attic or garage, and behind appliances, such as the refrigerator, washer, or drier. The balls can deter mice, but further measures may be required if the infestation is serious.

## 27.10 Advantages of Mixed Essential Oils [18]

1. The effect of mixtures of two or more essential oils is more than that of individual oils, as well as greater than the additive effective of individual oils. It has been found to be true in the mix of lavender oil and chamomile oil.
2. A mixture of two to four essential oils has the same therapeutic properties, and the mixture works synergistically and with more effectiveness than any oil alone.
3. The order of the mixture can be made pleasant by mixing pleasant oil in less pleasant or bitter oil.
4. A costly essential oil can be made affordable by diluting it with less expensive oil.
5. Mixed into the correct concentration, two or more mixed oils may remain safe to use and stable over a long duration.
6. Old products of daily use can be made attractive to customers by adding a suitable essential oil as a blend.
7. Mixed essential oils are ready for immediate use—no fuss and no mess.

## 27.11 Disadvantages of Mixing Essential Oils Together

1. Pure oil or concentrated oil usage can produce the reverse effect; that is, concentrated lavender oil can cause restlessness, agitation, and insomnia.
2. Little work has been carried out to understand the end properties of the mixed oil.

## References

1. Sharma, B.K. 1944. *Industrial Chemistry: Including Chemical Engineering.* Meerut, India: Goel Publishing House.
2. Chokechaijaroenporn, O., Bunyapraphatsara, N., and Kongchuensin, S. 1994. Mosquito repellant activities of *Ocimum* volatile oils. *Phytomedicine* 135–139.

3. Perumalsamy, H., Kim, J.Y., Kim, J.R., Hwang, K.N., and Ahn, Y.J. 2014. Toxicity of basil oil constituents and related compounds and the efficacy of spray formulations to *Dermatophagoides farinae* (Acari: Pyroglyphidae). *J. Med. Entomol.* 51(3):650–657.

4. Park, B.S., Choi, W.S., Kim, J.H., Kim, K.H., and Lee, S.E. 2005. Monoterpenes from thyme (*Thymus vulgaris*) as potential mosquito repellents. *J. Am. Mosq. Control Assoc.* 21(1):80–83.

5. Ansari, M.A., Vasudevan, P., Tandon, M., and Razdan, R.K. 2000. Larvicidal and mosquito repellent action of peppermint (*Mentha piperita*) oil. *Bioresour. Technol.* 71:267–271.

6. Maciel, M.V., Morais, S.M., Bevilaqua, C.M., Silva, R.A., Barros, R.S. et al. 2010. Chemical composition of *Eucalyptus* spp. essential oils and their insecticidal effects on *Lutzomyia longipalpis*. *Vet. Parasitol.* 167(1):1–7.

7. Maia, M.F., and Moore, S.J. 2011. Plant-based insect repellents: A review of their efficacy, development and testing. *Malaria J.* 10(1):S11. Retrieved from http://www.malariajournal.com/content/10/S1/S11.

8. Tripathi, A.K., Upadhyay, S., Bhuiyan, M., and Bhattacharya, P.R. 2009. A review on prospects of essential oils and biopesticides in insect-pest management. *J. Pharmacogn. Phytother.* 1(5):52–62.

9. Singh, O., Rathore, H.S., and Nollet, L.M.L. 2015. Biochemical pesticides: Oil pesticides. In L.M.L. Nollet, H.S. Rathore (eds.), *Biopesticides Handbook*. Boca Raton, FL: CRC Press/Taylor & Francis Group, 183–226.

10. 101 essential oils uses & benefits: 9 clever ways to use essential oils in the garden. http://www.naturallivingideas.com/9-clever-ways-use-essential-oils-garden/.

11. Green cleaning: 10 essential oils that naturally repel insects. http://info.achs.edu/blog/green-cleaning-10-essential-oils-that-naturally-repel-insects.

12. All natural homemade bug spray recipes that work! http://wellnessmama.com/2565/homemade-bug-spray//.

13. Aromatherapy and pest control: Essential oils to the rescue in the garden. http://www.suzannebovenizer.com/aromatherapy-essential-oils/aromatherapy-and-pest-control.

14. 10 homemade organic pesticides. http://www.globalhealingcenter.com/natural-health/organic-pesticides/.

15. Healthy living: DIY peppermint spider repellent. http://www.stepin2mygreenworld.com/healthyliving/around-the-home/spider-repellent-recipe/.

16. Organic pesticide and fungicide spray. http://www.instructables.com/id/Organic-pesticide-and-fungicide-spray/.

17. Pest control using essential oils for ants, mice, cockroaches and more. http://www.essentialoilspedia.com/pest-control/.

18. How to mix (or blend) essential oils + ways to benefit from aromatherapy at home. https://household-tips.thefuntimesguide.com/2007/03/mixessentialoil.php.

# Index

Page numbers followed by f and t indicate figures and tables, respectively.